Lecture Notes in Networks and Systems

Volume 437

The series “Lecture Notes in Networks and Systems” publishes the latest developments in Networks and Systems—quickly, informally and with high quality. Original research reported in proceedings and post-proceedings represents the core of LNNS.

Volumes published in LNNS embrace all aspects and subfields of, as well as new challenges in, Networks and Systems.

The series contains proceedings and edited volumes in systems and networks, spanning the areas of Cyber-Physical Systems, Autonomous Systems, Sensor Networks, Control Systems, Energy Systems, Automotive Systems, Biological Systems, Vehicular Networking and Connected Vehicles, Aerospace Systems, Automation, Manufacturing, Smart Grids, Nonlinear Systems, Power Systems, Robotics, Social Systems, Economic Systems and other. Of particular value to both the contributors and the readership are the short publication timeframe and the world-wide distribution and exposure which enable both a wide and rapid dissemination of research output.

The series covers the theory, applications, and perspectives on the state of the art and future developments relevant to systems and networks, decision making, control, complex processes and related areas, as embedded in the fields of interdisciplinary and applied sciences, engineering, computer science, physics, economics, social, and life sciences, as well as the paradigms and methodologies behind them.

Indexed by SCOPUS, INSPEC, WTI Frankfurt eG, zbMATH, SCImago.

All books published in the series are submitted for consideration in Web of Science.

For proposals from Asia please contact Aninda Bose (aninda.bose@springer.com).

Sazzad Hossain · Md. Shahadat Hossain ·
M. Shamim Kaiser · Satya Prasad Majumder ·
Kanad Ray
Editors

Proceedings of International Conference on Fourth Industrial Revolution and Beyond 2021

Springer

Editors
Sazzad Hossain
CSE
University Grants Commission of Banglade
Dhaka, Bangladesh

M. Shamim Kaiser
Institute of Information Technology
Jahangirnagar University
Dhaka, Bangladesh

Kanad Ray
Department of Physics
AMITY University Rajasthan
Jaipur, India

Md. Shahadat Hossain
Department of Computer Science
and Engineering
University of Chittagong
Chittagong, Bangladesh

Satya Prasad Majumder
Department of EEE
Bangladesh University of Engineering &
Technology
Dhaka, Bangladesh

ISSN 2367-3370 ISSN 2367-3389 (electronic)
Lecture Notes in Networks and Systems
ISBN 978-981-19-2444-6 ISBN 978-981-19-2445-3 (eBook)
https://doi.org/10.1007/978-981-19-2445-3

This Springer imprint is published by the registered company Springer Nature Singapore Pte Ltd.
The registered company address is: 152 Beach Road, #21-01/04 Gateway East, Singapore 189721, Singapore

Organization

Chief Patron

Prof. Datuk Ts. Dr. Wahid Bin Razzaly, Vice-chancellor, UTHM, Malaysia

Steering Committee

Kazi Shahidullah, Chairman, UGC of Bangladesh
Dil Afroza Begum, Member, UGC of Bangladesh
Md. Sazzad Hossain, Member, UGC of Bangladesh
Muhammad Alamgir, Member, UGC of Bangladesh
Biswajit Chanda, Member, UGC of Bangladesh
Md. Abu Taher, Member, UGC of Bangladesh

Local Advisory Committee

Dil Afroza Begum, Member, UGC
A. B. M. Siddique Hossain, AIUB
A. Z. M. Touhidul Islam, RU
Mohammad Kaykobad, BRACU
M. Rizwan Khan, UIU
M. M. A. Hashem, KUET
M. Lutfar Rahman, DIU
Mohammod Abul Kashem, DUET
Mozammel Huq Azad Khan, EWU
Nasim Akhtar, CUST
S. I. Khan, BRACU

Subrata Kumar Aditya, SHU
Syed Akhter Hossain, ULAB
Suraiya Pervin, DU

International Advisory Committee

Muhammad Alamgir, Member, UGC
Alamgir Hossain, Teesside University, UK
Anirban Bandyopadhyay, NIMS, Japan
Ayman Alfalou, University of Brittany, France
Ekram Hossain, University of Manitoba, Canada
Latifur Khan, University of Texas, USA
M. A. Karim, University of Massachusetts, USA
M. Ashraful Alam, Purdue University, USA
Marek Perkowski, Portland State University, Oregon, USA
Mohammad S. Alam, Texas A&M University, USA
Mustafizur Rahman, Intel Co., USA
M. Zahid Hasan, Princeton University, USA
N. Kishi, University of Electro-communication, Japan
Ramjee Prasad, Aarhus University, Denmark
Ranjan Gangopadhy, LNMIIT, India
Vijayan Asari, University of Dayton, USA
Saifur Rahman, Virginia Tech, USA
Syed Ishtiaque Ahmed, University of Toronto, Canada
Taher Saif, Stanford University, USA
Kanad Ray, Amity University, India

Organizing Committee Chair

Md. Sazzad Hossain, Member, UGC

Organizing Committee

Md. Sarwar Morshed, AUST
M. Mahfuzul Islam, BUET
A. K. M. Zakir Hossain, BAU
Mamunur Rashid, DU
Md. Shah Azam, RU
Mohammad Tofayal Ahmed, JUST

Md. Roshidul Hasan, BSMRAU
Shamim Al Mamun, JU
Ferdous Zaman, UGC
Sultan Mahmud Bhuiyan, UGC
A. K. M. Shamsul Arefin, UGC
Md. Kamal Hossain, UGC
Md. Fakhrul Islam, UGC
Md. Omar Faruque, UGC
Durga Rani Sarker, UGC
Md. Shah Alam, UGC
Muhammad Jaminur Rahman, UGC
M. Maksudur Rahman Bhuiyan, UGC
Md. Rezaul Karim Hawlader, UGC
Nahid Sultana, UGC
Bishnu Mallick, UGC
Probir Chandra Das, UGC

Technical Program Committee Chair

Satya Prasad Majumder, VC, BUET

Technical Program Committee

Md. Shahadat Hossain, CU
M. Shamim Kaiser, JU
Md. Abdur Razzaque, DU
Mufti Mahmud, NTU, UK
Nazrul Islam, MIST
Md. Atiqur Rahman Ahad, DU
Nazmul Siddique, Ulster University, IR
Mohammad Shorif Uddin, JU
Celia Sahanaz, BUET

Track Committee

Abdullah-Al Nahid, Khulna University, BD
Ahmed Wasif Reza, EWU, BD
Saadia Alam, University of Hyogo, Japan
Fazlul Hasan Siddiqui, DUET, BD

M. M. A. Hasehm, KUET, BD
Md. Obaidur Rahman, DUET, BD
Md. Abu Layek, Jagannath University, BD
Md. Saiful Islam, BUET, BD
Muhammad Golam Kibria, ULAB, BD
Farhana Sarker, ULAB, BD
Fida Hasan, RMIT University, BD
Hussain Nyeem, MIST, BD
M. Shahidur Rahman, SUST, BD
M. M. A. Hashem, KUET, BD
Md. Atiqur Rahman Ahad, DU, BD
Md. Golam Mostafa, MIST, BD
Md. Hossam-E Haider, MIST, BD
Md. Kafiul Islam, IUB, BD
Md. Sazzadur Rahman, IIT, JU, BD
Md. Abdul Alim, KU, BD
Md. Abdur Rahman, AIUB
Md. Farhad Hossain, BUET
Md. Golam Rashed, University of Rajshahi
Md. Saiful Islam, BUET
Mohammad Abu Yousuf, JU
Mohammad Nurul Huda, UIU
Mohammad Shahadat Hossain, University of Chittagong
Mohammad Shamsul Arefin, CUET
Mohammed Hoque, CUET
Mohammed Eunus Ali, BUET
Mufti Mahmud, NTU, UK
Muhammad Nazrul Islam, MIST
Nabeel Mohammed, North South University
Nasima Begum, University of Asia Pacific
A. K. M. Fazlul Haque, DIU
Shaikh Anowarul Fattah, BUET
S. M. Abdur Razzak, RUET
Rameswar Debnath, Khulna University
Rashed Mustafa, University of Chittagong
Risala Tahsin Khan, JU
S. M. Riazul Islam, Sejong University, Korea
Sejutib Rahman, University of Dhaka
M. Shamim Kaiser, JU
Shamim Al Mamun, JU
Syed Galib, JUST

Technical Review Committee

A. F. M. Shahab Uddin, KyungHee University
A. N. M. Shahebul Hasan, AIUB, BD
A. S. M. Touhidul Hasan, UAP PacificBD
A. B. M. Alim Al Islam, BUET, BD
A. B. M. Aowlad Hossain, KUET, BD
A. H. M. Hoque, University of Chittagong, BD
A. K. M. Fazlul Haque, DIU, BD
A. S. M. Sanwar Hosen, JNU, South Korea
Abdullah-Al Nahid, Khulna University, BD
Abir Ahmed, AIUB, BD
ABM Rashid, BUET, BD
Abu Hena Shatil, AIUB, BD
Abu Nowshed Chy, CU, BD
Abu Shamim Mohammad Arif, Khulna University, BD
Abul Kalam Azad, ULAB, BD
Afsana Begum, DIU, BD
Ahsanur Rahman, NSU, BD
AKM Baki, AUST, BD
Aminul Islam, MIST, BD
Aminul Islam, Khulna University, BD
Anas Abdullah Alvi, AIUB, BD
Anindya Iqbal, BUET, BD
Anupam Bairagi, Khulna University, BD
Arif Ahmad, SUST, BD
Ashikur Rahman, BUET, BD
Asif Mahfuz, AIUB, BD
Atanu Shome, Khulna University, BD
Alessandra Pedrocchi, Politechnico di Milano, Italy
Alessandro Gozzi, IIT, Italy
Amir Hussain, ENU, UK
Anirban-Bandyopadhyay, NIMS, Japan
Antesar Shabut, LTU, UK
Atik Mahabub, Concordia University, Canada
Bishwajit Banik Pathik, AIUB, BD
Bobby Barua, AUST, BD
Belayat Hossain, LU, UK
Bernd Kuhn, OIST, Japan
Bo Song, USQ, Australia
Chinmay Bepery, PSTU, BD
Cosimo Ieracitano, UNIRC, Italy
Cris Calude, University of Auckland, New Zealand
Daniel K. Wójcik, Nencki Institute of Experimental Biology, Poland

Daniel Marcus, UW, USA
David Brown, NTU, UK
Davide Zoccolan, SISSA, Italy
Derong Liu, UIC, USA
Dipankar Das, University of Rajshahi, BD
Effat Jahan, AIUB, BD
Egidio D'Angelo, University of Pavia, Italy
Eleni Vasilaki, The University of Sheffield, UK
Francesco Carlo Morabito, UNIRC, Italy
Francesco Papaleo, University of Padova and IIT, Italy
Fahim Alam, University of Chittagong, BD
Farhadur Arifin, AIUB, BD
Farhana Sarker, ULAB, BD
Farhana Tazmim Pinki, Khulna University, BD
Fazlul Hasan Siddiqui, DUET, BD
Ferdous Shaun, AIUB, BD
Feroz Ahmed, IUB, BD
Fida Hasan, RMIT University
Gabriella Panuccio, IIT, Italy
Gaute Einevoll, NMBU, Norway
Giacomo Indiveri, UZH, Switzerland
Giancarlo Ferregno, Politechnico di Milano, Italy
Giorgio A. Ascoli, GMU, USA
Golam Dastoger Bashar, Boise State University, USA
Guenther Zeck, NMI, Germany
Gustavo Deco, UPF, Spain
Hussain Nyeem, MIST, BD
Imtiaz Mahmud, Kyungpook National University
Iqbal Sarker, CUET, BD
Iqbal Aziz Khan, RU, BD
Jabed Faysal, KU, BD
Joyprokash Chakrabartty, CUET, BD
Imtiaz Mahmud, KNU, Korea
Joarder Kamruzzaman, Federation University, Australia
Jonathan Mappelli, University of Modena, Italy
Kafi Rahman, Truman State University, BD
Kamruddin Nur, AIUB, BD
Kawshik Shikder, AIUB, BD
Kazi Firoz Ahmed, AIUB, BD
Kazi Masudul Alam, KU, BD
Kazi Md. Rokibul Alam, KUET, BD
Kazi Sakib, University of Dhaka, BD
Khaleda Ali, Dhaka University, BD
Khandaker Tabin Hasan, AIUB, BD
KC Santosh, USD, USA

Lu Cao, Saitama University, Japan
Luca Benini, ETH, Switzerland
Luca Berdondini, IIT, Italy
Luciano Gamberini, University of Padova, Italy
Lafifa Jamal, University of Dhaka, BD
Lasker Ershad Ali, Khulna University, BD
Lutfa Akter, BUET, BD
M. Akhtaruzzaman, MIST, BD
M. Kabiruzzaman, AIUB, BD
M. M. Manjurul Islam, AIUB, BD
M. Raihan, North Western University, Khulna, BD
M. A. H. Akhand, KUET, BD
M. A. H. AKHAND, KUET, BD
M. Shahidur Rahman, SUST, BD
M. M. A. Hashem, KUET, BD
Mahfuzul Hoq Chowdhury, CUET, BD
Manan Noor, Jahangirnagar University, BD
Manishankar Mondal, Khulna University, BD
Manowarul Islam, JNU, BD
Marium-E -Jannat, SUST, BD
Md. Abdul Masud, Patuakhali Science and Tech. University, BD
Md. Aktaruzzaman, Islamic University, BD
Md. Arefin, BUBT, BD
Md. Al Mamun, RUET, BD
Md. Ashif Islam Oni, AIUB, BD
Md. Atiqur Rahman Ahad, University of Dhaka, BD
Md. Awal, Khulna University, BD
Md. Dulal Haque, HSTU, BD
Md. Ezharul Islam, JU, BD
Md. Forhad Rabbi, SUST, BD
MD. FOYSAL HAQUE, Dong-A University
MD. HALIM, MIST, BD
Md. Islam, University of Rajshahi, BD
Md. Jaber Nahian, BUP, BD
Md. Mahadi Hasan Nahid, SUST, BD
Md. Mehedi Hasan, University of Dhaka, BD
Md. Mostofa Kamal Rasel, East West University, BD
Md. Nasim Adnan, JUST, BD
MD. Nazmul Hasan, University of Ulsan
Md. Obaidur Rahman, DUET, BD
Md. Rahman, UITS, BD
Md. Rajib Hossain, CUET, BD
Md. Rakibul Islam, University of Wisconsin—Eau Claire
Md. Rifat Ahmmd Rashid, ULAB, BD
Md. Samsuzzaman, PSTU, BD

Md. Sazzadur Rahman, JU, BD
Md. Shahidul Islam, JU, BD
Md. Tanzil Shahria, University of Wisconsin-Milwaukee
Md. Zahidul Islam, Islamic University
Md. Zulfiker Mahmud, Jagannath University, BD
Md. Islam, University of Dhaka, BD
Md. Abdul Alim, Khulna University, BD
Md. Abu Yousuf, JU, BD
Md. Abu Layek, Jagannath University, BD
Md. Ahsan Habib, MBSTU, BD
Md. Alamgir Hossain, Islamic University, BD
Md. Anwarul Abedin, DUET, BD
Md. Arafat Hossain, KUET, BD
Md. Ashraful Islam, Rajshahi University, BD
Md. Delowar Hossain, HSTU, BD
Md. Eamin Rahmn, SUST, BD
Md. Farhad Hossain, BUET, BD
Md. Faruque Hossain, KUET, BD
Md. Foisal Hossain, KUET, BD
Md. Forkan Uddin, BUET, BD
Md. Golam Rashed, RU,BD
Md. Habibur Rahman, BDU
Md. Hasanul Kabir, IUT, BD
Md. Hossain, BUBT, BD
Md. Humayun Kabir, AIUB
Md. Humayun Kabir, JU, BD
Md. Iqbal Bahar Chowdhury, UIU, BD
Md. Jahedul Islam, KUET, BD
Md. Kafiul Islam, Independent University, Bangladesh
Md. Kamrul Hassan, AIUB
Md. Liakot Ali, BUET, BD
Md. Mahabub Hossain, HSTU, BD
Md. Mahbub Hasan, KUET, BD
Md. Mahbubur Rahman, MIST, BD
Md. Mahbubur Rahman Rahman, MIST, BD
Md. Meganur Rhaman, AUST, BD
Md. Mizanur Rahman, KU, BD
Md. Motaharul Islam, UIU, BD
Md. Motiur Rahman, MBSTU, BD
Md. Musfique Anwar, JU, BD
Md. Nahiyan Uddin, UAP, BD
Md. Nawab Yousuf Ali, EWU, BD
Md. Rabiul Islam, AIUB, BD
Md. Rafiqul Islam, KUET, BD
Md. Ragib Amin Nihal, DU, BD

Md. Rifat Hazari, AIUB, BD
Md. Saddam Hossain Mukta, UIU, BD
Md. Saiful Islam, BUET, BD
Md. Salah Uddin Yusuf, KUET, BD
Md. Samsuzzaman Sobuz, PSTU, BD
Md. Sanaullah Chowdhury, CU, BD
Md. Saniat Rahman Zishan, AIUB, BD
Md. Shahariar Parvez, AIUB, BD
Md. Shahjahan, KUET, BD
Md. Shahriar Mahbub, AUST, BD
Md. Shamim Ahsan, KU, BD
Md. Sheikh Sadi, KUET
Md. Sohel Mahmud Sher, KU, BD
Md. Tanvir Hasan, JUST, BD
Md. Tariq Hasan, KU, BD
Mehedi Hasan, AIUB, BD
Milon Biswas , BUBT
Mir Md. Jahangir Kabir, RUET, BD
M. M. Shahriar Maswood, KUET, BD
Mohammad Shidujaman, AIUB
M. Abdullah Al Mumin, SUST, BD
Mohammad Abu Yousuf, JU, BD
Mohammad Farhad Bulbul, JUST, BD
Mohammad Khairul Islam, CU, BD
Mohammad Mamun Elahi, UIU, BD
Mohammad Nasir Uddin, AIUB
Mohammad Nurul Huda, UIU, BD
M. Rashedur Rahman, NSU, BD
M. Shamsul Arefin, CUET, BD
M. Moshiul Hoque, CUET, BD
Mohidul Islam, KU, BD
Molla Hussein, UAP, BD
Monir Hossen, KUET, BD
Monir Morshed, MBSTU, BD
Mosabber Uddin Ahmed, DU, BD
Mostafa Zaman Chowdhury, KUET, BD
Mostofa Nasir, MBSTU, BD
Muammer Arif, MIST, BD
Mufti Mahmud, NTU, UK
Muhammad Abdullah Adnan, BUET, BD
Muhammad Asad Rahman, CUET, BD
Muhammad Golam Kibria, ULAB, BD
Muhammad Hasan, ULAB, BD
Muhammad Nazrul Islam, MIST, BD
Muhammad Sajjadur Rahim, RU, BD

Muhammad Shahin Uddin, MBSTU, BD
Muhammed Patwary, IIUC, BD
Mustafa Chowdhury, Independent University, Bangladesh
Nabeel Mohammed, North South University
M. Ali Akber Dewan, Athabasca University, Canada
Manjunath Aradhya, JSS STU, India
Manohar Das, Oakland University, USA
Marco Mongillo, University of Padova, Italy
Martin McGinnity, Ulster University, UK
Marzia Hoque-Tania, Oxford University, UK
Mathew Diamond, SISSA, Italy
Nadia Nowshin, AIUB, BD
Nafees Mansoor, ULAB, BD
Nafisa Noor, NSU, BD
Nafiz Ahmed Chisty, AIUB, BD
Nafiz Imtiaz Khan, MIST, BD
Naruttam Kumar Roy, KUET, BD
Nasrin Akhter, KUET, BD
Nasrin Sultana, HSTU, BD
Nihad Karim Chowdhury, CU, BD
Niloy Sikder, KU, BD
Nirjhor Tahmidur Rouf, AIUB, BD
Norhidayah Azman, Universiti Sains Islam Malaysia
Nowshin Alam, AIUB, BD
Nusrat Sharmin, MIST, BD
Nelishia Pillay, UP, South Africa
Nikola Kasabov, AUT, New Zealand
Nilanjan Dey, JIS University, India
Ning Zhong, Maebashi Institute of Technology, Japan
Noushath Shaffi, CAS, Oman
Ofer Yizhar, Weizmann Institute of Science, Israel
Omprakash Kaiwartya, NTU, UK
Omar Sharif, CUET, BD
Pallab Kumar Choudhury, KUET, BD
Pintu Shill, KUET, BD
Pranab Kumar Dhar, CUET, BD
Paolo Del Giudice, ISS, Italy
Paolo Massobrio, UniGe, Italy
Patrick Ruther, University of Freiburg, Germany
Rafia Nishat Toma, university of ulsan
Raja Hasan, AIUB, BD
Rameswar Debnath, KUy, BD
Rashed Mazumder, JU, BD
Rashed Mustafa, CU, BD
Riasat Khan, NSU, BD

Rifat Shahriyar, BUET, BD
Risala Tahsin Khan, JU, BD
Ruhul Amin, Fordham University
S. M. Imrat Rahman, AIUB
Sajeeb Saha, JnU, BD
Sajjad Waheed, MBSTU, BD
Shahriar Rahman, UIU, BD
Shamim Ahmad, RU, BD
Shamim Al Mamun, JU, BD
Shazzad Hosain, NSU, BD
Sifat Momen, NSU, BD
Silvia Ahmed, NSU, BD
Sohrab Nizami, University of Queensland
Subrata Aditya , Dhaka University, Bangladesh
Sudipta Kar, Amazon
Sujan Sarker, DU
Syed Galib, JUST, BD
S. M. Riazul Islam, Sejong University, South Korea
Tahseen Asma Meem, AIUB
Tama Fouzder, ULAB, BD
Tanvir Rahman, MBSTU, BD
Tapotosh Ghosh, NUB, BD
Taslima Ferdaus, DIU, BD
Tasmiah Anannya, MIST, BD
Tawsin Uddin Ahmed, CU, BD
Tofael Ahmed, CUET, BD
Uzzal Acharjee, Jagannath University, BD
Zinat Tasneem, RUET, BD

Preface

The International Conference on Fourth Industrial Revolution and Beyond, IC4IR 2021, is being organized by the University Grants Commission (UGC) of Bangladesh to be held through both physical and virtual online formats from 10 and 11 December 2021. On the 100th Birth Anniversary of the Father of the Nation of Bangladesh, Bangabandhu Sheikh Mujibur Rahman and 50th anniversary of independence of the country, the UGC is planning to create a premier international forum for bringing together researchers and practitioners from diverse domains to share cutting-edge research results obtained through the application of artificial intelligence, the internet of things, data analytics, and cloud computing to solve problems in the industrial domain. IC4IR 2021 also promotes the sharing and dissemination of innovative and practical development of methodologies and technologies with industrial applications through the exchange and dissemination of best practices. The scope of the conference includes artificial intelligence; robotics and automation; IoT and smart agriculture; data analytics and cloud computing; communication and networks signal and natural language processing. Researchers of home and abroad are requested to submit their research paper that can resolve the local problems, to improve life and the environment utilizing state-of-the-art 4IR techniques, and to help to achieve the sustainable development goals.

The IC4IR 2021 attracted 525 full papers from 16 countries in six tracks. These tracks include—artificial intelligence; robotics and automation; IoT and smart agriculture; data analytics and cloud computing; signal and natural language processing; and communication and networking. The submitted papers underwent a double-blind review process, soliciting expert opinion from at least two experts: at least two independent reviewers, the track co-chair, and the respective track chair. Following rigorous review reports from the reviewers and the track chairs, the technical program committee has selected 59 high-quality full papers from nine countries that were accepted for presentation at the conference. Due to the COVID-19 pandemic, the Organizing Committee decided to host the event in hybrid mode. However, the research community reacted amazingly in this challenging time. More than 2000 people attended the events, which featured keynote speeches from three Nobel laureates and six distinguished professors.

The book series will be insightful and fascinating for those interested in learning about the tool and technique of 4IR that explores the dynamics of exponentially increasing knowledge in core and related fields. We are thankful to the authors who have made a significant contribution to the conference and have developed relevant research and literature in computation and cognitive engineering.

We would like to express our gratitude to the Organizing Committee and the Technical Committee members for their unconditional support, particularly the Chair, the Co-Chair and the Reviewers. Special thanks to the Chairman and Members of UGC of Bangladesh for their thorough support. IC4IR 2021 could not have taken place without the tremendous work of the team and the gracious assistance. We are grateful to Mr. Aninda Bose, Ms. Sharmila Mary Panner Selvam, and other term members of Springer Nature for their continuous support in coordinating this volume publication. We would also like to thank vice chancellor of the public and private universities for continuous support. Last but not the least, we thank all of our contributors and volunteers for their support during this challenging time to make IC4IR 2021 a success.

Dhaka, Bangladesh — Md. Sazzad Hossain
Chittagong, Bangladesh — Md. Shahadat Hossain
Dhaka, Bangladesh — M. Shamim Kaiser
Dhaka, Bangladesh — Satya Prasad Majumder
Jaipur, India — Kanad Ray
October 2021

Contents

Emerging Applications

Healthcare

Editors and Contributors

About the Editors

Dr. Md. Sazzad Hossain is a distinguished professor, a researcher, an academic teaching advisor with extensive experience in ICT sector, and a writer. Currently, he is working as a full-time member of University Grants Commission of Bangladesh. He is also a director of the Board of Directors of Bangladesh Communication Satellite Company Ltd (BCSCL). Dr. Sazzad is a masterly instructor in ICT field with enormous knowledge on subject matter. He joined as a professor as well as the head of the Department of Computer Science and Engineering at University of Liberal Arts Bangladesh in December 2009. He received his Ph.D. in Electrical and Computer Engineering from Portland State University, Oregon, USA, which is one of the top-ranking universities of the world. His doctoral dissertation in quantum computing takes background and ideas from several fields of physics, mathematics, computer science, computer engineering, and biology. He completed his Master's in Electrical and Computer Engineering from the same university in the USA. He received his Bachelor of Science in Electrical System Network Engineering from Moscow Technical University (popularly known as Moscow Power Institute), Moscow, Russia. He also lent his teaching and research experiences in many universities and institutes as a visiting faculty/researcher/guest speaker. He has twenty-six (26) years of experience in information technology. He has served as a consultant to several nationally and internationally renowned ICT companies, government agencies, and organizations. His research interest includes quantum computing, quantum algorithms, quantum entanglement, de-coherence, natural computing, fault-tolerant computing, theoretical computer science, nanotechnology, low-power VLSI design, reversible logic, software engineering, Internet of things (IoT), artificial intelligence, robotics, and human–robot interaction. He has authored more than 50 research articles in peer-reviewed international journals and conference proceedings.

Prof. Dr. Md. Shahadat Hossain is serving as the professor of Computer Science and Engineering at the University of Chittagong, Bangladesh. He did both his M.Phil.

and Ph.D. in Computation from the University of Manchester Institute of Science and Technology, UK. He is the awardee of a number of fellowships including Commonwealth, Tyndall, and Erasmus Mundus. He is also the holder of Erasmus Mundus Scholar Visiting Professor in Europe. He successfully completed a number of research projects. Recently, he was awarded a prestigious Swedish Research Council grant for the project entitled "A Belief Rule Based DSS to Assess Flood Risks using Wireless Sensor Networks". His interest of research is in the areas of artificial intelligence, expert systems, pervasive computing, health informatics, deep learning, Internet of things (IoT), big data, e-government, hyper-optimization, and philosophy of computing. He has published 170+ scholarly articles in the reputed international journals and conferences. Many of his papers obtained the best paper award. Professor Hossain is usually invited to deliver talks in various countries of the world. He is the author of a number of books. He is the senior member of IEEE, USA, and the member of British Computer Society.

Dr. M. Shamim Kaiser is currently working as a professor at the Institute of Information Technology of Jahangirnagar University, Savar, Dhaka-1342, Bangladesh. He received his Bachelor's and Master's degrees in Applied Physics Electronics and Communication Engineering from the University of Dhaka, Bangladesh, in 2002 and 2004 respectively, and the Ph.D. degree in Telecommunication Engineering from the Asian Institute of Technology (AIT), Pathum Thani, Thailand, in 2010. He worked as a postdoc fellow in the Big Data and Cyber Security Lab of Anglia Ruskin University, UK, from 2017–2018. He also worked as a special research student at the Wireless Signal Processing and Networking Lab (Adachi Lab) of Tohoku University, Japan, in 2008. His current research interests include data analytics, machine learning, wireless network & signal processing, cognitive radio networks, big IoT data, health care, neuroinformatics, and cyber security. He has authored more than 170 papers in different peer-reviewed journals and conferences. He is an academic editor of Plos One Journal, an associate editor of the IEEE Access and Cognitive Computation Journal,and the guest editor of Brain Informatics Journal, IJACI (IGI Global), Electronics MDPI, Frontiers in Neuroinformatics, and Cognitive Computation Journal. Dr. Kaiser is a life member of Bangladesh Electronic Society, Bangladesh Physical Society, and NOAMI. He is also a senior member of IEEE, USA, and IEICE, Japan, and an active volunteer of the IEEE Bangladesh Section. He is the founding chapter chair of the IEEE Bangladesh Section Computer Society Chapter.

Dr. Satya Prasad Majumder joined the position of the vice chancellor of Bangladesh University of Engineering and Technology (BUET) on June 2020. He is a professor of the Department of Electrical and Electronic Engineering (EEE), BUET. Since joining BUET in 1981, he served in various positions in the department, faculty, and university levels, including the head of the EEE department at BUET from June 2006–May 2008. He was the dean of the Faculty of EEE, BUET, during June 2010 to June 2012. He served as a visiting professor in the Faculty of Engineering (FoE) of Multimedia University (MMU), Cyberjaya, Selangor, Malaysia, during January

1999 to July 2002. He was a visiting research professor in the department of Electrical Engineering of Gunma University (Japan) during December 2001 to January 2002. He was a visiting researcher in the department of Information Engineering of University of Parma (Italy) during 1996 and 1997. He worked as a consultant to the South Asian Association for Regional Co-operations (SAARC) & Dhaka Chamber of Commerce (DCC). He received the B.Sc. (1st class, 4th) and M.Sc. in Electrical & Electronic Engineering from BUET, Bangladesh, in 1981 and 1985, respectively. In 1993, he obtained his Ph.D. degree in Optical Fiber Communication Engineering from the Indian Institute of Technology (IIT), Kharagpur, India. His research interests include opto-electronics & photonics, optical fiber communication systems, DWDM optical networks, soliton propagation, broadband wireless communications, and mobile and infrared networks. He supervised more than 50 M.Sc. theses and 6 Ph.D. dissertations and published more than 200 research papers in internationally reputed referred journals and conference proceedings.

Kanad Ray (Senior Member, IEEE) received the M.Sc. degree in physics from Calcutta University and the Ph.D. degree in physics from Jadavpur University, West Bengal, India. He has been a professor of Physics and Electronics and Communication and is presently working as the head of the Department of Physics, Amity School of Applied Sciences, Amity University Rajasthan (AUR), Jaipur, India. His current research areas of interest include cognition, communication, electromagnetic field theory, antenna and wave propagation, microwave, computational biology, and applied physics. He has been serving as an editor for various Springer book series. He was an associate editor of the Journal of Integrative Neuroscience (The Netherlands: IOS Press). He has been a visiting Professor to UTM & UTeM, Malaysia, and a visiting scientist to NIMS, Japan. He has established MOU with UTeM, Malaysia; NIMS, Japan; and University of Montreal, Canada. He has visited several countries such as the Netherlands, Turkey, China, Czechoslovakia, Russia, Portugal, Finland, Belgium, South Africa, Japan, Singapore, Thailand, and Malaysia for various academic missions. He has organized various conferences such as SoCPROS, SoCTA, ICOEVCI, and TCCE as the general chair and a steering committee member.

Contributors

Abrar Ridwan Department of Electrical and Electronic Engineering (EEE), Bangladesh University of Engineering and Technology (BUET), Dhaka, Bangladesh

Adnan Palash Md. Istakiak Department of Information and Communication Technology, Bangladesh University of Professionals, Dhaka, Bangladesh

Adnan Nasim Jashore University of Science and Technology, Jashore, Bangladesh

Afrin Faria Pioneer Alpha, Dhaka, Bangladesh

Afrin Rumana Department of Computer Science and Engineering, Khulna University of Engineering and Technology, Khulna, Bangladesh

Ahmed Kazi Sadman InNeed Intelligent Cloud, Dallas, TX, United States

Ahmed Khandakar Razoan Associate System Administrator Therap BD Limited, Dhaka, Bangladesh

Ahmed Md. Rasel American International University-Bangladesh, Dhaka, Bangladesh

Ahmed Md. Rayhan United International University, United City, Dhaka, Bangladesh

Ahmed Md. Sojib United International University, United City, Dhaka, Bangladesh

Ahmed R. B. University Malaysia Perlis, Perlis, Malaysia

Ahmed Sabbir Institute of Information Technology, Jahangirnagar University, Savar, Dhaka, Bangladesh

Akash Ahsan Habib Department of Electrical and Electronic Engineering (EEE), Bangladesh University of Engineering and Technology (BUET), Dhaka, Bangladesh

Akhtar Md. Nasim Dhaka University of Engineering and Technology, Gazipur, Bangladesh;
Chandpur Science and Technology University, Chandpur, Bangladesh

Akter Nasrin BGC Trust University Bangladesh, Chandanaish, Bangladesh

Al Asad Nafiz Department of Information and Communication Technology, Bangladesh University of Professionals, Dhaka, Bangladesh

Al Bhuyain Naim Ibna Khadem Department of Computer Science and Engineering, Military Institute of Science and Technology (MIST), Dhaka, Bangladesh

Al Nahian Atik Ishrak Department of Mechanical Engineering, Bangladesh University of Engineering and Technology, Dhaka, Bangladesh

Al Razi Alim University of Asia Pacific, Dhaka, Bangladesh

Alam Dr. Md. Golam Rabiul Brac University, Dhaka, Bangladesh

Alam Md. Golam Rabiul BRAC University, Dhaka, Bangladesh

Alam Md. Samiul American International University-Bangladesh (AIUB), Dhaka, Bangladesh

Alam Shaikh Md. Mominul Professor, Department of Textile Machinery Design and Maintenance, Bangladesh University of Textiles, Dhaka, Bangladesh

Alfaz Nazia Department of Computer Science, American International University-Bangladesh(AIUB), Dhaka, Bangladesh

Ali Md. Liakot Institute of Information and Communication Technology (IICT), Bangladesh University of Engineering Technology (BUET), Dhaka, Bangladesh

Alvy Abu Saleh Muhammod American International University-Bangladesh, Dhaka, Bangladesh

Andersson Karl Lulea University of Technology, Skelleftea, Sweden

Ashrafi Shamim InNeed Intelligent Cloud, Dallas, TX, United States

Awal Md. Abdul Electronics and Communication Engineering Discipline, Khulna University, Khulna, Bangladesh

Awosaf Zahin American International University-Bangladesh, Dhaka, Bangladesh

Ayub Fahad Bin Department of Electrical and Electronic Engineering, Leading University, Sylhet, Bangladesh

Bakul Mehedi H. American International University-Bangladesh, Dhaka, Bangladesh

Banna Md. Hasan Al Bangladesh University of Professionals, Dhaka, Bangladesh

Baromon Rathindranath Dhaka University of Engineering and Technology, Gazipur, Bangladesh

Barros Alistair School of Information Systems, Queensland University of Technology, Brisbane, QLD, Australia

Bhowmik Moumita Department of Computer Science and Engineering, Military Institute of Science and Technology (MIST), Dhaka, Bangladesh

Bhowmik Pankaj Department of Computer Science and Engineering, Hajee Mohammad Danesh Science and Technology University (HSTU), Dinajpur, Bangladesh

Bhowmik Pulak Chandra Department of Computer Science and Engineering, Stamford University Bangladesh, Dhaka, Bangladesh

Bhuiyan Md. Harunur Rashid Bangladesh University of Textiles, Dhaka, Bangladesh

Bhuiyan Miyad Department of Computer Science and Engineering, Ahsanullah University of Science and Technology, Dhaka, Bangladesh

Bijoy Md. Hasan Imam Department of Computer Science and Engineering, Daffodil International University, Dhaka, Bangladesh

Buyya Rajkumar School of Computing and Information Systems, University of Melbourne, Parkville, VIC, Australia

Chakraborty Chinmay Department of Electronics and Communication Engineering, Birla Institute of Technology, Mesra, Jharkhand, India

Chang Vincent Brac University, Dhaka, Bangladesh

Chowdhury Arshad M. Brac University, Dhaka, Bangladesh

Chowdhury Ashraf Alam Khan Northern University Bangladesh, Dhaka, Bangladesh

Chowdhury Md. Solaiman Rajshahi University, Rajshahi, Bangladesh

Chowdhury Moinul H. CMED Health Limited, Dhaka, Bangladesh

Chowdhury Shadman Islam InNeed Intelligent Cloud, Dallas, TX, United States

Das Avijit Department of Computer Science and Engineering, Shahjalal University of Science and Technology, Sylhet, Bangladesh

Das Dipankar Department of Information and Communication Engineering, University of Rajshahi, Rajshahi, Bangladesh

Dastider Ankan Ghosh Department of Electrical and Electronic Engineering (EEE), Bangladesh University of Engineering and Technology (BUET), Dhaka, Bangladesh

Dey Samrat Kumar School of Science and Technology, Bangladesh Open University, Gazipur, Bangladesh

Doumanidis Charalabos C. Brac University, Dhaka, Bangladesh

Dowla Nozib Ud American International University-Bangladesh (AIUB), Dhaka, Bangladesh

Dowland Dave Brac University, Dhaka, Bangladesh

Dumitrescu Raluca TU Berlin, Berlin, Germany

Faisal Sekh Department of Electrical and Electronic Engineering, Leading University, Sylhet, Bangladesh

Faruque Mithila CMED Health Limited, Dhaka, Bangladesh;
Department of NCD, Bangladesh University of Health Sciences, Dhaka, Bangladesh

Fidge Colin School of Computer Science, Queensland University of Technology, Brisbane, QLD, Australia

Ghosh Partho Department of Electrical and Electronic Engineering (EEE), Bangladesh University of Engineering and Technology (BUET), Dhaka, Bangladesh

Ghosh Tapotosh Northern University Bangladesh, Dhaka, Bangladesh

Goswami Pritom Bangladesh University of Textiles, Dhaka, Bangladesh

Groh Sebastian Brac University, Dhaka, Bangladesh

Gupta Akash Kumar Department of Electronics and Communication Engineering, Birla Institute of Technology, Mesra, Jharkhand, India

Gupta Atanu Sen Department of Mechanical Engineering, Bangladesh University of Engineering and Technology, Dhaka, Bangladesh

Gupta Bharat Department of Electronics and Communication Engineering, National Institute of Technology Patna, Patna, Bihar, India

Hai Md. Abdul Department of Computer Science and Engineering, Brac University, Dhaka, Bangladesh

Hanif Md. Abu American International University-Bangladesh, Dhaka, Bangladesh

Haque Summit Department of Computer Science and Engineering, Shahjalal University of Science and Technology, Sylhet, Bangladesh

Hasan Mahmudul Military Institute of Science and Technology (MIST), Dhaka, Bangladesh

Hasan Md. Kamrul Systems and Software Lab (SSL), Department of Computer Science and Engineering (CSE), Islamic University of Technology (IUT), Dhaka, Bangladesh

Hasan Md. Manzurul American International University-Bangladesh (AIUB), Dhaka, Bangladesh

Hasan Taufiq mHealth Research Group, Department of Biomedical Engineering (BME), Bangladesh University of Engineering and Technology (BUET), Dhaka, Bangladesh

hasan Md. Mynul United International University, United City, Dhaka, Bangladesh

Hasnat Abul Civil Engineering, Faculty of Engineering, American International University-Bangladesh(AIUB), Dhaka, Bangladesh

Hassan Onim Md. Saif Military Institute of Science and Technology (MIST), Dhaka, Bangladesh

Himel Mahammudul Hassan United International University, United City, Dhaka, Bangladesh

Hoassain Md. Sazzad University of Liberal Arts Bangladesh, Dhaka, Bangladesh

Hoque Akif Md. Akiful Military Institute of Science and Technology (MIST), Dhaka, Bangladesh

Hossain Chowdhury Akram American International University-Bangladesh, Dhaka, Bangladesh

Hossain Jobair American International University-Bangladesh (AIUB), Dhaka, Bangladesh

Hossain Md. Akbar Bangladesh University of Business and Technology (BUBT), Dhaka, Bangladesh

Hossain Md. Muzakker American International University-Bangladesh, Dhaka, Bangladesh

Hossain Md. Sakir American International University-Bangladesh, Dhaka, Bangladesh

Hossain Md. Sazzad University of Liberal Arts Bangladesh, Dhaka, Bangladesh

Hossain Md. Shakhawat American International University-Bangladesh, Dhaka, Bangladesh

Hossain Mohammad Shahadat University of Chittagong, Chittagong, Bangladesh

Hossain Shahadat City University, Dhaka, Bangladesh

Hossen Imran Department of Information and Communication Engineering, University of Rajshahi, Rajshahi, Bangladesh

Hridhee Rubaiyat Alim CMED Health Limited, Dhaka, Bangladesh

Ishmam Abtahi Military Institute of Science and Technology (MIST), Dhaka, Bangladesh

Islam A. K. M. Mohaiminul Department of Information and Communication Technology, Bangladesh University of Professionals, Dhaka, Bangladesh

Islam Litan Dhaka University of Engineering and Technology, Gazipur, Bangladesh

Islam Md. Imranul InNeed Intelligent Cloud, Dallas, TX, United States

Islam Md. Mynul Bangladesh University of Textiles, Dhaka, Bangladesh

Islam Md. Rafiqul Bangladesh University of Textiles, Dhaka, Bangladesh

Islam Motaharul Department of Computer Science and Engineering, United International University, Dhaka, Bangladesh

Islam Muhammad Nazrul Department of Computer Science and Engineering, Military Institute of Science and Technology (MIST), Dhaka, Bangladesh

Islam S. M. Yeaminul Robi Axiata Limited, Dhaka, Bangladesh

Islam Salekul United International University, United City, Dhaka, Bangladesh

Islam Sihabul Institute of Information and Communication Technology (IICT), Bangladesh University of Engineering Technology (BUET), Dhaka, Bangladesh

Islam Tanvir CMED Health Limited, Dhaka, Bangladesh

Islam Tohidul Department of Information and Communication Engineering, University of Rajshahi, Rajshahi, Bangladesh

Istiak Md. Abrar Department of Electrical and Electronic Engineering (EEE), Bangladesh University of Engineering and Technology (BUET), Dhaka, Bangladesh

Jackson Majwega Brac University, Dhaka, Bangladesh

Jahan Nusrat International Islamic University Chittagong, Chittagong, Bangladesh;
Department of Computer Science and Engineering, International Islamic University Chittagong (IIUC), Chattogram, Bangladesh;
Department of Computer Science and Engineering, Military Institute of Science and Technology, Dhaka, Bangladesh

Jahan Sobhana Information and Communication Technology, Bangladesh University of Professionals, Dhaka, Bangladesh

Junjun Jubair Ahmed BGC Trust University Bangladesh, Chandanaish, Bangladesh

Kaiser M. Shamim Institute of Information Technology, Jahangirnagar University, Savar, Dhaka, Bangladesh

Kalam Rakiba Benta Department of Mechanical Engineering, Bangladesh University of Engineering and Technology, Dhaka, Bangladesh

Kalpo Kallul R. American International University-Bangladesh, Dhaka, Bangladesh

Khaleque Muntasir InNeed Intelligent Cloud, Dallas, TX, United States

Khaliluzzaman Md. Department of Computer Science and Engineering, International Islamic University Chittagong (IIUC), Chattogram, Bangladesh

Khan Alvi Md. Ragib Nihal Department of Architecture, American International University-Bangladesh(AIUB), Dhaka, Bangladesh

Khan Nafiz Imtiaz Department of Computer Science and Engineering, Military Institute of Science and Technology (MIST), Dhaka, Bangladesh

Leon Mazharul Islam CMED Health Limited, Dhaka, Bangladesh

Mahfuza Rezwana Department of Computer Science and Engineering, Brac University, Dhaka, Bangladesh

Mahmud Hasan Systems and Software Lab (SSL), Department of Computer Science and Engineering (CSE), Islamic University of Technology (IUT), Dhaka, Bangladesh

Mahmud Mufti Nottingham Trent University, Nottingham, UK

Mahmud Pranto Md. Appel Department of Information and Communication Technology, Bangladesh University of Professionals, Dhaka, Bangladesh

Mahzabin Mayeesha Department of Information and Communication Technology, Bangladesh University of Professionals, Dhaka, Bangladesh

Mallik Raisa Pioneer Alpha, Dhaka, Bangladesh

Mamun Khondaker A. CMED Health Limited, Dhaka, Bangladesh;
AIMS Lab, Department of CSE, United International University, Dhaka, Bangladesh

Mazumder Faria Tashnim Bangladesh University of Textiles, Dhaka, Bangladesh

Meghla Tamara Islam Faculty of Information Technology and Communication Sciences, Tampere University, Tampere, Finland

Mia Mohammad Badruddozza School of Business and Economics, United International University, Dhaka, Bangladesh

Mim Ahsan Habib United International University, United City, Dhaka, Bangladesh

Mim Kanis R. American International University-Bangladesh, Dhaka, Bangladesh

Mondal Proloy Kumar Electronics and Communication Engineering Discipline, Khulna University, Khulna, Bangladesh

Mondol Anupom Assistant Professor, Department of Textile Machinery Design and Maintenance, Bangladesh University of Textiles, Dhaka, Bangladesh

Moon Mohasina Jannat Bangladesh University of Business and Technology (BUBT), Dhaka, Bangladesh

Moon Nazmun Nessa Daffodil International University, Dhaka, Bangladesh

Moral Iqbal Hossain Northern University of Business and Technology Khulna, Khulna, Bangladesh

Morol Md. Kishor American International University-Bangladesh, Dhaka, Bangladesh

Mostafa Javed University of North Carolina at Chapel Hill, Chapel Hill, North Carolina, United States

Mozumder Samsil Arefin East Delta University, Chattogram, Bangladesh

Mukta Md. Saddam Hossain United International University, United City, Dhaka, Bangladesh

Muntasir Fahim American International University-Bangladesh (AIUB), Dhaka, Bangladesh

Murad Hasan University of Asia Pacific, Dhaka, Bangladesh

Nahar Lutfun International Islamic University Chittagong, Chittagong, Bangladesh

Nahar Nazmun BGC Trust University Bangladesh, Chandanaish, Bangladesh

Nahian Md. Jaber Al Bangladesh University of Professionals, Dhaka, Bangladesh

Nahid Abdullah-Al Electronics and Communication Engineering Discipline, Khulna University, Khulna, Bangladesh

Nasim Md. Abdullah Al Pioneer Alpha, Dhaka, Bangladesh

Nayeen Mahi Md. Julkar Institute of Information Technology, Jahangirnagar University, Savar, Bangladesh

Nazmin Tamanna Department of Computer Science and Engineering, Ahsanullah University of Science and Technology, Dhaka, Bangladesh

Niam Arefin Department of Computer Science and Engineering, Shahjalal University of Science and Technology, Sylhet, Bangladesh

Niloy Fardin Ahmed American International University-Bangladesh (AIUB), Dhaka, Bangladesh

Nizam Nusrat Binta Department of Biomedical Engineering, Bangladesh University of Engineering and Technology, Dhaka, Bangladesh

Nyeem Hussain Military Institute of Science and Technology (MIST), Dhaka, Bangladesh

Palash Md. Aminul Haque Pioneer Alpha, Dhaka, Bangladesh

Philipp Daniel MicroEnergy International, Berlin, Germany

Pramanik Anik Department of Computer Science and Engineering, Daffodil International University, Dhaka, Bangladesh

Prottoy Md. Salman Hossan American International University-Bangladesh, Dhaka, Bangladesh

Rafew S. M. Department of Civil Engineering, Khulna University of Engineering and Technology (KUET), Khulna, Bangladesh

Rafizul Islam M. Department of Civil Engineering, Khulna University of Engineering and Technology (KUET), Khulna, Bangladesh

Rahaman Md. Saidur Metropolitan University Sylhet, Sylhet, Bangladesh

Rahi Md. Muzahidul Islam Brac University, Dhaka, Bangladesh

Rahim Rafia Rizwana Department of Mechanical Engineering, Bangladesh University of Engineering and Technology, Dhaka, Bangladesh

Rahman Ashifur Bangladesh University of Business and Technology (BUBT), Dhaka, Bangladesh

Rahman Mafi Department of Business Administration-General, Faculty of Business Studies, Bangladesh University of Professionals (BUP), Dhaka, Bangladesh

Rahman Md. Mashiar Palli Karma-Sahayak Foundation, Dhaka, Bangladesh

Rahman Md. Mahbubur Department of Computer Science and Engineering, Military Institute of Science and Technology, Dhaka, Bangladesh

Rahman Md. Mizanur BRAC University, Dhaka, Bangladesh

Rahman Md. Mosfikur Daffodil International University, Dhaka, Bangladesh

Rahman Md. Sadekur Department of Computer Science and Engineering, Daffodil International University, Dhaka, Bangladesh

Rahman Md. Sazzadur Institute of Information Technology, Jahangirnagar University, Savar, Dhaka, Bangladesh

Rahman Mehrab Mustafy Systems and Software Lab (SSL), Department of Computer Science and Engineering (CSE), Islamic University of Technology (IUT), Dhaka, Bangladesh

Rahman Mostafijur Green University of Bangladesh, Dhaka, Bangladesh

Rahman Noor Nafeur Department of Computer Science and Engineering, Ahsanullah University of Science and Technology, Dhaka, Bangladesh

Rahman Rafsan Department of Computer Science and Engineering, Ahsanullah University of Science and Technology, Dhaka, Bangladesh

Rahman Rashik University of Asia Pacific, Dhaka, Bangladesh

Rahman Yeaminur Department of Computer Science and Engineering, Brac University, Dhaka, Bangladesh

Raisa Jasiya Fairiz Information and Communication Technology, Bangladesh University of Professionals, Dhaka, Bangladesh

Rashed Md. Golam Department of Information and Communication Engineering, University of Rajshahi, Rajshahi, Bangladesh

Rashid Md. Moontasir Department of Electrical and Electronic Engineering, Leading University, Sylhet, Bangladesh

Rashid Nayeeb Department of Electrical and Electronic Engineering (EEE), Bangladesh University of Engineering and Technology (BUET), Dhaka, Bangladesh

Reza Md. Rokonuzzaman Department of Computer Science and Engineering, Military Institute of Science and Technology (MIST), Dhaka, Bangladesh

Riad Sumsuttibriz United International University, United City, Dhaka, Bangladesh

Rimi Sazeda Akter Department of Computer Science and Engineering, International Islamic University Chittagong (IIUC), Chattogram, Bangladesh

Roy Koushik Military Institute of Science and Technology (MIST), Dhaka, Bangladesh

Sadi Muhammad Sheikh Institute of Information and Communication Technology (IICT), Bangladesh University of Engineering Technology (BUET), Dhaka, Bangladesh;
Department of Computer Science and Engineering, Khulna University of Engineering and Technology, Khulna, Bangladesh

Saha Sourav Pioneer Alpha, Dhaka, Bangladesh

Saif Adib Md. Rawnak Department of ICT, Bangladesh University of Professionals, Dhaka, Bangladesh

Saif Md. Zilan Uddin American International University-Bangladesh, Dhaka, Bangladesh

Saifuzzaman Mohd. Daffodil International University, Dhaka, Bangladesh

Sakib Md. Nazmus Systems and Software Lab (SSL), Department of Computer Science and Engineering (CSE), Islamic University of Technology (IUT), Dhaka, Bangladesh

Saleh Abu Jafar Mohammad Bangladesh University of Professionals, Dhaka, Bangladesh

Samiappan Sathishkumar Mississippi State University, Starkville, USA

Sarker Farhana CMED Health Limited, Dhaka, Bangladesh;
Department of CSE, University of Liberal Arts Bangladesh, Dhaka, Bangladesh

Sayom Nazmus Shakib Department of Computer Science, American International University-Bangladesh(AIUB), Dhaka, Bangladesh

Shah Faisal Muhammad Department of Computer Science and Engineering, Ahsanullah University of Science and Technology, Dhaka, Bangladesh

Shahariar G. M. Department of Computer Science and Engineering, Ahsanullah University of Science and Technology, Dhaka, Bangladesh

Shahpar Shahrin Department of Information and Communication Technology, Bangladesh University of Professionals, Dhaka, Bangladesh

Shakil MD. Dhaka University of Engineering and Technology, Gazipur, Bangladesh

Sharifuzzaman Sagar A. S. M. Sejong University, Seoul, South Korea

Shartaj Uddin Rafsan Department of Computer Science and Engineering, Brac University, Dhaka, Bangladesh

Shikdar Tareq Anwar Department of Electrical and Electronic Engineering, Leading University, Sylhet, Bangladesh

Shikder Kawshik American International University-Bangladesh, Dhaka, Bangladesh

Sookhak Mehdi Department of Computer Science, Texas A &M University, Corpus Christ, TX, USA

Sushmit Asif Shahriyar mHealth Research Group, Department of Biomedical Engineering (BME), Bangladesh University of Engineering and Technology (BUET), Dhaka, Bangladesh

Taher Abu University Grants Commission of Bangladesh (UGC), Dhaka, Bangladesh

Tasnim Farhan American International University-Bangladesh, Dhaka, Bangladesh

Tasnim Mayesha International Islamic University Chittagong, Chittagong, Bangladesh

Tasnim Noshin Department of Computer Science and Engineering, Military Institute of Science and Technology, Dhaka, Bangladesh

Tisha Sumona A. American International University-Bangladesh, Dhaka, Bangladesh

Toha Tarik Reza Lecturer, Department of Textile Machinery Design and Maintenance, Bangladesh University of Textiles, Dhaka, Bangladesh

Tonni Surraiya Islam Department of Information and Communication Technology, Bangladesh University of Professionals, Dhaka, Bangladesh

Uddin Md. Jashim Palli Karma-Sahayak Foundation, Dhaka, Bangladesh

Ullah A. K. M. Amanat University of British, Okanagan, BC, Canada

Whaiduzzaman Md School of Information Systems, Queensland University of Technology, Brisbane, QLD, Australia;
Institute of Information Technology, Jahangirnagar University, Savar, Bangladesh

Yesmin Farzana Daffodil International University, Dhaka, Bangladesh

Zaman Akib United International University, United City, Dhaka, Bangladesh

Zaman Sifat Bin Department of Computer Science and Engineering, Military Institute of Science and Technology, Dhaka, Bangladesh

Artificial Intelligence for Detection and Prediction

ColorNet: A Deep Learning-Based Approach to Colorize the Historical Documentaries of Bangladesh

Hasan Murad, Alim Al Razi, and Rashik Rahman

Abstract Image colorization is a process where each pixel in images or videos is given a suitable amount of red, green, and blue (RGB) main color components in order to make it appear more vivid. The process of colorization has significant values while preserving historical documentaries or investigating cosmological photography. However, the colorization of historically significant documentaries requires real witnesses, and a significant amount of human labor is needed to colorize even a single image frame. Recently, deep learning-based approaches have become one of the most reliable techniques to automate challenging computer vision tasks. Various deep learning methods are applied to automate the colorization process. However, we have found that most of the colorization models show poor performance while colorizing Bangladeshi historical documentaries. In our work, we developed ColorNet: a deep learning-based model using an encoder and a pre-trained feature extractor named InceptionV3, also, a fusion layer and a decoder. We have built a new dataset from Bangladeshi historical documentaries. After training our colorization model, it shows a significant result while colorizing Bangladeshi historical documentaries.

Keywords Colorization · Encoder network · Decoder network · Fusion network

H. Murad · A. Al Razi · R. Rahman (✉)
University of Asia Pacific, Dhaka, Bangladesh
e-mail: 17201012@uap-bd.edu

H. Murad
e-mail: hasan.murad.cse@uap-bd.edu

A. Al Razi
e-mail: 17201006@uap-bd.edu

S. Hossain et al. (eds.), *Proceedings of International Conference on Fourth Industrial Revolution and Beyond 2021*, Lecture Notes in Networks and Systems 437,
https://doi.org/10.1007/978-981-19-2445-3_1

1 Introduction

With the advancement of camera technology, historical events have been recorded and preserved using photography and video documentaries. Most of the photography and video documentaries before 1980 are in black and white or grayscale format.

Nowadays, the Colorization of historically valuable documentaries has become a common trend so that new generations can explore those historical events with excitement. However, the colorization of historically significant documentaries requires real witnesses which are not available most of the time. Moreover, a significant amount of human labor is needed to colorize even a single image frame. Therefore, automatic colorization techniques have been more relevant.

Many traditional image processing and machine learning methods for automatic colorization have been found in literature [3–5, 7, 9, 10, 13, 21]. However, most of the previous works have limitations as traditional image processing and machine learning techniques cannot extract meaningful features from unstructured data like image or video.

Recently, deep learning-based approaches have become one of most reliable techniques to automate challenging computer vision tasks [18]. In the literature, different deep learning-based model architectures have been devoted to automate the colorization process [2, 6, 8, 11, 12, 17].

Cheng et al. [6] have designed a deep neural networks (DNNs)-based model architecture for image colorization. However, DNN-based models cannot generate important features from image-like data. To resolve the problem, diverse convolutional neural networks (CNNs)-based models have been put forth for image colorization [1, 8, 14, 17]. CNN network can extract more meaningful features from image data and utilize these features in colorization. Encoder-decoder networks have been devoted to automatically colorizing images in [11]. Baldassarre et al. [2] introduce a fusion layer with a pre-trained feature extractor along with an encoder-decoder network to improve the result of automatic colorization. Recently, generative adversarial networks (GANs) have been deployed to automatically colorize images in [12].

However, we have found that most of the colorization models show poor performance while colorizing Bangladeshi historical documentaries. These models are not generalized colorization models and are mostly based on the dataset used during the training process.

Our research's primary purpose is to develop an essential deep learning-based model that can colorize historically significant documentaries in Bangladesh.

We have designed a deep learning-based model using an encoder, and feature extraction using transfer learning, fusion layer, and decoder. A high-level preview of model architecture is demonstrated in Fig. 1. The encoder network consists of a sequence of stack CNN layers. The encoder network takes grayscale images as input and generates high-level features using the CNN layers. A pre-trained feature extractor called InceptionV3 extracts high-level features from the grayscale image. InceptionV3 was originally trained on a large ImageNet dataset [16]. Finally, the fusion layer concatenates the extracted features from the encoder network and InceptionV3. The decoder network also consists of a sequence of stack CNN layers.

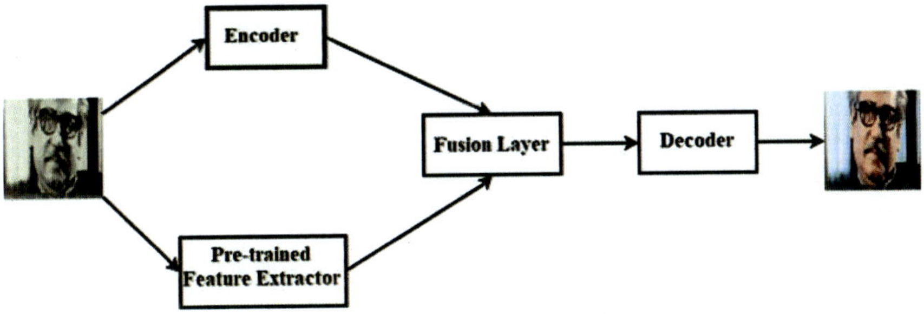

Fig. 1 An overview of our model architecture

The decoder network takes the concatenated features as input and generates the color components of the input image.

We have collected colorized Bangladeshi historical video documentaries from different online sources like Facebook and Youtube. We have created a new dataset by extracting image frames from video documentaries and making pairs of grayscale images and their corresponding color image. Section 4 provides a full description of the dataset creation process.

Finally, we have trained our model using our dataset. After training our colorization model, it shows a significant result while colorizing Bangladeshi historical documentaries.

Major contributions of our research work are given below

- We developed a deep learning-based model using an encoder, a pre-trained feature extractor named Inception V3, a fusion layer, and a decoder.
- We have created a new dataset with pairs of grayscale and color images collected from Bangladeshi historical documentaries.
- Our trained model shows a significant result while colorizing Bangladeshi historical documentaries.

In the following order, we arrange the remaining sections of our paper. In Sect. 2, we discuss previous related research works and their limitations. Section 3 describes the design of our model's architectural structure, as well as its strength. In Sect. 4, we will discuss the specifics of implementation. Section 5 will show the experimental results with analysis. In Sect. 6, we come to a conclusion and offer some ideas for future work.

2 Literature Review

Here, we present the relevant works on automatic colorization in the literature and their limitations. A significant number of research works have been devoted to automatically creating colorized images from grayscale images. We can divide them into three categories—user-guided colorization with image processing approaches, traditional machine learning approaches, and deep learning-based approaches.

2.1 User-Guided Colorization with Image Processing

Levin et al. [9] design a user-guided colorization technique where the user needs to provide color components without considering the object boundaries. This approach introduces efficiency in user-guided colorization by eliminating the use of unnecessary colors.

Yatziv et al. [21] suggested a technique for colorization based on luminance-weighted chrominance mixing and a short intrinsic distance. Qu et al. [13] present a colorization method that spreads color over diverse locations while maintaining pattern and intensity consistency.

Li et al. [10] have an efficient colorization method where information from grayscale photos such as the slopes or gradients, and the edges of the items in the image are obtained and the user scribbles to color the images.

2.2 Traditional Machine Learning-Based Colorization

Charpiat et al. [5], rather than selecting the most likely color, developed a colorization process that was a posterior distribution of each pixel in the images.

Bugeau et al. [3] propose a patch-based colorization technique where a distance-based technique is utilized for predicting color components. However, the resultant image looks unrealistic and desaturated. To enhance the colorization method's performance, Bugeau et al. [4] utilize variation formulations, energy reduction, and patch-based to choose a group of optimal colors. Gupta et al. [7] develop a colorization method by exploiting the replacement example-based technique for coloring gray images pixel levels.

2.3 Deep Learning-Based Colorization

Cheng et al. [6] design a deep neural networks (DNNs)-based model architecture for image colorization. Their model architecture has an input layer, various hidden layers, and output layer. However, DNN-based models cannot generate important features from an image-like data.

To resolve the problem, different CNN-based models are being put forth for image colorization [8, 17]. Hwang et al. [8] show CNN networks can extract more meaningful features from image data and utilize these features in colorization.

Baldassarre et al. [2] introduce a fusion layer with a pre-trained feature extractor along with an encoder-decoder network to improve the result of automatic colorization. Their model provides a significant improvement and the generated images look more realistic.

Yamini et al. [20] introduce a convolutional neural network-based system that incorporates OpenCV and accurately colorizes images without human intervention. Recently, generative adversarial networks (GANs) have been deployed to automatically colorize images in [12, 15, 19]. However, GAN models are heavy and cannot be quantized for embedded systems.

However, we have found that most of the deep learning-based colorization models show poor performance while colorizing Bangladeshi historical documentaries. These models are not generalized colorization models and are mostly based on the dataset used during the training process. Therefore, in our research work, we developed a deep learning-based model that can colorize historically significant documentaries in Bangladesh.

3 Proposed Approach

We explain the architecture of our colorization model with necessary diagrams. Moreover, we discuss the strength of our model architecture.

3.1 Model Architecture

We have designed our model architecture using encoder, transfer learning pre-trained model InceptionV3, fusion layer, and decoder. Figure 2 shows the architecture of our colorization model. The encoder extracts features from the grayscale input images. The pre-trained feature extractor also extracts features from the input image. Then fusion layer concatenates both feature vectors. Finally, the decoder utilizes the concatenated feature vectors to generate the output image. Figure 3 demonstrates how the encoder, decoder, and fusion layers are integrated into the model.

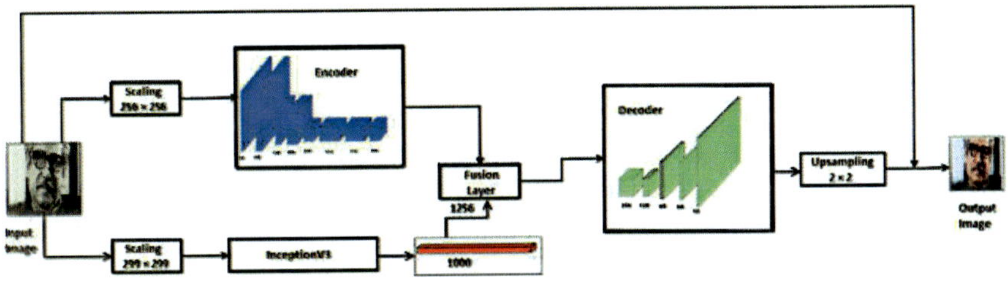

Fig. 2 Architecture of our colorization model

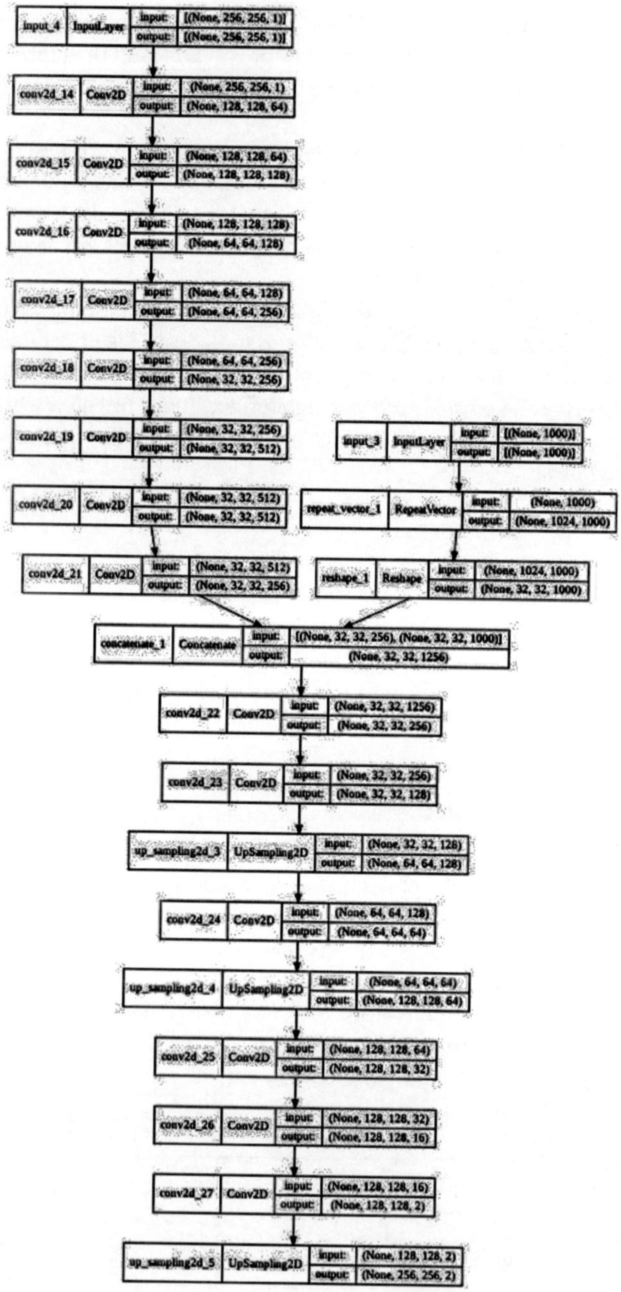

Fig. 3 Encoder-decoder architecture

Encoder Encoder is formed by a sequence of stack CNN layers. individually, the CNN layer is formed by a convolution layer, ReLE layer, and pooling layer. The primordial layers learn features like edges and corners, middle layers learn more advanced features as part of an object, and finally, the last layers learn features to understand the input image properly for colorization.

Pre-trained Feature Extractor We deploy a pre-trained feature extractor named InceptionV3 which was originally trained on a large dataset called ImageNet[1] and is available as a pre-trained model with model weight in the Keras library. InceptionV3 takes the grayscale image as input and creates a high-level feature vector.

Fusion Layer The fusion layer applies a repeat vector function to reshape the feature vector from the pre-trained model to match with the feature vector from the encoder. Finally, the fusion layer concatenates the reshaped feature vector with the feature vector created by the encoder.

Decoder The decoder consists of a sequence of stack CNN layers along with 2D Upsamping layers. The decoder takes the concatenated feature vector as input and finally generates the color components of the input image.

3.2 Strength of Our Model Architecture

We have collected high-level features by utilizing the advantages of both the encoder and the pre-trained model. Therefore, the decoder gets better feature vectors to generate color components of input image. Also, the model is lighter compared to other available GAN [15, 19] solutions.

4 Implementation

We will discuss the specifics of implementation, with dataset creation, parameters, and environment settings.

4.1 Dataset Creation

We have collected colorized Bangladeshi historical video documentaries from different online sources like Facebook and Youtube. All the collected video documentaries were created from 1960 to 1980. From the video documentaries, we have extracted one image frame from each 50 consecutive image frames. We convert each extracted image frame into the grayscale format and make pairs of grayscale images and their corresponding color image. By using this technique, we have created a dataset with 6780 pairs of a grayscale image and its original color image.

---

[1] https://www.image-net.org/.

4.2 Dataset

Our dataset is separated into three components: a training set (80%), a validation set (10%), and a test set (10%). The training, validation, and test sets each comprise 5444, 668, and 668 pairings of grayscale images and their corresponding color images.

4.3 Parameter Settings

Before training our model, we compile it with Adam as the optimizer. We set the mean squared error (MSE) as the loss function. Finally, the batch size is set to 32.

4.4 Environment Settings

Throughout the training, validation, and testing of our model, we utilize resources available in Google Colaboratory.[2] Google Colaboratory currently uses Python 3.7.12 and TensorFlow 2.6.0. The hardware configuration of Google Colaboratory is Intel(R) Xeon(R) CPU @ 2.20 GHz with 12.69 GB of RAM and also a Tesla p100 16 GB graphics card.

5 Experiment Results and Analysis

This section shows the experimental results and analysis. Moreover, we summarize the findings of our experiment.

5.1 Performance Metrics

Peak Signal-to-Noise Ratio (PSNR): PSNR is a commonly used metric to assess the reconstruction quality of images and videos that have undergone lossy compression. The most straightforward way to define PSNR is to use the mean squared error (MSE), which is defined as

$$MSE = \frac{1}{mn} \sum_{p=0}^{m-1} \sum_{q=1}^{n-1} [I(p,q) - K(p,q)]^2 \tag{1}$$

[2] https://colab.research.google.com/.

The PSNR (measured in decibels) is defined as

$$PSNR = 20.log_{10}(MAX_I) - 10.log_{10}(MSE) \quad (2)$$

In this case, MAX_I represents the image's maximum possible pixel value. With 8 bits per sample, this is equal to 255 when the pixels are represented as a single number.

Structural Similarity Index Measure (SSIM): The SSIM algorithm is used to determine the degree of similarity between the two images. While the SSIM model accounts for image deterioration as a perceived change in structural information, it also contains significant perceptual phenomena such as brightness and contrast masking to account for the overall experience of picture degradation. Other approaches, such as MSE or PSNR, estimate relative error, whereas this approach does not. Structural information relates to the idea that pixels are highly interconnected, even more so when they are spatially close together. These relationships transmit crucial information about the structure of the objects in the visual scene.

The SSIM formula measures difference between the input image(p) and the ground truth image(q) based on three parameters which are luminance (l), contrast (c), and structure (s).

Functions for these parameters are as follows:

$$l(p,q) = \frac{2\mu_p\mu_q + c_1}{\mu_p^2 + \mu_q^2 + c_1}$$

$$c(p,q) = \frac{2\sigma_p\sigma_q + c_2}{\sigma_p^2 + \sigma_q^2 + c_2}$$

$$s(p,q) = \frac{\sigma_{pq} + c_3}{\sigma_p\sigma_q + c_3}$$

Here, μ_p, μ_q is the average of p, q, and σ_p^2, σ_q^2 is the variance of p, q, respectively. σ_{pq} is the covariance of p and q. $c_1 = (K_1 * L)^2$, $c_2 = (K_2 * L)^2$, and $c_3 = \frac{c_2}{2}$. Here, L is the dynamic range of pixel values and $k_1 = 0.01$, $k_2 = 0.03$ by default. SSIM is then calculated as a weighted average of these parameters.

$$SSIM = [l(p,q)^\alpha.c(p,q)^\beta.s(p,q)^\gamma] \quad (3)$$

5.2 *Training Results*

Using the parameter setting given in Sect. 4, we train our model with the training dataset. After training our model for 100 epochs, our model reaches convergence. Figure 4 shows how the loss of our model decreases after each epoch.

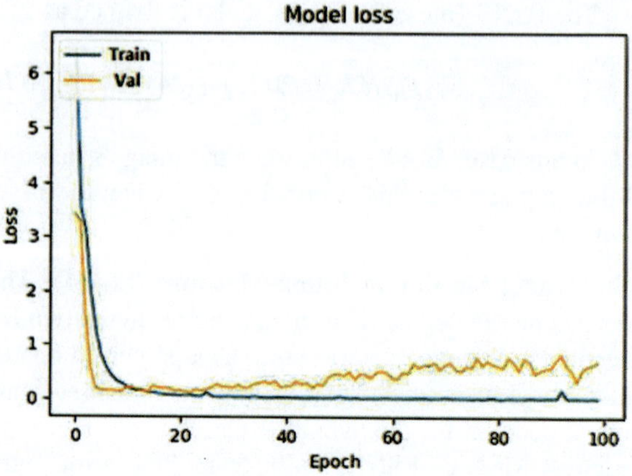

Fig. 4 Loss versus epoch curve

The last training loss of our model is 0.2 and the validation loss is 0.7. We evaluate our model on the test set and the testing loss is 0.8.

5.3 Evaluation

In Fig. 5, we demonstrate the performance of our model by providing some predicted color image and its original color image. We have found that the predicted color image is almost similar to its original color image. PSNR and SSIM are calculated using Eqs. 2 and 3, respectively, and from Table 1 it is observed that InceptionV3 incorporated with our proposed encoder-decoder model performs best having PSNR value of 36.10 and SSIM value of 0.9768.

6 Conclusion and Future Work

In this research work, we have developed a deep learning-based model to automatically colorize historically significant documentaries of Bangladesh. We have designed our model architecture using an encoder, a pre-trained feature extractor called InceptionV3, a fusion layer, and a decoder. We have created a new dataset with pairs of grayscale and color images collected from Bangladeshi historical documentaries from different online sources. After training our deep learning-based colorization model, it shows a significant result while colorizing Bangladeshi historical documentaries. In the future, we can train a generalized model which can colorize both new and old images or video content.

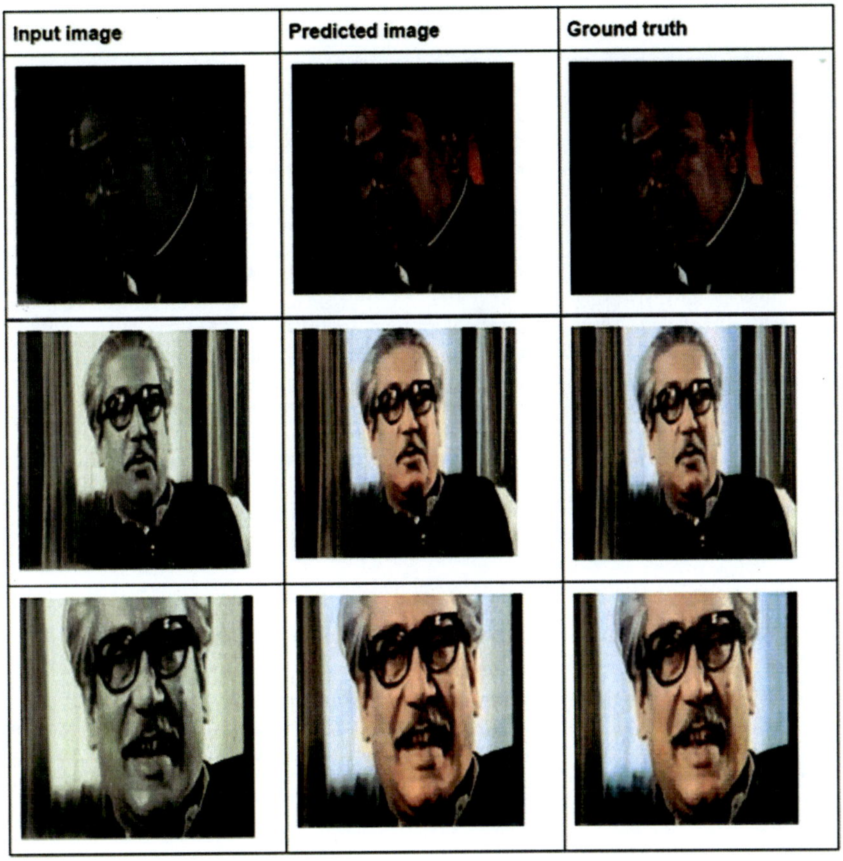

Fig. 5 Sample predicted color images

Table 1 PSNR and SSIM results

Method	PSNR	SSIM
VGG16 + proposed encoder-decoder model	27.52	0.8593
ResNet50V2 + proposed encoder-decoder model	30.83	0.9128
InceptionV3 + proposed encoder-decoder model	36.10	0.9768

References

1. An, J., Gagnon, K.K., Shi, Q., Xie, H., Cao, R.: Image colorization with convolutional neural networks. In: 12th CISP-BMEI. IEEE (2019)
2. Baldassarre, F., Morín, D.G., Rodés-Guirao, L.: Deep koalarization: image colorization using cnns and inception-resnet-v2. arXiv preprint arXiv:1712.03400 (2017)
3. Bugeau, A., Ta, V.T.: Patch-based image colorization. In: 21st ICPR. IEEE (2012)
4. Bugeau, A., Ta, V.T., Papadakis, N.: Variational exemplar-based image colorization. IEEE TIP (2013)
5. Charpiat, G., Hofmann, M., Schölkopf, B.: Automatic image colorization via multimodal predictions. In: ECCV. Springer (2008)
6. Cheng, Z., Yang, Q., Sheng, B.: Deep colorization. In: IEEE ICCV (2015)
7. Gupta, R.K., Chia, A.Y.S., Rajan, D., Ng, E.S., Zhiyong, H.: Image colorization using similar images. In: 20th ACM ICM (2012)
8. Hwang, J., Zhou, Y.: Image colorization with deep convolutional neural networks. In: Technical Report (2016)
9. Levin, A., Lischinski, D., Weiss, Y.: Colorization using optimization. In: ACM SIGGRAPH Papers (2004)
10. Li, Y., Lizhuang, M., Di, W.: Fast colorization using edge and gradient constrains (2007)
11. Limmer, M., Lensch, H.P.: Infrared colorization using deep convolutional neural networks. In: 15th IEEE ICMLA. IEEE (2016)
12. Nazeri, K., Ng, E., Ebrahimi, M.: Image colorization using generative adversarial networks. In: ICAMDO. Springer (2018)
13. Qu, Y., Wong, T.T., Heng, P.A.: Manga colorization. ACM TOG (2006)
14. Quan, W., Wang, K., Yan, D.M., Pellerin, D., Zhang, X.: Improving the generalization of colorized image detection with enhanced training of CNN. In: 11th ISPA. IEEE (2019)
15. Sun, Q., Chen, Y., Tao, W., Jiang, H., Zhang, M., Chen, K., Erdt, M.: A GAN-based approach toward architectural line drawing colorization prototyping. TVC (2021)
16. Szegedy, C., Vanhoucke, V., Ioffe, S., Shlens, J., Wojna, Z.: Rethinking the inception architecture for computer vision. In: IEEE CCVPR (2016)
17. Titus, S., NM, J.R.: Fast colorization of grayscale images by convolutional neural network. In: ICCSDET. IEEE (2018)
18. Voulodimos, A., Doulamis, N., Doulamis, A., Protopapadakis, E.: Deep learning for computer vision: a brief review. CIN (2018)
19. Wu, Y., Wang, X., Li, Y., Zhang, H., Zhao, X., Shan, Y.: Towards vivid and diverse image colorization with generative color prior. In: IEEE/CVF (2021)
20. Yamini, K., Swetha, K.S., Prasanna, P.L., Swathi, M.R.V., Maddumala, V.R.: Image colorization with deep convolutional open CV. J. Eng. Sci. (2020)
21. Yatziv, L., Sapiro, G.: Fast image and video colorization using chrominance blending. IEEE TIP (2006)

A Methodology for Recommending In-Vehicle Coupons Incorporating Machine Learning Algorithms for Efficient Financial Schemes

Md. Abdul Hai, Rafsan Shartaj Uddin, Yeaminur Rahman, and Rezwana Mahfuza

Abstract Coupon marketing is one of the oldest and most effective business marketing methods, in which special offers on products or services entice clients. This paper proposes a framework that acts as a coupon recommender system for in-vehicle users collaborating Machine Learning techniques to suggest the most suitable coupon for the user based on their attributes. ANN, CNN, LSTM, Random Forest, XGBoost, and Stacking Machine Learning Algorithms are applied to train and test the dataset to determine the most efficient and accurate algorithm that can predict the desirable coupon for the vehicle user. Accuracy, Precision, Recall, and F1-scores were calculated and compared of these algorithm models to obtain the most efficient model for the framework. Lastly, the framework determines the possibilities of accepting the available coupons based on the user features using the best-performing algorithm and recommends the coupon, thus providing a highly beneficial and effective system for the in-vehicle users and business owners.

Keywords Coupon · ANN · CNN-LSTM · Stacking · XGBoost · In-vehicle coupon · Recommendation · Machine learning · Framework

Md. A. Hai · R. Shartaj Uddin · Y. Rahman (✉) · R. Mahfuza
Department of Computer Science and Engineering, Brac University, Dhaka, Bangladesh
e-mail: yeaminur.rahman@g.bracu.ac.bd

Md. A. Hai
e-mail: md.abdul.hai@g.bracu.ac.bd

R. Shartaj Uddin
e-mail: rafsan.shartaj.uddin@g.bracu.ac.bd

R. Mahfuza
e-mail: rezwana.mahfuza@g.bracu.ac.bd

S. Hossain et al. (eds.), *Proceedings of International Conference on Fourth Industrial Revolution and Beyond 2021*, Lecture Notes in Networks and Systems 437,
https://doi.org/10.1007/978-981-19-2445-3_2

1 Introduction

Marketing is defined as an endeavor to elicit the desired response from clients by producing and providing value in the market. An effective marketing strategy should focus on its business model, which tells customers what it stands for, how it runs, and why they should do business with it. As a result, business owners gather data from market research and use different methods to create an organizational plan that appeals to client wants.

One of the most prevalent marketing tactics is coupon marketing, which takes advantage of consumers' desire to save money on purchases by offering products or services lower than the standard price with certain redemption conditions applied. Coupons are essentially codes or vouchers used to entice and retain customers, as they have a significant influence on consumers' loyalty and contentment. According to Juniper Research, the valuation of coupon redemptions is expected to be $91 billion by 2022, whereas it was $47 billion in 2017 [1]. Moreover, according to a Claremont University study, acquiring and then using coupons reduces stress and boosts oxytocin levels by 38%, which directly links to happiness [12]. Coupons provide value to the customer experience by delivering savings and fostering long-term relationships as they become frequent users. Although the same coupons are not applicable to all customers, choosing which one to provide is complicated because each customer profile reacts differently to each offer. Repeatedly offering them undesirable offers may cause them not to obtain those coupons, which may hamper customer relations. Advanced machine learning approaches can help implement personalized coupon filtering based on recommendation systems to surpass the obstacle, benefiting customers and coupon providers. According to the user's characteristics and features, a competent coupon recommendation system using machine learning can properly detect which coupon the user requires and intends to use.

The main objective of this paper is to propose a framework for a coupon recommendation system for real-life vehicle users using machine learning techniques. Algorithms such as Artificial Neural Networks (ANN), Convolutional Neural Networks (CNN), Long Short-Term Memory (LSTM), Random Forest, Extreme gradient boosting (XGBoost), and Stacking are experimented and compared to determine the most efficient algorithm that can accurately predict whether a user will accept the coupon or not based on their attributes. The dataset used contains data of 12685 distinct cases where each driver has 25 features, and five types of coupons are provided, which are Restaurant (less than $20), Coffee House, Carryout & Takeaway, Bar, and Restaurant ($20–$50). The framework can obtain data from the user, check the number of coupons available, and predict the probability of getting more accurate results using the optimal machine learning algorithm. Following the prediction of probabilities, the framework will select the coupon with the highest chance of being accepted and recommend it to the user. The key contributions of this paper are

- A novel framework for identifying and proposing coupons for in-vehicle users collaborating effective Machine Learning techniques to surge coupon acceptance rates.
- To discern the most efficient technique for the coupon recommendation framework, and a thorough comparison and study of several Machine Learning algorithms, including ANN, CNN, LSTM, Random Forest, XGBoost, and Stacking.

2 Literature Review

Several academics have explored the market profitability of delivering coupons, but there hasn't been significant analysis of in-vehicle coupons. For instance, Pandey. N analyzed coupon's involvement in increasing market efficiency from 1976 to 2015 [8]. The study conducted by Chen et al. [3] showed how successful discount promotions influence a customer's intention while purchasing a product. Furthermore, Khajehzadeh et al. [5] said that in the case of multipurpose items, the geographical location and the type of product have a role in coupon redeeming. Besides, while assessing profit in market strategies, Osuna et al. proposed factoring existing purchasers when clients redeem the brand's coupon [7].

Rahman proposed [11] a wearable sensor-based physical activity classification system applying a particular class of supervised machine learning. The study suggested a performance analysis of several boosting algorithms, including extreme gradient boosting, light gradient boosting machine, cat boosting, and gradient boosting. Among these algorithms, XGB, LGBM, and ADA (RF) classifiers have shown the most promising result.

Su et al. [14] focused on coupon trading and its contribution to market gains. The authors concentrated on three situations (without a coupon, coupon without trading, and coupon with trading). They observed that coupon trading could help merchants and manufacturers make increased profits over extended periods of time. In addition, the study contrasted coupon targeting with coupon trading, concluding that coupon trading outperforms coupon targeting in terms of revenues. Finally, the authors employed numerical analysis to determine the best coupon face values and recommended coupon trading with impulse schemes. Nevertheless, alternative hypotheses can be established using more straightforward approaches to assess how well they function in terms of increasing revenue for both retailers and manufacturers.

Lee [6] suggested a real-time genetic algorithm-based coupon recommendation service for moving cars that were filtered based on location, diversity, and proximity criteria. To guarantee that the most significant economic benefit is maintained, the maximum gain fitness function is employed to develop subsets consisting of substantial aspects. Furthermore, evolutionary algorithms solve the problem of temporal complexity. However, a comparative study with other algorithms should be included while evaluating the recommended algorithm's performance. Consequently, it is unknown how effectively the algorithm will handle the complex and diverging variables.

For building classifiers with a minimal number of brief rules, Wang [16] presented two probabilistic models with prior parameters. With demographic data and preferences, Bayesian Rule Sets in disjunctive standard form were included in the study to describe and forecast user behavior in in-vehicle recommender systems. The ROC curve was used to assess predictive performance, and BSR classifiers surpassed decision trees and RIPPER in terms of accuracy. The authors also asserted that it outperforms standard linear models in numerous respects. Indeed, the algorithm's viability with larger and highly complex datasets is still uncertain.

Rahahleh provides [10] in their research about the assessment of clients' purchasing behavior utilizing sustainable electronic vehicle sales to diversify into new markets, with an emphasis on electric and hybrid cars. Their goal is to figure out how green electronic marketing affects customer purchase decisions. Apart from that, they also investigate various factors such as the relationship between stimuli factors and response purchase decisions. After analyzing, they found that Consumers of Hybrid and Electric vehicles are affected by psychological factors, and the findings are consistent with previous research.

3 Work Flow Layout of the Proposed System

Here, Fig. 1 represents the layout for the working procedures of the proposed system. Firstly, a suitable, as well as a reliable real-life dataset is collected for coupon recommendation. Several approaches such as data cleaning, label encoding, imputing null values, and min-max scaling are used to pre-process the data. Afterward,

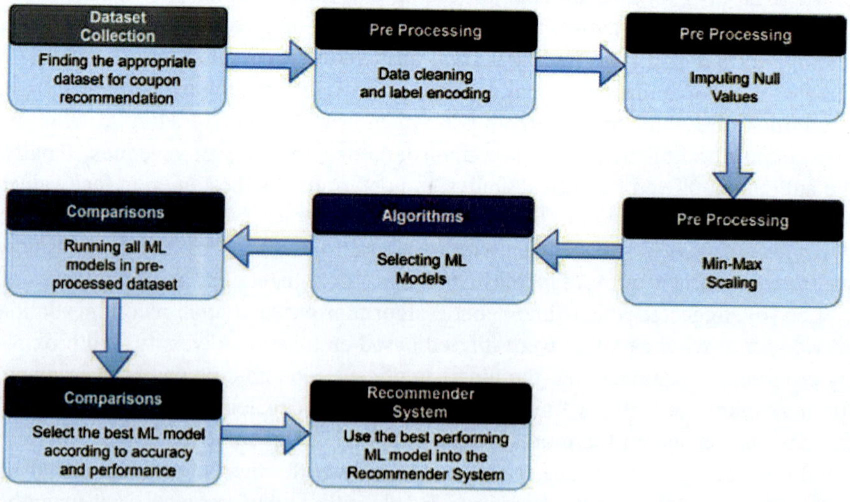

Fig. 1 Work flow layout of the proposed system

relevant machine learning algorithms are chosen and executed on the pre-processed dataset. The best-suited algorithm is then selected and employed based on accuracy and performance in the proposed recommender system.

4 Algorithms

The performance of three bagging and boosting algorithms were evaluated in the study of Singhal et al. [13] where gradient boosting and random forest were found to be the best-performing algorithms in unbalanced binary classification. Additionally, in terms of different boosting algorithms, XGBoost and Gradient boost were found most effective among the boosting algorithms [11]. On the other hand, the dataset used for the proposed system consists of a large number of user data, and neural networks tend to perform better in large datasets. So, to get the best result out of the dataset, boosting, bagging algorithms, and neural networks have been used. Moreover, to evaluate the combined performance of algorithms, stacking with boosting algorithms and Neural Networks has been employed.

Additionally, to optimize the hyperparameters of the employed algorithms, grid search has been used. Grid search creates and assesses a model for each combination of hyperparameters provided. Based on the result produced by grid search, the best combination of hyperparameters has been used to evaluate every model. Moreover, in the case of neural network models, all the algorithms were run for 100 epochs with Adam optimizer. For the evaluation of the loss function, binary cross-entropy has been used. The model with the least prediction error has been saved for further evaluation to overcome the overfitting situation.

4.1 Artificial Neural Networks

The name neural network comes from the biological term neurons. Artificial Neural Network is mainly motivated by how neurons carry signals from the human brain [15]. Neural Network is a collection of input layers, hidden layers, and output layers containing different neurons. Each neuron from the input layer takes input, and neurons of the first layers are connected to neurons of hidden layers through the channel. Moreover, each channel is assigned weight. In addition, there is an activation function that decides if a neuron of a particular layer will be activated or not. After passing through all the layers, the output layers have neurons according to the classes needed to predict. Furthermore, to train a neural network, the model has to be run for several epochs. In each epoch, the model generates a loss function, and after each epoch, the model tries to minimize the loss function that significantly improves the prediction's accuracy.

4.2 CNN LSTM

The CNN LSTM uses CNN layers to retrieve in-depth features from input data, while LSTMs are used to assist in sequence prediction. Nowadays, CNN LSTM is widely used in machine learning techniques, including image and video processing, activity recognition, etc. Islam et al. [4] proposed CNN LSTM to detect new COVID-19 dynamically from X-ray pictures with better accuracy. Additionally, the convolution frames comprise 12 convolutional layers, one FC layer, five max-pooling layers, one LSTM layer, and one softmax activation overlay. The design uses a fully connected layer to classify the X-ray images and identify which of the three categories they correspond to.

4.3 Random Forest

Random Forest is a robust machine learning algorithm that was developed by Breiman in 2001 [2]. It can process a massive quantity of data quickly and effectively, and it has high performance since it can handle thousands of input variables without deleting them. Besides, it requires less tuning of its hyperparameters compared to other machine learning algorithms.

Random forest algorithm begins with a root node where it constructs several shorter decision trees training with a random subset of features. It can handle thousands of variables at once and can efficiently run on enormous datasets.

4.4 Extreme Gradient Boosting

Extreme gradient boosting (XGBoost) is a supervised machine learning algorithm that was initially Chen and Guestrin in 2016 [9, 17]. It's been widely used for forecasting applications, such as predicting power load and store sales. It combines the estimates of a collection of more straightforward, weaker models to attempt to forecast a target variable correctly. After multiple boosting iterations, it can update the previous predictor so that the specified loss function can be optimized.

4.5 Stacking

Instead of choosing one best algorithm for machine learning predictions, stacking allows researchers to use the effectiveness of two or more well-performing algorithms to produce a superior result. A collection of ground classifiers is created in the first phase, wherein a meta-level classifier is trained in the second step. Stacking is highly preferable when many distinct machine learning models with low correlation make predictions in different but better ways. Diverse Ensemble Members, Member Assessment, and Combine With Model make up the framework of the stacking method. For the proposed system, Gradient boosting, XGBoosting, and ANN are used as the estimators for stacking. Moreover, ANN is used as a final estimator for the stacking model.

5 Dataset and Data Pre-processing

5.1 Dataset Description

The dataset used for the proposed paper was collected from the UCI Machine Learning repository [16]. The dataset was primarily collected using a survey conducted by Amazon Mechanical Turk. In the survey, drivers were given different scenarios and were asked to choose one coupon among other types. The data of the scenarios varied with different temperatures, car sitting passengers, coupon expiration time, etc., including the selected coupon noted in the dataset.

The total dataset consists of data of 12685 different scenarios and drivers with 25 features. The columns contain data of destination, coupon asked to take, weather, passenger number, expiration, marital status, distance from the restaurant, coffee house, or bar from the location coupon is taken, the direction of the driver (whether the car was going to the order of the taken coupon or to the opposite direction), and coupon taken. In the proposed system, the target feature is coupon has taken, which has binary attributes yes or no. In the coupon column, five types of coupons were offered to the drivers, the first one was restaurant coupon—less than 20$, another was between 20$ and 50$, the third coupon was for the coffee house, the fourth one was for carrying out and take away, and the last one was a coupon of a bar.

5.2 Dataset Pre-processing

The dataset used for the proposed system consists of 25 features where the attribute 'car' contains 99.14% null values; as a result, the attribute was dropped from the dataset. After that, most of the columns have string values in the dataset, and to convert the dataset to trainable, the string values were encoded with label encoding. Every unique string value in the dataset is converted into integer values. In addition to that, 0.84% data of attribute Bar, 1.7% data of attribute CoffeeHouse, 1.19% data of Carrayway, 1.02% data of Restaurant less than 20 km, and 1.49% data of Restaurant under 20–50 km were null values. The mean value of each column has been used to impute those data with dummy variables. Moreover, to convert the data into a standardized format, min-max scaling was used. Afterward, 80% of data were selected randomly to train the data, and the rest of the 20% data were used for testing.

Table 1 Performance evaluation of algorithms used

Algorithm	Accuracy (%)	Precision (%)	Recall (%)	F1-Score (%)
Stacking	76.42	76.00	76.00	76.00
XGBoost	76.38	76.00	76.00	76.00
Random forest	74.93	75.00	75.00	75.00
CNN + LSTM	71.46	71.00	71.00	71.00
ANN	70.87	71.00	71.00	71.00

6 Comparative Analysis and Result Discussions

To compare the achieved result, Accuracy, Precision, Recall, and F1-scores were used.

From Table 1, it is visible that the stacking algorithm with Gradient boosting, XGBoost, and ANN delivers the best accuracy score of all the algorithms, with a score of 76.42%. On the other hand, XGBoost has the closest accuracy following that, which is 0.004% lower than stacking accuracy. However, stacking and XGBoost perform identically in terms of Precision, Recall, and F1-score.

The confusion matrix of the stacking algorithm shown in Fig. 2 reveals that the algorithm does a better job of predicting rejected coupons.

In Fig. 3, the comparison of accepted and rejected coupons among stacking, XGBoost, and random forest algorithms is demonstrated. When it comes to forecasting approved coupons, the stacking algorithm has a 67.70% accuracy rate, but it outperforms both XGBoost and random forest in predicting rejected coupons (83.49%). However, for predicting accepted coupons, XGBoost surpasses both stacking and random forest algorithms (70.52%).

A ROC curve of Stacking, XGBoost, Random Forest, and ANN has been illustrated in Fig. 4 to compare the models. From the curve, it is evident that the stacking with the neural network model has the highest specificities associated with the highest sensitivities. Moreover, XGBoost comes right after the stacking model.

7 Proposed Coupon Recommendation Framework

Coupon Marketing systems can be a wonderful method of expanding and marketing in a business sector. An efficient framework can be built to extract the best out of the coupon recommendation system. In our paper, a proposed framework has been suggested to recommend coupons to vehicle users in real life. A flowchart of the proposed framework has been illustrated in Fig. 5.

Whenever a car arrives at a certain coupon collection point that will be obtained related to the vehicle, the data can be collected manually, or different modern technology can be used to collect data even faster. After collecting data, the framework

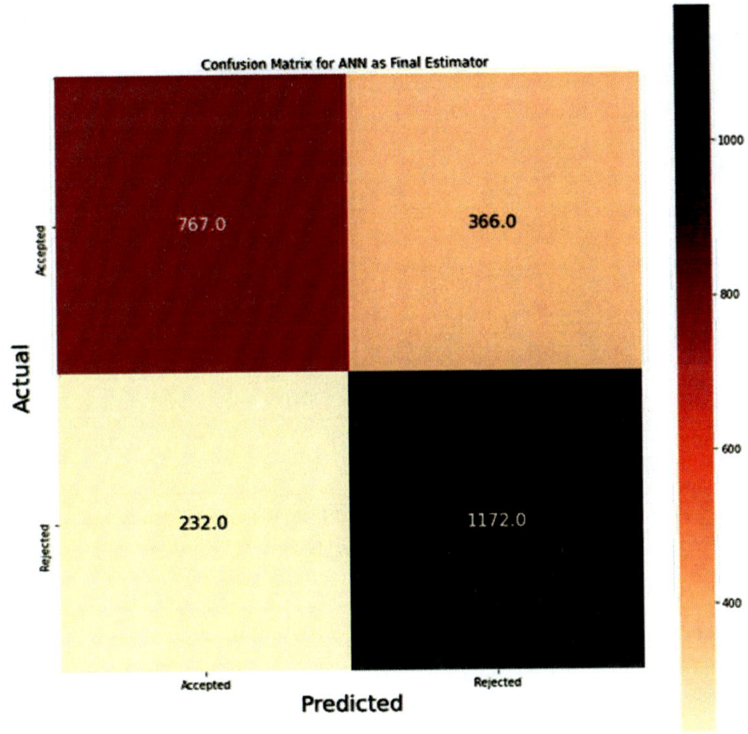

Fig. 2 Confusion matrix for ANN as final estimator

will check the number of coupons and predict the probability of accepting the coupon for every different type of coupon in its inventory. According to our research, the best-performing machine learning algorithm can be used to predict the probability of getting more accurate results. After predicting probabilities, the framework will choose the coupon with the best likelihood of accepting and recommending the coupon to the user.

8 Conclusion and Future Works

Coupon marketing is a strategy to boost the advertisement of an organization, and recommending suitable coupons can elevate an organization's marketing performance. This study provides a framework for marketing coupons to real-world car users in order to propose the best coupon to the user. Instead of applying conventional machine learning approaches, some well-established ML algorithms such as Stacking, XGBoost, ANN, and other algorithms have been used, and among these algorithms, Stacking has shown the most promising result (76.42%). This study aims to

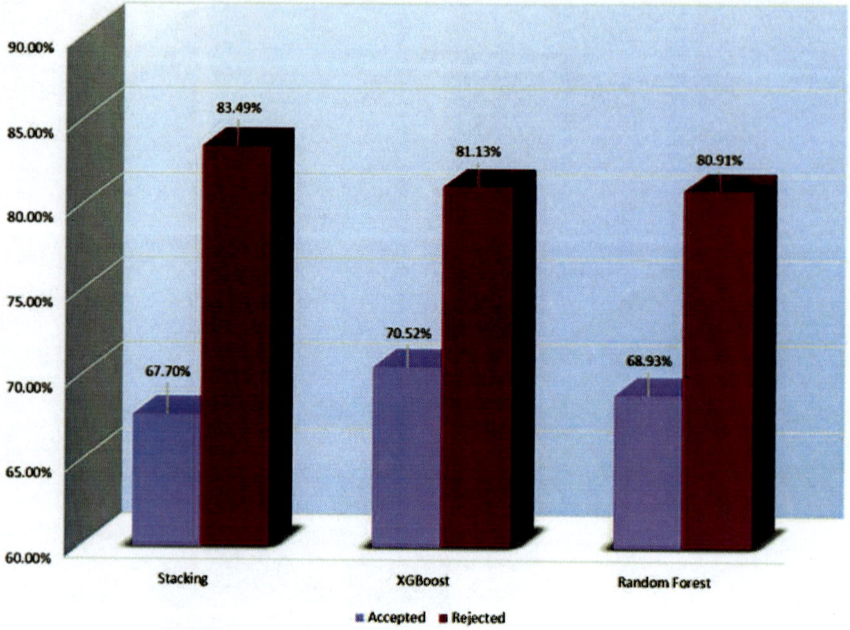

Fig. 3 Comparison of accepted and rejected coupons (top 3 performing algorithms)

Fig. 4 ROC curve with the selected ML algorithms

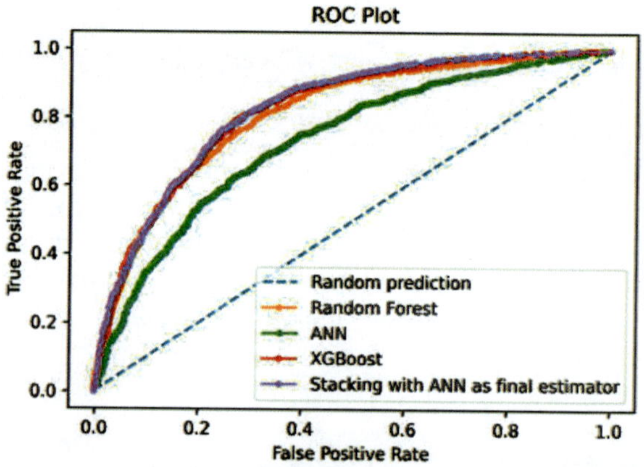

predict the best probability of accepting and recommending the coupon to the user, so it will surely improve the marketing strategy.

We intend to bring out further improvements and additions to our framework to propose a more robust system. In the future, the proposed framework will be experimented with a sizeable real-life dataset where there are various coupon types and determine how well it performs to recommend the coupon to the users. Further-

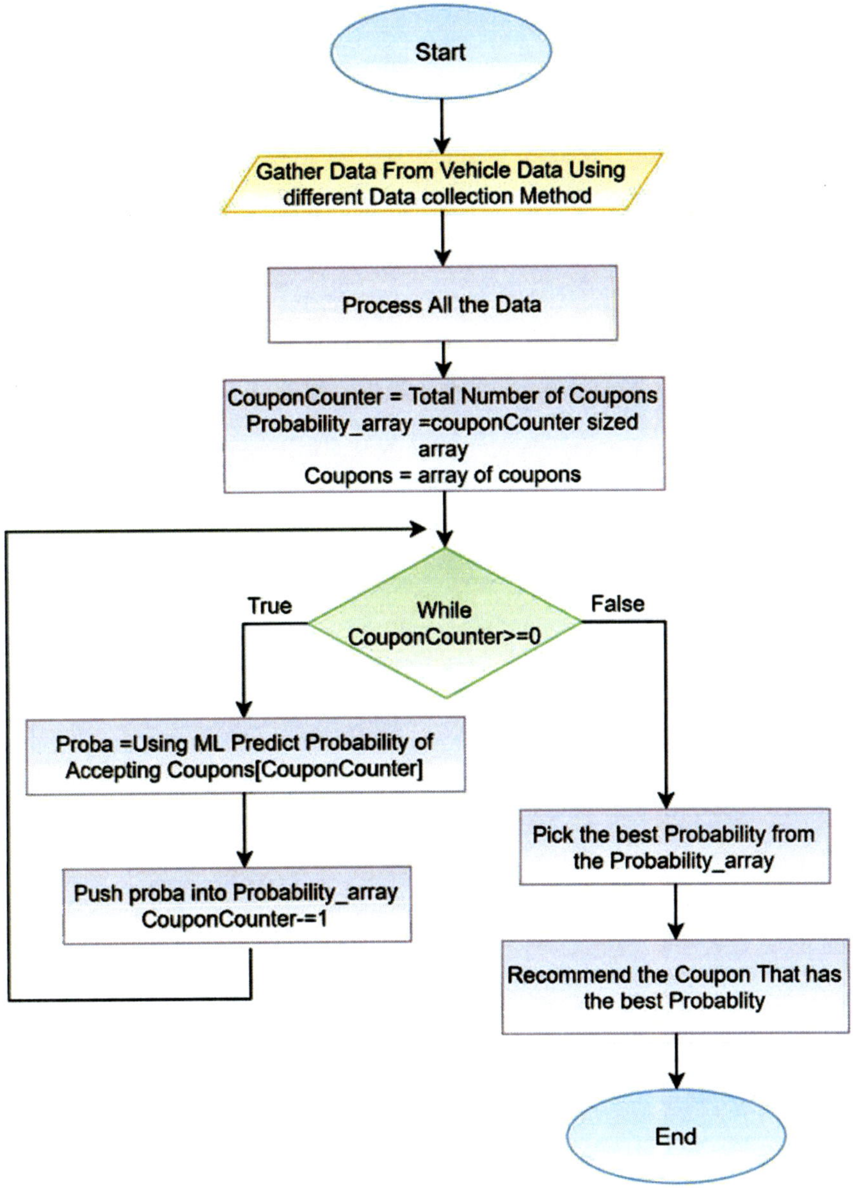

Fig. 5 Top-level approach for the proposed coupon recommendation framework

more, developing a user-friendly application to assist vehicle users in recommending coupons in real time can be highly beneficial.

References

1. Digital coupon redemptions to surpass $90 billion by 2022. https://www.juniperresearch.com/press/digital-coupon-redemptions-surpass-90-billion-2022. Accessed 21 Oct 2021
2. Breiman, L.: Random Forests. https://www.stat.berkeley.edu/~breiman/randomforest2001.pdf. Accessed 21 Oct 2021
3. Chen, S.F.S., Monroe, K.B., Lou, Y.C.: The effects of framing price promotion messages on consumers' perceptions and purchase intentions. J. Retail. **74**(3), 353–372 (1998)
4. Islam, M.Z., Islam, M.M., Asraf, A.: A combined deep CNN-LSTM network for the detection of novel coronavirus (COVID-19) using x-ray images. Inform. Med. Unlocked **20**(100412), 100412 (2020)
5. Khajehzadeh, S., Oppewal, H., Tojib, D.: Mobile coupons: what to offer, to whom, and where? Eur. J. Mark. **49**(5/6), 851–873 (2015)
6. Lee, J., Hyun, Y., Yun, Y., Lee, J., Lee, J.S., Park, G.L.: Design of a genetic algorithm-based coupon advertisement for electric vehicles. http://www.iraj.in/journal/journal_file/journal_pdf/12-335-15435534359-10.pdf. Accessed 21 Oct 2021
7. Osuna, I., González, J., Capizzani, M.: Which categories and brands to promote with targeted coupons to reward and to develop customers in supermarkets. J. Retail. **92**(2), 236–251 (2016)
8. Pandey, N., Maheshwari, V.: Four decades of coupon research in pricing: evolution, development, and practice. J. Revenue Pricing Manag. **16**(4), 397–416 (2017)
9. Park, S., Moon, J., Hwang, E.: 2-stage electric load forecasting scheme for day-ahead CCHP scheduling. In: 2019 IEEE 13th International Conference on Power Electronics and Drive Systems (PEDS). pp. 1–4. IEEE (2019)
10. Rahahleh, A.H., Moflih, M.A., Alabaddi, Z.A., AL-Nsour, S.N.: The moderating effect of psychological factors on consumer of electric and hybrid vehicles' response purchase decisions. Manag. Sci. Lett. **10**(8), 1649–1658 (2020)
11. Rahman, S., Irfan, M., Raza, M., Moyeezullah Ghori, K., Yaqoob, S., Awais, M.: Performance analysis of boosting classifiers in recognizing activities of daily living. Int. J. Environ. Res. Public Health **17**(3) (2020). https://doi.org/10.3390/ijerph17031082
12. Ruby, A.: The psychology of coupons: why do customers still use them? https://www.indoormedia.com/blog/the-psychology-of-coupons-why-do-customers-still-use-them (2017). Accessed 21 Oct 2021
13. Singhal, Y., Jain, A., Batra, S., Varshney, Y., Rathi, M.: Review of bagging and boosting classification performance on unbalanced binary classification. In: 2018 IEEE 8th International Advance Computing Conference (IACC). pp. 338–343 (2018). https://doi.org/10.1109/IADCC.2018.8692138
14. Su, M., Zheng, X., Sun, L.: Coupon trading and its impacts on consumer purchase and firm profits. J. Retail. **90**(1), 40–61 (2014)
15. Turjo, A.A., Rahman, Y., Karim, S.M., Biswas, T.H., Dewan, I., Hossain, M.I.: Cram: a credit risk assessment model by analyzing different machine learning algorithms. In: 2021 4th International Conference on Information and Communications Technology (ICOIACT). pp. 125–130 (2021). https://doi.org/10.1109/ICOIACT53268.2021.9563995

16. Wang, T., Rudin, C., Doshi-Velez, F., Liu, Y., Klampfl, E., MacNeille, P.: A Bayesian framework for learning rule sets for interpretable classification. J. Mach. Learn. Res. **18**(70), 1–37 (2017)
17. Zhang, W., Wu, C., Zhong, H., Li, Y., Wang, L.: Prediction of undrained shear strength using extreme gradient boosting and random forest based on Bayesian optimization. Geosci. front. **12**(1), 469–477 (2021)

A Deep Convolutional Neural Network Based Approach to Classify and Detect Crack in Concrete Surface Using Xception

Nazia Alfaz, Abul Hasnat, Alvi Md. Ragib Nihal Khan, and Nazmus Shakib Sayom

Abstract Deep learning has become a widely practiced approach in research arenas related to civil infrastructures. Monitoring concrete structures is time-consuming, costly, unsafe, and laborious. Instead of manual inspection, the deep learning approach increases more possibility to automate this inspection process helping to mitigate future risk. This study introduces an automatic concrete surface crack detection and classification technique using a deep learning architecture, namely Xception to alleviate the risks due to deteriorating structure conditions. At first, the Xception model was trained and tested on a public dataset consisting of cracked and non-cracked images, and the model has shown superior accuracy in two-class classification. Afterward, the cracked sub-dataset was split into two classes–horizontally cracked and vertically cracked using a traditional computer vision approach to determine the inclination angle of a crack. The proposed deep learning model was trained on the newly formed dataset and performed remarkably in three-class classification as well. This paper demonstrates the proposed model's effectiveness, performance, and findings, providing a reference for concrete surface crack detection and classification for related domains.

N. Alfaz (✉) · N. S. Sayom
Department of Computer Science, American International University-Bangladesh(AIUB), Dhaka, Bangladesh
e-mail: nazia.alfaz@aiub.edu

N. S. Sayom
e-mail: nsssayom@gmail.com

A. Hasnat
Civil Engineering, Faculty of Engineering, American International University-Bangladesh(AIUB), Dhaka, Bangladesh
e-mail: abul.hasnat@aiub.edu

A. Md. R. N. Khan
Department of Architecture, American International University-Bangladesh(AIUB), Dhaka, Bangladesh
e-mail: khan.alvi.md.ragib.nihal@gmail.com

S. Hossain et al. (eds.), *Proceedings of International Conference on Fourth Industrial Revolution and Beyond 2021*, Lecture Notes in Networks and Systems 437,
https://doi.org/10.1007/978-981-19-2445-3_3

Keywords Concrete surface · Crack detection · Crack classification · Xception · Deep learning

1 Introduction

The increasing number of deteriorating infrastructures with time throughout the world has raised concerns regarding their safety and functional efficiency [1]. Concrete infrastructures are very often subjected to severe environmental conditions, fatigue loads, and cyclic loading, all of which cause surface cracking [2, 3]. Cracks in structure have a substantial effect on its durability and make it much easier for external abrasive chemicals to contact the reinforcing bars and lead to corrosion [4, 5]. Because of this, cracks must be detected and repaired in order to preserve the structural integrity and performance of the building.

The most common approach to identify crack is visual inspection, which is very time-consuming, subjective to evaluation technique, moreover highly dependent on the level of expertise [6–8]. There are several nondestructive tests as well to detect a crack in structure [9]. Though these test methods intend to reliable results, expensive labor-intensive instruments are required to perform those tests [10].

In order to resolve the limitations of the above-mentioned approaches, several image processing algorithms have been developed to detect a crack in concrete buildings [11]. To increase the performance of image processing methods, machine learning algorithms were introduced using pattern identification and extraction [12]. Deep learning is one of the most remarkably operative research domains of computer vision and image classification. Many researchers have applied different deep learning approaches like AlexNet [13], ResNet, Inception V3, VGG16 [14], U-Net [6], DenseNet [15], to detect and classify crack in concrete structure [16, 17]. In this study, a deep learning based architecture, namely Xception is designed and employed using a public data repository to identify a crack in the concrete surface. Unlike the traditional deep learning architectures [18, 19], a different approach to performing the convolution operation is included in Xception to reduce the computational cost and training time [20].

2 Methodology

2.1 Data Acquisition and Cleaning

Version 2 of the "Concrete Crack Images for Classification" dataset [21] consists of 40,000 RBG images evenly distributed into 2 classes as negative and positive crack images, with each image having the dimension of 227 × 227 pixels generated from 458 high-resolution images with a method proposed in [22] contributed by Çağlar Fırat Özgenel into Mendeley Data, was used in this study for training and validating

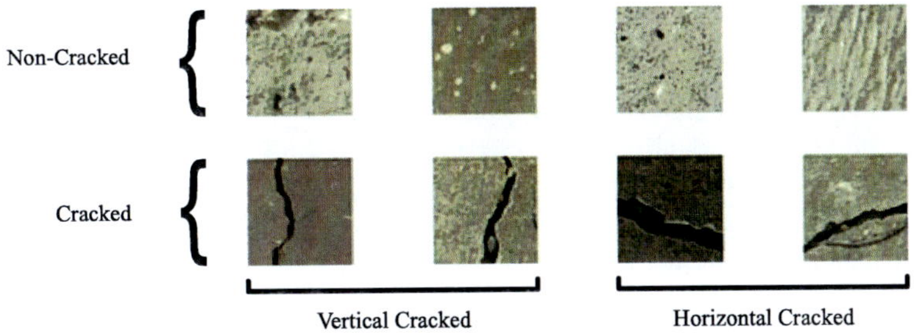

Fig. 1 Examples of the images with horizontal cracks, vertical cracks, and without cracks in the dataset

the network. However, distorted, blurry, and low-resolution images were encountered in the dataset, and thus the need for manual cleaning was evident because a classifier's performance is mainly determined by visual information and image resolution [23]. During the cleaning, we also checked for the cracks near the terminal pixels of the sides of images because passing through a network resizes the images to be smaller and renders it difficult to extract features from the edges [24]. As the training was performed in an Nvidia GeForce 1050 Ti GPU, the original dataset images were resized to 64 × 64 pixels due to its memory limitation. It also reduces computational cost and training time. Some samples from the final data repository of each class have been displayed in Fig. 1.

2.2 Data Distribution

For Two-Class Classification. After dataset acquisition and cleaning, 10,000 images were selected for the two-class classifier, and the whole dataset was split into a training set and a validation set with a proportion of 80:20. Each set contained an equal number of cracked and non-cracked images. The data distribution for the two-class classification is represented in Table 1.

An extra step was required for the there-class classifier to split the images of the cracked class into two new classes: Horizontally Cracked and Vertically Cracked.

Table 1 Data distribution for two-class classification

Class	Training	Validation
Non-cracked	4000	1000
Cracked	4000	1000

For Three-Class Classification. A Python script was used to automate this data labeling using the OpenCV library. The images were read as grayscale, and Gaussian filtering was applied with a 5 × 5 kernel to remove Gaussian noises from the image. Then Canny Edge Detection [25] is performed on the image to extract structural information from the image so that the amount of data that needs to be processed can be reduced. It detects a wide range of edges in images using a multistage method, and the major steps are the following:

i. Finding intensity gradient of the image: The scale of the gradient vector is determined for each pixel in this phase.
ii. Non-maximum suppression: The objective of this method is to "thin" the edge such that it is only one pixel wide.
iii. Threshold hysteresis: Finally, to reduce the false edges, a two-step threshold hysteresis is performed.

After this step, Morphological Closing Transformation [26] transformed the potential crack spots into continuous shapes. The largest continuous shape in an image can be assumed as the crack, which was determined by detecting the contour enclosing the maximum area. A straight line is taken afterward from the extreme top point to the extreme bottom point. The slope, M was calculated for that line, and the slope's arctangent value was the line's inclination angle, θ. If the absolute value of that angle was equal to or greater than 45°, the crack was classified as vertical, and if it was less than 45°, the crack was classified as horizontal. This algorithm worked fine for most of the cracked images in the cleaned dataset used for the two-class classification in successfully classifying those into two classes (Fig. 2).

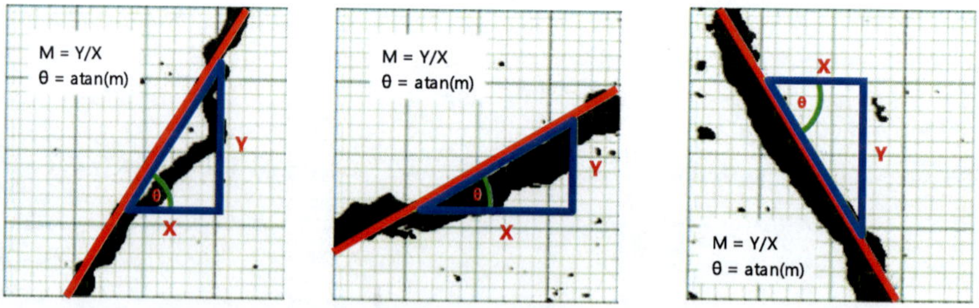

Fig. 2 Slope calculation of the crack to find the inclination angle

Table 2 Data distribution for three-class classification

Class	Training	Validation
Non-cracked	960	240
Horizontally cracked	1920	480
Vertically cracked	1930	480

However, some mislabeled images were noticed in the newly constructed three-class dataset because traditional computer vision techniques used here are not as accurate as deep learning approaches for image classification [27], and not much effort was made to make this algorithm handle all the edge cases because, after all, our goal was to develop a deep learning based classifier. At that point, we manually checked the dataset and corrected the labels. The data distribution for the three-class classification is demonstrated in Table 2.

2.3 *Xception Architecture*

Xception, also known as extreme inception [20], is a deep convolutional neural network inaugurated inception layers having depth-wise convolution followed by pointwise convolution, unlike traditional operations. In depth-wise convolution operation, each filter convolves only one channel of the input image at a time, and in pointwise convolution operation, the kernel of size 1×1, namely filter, iterates throughout every individual point of the input image. The traditional convolution operation used in deep learning models is expensive; therefore, to overcome the cost of convolution operation, separable depth-wise convolution [28] is introduced in the Xception model. Moreover, the depth-wise and pointwise convolution used in Xception provides faster training. In original depth-wise separable convolutions, a standard 3×3 convolution is factorized into a 3×3 depth-wise convolution that runs per input channels, and a 1×1 pointwise convolution is used to combine the outputs of the depthwise convolution into a linear combination. In the case of Xception, the 1×1 pointwise convolution takes place first, followed by a channel-wise spatial convolution. Another difference between Xception and depth-wise separable convolution is the usage of ReLU [29] nonlinearity. The Xception architecture mainly comprises three structures that are entry flow, middle flow, and exit flow. The entry flow consists of multiple convolution layers, activation layer, batch normalization, and max pooling layer. The middle flow, which repeats eight times, includes several convolution and activation layers. Finally, the exit flow is established with multiple convolution layers, activation layers, a max pooling layer, and global average pooling is operated at the final layer of exit flow. Lastly, the Softmax classifier presents exit flow plays the role of probability distributor in the output layer of the architecture (Figs. 3 and 4).

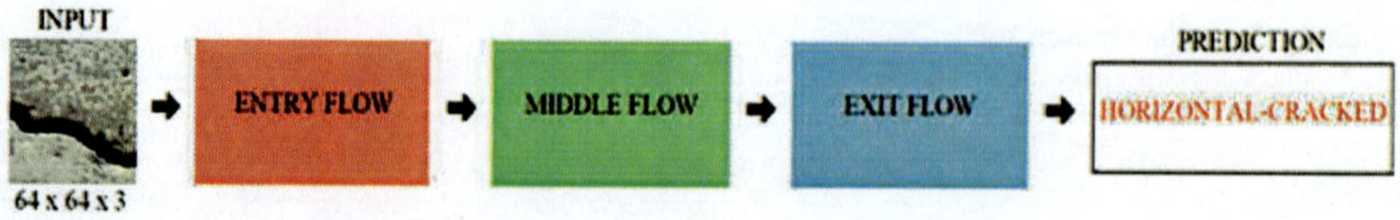

Fig. 3 Crack detection using Xception

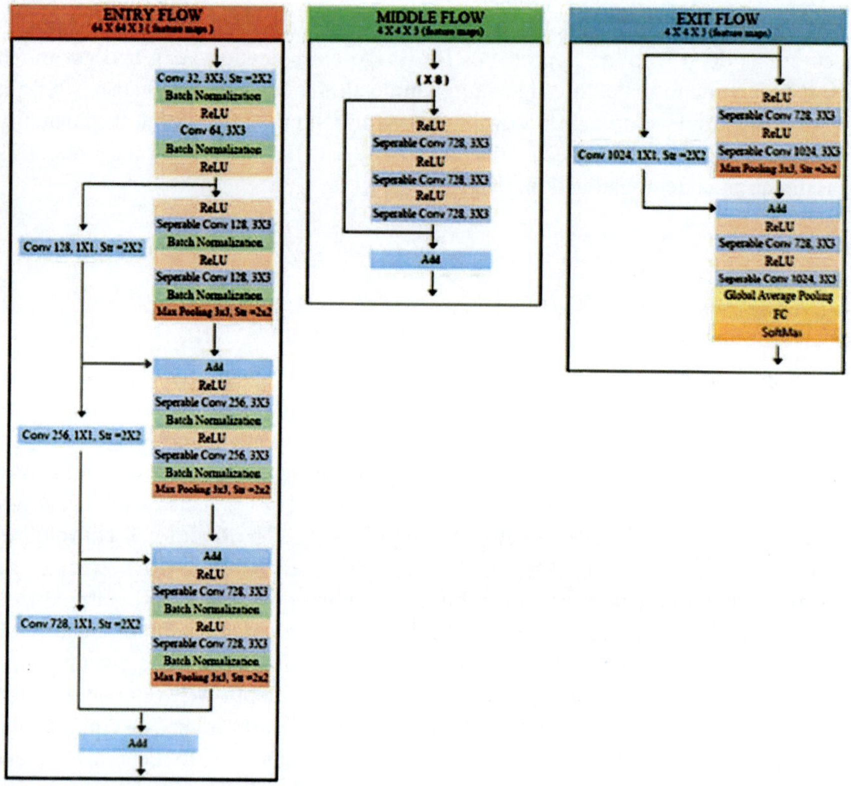

Fig. 4 Proposed Xception architecture

Rectified Linear Unit (ReLU). ReLU is a non-linear activation function first proposed in 2010 by Nair and Hinton [29]. It is heavily used in deep learning applications as it preserves the linear model properties and is comparatively easy to optimize [30]. ReLU activation function performs a thresholding operation that sets the value zero for the inputs, which are less than zero and preserves the values equal to or greater than zero. This rectification solves the vanishing gradient problem. The ReLU function is presented in Eq. (1).

$$f(x) = \max(0, x) = \begin{cases} x_i, & \text{if } x_i \geq 0 \\ 0, & \text{if } x_i < 0 \end{cases} \quad (1)$$

ReLU does not require any exponentials or division calculations, and thus it delivers faster computations which is a major advantage when deep learning is concerned [31].

Batch Normalization. If training data is limited in deep learning, the network may learn sampling noises in the training set but not in real test data [32]. Many methods were developed to reduce this problem by regularizing a fixed-sized model. Batch Normalization is a data regularization technique proposed by Ioffe and Szegedy in 2015 [33]. Batch Normalization adds extra layers to perform standardization and normalization operations on the input that is returned by the previous layer. The paper also showed that Batch Normalization accelerates training speed with no side effects and decreases learning rate decay.

Max Pooling. Max Pooling is a pooling operation that down-samples a feature map passing forward the maximum value within a group. Max Pooling usually comes after a convolution layer. It adds a small degree of translation invariance, which means that slight changes in the image do not substantially impact the values of most pooled outputs.

Softmax Layer. Bridle first introduced the idea of Softmax back in 1989 [34]. It is an output function that produces a vector of possibilities transforming the previous layer's output. Our work uses Softmax as the last activation function for the network for multinominal logistic regression. Equation (2) represents the Softmax classifier for an input vector x and a weighting vector w:

$$P\,(y = j\,|x) = \frac{e^{x^T w_j}}{\sum_{k=1}^{K} e^{x^T wk}} \quad (2)$$

3 Optimization and Loss Estimation

Adaptive moment estimation or Adam is an efficient algorithm for gradient-based optimization of stochastic objective functions requiring only a first-order gradient and has a very small memory footprint [35]. The optimizer was created with nonstationary objectives and situations with highly noisy and/or sparse gradients. The default learning rate of 0.001 for Adam is used. Loss is another significant measure for performance evaluation; therefore, categorical cross-entropy is used in this regard.

4 Result and Analysis

4.1 Training and Validation of Xception Model

In this work, the proposed Xception model for crack identification of two and three classes has been verified using two validation sets consisting of 2000 images and 1200 images, respectively. The Xception model was trained for 30 epochs using batch size 64, and the model achieved 99.13% validation accuracy at the 29th epoch for three-class classification and 99.90% validation accuracy at 19th epoch for two-class classification.

The progress of training accuracy and validation accuracy for two-class classification and three-class classification during training is represented in Figs. 5 and 7, respectively, which demonstrates excellent performance of the model for the identification of cracks on the concrete surface. It is visible in Figs. 6 and 8 that the Xception model hits desirable convergence as the training and validation loss reduce with the increasing number of epochs for two-class classification and three-class

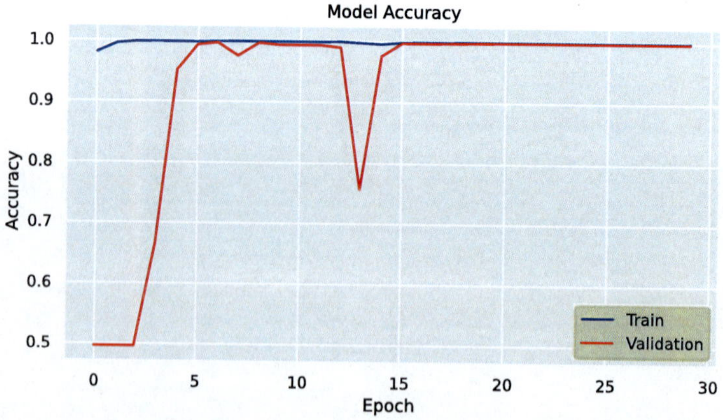

Fig. 5 Training and validation accuracy for two-class classification

classification severally. Besides, the superior training accuracy and validation accuracy exhibit remarkable performance of the model in both seen and unseen data along with the small gap between the training accuracy and validation accuracy manifested in Figs. 5 and 7, recognizing the model as an unstirred model toward the overfitting and underfitting problem. The training and validation procedure was conducted in a system equipped with Intel® Core™ i5-6600 K CPU, 16 GB Memory, and Nvidia GeForce GTX 1050 Ti running EndeavourOS based on Linux kernel 5.13.12.arch1-1.

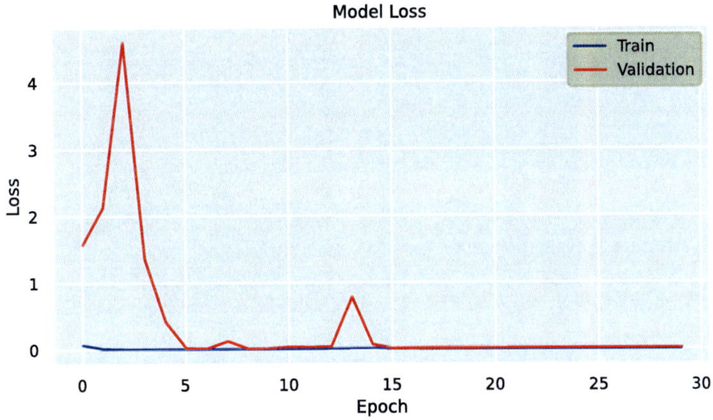

Fig. 6 Training and validation loss for two-class classification

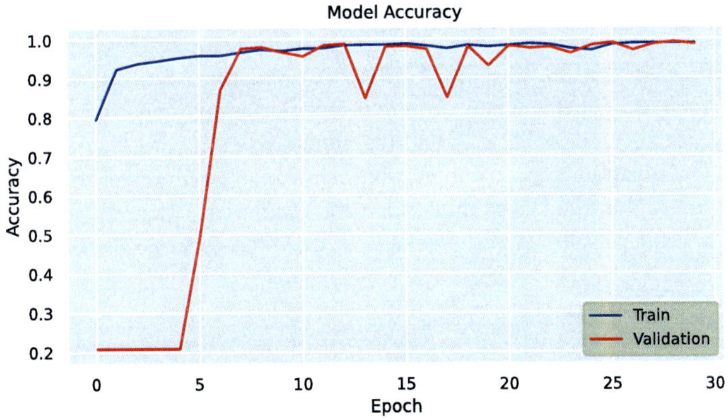

Fig. 7 Training and validation accuracy for three-class classification

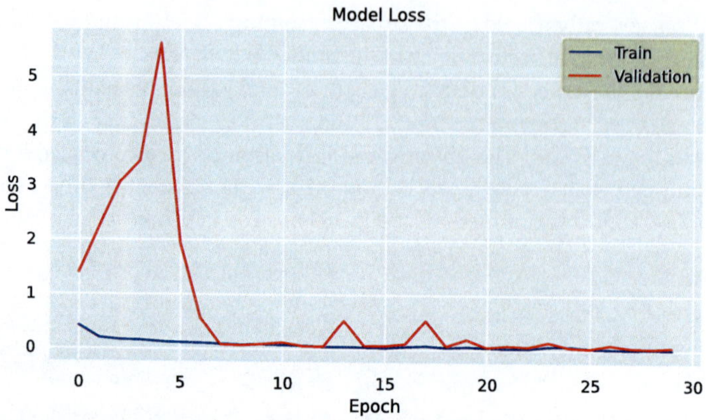

Fig. 8 Training and validation loss for three-class classification

Table 3 Performance evaluation report for two-class classification

Class	Precision	Recall	F1-score	Support
Non-Cracked	1.000	0.998	0.999	1000
Cracked	0.998	1.000	0.999	1000
Accuracy			0.999	2000

4.2 *Performance Evaluation*

Precision, recall, accuracy, and F1-score metrics are measured by generating confusion or error matrix using Eqs. (3)–(6) for further performance evaluation and visualization of the model. The results of the performance matrices for two-class classification are delineated in Table 3, and the generated confusion matrix is illustrated in Fig. 9. The illustrated confusion matrix for two-class classification shows that the model correctly classified 998 samples of 1000 samples as cracked samples and 1000 samples of 1000 samples as non-cracked samples, representing 99.9% accuracy.

$$\text{Accuracy} = \frac{\text{True Positive} + \text{True Negative}}{\text{True Positive} + \text{True Negative} + \text{False Positive} + \text{False Negative}} \tag{3}$$

$$\text{Precision} = \frac{\text{True Positive}}{\text{True Positive} + \text{False Positive}} \tag{4}$$

$$\text{Recall} = \frac{\text{True Positive}}{\text{True Positive} + \text{False Negative}} \tag{5}$$

$$\text{F1} - \text{score} = 2 \times \frac{\text{Precision} \times \text{Recall}}{\text{Precision} + \text{Recall}} \tag{6}$$

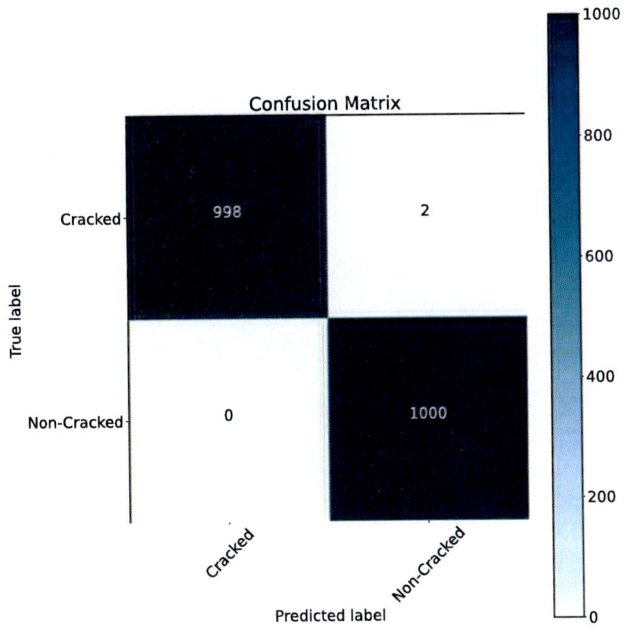

Fig. 9 Confusion matrix of validation data for two-class classification

The precision, recall, and F1-score achieved from the proposed Xception model for the cracked class are 0.998, 1.000, and 0.999, respectively, and for the non-cracked class are 1.000, 0.998, and 0.999, which signifies a superior performance of the model in identifying both cracked and non-cracked samples of concrete surface.

The generated confusion matrix for three-class classification is displayed in Fig. 10. The proposed architecture performed remarkably on the validation dataset for three-class and predicted 1184 samples accurately of 1200 samples. The precision, recall, and F1-score on unseen data obtained by the model are demonstrated in Table 4.

5 Conclusion

In this study, a deep learning model, namely Xception, is approached to detect cracks on a concrete structure to sustain the reliability of the concrete structure. The Xception model has been trained independently for two-class and three-class classification, and the training progress is recorded using accuracy and cross-entropy loss. Afterward, the validation data portion is validated using the Xception architecture, and the architecture has shown superior crack detection accuracy exceeding 99%. Furthermore, a confusion matrix was generated to calculate precision, recall, and F1-score for the quantification and performance visualization of the model. All the

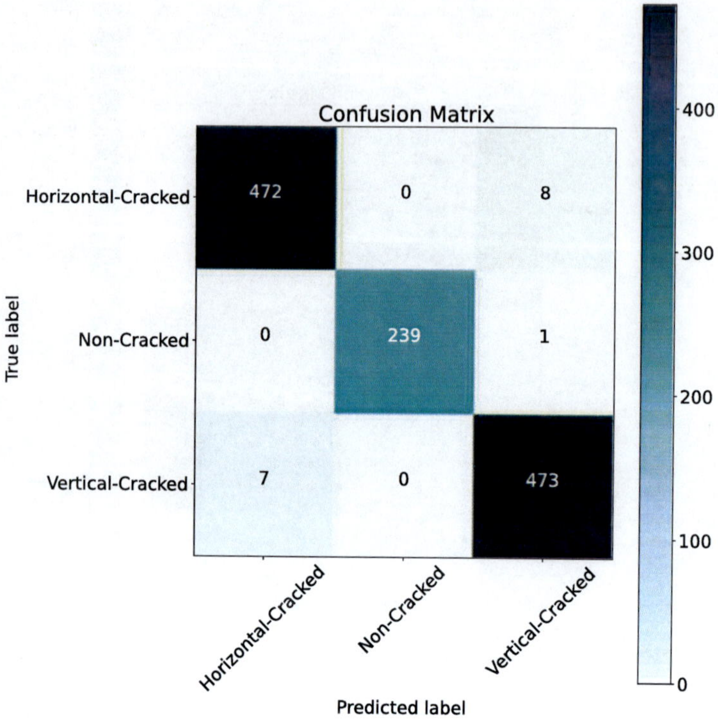

Fig. 10 Confusion matrix for three-class classification

Table 4 Performance evaluation report for three-class classification

Class	Precision	Recall	F1-score	Support
Horizontally cracked	0.985	0.983	0.984	480
Non-cracked	1.000	0.996	0.998	240
Vertically cracked	0.981	0.985	0.983	480
Accuracy			0.987	1200

values of performance evaluation matrices exceeded 99% in the two-class classification of cracks and 98% in the three-class classification of cracks. It is concluded from this study that the proposed model is capable of detecting and classifying cracks on concrete surfaces successfully without any overfitting and reduced training time.

References

1. Madanat, S.M., Karlaftis, M.G., McCarthy, P.S.: probabilistic infrastructure deterioration models with panel data. J. Infrastruct. Syst. **3**, 4–9 (1997). https://doi.org/10.1061/(ASCE)1076-0342(1997)3:1(4)
2. Huang, G., Liu, Z., Van Der Maaten, L., Weinberger, K.Q.: Densely Connected Convolutional Networks. In: 2017 IEEE Conference on Computer Vision and Pattern Recognition (CVPR). pp. 2261–2269. IEEE (2017). https://doi.org/10.1109/CVPR.2017.243
3. Kim, H., Ahn, E., Cho, S., Shin, M., Sim, S.H.: Comparative analysis of image binarization methods for crack identification in concrete structures. Cem. Concr. Res. **99**, 53–61 (2017). https://doi.org/10.1016/J.CEMCONRES.2017.04.018
4. Song, J., Kim, S., Liu, Z., Quang, N.N., Bien, F.: A real time nondestructive crack detection system for the automotive stamping process. IEEE Trans. Instrum. Meas. **65**, 2434–2441 (2016). https://doi.org/10.1109/TIM.2016.2583218
5. Bas, P.L., Anderson, B.E., Remillieux, M., Pieczonka, L., Ulrich, T.J.: Elasticity Nonlinear Diagnostic method for crack detection and depth estimation. J. Acoust. Soc. Am. **138**, 1836 (2015). https://doi.org/10.1121/1.4933844
6. Liu, Z., Cao, Y., Wang, Y., Wang, W.: Computer vision-based concrete crack detection using U-net fully convolutional networks. Autom. Constr. **104**, 129–139 (2019). https://doi.org/10.1016/J.AUTCON.2019.04.005
7. Mujeeb, A., Dai, W., Erdt, M., Sourin, A.: One class based feature learning approach for defect detection using deep autoencoders. Adv. Eng. Inf. **42**, (2019). https://doi.org/10.1016/J.AEI.2019.100933
8. Lynch, J.P., Farrar, C.R., Michaels, J.E.: Structural health monitoring: technological advances to practical implementations. Proc. IEEE. **104**, 1508–1512 (2016). https://doi.org/10.1109/JPROC.2016.2588818
9. Jahanshahi, M.R., Jazizadeh, F., Masri, S.F., Becerik-Gerber, B.: Unsupervised approach for autonomous pavement-defect detection and quantification using an inexpensive depth sensor. J. Comput. Civ. Eng. **27**, 743–754 (2012). https://doi.org/10.1061/(ASCE)CP.1943-5487.0000245
10. Fujita, Y., Hamamoto, Y.: A robust automatic crack detection method from noisy concrete surfaces. Mach. Vis. Appl. **222**(22), 245–254 (2010). https://doi.org/10.1007/S00138-009-0244-5
11. Zakeri, H., Nejad, F.M., Fahimifar, A.: Image Based Techniques for Crack Detection, Classification and Quantification in Asphalt Pavement: A Review. Arch. Comput. Methods Eng. 2016 244. 24, 935–977 (2016). https://doi.org/10.1007/S11831-016-9194-Z.
12. Zhang, W., Zhang, Z., Qi, D., Liu, Y.: Automatic crack detection and classification method for subway tunnel safety monitoring. Sensors, **14**, 19307–19328 (2014). https://doi.org/10.3390/S141019307
13. Li, S., Zhao, X.: Image-based concrete crack detection using convolutional neural network and exhaustive search technique. Adv. Civ. Eng. (2019). https://doi.org/10.1155/2019/6520620
14. Vu Dung, C., Duc Anh, L.: Autonomous concrete crack detection using deep fully convolutional neural network (2018). https://doi.org/10.1016/j.autcon.2018.11.028
15. Droguett, E.L., Tapia, J., Yanez, C., Boroschek, R.: Semantic segmentation model for crack images from concrete bridges for mobile devices (2020) https://doi.org/10.1177/1748006X20965111

16. Prasanna, P., Dana, K.J., Gucunski, N., Basily, B.B., La, H.M., Lim, R.S., Parvardeh, H.: Automated crack detection on concrete bridges. IEEE Trans. Autom. Sci. Eng. **13**, 591–599 (2016). https://doi.org/10.1109/TASE.2014.2354314
17. Hsieh, Y.-A., Tsai, Y.J.: Machine learning for crack detection: review and model performance comparison. J. Comput. Civ. Eng. **34**, 04020038 (2020). https://doi.org/10.1061/(ASCE)CP.1943-5487.0000918
18. Pauly, L., Peel, H., Luo, S., Hogg, D., Fuentes, R.: deeper networks for pavement crack detection. In: ISARC 2017 Proceedings of the 34th International Symposium on Automation and Robotics in Construction, pp. 479–485 (2017). https://doi.org/10.22260/ISARC2017/0066
19. Alipour, M., Harris, D.K.: Increasing the robustness of material-specific deep learning models for crack detection across different materials. Eng. Struct. **206**, 110157 (2020). https://doi.org/10.1016/j.engstruct.2019.110157
20. Chollet, F.: Xception: deep learning with depthwise separable convolutions. In: 2017 IEEE Conference on Computer Vision and Pattern Recognition (CVPR), pp. 1800–1807. IEEE (2017). https://doi.org/10.1109/CVPR.2017.195
21. Özgenel, Ç.F.: Concrete crack images for classification. Mendeley Data. V2. https://doi.org/10.17632/5y9wdsg2zt.2
22. Zhang, L., Yang, F., Daniel Zhang, Y., Zhu, Y.J.: Road crack detection using deep convolutional neural network. In: 2016 IEEE International Conference on Image Processing (ICIP), pp. 3708–3712. IEEE (2016). https://doi.org/10.1109/ICIP.2016.7533052
23. Kannojia, S.P., Jaiswal, G.: Effects of varying resolution on performance of cnn based image classification an experimental study. Int. J. Comput. Sci. Eng. **6**, 451–456 (2018). https://doi.org/10.26438/ijcse/v6i9.451456
24. Flah, M., Suleiman, A.R., Nehdi, M.L.: Classification and quantification of cracks in concrete structures using deep learning image-based techniques. Cem. Concr. Compos. **114**, 103781 (2020). https://doi.org/10.1016/j.cemconcomp.2020.103781
25. Canny, J.: A computational approach to edge detection. IEEE Trans. Pattern Anal. Mach. Intell. PAMI **8**, 679–698 (1986). https://doi.org/10.1109/TPAMI.1986.4767851
26. Vincent, L.: Morphological area openings and closings for grey-scale images. In: Toet, O,Y.-L., Foster, A., Heijmans, D., Meer, P. (eds.) *Shape in Picture*, pp. 197–208. Springer, Berlin, Heidelberg (1994). https://doi.org/10.1007/978-3-662-03039-4_13
27. O'Mahony, N., Campbell, S., Carvalho, A., Harapanahalli, S., Hernandez, G.V., Krpalkova, L., Riordan, D., Walsh, J.: Advances in Computer Vision. Springer International Publishing, Cham (2020). https://doi.org/10.1007/978-3-030-17795-9
28. Guo, Y., Li, Y., Wang, L., Rosing, T.: Depthwise convolution is all you need for learning multiple visual domains. Proc. AAAI Conf. Artif. Intell. **33**, 8368–8375 (2019). https://doi.org/10.1609/aaai.v33i01.33018368
29. Nair, V., Hinton, G.E.: Rectified linear units improve restricted boltzmann machines. In: Fürnkranz, J., Joachims, T. (eds.) Proceedings of the 27th International Conference on Machine Learning (ICML-10), June 21–24, pp. 807–814. Omnipress, Haifa, Israel, (2010)
30. Goodfellow, I., Bengio, Y., Courville, A.: Deep Learning. MIT Press (2016)
31. Zeiler, M.D., Ranzato, M., Monga, R., Mao, M., Yang, K., Le, Q.V., Nguyen, P., Senior, A., Vanhoucke, V., Dean, J., Hinton, G.E.: On rectified linear units for speech processing. In: 2013 IEEE International Conference on Acoustics, Speech and Signal Processing, pp. 3517–3521. IEEE (2013). https://doi.org/10.1109/ICASSP.2013.6638312
32. Srivastava, N., Hinton, G., Krizhevsky, A., Sutskever, I., Salakhutdinov, R.: Dropout: a simple way to prevent neural networks from overfitting. J. Mach. Learn. Res. **15**, 1929–1958 (2014)
33. Ioffe, S., Szegedy, C.: Batch normalization: accelerating deep network training by reducing internal covariate shift. In: 32nd International Conferences Machine Learning ICML, vol. 1, pp. 448–456 (2015)

34. Bridle, J.S.: Training stochastic model recognition algorithms as networks can lead to maximum mutual information estimation of parameters. In: Proceedings of the 2nd International Conference on Neural Information Processing Systems, pp. 211–217. MIT Press, Cambridge, MA, USA (1989)
35. Kingma, D.P., Ba, J.: Adam: a method for stochastic optimization. In: 3rd International Conferences Learning Representation ICLR 2015 Conferences Track Proceeding, pp. 1–15 (2014)

Machine-Learning Based Physical Exercise Identification with Heuristic Optimized Features Prioritization

Proloy Kumar Mondal, Md. Abdul Awal, and Abdullah-Al Nahid

Abstract Human Activity Recognition (HAR) is a sensor-based observation of human action. HAR is used to monitor the health issues associated with human activities. HAR is used to monitor human health, observation of elderly people, and fitness tracking activity. HAR is an advanced application of Machine Learning (ML), and Artificial Intelligence is used to detect various human-related activities. In this study, we have described a few physical exercises, using Random Forest (RF) classifier. In that case, we have obtained an accuracy of 99%. We have also used a few feature selection algorithms such as Equilibrium Algorithm (EQ), Marine Predators algorithm (MPA), Tree Growth Algorithm (TGA), Artificial Butterfly Algorithm (ABO), and Bat Algorithm (BA) to reduce computational complexity and time. These algorithms were tested using a KU-HAR dataset, which provided 97%, 98%, 99.8%, 97%, and 97% accuracy in classification for the dataset used for testing. This study aims to understand the extent to which supervised machine learning algorithms can achieve well and how it affects the accuracy of sensor recognition. This approach will help researchers conduct more research in the future on the recognition of human exercise activities.

Keywords Exercise recognition · Heuristic algorithms · Random forest · Feature selection

P. K. Mondal · Md. A. Awal · A.-A. Nahid (✉)
Electronics and Communication Engineering Discipline, Khulna University, Khulna 9208, Bangladesh
e-mail: nahid.ece.ku@gmail.com

P. K. Mondal
e-mail: proloykumar1996@gmail.com

Md. A. Awal
e-mail: m.awal@ece.ku.ac.bd

S. Hossain et al. (eds.), *Proceedings of International Conference on Fourth Industrial Revolution and Beyond 2021*, Lecture Notes in Networks and Systems 437,
https://doi.org/10.1007/978-981-19-2445-3_4

1 Introduction

The ability to recognize human activity is the most effective method of studying human behavior by analyzing physical and mental aspects. Wearable smartphone devices have many types of sensors. Due to the perception of these sensors, human activity has become a dynamic and challenging field of activity recognition study. Since the development of sensors, any activity has been organized by collecting information from these sensors, such as daily life, work activities, eating activities, and exercise activities. The number of older people in the world is gradually increasing. It's becoming increasingly difficult to care for those who are getting older over the next few decades. The number of elderly people will increase day by day in society and this will continuously increase as the days go. It takes a lot of doctors to monitor these older people, which can be a lot more challenging. If we can bring the older people into a systematic institute, this process will be comfortable for everyone and easier to manage. However, monitoring older people's daily activities considering their safety and well-being is a challenging task. Considering these issues, wearable sensors technology can be a reasonable solution to monitor the unexpected health conditions of older people. Al most every adult person in the world has a smartphone. Figure 1 shows the exploration of raw data extracted from wearable sensors.

Have you ever thought, how does your smartphone know when you are walking, when you are running, when you are sleeping and everything? Your smartphone has lots of sensors that give various information, such as Wi-Fi, audio, image, direction, and acceleration; sensors are very common nowadays. We can use this smartphone sensor to collect data from these older people from different organizations. However, older people can be accurately monitored by applying ML and artificial intelligence technology. As the number of admitted people for checking their health condition can be reduced, which ultimately can reduce the doctor's activities. Michalis Vrigkas et al. used the HAR dataset to classify the activity as human along with the analysis of social behavior [1]. In Zebin et al. [2], they tested various algorithms of human activity recognition with respect to the time and frequency domain. They found these results: Support Vector Machine (SVM) with quadratic kernel classifier with an accuracy of 93.5 % and an ensemble classifier that incorporates bagging and boosting that has the

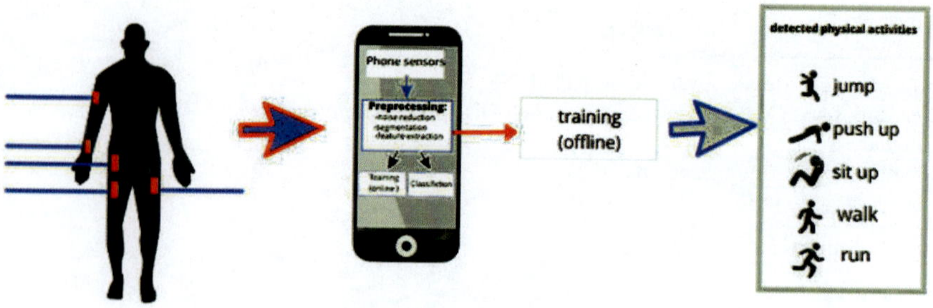

Fig. 1 Human activity monitoring based on wearable sensors

precision of 94.6 %. It outperforms other algorithms for classifying activities. The author in Ogbuabor and La [3] says that the accelerometer and the gyroscope both sensors perform together better than performing individually to recognize human activity. When the sensor performs individually then accelerometer accuracy was 92% but the accuracy became 95% when the sensors perform together. In Huang et al. [4], they collected data and classified them using RF, Decision Tree, and SVM algorithm. Through their work, they have achieved astounding accuracy levels of 81.00 % and a dominant hand sensor accuracy of 80 %, which is 7 % more than the dominant. The researchers in Wang and Zhang [5] utilized an Android-based application for data collection to create human-like speed. They are able to recognize the speed of three people. Then, they categorize the information using an algorithm called the Waikato Environment for Knowledge Analysis (WEKA) algorithm. In their case, they have obtained an accuracy of 86%. The authors in Kirchner et al. [6] have classified physical activity on the basis of the KNN algorithm that uses combination sensors of (accelerometer, gyroscope, and barometer) for accuracy in classification as well as power consumption and time computation. In Zebin et al. [7], the authors have used SVM and KNN-based ML algorithms. They have presented a system-recognized and general-purpose structure with six different activities. They collected data from several volunteers using accelerometer and gyroscope sensors. In Kańtoch [8], the authors have used wearable sensors to classify daily activities. They have applied neural network classification. Till the end, they have got 82% accuracy in implementation and recognized system activities and free-living for practice. This study aimed to conduct regular rehabilitation exercises for home visits, especially for long-term residents. Akram Bayat et al. proposed a recognition system designed for new digital low-pass filters and selected five classification models of machine learning to offer better results. They have achieved a recognition rate of 91.15% during daily tasks using only a single Axial-Accelerometer [9]. In Casale et al. [10], the authors developed a wearable system that is easy and comfortable to use. They have collected 20 computationally efficient features using a wearable system. In their work, they have applied the RF machine learning algorithm. In their instance, they obtained an accuracy of 94%. In 2017, Song-Mi Lee et al. showed that 1D convolutional neural network (CNN) method has 92 % classification accuracy, and the RF has 89.10% accuracy in ternary activity recognition [11]. In the next year, Chandini [12] proposed a method to recognize and classify activities using sensor data and this method employs the most popular algorithm in deep learning, namely LSTM. In that instance, they have obtained 97% classification accuracy. In this research, we will define a classification system built on an RF that can identify various human actions. In that case, the activities of different people are investigated using accelerometer and gyroscope sensors. The method is evaluated using the KU-HAR dataset. The structure of this paper is according to the following format Sect. 2 provides an explanation of the methodology used in this study, while Sect. 3 presents the research findings, and we conclude our conclusion in Sect. 4.

2 Methodology

In this study, we have used the KU-HAR dataset shared in Nahid et al. [13]. In this experiment, we have performed a five-class classification problem based on a supervised manner. Since it is supervised machine learning, we need labeled information. Though there are lots of statistical features available, we have made the work using only four statistical feature selections. They are mean, maximum, minimum, and standard deviation. And finally, results are presented based on the RF classification model. Figure 2 shows a graphical view of the proposed model.

2.1 Dataset Description

The KU-HAR dataset comprises 18 classes covering a range of behavior containing 90 people (75 males and 15 females). Five categories (Jump Sit-up, Push-up, Sit-up Walk, and Run) were selected for the study to create an additional sub-dataset that is based on physical activity. An accelerometer worn by a wearable sensor and the gyroscope were employed to gather this information. The dataset contains subsamples of 20,749 as well as a total of 1800 features. This dataset includes a total count of 3628 samples. It is split into two sets (i.e., one set of training and one test set). The first set comprises 2539 samples (69.99%). The remaining set includes 1021 samples (30.01%). The remainder of the data is presented in Table 1 as described below.

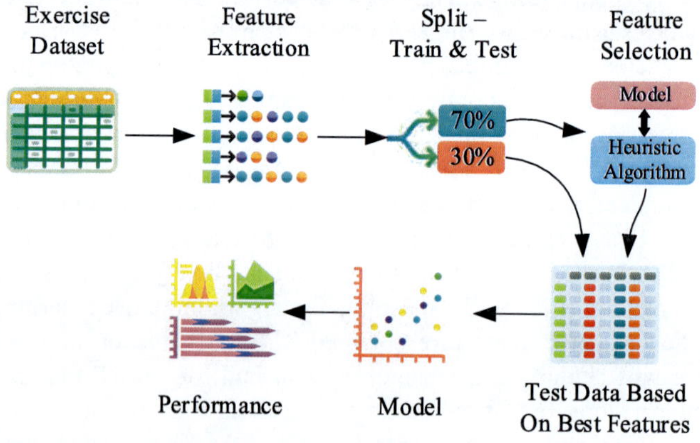

Fig. 2 Block diagram of the classification result

Table 1 Activity class, their labels, and samples

Activity					
Class label	Jump	Push-up	Sit-up	Walk	Run
Samples	666	480	1005	882	595

Table 2 Types of features for the HAR in this work

Domain	Time	Time	Time	Time
Feature	Mean Value	Maximum	Minimum	Standard Deviation
Formula	Mean = $\frac{1}{N}\sum_{i=1}^{N} S_i$	max = MAX(S_t)	min = min (S_t)	Std = $\sqrt{\frac{1}{N}\sum_{i=1}^{N}(si - men)^2}$

2.2 *Feature Explanation*

The extraction of features is a crucial component of the activity recognition systems. The purpose of feature extraction is to extract the common characteristics of acquired signals related to the same type of activity. Then, term feature describes data that is computed using the sensor's raw signal and any other method [14]. If we would use the raw data, then our dataset feature would become 1800 and create more complexity. That is why, we used four statistical features to reduce complexity, which are mean, minimum, maximum, and standard deviation. Now, we have 24 total features in our dataset. For example, the data of the X-axis of the accelerometer is identified by A (x, mean), A (x, max), A (x, min), A (x, std) the data of the Y-axis is identified by A (y, mean), A (y, max), A (y, min), A (y, std) and the axis of Z is identified by A (z, mean), A(z, max), A(z, min), A(z, std). Similarly, the data of the gyroscope has been identified following the same procedure. Assume that S_t is the data segment in a data Window, S_i stands for the ith data obtained from the accelerometer or gyroscope sensor at the time. All four function formulas in Table 2 were used to calculate the data in this paper.

2.3 *Classifier Model*

The paper describes RF classification algorithms employed to determine human activity recognition as described below. RF is a combination of tree estimates and random vectors. Each tree is constructed based on the probability of a random vector. The RF algorithm is an important modification to the bagging process that results in a variety of unbound trees [15]. Breiman invented it in 2001. RF excels in a variety

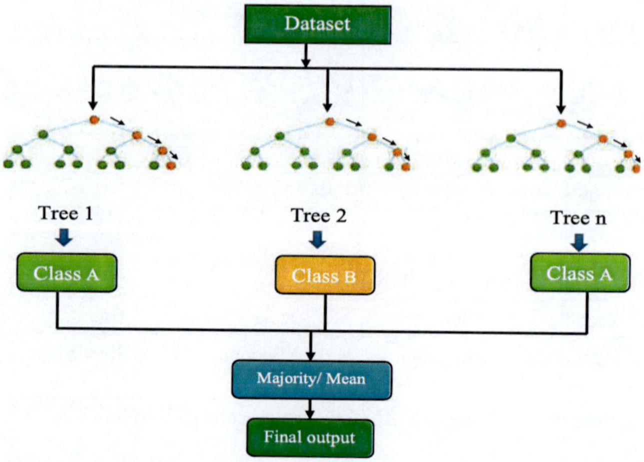

Fig. 3 Random forest representation

of classification tasks. It is a great choice for many classification problems. RF algorithms are also simpler to train and tune. It is a popular choice in ML algorithms and is present in numerous software. The best advantage of the random forest model is its sophistication. It is used for classification task, and it is much easier to show the importance of the features of the input. In the random forest, the classifier will not overfit. The hyperparameters are easy to understand and this model produced good prediction results. Figure 3 illustrates the operation of the random forest classification algorithm.

2.4 *Feature Selection*

Feature selection is a difficult and traditional optimization task in the fields of ML. This allows us to select the most important features from the actual features[16]. Although feature selection might be thought of as an exception to feature extraction, mathematically it is not the same thing [16]. The researchers are different in both these areas. An objective function is required to evaluate candidate features in feature selection. There are two types of objective functions: wrappers and filters. In this study, we have used five heuristic-based feature selection algorithms, which are described in the following.

2.4.1 Equilibrium Algorithm (EQ)

An equilibrium optimizer or (EA) [17] is an innovative physics-based algorithm for optimizing dynamic and equilibrium states. The EA is a new, physics-based optimization algorithm to solve continuous optimization issues. The EA can randomly alter the solution via high exploration and exploitation. Particles containing EA can have similar concentrations to those found in particle swarm optimizing (PSO). These are search agents. To reach the equilibrium state (optimal solution), the search agents will randomly adjust their concentrations to reflect the best solutions [17].

2.4.2 Marine Predators Algorithm (MPA)

The marine predators algorithm (MPA) is a recently developed metaheuristic algorithm that mimics the nature of predator and prey [18]. MPA simulates most animals' main aim of searching for food. However, a predator will continue to hunt for prey. The algorithm was based on both a predator and a prey perspective. The MPA is composed of three phases, each concerning the prey's rational velocity and the predator.

2.4.3 Bat Algorithm (BA)

Bats' fascinating behavior has drawn scientists from all disciplines. Their ability to locate echolocation has made them fascinating. Echolocation is a sonar-like phenomenon. Microscopic bats typically emit a loud swell of sound and then expect it to strike an object. The sound then comes back to them after a brief period of time [19]. An interesting metaheuristic optimization method basing itself on the bat's behavior.

2.4.4 Artificial Butterfly Algorithm (ABO)

Qi et al. were inspired by the mate-finding strategy in speckled woods and proposed artificial butterfly optimization, a metaheuristic algorithm [20]. Spoiled woods prefer to reside close to the edge of forests and in areas where the sun shines brilliantly on the trees and produces lots of sunspots [21]. The health of the population of butterflies is assessed and classified into two categories. Butterfly species with higher fitness are classified as sunspot butterflies, while those with lower fitness are called canopy butterflies. Each group is assigned a different flight strategy. ABO algorithms can be divided into three flight modes: sunspot mode, canopy mode, and free mode [21].

2.4.5 Tree Growth Algorithm (TGA)

The tree growth algorithm is a novel optimization approach and a brand metaheuristic optimization algorithm [22]. The jungle has four types of the tree population. The best tree group (or the best) will grow faster and have a slower growth rate. The second group, which is competitive for light tree groups, is the best. This group aims to move the tree toward the location that is closest to the nearest two trees. The third group is called the removal or substitute group. It takes out the most dangerous trees and plant the new ones. The reproduction group is the final group. The best plants in this group can be used to create new plants. TGA's algorithm can be as follows: First, a random population of trees will be generated. Next, each tree's fitness is assessed. The fitness score determines the order in which trees are sorted. The best N1 trees will be assigned to the first group.

3 Result and Discussion

In this part, we will discuss the findings of our research and present the various results of the study. However, our test subject was physical exercise categorized based on different human activities such as Jump, Push-up, Sit-up, Walk, and Run. In this segment, we have compared the execution of the RF model dependent on the confusion matrix and the time taken to prepare the model. RF models will be able to classify our human test activities efficiently after being trained with ML algorithms using smartphone sensor data. Our model has a precision–recall of 99% for the random forest algorithm. Analyzing a classification task's resultant confusion matrix is another abstinent method to determine the potency of a classifier. The confusion matrix is described as a matrix that is used to compare the test specimens to their actual class. Figure 4 illustrates how many samples have been properly indexed and how many of them were misclassified by random forest Fig. 4. It provides the confusion matrix for our model of HAR classification. We can see that the classification matrix gives 196 correct predictions and 4 wrong predictions for the jump class, 142 correct predictions and 2 wrong predictions for the push-up class, 176 correct predictions and 2 wrong predictions for the run class, 300 correct predictions and 2 wrong predictions for the sit-up class. Finally, classification matrix gives 264 correct predictions and 1 wrong prediction for the walking class. However, the one that was most wrongly classified was the jump class, and the class with the lowest miscalculation was the walking class.

In this study, we employed five heuristic algorithms. Data loss is certain to the feature of each algorithm. Figure 5 illustrates the loss curve of five algorithms at each period. Here, the blue line represents EQ, the gray line represents ABO, the green line represents BA, the yellow line represents TGA, and the red line represents MPA Algorithm. As you can see in the figure, the loss has slowly diminished with each passing interval. The amount of loss was low throughout the first 20 epochs, but later it reached the final epoch of loss. From the graph, while the epoch is minimum, BA

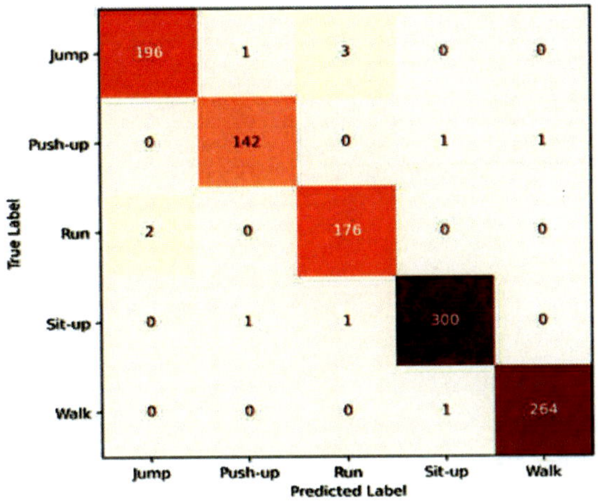

Fig. 4 Confusion matrix of our proposed model based on the test dataset

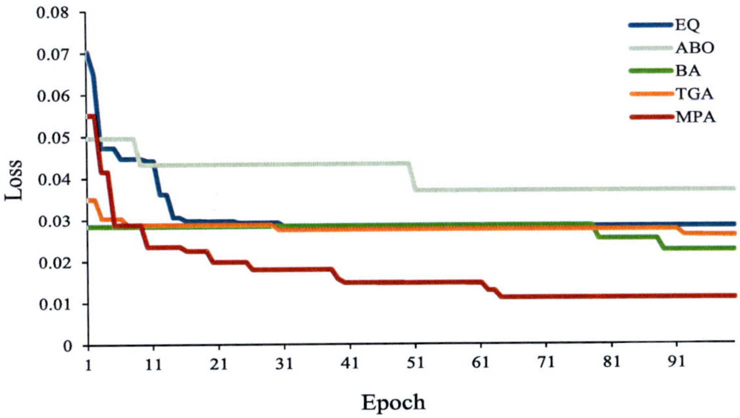

Fig. 5 Possible data loss at each epoch

shows minimum loss and EQ shows a maximum loss. With the change of epoch loss changes for every algorithm, when epoch is maximum, MPA shows minimum loss and ABO shows a maximum loss. Therefore, for the lowest epoch, BA accuracy is higher than others and at the maximum epoch, MPA accuracy is maximum.

F1 scores, recall, and precision are the three variables you can use to determine the classifier's effectiveness. The percentage of correctly classified samples that fall within positively classified models in the same class is precision [23]. The high precision implies a low false positivity rate and vice versa. The recall is the percentage of instances that are truly positive in an area that the classifier recognizes. A high level of recall could suggest that the classifier could correctly predict some false negative examples of the class and reverse them. Typically, the recall and precision from all class types are summed to determine the accuracy and recall of the model.

Table 3 Classification report (Precision, recall, and F1-score)

Class	Jump	Push-up	Run	Sit-up	Walk
Precision	0.99	0.99	0.98	0.99	1.00
Recall	0.98	0.99	0.99	0.99	1.00
F1-Score	0.98	0.99	0.98	0.99	1.00

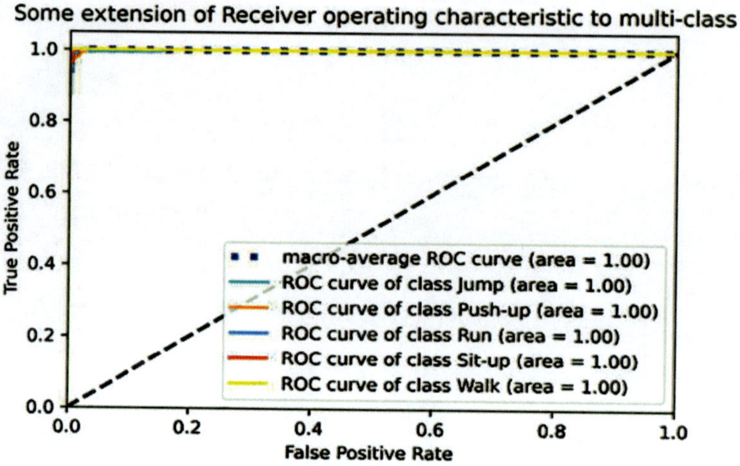

Fig. 6 ROC curves of various classes

F1-score is either a harmonic average of precision and recall values or the median weighted. Table 3 displays the precision–recall, precision as well as F1 score of the random forest model classified as a forest. The three matrices all have similar scores for the test and train classifications.

In Fig. 6, we present the Receiver Operating Characteristics (ROC) curves in order to show the model's capabilities. The ROC curve does not become affected by classes that are not balanced within the dataset and provides a more accurate measure of the accuracy of classification. It plots the actual rate of classifier recall versus an inaccurate rate for the fall-out of the classifier. The graph illustrates that the actual positive rate of all classes is very close to 1. This indicates that the algorithm used to classify is very accurate to categorize the various activities into different classes.

In this paper, we checked what would be the overall performance if any random feature fails. For this test, at first, we randomly set a single feature to null ten times and collected average accuracy and that was 98.8%. Secondly, we randomly set two features to null ten times and collected the average accuracy, which was 93%. We repeated this until we got the average accuracy while randomly five features were set to null. From this, we can understand that if for some reason our sensors fail to give any one feature then what will be the performance of our model.

In the random forest classifier, we see that we get very good accuracy using 24 features. However, now we wanted to see which features provide very good performance. For this, we apply five heuristic algorithms such as EQ, MPA, TGA,

Table 4 Results based on the heuristic algorithm

Algorithm	Class	Jump	Push-up	Run	Sit-up	Walk
EA	Precision	0.98	0.92	0.98	0.98	0.99
	Recall	0.99	0.95	0.97	0.97	0.98
	F1-score	0.98	0.93	0.98	0.97	0.98
MPA	Precision	0.97	0.94	0.99	1.00	0.97
	Recall	0.98	0.98	0.97	0.99	0.97
	F1-score	0.97	0.96	0.98	0.99	0.97
TGA	Precision	0.94	0.93	0.97	0.97	1.00
	Recall	0.98	0.95	0.97	0.95	0.98
	F1-score	0.96	0.94	0.97	0.96	0.99
ABO	Precision	0.98	0.95	0.98	0.99	0.96
	Recall	0.97	0.96	0.98	0.98	0.97
	F1-score	0.97	0.96	0.98	0.99	0.97
BA	Precision	0.97	0.95	0.96	1.00	0.98
	Recall	0.98	0.93	0.98	0.99	0.97
	F1-score	0.98	0.94	0.97	0.99	0.97

ABO, and BA. Among all the 24 features, few features perform well for EA A (x, std), A(y, std), A(z, mean), A(z, std), G(x, max), G(x, min), G(y, mean), G(y, max), G(y, min), G(z, max), G(z, min), G(z, std). We obtained 97% accuracy using this feature. For MPA, few features perform well, such as A (x, std), A (z, std), G (x, min), G (y, max), G (y, min), G (y, std), G (z, max), G (z, min), G (z, std) and we have gotten 98% accuracy. For TGA, few features perform well, such as A (x, min), A (y, mean), A (y, std), A (z, mean), A (z, std), G (x, mean), G (x, max), G (x, min), G (x, std), G (y, std), G (z, max), G (z, std). For ABO, few features perform very well, such as A (x, mean), A (x, max), A (y, mean), A (z, mean), A(z, std), G (x, mean), G (x, min), G(x, std), G (y, max), G (y, min), G (y, std), G(z, min), G (z, std) and we obtained 97% accuracy. Finally, for BA, few features perform well, such as A (x, std), A (z, mean), A (z, std), G (x, std), G (y, max), G (y, min), G (z, max), G (z, min) and we obtained 97% accuracy. Table 4 presents the accuracy as well as recall and F1-scores of the five algorithmic methods for selecting features.

4 Conclusions

In this paper, we suggest ML to carry out human-based activity detection (HAR) with the help of the accelerometer and gyroscope of the smartphone. The paper presents the physical- exercise-based HAR classification model and then tests it on the KU-HAR dataset. We have been able to obtain results of 99.00% accuracy for classification. However, the accuracy of the dataset is measured by using five heuristic methods to assess the reduced computational complexity and time. The five algorithms have yielded good results. The method is able to be applied to an environment that is smart, especially for people who are older and need to carry out and monitor particular physical activities frequently. The method developed is employed to meet as a theoretical approach to monitoring physical rehabilitation.

References

1. Vrigkas, M., Nikou, C., Kakadiaris, I.A.: A review of human activity recognition methods. Front. Robot. AI **2**, 28 (2015)
2. Zebin, T., Scully, P.J., Ozanyan, K.B.: Evaluation of supervised classification algorithms for human activity recognition with inertial sensors. In: 2017 IEEE Sensors, pp. 1–3. https://doi.org/10.1109/ICSENS.2017.8234222
3. Ogbuabor, G., La, R.: Human activity recognition for healthcare using smartphones. In: Proceedings of the 2018 10th International Conference on Machine Learning and Computing, pp. 41–46 (2018)
4. Huang, Y.-C., Yi, C.-W., Peng, W.-C., Lin, H.-C., Huang, C.-Y.: A study on multiple wearable sensors for activity recognition. In: IEEE Conference on Dependable and Secure Computing, pp. 449–452 (2017)
5. Wang, C., Zhang, W.: Activity recognition based on smartphone and dual-tree complex wavelet transform. In: 2015 8th International Symposium on Computational Intelligence and Design (ISCID), vol. 2, pp. 267–270 (2015)
6. Kirchner, J., Faghih-Naini, S., Bisgin, P., Fischer, G.: Sensor selection for classification of physical activity in long-term wearable devices. In: IEEE Sensors, pp. 1–4 (2018)
7. Zebin, T., Scully, P.J., Ozanyan, K.B.: Inertial sensor-based modelling of human activity classes: Feature extraction and multi-sensor data fusion using machine learning algorithms. In: e-Health 360°. Springer, pp. 306–314 (2017)
8. Kańtoch, E.: Human activity recognition for physical rehabilitation using wearable sensors fusion and artificial neural networks. Comput. Cardiol. (CinC) **2017**, 1–4 (2017)
9. Bayat, A., Pomplun, M., Tran, D.A.: A study on human activity recognition using accelerometer data from smartphones. Procedia Comput. Sci. **34**, 450–457 (2014)
10. Casale, P., Pujol, O., Radeva, P.: Human activity recognition from accelerometer data using a wearable device, vol. 6669, p. 296. https://doi.org/10.1007/978-3-642-21257-4_36
11. Lee, S.-M., Yoon, S.M., Cho, H.: Human activity recognition from accelerometer data using convolutional neural network. In: 2017 IEEE International Conference on Big Data and Smart Computing (BigComp), pp. 131–134. https://doi.org/10.1109/BIGCOMP.2017.7881728
12. Chandini, U.: A machine learning based activity recognition for ambient assisted living. Int. J. Future Rev. Comput. Sci. Commun. Eng. **4**(3) (2018)
13. Nahid, A.-A., Sikder, N., Rafi, I.: KU-HAR: an open dataset for human activity recognition 5 (2021). https://doi.org/10.17632/45f952y38r.5
14. Dehkordi, M.B., Zaraki, A., Setchi, R.: Feature extraction and feature selection in smartphone-based activity recognition. Procedia Comput. Sci. **176**, 2655–2664 (2020). https://doi.org/10.1016/j.procs.2020.09.301
15. Nour, M., Polat, K.: automatic classification of hypertension types based on personal features by machine learning algorithms. Math. Probl. Eng. **2020**, e2742781 (2020). https://doi.org/10.1155/2020/2742781
16. Rasekh, A., Chen, C.-A., Lu, Y.: Human activity recognition using smartphone (2014). http://arxiv.org/abs/1401.8212. Accessed 2 Sept 2021
17. Gao, Y., Zhou, Y., Luo, Q.: An efficient binary equilibrium optimizer algorithm for feature selection. IEEE Access **8**, 140936–140963 (2020). https://doi.org/10.1109/ACCESS.2020.3013617

18. Abd Elminaam, D.S., Nabil, A., Ibraheem, S.A., Houssein, E.H.: An efficient marine predators algorithm for feature selection. IEEE Access **9**, 60136–60153 (2021)
19. Nakamura, R.Y., Pereira, L.A., Costa, K.A., Rodrigues, D., Papa, J.P., Yang, X.-S.: BBA: a binary bat algorithm for feature selection. In: 2012 25th SIBGRAPI conference on graphics, patterns and images, pp. 291–297 (2021)
20. Chand, S., Wagner, M.: Evolutionary many-objective optimization: a quick-start guide. Surv. Oper. Res. Manag. Sci. **20**(2), 35–42 (2015). https://doi.org/10.1016/j.sorms.2015.08.001
21. Rodrigues, D., de Albuquerque, V.H.C., Papa, J.P.: A multi-objective artificial butterfly optimization approach for feature selection. Appl. Soft Comput. **94**, 106442 (2020). https://doi.org/10.1016/j.asoc.2020.106442
22. Too, J., Abdullah, A.R., Mohd Saad, N., Mohd Ali, N.: Feature selection based on binary tree growth algorithm for the classification of myoelectric signals. Machines **6**(4), 65 (2018)
23. Sikder, N., Chowdhury, M.S., Arif, A.M.S., Nahid, A.-A.: Human activity recognition using multichannel convolutional neural network. In: 2019 5th International Conference on Advances in Electrical Engineering (ICAEE), pp. 560–565 (2019)

Modelling Lips State Detection Using CNN for Non-verbal Communications

Abtahi Ishmam, Mahmudul Hasan, Md. Saif Hassan Onim, Koushik Roy, Md. Akiful Hoque Akif, and Hussain Nyeem

Abstract Vision-based deep learning models can be promising for speech-and-hearing-impaired and secret communications. While such non-verbal communications are primarily investigated with hand gestures and facial expressions, no research endeavour is tracked so far for the lips state (*i.e.*, open/close)-based interpretation/translation system. In support of this development, this paper reports two new convolutional neural network (CNN) models for lips state detection. Building upon two prominent lips landmark detectors, DLIB and MediaPipe, we simplify the lips state model with a set of six key landmarks and use their distances for the lips state classification. Thereby, both the models are developed to count the opening and closing of lips and thus, they can classify a symbol with the total count. Varying frame rates, lips movements and face angles are investigated to determine the effectiveness of the models. Our early experimental results demonstrate that the model with DLIB is relatively slower in terms of an average of six frames per second (FPS) and higher average detection accuracy of 95.25%. In contrast, the model with MediaPipe offers faster landmark detection capability with an average FPS of 20 and detection accuracy of 94.4%. Both models thus could effectively interpret the lips state for non-verbal semantics into a natural language.

Keywords Lips state detection · DLIB · MediaPipe · CNN · Non-verbal communications · Human–robot interaction

A. Ishmam (✉) · M. Hasan · Md. S. Hassan Onim · K. Roy · Md. A. Hoque Akif · H. Nyeem
Military Institute of Science and Technology (MIST), Mirpur Cantonment, Dhaka 1216, Bangladesh
e-mail: abtahiishmam3@gmail.com

Md. S. Hassan Onim
e-mail: saif@eece.mist.ac.bd

H. Nyeem
e-mail: h.nyeem@eece.mist.ac.bd

S. Hossain et al. (eds.), *Proceedings of International Conference on Fourth Industrial Revolution and Beyond 2021*, Lecture Notes in Networks and Systems 437,
https://doi.org/10.1007/978-981-19-2445-3_5

1 Introduction

Lips state (*i.e.*, *open* or *close*) detection can be promising for vision-based non-verbal communication system, which has traditionally been investigated with the head movement and gestures, and facial expression. Lips state detection and interpretation is a key step in many security, surveillance and law-enforcement applications. For example, lips reading can be helpful in emergency hostage situation. Such communications also enable a speech-and-hearing-impaired person (with a disorder like *stuttering*, *apraxia*, *dysarthria* or *muteness*) to communicate using lips state with a minimal effort. In support of developing such a system for non-verbal communication for translating simple lip state combinations to a complete instruction as illustrated in Fig. 1, we aim to start with the development of a lip state detection model in this paper. The envisaged model with higher possible detection accuracy could thus be a promising, cost-effective and as simple solution as a user-friendly mobile application.

Despite having an obvious potential for non-verbal communication, no or a few research endeavours can be tracked in the literature on lips state detection. For example, lip reading was studied for verbal communications using convolutional neural network (CNN) models [6]. Besides, lips states and movements have been partially considered for facial detection and emotion recognition, both with and without the landmarks of lips. The case of without lips landmarks has a higher dependency on the image contrast and spatial resolution, and thus the performance of this approach significantly varies [2]. In contrast, the lips landmarks-based detection widely investigated facial detection and emotion recognition. For example, Sharma et al. [9] considered face alignment and feature extraction in their face recognition model using convolutional neural network (CNN) and DLIB face alignment.

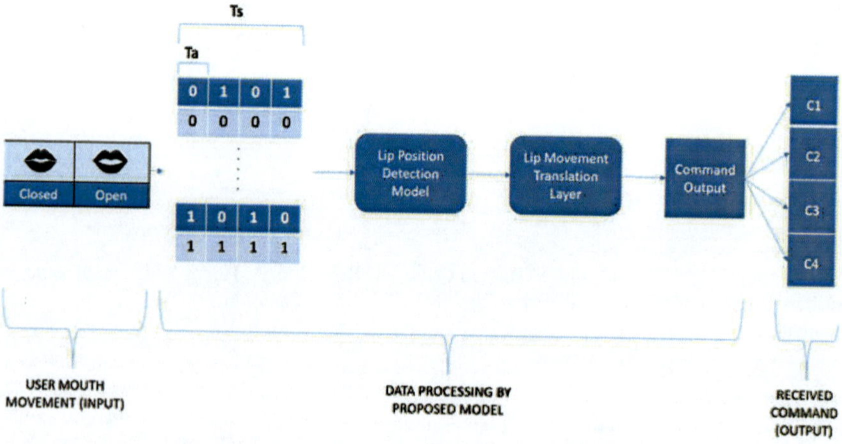

Fig. 1 A general framework of the proposed lips state detection-based interpretation/translation model for non-verbal communication system

To better tackle the challenges in lips detection, including varying skin colour, appearance and different lighting conditions, Juan et al. [11] considered the relative distances among faces, eyes and mouth to locate the mouth region. Their model can segment the lips more efficiently than some other prominent models. Amornpan et al. [1] applied transfer learning for feature extraction for a face recognition application using a pre-trained deep learning model and validated with the public face datasets like Extended Yale Face Database B (Cropped) and the Extended Cohn-Kanade Dataset (CK+).

Similarly, Singh et al. [10] proposed to use Viola Jones algorithm to localise the face and mouth in the image with an iterative and adaptive construction of merging threshold to make the model image quality invariant. Krause et al. [5] introduced a highly portable, and automatic solution for extracting oral posture from digital video using an existing face-tracking utility and OpenFace2. Later, Xu et al. [12] improved the facial landmark localisation across large poses using a split-and-aggregate strategy.

The above vision-based models focus on face tracking and have a partial consideration of lips detection. In addition to the contextual information extracted from the landmarks of eyes, nose and face shape, lips were considered to complement the features required for face detection and recognition and emotion classification. However, as mentioned at the beginning of this section, no lips state detection is considered so far for developing the envisaged translation model for non-verbal communication system.

In this paper, two new lips state detection models have been proposed for non-verbal communications. Both models use landmark detectors for the localisation of lips. Two highly accurate and robust facial landmark detectors: DLIB [4] and MediaPipe [7] are considered, which can gather facial landmark information from both real-time and static images. Building upon these detectors, we simplify modelling the lips state with a set of six key landmarks and their distances to detect relative variations in lips.

Thereby, we develop two models to accurately track the lips states. Particularly, these models start with the detection of human faces from the captured frames and isolate the face region followed by the extraction of the landmarks of lips. These identified landmarks and the distance among the landmark points are then used for the successive decision-making, approximation and classification of the lips state. The performance of these proposed models is finally analysed to learn their merits for the non-verbal communication system (see Sect. 3).

2 New Lips State Detection Models

Our research aims at developing an alternative interpretation or translation system for speech-and-hearing-impaired communication based on lips state. To this end, we have modelled lips states streamlining the landmarks, and thus, the developed lips states classifier model using two prominent landmark detectors, and thus, we construct two models as illustrated in Fig. 3. The first model that we call model-I uses the popular landmark detector *DLIB* to feed data into our neural network-based prediction model to predict the lips states. Similarly, we use the MediaPipe face-mesh landmark detector to develop the second model called model-II with support vector regression (SVR) prediction block. Having similar network architecture, the lips state classification accuracy and speed may primarily depend on their underlying detectors. Prior to analysis of these performances in Sect. 3, we now present below the necessary technical details of the proposed models.

2.1 Landmark Detection

DLIB. For the first model, pre-trained face feature point detector that comes with the DLIB library is used to obtain 68 Cartesian coordinate points corresponding to a specific area of the face. These 68 points shown in Fig. 2a come from the DLIB model which is trained on the iBUG 300-W dataset [8]. From those landmark points, six landmark points were selected for lip distance calculation.

2.1.1 MediaPipe

MediaPipe face mesh is a face geometry detector based on the blaze-face model. It can estimate 468 3D face landmarks as shown in Fig. 2b, from both recorded and real-time video streams even on mobile devices. The detector employs machine learning to infer the 3D surface geometry, requiring only a single camera input without the need for a dedicated depth sensor, a feature suitable for our purpose, accurate tracking of lips. Utilising lightweight model architectures together with GPU acceleration throughout the pipeline, the detector can also deliver real-time performance even on weaker processing machines such as *raspberry pi* or *jetson* devices. Out of the 468 landmarks, six were used for lips distance calculation (Fig. 3).

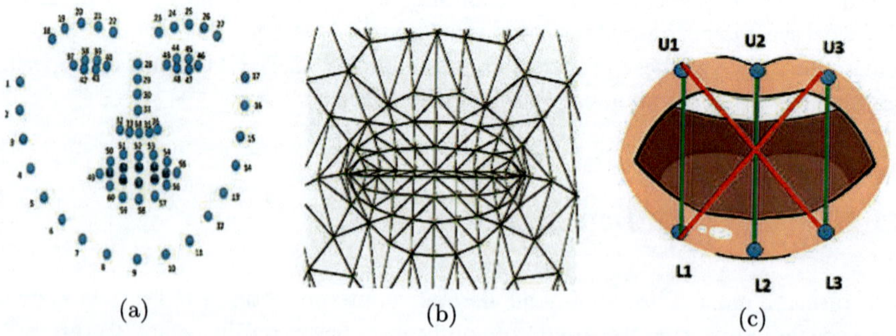

Fig. 2 Facial landmark points: **a** DLIB, **b** MediaPipe, and **c** simplified lips landmarks and distances of our model

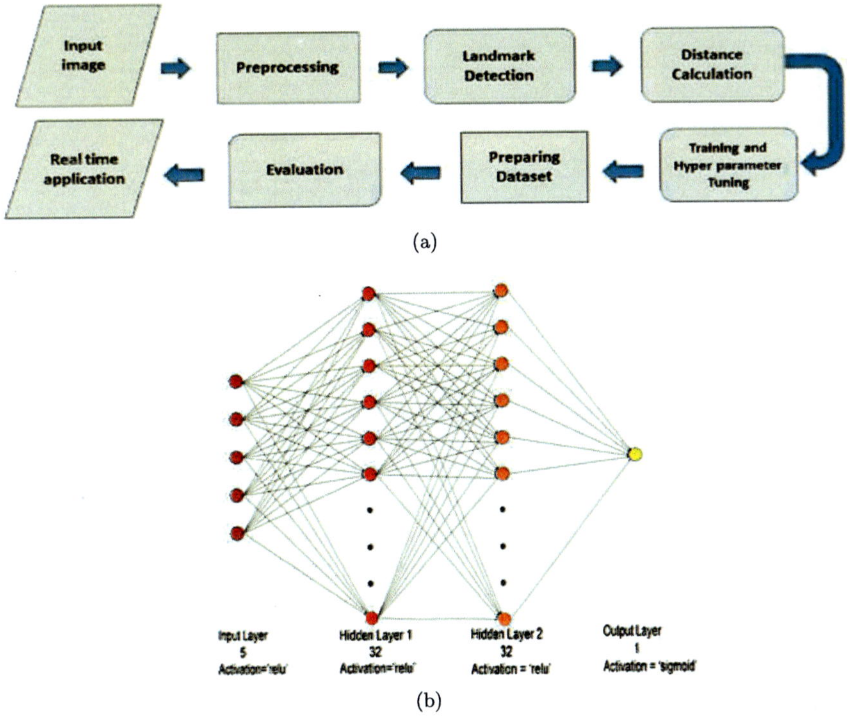

Fig. 3 Proposed model: **a** key processes and **b** neural network architecture

2.2 *Distance Calculation*

After detection of landmarks, five distances were calculated Fig. 2c from the (x_i, y_i) coordinates of the points. Six landmark points were selected for distance calculation shown in Table 1. These distances are computed using Eq. (1a)–(1e).

$$LD = \alpha \times \sqrt{(U_1[x_1] - L_1[x_2])^2 + (U_1[y_1] - L_1[y_2])^2} \tag{1a}$$

$$MD = \beta \times \sqrt{(U_2[x_1] - L_2[x_2])^2 + (U_2[y_1] - L_2[y_2])^2} \tag{1b}$$

$$RD = \gamma \times \sqrt{(U_3[x_1] - L3[x_2])^2 + (U_3[y_1] - L_3[y_2])^2} \tag{1c}$$

$$D_1 = \delta \times \sqrt{(U_1[x_1] - L_3[x_2])^2 + (U_1[y_1] - L_3[y_2])^2} \tag{1d}$$

$$D_2 = \epsilon \times \sqrt{(U_3[x_1] - L_1[x_2])^2 + (U_3[y_1] - L_1[y_2])^2} \tag{1e}$$

Table 1 Landmark used for distance calculation

Lips landmark point	Our model-I	Our model-II
U1	51	37
U2	52	267
U3	53	0
L1	59	84
L2	58	314
L3	57	17

Here, α, β, γ, δ, ϵ are the length control coefficient, which can be tuned for improved observation; U_1 = upper lip left point; U_2 = upper lip middle point; U_3 = upper lip right point; L_1 = lower lip left point; L_2 = lower lip middle point; L_3 = lower lip right point; LD = lip left distance; MD = lip middle distance; RD = lip right distance; $D1$ = lip diagonal 1 distance; $D2$ = lip diagonal 2 distance. These selected landmark points demonstrate the maximum variation upon lips state. Although the middle points U2 and L2 can be sufficient for our purpose, the other points were considered to tackle the variety of lips shapes during their states.

2.3 Dataset Collection

We collected video recordings of 15 individuals for training. For each person, the video was captured for both closed and open mouth positions. The lips training dataset consists of 23,412 frames. All of these frames were passed through the preprocessing stage to extract landmarks and calculate the five distances mentioned in distance calculation. Thus a dataset consisting of five lips landmark distances and mouth position, either open or close for all frames, was created and used to train the data on the train dataset. The validation data consists of 2325 frames of three people excluded from the training set.

2.4 Lips State Detection

2.4.1 Proposed Model-I Using DLIB Data

The proposed model-I utilises a typical and very lightweight dense neural net with one dense layer of 32 units and rectified linear unit (RELU) activation. Early stopping was employed for better optimisation of training time and to avoid over-fitting. For compilation of the model, the optimiser was set to Adam and loss was binary cross-entropy. After training and tuning, it was used in real-time application.

Even though the model is very lightweight, the underlying DLIB detector in the data preprocessing stage was adding enough complexity to turn the frames per second (FPS) down. We got about 5–9 FPS in real-time application, which is usable at best but not robust enough to work in every situation.

2.4.2 Proposed Model-II Using MediaPipe Data

The proposed model-II uses the landmark distance values obtained from MediaPipe and plugs it into a support vector classifier (SVC) model. The model was used with the default parameters as it was able to provide sufficiently accurate results.

2.5 Training Performance Evaluation

The mouth position detection model was trained for 350 epochs with early stopping employed to stop the model training if the accuracy improvement is negligible. It can be seen from the loss versus epoch curve that the model is very quick to get to a lower loss value (see Fig. 4). After a while the improvement as well as the learning rate gets saturated.

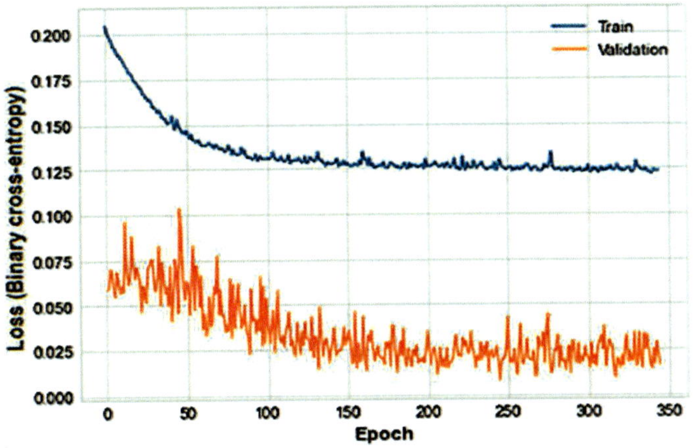

Fig. 4 Average loss versus epoch

2.6 Time Slots and Commands

Each of the detected actions is placed into a time slot for converting into commands as shown in Fig. 1. If the time slots are field with N number of actions, the command will be known as N actions per time slot window. The total number of actions per window will increase the maximum number of commands. This is limited by the maximum rate of the frame captured and the maximum number of actions taken by the user. Their overall relation can be expressed as follows:

$$T_a = T_s / N_a \tag{2}$$

T_s = time for N_a; T_a = time for action; N_a = total number of actions per time slot. Since it is binary action, $N_c = 2^{Na}$, where N_c = total number of commands. For example if $T_a = 0.6$ s, $N_a = 4$, then T_s will be $0.6 \times 4 = 2.4$ s.

3 Results

The model was tested on both recorded video footage with video resolution of 1920 × 1080 at 30 fps as shown in Fig. 5 and still images from Flickr-Faces-HQ Dataset (FFHQ) [3]. A model trained with a particular dataset may not give similar results when tested with data from the real world. The model is therefore evaluated for the cases unknown to it.

A step by step working procedure of our models for real-time application scenario is illustrated in Fig. 5. A few cases of visual detection of lips state are also illustrated for the considered datasets in Fig. 6. The images in the figure captured varying con-

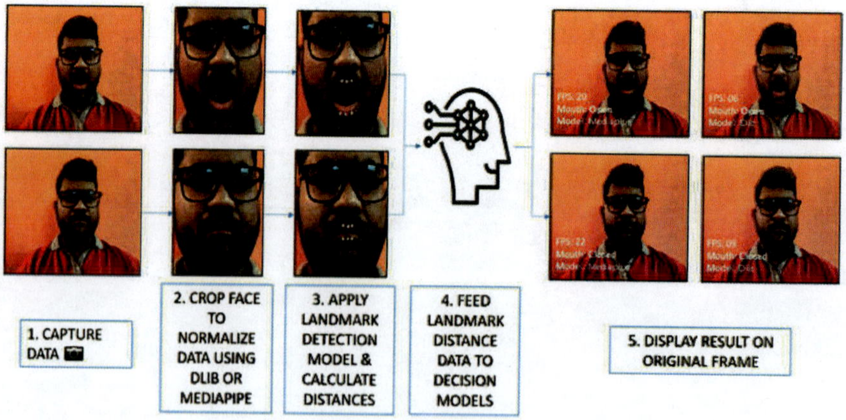

Fig. 5 Performance of the proposed model in a real-time application scenario

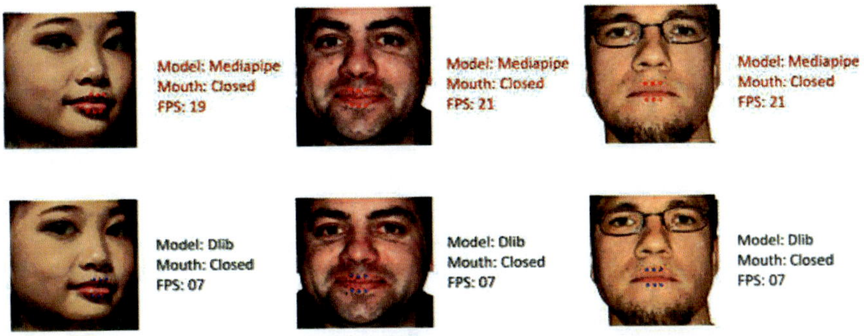

(a) Closed lips

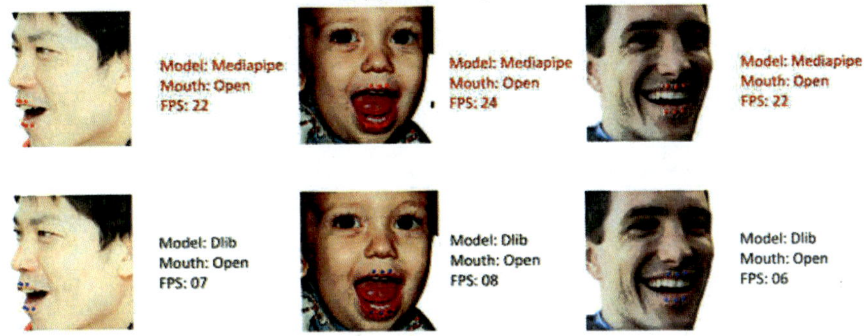

(b) Open lips

Fig. 6 Examples of lips state detection using proposed model-I (*bottom row*) and model-II (*top-row*) for opening and closing lips

ditions of the face, angle, appearance and lighting that illustrate the performance of the model for detecting the opening and closing of the lips.

Similarly, the performance of the models while applied to faces turned in different angles has been given in Table 2, which shows that the effect of face rotation can be quite significant and similar for both models. The detection accuracy is 100% from −40 °C to +40 °C angle. But the results are affected when the face is rotated beyond that. The accuracy starts declining at +-60°C still staying at a quite appreciable 90%. But if the face is rotated any further, both the models fail to detect the facial landmarks.

The models' accuracy at different speeds of changing the lips states is also evaluated and given in Table 3. We observed that the accuracy decreases for both models with the increase in the lips state per second. For instance, as in Table 3, both of the models can detect the lips states, while the lips open/close at a rate of two per second. For a different case, when the rate increases to six per second, nearly 40% of the state changes are not detected. However, compared to model-I, model-II is found more accurate for the states being changed at a higher rate.

Table 2 Detection accuracy of proposed model at varying face angles

Angle (degree)	Our model-I (%)	Our model-II (%)
+0	100	100
+20	100	100
+40	100	100
+60	95	95
+80	Not detected	Not detected
−0	100	100
−20	100	100
−40	100	100
−60	95	95
−80	Not detected	Not detected

Table 3 Detection accuracy at varying lips state detection rate

Movements per s	Our model-I (%)	Our model-II (%)
1	100	100
2	100	100
3	95	100
4	85	95
5	75	90
6	60	80

The overall training and test accuracy of the models is given in Table 4, where we observed that model-I is slightly more accurate, while model-II is much faster, almost three times than model-I. So, the proposed model-II can be a better choice for fast-moving scenarios. In contrast, model-I is more suited for situations where accuracy is the main priority. In other words, Table 4 indicates that with only six frames per second, model-I has an average accuracy of 95.25%, and with 20 frames per second, model-II has an average accuracy of 94.40%.

Additionally, some values of the dataset are given in Table 5 that suggest that the higher distance values correspond to mouth open position, whereas lower distance values normally indicate closed-mouth position. For example, in Table 5, we see that the maximum distance between the landmarks is for *diagonals* and the values are comparatively higher in the case of open lips than that of the closed lips. The shortest distances, in contrast, are between the landmarks pair of leftmost and rightmost sides, and they are relatively lower for the case of closed lips.

Finally, the lips state detection for varying conditions is illustrated in Fig. 6. The trends in accurately detecting the lips states also hold for the other test images. The capability of accurately detecting the lips opening and closing at different angles and lighting conditions means that the proposed model has the potential for a lips-state-based electronic translator of a non-verbal communication system.

Table 4 Overall performance of lips state detection

Parameters	Our model-I	Our model-II
Best validation accuracy	95%	94%
Best training loss	0.122%	0.130%
Best training accuracy	95.47%	94.808%
Average Accuracy	95.25%	94.40%
Average FPS	6	20

Table 5 Sample of training data

Left	Middle	Right	Diagonal-1	Diagonal-2	Output
30.06659	29.123	28.01785	31.38471	33.10589	1
30.06659	30.01666	29.01724	32.31099	33.54102	1
34.05877	34.0147	33.01515	36.05551	37.16181	1
17.02939	17.02939	17.02939	21.40093	21.40093	0
18.02776	17.03876	17.02939	21.63331	21.40093	0
18.02776	17.01345	16.03122	21.63331	21.40093	0

4 Conclusion

In support of developing an alternative interpretation/translation system for non-verbal communication system, we have introduced two new models for lips state detection. Building upon two popular facial landmark detectors, DLIB and MediaPipe, the proposed models have been investigated for the classification of opening and closing of lips for the standard datasets. Being faster, computationally efficient (in terms of FPS) and reasonably accurate, the proposed model-II with MediaPipe can be a promising candidate for the envisaged translation system. Our research continues to further develop the detection accuracy of the proposed models, particularly for extreme facial rotations and very poor lighting conditions.

References

1. Amornpan, P., Praisan, P.: Face recognition using transferred deep learning for feature extraction. In: 2019 Joint International Conference on Digital Arts, Media and Technology with ECTI Northern Section Conference on Electrical, Electronics, Computer and Telecommunications Engineering (2019)
2. Bouvier, C., Benoit, A., Caplier, A., Coulon, P.Y.: Open or closed mouth state detection: static supervised classification based on log-polar signature. In: ACIVS 2008—International Conference on Advanced Concepts for Intelligent Vision Systems. vol. Volume 5259/2008, pp. 1093–1102. Springer Berlin/Heidelberg, Juan-Les-Pins, France (Oct 2008), https://hal.archives-ouvertes.fr/hal-00372148

3. Karras, T., Laine, S., Aila, T.: A style-based generator architecture for generative adversarial networks (2019)
4. King, D.E.: Dlib-ml: a machine learning toolkit. J. Mach. Learn. Res. **10**, 1755–1758 (2009)
5. Krause, P.A., Kay, C.A., Kawamoto, A.H.: Automatic motion tracking of lips using digital video and openface 2.0. Lab. Phonol.: J. Assoc. Lab. Phonol.**11**(1) (2020)
6. Lu, Y., Li, H.: Automatic lip-reading system based on deep convolutional neural network and attention-based long short-term memory. Appl. Sci.**9**(8), 1599 (Apr 2019), https://doi.org/10.3390/app9081599
7. Lugaresi, C., Tang, J., Nash, H., McClanahan, C., Uboweja, E., Hays, M., Zhang, F., Chang, C.L., Yong, M.G., Lee, J., et al.: Mediapipe: A framework for building perception pipelines. arXiv preprint arXiv:1906.08172 (2019)
8. Sagonas, C., Antonakos, E., Tzimiropoulos, G., Zafeiriou, S., Pantic, M.: 300 faces in-the-wild challenge: database and results. Image Vis. Comput. **47**, 3–18 (2016)
9. Sharma, S., Shanmugasundaram, K., Ramasamy, S.K.: Farec-cnn based efficient face recognition technique using dlib. In: 2016 International Conference on Advanced Communication Control and Computing Technologies (ICACCCT). pp. 192–195. IEEE (2016)
10. Singh, B., Sahoo, S., Kumar, V., Issac, A., Dutta, M.K.: Improved lip contour extraction using k-means clustering and ellipse fitting. In: 2016 2nd International Conference on Communication Control and Intelligent Systems (CCIS), pp. 99–103. IEEE (2016)
11. WenJuan, Y., YaLing, L., MingHui, D.: A real-time lip localization and tacking for lip reading. In: 2010 3rd International Conference on Advanced Computer Theory and Engineering (ICACTE). vol. 6, pp. V6–363. IEEE (2010)
12. Xu, Z., Li, B., Geng, M., Yuan, Y., Yu, G.: Anchorface: An anchor-based facial landmark detector across large poses. arXiv preprint arXiv:2007.03221 (2020)

Depression-Level Prediction During COVID-19 Pandemic Among the People of Bangladesh Using Ensemble Technique: MIRF Stacking and MIRF Voting

Anik Pramanik, Md. Hasan Imam Bijoy, and Md. Sadekur Rahman

Abstract Machine learning is currently one of the most prominent approaches for the prediction of different diseases, conditions, and disorders in our human life. In Bangladesh, most people are unaware of their mental health. Only when they suffer from serious mental problems and trauma then they start to take treatment. But if they know about the features which are important to understand, realize, and have to figure out, then the total number of affected people will decrease. The main goal of this research is to predict the depression level of a person using machine learning and data analysis approaches by only filling up 30 basic questionnaires, which are related to depression collected through a public survey. We were able to collect responses of 1088 people from the participants for those 30 instances from all over Bangladesh. To achieve our aim of predicting depression levels, we used a total of ten classifiers, eight of which were based classifiers, which we combined with the best three top-scoring classifiers to build a novel ensemble approach called MIRF Stacking and MIRF Voting ensemble classifier. With 96.78% accuracy, the Random Forest (RF) classifier is the most accurate of the eight base classifiers. Then, our proposed ensemble MIRF Stacking and MIRF Voting classifiers achieve the supremacy performance of 96.81% and 97.18% accuracy, respectively. The proposed method would be used in a framework, where mental health counselors find the root causes and minor explanations for depression in people so that they can better understand all aspects of local psychology and provide them with the best advice and solutions to their problems.

Keywords Predict depression level · Depression analysis · COVID-19 impact · MIRF stacking · MIRF voting · Bangladesh

A. Pramanik · Md. H. I. Bijoy (✉) · Md. S. Rahman
Department of Computer Science and Engineering, Daffodil International University, Dhaka 1207, Bangladesh
e-mail: hasan15-11743@diu.edu.bd

A. Pramanik
e-mail: anik15-9495@diu.edu.bd

Md. S. Rahman
e-mail: sadekur.cse@daffodilvarsity.edu.bd

S. Hossain et al. (eds.), *Proceedings of International Conference on Fourth Industrial Revolution and Beyond 2021*, Lecture Notes in Networks and Systems 437,
https://doi.org/10.1007/978-981-19-2445-3_6

1 Introduction

With technologies and human expertise, the planet is advancing rapidly. Everybody is so distracted and materialistic that we fail to worry about our mental health. To keep pace with the speed of global development, people constantly exert pressure on physical and mental health, particularly on their mental health. Their health is adversely affected. The world is advancing at a quick rate with the assistance of innovation and human abilities. Everybody is turning out to be so occupied and materialistic that occasionally we neglect to give some time to consider our psychological well-being. To stay aware of the quick speed on the planet, individuals are continually propelling themselves by taking a ton of pressing factors both actually and intellectually which hurts their well-being, particularly their psychological well-being. Misery is a quiet executioner that can hurt a person in an extraordinary manner if not treated at the perfect time. It is a normal psychological instability and everybody eventually in their life is discouraged. Be that as it may, due to the absence of self-ignorance, society, and individuals' judgment. This sickness is thought of as a no–no in numerous spots across the world and individuals will in general ridicule this sickness and bother individuals who are determined to have this issue.

Depression is a common illness worldwide, with more than 264 million people affected [1]. In Bangladesh, the prevalence of depression, anxiety, and stress have been reported to be as high as 54.3%, 64.8%, and 59.0%, respectively [2–5]. Globally, it is estimated that 4.4% of the global population suffers from a depressive disorder, and 3.6% from an anxiety disorder. All over the world, only 87% offer some primary care health services to control mental illness, 30% do not have programs, on the other side, there are 28% who have no budget identified for mental health in a specific way [3]. Even there is no proper laboratory or diagnostic center for diagnosing all the stages of mental illness; typically, the diagnosis will be done based on the patient's self-reported experiences based on some incidents and behaviors reported by their surroundings. It accounts for more disability-adjusted life years (DALYs) than all other mental disorders [4], and it is projected to become the leading cause of disability in developed countries by 2030 [5].

The economic burden of depression is estimated at 83 billion dollars annually from reduced worker productivity, increased medical expenses, and suicide [6]. World Health Organization (WHO) figured out that the effect of spreading misinformation and rumor through social media and people are basically accounted for anxiety, stress, fears, depression and emotional trauma, and many other mental health problems during this pandemic [7]. To make a decision without anyone else, we should have classified data. Even though the world is progressing, and everyone is open to new experiences, psychological instabilities are not welcomed favorably in many social structures. It is indeed very little, particularly in many Asian countries, therefore those who are affected are not outspoken about their concerns and seek to conceal their sickness and overcome it without the help of others. As a result, their sickness may demolish, which may disintegrate their well-being both intellectually and genuinely and lead to major sadness, which will directly cause self-destructive behavior and self-hurt mindset in persons. In addition, many people are unaware of their downward condition and treat it as if it were any other illness, if it would spontaneously heal when given time. On the contrary, some individuals get disheartened because they believe they are discouraged, although this may not be the case. While leading this research, we discovered that most people act in this manner and aren't all that vocal about their issues. Some of them believe they are discouraged, which is influencing their performances. However, they were not discouraged at all; a few people are unaware that they have moderate and serious burdensome indications. Another issue we've noticed is that there is a bunch of people, who are outspoken about their economic slump. And there is another gathering of individuals, who are perplexed and need to know whether they are disheartened so they may seek expert help. Regardless, the complicated surveys in scaling techniques and feelings of discomfort in providing all of the individual data or instructions demotivated individuals to address their problematic behavior. Aside from that, there is no precise technique or research that has ever truly uncovered the core causes that are causing unhappiness in college-going, undergraduate, and graduate students alike.

In this approach, the remainder of the paper is in the following order: Section 2 of the paper is a review of the literature. The approach for predicting the depression level of Bangladeshis is discussed in Sect. 3. The experimental results are demonstrated and discussed in Sect. 4. Section 5 of the document, certainly, brings the whole thing to a conclusion.

2 Literature Review

We looked at some of the publications in this section to determine if there was anything missing from prior studies.

Bueno-Notivol et al. [8] proposed in their paper about the depression's appearance during the pandemic of COVID-19 based on about 12 paper analyses. They had reported a systematic approach of reviews and meta-analysis with PRISMA and looked in PubMed and Web of Science for cross-sectional, community-based studies. The prevalence rates of depression were observed to range from 7.45 to 48.30% in the selected studies. With substantial variation between studies, the total prevalence of depression was 25%. Ramalingam et al. [9] analyzed depression with the help of some machine learning techniques in their paper on the basis of various relevant datasets coming from social media platforms. To identify the dataset related to depression there must be a need for some machine learning algorithms and for classification of the rate of depression, there are so many techniques such as image processing with a 2D and 3D processing, semantic algorithms, 2D feature vector for classification, Support Vector Machines (SVM) techniques, etc. Hyland et al. [10] conducted a research study to estimate the probability rate of attacking Generalized Anxiety Disorder (GAD) and depression in Ireland. They tried to find out the sociodemographic risk factor and also tried to recognize some screening positives for GAD or depression. They collect their data from self-reported data of Irish citizens and were able to collect 1041 responses during the pandemic situation. In their findings, they got 20% GAD, 22.8% depression, and GAD or depression 27.7% in the people of Ireland. GAD or depression testing was linked to younger age, female sex, COVID-19-related income loss, COVID-19 infection, and a greater perceived risk of COVID-19 infection. COVID-19-related anxiety was shown to be considerably greater in citizens who are 65 years and older than in people 18–34 years old. Choudhury et al. [11] introduced depression prediction based on social media platforms. They had formed a classifier for analyzing the behavioral signals from social media like Twitter with the clinical report of the participants that they have depression. Crowdsourcing techniques had been used by them for analysis of behavioral data like Major Depressive Disorder (MDD). There is also a term they used named Amazon's Mechanical Turk interface to trail by a few inquiries on their downturn history. For 476 users, they had collected a total of 2,157,992 posts, 1,583 data were performed in crowd workers for Human Intelligence Tasks. They used the data to make an SVM classifier, which had given an accuracy of 70% for the dataset. Bijoy et al. [12] analyzed broadly COVID-19's impact on Bangladeshi people. According to their study, 57.76% of people have severe depression. According to their findings, 57.76% of people suffer from clinically significant depressive disorder (MDD). From Hasan et al. [13] 76.96% are suffering mental health problems and problems that are considered depression in the Covid scenario, and this has a negative impact on the education sector. Hasan et al. [14] illustrate the supervised approach, where students are exposed to depression and the overall influence of COVID-19 on schooling throughout the pandemic.

3 Methodology

The research study follows certain basic stages such as data collecting, preprocessing data, training the model for a specific application, testing the machine, and obtaining findings. As a basis, we proceeded through step-by-step working procedures to perform this research, and the workflow diagram is shown in Fig. 1.

3.1 Dataset Description

To perform an analysis, we will require a significant amount of data, which will aid us in training the model to provide a better result. So, this research is based on a public survey. During the pandemic, we conducted a public survey in which we created a Google Form with 30 questions. We were able to acquire 1088 responses from participants after conducting a public survey. To model attributes and label data, all of the questions are used as variables. In our early phases, there were 29 variables that were independent and 1 variable that was utilized as label data for the dependent

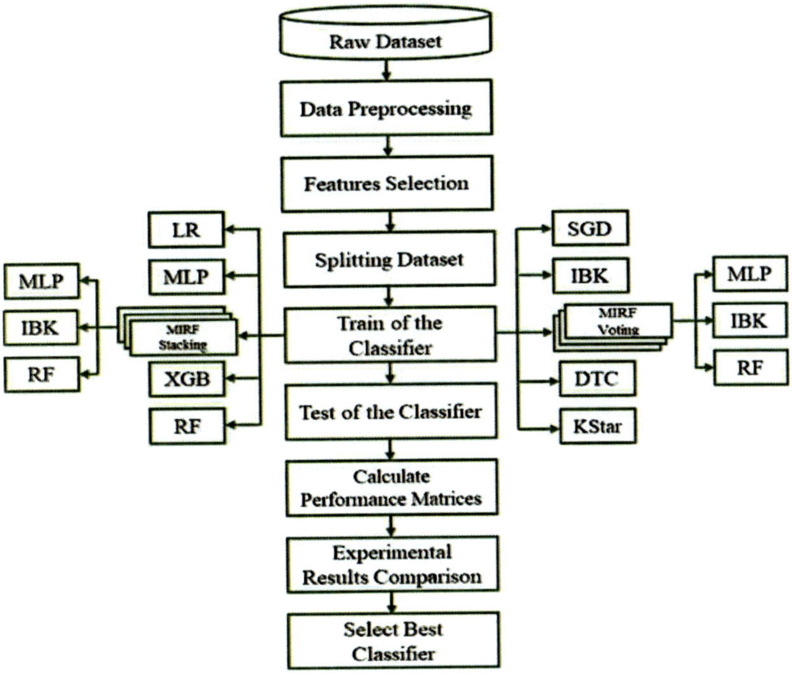

Fig. 1 Step-by-step working procedure to predict the depression level during the COVID-19 pandemic among the people of Bangladesh

variable. All variables, descriptions, variable types, and possible values are listed in Table 1.

3.2 Data Preprocessing

We require machine-readable data to model feeding, and then we preprocess the data to make it suitable for training the model. To begin, we remove any extraneous text from our dataset. Then, rather than dealing with lost data from our participants, we deal with missing data. To fill in the gaps in the data, we selected the most frequent value for categorical data and computed the mean value for numerical data. After that, we use Label Encoder to transform the category data into a machine-readable numeric number.

Table 1 All attributes with description, type, and possible values

Variable	Description	Variable type	Possible values
GT	Gender Type	Independent	Male (0), Female (1)
AG	Age Group	Independent	Less than 20 (0), 20–25 (1), 25–30 (2), 30 Above (3)
OT	Occupation Type	Independent	Student (0), Business (1), Job Holder (2), Housewife (3), Doctor (4), Engineer (5), Teacher (6), Others (7)
MTSI	More Time Spent on the Internet	Independent	Yes (1), No (0)
HMTSI	How Much Time Spent on the Internet	Independent	1 h (0), 2 h (2), 3 h (2), 4 h or More (3)
HMDOH	How Many Days have not been Out of the House	Independent	Go out Every day (0), 1–14 Days (1), 15–30 Days (2), 31–60 Days (3), More Than 2 Months (4)
HACSC	Have any Change in Sleep Cycle	Independent	No (0), Maybe (1), Yes (2)
MCLDQ	Mood Changed a Lot During Quarantine	Independent	Low (0), Medium (1), High (2)
HAOEF	Have Any Other Earner in Family	Independent	Yes (1), No (0)
TDA	Take Drug or Alcohol	Independent	Yes (1), No (0)
HFSD	Have any Fatal or Serious Disease	Independent	Yes (1), No (0)
EHSMA	Ever Had a Stroke, Meningitis, or Amputation	Independent	Yes (1), No (0)
RWF	Relationship With Family	Independent	Least Satisfied (0), Satisfied (1), very Satisfied (2)
SFDQ	Stay With Family During Quarantine	Independent	Yes (1), No (0)
DNISDQ	Disappointed by Any News, Incident, Social-Media During Lockdown	Independent	Yes (1), No (0)
HFET	Have Faced Emotional Trauma	Independent	Yes (1), No (0)
Variable	Description	Variable Type	Possible Values
HAMDF	Have Any Mental Disorder in Family	Independent	Yes (1), No (0)
HPSMDQ	Have attacked Personally on Social-Media During Quarantine	Independent	Yes (1), No (0)
HDQ	Have Been Harassed During Quarantine	Independent	Yes (1), No (0)

(continued)

Table 1 (continued)

Variable	Description	Variable type	Possible values
SMK	Smoker	Independent	Yes (1), No (0)
DIDC	Disrupted Income During COVID-19	Independent	No (0), Yes (1), No Job (2)
FEWTP	Feel Emotionally Weak While Talking to People	Independent	No (0), Maybe (1), Yes (2)
LIW	Losing Interest in Work	Independent	No (0), Maybe (1), Yes (2)
HADMP	Have the Ability to Deal with Mental Problem	Independent	Yes (1), No (0)
FFFTMDQ	Feel Frustrated First Two Months During Quarantine	Independent	No (0), Maybe (1), Yes (2)
ISDBD	Increase in Stress Day by Day	Independent	No (0), Maybe (1), Yes (2)
HFMPBL	Have Faced Mental Problem Before Lockdown	Independent	Yes (1), No (0)
HFPP	Have Faced Any Physical Problem	Independent	Yes (1), No (0)
HABCov	Have been Attacked by COVID-19	Independent	Yes (1), No (0)
HDL	Having Depression Level	Dependent	Low (0), Medium (1), High (2)

3.3 Features Selection

In the physical realm, it's infrequent that all of the variables in a dataset are relevant for creating a machine learning model. Adding superfluous variables decreases the model's classification performance and may also diminish a classifier's overall accuracy. Furthermore, adding more variables to a model enhances the model's total complexity. To select top features from our dataset, we analyze the correlation matrix and apply f_classif feature selection techniques to find the top features for our model implementation. Then, following the application of f_classif, we attempt to choose the top 20 features for our model. Then we compute score values and select the top 20 features to proceed to the model implementation stage. The top 20 features are depicted in Fig. 2 along with their scores.

3.4 Model Implementation

Following the selection of top features, we split our dataset into two parts using an 80:20 ratio, which means that 80% of the data is used to train the model and 20% is used to test the model. Several classifiers are used to predict the severity of depression, and the following is the relevant model theory.

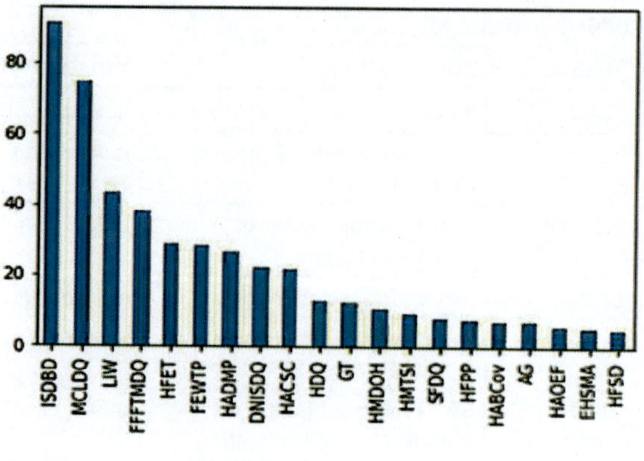

Fig. 2 Top 20 features in our working dataset

The supervised learning approach of logistic regression [15] is used to predict the categorical dependent variable using a collection of the independent variable. The logistic function is at the crux of the logistic regression approach. The logistic function, also known as the sigmoid function "(1)", was created by statisticians to characterize the features of the population. It is an S-shaped curve that can translate any real-valued number to a value between 0 and 1 but never exactly between those two points.

$$y = \frac{1}{1 + e^{-z}} \tag{1}$$

A Multilayer Perceptron (MLP) is a neural network with an input, an output, and one or more hidden layers [16]. A single-layer perceptron can only learn linear functions; however, a multilayer perceptron can learn both linear and nonlinear functions. MLP's learning method is known as the backpropagation algorithm. The input layer receives the signal, and the output layer predicts a decision based on the input. The hidden layers act as a computational engine to approximate continuous functions. In MLPs, the previous layer's output is utilized as the input to the next layer using the following formula "(2)".

$$y = f(WxT + b) \tag{2}$$

The lazy classifier category variant of the K-Nearest Neighbors (KNN) technique is Instance-Based k (IBK). Instead of creating a model, the IBk technique delivers a just-in-time forecast for a test case. The IBk method uses a distance metric to choose k "near" cases from the training data for each test case, then produces a prediction based on those instances using the function "(3)" [17].

KStar is a type of instance-based classifier that uses an entropy-based distance function, making it unique among instance-based classifiers. It is a spin-off of K-Nearest Neighbors (KNN), commonly known as the lazy learner. Instead of learning, this classifier memorizes the training data, conducts some preprocessing, and then waits for the test tuple, which it identifies and classifies based on its likeness to the preset training tuples.

$$\mathrm{Dist}(X, Y) = \sqrt{\sum_{i=1}^{D} (X_i Y_i)^2} \tag{3}$$

The nonparametric Decision Tree Classifier [18] is one of the most often used supervised methods. The goal is to learn effective decision rules from attributes to construct a model that predicts the value of a target variable. A tree is an estimate of a piecewise invariant. With a set of if–then-else decision rules, decision trees learn from data to approximate a sine curve which follows the formula "(4)". The decision criteria get more complicated as the tree grows deeper, and the model becomes more accurate.

$$H(X) = -\sum_{i=1}^{n} P(x_i)\log_2 P(x_i) \tag{4}$$

Random Forest [19] is a supervised learning technique, which is a basic machine learning algorithm that, in the majority of cases, produces great results even without hyperparameter modification. It generates a "forest" out of a group of decision trees that have been trained using "(5)" which is the "bagging" method. The essential concept of the bagging approach is that mixing several learning models enhances the final result. It is also one of the most often utilized algorithms due to its simplicity and adaptability (it can be used for both classification and regression tasks).

$$\mathrm{RFfi}_i = \frac{\sum_j \mathrm{normfi}_{ij}}{\sum_{j \in \mathrm{allfeatures}, k \in \mathrm{alltrees}} \mathrm{normfi}_{jk}} \tag{5}$$

eXtreme Gradient Boosting (XGB) [20] is a machine learning technique that has subsequently prevailed. The XGB method is a decision tree-based ensemble technique that uses a gradient boosting framework to improve speed. Gradient boosting is a supervised learning approach that attempts to properly predict a target variable by combining the estimates of a series of smaller, poorer algorithms using the formula "(6)".

$$L^{(t)} = \sum_{i=1}^{n} l(y_i, \hat{y}_i)^{(t-1)} + f_t(x_i)) + \Omega(f_t) \tag{6}$$

Stochastic Gradient Descent (SGD) is a quick and easy way to fit linear classifiers and regressors to convex loss functions like (linear) Support Vector Machines and Logistic Regression. At each iteration "(7)", SGD selects one data point randomly from the whole data collection, greatly reducing calculations. It is also usual to sample a small number of data points rather than simply one at each step, which is referred to as "mini-batch" gradient descent.

$$Q(w) = \frac{1}{n}\sum_{i=1}^{n} Q_i(w) \tag{7}$$

Staking or stacked generalization is an ensemble learning technique that uses meta-classifier in machine learning. Stacking uses the meta-classifier (level-1 classifier) idea to aggregate the separate outputs of the basic classifiers (level-0 classifiers). Though any classifier might very well be used as a level-1 classifier, so in our proposed stacking method we chose the RF as a meta classifier, and MLP, IBK, and RF were our three highest scoring base classifiers to get the final ensemble method as MIRF Stacking.

In order to avoid over-fitting, we employed cross-validation to create the level-1 classifier mode which is highly recommended. In our proposed MIRF Stacking, MPL generates 19 sigmoid nodes, IBK = IB1 (K = 1) used the 1 nearest neighbors classification, bagging with 100 iterations used in RF and meta-RF. The graphical presentation of MIRF stacking is shown in Fig. 3a. The voting classifier is a machine learning technique, where the model is trained based on multi-ensemble models and generates the prediction combining the highest probability distributions of these base learners as per chosen output classes. In our proposed MIRF voting, we conduct the soft voting methods using the average probability combination rule. In our proposed MIRF voting, MPL generates 19 sigmoid nodes, IBK = IB1 (K = 1) used the 1 nearest neighbors classification, bagging with 100 iterations used in RF. The graphical presentation of MIRF voting is shown in Fig. 3b.

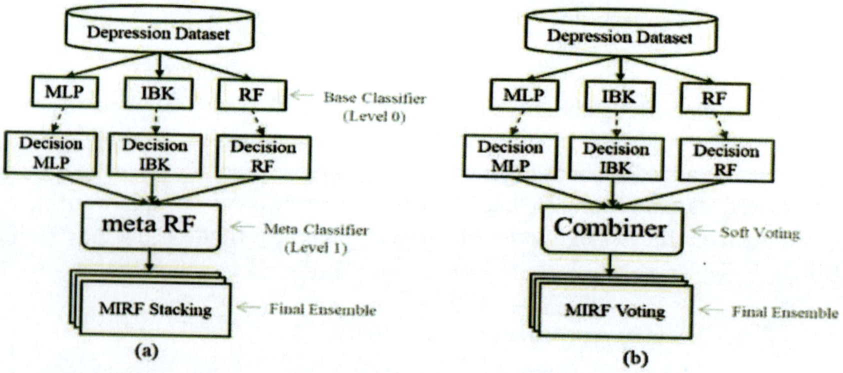

Fig. 3 Classifier Ensemble using: **a** MIRF Stacking (MLP, IBK, RF, meta-RF) and **b** MIRF Voting (MLP, IBK, RF)

3.5 Calculate Performance Metrics

Here are some of the performance evaluation metrics that were computed. Using these criteria, we found the best classifier to predict in this scenario. Many performance metrics in percent (%) have been calculated using Eqs. (8–14) based on the confusion matrix generated by the classifier.

$$Accuracy = \frac{TP + TN}{TP + FN + FP + TN} \times 100\% \quad (8)$$

$$TruePositiveRate(TPR) = \frac{TP}{TP + FN} \times 100\% \quad (9)$$

$$TrueNegativeRate(TNR) = \frac{TN}{FP + TN} \times 100\% \quad (10)$$

$$FalsePositiveRate(FPR) = \frac{FP}{FP + TN} \times 100\% \quad (11)$$

$$FalseNegativeRate(FNR) = \frac{FN}{FN + TP} \times 100\% \quad (12)$$

$$Precision = \frac{TP}{TP + FP} \times 100\% \quad (13)$$

$$F1Score = 2 \times \frac{Precision \times Recall}{Precision + Recall} \times 100\% \quad (14)$$

4 Results and Discussions

The following section contains a detailed empirical investigation of the classifier ensemble for depression levels. We show that classifier ensembles, which outperform single classifiers in the domain of depression detection, are promising approaches for prediction tasks besides our proposed model. In our model implementation section, we applied 10 classification algorithms, where 8 classifiers were base classifiers, and MIRF stacking and MIRF voting were combined with MLP, IBK, and RF classifiers. Low (0), Medium (1), and High (2) are the three classes in our label column; Having Depression Level (HDL) indicate that our work is a multiclass problem. Table 2 shows the result of the applied classifier, constructing a 3 × 3 confusion matrix as stated in [21].

Accuracy, TPR, TNR, FPR, FNR, Precision, and F1-score from the above confusion matrix are computed to determine the best model for our work and to assess this work. Table 3 shows the results of many performance evaluation measures. The

Table 2 Confusion matrices for all applied algorithms

Model	Class	TP	FN	FP	TN
MLP	High	345	41	43	659
	Medium	401	35	51	601
	Low	218	48	30	792
LR	High	213	173	174	528
	Medium	233	203	226	426
	Low	138	128	104	718
IBK	High	376	10	20	682
	Medium	418	18	18	634
	Low	246	24	13	801
KSTAR	High	374	12	26	676
	Medium	410	26	22	630
	Low	242	24	14	808
RF	High	376	10	24	678
	Medium	408	25	20	635
	Low	248	14	12	814
DTC	High	264	10	30	784
	Medium	364	60	30	634
	Low	349	40	50	649
XGB	High	265	10	40	774
	Medium	374	50	60	604
	Low	329	60	20	679
SGD	High	264	10	30	784
	Medium	364	60	30	634
	Low	349	41	50	648
MIRF stacking	High	378	8	26	676
	Medium	416	20	16	636
	Low	242	24	10	812
Model	Class	TP	FN	FP	TN
MIRF voting	High	374	12	18	684
	Medium	420	16	18	634
	Low	248	18	10	812

overall analysis is separated into two parts: evaluation results for eight base classifiers and results for the two proposed MIRF Stacking and MIRF voting techniques. Table 3 demonstrates that the random forest classifier outperforms the other eight base classifiers. The random forest classifier's accuracy for the Low (0), Medium (1), and High (2) classifications are 97.61%, 95.86%, and 96.88%, respectively. The random forest classifier's F1-score for the Low (0), Medium (1), and High (2) classes

Table 3 Performance evaluation of the applied base classifier

Model	Class	Accuracy (%)	TPR (%)	TNR (%)	FPR (%)	FNR (%)	Precision (%)	F1 score (%)
MLP	High	92.28	89.38	93.87	6.13	10.62	88.92	89.15
	Medium	92.10	91.97	92.18	7.82	8.02	88.72	90.32
	Low	92.83	81.95	96.35	3.65	18.05	87.90	84.82
LR	High	68.11	55.18	75.21	24.77	44.82	55.04	55.11
	Medium	60.57	53.44	65.34	34.66	46.56	50.76	52.07
	Low	78.68	51.88	87.35	12.65	48.12	57.02	54.33
IBK	High	92.24	97.41	97.15	2.85	2.60	94.95	96.16
	Medium	96.70	95.87	97.24	2.76	4.13	95.87	95.87
	Low	96.23	91.11	98.40	1.60	8.89	94.98	93.00
KStar	High	95.51	96.89	96.30	3.70	3.11	93.50	95.17
	Medium	95.59	94.03	96.63	3.37	5.96	94.91	94.47
	Low	96.51	90.98	98.30	1.70	9.02	94.53	92.72
RF	High	96.88	97.41	96.58	3.42	2.60	94.00	95.67
	Medium	95.86	94.23	96.95	3.05	5.77	95.33	94.77
	Low	97.61	94.66	98.55	1.45	5.34	95.39	95.02
DTC	High	96.32	96.35	96.31	3.69	3.65	89.80	92.96
	Medium	91.73	85.85	95.48	4.52	14.15	92.39	88.90
	Low	91.73	89.72	92.85	7.15	10.28	87.47	88.58
XGB	High	95.50	96.36	95.08	4.91	3.64	86.89	91.38
	Medium	89.89	88.21	90.96	9.03	11.79	86.18	87.18
	Low	92.65	84.58	97.14	2.86	15.42	94.27	89.16
SGD	High	96.32	96.35	96.31	3.69	3.65	89.80	92.96
	Medium	91.73	85.85	95.48	4.52	14.15	92.39	89.00
	Low	91.64	89.49	92.84	7.16	10.51	87.47	88.47

is 95.02%, 94.77%, and 95.67%, respectively, which is outrageous of all the classifiers. Furthermore, the result of other data in Table 3 corroborates the random forest classifier.

After eight base classifier results evaluation, we compute the same Accuracy, TPR, TNR, FPR, FNR, Precision, and F1-score from the above confusion matrix for our proposed two methods: MIRF stacking and MIRF voting techniques. Performance evaluation results are presented in Table 4.

By employing our proposed model, we are attempting to determine the true cause and impact of depression in people. We obtained the connection among the data from our dataset, and the primary reason for experiencing depression was Increase in Stress Day by Day (ISDBD), which means people's stress levels escalated because of the lockdown, as shown in Fig. 4a. Then, another depression factor that might influence their mood was Mood Changed a Lot During Quarantine (MCLDQ) in

Table 4 Performance evaluation of the proposed MIRF stacking MIRF voting techniques

Model	Class	Accuracy (%)	TPR (%)	TNR (%)	FPR (%)	FNR (%)	Precision (%)	F1 score (%)
MIRF stacking	High	96.88	97.93	96.30	3.70	2.07	93.56	95.70
	Medium	96.69	95.40	97.55	2.45	4.56	96.30	95.85
	Low	96.88	90.98	98.78	1.22	9.02	96.03	93.44
MIRF voting	High	97.24	96.89	97.44	2.56	3.11	95.41	96.14
	Medium	96.88	96.33	97.24	2.76	3.67	95.89	96.11
	Low	97.43	93.23	98.78	1.22	6.77	96.12	94.66

Fig. 4b, Loss of Interest in Workplace (LIW) in Fig. 4c and Feel Frustrated First Two Months During Quarantine (FFFTMDQ) in Fig. 4d, they Have Experienced Emotional Trauma (HFET) and Feel Emotionally Weak While Talking to People (FEWTP) in Fig. 4e, f, respectively.

5 Conclusion

The most common mental issue or disease, despondency is an expansive region of examination that has a ton of suggestions in medication and brain science. Youngsters are bound to experience the ill effects of this illness particularly inhabitants of lower center pay nations because of different socio-segment reasons. This examination looked to misuse this connection by utilizing some social and individual information to foresee wretchedness in people. There has not been a lot of work on dissecting wretchedness among Bangladesh residents. Nonetheless, this is one of the significant social and clinical issues that is persistently developing and consequently should be tended to the right away. One of the fundamental targets of this investigation was to distinguish early instances of gloom and request people in danger from creating wretchedness to seek counselling from a therapist or expert. Additionally, experts can utilize our framework to recognize reasons for discouragement in explicit people. MIRF voting ensemble classifier was utilized to discover pertinent highlights that add to gloom and the ten calculations were utilized to anticipate despondency in people groups of Bangladesh depending on the chosen highlights. The proposed model could be utilized by clinicians, advocates, and colleges to discover the downturn in understudies so that proper advances can be produced to decrease the results and effectiveness of the infection and help youngsters carry on with a better and a more joyful life. Studying supervised learning also helps us to know about unsupervised learning and we are all interested to work with different techniques and models in the future.

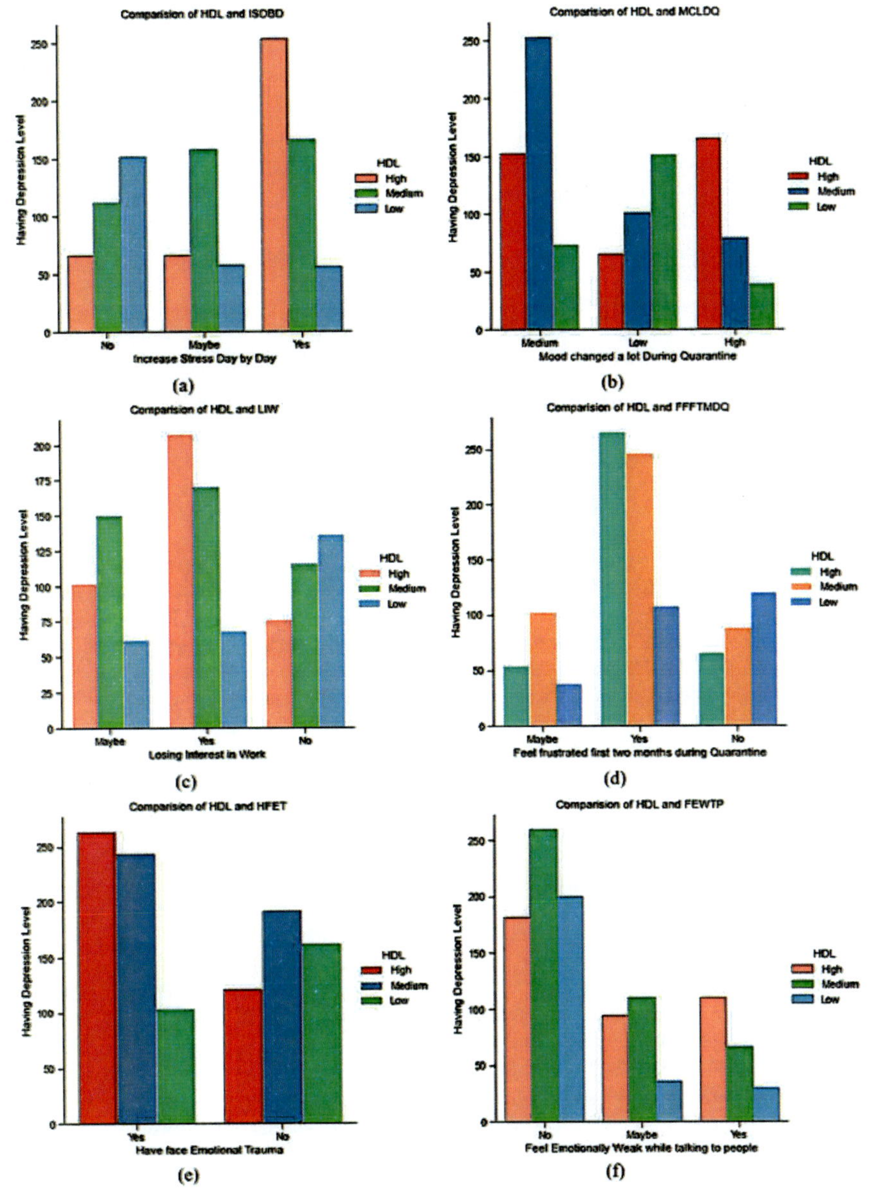

Fig. 4 Reasons behind depression among people of Bangladesh during the COVID-19 pandemic; **a** ISDBD **b** MCLDQ, **c** LIW, **d**, FFFTMDQ, **e** HFET **f** FEWTP with HDL

References

1. Depression, Who.int.: https://www.who.int/news-room/fact-sheets/detail/depression (2021). Accessed 20 Sep 2021
2. Mohammed, M.A., Rafi, M.A., Shamim Al Mamun, A.H.M., Hasan, M.D., Akter, K., Hsan, K., Griffiths, M.D.: Prevalence and psychiatric risk factors of excessive internet use among northern Bangladeshi job-seeking graduate students: a pilot study. Int. J. Mental Health Addiction 1–11 (2019)
3. Detels, R., Tan, C.C.: The scope and concerns of public health. Oxford Textbook Public Health **1**, 392 (2009)
4. Murray, C.J.L., Vos, T., Lozano, R., Naghavi, M., Flaxman, A.D., Michaud, C., Ezzati, E., et al.: Disability-adjusted life years (DALYs) for 291 diseases and injuries in 21 regions, 1990–2010: a systematic analysis for the global burden of disease study 2010. Lancet **380**(9859), 2197–2223 (2012)
5. Mathers, C.D., Loncar, D.: Projections of global mortality and burden of disease from 2002 to 2030. PLoS Med. **3**(11), e442 (2006)
6. Greenberg, P.E., Kessler, R.C., Birnbaum, H.G., Leong, S.A., Lowe, S.W., Berglund, P.A., Corey-Lisle, P.K.: The economic burden of depression in the United States: how did it change between 1990 and 2000? J. Clin. Psychiatry **64**(12), 1465–1475 (2003)
7. World experts and funders set priorities for COVID-19 research.: Who.int. https://www.who.int/news-room/detail/12-02-2020-world-experts-and-funders-set-priorities-for-covid-19-research (2021). Accessed 20 Sep 2021
8. Bueno-Notivol, J., Gracia-García, P., Olaya, B., Lasheras, I., López-Antón, R., Santabárbara, J.: Prevalence of depression during the COVID-19 outbreak: a meta-analysis of community-based studies. Int. J. Clin. Health Psychol. **21**(1), 100196 (2021)
9. Ramalingam, D., Sharma, V., Zar, P.: Study of depression analysis using machine learning techniques. Int. J. Innov. Technol. Explor. Eng **8**(7C2), 187–191 (2019)
10. Hyland, P., Shevlin, M., McBride, O., Murphy, J., Karatzias, T., Bentall, R.P., Martinez, A., Vallières, F.: Anxiety and depression in the Republic of Ireland during the COVID-19 pandemic. Acta Psychiatr. Scand. **142**(3), 249–256 (2020)
11. De Choudhury, M., Gamon, M., Counts, S., Horvitz, E.: Predicting depression via social media. In: Seventh International AAAI Conference on Weblogs and Social Media (2013)
12. Bijoy, M.H.I., Rabbani, M., Zarif, M.I.I., Rahman, M.M., Hasan, M.R., Bhowmik, T.: A stupendous statistics on the pandemic impact on all sectors of Bangladesh. In: The International Conference on Advances in Multi-Disciplinary Sciences and Engineering Research (ICAMSER-2021), Chitkara University, Himachal Pradesh, Jul. 2–3 (2021)
13. Hasan, M., Bijoy, M.H.I., Akhi, S.A.: Refute the decision of auto-promotion and real facts of digital online classes during the pandemic in Bangladesh. In:IEEE International Conference on Advent Trends in Multidisciplinary Research and Innovation (ICATMRI), pp. 1–6 (2020). https://doi.org/10.1109/ICATMRI51801.2020.9398326
14. Hasan, M.R., Imam Bijoy, M.H., Khushbu, S.A., Akter, S., Hossain, S.A.: Supervised method pursued for overall impact of online class during lockdown in Bangladesh. In: 2021 12th International Conference on Computing Communication and Networking Technologies (ICCCNT), pp. 1–5. https://doi.org/10.1109/ICCCNT51525.2021.9579827
15. Kleinbaum, D.G., Klein, M.: Introduction to logistic regression. In: *Logistic Regression*, pp. 1–39. Springer, New York (2010)
16. Murtagh, F.: Multilayer perceptrons for classification and regression. Neurocomputing **2**(5–6), 183–197 (1991)
17. Ayu, M.A., Ismail, S.A., Matin, A.F.A., Mantoro, T.: A comparison study of classifier algorithms for mobile-phone's accelerometer based activity recognition. Procedia Eng. **41**, 224–229 (2012)
18. Safavian, S.R., Landgrebe, D.: A survey of decision tree classifier methodology. IEEE Trans Syst. Man Cyber. **21**(3), 660–674 (1991)

19. Liaw, A., Wiener, M.: Classification and regression by randomForest. R News **2**(3), 18–22 (2002)
20. Chen, T., He, T., Benesty, M., Khotilovich, V., Tang, Y., Cho, H.: Xgboost: extreme gradient boosting. R Package Version **0.4–2.1**(4), 1–4 (2015)
21. Bijoy, M.H.I., Akhi, S.A., Nayeem, M.A.A., Rahman, M.M., Mia, M.J.: Prediction of internet user satisfaction levels in bangladesh using data mining and analysis of influential factors. Bulletin Electr. Eng. Inf. **10**(6) (2021)

HandFruitNet: A Deep Learning Based Model for Fruits Classification from Hand

Nusrat Jahan, Sazeda Akter Rimi, and Md. Khaliluzzaman

Abstract Detecting a hand and classifying fruits from an image is a significant research area in the field of computer vision. For many applications, detecting a hand and classifying fruits together from a single system is also significant. This type of system is also effective for shop owner to detect suspicious people, as well as for the disabled people to easily pick different fruits from fruits shop without any other help. In this regard, in this paper, a system is proposed where hand detection and fruit classification are performed together when the fruit is in a person's hand. For that, initially, a hand is detected by using two methods. The first method is developed based on the hand features extraction, i.e., contour feature of the hand. Another method is developed based on deep learning, i.e., YOLOv3. After that, a convolutional neural network model based on deep learning is developed to classify the fruits. Finally, both models are combined to detect hands and classify the fruits from a single system. For training the model, 11 K and Fruits_360 datasets are used for hand detection and fruits classification, respectively. Beyond that, two different own datasets are created for hand detection and fruit classification. These datasets are utilized during testing the models. Among the models, the feature extraction and deep learning based hand detection model show 95.17% and 99.90% accuracy, whereas deep learning based fruits classification model shows 99.99% accuracy.

Keywords YOLOv3 · Computer vision · Deep learning · Hand detection · Fruits classification

1 Introduction

Throughout deep learning's research, detection and classification have spread as an important research area in computer vision and human–computer interaction. Detecting a hand and classifying fruits together is an effective system because of

N. Jahan · S. A. Rimi · Md. Khaliluzzaman (✉)
Department of Computer Science and Engineering, International Islamic University Chittagong (IIUC), Chattogram 4318, Bangladesh
e-mail: khalilcse021@gmail.com

S. Hossain et al. (eds.), *Proceedings of International Conference on Fourth Industrial Revolution and Beyond 2021*, Lecture Notes in Networks and Systems 437,
https://doi.org/10.1007/978-981-19-2445-3_7

its various applications such as security systems, market analysis, and robotics. This system is also efficient for the person who goes to the fruit shop and picks fruits that he does not know. In this situation, this system detects the hand and then automatically recognizes the type of fruit in his hand. This system is also effective in detecting the suspicious persons, who want to steal the fruit and to help the blind person to pick the different types of fruits from the fruit shop.

This type of system is quite a bit complicated due to the hands and fruits patterns' similarity and variability. Strawberry and lychee are pretty similar in terms of color. Many researchers use many hand-crafted feature methods to detect hands [1, 2] and classify the different types of fruits [3–5]. However, at the present, many researchers apply artificial neural networks (ANN) and convolutional neural networks (CNN) to detect a hand and classify the fruits and found better performance with respect to accuracy and time complexity.

In this regard, in this paper, we have applied a deep convolutional neural network (DCNN) to detect a hand and classify fruits together in a single system. We have also used a hand feature-based model to detect a hand from an image. Here, the contour feature of a hand is utilized. The DCNN model used for hand detection is YOLOv3, a new network used for redacting feature extraction. YOLO is extremely fast compared to its predecessors. YOLOv3 is the latest and one of the popular object detection algorithms. For fruit classification, a convolutional neural network model based on deep learning is developed. Finally, a comparative analysis is presented to show the performance of the hand feature extraction method and DCNN model.

The remaining part of our paper will be arranged in this way. In Sect. 2, we have shortly described some of the papers related to our topic. We have described the hand detection and fruits classification framework in the Sect. 3. The experimental analysis is presented in Sect. 4. Finally, the paper is concluded in Sect. 5.

2 Related Works

Hand detection and fruit classification have spread as an important research area. Hand detection is challenging because of the structural variation and the different positions of the hand. There are many works held in previous times for hand detection. Many researchers used various effective methods to improve performance and accuracy. For example, in Khan and Pathan[6], a prototype system is presented for a group of deaf people. The purpose of this system is to recognize gestures of a set of 24 letters from the American Sign Language. The digital image processing technique is used to solve the problem using color segmentation, skin detection, image filtering, and template matching. For gender recognition and biometric identification, the authors proposed a new dataset containing more than 11,000 hand images in Afifi [7]. The authors proposed a two-stream CNN to classify gender. The system shows better accuracy when compared to the other CNN models such as AlexNet and GoggleNet. For biometric identification, the proposed convolutional layers are used as feature extractor and send these features to a support vector machine (SVM) for classification.

Vision-based hand gesture recognition for human–computer interaction is becoming popular day by day. To recognize hand gesture recognition, the author presented a method in Xu et al. [8] for convex defect detection, which is used to extract a series of features. The relationship between convex defects and fingertips made this procedure much easier. In Mujahid et al. [9], a lightweight model based on YOLOv3 and DarkNet53 convolutional neural network is used for gesture recognition from static hand images and video feed. The authors showed that the YOLOv3 model gives better performance than Single Shot Detector (SSD) and Visual Geometry Group (VGG16). In Il Joo et al. [10], depth information is used to detect the hand region. The difference between the depth value at the center of the input image and the depth value within a segmented block is calculated to extract the features. To predict the size of the hand region, the central depth value and the secondary linear model are used. To train and recognize the features from the hand region, the AdaBoost classifier is used. Scale-invariant features transform (SIFT) on depth images suggested in Mahmud et al. [11], to develop a method that enables a user to recognize hand gestures that contain a number of symbols. In order to minimize the number of features dimension, the authors have used k-means clustering and SIFT features. These features were fed into the Support Vector Machine (SVM) classifier for training.

Every fruit classification consists of unique characteristics. In such a manner, the feature extraction methods must be susceptible to extract the features from the essential parts of any fruits. Most researchers take advantage of these elusive attributes to recognize the fruits. Numerous researchers used deep learning based algorithms to identify the fruits in the last period of a decade. For example, in Mureşan and Oltean [12], a new high-quality dataset is introduced. The dataset is named Fruits_360, which contains 38,409 images of fruits and is updated day by day. A convolutional neural network is proposed in this paper to detect fruits. Another CNN-based model named MobileNetv2 is proposed in Rojas-Aranda et al. [13]. Here, the authors try to detect the fruits and decide whether the fruits are inside or outside the plastic bag. Here, the authors introduced a new dataset for three classes. Different input features such as RGB color, RGB histogram, and RGB centroid obtained from K-means clustering improves the classification accuracy. In Ashraf et al. [14], the authors proposed a system that detects whether the fruits are fresh or rotten from an input image. For which, the inception-v3 model was presented and trained the dataset through transfer learning. In Bargoti and Underwood [15], a system is presented for fruits detection in orchards including mangoes, apples, and almonds. A tiling approach is introduced for Faster R-CNN to apply a trained model over large images. In Patel and Prajapati [16], to extract the features of fruits, the authors presented a new method named wavelet transformation using the Haar filter. Wavelet transformation with Haar filter was applied to the region of interest in an image. Using the extracted features, an SVM classifier is used to classify the fruits.

3 The Proposed Framework

In this work, we have proposed a framework, where both hand detection and fruits classification are performed in a single system. For that, initially, two hand detection model is introduced, where one model is based on the hand feature extraction model, i.e., contour-based model and another one is deep learning based model, i.e., YOLOv3-based model. Finally, a convolution neural network based on deep learning is developed for fruits classification. The overall diagram of the proposed framework is presented in Fig. 1.

3.1 Hand Detection

In this work, two different methods are used for hand detection. One is a contour feature based model and another one is a deep learning based YOLOV3 model. The overall workflow diagram for hand detection is presented in Fig. 2.

3.1.1 Hand Detection using Hand Features Extraction

Hand detection is an important part of this work. By which we can specify whether the user holds a fruit or not. In this part, contour-based hand detection procedure is described. This procedure is divided into four primary steps. Those are: (1) background subtraction, (2) morphological transformation, (3) extracting contour feature of hand, and (4) detecting large convex hull and hand detection.

3.1.2 Background Subtraction and Morphological Transformations

In the first step, a normal camera is used to capture the hand pictures. After that, the input image is preprocessed to remove the noise using the gaussian filter. The filtered RGB image is remodeled into an HSV image. This operation is performed to extract

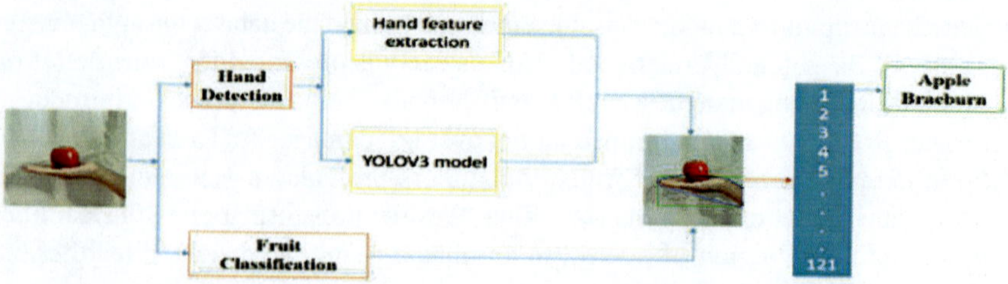

Fig. 1 Framework of the proposed model

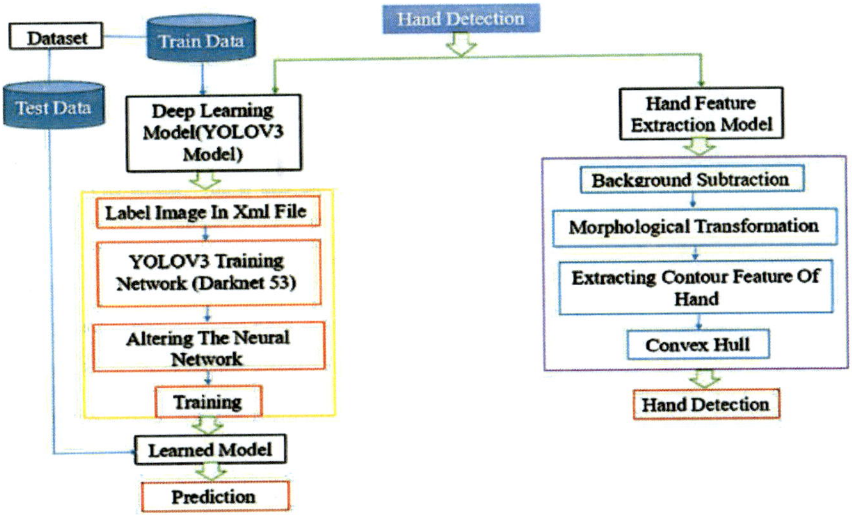

Fig. 2 Workflow diagram for hand detection models

the hand skin color region. The HSV image is used instead of RGB because BGR carries more information about the color's brightness. The hand skin color region extraction formula is presented in (1).

$$Hand = \begin{cases} 255if((50 \le R \le 255)and(50 \le B \le 255)and(2 \le G \le 15)) \\ 0otherwise \end{cases} \quad (1)$$

The skin color conversion image from a filtered image is shown in Fig. 3b. In Fig. 3b, there are some gaps due to the image shape. To recover these gaps, mathematical morphological operations such as dilation and erosion are applied. The result of a morphological operation is shown in Fig. 3c.

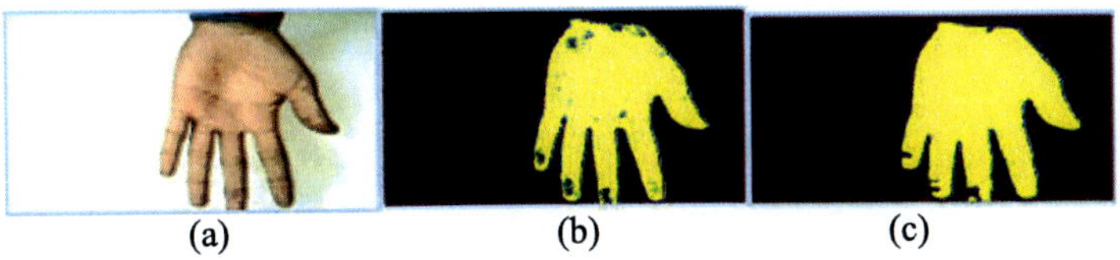

Fig. 3 Background subtraction: **a** input image, **b** subtraction image, and **c** image after morphological transformations

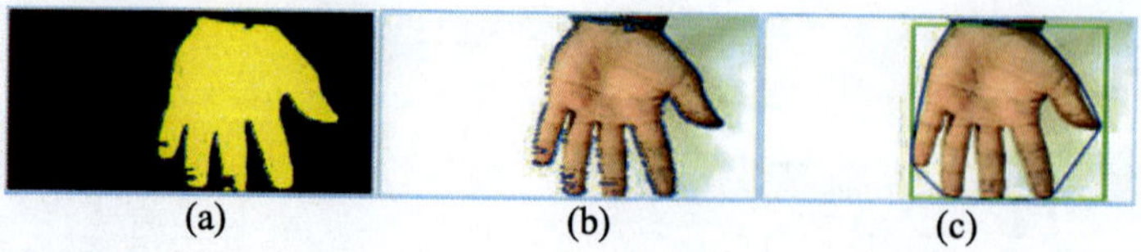

Fig. 4 Detected contours and convex hull: **a** morphological transformations, **b** contour of the hand, and **c** convex hull with a rectangle box

3.1.3 Extracting Contours Feature of Hand

The curves change of integrity for all the continual points on the boundary are measured by contours. Extracting contours is important for analysis and object detection. In this work, the contours are used to detect the hand boundary of the hand image. Here, the find Contours function helps to extract the contours. After that, the CHAIN_APPROX_SIMPLE function is used to remove the redundant points and help to compress diagonal, horizontal, and vertical segments. The contours with hand are presented in Fig. 4b. Finally, a boundary line is drawn around the hand.

3.1.4 Convex Hull and Hand Detection

The convex hull is explained in the way of characterizing the shape of an image. The convex hull is joined to contours. Around the contour of the hand, the convex hull is drawn because contour points within the convex hull. It uses the least number of points to build the hull. So that, all of the points of the contour are inside the hull. Convexity defects can be defined using the difference between contour and convex hull. A defect is located where the contour of the hand is away from the convex hull drawn around the same contour. Convexity defect offers the set of values for every defect. The convex hull with a rectangle box around the hand is presented in Fig. 4c.

3.2 *Deep Learning Based Model*

For hand detection using deep learning based model, 345 images from 11 K hand dataset [7] are used. To find effective learning accuracy from all environments, 100 real-time hand images are used in this process. The similarity between the images of the hand 11 k dataset and the real data set was ensured.

YOLOv3 uses a single convolutional network making it a fully convolutional layer. The convolution is performed on the input image only once, which is right only once to heal the predictions. YOLOv3 performs the whole image rather than the sections. Using multiscale training, the model learned to generalize and detect objects with different aspect ratios or configurations. This algorithm applies a single neural network to divide the image into the region and predicts bounding boxes.

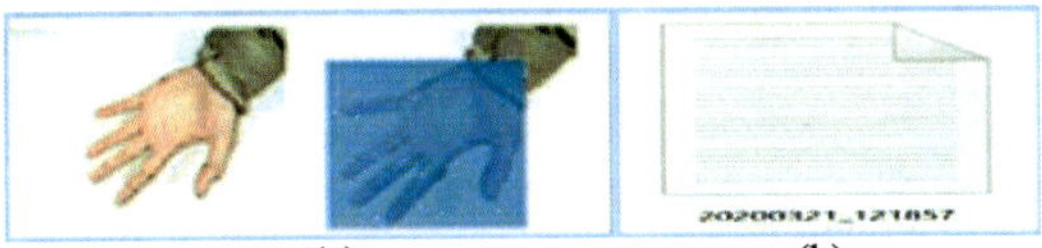
(a) (b)

Fig. 5 Labeling images: **a** sample label image, and **b** XML file

Table 1 The parameters of the model

Image's size	Batch	Learning rate	Momentum	Decay	Steps for training
1600*1200	64	0.0001	0.9	0.0005	4000

3.2.1 Labeling the Images

In the labeling of the image, the Labelimg is used to select where the hand is located. By this function, a rectangle box is created around the hand and names the label as 'hand'. Finally, the information is saved in the XML file. This information is used by the YOLO model during the training process to train the model. The label partition of the sample date is shown in Fig. 5(a).

3.2.2 Training Framework Setup

Darknet 53 is used for the training session. From the author's GitHub project Darknet, a configuration file for Darknet was retrieved [17]. YOLOv3.cfg is the most important file because it contains all the information of layers in the network and hyperparameters.

3.2.3 Altering the Neural Network

For all training sessions, the 64 batch size is used. The learning rate used in this work is 0.001. As the system focused on only one class, i.e., hand, for that reason, the neural network was altered to detect one class. These parameters are stored through editing the Darknet configuration file yolov3.cfg. The parameters of the model are shown in Table 1.

3.2.4 Training

In this paper, the feature of annotated custom images was extracted using the Darknet-53 framework, and Yolov3 was used for identification. Table 1 shows the changes made to the configuration file for training the models. Two new files with data and names have been created. The data file contains the path to train.txt, test.txt, name

Table 2 Average loss from 9 training sessions

Iteration	Loss	Average loss
3926	0.026	0.032
3929	0.030	0.031
3931	0.032	0.032
3936	0.030	0.032
3945	0.022	0.031
3950	0.021	0.030
3964	0.020	0.025
4000	0.030	0.024

file, and backup folder in Cloud VM via Google Drive. The backup folder is where all the trained weights will be stored after every 1000 iterations. The 'name file' contains the names of the class, such as 0. These two files were copied from Google Drive to Cloud VM. The training session was stopped after 4000 training iterations. During the training session, an average loss of each of the iteration is estimated. From 4000 iterations, 9 iterations are shown in Table 2.

3.3 Proposed Network for Fruits Classification

In this section, the fruit classification model is developed based on the convolutional neural network explained. The workflow diagram of the fruits classification is presented in Fig. 6.

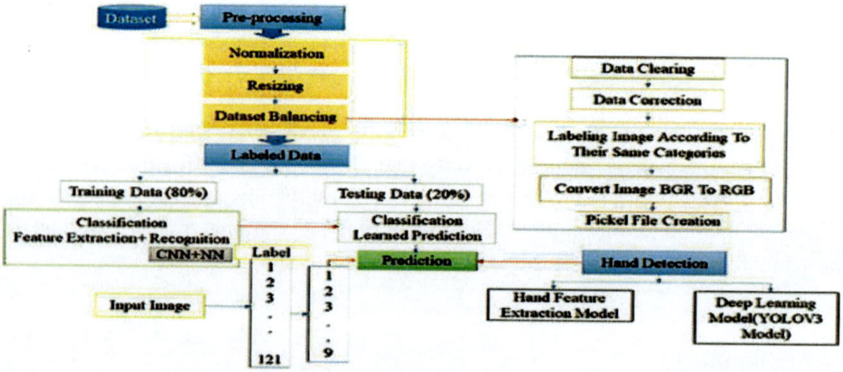

Fig. 6 The workflow diagram for fruits classification

3.3.1 Dataset

In this work, the "Fruits_360 dataset" is used. And the dataset is known as the "Fruits_360 dataset" [18]. Fruits_360 is made up of 82,213 images and the size of the images is 100 × 100 pixels. Where the size of the training set contains 61.488 images and the size of the test set is 20,622 images. The dataset contains a variety of vegetables and fruits.

We have collected 1000 images for our owned data. These images are captured in similar way to Fruits_360 dataset. In this dataset, one more label is added named 'No fruits on hand'. The total number of labels in this dataset is 121.

3.3.2 Model Preparation

To establish the final model, some models with different features are tested. Here, some models with different convolutional layers and hidden layers with different image formats such as BGR to RGB to GRAY, BGR to RGB to HSV, and BGB to RGB are tested to establish the final model. The transfer model MINVGGNET with different image formats is tested and the results are shown in Table 3. From the result, it is revealed that BGB to RGB model gives more strong accuracy and that model also gives more accurate results to identify different attributes. In the other formats, this model faces different problems to identify attributes accurately. BGR to RGB to GRAY model is not able to recognize banana labels accurately and also faces difficulties to recognize Apple Red labels.

MINIVGGNET model gives appropriate results during training and validation. However, during the phase of testing, it faces trouble recognizing real-time data. BGB to RGB model is more effective and accurate to recognize all of our given attributes' labels and it performed equally well when it tested any real data type.

To establish a suitable model for fruits classification, different convolutional layers with the same fully connected layer models are considered. These models' training and test accuracy are presented in Table 4. In that table, it is revealed that Model-6, i.e., 16_64_64_128 gives better train and test accuracy. Based on the results, this model is selected as the proposed model. The elaborate representation of the model is shown in Fig. 7. Here, the framework contains four convolutional layers by which

Table 3 Comparison between different model scenarios

Models	Accuracy on training set (%)	Accuracy on test set (%)
BGR to RGB	99.99	99.99
BGR to RGB to GRAY	99.96	99.87
BGR to RGB to HSV	97.98	90.93

Table 4 Training and test accuracy for different model configurations

Model	Filter configuration (Number of filters in each layer)	Fully connected layer configuration	Train accuracy (%)	Test accuracy (%)
Model-1	16_16_64_128	512	99.95	99.94
Model-2	16_32_32_128	512	99.97	99.91
Model-3	16_32_64_64	512	99.57	99.92
Model-4	16_32_64_128	512	99.96	99.95
Model-5	16_32_128_128	512	99.94	99.73
Model-6	**16_64_64_128**	**512**	**99.99**	**99.98**
Model-7	32_32_64_128	512	99.93	99.83
Model-8	8_32_64_128	512	99.99	99.94

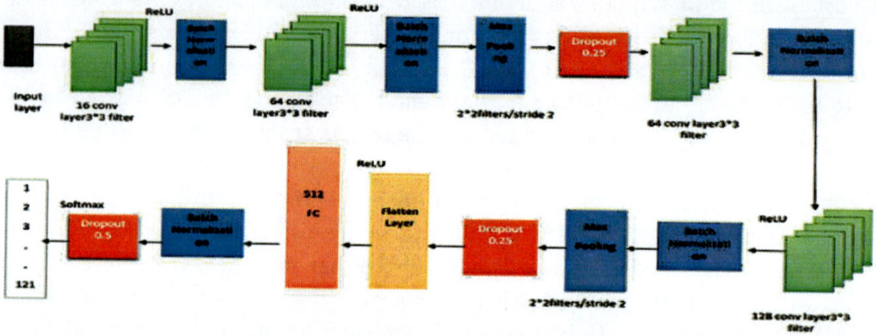

Fig. 7 Proposed convolutional neural network model's architecture for fruits classification

the important features of the object region are extracted and one fully connected layer by which objects are recognized.

The first layer of the convolutional is made up of 16 filters and the filter's size is 3 × 3. The size of the output features maps is the same as the same padding is used. For instance, the size of the features map is 48 × 48 for 16 channels, and the output of the feature maps are the same as before. The ReLU is used as an activation function to calculate the output of the nonlinearity. Batch normalization makes the network stable and also makes the network faster during the training session.

The second layer of the convolutional is made up of 64 filters and the filter's size is 3 × 3. And the size of the output feature maps is 48 × 48. The batch normalization is again used to standardize the model's performance. After that, the max-pooling is used which filter size is 2 × 2 and the number of strides is 2. After the max operation, the size of the output feature maps is 24 × 24. The value of dropout has been set to 0.25 to generalize the model and preventing from overfitting. We have used the value of dropout as an empirical value.

The third layer of the convolutional is made up of 64 filters and the filter's size is 3×3. In the section of nonlinearity, the ReLU activation function is used again. After applying the activation function, batch normalization is used for the model.

The fourth layer of the convolutional is made up of 128 filters and the filter's size is 3×3. Moreover, the size of the output feature maps is 24×24 for 128 channels. ReLU activation function is again used to update the value of output successfully. Again learning process is stabilized by batch normalization. After that, the filter size of the max-pooling is 2×2 and the number of strides is 2. After the max-pooling operation, the size of the output feature maps is 12×12 for 128 channels. In this layer, dropout has been set to 0.25.

For converting data into one-dimensional array, flattened layer is used. The fully connected layer is used to classify images into a class or labels. It contains $12 \times 12 \times 128 = 18{,}432$ elements, which are flattened in the last layer into the form of a vector. The fully connected layer contains 512 neurons. Batch normalization is applied again. Dropout value of 0.5 is used. Here, ReLU is used again for the activation function. Softmax is used as an output activation function for multilabel classification. Finally, in the output layer, the expected result represents 121 classes.

3.3.3 Loss Function and Optimizer

The loss function used in this work is categorical cross-entropy as our problem is multiclass classification. This loss function is used to assess the model's performance by evaluating the loss value. Based on these loss values, the parameters of the model is learned through optimization algorithm. In this work, Adam optimizer is used with momentum. The loss function is shown in (2).

$$Loss = -\text{logp} \tag{2}$$

where p is the probability distribution of the predicted class labels.

4 Experimental Results

The detailed experimental review of the model being proposed based on hand detection and numerous fruits classification recognition databases is presented in this section.

4.1 Hand Detection using Hand Features Extraction

The hand features extraction model works in four basic steps. The steps are as follows: in the first step, background subtraction is used through skin color conversion. After that, for dispelling the background's noise morphological transformation is performed. Furthermore, the contour feature of the hand is extracted. Finally, it finds the convex hull and detects the hand with a rectangle box. Some processing examples of hand detection based on the hand feature extraction are presented in Fig. 8. Here, images are taken from 11 K dataset.

Some processing examples for own dataset with different angles are presented in Fig. 9. Some images are captured where fruits are in the hand. From the experiments, the experiments confirm that the proposed method detects the hand from different angles efficiently and accurately.

Using hand features extraction, sometimes it detects the hands incorrectly when the hand is not shown properly. Some images with poor hand detection are given in Fig. 10.

We tested 80 images from the dataset. At the time of testing, 78 are correctly detected out of 80 images. The recognition accuracy is 97.50%. We also test 70 images where the fruits were in the hand and 65 images are correctly recognized.

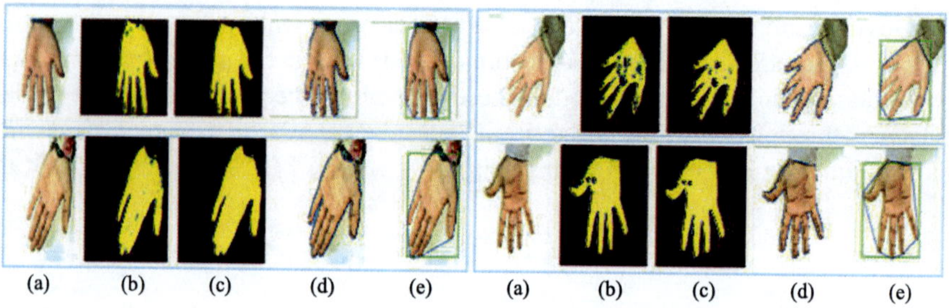

Fig. 8 Hand detection using hand features extraction: **a** input image, **b** background subtraction, **c** labeled image, **d** detected contour, and **e** detected convex hull with a rectangle box

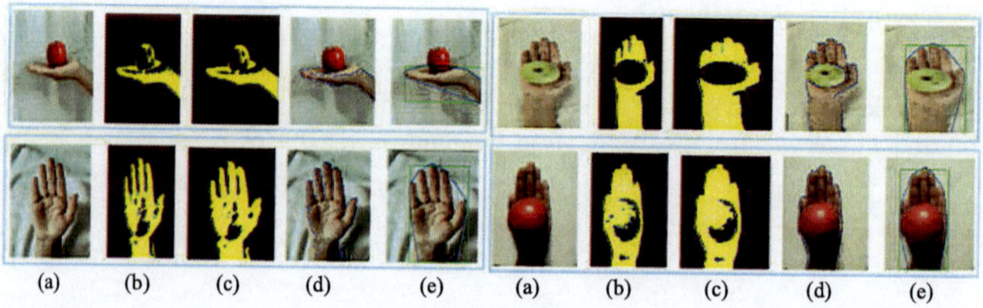

Fig. 9 Hand detection using hand features extraction using own dataset: **a** input image, **b** background subtraction, **c** labeled image, **d** detected contour, and **e** detected convex hull with a rectangle box

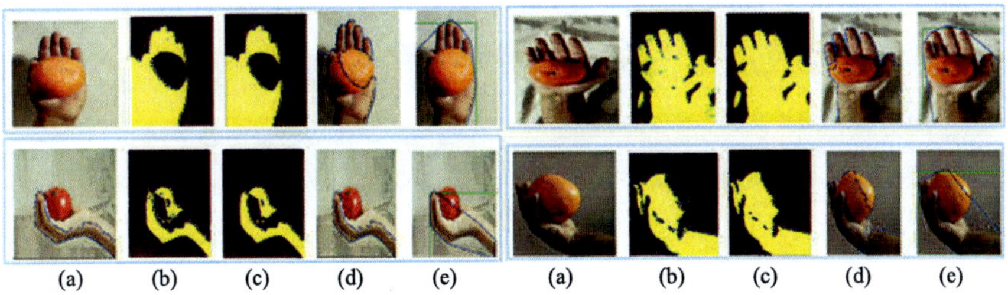

Fig. 10 Some images with poor hand detection: **a** input image, **b** background subtraction, **c** labeled image, **d** detected contour, and **e** detected convex hull with a rectangle box

Table 5 Accuracy for detecting hand using hand feature extraction

Types	Total image	Correctly recognized	Recognition accuracy (%)
Hand	80	78	97.50
Fruits in hand	70	65	92.85
Avg	150	143	95.17

The recognition accuracy in this time is 92.85%. The average accuracy of the hand feature extraction model is 95.17%. The accuracy is shown in Table 5.

4.2 Hand Detection using Deep Learning Based Model

The YOLOv3 method has several advantages as compared to other object detection algorithms. CNN, FCNN algorithm does not work with the image completely, however, the YOLO model looks at the image correctly and predicts the bounding box using the convolutional network. There we use 345 images from 11 K dataset and 100 images from our own dataset to train and test the YOLOv3 model. Some experimental results of the YOLOv3 model are given in Fig. 11. Experimental results of own dataset images that are captured in different orientation are presented in Fig. 12.

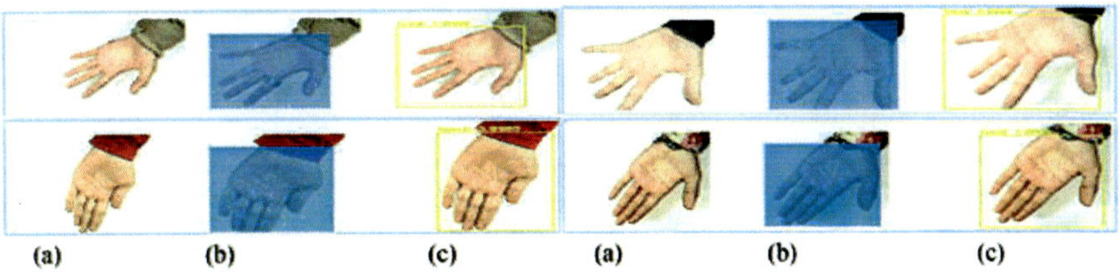

Fig. 11 Hand detection using deep learning Yolov3 model with11K dataset: **a** input image, **b** label image, and **c** output image

Fig. 12 Hand detection using deep learning Yolov3 model with real dataset: **a** input image, **b** label image, and **c** output image

The performance analysis of the YOLOv3 model is given in Table 6. Where TP is used to measure the percentage of actual positive features correctly identified, FP is used to measure the percentage of features which are not correctly identified. To calculate the overlap between two areas, intersection over union (IoU) is conducted. It also shows the precision, recall, F1-factor, average precision (AP), and average accuracy.

We clearly see that the Yolo model works better than the hand feature extraction model. Yolo model detected hands more effectively and the accuracy is better than the hand feature extraction model. So the YOLOv3 model is our proposed network.

4.3 *Experimental Analysis of Fruits Classification*

For fruits classification, two models are used in this work. One is miniVGGNet and another one is the HnadFruitNet, i.e., proposed model. To distinguish between the proposed model and miniVGGNet, historical curves are used. The performances of the models are also estimated based on the confusion matrix and classification report.

The training loss and validation loss curve of MiniVGGNet and the proposed model for the Fruit_360 dataset are shown in Fig. 13a, c, respectively. The curve for training and validation accuracy for MiniVGGNet and the proposed model are represented in Fig. 13b, d, respectively. From the curve, it is revealed that MiniVGGNet gives 99.93% accuracy, and the proposed model gives 99.99% accuracy on the Fruits_360 dataset.

From the confusion matrix, it can be seen that which label predicted how well. The confusion matrix without normalized and with normalized form is presented in Fig. 14a, b, respectively. From the figures, it is revealed that the prediction label of Apple Granny Smith and Pomegranate is 99%, and other label models predicted 100%.

Some processing examples of correctly predicted fruits from the Fruits_360 dataset and own dataset with prediction scores are shown in Figs. 15 and 16, respectively.

The proposed model is misclassifying some fruits images. The misclassifying images are shown in Fig. 17. The names of the class that have been properly classified

Table 6 TP, FP, IoU, and AP of each class in YOLOv3

Class	TP (%)	FP (%)	IoU (%)	Precision (%)	Recall (%)	F1-Score (%)	Accuracy (%)	Average accuracy (%)	AP (%)
Hand	99	1	99.10	99.12	98.89	99.20	99.90	99.90	99.90

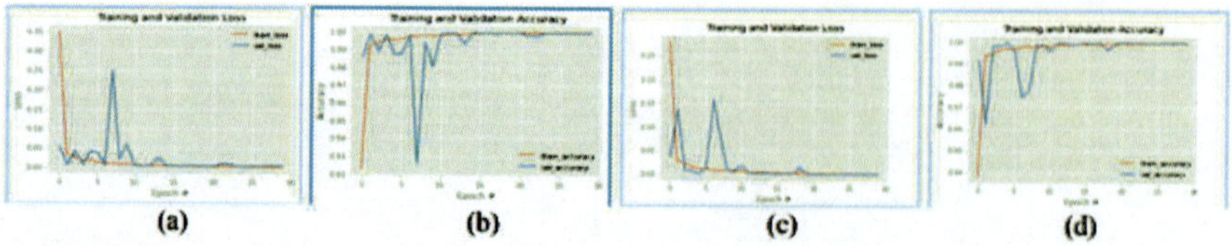

Fig. 13 Loss and accuracy curve: **a** training and validation loss curve for miniVGGNet, **b** training and validation accuracy curve for miniVGGNet, **c** training and validation loss curve for the proposed model, and **d** training and validation accuracy curve for the proposed model

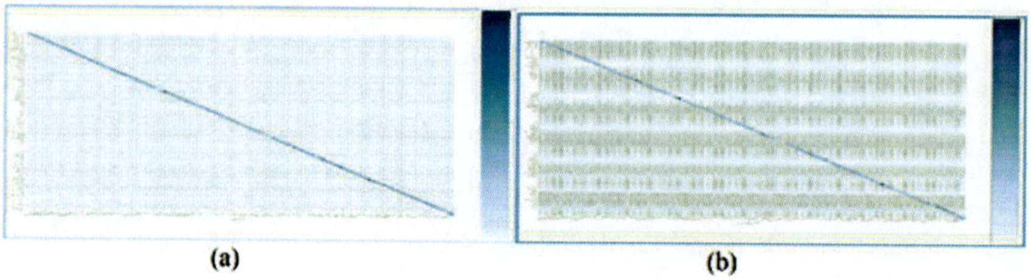

Fig. 14 Confusion matrix of our dataset for proposed model: **a** confusion matrix (without normalized), and **b** normalized confusion matrix

Fig. 15 Proposed model test accuracy using the Fruits_360 dataset

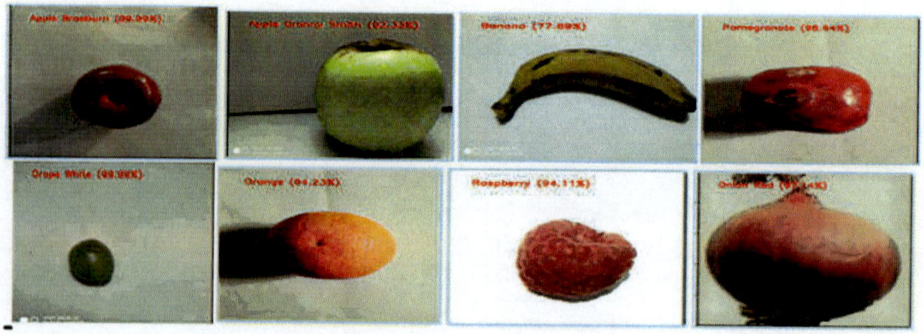

Fig. 16 Some processing examples of the proposed model with test accuracy for owned dataset

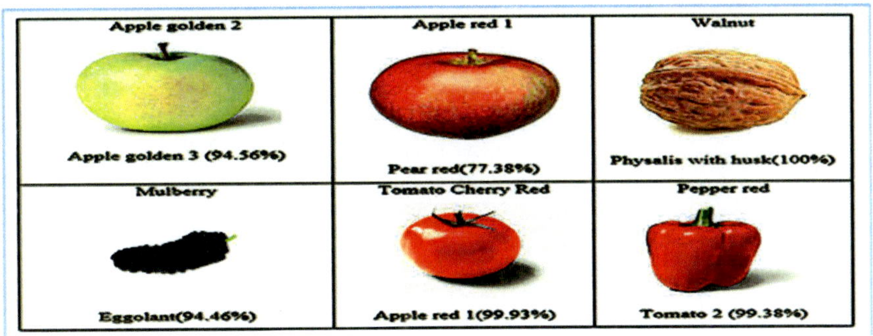

Fig. 17 Sample images with incorrect data

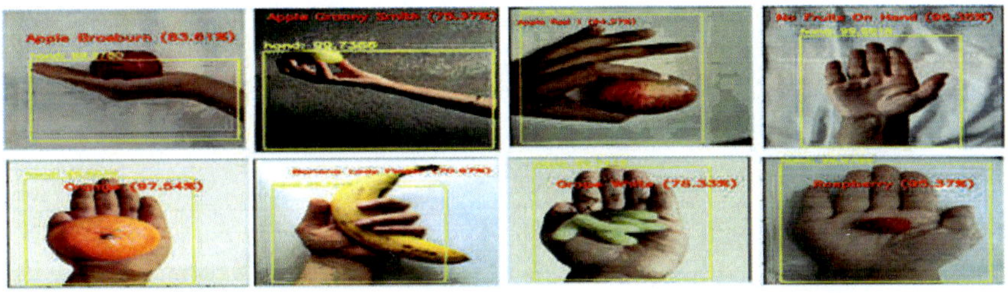

Fig.18 Experimental result using fruits classification and hand detection model together

are given on the top and the names of the class that have been classified incorrectly are given below.

Finally, the YOLOV3 is used for hand detection as this model gives better accuracy than the hand feature extraction model and HnadFruitNet proposed model is used for fruits classification together in a single system to recognize the fruits in a hand. Some processing examples of hand detection and fruits classification from a single system are presented in Fig. 18. From the results, it is shown that the proposed systems detect and classify both hand and fruits together efficiently and accurately.

The hand detection and fruits classification models are compared with the present state of the art. The comparisons of hand detection and fruits classification are shown in Tables 7 and 8, respectively. From the comparison, it is clear that both deep learning models give better results with compared to the present state of the art.

5 Conclusions

In this paper, hand detection and fruits classification models are introduced. For hand detection, two models are introduced. The first hand detection model is developed based on hand feature extraction. The feature used in this work is the contour feature

Table 7 Accuracy comparison of hand detection

Method	Accuracy (%)
Proposed YOLOv3	99.90
Hand feature extraction method	95.17
Image processing [6]	92.30
YOLOv3 [9]	97.68
SIFT + SVM[11]	99.00 (depth image) 96.00 (binary image)

Table 8 Accuracy comparison of fruit classification

Method	Accuracy (%)
Proposed Framework	99.99
MINIVGGNET	99.93
CNN + TensorFlow [12]	96.30
MobileNetV2 + Single color [13]	97.23
VGG16 Inception-v3 [14]	95.08 87.08
Faster R-CNN [15]	86.10

of a hand. The second model is developed based on deep learning, i.e., YOLOv3. The fruits are classified based on the convolutional neural network model. Finally, both models are combined to easily detect the hand and accurately classify the fruit in a person's hand together. 11 K and Fruits_360 datasets are used for hand detection and fruits classification to train and validate the models. For testing, own created datasets are utilized. Between the hand detection models, YOLOv3 performs better, which has the higher accuracy of 4.73%. The fruits classification model shows better accuracy with respect to the present state-of-the-art models, which is 99.99%. In future, the system will be developed for real-time applications and detecting the hand where there are multiple fruits.

References

1. Karbasi, M., Bhatti, Z., Nooralishahi, P., Mazloomnezhad, S.M.R., Shah, A.: Real-time hands detection in depth image by using distance with kinect camera. vol. 4, pp. 1–6 (2015). https://doi.org/10.5923/c.ijit.201501.01
2. Dhawan, A., Honrao, V.: Implementation of hand detection based techniques for human computer interaction (2013). https://doi.org/10.5120/12632-9151
3. Song, Y., Glasbey, C.A., Horgan, G.W., Polder, G., Dieleman, J.A., van der Heijden, G.W.A.M.: Automatic fruit recognition and counting from multiple images. Biosyst. Eng. **118**(1), 203–215 (2014). https://doi.org/10.1016/j.biosystemseng.2013.12.008
4. Mia, M.R., Mia, M.J., Majumder, A., Supriya, S., Habib, M.T.: Computer vision based local fruit recognition. Int. J. Eng. Adv. Technol. **9**(1), 2810–2820 (2019). https://doi.org/10.35940/ijeat.A9789.109119

5. Rahnemoonfar, M., Sheppard, C.: Deep count: fruit counting based on deep simulated learning. Sensors (Switzerland) **17**(4) (2017). https://doi.org/10.3390/s17040905
6. Khan, T., Pathan, A.H.: Hand gesture recognition based on digital image processing using MATLAB. Int. J. Sci. Eng. Res. **6**(9) 338 (2015). http://www.ijser.org
7. Afifi, M.: 11K Hands: Gender recognition and biometric identification using a large dataset of hand images. Multimed. Tools Appl. **78**(15), 20835–20854 (2019). https://doi.org/10.1007/s11042-019-7424-8
8. Xu, Y., Park, D.W., Pok, G.: Hand gesture recognition based on convex defect detection. Int. J. Appl. Eng. Res. **12**(18), 7075–7079 (2017). http://www.ripublication.com
9. Mujahid, A., et al.: Real-time hand gesture recognition based on deep learning YOLOv3 model. Appl. Sci. **11**(9) (2021). https://doi.org/10.3390/app11094164
10. Il Joo, S., Weon, S.H., Hong, J.M., Il Choi, H.: Hand detection in depth images using features of depth difference. In: Proceeding 2013 International Conferences Image Processer Computer Vision, Pattern Recognition, IPCV 2013, vol. 2, pp. 823–824 (2013)
11. Mahmud, H., Hasan, M.K., Al-Tariq, A., Mottalib, M.A.: Hand gesture recognition using SIFT features on depth image. In: Ninth International Conferences Advances Computer Interaction Hand, pp. 359–365 (2016)
12. Mureşan, H., Oltean, M.: "Fruit recognition from images using deep learning. Acta Univ. Sapientiae Inform. **10**(1), 26–42 (2018). https://doi.org/10.2478/ausi-2018-0002
13. Rojas-Aranda, J.L., Nunez-Varela, J.I., Cuevas-Tello, J.C., Rangel-Ramirez, G.: Fruit classification for retail stores using deep learning. In: Lecture Notes Computer Science (including Subser. Lecture Notes Artificial Intelligent Lecture Notes Bioinformatics), vol. 12088, pp. 3–13. LNCS (2020). https://doi.org/10.1007/978-3-030-49076-8_1
14. Ashraf, S., Kadery, I., Chowdhury, A.A., Mahbub, T.Z., Rahman, R.M.: Fruit image classification using convolutional neural networks. Int. J. Softw. Innov. **7**(4), 51–70 (2019). https://doi.org/10.4018/IJSI.2019100103
15. Bargoti, S., Underwood, J.: Deep fruit detection in orchards. In: Proceeding of IEEE International Conference Robotics Automation, pp. 3626–3633 (2017). https://doi.org/10.1109/ICRA.2017.7989417
16. Patel, H., Prajapati, P.: Fruits classification using image processing techniques. Int. J. Computer. Sci. Eng. **6**(10), 628–632 (2018). https://doi.org/10.26438/ijcse/v7si5.131135
17. Bochkovskiy, A.: AlexeyAB/darknet: YOLOv4v / Scaled-YOLOv4–Neural Networks for Object Detection (Windows and Linux version of Darknet). GitHub (2020). https://github.com/AlexeyAB/darknet. Accessed 10 Mar 2020
18. Mureşan, H., Oltean, M.: Fruits 360 dataset on kaggle. Kaggle (2021). https://www.kaggle.com/moltean/fruits. Accessed 03 Jan 2020

Entrepreneurship Education and Entrepreneurial Intention of Business Graduates: Does Artificial Intelligence Matter?

Md. Mizanur Rahman, Md. Saidur Rahaman, Iqbal Hossain Moral, and Md. Solaiman Chowdhury

Abstract The main goal of this research is to investigate the mediating role of Artificial Intelligence (AI) in the relationship between Entrepreneurship Education (EE) and Entrepreneurial Intention (EI) among business graduates. A well-structured questionnaire in "Google Form" was created and distributed to respondents via various social media sites, including WhatsApp, LinkedIn, Messenger, etc. A snowball sampling technique was used to distribute the questionnaire to the respondents for data collection in this study. However, the sample size was set at 401 business graduates from different universities in Bangladesh. The analysis was carried out using SPSS and AMOS software. The findings revealed that entrepreneurship education has a significant positive impact on entrepreneurial intention. Besides, artificial intelligence has substantial positive impacts on entrepreneurial intention. Moreover, artificial intelligence partially mediates the relationship between entrepreneurship education and entrepreneurial intention. This study incorporates artificial intelligence into entrepreneurial education for the purpose of establishing an entrepreneurial mindset among business graduates. Only business students from Bangladeshi universities took part in the study, thus, it is recommended that future studies include students from other disciplines or fields such as Computer Science and Engineering, information technology (IT), English, Law, and Engineering to make the findings more acceptable. To the best of the authors' knowledge, this is one of the leading studies relating to artificial intelligence and entrepreneurship in Bangladesh.

Md. M. Rahman (✉)
BRAC University, Dhaka, Bangladesh
e-mail: mizanur.rahman@bracu.ac.bd

Md. S. Rahaman
Metropolitan University Sylhet, Sylhet, Bangladesh

I. H. Moral
Northern University of Business and Technology Khulna, Khulna, Bangladesh

Md. S. Chowdhury
Rajshahi University, Rajshahi, Bangladesh
e-mail: schowdhury@ru.ac.bd

S. Hossain et al. (eds.), *Proceedings of International Conference on Fourth Industrial Revolution and Beyond 2021*, Lecture Notes in Networks and Systems 437,
https://doi.org/10.1007/978-981-19-2445-3_8

Keywords Artificial intelligence · Business graduates · Entrepreneurship education · Entrepreneurial intention

1 Introduction

As a developing country, Bangladesh is continuously trying to increase the number of entrepreneurs for the development of its economy. According to the Global Entrepreneurship Index [1], among 137 studied countries, the position of Bangladesh is 134th, whereas the ranks of Singapore, Brunei Darussalam, and Malaysia are consecutively 27, 53, and 58 [2]. To deal with this issue, the government of Bangladesh has taken several initiatives. Bangladesh is such a country where most people are young; half of its population is under the age of 25. Bangladesh aimed at becoming a middle-income country by 2021, focusing on an innovation-based economy. It has been concentrating more on developing entrepreneurship by analyzing its necessity for economic development and poverty reduction. In this regard, the Bangladesh government has launched a national-level entrepreneurship platform to facilitate Innovation Design and Entrepreneurship Academy (IDEA) project, which the ICT division governs. Furthermore, the central bank of Bangladesh has inaugurated a project worth one billion takas, Entrepreneurship—Support Fund (ESF), to help entrepreneurship development. Many private and public universities, development institutions, and organizations offer *entrepreneurship* courses to facilitate entrepreneurship education and training. Noticeably, currently, the insufficient infrastructure and unemployed graduates indicate that Bangladesh has lagged in entrepreneurship development. One of the researchers agrees that entrepreneurial education can play a vital role in increasing entrepreneurial passion, spirit, and attitude among the young generation [3]. Additionally, Cui et al. [4] stated that if entrepreneurship education can be promoted, it can help to create a positive mindset, attitude, and behavior among the graduates to engage themselves with entrepreneurship as a career. Besides, the family can play an important role as a social institution to encourage entrepreneurship along with entrepreneurship education [5].

Based on the report provided by World Vision, around 4.4 million young people in Bangladesh are unemployed [6]. Currently, the Bangladesh Bureau of Statistics (BBS) reported that the unemployment rate of university graduates aged between 15 and 29 years is 11.2% [7]. This statement raises a question about entrepreneurship education (EE) and creation in Bangladesh. Contrary to this situation, this study tries to identify the contributions of BBA and MBA programs to develop traits, skills, and knowledge among the graduates that can help them be entrepreneurs. The key point of Bangladesh's education policy, according to the Academic Policy 2010, is that all levels of education, including the tertiary level, have been designed to focus on entrepreneurship development. As per that policy, at tertiary-level education, the main focus of business studies education is to develop the required knowledge, skills, and attitudes to be a successful entrepreneur. Currently, business administration departments, faculty of business studies, and business administration institutions are providing entrepreneurial education (EE) among the students. Entrepreneurship education can be taught by providing knowledge regarding entrepreneurship theory, challenges of being entrepreneurs, motivation toward entrepreneurial mindset, entrepreneurship as a profession, circumstances of doing business, the contribution of entrepreneurs in the economy, how to generate an idea, ways of selecting a suitable project, project management, how to prepare business plans, corporate social responsibility and ethical issues of entrepreneurs. The number of private and public universities in Bangladesh has been growing because of the progression of the increasing number of students [8]. Currently, Bangladesh has 47 public universities and 107 private universities [9]. The total number of students studying business or business-related courses from private universities is 87695 and from public universities is 41090 [9]. After completing graduation, the majority of the students try to involve themselves with jobs, but it is a matter of great regret that many of them fail to manage it. In Bangladesh, the unemployment condition of graduate students is a topped of topic. However, the unemployment rate of graduate students is 26.1%. Comparatively, the unemployment rate of primary-level education completed youths is 6.0%, which is four times lower than those who have completed graduation [10]. Furthermore, as the number of unemployed graduate students is increasing, this situation can be a source of economic improvement if the entrepreneurship mindset and attitude can be improved [11].

The former chairman of UGC, Abdul Mannan stated that till now, the students of Bangladesh had not had a satisfactory level of knowledge about Information Technology (IT) and Artificial Intelligence (AI) (2017, www.thedailystar.net). World Economic Forum (WEF) published a report on Global Competitiveness Index, which divulges that the score of Bangladesh about global IT competitiveness is 106th among 138 countries. But, it is a matter of great concern for digital Bangladesh, most the young internet users keep themselves busy with different social media such as Facebook, Messenger, WhatsApp, TikTok, Likee, and so on. The majority of the top positions related to IT are employed by foreign experts, who can create serious security and control problems. This circumstance denotes the lack of concern about the implementation of AI in our education system. Currently, AI takes place as one of the most discussed topics and is widely recognized because of its close connection with our daily lives. Artificial Intelligence has been widely used in many sectors like science fiction, health care, medicine, transportation, retail, banking, and education [12]. Noticeably, around the world, for the improvement of education regarding AI, about $1047 billion was financed between the years 2008 and 2017 [12]. According to recent studies, it can be concluded that a lot of attention has been given to Artificial Intelligence as it influences becoming an entrepreneur, and AI helps to assure the better performance and economic growth of a country [13]. However, the relationship between AI and entrepreneurship education and intentions to be an entrepreneur is under development.

It has been found from previous studies that there is a direct effect of entrepreneurship education (EE), artificial intelligence, and entrepreneurial intention (EI) [14, 15]. But most of the earlier investigations had a limitation that they failed to identify the intervening effect of artificial intelligence (AI) on EE and EI. Therefore, this study identifies artificial intelligence as the mediating variable between entrepreneurship education and entrepreneurial intention among the students, especially business graduates in Bangladesh.

2 Hypotheses Development Based on Theory

In recent days, continuous development in higher education emerges the extreme need for entrepreneurship to alleviate unemployment and poverty reduction. Since, higher education performs a crucial role in forming students' high attitude, stable mindset, and high intention to become entrepreneurs [16]. Besides, entrepreneurship in higher education supports a lot to prepare the students to initiate new ideas and convert the ideas into an established business by integrating knowledge, skills, abilities (KSA), and experiences [17]. Therefore, Din et al. [18] extended that entrepreneurship education should be particular; it might be more effective for business-centered institutions than multidisciplinary schools. It is a universal truth that creative, innovative, and sustainable business ideas may come from any department. According to Ajzen [19], there is a significant association between entrepreneurship education and intention, which is theoretically tested. In this case, Ajzen's TPB is essentially effective in analyzing entrepreneurs' behavior by attaining entrepreneurship education. This is not always true that participants can make the business sustainable after getting the entrepreneurship education for a lifetime. As a result, sometimes entrepreneurial behavior might not be the only predictor. But, promoting students to become entrepreneurs in higher education must ensure that they have grown with high entrepreneurial intention. This study is a new avenue of using the theory of planned behavior, where the intention can use as a substitute for behavior. With the adoption of this concept, it is possible to develop the students' intention to become an entrepreneur through entrepreneurship education. Because knowledge acquisition with formal education can divert the individuals' behavior, whereas learning outcomes can influence entrepreneurial intention (Fig. 1).

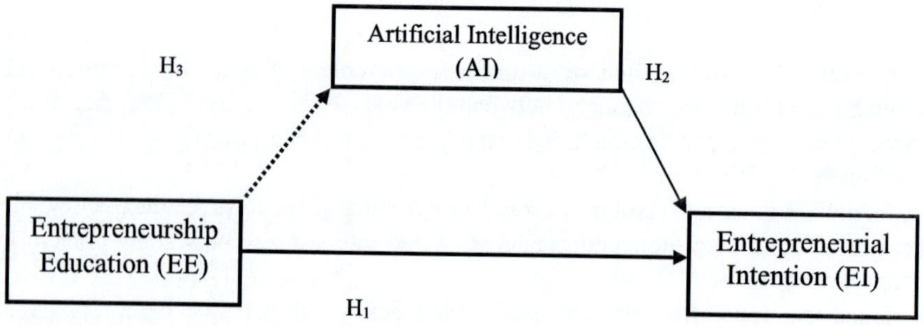

Fig. 1 A research framework

Indirect (mediating) Effects and Direct Effects.

According to Ferreira et al. [20], continuous learning behavior can change the students' attitude, which also might affect entrepreneurial intention. Not only that, specific entrepreneurship education can change individuals' knowledge, skills, spirits, and abilities and turn around this change to entrepreneurial intention. Some components are related to the educational activities that lead the entrepreneurs' intention to behavior [21]. The findings of Saptono et al. [2] support the entrepreneurial human capital (EHC) theory, which asserts that entrepreneurial education and intentions are linked. According to this theory, human capital is a vital component in reaching individual entrepreneurial intention [22]. Brush et al. [23] discovered a strong link between entrepreneurial education and human capital achievement. Similarly, Gurbuz and Aykol [24] asserted that entrepreneurial education is also measured by some indicators such as entrepreneurial knowledge, values, motives, social interaction, and entrepreneurial competencies.

H_1: Entrepreneurial education has a positive impact on entrepreneurial intention.

In academic research, the TAM model association with TPB is helpful to determine the applicability, accessibility, adaptability, and acceptance of updated technology in many circumstances. TAM theory helps people know the benefits of using AI and how the behavior affects adopting the information technology [25]. Therefore, it is clear that there is a robust association between TPB and TAM model [19]. But, TAM is specifically used to determine the internal variables that direct individuals to use technology. At present, the impact of AI does not limit to any sector, and entrepreneurship is impacted at large. AI influences people's significant role and places them in this new AI-infused world of work in various ways [26]. Artificial intelligence integrated with entrepreneurship in the start-up of new firms has the potential to boost the economy. Artificial intelligence is being employed in almost every industry, including heavy industries [27]. Among many, AI is becoming a buzzword in education, medical, finance, marketing, R&D, T&D, and agriculture to solve and face unforeseen critical problems. It is proven from the previously published articles that universities and higher education institutions are in advance due to adopting AI. So, the application of AI can be segmented into three categories: institutions, students, and instructional [28]. Consequently, it is also found that AI significantly penetrates EI in higher education for entrepreneurship development. Artificial intelligence (AI) is a machine learning concept to develop commercial solutions that employ robots and algorithms to solve problems.

H_2: Artificial intelligence has a positive impact on entrepreneurial intention.

According to Baron and Kenny [29], there is a requirement to investigate the relationship between independent and dependent variables to detect mediating factors. Furthermore, a requirement must be met to develop mediating variables: the mediator and dependent variables have a strong association. Finally, if we focus on the variables discussed in this study, we can expect AI to play a mediating variable. As a result, based on the conditions, the following hypothesis can be formed.

H3: Artificial intelligence might mediate the relationship between EE and EI.

Table 1 Measurement of scale

Variables	Entrepreneurial intention (EI)	Entrepreneurship education (EE)	Artificial intelligence (AI)
No. of items	10	6	8
Sources	[19]	[32]	Self-developed

3 Methodology

3.1 *Study Design and Subjects' Selection*

Since the virus originated, the Bangladesh Government has tried its best to stop the mass spreading of COVID-19. But, WHO declared the COVID-19 pandemic in March 2020, the government announced to close to all the educational institutions and request students to stay at home and continue their study online[30, 31]. Therefore, researchers also took this opportunity to collect the data through an electronic platform. They adopted the snowball sampling technique for this study. First, a structured questionnaire has been developed using Google Form and sent it to the respondents to collect their opinions. Initially, the questionnaire link was sent to the business faculty members in various institutions across the country to disburse the Google Form (questionnaire) link to their respective students. Apart from that, students were also requested to disseminate the questionnaire to their friends and relatives studying in the business department. In most cases, researchers used some social media to forward the link to others, such as Email, WhatsApp, Messenger, LinkedIn, etc. The questionnaire link was accessible to the students until researchers found the expected sample size (Table 1).

3.2 *Justification of Number of Sample Size*

After a certain period, researchers received 415 responses. According to Kline [33], it is large sample size for covariance-based structural equation model (CB-SEM) if researchers want to use AMOS as a statistical tool to analyze the data. Also, Kline [33] suggests that, for doing CB-SEM, this sample size is perfect for all. Therefore, during data mining, researchers found some inconsistent and distorted numerical values, which may significantly impact the analysis and results. Finally, we have selected the actual sample size of this research, 401. Its needs to be mentioned that after disabusing the questionnaire and getting 75 responses, we have done a pilot test where it shows good data consistency. The reliability values are pointed out in Table 2.

Table 2 Reliability and validity

	Original Items	Reliability	Items after CFA	Composite reliability	Convergent validity (AVE)	Discriminant validity r ≤ 0.90
EI	10	0.834	4	0.806	0.512	r ≤ 0.90
EE	6	0.904	5	0.900	0.646	
AI	8	0.874	5	0.892	0.628	

3.3 Demographic Detail of the Respondents

In this study, business students are respondents; 61% are male, 84% are unmarried, and 44% have completed their third year. Regarding job experience, 69% of the respondents have no experience, but 58% of the respondents are from business background families.

3.4 Instrument Development and Data Analysis

Researchers used some measurements for this study collected from the previously published literature aligning with the Bangladeshi context considering slight changes and adjustments. In this questionnaire, we have used 32 questions with the demographic profile of the respondents. First, we employ 6 popular items to measure entrepreneurial education (EE) [32]. Again, 10 instruments have been used to know the business students' entrepreneurial intention (EI) from Ajzen (1991). Researchers used self-developed 8 Artificial Intelligence (AI) constructs to accumulate their perceptions on using AI in entrepreneurship education and entrepreneurial intention. Each of the constructs maintains the five-point Likert Scale from "strongly disagree" (1) to "strongly agree" (5). Finally, AMOS has been used to employ covariance-based structural equation model (CB-SEM) to generalize the findings of the hypotheses.

4 Data Analysis and Findings

The study is contemporary, focusing on a new research area on entrepreneurship through AI. For this study, we have conducted the data analysis process into two standard sections, primary and inferential analysis. Researchers enclosed the reliability, validity, and measurement model with CFA for the first part of the analysis. Later on, for the inferential analysis, correlation, structural, and mediation models with bootstrapping have used in the research.

A. Preliminary analysis

Table 2 in this study shows the reliability and composite reliability. The value of both reliability conforms to general acceptance criteria. Apart from the reliability, Table 2 also illustrates the convergent and discriminant validity for excellent construct validity in this study (Fig. 2).

Table 3 using individual CFA reveals the model fit value of each variable with the measurement model (CFA). And significantly, all the measurement model fit values met the primary acceptance criteria. Therefore, Table 3 values persuade researchers to continue the further analyses (Fig. 3 and Table 4).

EE and AI together explained 83 percent of the variance in EI (R2 = 0.83). The researchers discovered that EE has a substantial positive effect on EI (EE = 0.81, p = 000). Furthermore, AI has a significant (positive) impact on EE (AI = 0.18, p = 000). These results indicate that H1 and H2 are acceptable with a 95% confidence interval (Fig. 4).

The researchers discovered a significant positive effect of EE on EI ($\beta_{EE} = 0.89$, $p < 0.000$ for the direct model) and ($\beta_{EE} = 0.84$, $p < 0.000$ for the mediation model) based on the results of the standardized direct effect from the direct and mediation

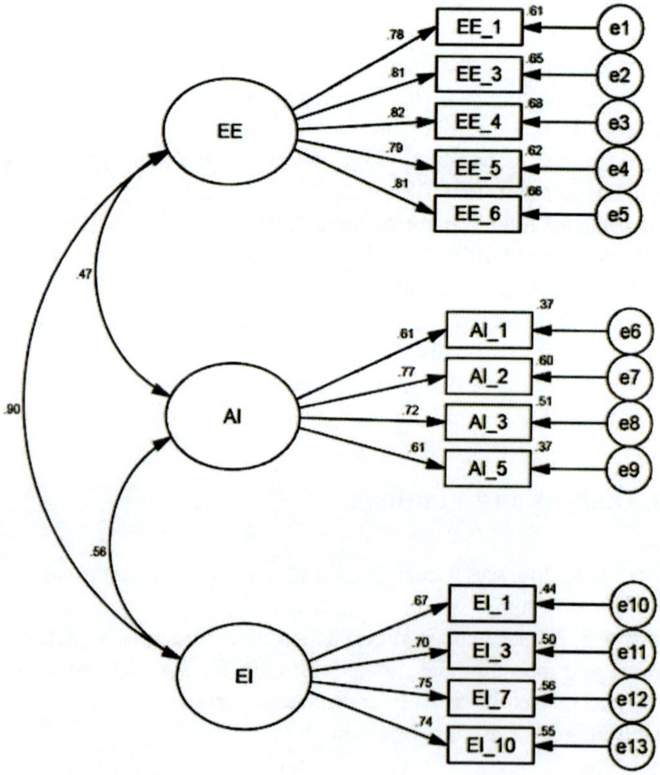

Fig. 2 Measurement model

Table 3 Results of model fit

	RMSEA	GFI	AGFI	CFI	Chisq/df
EI	0.059	0.998	0.975	0.995	2.911
EE	0.061	0.966	0.959	0.998	3.801
AI	0.058	0.954	0.961	0.999	3.545
Measurement model	0.076	0.933	0.902	0.945	3.323

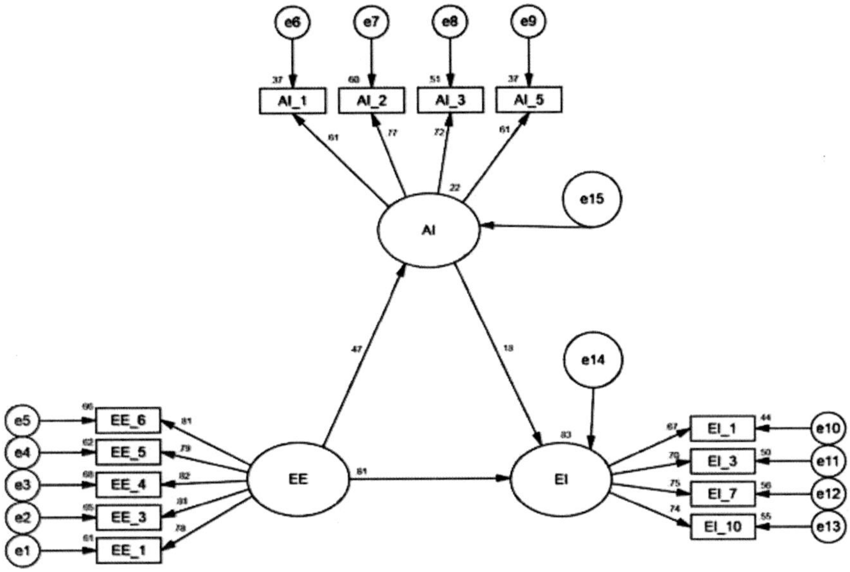

Fig. 3 Structural model (hypotheses testing)

Table 4 Results of the direct standardized effect

Variables	Variables	Estimate	S.E	C.R	P
EE	EI	0.81	0.064	11.695	0.000
AI	EI	0.18	0.047	3.778	0.000

models (Table 5). Furthermore, the standardized indirect effect reveals that EE has a substantial beneficial impact on EI $\beta_{EE} = 0.80$, $p < 0.000$ for standardized indirect effect). As a result, AI plays a role in mediating the relationship between EE and EI (H3 accepted) with a 95% confidence interval.

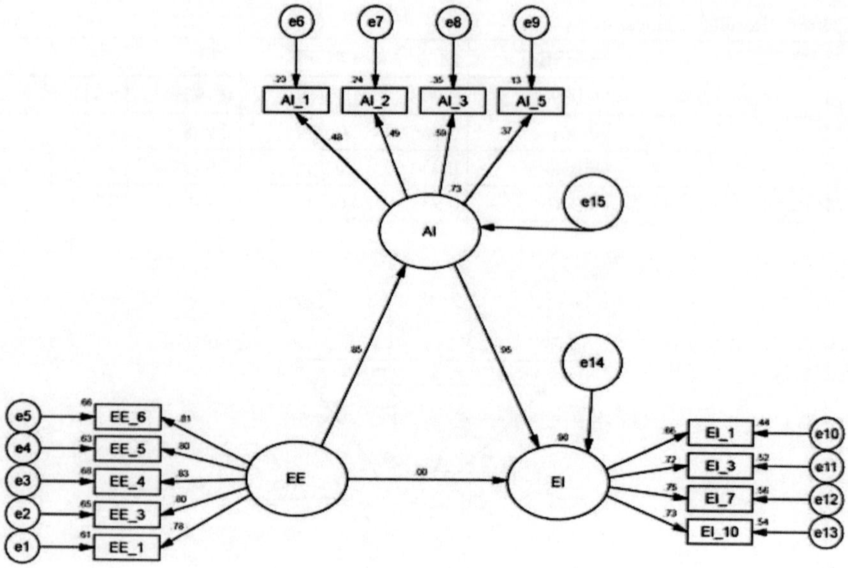

Fig. 4 Mediation model (hypotheses testing)

Table 5 Results of mediation

Model/Hypothesized paths			Beta	P	95% CL Bootstrap BC	
					LB	UB
Direct Model (SDF)						
EE	---- AI ----	EI	0.89	0.000	0.692	0.973
Mediation Model (SDF)						
EE	---- AI ----	EI	0.84	0.000	0.602	0.894
SIE						
EE	---- AI ----	EI	0.80	0.000	0.315	0.915

SDF = standardized direct effect, SIE = Standardized indirect effect

5 Discussions of the Findings

The fundamental objective of this study was to know the entrepreneurial intention of business graduates in Bangladesh, including two important predicted variables, i.e., entrepreneurship education and artificial intelligence. In order to reach the decision, the researchers formulated three hypotheses. In the following section, we have designed our discussion part based on our formulated hypotheses. The first finding was that entrepreneurship education has a significant positive effect on the entrepreneurial intention of business graduates in Bangladesh. This finding is consistent with the findings of previous scholars from Asian and European countries [2,

3, 34, 35]. This finding proves that students who receive entrepreneurship education can easily understand that starting a new business is not easy. It is also high risk and uncertain. Besides, entrepreneurship education can enhance the students' knowledge, skills, norms, and values that increase their self-efficacy, which help to increase the students' entrepreneurial intention.

The second finding showed that artificial intelligence has a significant positive impact on the entrepreneurial intention of business graduates in Bangladesh. Recent studies from developed countries found the same findings [36]. In one of the major findings found by Nawaz [13], it was mentioned that innovation and technology, including artificial intelligence, can positively influence entrepreneurial intention and positively contribute to a country's economic growth.

The major finding highlighted that artificial intelligence positively mediates the relationship between entrepreneurship education and entrepreneurial intention of business graduates in Bangladesh. It indicates that if the current entrepreneurship courses or programs are being designed with new technological support, such as artificial intelligence, it will enhance entrepreneurship education effectively with efficiency. In Bangladesh, both public and private universities run their entrepreneurship courses following the traditional teaching methods. Usually, the course teachers of entrepreneurship development focus more on theoretical perspectives, including advantages, disadvantages, characteristics of entrepreneurship, role of entrepreneurs, social and ethical responsibility of entrepreneurs, etc. Besides this technique, if the university authority and the teachers add some emerging issues such as introducing statistical software for data analysis, arranging virtual competition, online quizzes, and online participation in different entrepreneurship programs, it will enhance the quality of entrepreneurship education. In addition, Han [37] stated that by introducing an intelligent teaching expert system, using virtual reality or human–computer interaction, the university authority, including the schools or departments, can help to estimate the statistical distribution of students' learning content efficiency and learning habit. Traditional network education relies on internet resources, which are characterized by a large number of low-quality materials. As a result, learners struggle to maintain an adequate concentration on the learning process, which is necessary for increased learning efficiency. Thus, the network teaching resources need to be further restructured after combining expert systems for the operation of the network education system, students' cognitive situation, build a unified study of the cognitive model that helps students learn better form adaptive learning. Thus, artificial intelligence should inject a predictive variable between entrepreneurship education and entrepreneurial intention for business students that will boost their ability to make judgments and identify mistakes as an intrinsic element of learning.

6 Contributions of the Study

The researchers have designed the contribution part including both theoretical and practical perspectives.

6.1 Theoretical Contributions

In terms of theoretical contribution, this study's primary contribution was to examine the two predicted variables (entrepreneurship education and artificial intelligence) concerning the assumption of TPB and TAM to foster entrepreneurial intent in a developing country. Another contribution would be to test these theories in higher education institutions (business graduates). Additionally, the findings of this study represent an empirical attempt to supplement existing conceptual and literature on the entrepreneurial intentions of business graduates. This study is possibly the first to examine entrepreneurship education, artificial intelligence, and entrepreneurial intentions in developing countries. The theoretical framework for this study is based on TPB and TAM, which has proved that artificial intelligence serves as a bridge between entrepreneurship education and entrepreneurial intention.

6.2 Practical Contributions

Students, academicians, entrepreneurs, the government, and the state are all involved in this research regarding its practical ramifications. For the purpose of establishing an entrepreneurial mindset among Bangladeshi business graduates, this study incorporates artificial intelligence into entrepreneurial education. As outlined in this paper, academics from business disciplines should reorganize the transforming educational system to improve students' willingness to be entrepreneurs. This paper draws the attention of Bangladeshi policy-makers to entrepreneurial education, artificial intelligence, and entrepreneurship.

7 Conclusion

This work has mainly focused on checking the mediating influence of artificial intelligence in the relationship between entrepreneurship education and entrepreneurial intention among students in Bangladesh, particularly business graduates, using the assumptions of TPB theory and TAM model. Clear evidence found that artificial intelligence partially mediates the association between entrepreneurship education and entrepreneurial intention. As a result, this study contributes significantly to the literature on TPB theory and the TAM model assumptions. Students, academicians, entrepreneurs, the government, and the state are all involved in this research regarding its practical ramifications. Finally, artificial intelligence can be used as an intervening variable (mediator) between entrepreneurship education and entrepreneurial intent in business students, enhancing their ability to make decisions and recognize failures as a natural part of learning.

8 Limitations and Future Research

Only business students from Bangladeshi universities took part in the survey, which proved that the sample size was relatively small. It is recommended that future studies include students from other disciplines or fields such as CSE, information technology (IT), English, Law, and other disciplines. The sample size needs to be increased for the findings to be more broadly applicable. In this work, only quantitative methodologies have been used, thus, a mixed-method research is suggested to get a better understanding of entrepreneurship education, artificial intelligence, and entrepreneurial intention.

Funding The researchers did not receive any funding.

Conflict of Interest There is no conflict of interest.

References

1. GEI.: https://thefinancialexpress.com.bd/public/economy/bangladesh-slips-one-step-in-global-entrepreneurship-index-1512185778 (2019)
2. Saptono, A., Wibowo, A., Narmaditya, B.S., Karyaningsih, R.P.D., Yanto, H.: Does entrepreneurial education matter for Indonesian students' entrepreneurial preparation: the mediating role of entrepreneurial mindset and knowledge. Cogent Edu. **7**(1), 1836728 (2020)
3. Karyaningsih, R.P.D., Wibowo, A., Saptono, A., Narmaditya, B.S.: Does entrepreneurial knowledge influence vocational students' intention? lessons from Indonesia. Entrepr. Bus. Econ. Rev. **8**(4), 138–155 (2020)
4. Cui, J., Sun, J., Bell, R.: The impact of entrepreneurship education on the entrepreneurship mindset of college students in China: the mediating role of inspiration and the role of educational attributes. Int. J. Manag. Edu. **100296** (2019)
5. Lingappa, A.K., Shah, A., Mathew, A.O.: Academic, family, and peer influence on entrepreneurship intention of engineering students. SAGE Open **10**(3–4), 2158244020933877 (2020)
6. Tribune, D.: 4.4 million Youths face unemployment in Bangladesh. Retrieved from https://www.dhakatribune.com/bangladesh/2018/09/25/4-4-million-youths-face-unemployment-in-Bangladesh (2018)
7. New Age Bangladesh.: Unemployment among university graduates rises. Retrieved from http://www.newagebd.net/article/37148/unemployment-among-univ-graduates-rises (2018)
8. Rahman, M.M., Ali, N.A.: Does work–family balance mediate the relationship between work–family conflicts and job satisfaction of academicians? J. Commun. Psychol. **49**(2), 361–374 (2021)
9. UGC.: http://universities.gov.bd/public-universities (2019)
10. Hossain, M.: Jobless rate of young varsity gradates 4 times more than those with pri level edn: study. The Financial Express. Retrieved from http://www.thefinancialexpress-bd.com/2014/04/11/28273/ (2014). Accessed 25 Nov 2014
11. Huq, S.M.M., Huque, S.M.R., Rana, M.B.: Entrepreneurship education and university students' entrepreneurial intentions in Bangladesh. In: Entrepreneurship: Concepts, Methodologies, Tools, and Applications, pp. 221–246. IGI Global (2017)
12. Mou, X.: Artificial intelligence: investment trends and selected industry uses. 1–8 (2019)
13. Nawaz, N.: Entrepreneurship in the age of artificial intelligence and robotics. SSRN **9**(S), 28–33 (2020)

14. Haque, S.: How to build an entrepreneurial mind. https://cutt.ly/JQZMfbl (2020)
15. Sansone, G., Battaglia, D., Landoni, P., Paolucci, E.: Academic spinoffs: the role of entrepreneurship education. Int. Entrepr. Manag. J. **17**(1), 369–399 (2021)
16. Alshebami, A., Al-Jubari, I., Alyoussef, I., Raza, M.: Entrepreneurship education as a predicator of community college of abqaiq students' entrepreneurship intention. Manag. Sci. Lett. **10**(15) (2020)
17. Bazkiaei, H.A., Heng, L.H., Khan, N.U., Saufi, R.B.A., Kasim, R.S.R.: Do entrepreneurship education and big-five personality traits predict entrepreneurship intention among universities students? Cogent Bus. Manag. **7**(1), 1801217 (2020)
18. Din, B.H., Anuar, A.R., Usman, M.: The effectiveness of the entrepreneurship education program in upgrading entrepreneurship skills among public university students. Procedia Soc. Behav. Sci. **224**, 117–123 (2016)
19. Ajzen, I.: The theory of planned behavior. Organ. Behav. Hum. Decis. Process. **50**(2), 179–211 (1991)
20. Ferreira, J. J., Ratten, V., Dana, L. P.: Knowledge spillover-based strategic entrepreneurship. Int. Entrepreneurship Manage. J. **13**(1), 161–167 (2017)
21. Ahmed, T., Chandran, V.G.R., Klobas, J.: Specialized entrepreneurship education: Does it really matter? fresh evidence from Pakistan. Int. J. Entrep. Behav. Res. **23**(1), 4–19 (2017)
22. Jogaratnam, G.: The effect of market orientation, entrepreneurial orientation and human capital on positional advantage: evidence from the restaurant industry. Int. J. Hosp. Manag. **60**(1), 104–113 (2017)
23. Brush, C., Ali, A., Kelley, D., Greene, P.: The influence of human capital factors and context on women's entrepreneurship: which matters more? J. Bus. Ventur. Insights **8**, 105–113 (2017)
24. Gurbuz, G., Aykol, S.: Entrepreneurial intentions of young educated public in Turkey. J. Glob. Strat. Manag. **4**(2), 47–56 (2008)
25. Davis, F.D.: Perceived usefulness, perceived ease of use, and user acceptance of information technology. MIS Quart. 319–340 (1989)
26. Frank, M.R., Wang, D., Cebrian, M., Rahwan, I.: The evolution of citation graphs in artificial intelligence research. Nature Mach. Intell. **1**(2), 79–85 (2019)
27. Shah, P., Kendall, F., Khozin, S., Goosen, R., Hu, J., Laramie, J., Schork, N.: Artificial intelligence and machine learning in clinical development: a translational perspective. NPJ Digital Med. **2**(1), 1–5 (2019)
28. Zeide, E.: Artificial intelligence in higher education: applications, promise and perils, and ethical questions, Edu. Rev. **54**(3), 31–39 (2019)
29. Baron, R.M., Kenny, D.A.: The moderator–mediator variable distinction in social psychological research: conceptual, strategic, and statistical considerations. J. Pers. Soc. Psychol. **51**(6), 1173 (1986)
30. Rahaman, M.S., Rahman, M.M., Moral, I.H.: Informative knowledge and challenges of home quarantine during COVID-19 in Bangladesh. Int. J. Technol. Transf. Commer. **17**(4), 354–367 (2020)
31. Rahman, M., Rahaman, S.: Home Quarantine Challenges and Psychological Status of Bangladeshi University Students during COVID-19. J. Edu. Manag. Devel. Stud. **1**(2), 1–9 (2021)
32. Hasan, S.M., Khan, E.A., Nabi, M.N.U.: Entrepreneurial education at university level and entrepreneurship development. Edu. Train. **59**(7/8) (2017)
33. Kline, R.B.: Structural equation modeling (2006)
34. Wardana, L.W., Narmaditya, B.S., Wibowo, A., Mahendra, A.M., Wibowo, N.A., Harwida, G., Rohman, A.N.: The impact of entrepreneurship education and students' entrepreneurial mindset: the mediating role of attitude and self-efficacy. Heliyon **6**(9), e04922 (2020)
35. Li, L., Wu, D.: Entrepreneurship education and students' entrepreneurship intention: does team cooperation matter? J. Glob. Entr. Res. **9**(1), 35 (2019)
36. Desai, S.: Artificial intelligence and entrepreneurship: some thoughts for entrepreneurship researchers. In: Audretsch, D., Lehmann, E., Link, A. (eds.) A Research Agenda for Entrepreneurship and Innovation, pp. 197–207 Edward Elgar Publishing (2019)

37. Han, L.: New advances in the application of AI to education system. In: Proceedings of the 2018 3rd International Conference on Education, E-Learning and Management Technology (EEMT 2018), pp. 608–611 (2018)

COVID-19 Chest X-Ray Classification with Augmented GAN

Md. Appel Mahmud Pranto, Nafiz Al Asad, Md. Istakiak Adnan Palash, A. K. M. Mohaiminul Islam, and M. Shamim Kaiser

Abstract Coronavirus disease (COVID-19) is a major concern now. According to the Globe Health Organization, the coronavirus (COVID-19) epidemic is straining healthcare systems worldwide (WHO). Early-stage detection using artificial intelligence of this virus will help in the fast recovery. Early identification of this infection utilizing artificial intelligence will aid in its quick recovery. In the fight against COVID-19, it's critical to have a positive chest X-ray for infected patients. Early research suggests that chest X-ray abnormalities in COVID-19 patients are common. Using augmented chest X-ray images, this research proposes a novel model for identifying the presence of COVID-19. As medical images are sensitive, GAN (Generative Adversarial Networks) augments chest X-Ray images. Augmented images are classified through our proposed model. Then classified images are segmented for validation. GAN augmentation model consists of a generator and a discriminator. Convolutional neural networks (CNNs) based classification model needs a substantial amount of training data. Augmentation increased dataset into 10x amount. Before augmentation, our initial model received 94.42% accuracy, and after augmentation, our final model accuracy raised to 98.58%. We hope our model will detect COVID-19 presence accurately through X-Ray images.

Keywords Chest X-ray · Classification · Augmentation · GAN · Segmentation · COVID-19 · Artificial intelligence

Md. A. Mahmud Pranto · N. Al Asad · Md. I. Adnan Palash · A. K. M. M. Islam
Department of Information and Communication Technology, Bangladesh University of Professionals, Dhaka, Bangladesh

M. Shamim Kaiser (✉)
Institute of Information Technology, Jahangirnagar University, Dhaka 1342, Bangladesh
e-mail: mskaiser@juniv.edu

S. Hossain et al. (eds.), *Proceedings of International Conference on Fourth Industrial Revolution and Beyond 2021*, Lecture Notes in Networks and Systems 437,
https://doi.org/10.1007/978-981-19-2445-3_9

1 Introduction

Coronavirus disease (COVID-19) is known as a lethal infectious virus. Later, the World Health Organization (WHO) proclaimed the epidemic a pandemic (global health emergency). COVID-19 epidemic disorder is produced through Severe Acute Respiration Syndrome (SARS-CoV-2). Developing countries like Bangladesh have also been severely affected by the virus. Initially, adults were more likely to be infected and die, but now young people are also infected. In the beginning, cough, sneezing, shortness of breath, fever, headache were found to be the symptoms of being infected with this virus, but now in many cases, these symptoms are not manifested [18]. As a result, many are unaware that they have been infected with the coronavirus. As a result, they are suffering from various physical and mental complications [2]. The diagnosis of covid-19 is an important criterion for solving this problem. A quick and rapid analysis of COVID-19 can lessen the damage of this virus [17]. There are several ways for distinguishing COVID-19. A chest X-ray can play an important role in the diagnosis of this deadly virus [1].

Chest X-ray produces images using less radiation that does not negatively affect the patient's health. On the other hand, this method is relatively simple and takes less time [4]. An automated computer-aided diagnosis(CAD) framework can effortlessly assist us to identify COVID-19 utilizing chest X-ray images exceptionally effectively [6].In this case, by applying Deep Learning algorithms in chest X-ray images, we can easily find out if a person has Covid-19 or not. Among all existing algorithms, the performance of the Convolutional Neural Network(CNN) is better than others [7]. Nowadays, this algorithm is also being widely used in radiography images, and it is showing very good results.

The Convolutional Neural Network (CNN) is the most widely used computer vision algorithm.. An important characteristic of this algorithm is that it can learn about various features from the picture on its own, which makes this algorithm more powerful than others. Even the performance of CNN in image classification is much higher than other algorithms. There are primarily three levels to this structure. CNN has three layers: a fully connected layer, a convolutional layer, and a pooling layer. These layers learn about various characteristics of the images [3]. A major drawback of CNN is that it performs better when the size of the data is large. That means the higher the amount of data, the better CNN will perform. For the low amount of data CNN can overfit easily. Gathering enough images in the medical domain is a challenging task. This is because protecting the security and privacy of the patient's personal information becomes a major issue here.

Data augmentation is a method by which we can easily increase the number of images automatically [16]. Since we do not get good results with fewer data in Deep Neural Network, data augmentation can help us eliminate this problem. Generative Adversarial Network(GAN) is one way of augmentation.GAN helps us create new images while maintaining the originality of the image. In this way, the problem that we had before is largely removed.GAN is usually used to synthesize images.GAN has two parts one is called Generator, and the other is called Discriminator. The job of the Generator is to create a fake image that looks exactly like a real image. And the job of the Discriminator is to identify this fake image. After a while, the Generator can create an image like the real image, and the Discriminator is not able to classify this image as a fake image. This is the basic working procedure of GAN. Another extraordinary advantage of GAN is that it secures the security of the patient's information, and this strategy is much more secure [5].

Identifying infected regions of the chest using COVID-19 chest X-ray pictures is an essential element of the research since it may assist in slowing the spread of the virus. For this, a semantic segmentation strategy can be applied. Semantic segmentation is a method by which we can label an image in a specific region according to its properties [15].

1.1 Related Works

In recent times, many researchers, scientists have worked in medical image augmentation, identification of COVID-19 using chest X-ray images and image segmentation. Saman Motamed et al. [12] used a generative model to develop semi-supervised pneumonia and COVID-19 chest X-ray image categorization system. For the image synthesis task, they suggested Inception Augmentation Generative Adversarial Network (IGGAN). They used two different chest X-ray datasets for training purposes. Finally, they showed that the traditional data augmentation method failed to improve the overall model performance. The IAGAN model improved the AUC by 3 percent and 2 percent compared to non-augmentation for both datasets.

Based on Generative Adversarial Network(GAN), Nour Eldeen M. Khalifa et al. [8] proposed a pneumonia chest X-ray detection model. For the detection of pneumonia, they used different transfer learning models. By applying GAN and transfer learning, they showed that considering the testing accuracy metric, the efficiency of the model increased much.

Mohamed Loey et al. [10] proposed a Generative Adversarial Network(GAN) with a transfer-learning-based COVID-19 detection work. Their dataset contained four different types of classes. For tested three different case scenarios every time they selected different deep transfer models, and GoogleNet achieved the best accuracy.

Keeping the low data volume of COVID-19 X-ray images in mind, Sumeet Menon et al. [11] proposed Mean Teacher + Transfer GAN (MTT-GAN) that generated high-quality COVID-19 X-ray images. For improved stability of training, they employed the Mean Teacher algorithm as a constrain. They used two different classifier models evaluation. Their model provides a better improvement for image generation.

Saman Motamed et al. [13] proposed a semi-supervised model for the detection of COVID-19 is called Randomized Generative Adversarial Network (RANDGAN). They evaluated how segmentation improved the task of classification. Finally, they showed a comparison between conventional generative network and their proposed model.

An automated machine learning model proposed by Abolfazl Zargari Khuzani et al. [9] classified COVID-19 infection in the chest X-ray images. To improve the efficiency of the model, they used a dimensionality reduction method.

Ali Narin et al. [14] developed a method for automatically detecting coronavirus illness using chest pictures. They used five pre-trained CNN models and implemented three binary classifications with four classes. Tulin Ozturk et al. [10] proposed a fully binary classification and multi-class classifier to better diagnose COVID-19 detection using chest X-ray.

1.2 Contributions

In this paper, we state a model that can identify COVID-19 presence from chest X-ray images. In our experiment, different COVID-19 chest X-ray datasets have been used to train and test the software. To compensate for the lack of good photos in datasets, we used Generative Adversarial Network (GAN) to enrich the data. As a result, the performance of our model increases a lot. Our paper makes the following contributions:

(1) Augmentation using a Generative Adversarial Network (GAN) can produce sufficient image for training a model

(2) Detection of positive Covid-19 using a Deep-Convolutional-Neural-Network and those who can only be seen as undesirable in enhanced photographs.

(3) Semantic segmentation strategy to label an image in a specific region for validation between normal image and image carrying COVID-19.

2 Methodology

In our proposed model, we demonstrated our model in three sub-models: Augmentation (to increase the amount of image data) with GAN, Binary classification (to classify the image data into two classes-Covid Normal) with 2D Convolution, and finally, Segmentation (to identify the covid affected region in the chest X-Ray) with Encoder-Decoder based Convolution. This demonstration drawn in Fig. 1.

2.1 Dataset Exploration

All of the sub-models were trained and tested using the publicly accessible dataset COVID-19 Radiography Dataset. There are two courses in it (Covid and Normal). Both have 3500 pictures in them. At first, for training, 2900 number of pictures from both classes were input, and for validation purposes, 600 images were used.

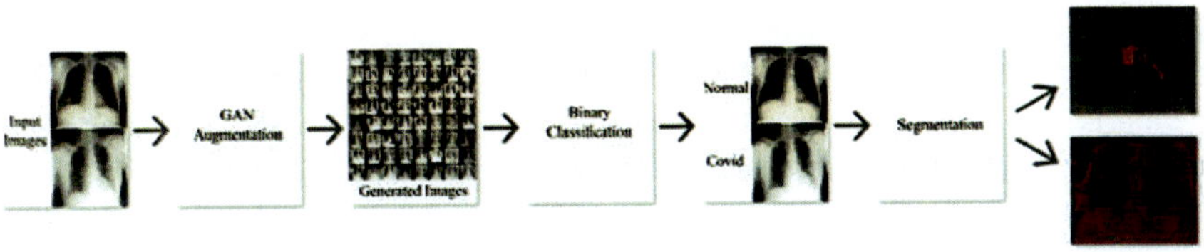

Fig. 1 Proposed model architecture. Model consists of chest X-ray image augmentation, classification of augmented images, and segmentation of classified images

2.2 *Sub-model 1: Dataset Enhancement*

The quantity of data points plays an important role in achieving better accuracy and reducing loss. That's why we utilized GAN to create fictitious pictures. There are primarily two causes for this: First, increase the dataset size; second while testing with real-world data, safeguard privacy. In this model, we utilized the Deep-Convolution Generative Adversarial Network (DCGAN), which is a GAN extension architecture that produces durable generator model training by using deep-convolutional models, as well as model and training settings. This image augmentation procedure drawn in Fig. 2.

2.2.1 Generator

To deceive the discriminator network, the generator model is trained to create false pictures. A single $64 \times 64 \times 3$ picture is required. A latent vector of noise (a 2D sample with a value taken from a uniform or Gaussian distribution) and a class

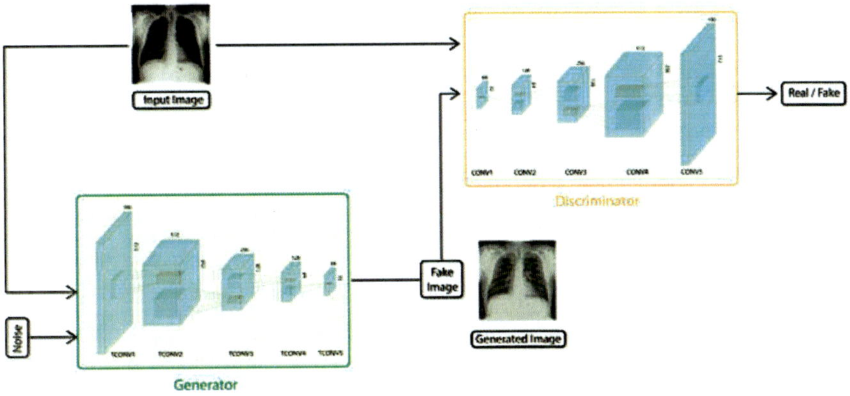

Fig. 2 Augmentation of chest X-ray images using Deep-Convolution Generative Adversarial Network (DCGAN)

label are required to input the generator. In reality, this is accomplished via the use of a series of strided two-dimensional convolutional transpose layers, each of which is connected to a 2d batch normalization layer and a relu activation. Each transpose convolutional layer, with the exception of the last one, is followed by batch normalization and activation layers. The output is sent into the tanh function, which returns to the [-1,1] [1, 1] input data range. The feature maps are then down sampled using five transpose convolutional layers. To 100 × 512, then 512 × 256, then 256, 128, 128 × 64, and ultimately 64 × 3. The first transpose conv layer has a kernel size of 4 × 4 and a stride of 1, while the others have kernel of 4 × 4 and stride of 2.

2.2.2 Discriminator

Strided-convolution layers, batch-normalization layers, and Leaky ReLU-activations make up the discriminator. Before delivering the final probability, the Sigmoid activation function processes a 3 × 64 × 64 input picture via a sequence of Conv2d, BatchNorm2d, and LeakyReLU layers. A CNN architecture with two output layers (Real/Fake) makes up the discriminator model. Each convolutional layer, with the exception of the last, is followed by a batch normalization layer and an activation layer (LeakyReLU). The output of the last convolution layer is sent into the sigmoid function. To up-sample the feature maps, the input tensor is processed through five convolutional layers. First, 64 × 3, then 128 × 64, 256, 128, 512 × 256, and ultimately 512 × 3. The first transpose convolutional layer has a kernel of 4 × 4 and stride of 2, whereas the other layers have a stride of 1. The first conv layer has a padding of 1.

2.3 *Sub-model 2: Binary Classification*

This model's two convolution layers and two max-pooling layers are used for feature extraction and two dense layers and one dropout layer are used for classification. Binary Classification architecture drawn in Fig. 3.

2.3.1 First Convolution Layer

The RGB picture format is compatible with the first convolution layer since it has an input channel with a size of three. This layer makes use of a 3 × 3 kernel with a stride of 1. To maintain consistency between the input and output feature maps, a padding of image size 1 has been added. The output feature map for this layer is 178 × 178 × 32. After the ReLU activation function, a max-pooling layer with a 4 × 4 kernel and a stride of 4 was added to decrease the number of training parameters

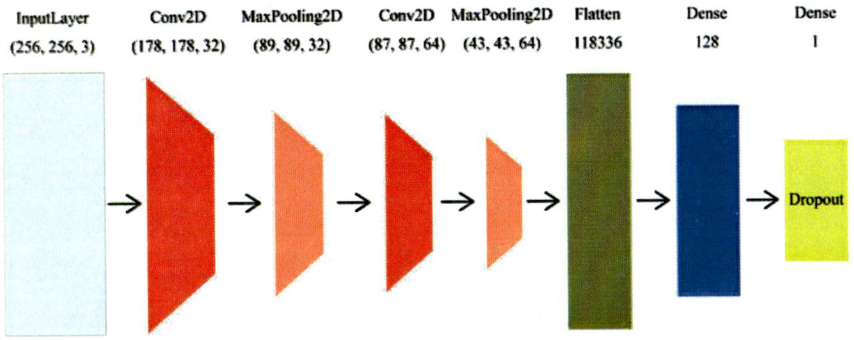

Fig. 3 Binary classification for augmented images

and address the overfitting problem. The max-pooling layer was used to reduce the feature map's resolution to 89 × 89x32.

2.3.2 Second Convolution Layer

In this layer, the kernel size is also 3 × 3 with stride of 1. This layer's feature map has a resolution of 87 × 87 × 64. Another max-pooling layer is added with a kernel which consist of a size of 2 × 2 and a stride which size of 2, the feature map was down sampled to 43 × 43 × 64.

2.3.3 Flatten Layer

Converting the feature map from second max-pooling layer into a 1-dimensional array for inputting it to the dense layer

2.3.4 Dense Layer

The dense layer (neural network layer) is highly linked. This is the final classification layer. Transfer data to the dropout layer after processing this layer. This model avoids overfitting thanks to the dropout layer.

2.4 Sub-model 3: Segmentation

Segmentation is utilized in this model to locate the covid afflicted area in the chest X-Ray. For this, we utilized semantic segmentation. This model was used to segment

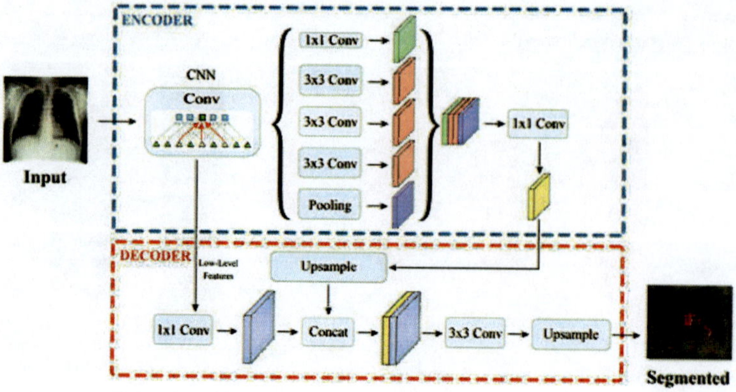

Fig. 4 Segmentation of classified images

both normal and covid pictures (Shown in Result Section). We utilized an Encoder-Decoder network to segment the pictures. Segmentation architecture drawn in Fig. 4.

2.4.1 Encoder

It is made out of four Conv blocks, the first of which is 1×1, and the others are 3×3. Finally, there is a spatial pyramid pooling layer. All Conv-blocks are given input individually, and their outputs are concatenated and sent to the Decoder Network. The encoder block's output feature map has 256-channels and a wealth of semantic-information.

2.4.2 Decoder

single 1×1 Conv block, 3×3 Conv block, and one up-sampling block constitute the Decoder network. We utilize 1×1 convolution to the number of layers, since comparable low-level features often have a large number of available networks (256 or 512), which may overpower the importance of the abundant encoder features (256 in the model) and make training more difficult. Following concatenation, we enhance the features using a pair of 3×3 convolutions, followed by a factor of four basic bilinear up-sampling.

3 Result Discussion

In this section, we optimize our model and evaluate result.

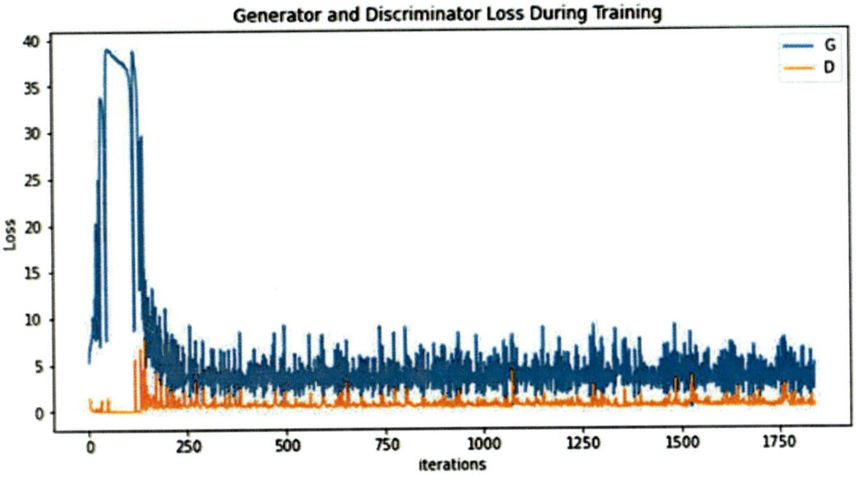

Fig. 5 Loss graph for generator and discriminator of GAN augmentation model during training over 1750 iterations

3.1 *Model Optimization*

3.1.1 GAN Augmentation Model

In the augmentation model, we determined hyperparameters: optimal learning rate as 0.002, 40 number of training epochs, beta1 for Optimizers as 0.5 for training generator and discriminator.

Figure 5 shows generator and discriminator loss during training of our proposed model's augmentation part using learning rate of 0.002 and 40 epochs. The line of generator and discriminator shows that it decreases gradually with respect to the increase of iterations. The figure shows that high loss occurred in iterations between 1 and 250. After the 250th iteration, loss was kind of stable.

3.1.2 Initial Classification Model

Classification Model was trained with both real images of datasets and augmented images.

After 30 Epochs, the classification model loss for both training and validation set is shown in Fig. 6. This was done with 3500 covid and 3500 normal images (Before Augmentation), as we can see that the losses were too great at the beginning but gradually decreased. But even after 30 epochs, there is still some loss.

The same goes for accuracy, as shown in Fig. 7. This is the training and validation accuracy with 3000 and 600 images for training and validation, respectively, for each

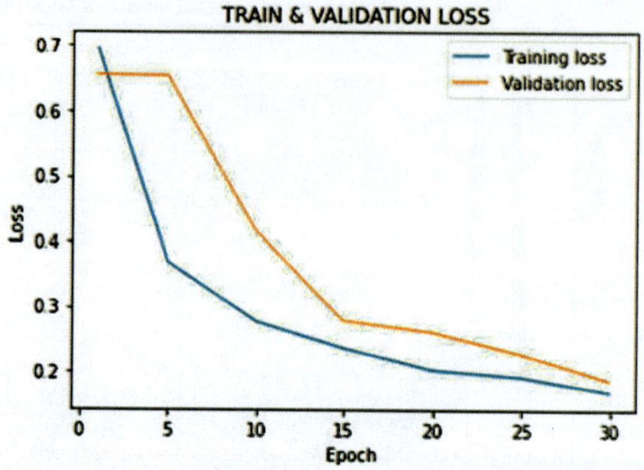

Fig. 6 Train and validation loss over 30 epochs

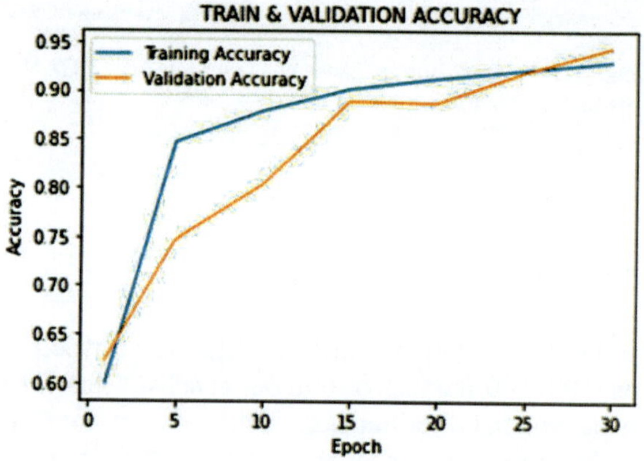

Fig. 7 Train and validation accuracy over 30 epochs

covid and normal class. After 30 epochs there the accuracies are 0.9297 and 0.9442 for training and validation, respectively.

3.1.3 Final Classification Model

After augmentation, we started with 7000 images in total. But this number is not enough for feeding into any CNN model. That's why we performed Image augmentation, to increase the dataset size as well as maintain the privacy in case of real-world data. After augmentation the 35000 images were created for each class. 29000 images from each class were used for training and rest for validation. As we can see that the loss was reduced drastically after augmentation. After 30 epochs it was 0.0068 and 0.0864 for training and validation respectively. This demonstration is drawn in Fig. 8.

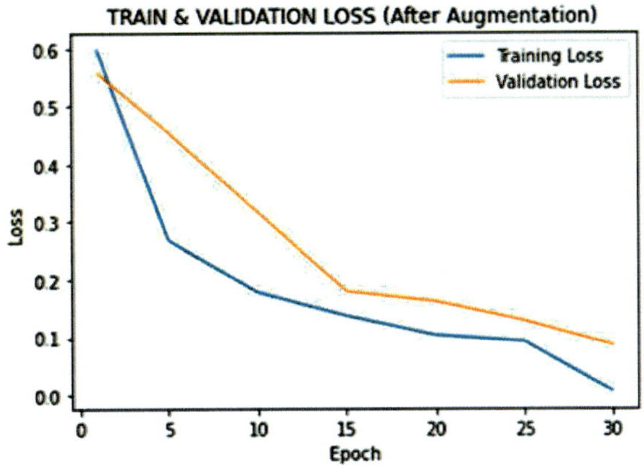

Fig. 8 Loss of train and validation using augmented images over 30 epochs

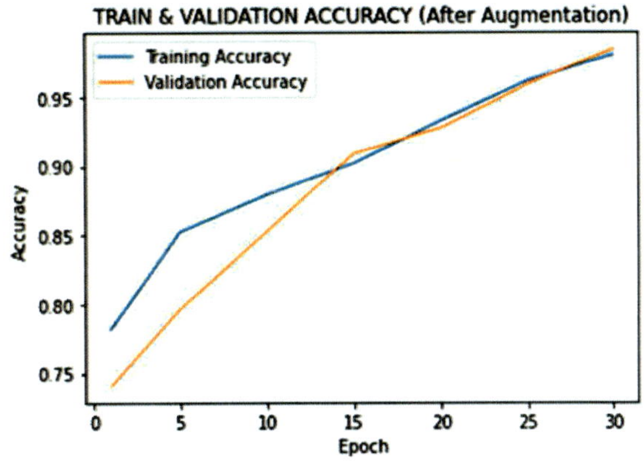

Fig. 9 Train and validation accuracy using augmented images over 30 epochs

Same goes for the accuracy. It is 0.9797 and 0.9858 for training and validation respectively. Both training and validation accuracy increased after our final classification model run with augmented images. From Fig. 9. we can see this increase.

3.2 *Result Discussion*

In our initial model, accuracy 94.42% is gained using real images of dataset. Accuracy of our final model increased to 98.58% using augmented images. Both the COVID and Normal classes show an improvement in accuracy and recall. This implies that the synthetic augments created have significant characteristics that aid in the improvement of CNN performance These details are demonstrated in Table 1.

Parameters: precision, recall, f1-score, and accuracy are used to evaluate the performance of our initial and final model. The formulae for the measurements of precision, recall, F1-score, and accuracy are as follows:

Table 1 Performance evaluation

Model	Data type	Precision	Recall	F1-Score	Accuracy
Initial	Real images	0.95	0.93	0.94	94.42%
Final	Augmented images	0.98	0.94	0.96	98.58%

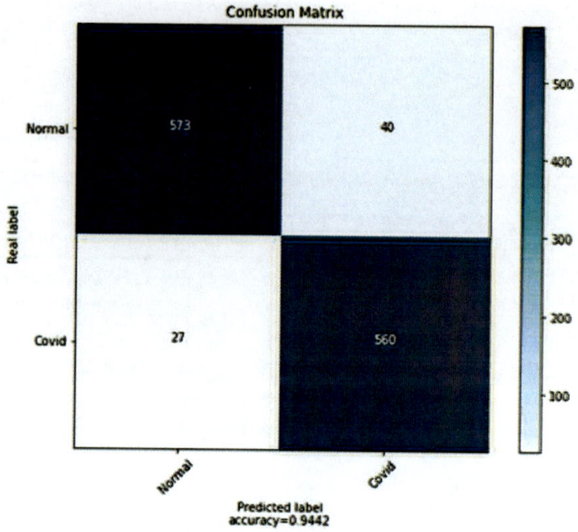

Fig. 10 Confusion matrix for validation before augmentation

$$Precision = TruePositive/(TruePositive + FalsePositive) \quad (1)$$

$$Recall = TruePositive/(TruePositive + FalseNegative) \quad (2)$$

$$F1 - Score = 2 * (Recall * Precision/Recall + Precision) \quad (3)$$

$$Accuracy = (TruePositive + TrueNegative)/Total \quad (4)$$

3.2.1 Performance Evaluation Using Confusion Matrix

From Fig. 10. we can see that for validation True Positive(TP) = 573 and True Negative(TN) = 560, and the total number of images in two classes is 1200. So by calculating the accuracy is we get 0.9442.

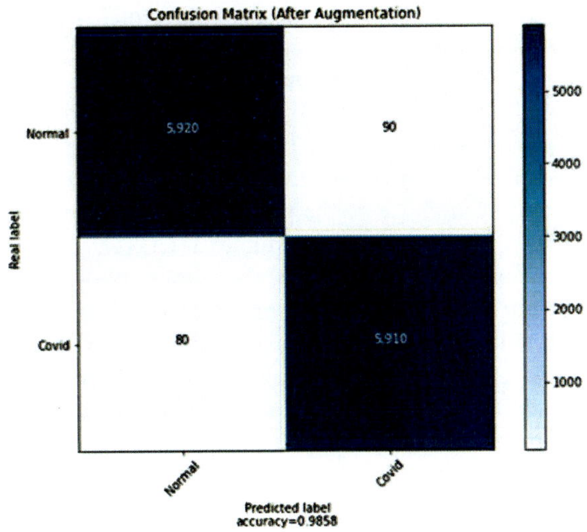

Fig. 11 Confusion matrix for validation after augmentation

Table 2 Performance evaluation with related works

Model	Precision	Recall	F1-Score	Accuracy
RandomForest	0.84	0.75	0.79	90.42%
VGG16	0.91	0.92	0.91	92.73%
VGG19	0.93	0.92	0.92	93.30%
Inception	0.96	0.95	0.96	96.57%
Proposed	0.98	0.94	0.96	98.58%

From Fig. 11. we can see that for validation True Positive(TP) = 5920 and True Negative(TN) = 5910, and the total number of images in two classes is 12,000 (after augmentation). So by calculating the accuracy is we get 0.9858.

As we can see from Table 2. our proposed model outperforms the existing state-of-the-art models. None of the pre-trained models could exceed 98.58% accuracy. Comparing all the evaluation metrics, we can say that our model can keep pace with current technology.

4 Conclusion

COVID-19 is a virus that eventually kills us. Numerous individuals have perished prematurely as a result of this infection. As a result, accurate and rapid diagnosis has become critical. This article discussed a strategy for generating fresh pictures using a Generative Adversarial Network (GAN). Following that, we classified COVID-19 using a Deep Neural Network (DNN) based on chest X-ray pictures. With GAN, we improved the model's accuracy significantly. We also included several dropout layers to mitigate the problem. Our testing results indicated that our method classified COVID-19 and conventional chest X-ray pictures from the dataset with an accuracy of 98.58 percent. Following that, we demonstrated how picture segmentation impacts the chest of a COVID-19 positive individual. It identifies sick areas in the chest of COVID-19 positive individuals. For segmentation objectives, we adopted a semantic technique. One of the most significant drawbacks of our investigation was the dataset's modest size. In the future, if COVID-19 can be classified using chest X-ray pictures from a bigger dataset, it will be simpler to detect and diagnose COVID-19 infection. Additionally, we may employ several sorts of data augmentation methods to improve performance and accuracy. As shown in Sect. 3, our model achieves 94.42 percent, which is an excellent result for any classification and segmentation model. However, since this model would be used for medical diagnosis, we made every effort to achieve the highest possible accuracy. That is why we employed GAN image augmentation to expand the dataset while maintaining the confidentiality of patient data. Finally, we obtained an accuracy of 98.58 percent, which is the greatest among currently available state-of-the-art models. Additionally, our model achieved this result in just 30 epochs, which significantly reduces computing time. After 30 periods, the loss in training and validation is around 0.1. This model is capable of identifying even the tiniest covid traces.

References

1. Abbas, A., Abdelsamea, M.M., Medhat Gaber, M.: Classification of Covid-19 in chest X-ray images using detrac deep convolutional neural network. Appl. Intell. **51**(2), 854–864 (2021)
2. Al Asad, N., Appel Mahmud Pranto, Md., Afreen, S., Maynul Islam, Md.: Depression detection by analyzing social media posts of user. In: 2019 IEEE International Conference on Signal Processing, Information, Communication & Systems (SPICSCON), pp. 13–17. IEEE (2019)
3. Moontari Ali, H., Shamim Kaiser, M., Mahmud, M.: Application of convolutional neural network in segmenting brain regions from mri data. In: International Conference on Brain Informatics, pp. 136–146. Springer (2019)
4. Antin, B., Kravitz, J., Martayan, E.: Detecting pneumonia in chest X-rays with supervised learning. Semanticscholar. org (2017)
5. Cai, Z., Xiong, Z., Xu, H., Wang, P., Li, W., Pan, Y.: Generative adversarial networks: a survey towards private and secure applications. arXiv:2106.03785 (2021)
6. Bahadur Chandra, T., Verma, K., Kumar Singh, B., Jain, D., Singh Netam, S.: Coronavirus disease (Covid-19) detection in chest X-ray images using majority voting based classifier ensemble. Expert. Syst. Appl. **165**, 113909 (2021)
7. Kalchbrenner, N., Grefenstette, E., Blunsom, P.: A convolutional neural network for modelling sentences. arXiv:1404.2188 (2014)
8. Khalifa, N.E.M, Taha, M.N., Hassanien, A.E., Elghamrawy, S.: Detection of coronavirus (Covid-19) associated pneumonia based on generative adversarial networks and a fine-tuned deep transfer learning model using chest X-ray dataset. arXiv:2004.01184 (2020)
9. Khuzani, A.Z., Heidari, M., Ali Shariati, S.: Covid-classifier: an automated machine learning model to assist in the diagnosis of Covid-19 infection in chest X-ray images. Sci. Rep. **11**(1), 1–6 (2021)

10. Loey, M., Smarandache, F., Khalifa, N.E.M.: Within the lack of chest Covid-19 X-ray dataset: a novel detection model based on gan and deep transfer learning. Symmetry **12**(4), 651 (2020)
11. Menon, S., Galita, J., Chapman, D., Gangopadhyay, A., Mangalagiri, J., Nguyen, P., Yesha, Y., Yesha, Y., Saboury, B., Morris, M.: Generating realistic Covid-19 x-rays with a mean teacher+ transfer learning gan. In: 2020 IEEE International Conference on Big Data (Big Data), pp. 1216–1225. IEEE (2020)
12. Motamed, S., Rogalla, P., Khalvati, F.: Data augmentation using generative adversarial networks (gans) for gan-based detection of pneumonia and Covid-19 in chest X-ray images. arXiv:2006.03622 (2020)
13. Motamed, Saman, Rogalla, Patrik, Khalvati, Farzad: Randgan: randomized generative adversarial network for detection of Covid-19 in chest X-ray. Sci. Rep. **11**(1), 1–10 (2021)
14. Narin, A., Kaya, C., Pamuk, Z.: Automatic detection of coronavirus disease (Covid-19) using X-ray images and deep convolutional neural networks. Pattern Anal. Appl., 1–14 (2021)
15. Novikov, A.A. et al.: Fully convolutional architectures for multiclass segmentation in chest radiographs. IEEE Trans. Med. Imaging **37**(8), 1865–1876 (2018)
16. Shorten, C., Khoshgoftaar, T.M.: A survey on image data augmentation for deep learning. J. Big Data 6(1), 1–48 (2019)
17. Kai-Wang To, K., Tak-Yin Tsang, O., Chik-Yan Yip, C., Chan, K.-H., Wu, T.-C., Man-Chun Chan, J., Leung, W.-S., Shiu-Hong Chik, T., Yau-Chung Choi, C., Kandamby, D.H. et al.: Consistent detection of 2019 novel coronavirus in saliva. Clin. Infect. Dis. **71**(15), 841–843 (2020)
18. Wang, Hai-Yang., Li, Xue-Lin., Yan, Zhong-Rui., Sun, Xiao-Pei., Han, Jie, Zhang, Bing-Wei.: Potential neurological symptoms of Covid-19. Ther. Adv. Neurol. Disord. **13**, 1756286420917830 (2020)

Artifical Intelligence for Smart City Applications

Computer Vision-based Waste Detection and Classification for Garbage Management and Recycling

S. M. Yeaminul Islam and Md. Golam Rabiul Alam

Abstract Waste management systems and their inherent problems are still a matter of great concern amid this cutting edge of science and technology. Piles of untreated waste in open locations contribute to a variety of issues, including air and water pollution, messy and unpleasant workplaces, waste of recyclable products, potential health risks for waste management workers, and so on. The root cause of this problem points to a single fact—too much manual labour involved in the garbage treatment process (collection, separation and recycling)—that can't keep up to the pace with which garbage generation happens. An efficient recycling method is imperative to solve this problem, which can be achieved by a fast, real-time garbage detection and classification system. In this research, we will propose a novel Deep Learning-based approach for automatic detection and classification of five kinds of waste materials, namely *Kitchen Waste, Glass Waste, Metal Waste, Paper Waste and Plastic Waste*, from the garbage dump for an efficient recycling process, which not only improves the efficiency of the current manual approach but also provides a scalable solution to the problem. The contributions of this paper include a fully human labelled dataset consisting of 2,200 images of a garbage dump with 135,541 annotated objects from the aforementioned categories and a real-time garbage object localisation and classification framework based on a lightning-fast, fairly accurate and end-to-end deep learning algorithm. For the baseline, we have articulated a single-stage object detection and classification framework and initialised the detector training process with pre-trained weights (trained on MS COCO Dataset) of the feature extractor. Then, with some fine-tuning and employing a few transfer learning tricks, we have proposed a waste object detection framework that yields a mAP of 66.08% at an IoU threshold of 0.35 with an inference speed of roughly 55–58 milliseconds in a single GPU environment on both images and videos. It surpasses the performances of all the contemporary frameworks which deal with waste separation tasks.

S. M. Y. Islam (✉)
Robi Axiata Limited, Dhaka, Bangladesh
e-mail: yeaminul1009@gmail.com

Md. G. R. Alam
BRAC University, Dhaka, Bangladesh
e-mail: rabiul.alam@bracu.ac.bd

S. Hossain et al. (eds.), *Proceedings of International Conference on Fourth Industrial Revolution and Beyond 2021*, Lecture Notes in Networks and Systems 437,
https://doi.org/10.1007/978-981-19-2445-3_10

Keywords Waste classification · Solid waste separation · Garbage recycling · Garbage waste dataset · Object detection · Computer vision · Deep learning · Urban environment

1 Introduction

Bangladesh is a densely populated country with its population being on course to be around 17 crores by the end of this year. This huge population comes with a lot of human-made hazards, and 'solid waste' is surely a major part of it. An enormous volume of garbage waste is getting generated each day, particularly in urban areas, and regrettably, their treatment process is getting deteriorated as the days pass by. The major cities of Bangladesh generate around 16 thousand tons of waste a day, which sum up to more than 5.84 million tons a year. This amount is on the rise and will reach 47 thousand tons per day and approximately 17.16 million tons per year due to the rapid expansion of urban areas, population growth and hike in per capita waste generation rate. With the current amount of urban inhabitants, the waste generation rate is 0.41 kg per capita per day, and the garbage collection efficiency ranges from 37% to 77% in different cities [2]. So, a big difference has been observed between the wastes being generated and the wastes being treated here. A considerable fraction of wastes are not properly processed, gathered, disposed of, or recycled due to a lack of understanding, technological limits, severe law, and budgetary restraints. Consequently, it's a growing headache for the municipalities—how to manage these solid waste efficiently and not only ensure a healthy environment for the residents but also make effective use of the litter as well through recycling (Fig. 1).

Fig. 1 The challenges of vision-based garbage recognition system for recycling

1.1 Research Problem

The entire waste management process, which includes proper recycling and disposal, takes a significant amount of time and manual labor. Recycling of solid waste is mostly dependent upon the garbage separation process, as different types of waste require different treatment. An automated solution to this problem involves computer vision- and sensor-based approaches, which can help an automatic machine to filter out the waste objects as per their predefined classes and tag them accordingly. It imposes a wider range of challenges as well. The waste objects in nature can be obscured, twisted, camouflaged, transparent, aged and transformed in different forms, which makes it extremely difficult for such systems to generalise well in the production environment. In spite of several attempts like letsdoitworld, the problems still persist in the absence of an enriched dataset with contextual images and a solid framework to detect and classify major categories of waste. So, our research problem is to find a fast and efficient waste object detection and classification system which has to work well in real-world multiple overlapping garbage objects and multi-class scenarios that impose the aforementioned challenges.

1.2 Aims and Objectives

A robotic agent working inside a dumping station and purposefully separating rubbish into specified categories would certainly qualify for a science fiction film a few decades ago! Fortunately, with the advent of IoT-enabled smart devices and robotics, fast and remote maintenance of mechanical activities can be facilitated, and waste management can surely be a part of it. But a problem arises that those systems require accurate representation of the kinds of waste to be treated. Some aspiring researchers addressed this problem and their solutions [10, 21, 23, 26, 29] but lack of contextual representation of garbage objects in their datasets, high inference cost, etc. make them questionable for edge device deployment. Now, detecting trash in an open environment is a daunting task, as it can be twisted, pellucid, shattered and camouflaged and adding to that, the vast feature diversity of our natural world has to be taken into consideration as well. In this work, we will propose a solution of an automated garbage separation system using Computer Vision and end-to-end Deep Learning techniques, to encounter these issues. Our proposed system will be able to detect, localise and classify garbage objects from piles of waste in the natural environment all in real time through pictures/videos/cameras. This information then can be passed to a robotic agent immediately to automate the process using a microcontroller-based embedded system. So, in summary our propositions are

1. A fully human labelled dataset consisting of 2200 images of garbage dump. The dataset should be evenly balanced to avoid data-driven bias, represent contextual images to fight against concept drift and data drift problems and annotate up to object-level granularity to detect and classify individual objects from images.

2. A real-time garbage separation framework based on a fast and accurate object recognition algorithm, which can detect and classify multiple overlapping waste objects across three different scales (small, medium, large) from a variety of backgrounds.

1.3 Organisation of the Report

The report is structured in the following manner—in Sect. 2, we shall try to explain the concepts related to the subject matter in detail. In Sect. 3, a comprehensive walk-through of the previous studies will be provided. Then, in Sect. 4, we shall briefly explain the data collection procedures and considerations of our proposed garbage detection dataset. Afterwards, in Sect. 5, the framework architecture and design choices will be reviewed. Finally, we'll wrap up with the experimental results and some future scopes of this research in Sects. 6 and 7, respectively.

2 Background Study

2.1 Waste Generation

Waste generation is something that can't be avoided as long as we, humans, live on this beautiful planet Earth. Garbage has been the by-product of this ecosystem since the outset of human civilisation. But during recent years, there is a significant amount of artificial materials prevailing around us, which, when converted to waste, result in dangerous non-bio-degradable objects—which is the most concerning thing about waste generation nowadays. As types and quantities of waste vary in different localities, we will try to depict the scenario from a Bangladesh perspective. In city areas, the MSW generation rate is increasing with the growth of the population. As reported in [4], a total of 7.7 thousand tons of solid waste is being generated from the six divisional cities of Bangladesh on a daily basis, while the capital itself has a 69% share of the total waste volume. The formation of the total stream of waste is approximately 74.4% of organic matter, 9.1% of paper, 3.5% of plastic, 1.9% of textile and wood, 0.8% of leather and rubber, 1.5% of metal, 0.8% of glass and 8% of other wastes [3]. The constituents behind this waste composition are population growth, advanced lifestyles, economic conditions, seasonality, weather, amount of recycling and waste management programs. The waste generation rate is linearly evolving with respect to the GDP growth of this nation as well.

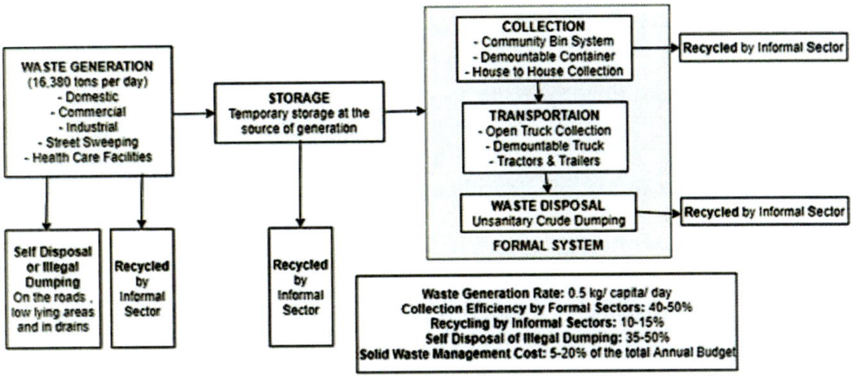

Fig. 2 Waste management framework in Bangladesh adopted from [2]

2.2 *Waste Management Systems*

The overall Waste handling process here is not structured in a profound manner. Figure 2 is a clear illustration of the existing garbage handling framework in Bangladesh. There are three roughly established processes of waste management in Bangladesh, and the recycling process is completely ignored in almost all the cases (only 10–15% of total waste, that too by informal private channels), as the prerequisite waste separation task is a tedious thing to do.

2.3 *Sensor-Based Robotic Agent*

An autonomous robotic agent can have a vision sensor, to capture real-time scenes while surfing around the designated places. The captured images will then be analysed and waste objects can be detected and classified via a waste object recognition framework. Then the robotic agent will auto-separate the garbage objects in their respective categories for recycling or permanent disposal.

2.4 *Computer Vision*

Now, people, who use cameras as the sensors of the micro-controllers, usually rely on images to give them information about the objects that they were looking at. In general, they use computer vision to identify objects within an image. The name computer vision is kind of self-explanatory which enables a computer to identify objects autonomously.

2.5 *Object Detection and Classification*

Image classification Image classification algorithms take an image as input, transform it into a giant one-dimensional vector, process the input and yield the predicted label of that image with some metric like loss, probability, accuracy, etc. **Object detection—**Object Detection algorithms take an image as input, transform it into a giant one-dimensional vector, process it and output one or more bounding boxes with the class labels, depicted on top of each bounding box.

Now, the object detection algorithms can be further classified into 2 major categories, namely **Single-Stage Detectors** which refer to the family of algorithms that detects and classifies the object using a single network and one forward pass. This has attracted many researchers of late, mainly because of its faster inference speed and practical applicability. These models can also be optimised in such a way that they can be used on mobile devices. Examples—YOLO models [11, 18, 22, 25], RetinaNet [16], SSD [9], etc.; **Multi-Stage Detectors**, which employ two networks—one for predicting bounding boxes around detected objects and another for classifying the object in the box. These algorithms are based on region proposals and are comparatively slow to make predictions than their single-stage counterparts. But they are more accurate than the single-stage detectors. Examples—R-CNN [5], Masked R-CNN [15], Fast R-CNN [8], etc.

2.6 *YOLO Models*

YOLO is a CNN-based object detection algorithm. This algorithm uses a single neural network, and within a forward pass the network breaks down the total image area into multiple local regions and predicts the bounding boxes and probabilities of those regions. The first version of this model [11] yielded greater localisation errors, although it operates at an impressive 45 FPS. In YOLOv2 [18], the authors adopted the idea of the anchor boxes from the R-CNN paper [8]. The representation of the bounding boxes is modified a bit (instead of estimating the bounding box position and size directly, offset values are predicted for moving and reshaping the predefined anchor boxes relative to a grid cell) to allow small changes which are eventually expected to have a less dramatic effect on the predictions, resulting in a more stable model. The latest improvement, YOLOv4, tried to combine all the common recipes of success from the rich vision recognition research domain. It introduced techniques called Bag of Specials which implied change in activation function, better backbone, better feature aggregation and propagation path, etc. and Bag of Free bees which introduced lots of data augmentation strategies, modification in the loss function, using a better regularization method. All of them together resulting in a fast, accurate and robust object detector which achieved state of the earth performance running on a single GPU environment (Fig. 3).

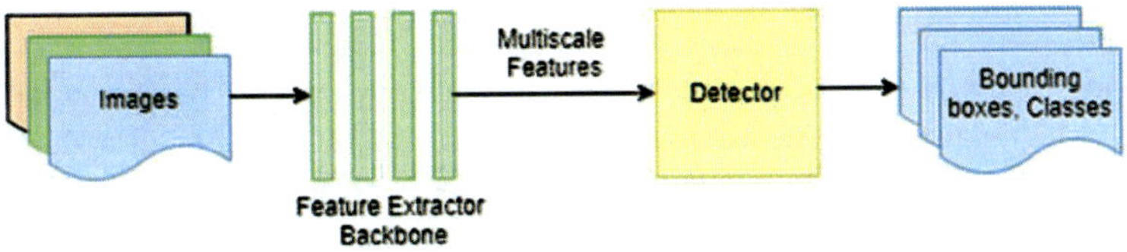

Fig. 3 High-level architecture of a vision based object detection framework

3 Related Works

Our local researchers pointed out the problem and its severity in [2, 7]. In [7], the authors proposed a high-level framework combining legal bindings, economic instruments and public awareness to tackle this problem. And in [2], the authors portray the existing process of garbage collection to separation and processing scheme and expressed their proposition to haunt them down (mostly high-level proposition demanding process changes). In [3], the authors proposed an electro-mechanical sensor-based system using micro-controllers and operational amplifiers to auto-separate common forms of wastes. But the practical applicability of this system is in question mainly because of the diversity of the waste objects possess which is difficult to capture with the amount of information that these sensors provide to make a prediction.

In [19], the authors dealt with a slightly easier problem which is tagging street cleanliness. They have used street cameras from different localities and tried to develop an array of localised models for area-specific inference. That indicates training a lot of discrete models which doesn't seem like a scalable solution to the problem. Moreover, they used Naïve Bayes and AdaBoost classifiers to make the prediction observing the whole image, which missed the detection granularity down to object level that is subjective for the object separator to work. Recently, deep learning techniques have become increasingly popular because of their capabilities to extract rich features automatically from images, highly efficient parallel processing on GPUs to satisfy these resource-hungry algorithms, fast and efficient optimising techniques and a wonderful research community. That is why most of the modern literature tried to address the problem using the same. In [17], the authors used an end-to-end deep learning-based system to do a similar task. For training efficiency, they have employed a pre-trained Overfeat model as the feature extractor backbone and GoogleNet model as the classification head. Moreover, this paper addressed the detection granularity down to object level which makes it a better candidate for the waste separator model. But the dataset they have developed is heavily biased towards leaves and cigarette buts, didn't handle overlapping objects which is a common phenomenon in practical scenario and didn't use techniques like non-max suppression to choose the most relevant bounding box from multiple predicted bounding boxes at a specific IoU threshold.

In [12], the authors adopted a deep learning-based object localiser and classifier approach to figure out waste objects of 2 categories, namely bio-degradable and non-bio-degradable. But the authors didn't portray any experimental results in this paper. No specific network architecture, loss function, result analysis or information regarding the dataset is present here. In [20], the authors adopted state of the earth networks like mobilenet, densenet and inception for waste object classification and portrayed good level of accuracy metric. But the waste object is not represented contextually (Fig. 4c) in the dataset as they adopted the TrashNet dataset and they have not taken the detection granularity to the object level as well. In [14], the authors used transfer learning to train a Faster R-CNN model pre-trained on PASCAL VOC Dataset which detects and classifies three categories of waste objects, namely landfill, recycling and paper. They have also adopted the Trashnet Dataset and engaged lots of augmentation techniques to artificially generate contextual images. This yields significant improvement over the last works involving this dataset with a mAP of 68.3% but the algorithm does poorly in paper class, and it also ignored some other waste categories like glass, metal, plastic, etc.

In [21], the authors trained a multi-layer hybrid deep learning network. This system involves both a high-resolution camera and sensors to extract useful features and finally processes the image features using a CNN-based algorithm and deploys a Fully Connected MLP to blend image features with other information to classify waste as recyclable or others. This system tried to mimic human sensory organs like eyes with a pre-trained AlexNet model and ears/noise with a MLP and achieved a better classification accuracy than other systems at that time. But 2 problems still persist—the model fails to pinpoint the waste object but rather classifies the whole image frame which can contain multiple waste objects and although they employed image augmentation to present different forms of the image to the model, the background and natural diversity of the real world is still absent here as seen from Fig. 4a. In [13], the authors tried classifiers like Support Vector Machine and Neural network models like AlexNet, with applied data augmentation techniques. But the context is missing here along with multiple overlapping object detection and classification which can be seen from Fig. 4b. An improved work using this dataset has been published in [23], where the authors tried data augmentation techniques like building a collage of images, employing GANs, etc. to artificially manipulate the real-world scenario and yield some good results but no inference test has been provided involving real-world waste image, which raises the question of their practical applicability (Fig. 5).

In [24], the authors used an end-to-end deep learning approach in a hierarchical manner to detect and classify waste in food trays. They have published their developed dataset (Sample: Fig. 5) and adopted a multi-label multi-class classification approach to annotate the dataset to improve the accuracy. But the usage of 7 individual CNNs to classify the waste objects significantly increases the computational complexity and inference time, making it hard to deploy the system in edge devices for practical applications. In [10, 26], the authors used more realistic images to deal with context problems and both contributed their own datasets GINI and TACO but both of them represented a very small amount of annotated objects (*1494* and *4784*, respectively)

(a) (b) (c)

Fig. 4 Representative of waste images from [13, 20, 21]

(a) Example 1. (b) Example 2. (c) Example 3.

Fig. 5 Representative of waste images from [24]

and unbalanced class distribution. Architecturally, [10] offers only binary detection—garbage is present or not which takes out the recycling application, and the inference time required is not feasible for practical applications as well. Proença and Simões [26] detects only litters leaving the bio-degradable wastes out of the equation. It doesn't support multi-scale detection; hence small-sized objects are often ignored. Moreover, both of them used a region proposal-based 2-stage object detector which is slower to make inferences on real-time data.

4 The Garbage Annotated in Context Dataset

Building the dataset is one of the more challenging tasks of any machine learning project. The struggle becomes scarier when we don't have any benchmark dataset for our problems—which actually happened in our case. Now, building a dataset can be done in a few different ways as follows. A brief overview of them is illustrated below.

4.1 Publicly Available Open Labelled Datasets

The existing open-source datasets regarding the waste management domain use mostly out of context images, portray unbalanced class ratios, don't provide object-level annotations, etc. (as illustrated in the previous section). Moreover, our required classes were absent in those datasets as well. That's why we were unable to adopt any of them for our work.

4.2 Web Scraping

The next option is writing an automated web scrapper script to search keywords of required images and download them accordingly. It's a proven method and has been adopted in various works like Web Scraping of COVID-19 News Stories to Create Datasets. Now, images found on the Internet are of lots of different sizes and pixel resolutions. Some images are of pretty low resolution and our target task can contain 50–100 objects per frame which can make the job difficult for the detector on very low-resolution images. So, we have written our own image-scrapper-from-web script. After accumulating all the downloaded images, we also manually inspected each image and took a decision as per the merit of the image to portray the real-world waste scenario. There is a demonstration of a few subcategories of waste images that we've downloaded in Table 1. We have tried to cover the common categories of waste that are seen in the garbage dump in Bangladesh.

4.3 Capturing Images

Taking videos or capturing images of the subject can also be considered as an option to develop a dataset of this kind. This can be achievable by (a) manually clicking each image or capture a video by ourselves or (b) crowdsourcing, which is employing people to take images of the subject matter for us. (c) Installing a programmed camera in our environment of choice. Due to a lack of financial support and other endorsements, we were unable to take (b) and (c) as an option. So, we had to proceed with manually clicked photographs by ourselves.

4.4 Using Data Augmentation

As deep learning algorithms are data-hungry ones, when we have a relatively small dataset, it becomes difficult to train a model and generalise it well on unseen examples. In these cases, different kinds of data augmentation can be adopted to extrapolate

Table 1 Search Category granularity to accumulate garbage photos through web scrapping

Metal	Glass	Kitchen Waste	Plastic	Paper
Alloy metal scrap	Broken glass jar	Bio-degradable waste	Coke plastic bottle waste	Cardboard waste
Crushed coke cans	Broken glass waste	Chicken bones waste	Plastic waster bottle waste	Craft paper waste
Electronic scrap	Broken plate waste	egg waste	industrial plastic waste	magazine paper waste
Ferrous metal scrap	Broken glass waste	fish waste	laboratory plastic waste	newspaper waste
Household metal scrap	Crushed glass waste	food waste	medical plastic waste	paper waste
Metal rod scrap	Glass bottle waste	food waste closeup	plastic bottle waste	pizza box waste
Metal waste	Glass waste	fruit waste	plastic container waste	paper scrap closeup
Non-ferrous scrap	Light bulb waste	fruit waste closeup	polythene waste	shredded paper waste
Crushed steel drums	Recycled glass waste	kitchen fish waste	7up plastic bottle waste	soap packet waste
Tin canes waste	Scrap glass waste	kitchen waste	onetime coffee cup waste	tissue waste
Used nuts and Bolts	Shattered glass waste	organic waste	onetime plastic cup waste	torn paper bag
	Broken window glass	tea bag waste	plastic bag waste	used diapers
		vegetable waste		used tissue waste
				waste paper junk

more training data. Geometric transformations like flipping, cropping, rotation and translation are the most common data augmentation techniques. Using data augmentation in training images not only enriches our dataset by creating variation but also reduces the model over-fitting problem. We have adopted a wide range of data augmentation techniques, which not only are limited to the image transformations stated above, but also are expanded towards some modern techniques like photo-metric distortion, cut mix, mix up and mosaic data augmentation as described in [25].

As we can see from Figs. 6 and 7, we have made our efforts clear to leave minimal distribution difference between the two sources of image collection, in easy words, we have tried to carefully pick up photos from the scrapped images which can better mimic the real-world scenario.

(a) (b) (c)

Fig. 6 Annotated Images of Garbage Waste—Self-Taken Photographs

(a) (b) (c)

Fig. 7 Annotated Images of Garbage Waste—Web Scrapped Images

Fig. 8 Annotation count and percentage per category

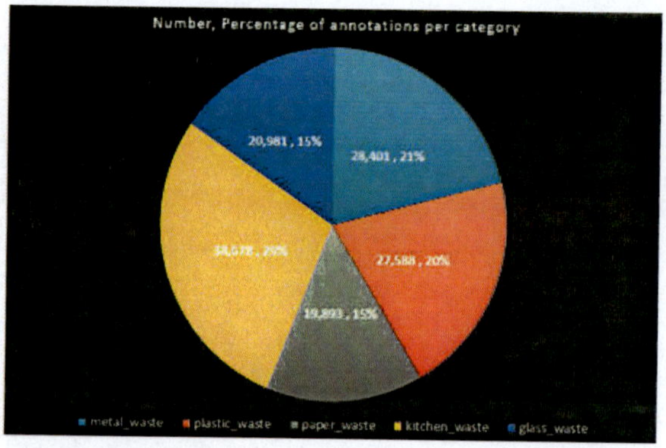

4.5 *Dataset Preparation Steps*

Finally, if we summarise the steps that we had taken for building the dataset:

1. First, we've visited 10 different dumping stations in Dhaka City, captured the spotted trash using mobile phones and accumulated more than 1200 images. We have taken images from different angles and verticals trying to mimic a moving agent which will automate the process. All the images are captured in different lighting conditions (Haze, Cloudy, Sunny, etc.) during daytime at an optimum resolution.

2. Then, we have carefully investigated the images and identified the varieties and different orientations with which waste appears in an open environment.
3. After that, we have written a web scrapper in Python to download photos of a given category (refer to Table 1) from Google images automatically. We have accumulated photos of more than 70 different subcategories of waste, which belong to our main five categories.
4. From there, we have carefully investigated and picked 1000 photos from the collection that matches the real scenarios by hand.
5. We have annotated the final images using the labelImg tool [28] in PASCAL VOC format.
6. The annotated images are then further verified by two other human individuals and finally used for model development purposes. The primary objective had been to keep the contextual images only to guard against the data drift and concept drift problem.

We have also tried to make our dataset evenly balanced so that no data-driven bias exists in the trained model as illustrated in Fig. 8.

5 Model Overview

5.1 Problem Formulation

Considering an image of garbage waste, our goal is to predict and draw rectangular bounding boxes wrapped around each and every waste object present in that image and assign each detected object to one of the considered classes. Consequently, this can be formulated as a multi-label K-way classification problem. We employ an end-to-end deep learning method which uses a proven (1) feature extractor backbone to project rich image features to latent space; (2) leverage the idea of Spatial Pyramid Pooling and Path Aggregation Networks to significantly increase receptive field with almost no incremental time cost and collect image features from different levels of the backbone network for different detector levels. We have also used a single objective function inclusive of all the losses and optimise that using Adam or RMSProp optimiser for dense prediction of waste objects. (3) Finally, we have used the YOLOv3 (anchor-based) head, to detect and classify waste objects on three different scales. We have also used post-processing techniques like Non-Max suppression to filter out the most ideally fitted bounding box around the subject (Figs. 9 and 10).

5.2 Feature Extractor Backbone

While choosing the backbone of this framework, we had to make an optimum choice between maintaining both classification accuracy and detector efficiency simulta-

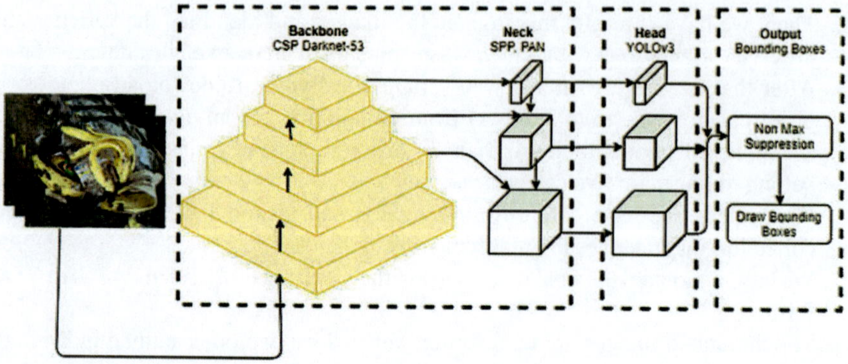

Fig. 9 CNN Architecture used in our waste detection and classification framework

neously. The criteria of choice are (1) Higher input size—for multiple small object detection, (2) Deep Layers—for higher receptive field size to deal with the increased input size and (3) More parameters—for greater learning capacity of the model. The Darknet53 network used in [22] which provided a significant improvement over the detection efficiency while maintaining good enough classification accuracy than the contemporary models. It has shown promising results in cases of training the detector using the pre-trained backbone weights, which is very convenient for the transfer learning strategy which we have adopted here. The CSPDarknet53 [27] that we have used here further optimises the framework by reducing the computations up to 20% hence decreasing the inference cost.

5.3 SPP-PAN Neck

Spatial Pyramid Pooling block, after the CSPDarknet53 block, significantly improves the receptive field size, highlighting the most significant contextual features without adding almost any computational overhead. Receptive field size is an important property of the convolutional neural nets which enables a network to detect small to largest objects on multiple scales. We have also used *Path Aggregation Neck* as a parameter aggregator from different backbone levels for different detection heads, to enable effective propagation of low-level image features towards the detector layers.

5.4 YOLOv3 Detection Head

This is the last block of the system and is responsible for detecting and classifying waste objects in images across 3 different scales having strides 32, 16 and 8, respectively. Consequently, an input image of size 416*416 will be able to make detections across scales 13*13, 26*26 and 52*52. Starting from input, up to the first detection layer, the network only downsamples the input, where feature maps of a layer with stride 32 are used to perform the detection. Afterwards, layers are upsampled by a factor of 2 and added with feature maps of a previous layer. Consequently, detections were made at layers with stride 16. A similar strategy is repeated to yield the final detection at the layer with stride 8. Output processing techniques like *Non-max suppression* also have been implemented in this block to filter the most accurate bounding boxes wrapped around an object from the predicted ones.

5.5 Loss Function

With the final detection result at hand, we will be able to calculate the loss with respect to the true labels now. The loss function consists of four parts: (1) centroid (xy) loss, (2) width and height (wh) loss, (3) objectness (object and no object) loss and (4) classification loss. Putting everything together in perspective, the final equation looks like below

$$
\begin{aligned}
TotalLoss = \lambda * \sum(CIoU((x, y), (x', y') * obj\_mask) + \lambda * \sum(CIoU((w, h), (w', h') * obj\_mask) \\
+ \sum(BCE(class, class') * obj\_mask) + \sum(BCE(obj, obj') * obj\_mask) \\
+ \lambda_{noobj} * \sum(BCE(obj, obj') * (1 - obj\_mask) * ignore\_mask)
\end{aligned}
$$

The first two components of the above function calculate the loss of the bounding box regression problem. Previously, RMSE had been used in this case, but it doesn't converge well (when the large-sized objects cause large loss values). IoU loss (refers to the ratio of the overlapping area of the predicted Bounding Box and union of them), being a ratio by nature, solves this problem. We have used a more recent advancement named CIoU loss which brought the distance between centre points, and the aspect ratio into the loss calculation. Then the usage of Binary Cross Entropy loss to calculate *objectness* and *noobjectness* loss is self-explanatory being a binary classification problem. By estimating this objectness loss, we can steadily show the model to distinguish an area of intrigue. At the same time, we don't need the algorithm to cheat by making wrong region proposals. Henceforth, we need noobjectness loss to punish those false positive assumptions. Finally, we used independent logistic classifiers instead of Softmax as suggested in [22] to calculate the class loss.

6 Implementation and Ablation Study

6.1 Implementation Details

Below are the key highlights of our training process:

- The weights of our feature extractor backbone has been initialised using the pre-trained model, CSPDarknet-53 [27], which has been trained on huge MS COCO Dataset [6]. As a result, our model was able to exploit the rich hierarchy of already learned representations, helping it to reduce over-fitting and achieve a better generalisation. The original architecture has been tweaked to perform a 5-way classification task as per our requirement. We have observed a significant uptake in performance with transfer learning through qualitative and quantitative experiments.
- We have adopted a fivefold cross-validation method for all the experiments and trained the detection head, freezing the weights of the feature extractor layer, to iteratively search for the optimum set of hyperparameters which yield the best training and validation results. For each fold, the dataset is split randomly into 80% training set, 10% development set and 10% test set.
- As we have used an anchor-based method for the object detector, the dimensions of those prior boxes are vital. We have run a K-Means algorithm on our whole dataset images, to figure out the clusters of bounding box dimension distribution and found 9 optimal anchor box coordinates (3 per each scale). Hypothetically, we can say, it would increase the accuracy of the detector as we have considered IoU loss for the bounding box regression task.
- We have used TensorFlow 2.0 for model building and training purposes. We have also exported the model using TFlite and TFServing for edge device deployment in real-world use cases.
- The hyperparameters are as follows: the training steps is 10,000; the step decay learning rate scheduling strategy is adopted with initial learning rate 0.01 and multiplied with a factor 0.1 at the 8000th steps and the 9000th steps, respectively; the network is optimised using RMSprop with momentum and weight decay respectively set as 0.9 and 0.0005. Images of input resolution 416*416 have been fed to the network during the forward passes. Because of the large architecture and GPU memory limitation, all of the experiments were run on a single GPU with a batch size of 64 and a mini-batch size of 4 (studies from [25] revealed that mini-batch size has a pretty limited effect on detectors performance).
- Different experiments have been carried out borrowing ideas from [25] which involves using a lot of image augmentation techniques, cross mini-batch normalisation, varying the IoU threshold for assigning ground truth, etc.
- All the training and experiments are run on a single GPU environment using 4GB GeForce GTX 1050 Ti GPU.

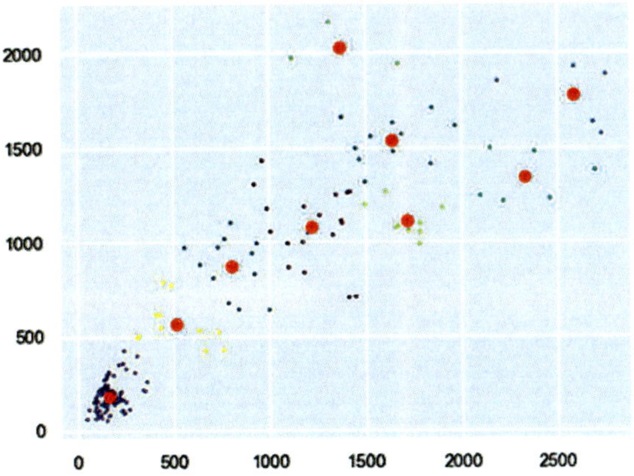

Fig. 10 Customised Anchor Box coordinates determined by running K-Means Algorithm over the whole dataset images

6.2 Overall Performance

Mean Average Precision (mAP) [1] has been used for model performance evaluation along with sensitivity (recall), F1 Score and inference speed. The mAP score has been calculated by taking the mean AP over all the classes and compared at different IoU thresholds to extract the best setting. Finally, the IoU threshold has been set as 0.35 which achieved a good balance between sensitivity (recall) and specificity (precision). The defined metrics have been monitored via the TensorBoard tool. Table 2 shows the performance of 2 Single-Stage Anchor box-based object recognition frameworks, fine-tuned for our use case.

It's evident that the modified YOLOv4 model exceeds the earlier one in almost all the categories as it embarks upon all the bag of tricks from the rich computer vision research domain. The 3-scaled approach makes the objects across various scales detectable. The deep architecture of the feature extractor has increased the size of the receptive field along with more parameters and robust learning. In both cases, the transfer learning strategy worked out well as the relatively small dataset yielded considerable results. We have exported the model via TF Serving for edge device deployment and test sample inference. It requires 52 MB of additional memory allocation for processing a single example. The relatively poor performance of glass waste and paper waste can be overcome by hard negative mining of the falsely detected examples. The overall performance of the model can be improved by enriching the dataset to deal with more distribution differences of waste in wild.

6.3 Ablation Study

We have collected 50 more images outside the main dataset from different places and conducted the ablation study. Firstly, we can see the effective low-level feature learning from the backbone in Fig. 11. This indicates the model generalises well on

Table 2 Comparison of the speed and accuracy of 2 experimented frameworks on our waste dataset

Method	Network	Inference Speed	Sensitivity	F1 Score	AP	AP (Paper)	AP (plastic)	AP (glass)	AP (kwaste)	AP (metal)
YOLOv3	Darknet53-FPN	70 ms	40.47%	0.47	57.3%	47.30%	47.02%	63.40%	60.06%	68.70%
YOLOv4	CSPDarknet53-SPP-PAN	58 ms	62.25%	0.65	66.08%	57.55%	65.67%	57.32%	73.06%	76.80%

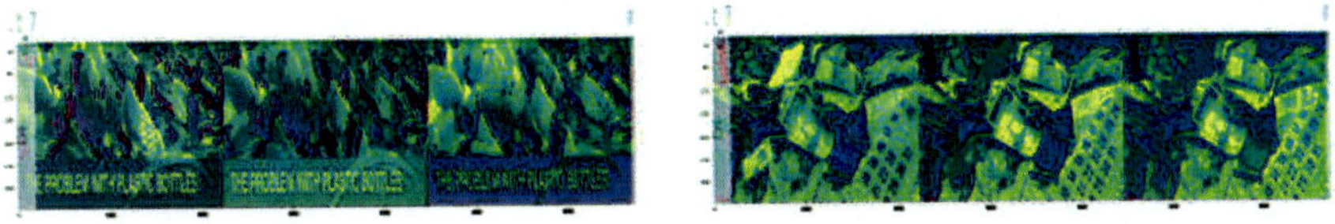

Fig. 11 Feature visualisation from feature extractor backbone

Fig. 12 Sample inference on real-world Garbage Data

unseen examples even from slightly different data distribution which points towards effective **transfer of knowledge** from pre-trained weights. The use of **SPP-PAN Neck** instead of Feature Pyramid Neck results in significant improvement over mAP score backed by better propagation of low-level features to high-level latent space and rise in receptive field resolution. The choice of **YOLOv3 head** as the Single-Stage Anchor-based detector makes the real-time inference possible by making all the estimations in one forward pass. The choice of **CIoU** loss over RMSE significantly speeded up the convergence and reduced the training time. Our proposed **GACO Dataset** featuring multiple overlapping objects in contextual waste images annotated up to object-level granularity makes the model operate well in real-life scenarios as well (Fig. 12).

6.4 Result Comparison

In Table 4, we evaluate the proposed method and other prevailing waste object recognition methods with respect to mAP score, the type of problem addressed, sensitivity, inference time and deployment readiness in edge devices. All the data of other methods are directly obtained from their original papers. We have only considered the works which addressed waste recognition up to object-level granularity as hypothetically we can say that garbage can't be separated if not detected properly. Moreover, most of the models are trained and evaluated on imbalanced datasets containing manipulated waste images as described in Related Works, which arise the question

Table 3 Benchmarking our 'Garbage Annotated in Context (GACO)' dataset w.r.t. other datasets

Dataset	#Images –#Objects	#Classes	Class distribution	Context
Garbage In Images (GINI) [10]	1,484 –1,484	2	Unbalanced	Real World
TrashNet [13]	3,000 –3,000	6	Balanced	Not Real World
Trash Annotated In Context (TACO) [26]	1,500 –4,784	25	Unbalanced	Real World
Labelled Waste in Wild (LWW) [24]	1,002 –7,200	19	Unbalanced	Real World
Garbage Annotated In Context (GACO)-ours	2,200 –135,541	5	Balanced	Real World

Table 4 Benchmarking the performance of our framework w.r.t. the contemporary works. All the data is taken from the published papers provided by the authors

Model	Dataset	#Annotation	Multi-label Multi-class	Sensitivity	mAP	Inference speed	Deployment readiness
OverFeat-GoogleNet [17]	Own	4,338	✓–✓	83%	63%		
Faster R-CNN-CNN [24]	LWW	7,200	✓–✓		73.6%		
Masked R-CNN [26]	TACO	4,784	✓–✓		26.10%		
Faster R-CNN [23]	TrashNet	8,000	✓–✓	87%	81%		
CSPDarknet53-SPP-PAN-YOLOv3 (ours)	GACO	135,541	✓–✓	73%	66.08%	58 ms	✓

of the general applicability of those methods. So, we can say, our objective of delivering a fast, accurate object detection model capable of recognising common categories of wastes from messy real-world scenarios has been served through our proposed method. Table 3 shows the high level comparison of our dataset with respect to others.

7 Conclusion

So, here we have proposed a deep learning-based single stage waste recognition framework which can make real-time predictions of garbage objects in contextual images both in images and videos. We have taken a data-centric approach to build up a relevant dataset and leverage the fine-grained ideas and recipes to build a decent framework for waste object detection and classification task. It's pretty lightweight, computationally cheap and easily deployable in edge devices which makes it convenient to use in practise. There is still huge room for improvement in this work. The diverse distribution difference of the domain along with limited number of training examples has been tried to cover up by transfer learning, data augmentation techniques up to some extent. Further research can be directed towards enriching the dataset using the techniques that have been elaborated in The Garbage Annotated In Context Dataset and optimising the model further to enhance the learning capability with fewer parameters involved. However, we have proposed a novel deep learning-based approach to fight against one of the most prominent problems that prevails in our country at the moment. So, we believe it's s solid baseline to take the work forward and enrich further.

References

1. Everingham, M., Van Gool, L., Williams, C.K., Winn, J., Zisserman, A.: The pascal visual object classes (voc) challenge. Int. J. Comput. Vision **88**(2), 303–338 (2010)
2. Bahauddin, K., Uddin, M.: Prospect of solid waste situation and an approach of environ-mental management measure (emm) model for sustainable solid waste management: Case study of dhaka city. J. Environ. Sci. Nat. Resour. **5**(1), 99–111 (2012)
3. Russel, M.M., Chowdhury, M., Shekh, M., Uddin, N., Newaz, A., Mehdi, M., Talukder, M.M.M.: Development of automatic smart waste sorter machine (2013)
4. Concern, W.: Bangladesh waste database 2014. Dhaka: Waste Concern **29**, 2019 (2014). Accessed June
5. Girshick, R., Donahue, J., Darrell, T., Malik, J.: Rich feature hierarchies for accurate object detection and semantic segmentation. In: Proceedings of the IEEE Conference on Computer Vision and Pattern Recognition, pp. 580–587 (2014)
6. Lin, T.Y., Maire, M., Belongie, S., Hays, J., Perona, P., Ramanan, D., Dollár, P., Zitnick, C.L.: Microsoft coco: common objects in context. In: European Conference on Computer Vision, pp. 740–755. Springer (2014)
7. Abedin, M.A., Jahiruddin, M.: Waste generation and management in Bangladesh: an overview. Asian J. Med. Biol. Res. **1**(1), 114–120 (2015)
8. Girshick, R.: Fast r-cnn. In: Proceedings of the IEEE International Conference on Computer Vision, pp. 1440–1448 (2015)
9. Liu, W., Anguelov, D., Erhan, D., Szegedy, C., Reed, S., Fu, C.Y., Berg, A.C.: Ssd: single shot multibox detector. In: European Conference on Computer Vision, pp. 21–37. Springer (2016)
10. Mittal, G., Yagnik, K.B., Garg, M., Krishnan, N.C.: Spotgarbage: smartphone app to detect garbage using deep learning. In: Proceedings of the 2016 ACM International Joint Conference on Pervasive and Ubiquitous Computing, pp. 940–945 (2016)
11. Redmon, J., Divvala, S., Girshick, R., Farhadi, A.: You only look once: United, real-time object detection. In: Proceedings of the IEEE Conference on Computer Vision and Pattern Recognition, pp. 779–788 (2016)
12. Sudha, S., Vidhyalakshmi, M., Pavithra, K., Sangeetha, K., Swaathi, V.: An automatic classification method for environment: friendly waste segregation using deep learning. In: IEEE Technological Innovations in ICT for Agriculture and Rural Development (TIAR), vol 2016, pp. 65–70 . IEEE (2016)
13. Yang, M., Thung, G.: Classification of trash for recyclability status, CS229 Project Report, vol. 2016 (2016)

14. Awe, O., Mengistu, R., Sreedhar, V.: Smart trash net: waste localization and classifica-tion (2017)
15. He, K., Gkioxari, G., Dollár, P., Girshick, R.: Mask r-cnn. In: Proceedings of the IEEE International Conference on Computer Vision, pp. 2961–2969 (2017)
16. Lin, T.-Y., Goyal, P., Girshick, R., He, K., Dollár, P.: Focal loss for dense object detection. In: Proceedings of the IEEE International Conference on Computer Vision, pp. 2980–2988 (2017)
17. Rad, M.S., von Kaenel, A., Droux, A., Tieche, F., Ouerhani, N., Ekenel, H.K., Thiran, J.-P.: A computer vision system to localize and classify wastes on the streets. In: International Conference on Computer Vision Systems, pp. 195–204. Springer (2017)
18. Redmon, J., Farhadi, A.: Yolo9000: better, faster, stronger. In: Proceedings of the IEEE Conference on Computer Vision and Pattern Recognition, pp. 7263–7271 (2017)
19. Alfarrarjeh, A., Kim, S.H., Agrawal, S., Ashok, M., Kim, S.Y., Shahabi, C.: Image classification to determine the level of street cleanliness: a case study. In: 2018 IEEE Fourth International Conference on Multimedia Big Data (BigMM), pp. 1–5. IEEE (2018)
20. Aral, R.A., Keskin, Ş.R., Kaya, M., Hacıömeroğlu, M.: Classification of trashnet dataset based on deep learning models. In: 2018 IEEE International Conference on Big Data (Big Data), pp. 2058–2062. IEEE (2018)
21. Chu, Y., Huang, C., Xie, X., Tan, B., Kamal, S., Xiong, X.: Multilayer hybrid deep-learning method for waste classification and recycling. Comput. Intell. Neurosci. **2018** (2018)
22. Redmon, J., Farhadi, A.: Yolov3: an incremental improvement (2018). arXiv preprint arXiv:1804.02767
23. Kulkarni, H.N., Raman, N.K.S.: Waste object detection and classification (2019)
24. Sousa, J., Rebelo, A., Cardoso, J.S.: Automation of waste sorting with deep learning. In: XV Workshop de Visão Computacional (WVC), vol 2019, pp. 43–48. IEEE (2019)
25. Bochkovskiy, A., Wang, C.-Y., Liao, H.-Y.M.: Yolov4: optimal speed and accuracy of object detection (2020). arXiv preprint arXiv:2004.10934
26. Proença, P.F., Simões, P.: Taco: trash annotations in context for litter detection (2020). arXiv preprint arXiv:2003.06975
27. Wang, C.Y., Liao, H.Y.M., Wu, Y.H., Chen, P.Y., Hsieh, J.W., Yeh, I.H.: Cspnet: a new backbone that can enhance learning capability of cnn. In: Proceedings of the IEEE/CVF Conference on Computer Vision and Pattern Recognition Workshops, pp. 390–391 (2020)
28. Tzutalin, Labelimg. https://github.com/tzutalin/labelImg
29. Vinodha, D., Sangeetha. J., Sherin, B.C., Renukadevi, M.: Smart garbage system with garbage separation using object detection

A System Dynamics Approach of Decisive Policy Measures for Sustainable Municipal Solid Waste Management in Khulna City of Bangladesh

S. M. Rafew and Islam M. Rafizul

Abstract A persisting municipal solid waste (MSW) management plan for any city in Bangladesh has to be focused on the understanding of dynamic interactions and ionized effect of all elements of the current MSW management system. Municipalities are the responsible authority for managing the generated MSW including the food and bio-degradable waste which is the major discarded constituent all over the country in terms of weight percentage. This study utilized system dynamics (SD) approach to not only understand the MSW management system of Khulna city with numerous interconnected and independent parameters but also investigated a number of policy scenario emphasized on the collection and separate industrial scale treatment for food and bio-degradable MSW of the city. The quantity of CH_4 emission potential for a certain time period has been found to increase from 6.71 Gg in year 2020 to 51.969 Gg in 2050. Three different policy scenario has been implemented in the developed SD model to reach to the dynamic conclusion that "Policy 3" is the most effective on the criteria of the illustrated sustainable development goal (SDG 12.3) with a less fluctuating composite index value of 6.05 until year 2050. To achieve the 71% composting scenario illustrated in "Policy 3" there is no alternative to industrial or commercial composting for the near future. Finally, the illustrated results represents that this developed SD model can be used as an effective simulation platform for MSW management system's evaluation as well as to determine the methods of achieving sustainable condition in the long run.

Keywords System dynamics · MSW management · Industrial composting · Policy scenario · Sustainable development

S. M. Rafew (✉) · I. M. Rafizul
Department of Civil Engineering, Khulna University of Engineering and Technology (KUET), Khulna 9203, Bangladesh
e-mail: rafew1801552@stud.kuet.ac.bd

S. Hossain et al. (eds.), *Proceedings of International Conference on Fourth Industrial Revolution and Beyond 2021*, Lecture Notes in Networks and Systems 437,
https://doi.org/10.1007/978-981-19-2445-3_11

1 Introduction

Civilization and development has been proportionally as well as expeditiously associated with the generation of municipal solid waste (MSW). The beginning phase of the 21st century has witnessed science fictional concepts turning into reality, resulting a drastically changing society, economy and consumer patterns. Cumulative piling of MSW in the designated landfill sites around the world for decades exhibits the evidence of insufficient visionary management along with resource limitations of the previous generation, authorized for the task [22]. Historically, MSW generation and its management opportunities has been considered extensively from both theoretical and implementation perspectives. The utmost thrifty solution is seen in the nature's cyclic pattern which involves utilization of refused materials and turning it back to the source it came from. For example, involvement of community citizens in the source separation process can be increased through frequents mainstream media and press release along with a progressive and already existing model for developed countries to provide separate colored bins and chambers for compostable MSW, paper products, recyclable and hazardous MSW to both community houses and collecting vehicles. This separation policy initiative, if implemented systematically, can significantly improve the MSW collection quality and quantity of Khulna city.

A sustainable development process in MSW system of Khulna city is a prime consideration for Bangladesh Delta Plan (BDP), 2100. Sustainable MSW management system contributes to improved efficiency in natural resources use and to mitigate environmental burdens. To be sustainable, MSW management cost should be reasonable and widely accepted by the general mass as well as by leading organizations [17]. Investment plan for Bangladesh delta plan envisage a tidy sum of 73,226 million BDT or 923 million USD capital formation in the improvement of flood control, drainage system and MSW system of Khulna city under the project code of "UA 3.1" [8]. Though this capital expenditure portfolio had a time period until 2020, the implication phase of the project is advancing with possible time extension till date. Sustainability with provision based on public service is considered in today's technologically advanced world with limited natural resources. The annual production of food waste for Bangladesh is estimated to be 10,618,233 tons/year with a per capita household waste of 65 kg/year [22]. Sustainable development goal (SDG 12.3) illustrates a commitment towards halving bio degradable food waste at retail and consumer level with the ultimate target of reducing squander nature towards this particular type of waste. A conventionally constant nature of MSW management system practiced throughout the country along with the deficiency of advanced infrastructure, industrial composting as well as recovery of existing CH_4 from landfills in the context of Bangladesh has deliberately being disregarded. In this study, precisely in the CH_4 emission sector of the developed SD model for Khulna city, IPCC guideline and methods were programmed along with implication of delay time converter for methane emission for time period until year 2050 [21]. The previous two sectors of MSW generation and management of Khulna city is based on the publication in the prospective journal authored by Rafew and Rafizul [17], except some obvious technical modification that has been conducted for this respective study. Finally, possible policy implementation scopes are analyzed with the prognostic dataset simulation of MSW of Khulna city from 2020 to 2050. The programming development of MSW generation quantification in yearly basis is the basic information required to design a sustainable MSW management system through SD model. Economically efficient design for collection, recycling, on site storage, MSW processing, transportation, budgeting, public concern, composite index value are based on such data. Moreover, all the related parameters that effect the overall MSW management system, is necessary to consider in order to replicate an actual scenario of the MSW management and to implement the long term policy measures for achieving the ultimate goal of a sustainable and efficient environmental system [2]. SD is considered to be an ideal choice for replicating the dynamically knitted complex parameters while considering the methodized parameters ranging from problem definition, hypothesis and model development to implementation of theoretical equation as well as model validation through sensitivity analysis and historical perspective options.

2 Synopsis and Applicability of System Dynamics

The basic advantage of system dynamics (SD) is that it consider the whole system in a rational and rather intuitive approach than isolated mechanisms of the system [2]. SD was developed by J. Wright Forrester in Massachusetts Institute of Technology (MIT) in the 1950s. Application of SD model in computing MSW generation has diversity options over models based on regression analysis, correlation based models and fuzzy logic system [12]. Due to the non-linearity as well as complex uncertain behavior, SD model has proven vital for the policy formulation [15]. Among recent studies Chen et al. [5] applied SD model to determine whether the present MSW disposal capacity is capable enough to cope up with the ever increasing MSW generation in Singapore. Dyson and Chang [6] emphasized on the capability of SD forecasting MSW generation. Plai et al. [16] used SD methodology to identify impact of existing population, population growth, total MSW generated and collected along with the collection volume of hospital waste on the amount of generated MSW.

Talyan et al. [21] used SD model to quantify CH_4 emitted from MSW in Delhi, India. The key featured they used is delay function in their model for methane generation from MSW to project the actual scenario. They proposed that energy recovery including various treatment and disposal methods as an alternative to fossil fuel will significantly contribute to the reduction of greenhouse effect due to electricity production. Manasakunkit and Chinda [13] has developed a dynamic model of MSW for Bangkok, Thailand and tested it on three validity methods covering dimensional consistency, behavior sensitivity and model scenario on extreme conditions. System dynamics had been applied for long-term integrated MSW management in Bandung city, Indonesia which emphasized on population and economic growth [18]. Federica et al. [7] applied SD methods for solution of MSW crisis in Campania, Italy and they also provided analytical framework to explore the effect of alternative policies in this regard. Bala et al. [2] applied SD method for predicting electric energy generation in case of Dhaka city of Bangladesh.

3 Research Methodology

3.1 Study Area

Khulna city is located in the tropic of cancer region with longitude and latitude of 89.4° East and 22.49° North respectively. It is divided in to 31 wards with a population of around 1.5 million who are responsible for most of the MSW generated around the city. The existing average total annual rainfall and temperature in Khulna are 1630 mm and 26.37 °C with an increment rate of 4.960 mm/year and 0.005 °C respectively [14]. Though the city had a wide reputation in the past decade as an industrial city, it deemed eventually due to the scarcity of investment and work opportunity. Collaboration of two independent department namely conservancy and engineering are responsible for the management of MSW generates throughout city area. Compositional analysis has not been conducted during any phase of MSW collection and dumping procedure which is a major concern to consider in case of achieving an efficient MSW management system. The budget for the financial year 2019–2020 for Khulna city was 3,350 million BDT with an implementation period until 2021. The geographic location with coordinates of Khulna city is shown in Fig. 1.

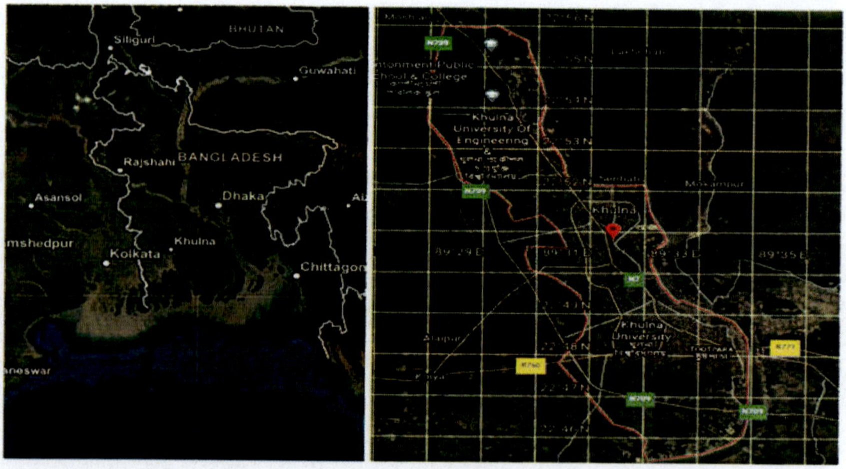

Fig. 1 Geographic location and boundary area of Khulna City. (*Source* Google Earth)

3.2 Development of System Dynamics Model

In this study, for the prediction of MSW generation and management of Khulna city through SD model, lots of independent factors such as GDP, population, population growth and death rate, income multiplier, incineration capacity, composting capacity, recyclable fractions of MSW, MSW stock, landfill development, number of trucks, collection capacity of trucks, compacted density of MSW for landfill, dumping height for landfill avg. time for depreciation of trucks and treatment plants, average yearly rainfall, annual temperature etc. are considered to ensure more precision. To avoid ambiguity of prediction data from conventional approach, SD model is being considered. Moreover, the research work is done mostly with orientation around a quantitative simulation approach in order to evaluate the performance of previously adopted policies as well as policies needed for sustainable development goal (SDG) for Khulna city. Two basic structure of a SD model namely casual loop for representing the dynamic hypothesis of the research and systematically arranged differential equation representing cause and effect through stock-flow diagram are being developed where the present management system of Khulna city is hypothesized in the base year 2020's reference mode.

Casual loop diagram: The casual loop is based on a theoretical framework of MSW generation and management along with the determination of possible CH_4 generation and emission from landfill sites of Khulna city. From a study conducted by Ahsan et al. [1] it has been extracted that the food and vegetable waste of Khulna city is 78.9% of the total generated waste which has an extensive potential of composting in an industrial scale. Moreover it will minimize the methane emission from landfills of Khulna city which is mainly known as Rajbandh. The model calculated the CH_4 emission from the landfill until year 2050, which will pose a serious adverse effect. For valuation of the environmental factors and other interconnected factors such as public concerns and awareness composite index is calculated for the generated model of Khulna city. The composite index value specifies the public health and environmental condition of a given situation. If composite index value rises, it denotes worsening in the environmental and public concern conditions in an incremental manner. The polarity of the connectors represents either the aggregative or the decretive interconnections with the parameters assigned for the MSW management of Khulna city. The composite index is being calculated with the following Eq. (1) and the casual loop diagram is shown in Fig. 2.

$$Composite\,index = (w_1 \times UNCR + w_2 \times UNTR + w_3 \times CH_4Emission) \times POR \tag{1}$$

Stock-flow diagram: In order to simulate and run the Stock-flow diagrams of three sectors representing MSW generation, management and CH_4 emissions of KCC, Euler's integration method has been used. Because of the non-linearity in the relation of datasets, mostly exponential graphical interpolation functions have been fabricated by inflowing the intermediate values. Moreover, complex computational procedures

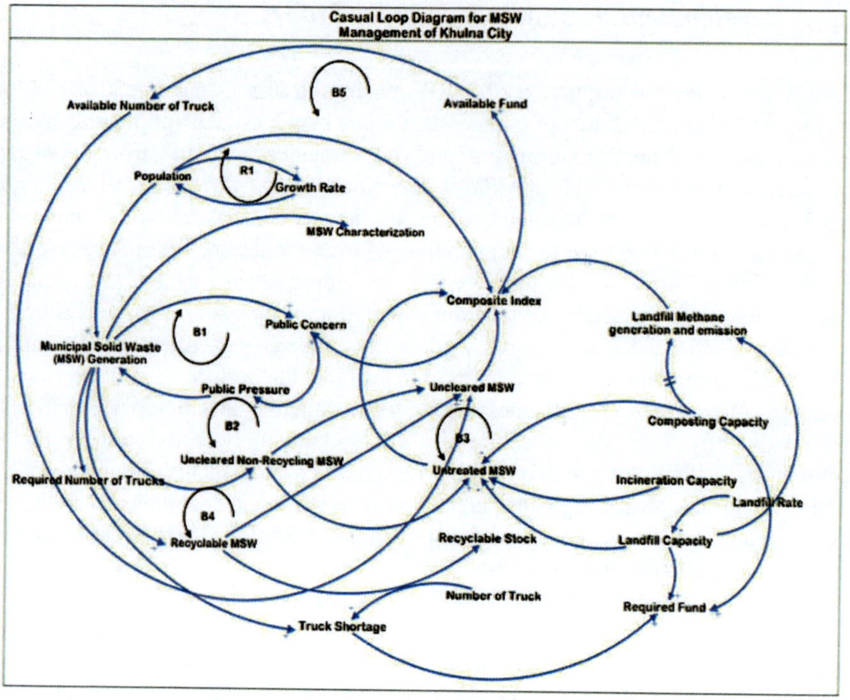

Fig. 2 Casual loop diagram of MSW management of Khulna City

containing delay functions and non-integrated function are used. The purpose of these specific functions are to alter the directive amplitude of the signal passing through the integration procedure. The system dynamics software used for this research is STELLA version 1.9.4 for programming the MSW generation, management and CH_4 emission potential sectors. The visual representation of stock-flow diagram is sown in Fig. 3a–c respectively along with a few number of vital equations such as the IPCC (1996) and IPCC (2006) default method equation for CH_4 emissions from landfill used during the programming procedure Eqs. (2)–(6). Moreover, the degradable organic carbon value (DOC) has been calculated by following IPCC guideline with MSW characterization data of Khulna city. The value obtained through extensive procedural calculation is 0.16578 which is proportionally identical to the default value of 0.156, used for calculating CH_4 emission for Matuail landfill site in Dhaka [9].

$$\begin{aligned}&Composite\_Index =\\&((Weight\_1 * (Uncleared\_MSW/Base\_Value\_for\_Uncleared\_MSW))\\&(Weight\_2 * (Untreated\_Waste/Base\_Value\_for\_Untreated\_MSW))+\\&(Weight\_3 * (CH_4\_Emission/CH_4\_emission\_\ Base\_\ year)))*Population\_Ratio\end{aligned} \quad (2)$$

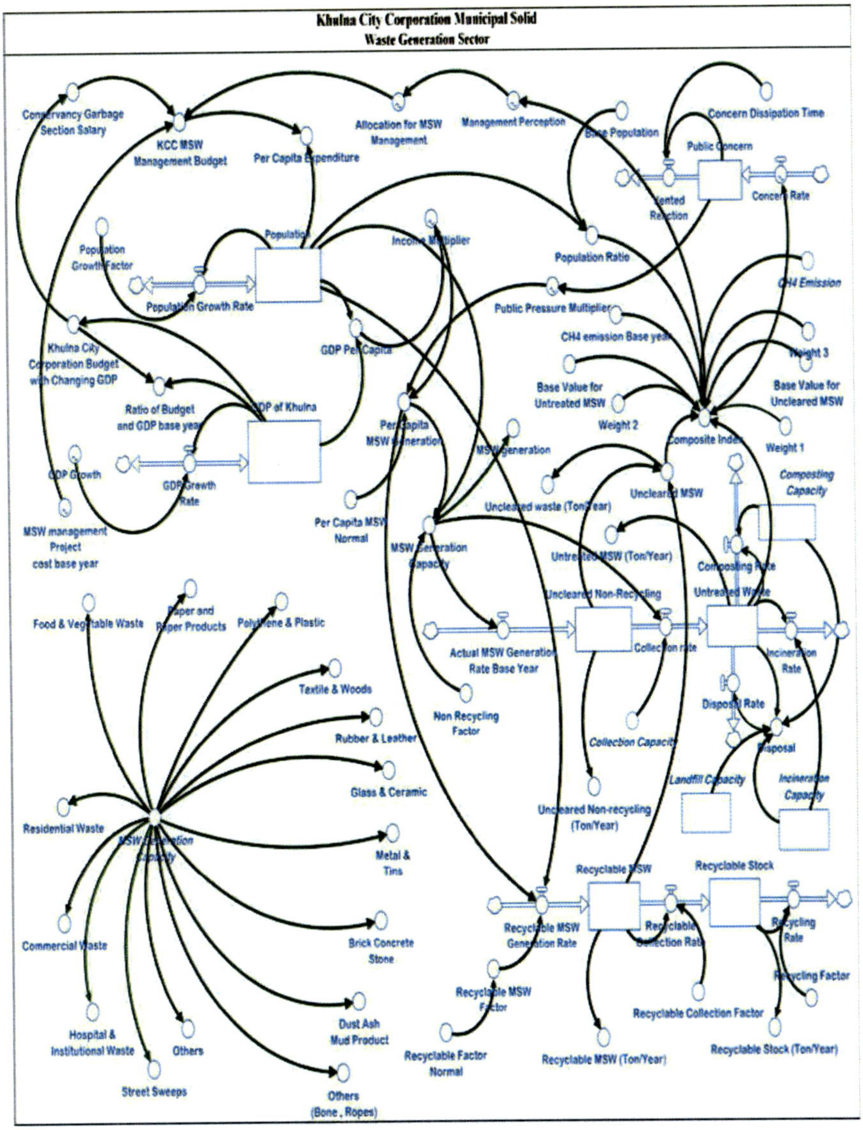

Fig. 3 **a** SD model for MSW Generation Sector of Khulna City. **b** SD model for MSW management sector of Khulna City. **c** SD model for CH4 emission quantification from discarded MSW of Khulna city

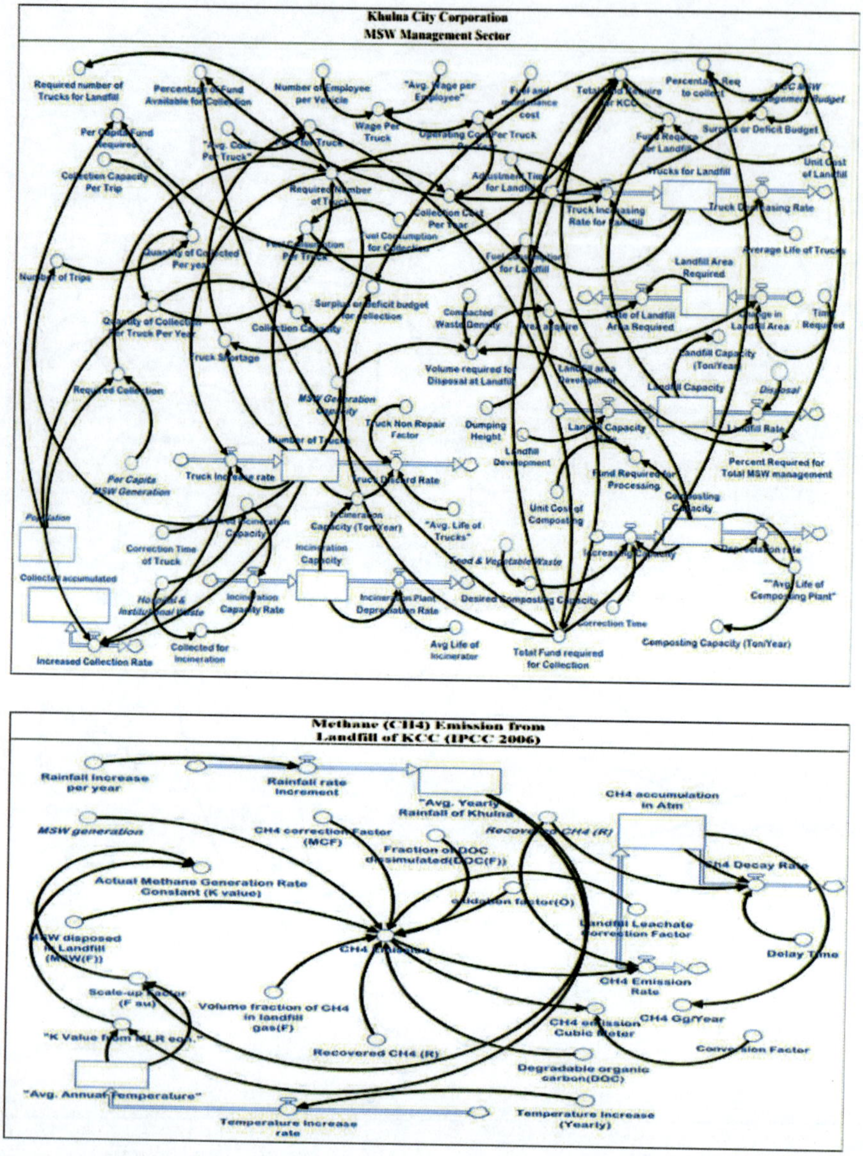

Fig. 3 (continued)

$$Composting\_Capacity(t) = Composting\_Capacity(t - dt)$$
$$+(Increasing\_Capacity - Depreciation\_rate) * dt\{NON - NEGATIVE\} \quad (3)$$

$$CH_4\_Emission = ((MSW\_generation * MSW\_disposed\_in\_Landfill\_(MSW(F))*$$

$$CH_4\_correction\_Factor\_(MCF) * Fraction\_of\_DOC\_dissimulated(DOC(F))*$$
$$Degradable\_organic\_carbon(DOC)*$$
$$Volume\_fraction\_of\_CH_4\_in\_landfill\_gas(F) * (16/12))-$$
$$Recovered\_CH_4\_(R)) * (1 - oxidation\_factor(O)) + (((MSW\_generation*$$
$$MSW\_disposed\_in\_Landfill\_(MSW(F)) * CH_4\_correction\_Factor\_(MCF)*$$
$$Fraction\_of\_DOC\_dissimulated(DOC(F)) * Degradable\_organic\_carbon(DOC)*$$
$$Volume\_fraction\_of\_CH_4\_in\_landfill\_gas(F) * (16/12))-$$
$$Recovered\_CH_4\_(R)) * (1 - oxidation\_factor(O))*$$
$$Landfill\_Leachate\_Correction\_Factor) \quad (4)$$

$$CH_4\_accumulation\_in\_Atm(t) = CH_4\_accumulation\_in\_Atm(t - dt)$$
$$+ (CH_4\_Emission\_Rate - CH_4\_Decay\_Rate) * dt\{NON - NEGATIVE\}$$
$$INITCH_4\_accumulation\_in\_Atm = CH_4\_Emission$$
$$INFLOWS:$$
$$CH_4\_Emission\_Rate = ((CH_4\_emission) - Recovered\_CH_4\_(R))$$
$$OUTFLOWS:$$
$$Ch4\_Decay\_Rate =$$
$$DELAY1((CH_4\_accumulation\_in\_Atm), Delay\_Time, Recovered\_CH_4\_(R))$$
$$\{UNIFLOW\} \quad (5)$$

$$Disposal = IF(((Untreated\_Waste + Incineration\_Capacity) >$$
$$(Composting\_Capacity)))THEN(MIN((Untreated\_Waste+$$
$$Incineration\_Capacity - Composting\_Capacity), Landfill\_Capacity))$$
$$ELSE(0) \quad (6)$$

The polarity of the connectors represents either the aggregative or the decretive interconnections with the parameters assigned for the MSW management of Khulna city. The composite index is calculated with Eq. (1) and the casual loop diagram (Fig. 2).

3.3 Development of Policy Measures for Implementation

Experimenting with the existing system would be impractical due to the economic factors as well as time consuming phenomena. Rather, in these regards, system's simulation is the most logical and efficient approach to consider. Moreover, while experimenting with different pathways, it will almost be impossible to turn all the way back and start from the very beginning without any devastating impact. Therefore, due to the MSW generation nature and constituents of Khulna city, industrial composting is considered to be the most efficient solution for achieving the ultimate sustainable development goal. The scenarios considered as policy measures are discussed below.

Policy 1: Increase collection capacity by increasing number of trucks with increased fund and evaluating the effect on uncleared MSW, untreated MSW, composite index, number of trucks, required number of trucks and public concern.

Policy 2: Increase project based allocation on composting capacity to 75% with a small scale increase of new collection vehicles to 33 nos. and evaluate the effect on uncleared MSW, untreated MSW, composite index, number of trucks, required number of trucks and public concern.

Policy 3: Increase the fund allocation and future funds for improving collection capacity by buying 32% new trucks with respect to the required number of trucks and improving the composting capacity to a mere 71% from the baseline scenario of 4.4% of the total MSW generated and evaluate the effect on uncleared MSW, untreated MSW, composite index, number of trucks, required number of trucks and public concern.

4 Model Validation and Optimization

Historical raw dataset is extracted from prospective published journals from Ahsan et al. [1, 4, 10, 11, 19, 20]. From the built model, during the extensive process run of data validation and optimization, the mean value of per capita MSW generation obtained for the year 2017 is 103.437 kg/year and the actual value was 106.59 kg/year with a MAEP value of 1.71%, indicating the validation precision of the built model. Similarly, the MSW generation for Khulna city observed for year 2019 is 160634395.4 kg and the model simulated results showed 156288502.3 kg where the MAPE value is 3.73, representing the robustness of the programmed MSW model for Khulna city. Moreover, the obtained simulated results represented a persistent behavior with the historical data from 2013 to 2020 with the error percentile of less than 5%.

5 Results and Discussion

The three different policy measures considered are simulated and then evaluated on the basis of composite index, public concern, uncleared waste and untreated waste and the lower composite index value within a period of thirty years will be considered as the best policy measures for MSW management of Khulna city. The CH_4 emitted from the simulation of baseline scenario is 6710.40 mg or 6.710 Gg for year 2020 which is quite identical with the result 25.06 Gg/year obtained from Matuail landfill site in Dhaka city [9]. Furthermore, with the landfill capacity it is estimated that the closing year for Rajbandh landfill in Khulna will be 2045. Because of that factor, an exponential rise in methane emission to 51.9 Gg in year 2050 from 34.8 Gg in year 2025 and a sudden drop in the disposal graph is visible while simulating the SD model.

5.1 Base Line Scenario

The baseline graph for MSW management of Khulna city is illustrated in Fig. 4. For the baseline scenario the composite index of 57 with a gradually increasing public concern of 14.7 until year 2050 is being observed. The untreated MSW piled from 84229.169 ton/year in 2020 to 153128.70 ton/year by 2050 and the number of collection trucks increased slightly to 83 nos. by 2050 from an initial 73 nos. in 2020. A merely small scale development of acquiring and repairing trucks with presently established policy and budget allocation of KCC is considered. The required number of collection trucks increased to 546 nos. in 2050 from an initial 84 nos. in 2020. Moreover, due to inexistence of visionary policy pattern of setting up a number of central collection systems in the city, the vehicle collection system has been considered.

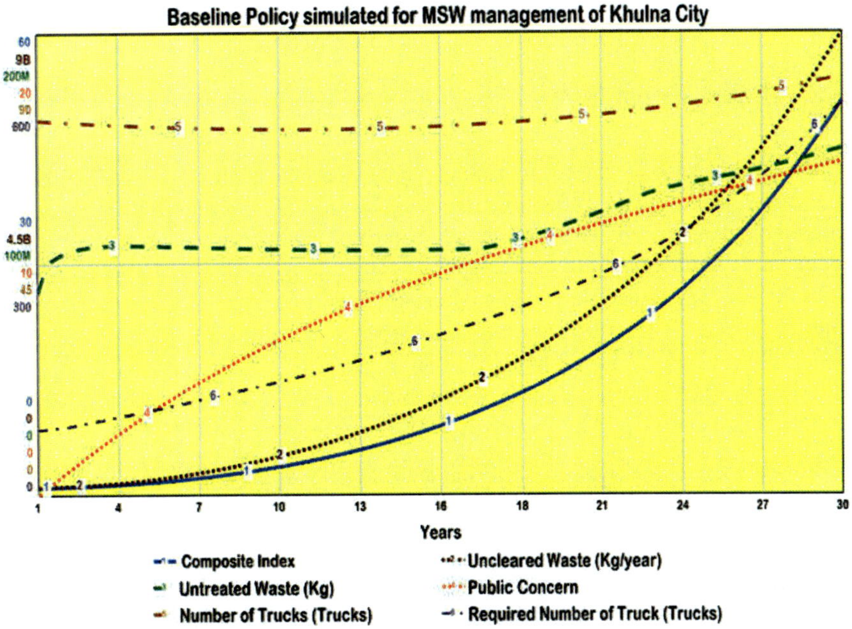

Fig. 4 Baseline scenario for MSW management of Khulna City

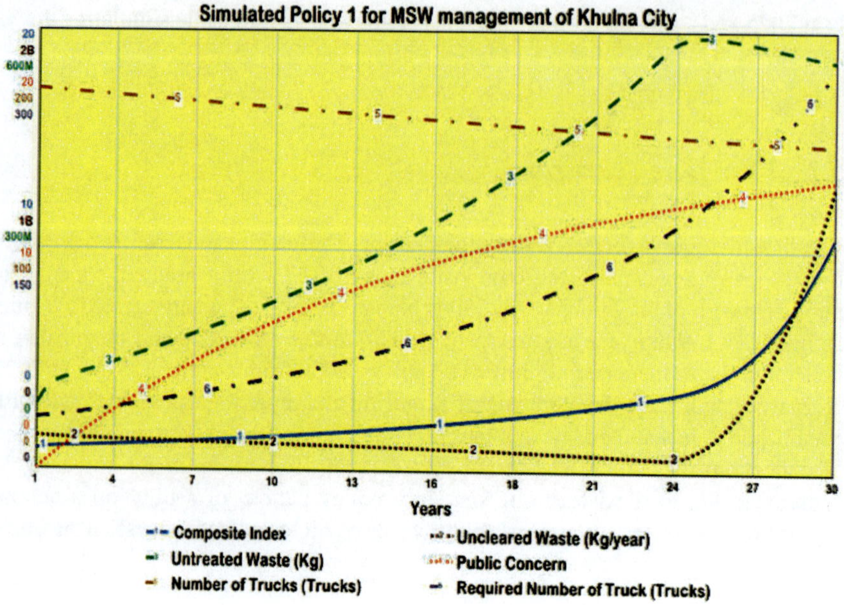

Fig. 5 Implemented Policy 1 for MSW management of Khulna city

5.2 Policy Scenario 1

The policy scenario 1 is based on existing policy measures taken by the KCC authority of initially increasing the collection system is shown in Fig. 5. Due to this purpose, the quantity of 100 nos. trucks are included for the collection system and the composite index value quite amazingly remained in sustainable condition to a highest of 4.34 until 2045 which is also the calculated closure year of the landfill site based on the landfill capacity. This is because the initial satisfaction of the inhabitants due to less or no piling of MSW in the SDS and community bins but index value jumped to 13.5 within next five years due to no improvement in MSW treatment facility. The untreated and uncleared MSW raised to 548995.5 ton/year and 1787775.24 ton/year respectively by 2050. The number of trucks remained at 147 nos. due to depreciation rate along with increased required number to 299 nos. by 2050 from an initial 173 nos. and 36 nos. in 2020 respectively.

5.3 Policy Scenario 2

An initial increase in MSW collection system in a pilot scale with added amount of 33 nos. collection vehicles and an improved industrial composting efficiency of 75% of the total accumulation of collected MSW is being imposed for evaluation of

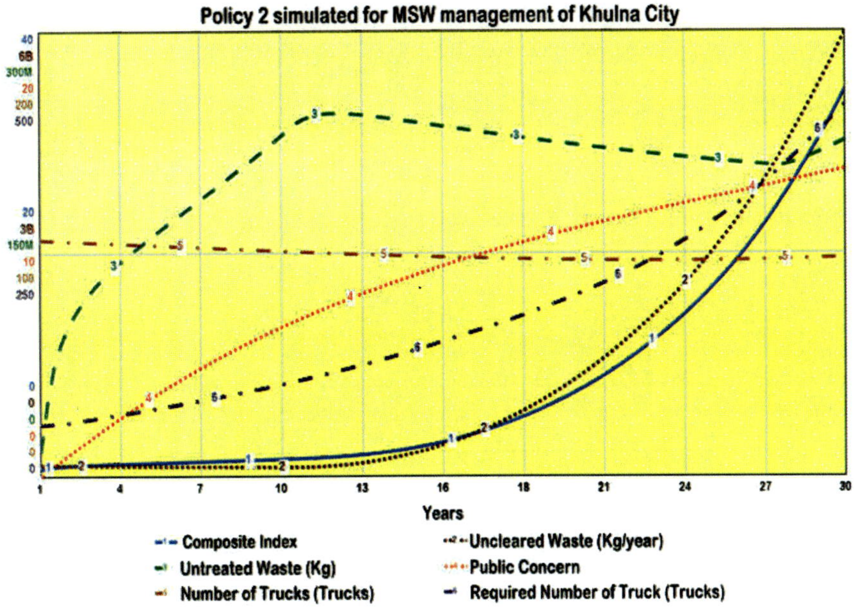

Fig. 6 Policy 2 implemented for MSW management of Khulna City

policy 2 is shown in Fig. 6. It indicates a composite index value of 40 which denotes a deteriorating environmental sustainability. Though the sustainability prevailed until 2041 with an index value of 10.1 due to the initial improvement in composting capacity but it mounted exponentially due to the upsurge of uncleared waste to 6823763.920 ton/year by 2050. The number of trucks during this period didn't show any significant improvement as it remained at 98 nos. with a required amount of truck reaching to the second highest quantity of 417 nos. during this duration of thirty years. The public concern remained at a significant value of 14.1 indicating the prerequisite of compulsory emphasis on the MSW treatment and collection process by introducing ground breaking as well as spatially adaptable technologies for further policy implementation in an efficient manner.

5.4 Policy Scenario 3

For policy 3, an industrial or commercial composting plant/ Bio-char reactor with a composting capacity of 71% of the total collected MSW has been considered with an initial fund allocation for buying 175 nos. trucks for MSW collection purpose. The SD model is then simulated with duration of 1000 second and the DT has been set to 0.25 for precision accuracy. The values obtained while simulating the SD model from 2020 to 2050 is shown in Fig. 7. The composite index value remained at stable

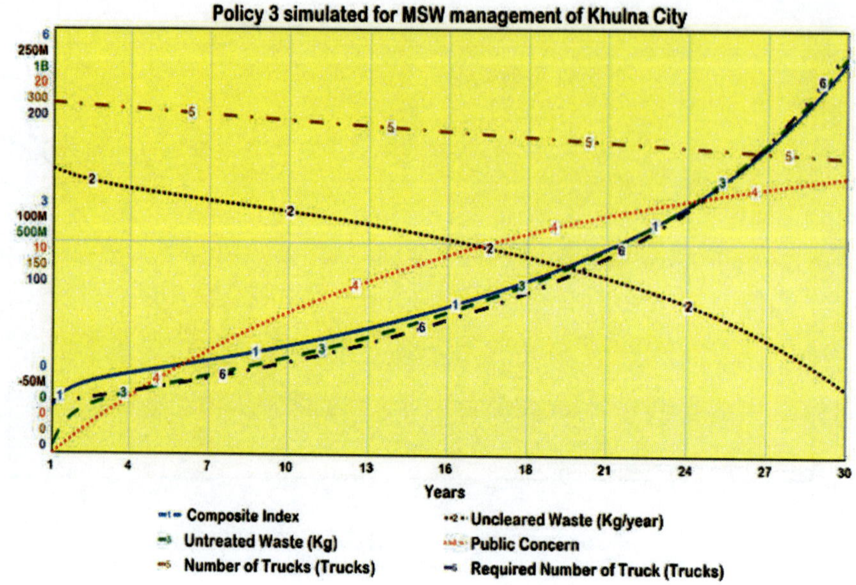

Fig. 7 Policy 3 implementation for MSW management of Khulna City

situation at 6.05 until year 2050. The increased composting capacity and collecting capacity interestingly reduced the uncleared MSW to 10607 ton/year by 2048 and to 0 by year 2050 but the untreated MSW stock rises to 1022563.14 Ton/year which indicates not only industrial composting but also a hybrid treatment solution involving effective full capacity recycling plants as well as energy recovery system through pilot incineration for hazardous hospital waste and landfill gas (LFG) burning is required in order to get rid of the pile of waste stocked throughout the initial three decades since the establishment of Khulna city corporation. The Number of trucks are observed to be 210 nos. for 2050 and also the required number of truck also reduced to 210 nos. by 2050 because of the improved composting capacity. The public concern has been represented as a logarithmic growth from initial value of 0 in 2020 to a prospective value of 13.3 by year 2050 due to the scarcity of implementation efforts with advanced technology to cope up with the continuous piling of MSW stocks throughout the course of these thirty years.

6 Industrial Composting Technologies

Achieving actual progress in the process of MSW management is a highly synchronized as well as knitted effort from all the sectors responsible. While taking in to consideration the 78.9% of food waste out of the total MSW generation of Khulna city, composting has been considered to be a first generation based advantageous method for the ultimate goal for environmental as well as economic geographical sustainability. There are basically three frequently used techniques for existing industrial composting present in today's world which are windrow, in-vessel and aerated static pile composting along with another exiting and till date non-industrial technique of vermin-composting. Windrows are mainly constructed with long narrow pile (approximately 5 feet) with the mechanism of temperature and oxidization requirement based controlled initiation process. Initiation of the compost can be labor-intensive and might require expensive machinery. The main constrains are the requirement of large area, odor problems and leachate concerns if exposed to rainfall. In-vessel systems is mainly an enclosed process in specifically manufactured containers, perforated barrels or drums etc. In-spite of the advantage of efficient control of environmental condition and low requirement of surface area, the requirement of high initial investment, time consuming nature and lower handling volume of MSW made this process quite an inefficient one in the context of present perspective. This process also requires to reach an in-vessel temperature of 140°F for at least two consecutive days in order to get rid of the harmful bacteria present in the final compost. The aerated static pile composting is relatively suitable in producing compost from homogenous mixture of organic waste and an advantage of function ability for generation of compost in larger quantity which is mainly conducted by government agencies and industries. The method has a disadvantage on the mechanism regarding pathogen reduction. Vermi-composting mainly follows the mechanism of feeding the waste to the worms inside a container, bin or greenhouse to produce high quality compost. Typically a pound of worm can consume four times more waste to its respective weight. The initial investment is high for this type of composting technique with an added disadvantage of occurring anaerobic condition with excessive amount of waste. Though the application of this process is slowly making its appearance in the agricultural sector on a pilot basis, a visionary based improvisation of the process will illuminate the potential of industrial scale implication in near future [3].

7 Conclusion

This study aimed to assess efficient MSW management policy decisions for the case of Khulna city in Bangladesh. The results obtained through the extensive research and experiments brought us to the concluding remarks that, system dynamics is definitely an impact assessment tool for the industrial corporation as it offers a wide spectrum of results in varied condition and is able to produce most effective decisions through result interpretations. However, it has to be kept in mind that, system dynamics models are knitted together with numerous influencing factors that reflects concrete reality, subjectivity and specificity which can be complex if applied to different contexts. Due to this phenomena, the construction process of the model has to be concentrated extensively on detail parameters for sustainable development in urban planning and policy implication for reaching the ultimate goal of a green economic future.

References

1. Ahsan, A., Alamgir, M., Imteaz, M., Shams, S., Rowshon, M.K., Aziz, M.G., Idrus, S.: Municipal solid waste generation, composition and management: Issues and challenges a case study. Environ. Prot. Eng. **41**, 43–59 (2015). https://doi.org/10.5277/epel50304
2. Bala, B.K., Arshad, F.M., Noh, K.M.: System Dynamics: Modelling and Simulation. Springer (2017). https://doi.org/10.1007/978-3-8349-8074-8_6
3. Barrena, R., Font, X., Gabarrell, X., Sánchez, A.: Home composting versus industrial composting: influence of composting system on compost quality with focus on compost stability. Waste Manag. **34**, 1109–1116 (2014). https://doi.org/10.1016/j.wasman.2014.02.008
4. BBS, SID, MoF.: Bangladesh Statistics 2016. Bangladesh Bur. Stat. Stat. Inf. Div. Minist. Plan. **73** (2017)
5. Chen, M., Giannis, A., Wang, J.-Y.: Application of system dynamics model for municipal solid waste generation and landfill capacity evaluation in Singapore, p. 1 (2012)
6. Dyson, B., Chang, N.B.: Forecasting municipal solid waste generation in a fast-growing urban region with system dynamics modeling. Waste Manag. **25**, 669–679 (2005). https://doi.org/10.1016/j.wasman.2004.10.005
7. Federica, M., Nola, D., Escapa, M., Ansah, J.P.: Modelling solid waste management solutions: the case of Campania Italy. Waste Manag. **78**, 717–729 (2018). https://doi.org/10.1016/j.wasman.2018.06.006
8. Godoi, F.C., Prakash, S., Bhandari, B.R.: Investment plan for the Bangladesh delta plan 2100, People's Republic of Bangladesh multisector approaches to delta management (2017)
9. Hoque, M.M., Rahman, M.T.U.: Estimation of greenhouse gas emissions from matuail landfill site. In: Urban Mining and Sustainable Waste Management, pp. 149–159 (2020). https://doi.org/10.1007/978-981-15-0532-4_22
10. Islam, M.S., Moniruzzaman, S.M.: Simulation of sustainable solid waste management system in Khulna city. Sustain. Environ. Res. **1**, 1–8 (2019). https://doi.org/10.1186/s42834-019-0013-8
11. KCC.: KCC Budget Book, 2019–2020 (2019)
12. Kolekar, K.A., Hazra, T., Chakrabarty, S.N.: A review on prediction of municipal solid waste generation models. Procedia Environ. Sci. **35**, 238–244 (2016). https://doi.org/10.1016/j.proenv.2016.07.087
13. Manasakunkit, C., Chinda, T.: Development of a municipal solid waste dynamic model in Bangkok, Thailand. Songklanakarin J. Sci. Technol. **39**, 685–695 (2017). https://doi.org/10.14456/sjst-psu.2017.84
14. Mondal, K.K., Akhter, A.E., Mallik, M.A.K., Hassan, S.M.Q.: Study on rainfall and temperature trend of Khulna division in Bangladesh study on rainfall and temperature trend of Khulna Division in Bangladesh. Dew-Drop (2017)
15. Mutingi, M., Mbohwa, C., Kommula, V.P.: System dynamics approaches to energy policy modelling and simulation. Energy Procedia **141**, 532–539 (2017). https://doi.org/10.1016/j.egypro.2017.11.071
16. Plai, R.R., Roodrigues, L.L.R., Mathew, A.O., Hebbar, S.: Impact of urbanization on municipal solid waste management : a system dynamics approach. Int. J. Renew. Energy Environ. Eng. **02**, 31–37 (2014)
17. Rafew, S.M., Rafizul, I.M.: Application of system dynamics model for municipal solid waste management in Khulna city of Bangladesh. Waste Manag. **129**, 1–19 (2021). https://doi.org/10.1016/j.wasman.2021.04.059
18. Rahayu, N., Arai, T., Yudoko, G., Morimoto, H.: System dynamics models for planning long-term integrated municipal solid waste management in Bandung city. WIT Trans. Ecol. Environ. **179**, 1153–1168 (2017). https://doi.org/10.2495/SC130982
19. Roy, S., Sowgat, T., Uddin, M., Islam, A.S.M.T., Anjum, N., Mondal, J., Rahman, M.M.: Bangladesh: National Urban Policies and City Profiles for Dhaka and Khulna, p. 156 (2018)
20. Shams, S., Sahu, J.N., Rahman, S.M.S., Ahsan, A.: Sustainable waste management policy in Bangladesh for reduction of greenhouse gases. Sustain. Cities Soc. **33**, 18–26 (2017). https://doi.org/10.1016/j.scs.2017.05.008

21. Talyan, V., Dahiya, R.P., Anand, S., Sreekrishnan, T.R.: Quantification of methane emission from municipal solid waste disposal in Delhi. Resour. Conserv. Recycl. **50**, 240–259 (2007). https://doi.org/10.1016/j.resconrec.2006.06.002
22. United Nations Environment Programme.: Food Waste Index Report 2021, Unep. (2021)

Prediction of Soil Nutrients Using Hyperspectral Satellite Imaging

Sumsuttibriz Riad, Md. Sojib Ahmed, Mahammudul Hassan Himel, Md. Rayhan Ahmed, Md. Mynul hasan, Ahsan Habib Mim, Akib Zaman, Salekul Islam, and Md. Saddam Hossain Mukta

Abstract Agricultural sector plays an important role in the overall development of Bangladesh. Bangladesh is a densely populated country where the production of high-yielding crops is necessary. In Bangladesh, approximately 50% of the population are primarily employed in agriculture and more than 70% of the land area is used for cultivation. However, due to the traditional agricultural system, the country still lags behind to the best use of its resources. In our research, we investigate the presence of major soil nutrients using the hyperspectral images from Landsat-8 satellite of the northern regions of Bangladesh. In these regions, people are still using the ancient agricultural system despite being mostly dependent on agriculture. They are less likely to understand the soil quality, which consequently impacts on the production of crops. In this research, we capture 1.5K+ hyperspectral satellite images

S. Riad · Md. S. Ahmed · M. H. Himel · Md. R. Ahmed · Md. M. hasan · A. H. Mim · A. Zaman · S. Islam · Md. S. H. Mukta (✉)
United International University, United City, Madani Ave, Dhaka 1212, Bangladesh
e-mail: saddam@cse.uiu.ac.bd

S. Riad
e-mail: sriad172076@bscse.uiu.ac.bd

Md. S. Ahmed
e-mail: mahmed171067@bscse.uiu.ac.bd

M. H. Himel
e-mail: mhimel171066@bscse.uiu.ac.bd

Md. R. Ahmed
e-mail: mahmed171063@bscse.uiu.ac.bd

Md. M. hasan
e-mail: mhasan171279@bscse.uiu.ac.bd

A. H. Mim
e-mail: amim171159@bscse.uiu.ac.bd

A. Zaman
e-mail: akib@cse.uiu.ac.bd

S. Islam
e-mail: salekul@cse.uiu.ac.bd

S. Hossain et al. (eds.), *Proceedings of International Conference on Fourth Industrial Revolution and Beyond 2021*, Lecture Notes in Networks and Systems 437,
https://doi.org/10.1007/978-981-19-2445-3_12

using *GloVis/Earth Explorer* from the northern regions of Bangladesh. Then, we apply the latest computer vision technologies, i.e., the *ArcMap* software to identify the major spatial properties. Later, we collect the actual soil ingredients of those locations from agriculture extension offices and government websites. Finally, we build a machine learning-based hybrid classification model where different features of satellite imagery are independent features and soil nutrients are dependent features. We also build a web tool for the agriculturists and government service providers to understand the soil quality well for better crop yielding.

Keywords Hyperspectral imaging · Digital number (dn) · Radiance · Reflectance · Image processing · Landsat-8 satellite · Arcmap software

1 Introduction

Bangladesh is a country where agriculture plays an essential role to gain the positive momentum of the sustainable development. Majority of the people of Bangladesh directly or indirectly depend on agricultural sub-sectors for their livelihood. Agriculture holds the lion's share in the production sectors, which comprises of 14% of the GDP and 43% of the labor force. Many factors including the types of soil, the amount of soil nutrients, the height of the cultivating land, the season, deluge, and aridity impact the fertility of the lands. Among these factors, soil nutrients play the pivotal role for the growth of crops through mineral nutrition and toxicity [10, 19]. In this paper, we analyze a significant amount of satellite imagery of the northern areas of Bangladesh and predict the soil nutrients from those imagery by calculating the reflectance from the hyper-spectrum of bands.

Hyperspectral Satellite Imaging (HSI) has numerous applications including meteorology, agriculture, fishing, oceanography, biodiversity conservation, forestry, landscape, geology, etc. In case of improvement of the agriculture, several works have been conducted from HSI. For example, rice model through the optimization of Normalized Difference Vegetation Index (NDVI) threshold [5, 27, 31], detection of agricultural land [2] and determination of land suitability [17], forecasting of crop production [18], etc. Successful results from these studies provide a scope of extending the use of HSI in unexplored sectors of agriculture. Sabbir et al. [28] show a theoretical hypothesis to the possibility of determining the amount of specific soil nutrients by using hyperspectral reflectance data. However, to the best of our knowledge, a comprehensive study to predict the soil nutrients using HSI is yet to be carried out in the existing literature.

Government and private agricultural organizations in Bangladesh do not have the capability to provide easily accessible agricultural information such as suitability of land, status of the soil nutrients, prediction of crop production, etc., to the farmers. Most importantly, farmers from underprivileged areas of the North-Bengal regions of Bangladesh such as Dinajpur, Panchogor, Nilfamari, etc., are the most affected people who are deprived of the necessary information regarding the soil nutrients.

In a similar note, the legacy of the agricultural system in Bangladesh does not keep record on variation of the soil nutrients and minerals on yearly basis. Thus, in the long run, the agricultural lands transform into barren fields though we use excess amount of fertilizers, bio-pesticides, and rinse. Unavailability of these information leads us toward the unreasonable usage of chemicals, which is injurious to the crop production. Moreover, incorrect land decision-making degrades the crop production and increases the farmers' production-related expenses. However, application of HSI data to create a land suitability evaluation system based on the prediction of soil nutrients can facilitate the information flow to the farmers especially in the underprivileged areas of Bangladesh.

To predict the soil nutrients from the satellite imagery, first we collect 1,560 satellite images by using *Glovis/Earth Explorer* for the Rangpur division of Bangladesh. We mostly collect images from the Landsat-8 satellite with a multi-spectral imagery at 30-meter resolution. From these multi-spectral images, we calculate the reflectance of the land, which is the key property to understand the soil nutrient in our study. Different bands of a multi-spectral image hold different characteristics. For example, the near-infrared of Bands 2, 3, and 4 reflect the light wave back to the sky. Similarly, from the Bands 6 and 7, we understand the quality of the wet to dry condition of the earth. Understanding the properties of these Bands and the previous study of Saleh et al. [28] motivate us to explore the relationship between the soil nutrients with the variation of range of Bands in a multi-spectral image. Therefore, we collect information regarding the soil nutrients from 8 districts of Rangpur division from the agriculture extension offices and extract the corresponding satellite images from the Glovis/Earth Explorer. Then, we find association between the numeric values of Bands 2 to 7 and the amount of soil nutrients. Later, we discretize the values of soil nutrients into three different class labels: *high, medium, and low*. Next, we develop a multi-class multi-label hybrid classifier to predict the soil nutrients from the Bands of the multi-spectral satellite imagery. Our model shows an outstanding performance with an average F1-score of 84%. To increase the usability of our model, we also develop a user friendly, interactive web tool to understand the soil nutrients from the satellite imagery.

Thus, the objectives of this research is to create a land suitability evaluation system that predicts the presence of different soil nutrient by using collected data from the satellite remote sensing. The findings of the research will facilitate the agricultural experts and enthusiasts to know the suitability of the land for a specific crop, which will enable higher production. We can summarize our contribution as follows:

- Build a satellite imagery data set for the investigating of the presence of soil nutrients in the context of Bangladesh.
- Develop a hybrid machine learning model to predict the soil nutrients from the collected imagery.
- Develop a web tool to understand the soil nutrients for the northern regions of Bangladesh.

2 Literature Review

Satellite imaging has been used in several studies [14, 33] to develop fire detection and monitoring applications. Forest fire detection is one of the popular applications of satellite imaging. The NOAA satellite is the widely used satellite sensors in fire detection. It provides daily data senses for each operational platform at approximately 1.1 km spatial resolution. Yoshiaki et al. [33] classify the snow area by using the NOAA AVHRR image dataset, which is based on Mongolia. Similarly in recent days, Mohammed et al. [20] used hyperspectral imagery from the Sentinel-2A satellite to assess and understand wildfires natural resources before they occur by extracting spectral highlights from satellite photos and analyzing the land-cover map of the targeted arer located in Athens, Greece. There are few more recent studies regrading the use of HSI in weather prediction and route loss distribution. Ahmedian et al. [1] described a novel method for predicting path loss distributions directly from 2D satellite images using deep convolutional neural networks which can correctly predicted route loss distribution for different communication frequencies and transmitter heights. Scheck et al. [29] used visible Meteosat SEVIRI pictures which are assimilated in a convective-scale data assimilation system based on a local ensemble transform Kalman filter (LETKF) to predict the weather.

We find several studies [5, 17, 31] that are related to agriculture and crop yielding from satellite imagery. Mostafiz et al. [5] propose a land suitability assessment system by using satellite remote sensing-derived soil-vegetation indicators. For suitability assessment, authors conduct a multi-criteria decision by integrating weighted linear combinations and fuzzy multi-criteria analyses in a GIS platform following eight criteria: elevation, slope, and LST vegetation indices (SAVI, ARVI, SARVI, MSAVI, and OSAVI). Turker et al. [31] describe a technique for detecting agricultural fields from satellite images based on field blocks. An agricultural field block is made up of one or more fields that the farmers own. The work proposes a method for detecting agricultural fields in satellite images by combining Otsu's thresholding algorithm with an MCW-based transform. Tello et al. [17] develops a method for recognizing crop patterns from the temporal sequence of satellite images using ANNs and propose a learning approach based on the neural networks, with training the pixels of the cultivated plots using satellite images over time as examples. Priyanka et al. [27] build a rule-based system for forecasting the crop yield production based on a pool of collected past data.

Mahboob et al. [18] proposed a model for estimating Boro rice regions in Bangladesh's Haor region by using the Normalized Difference Vegetation Index (NDVI). Alam et al. [2] implement an algorithm to determine the NDVI threshold values. Four regression models (Linear, Ridge, Lasso, and Decision Tree) are carried out to estimate the total Boro production of each district of Bangladesh. Among the regression models, maximum R^2 (co-efficient of determination) values of 0.492, 0.790, 0.899, 0.891, 0.848, 0.942, 0.777, and 0.848 are acquired for Barisal, Chittagong, Dhaka, Khulna, Mymensingh, Rajshahi, Rangpur, and Sylhet divisions, respectively.

There have been numerous successful applications of the satellite imaging such as forest fire monitoring system [11], snow area detection [33], dust index determination [14], etc., along with the continuous development of the quality of the hyperspectral image dataset [12, 13, 15, 23, 24]. These studies show the credibility of the satellite imaging dataset for creating ground based applications. Moreover, several successful recent studies [1, 20, 29] in the precise prediction of weather and route loss distribution present a viable propspect of HSI in groud based applications. Secondly, several works related to the agricultural applications such as land suitability prediction [5], crop production forecasting [31], crop pattern recognition [17], detection of agricultural field [27], etc., have been conducted successfully by applying machine learning classification and regression-based models by using the satellite imaging dataset. These studies provide the scope for extending the success in newer agricultural applications. Thirdly, though some of the works [2, 18] have been conducted on agricultural sector of Bangladesh using the satellite imaging, they are very limited in numbers and mostly neglecting the underprivileged areas of North-Bengal region of Bangladesh. For addressing the mentioned research gaps, the main objective of this study is to develop a land suitability model through soil nutrient prediction for the underprivileged Northern regions of Bangladesh. An overview of the research methodology is illustrated in Fig. 1 and described in the following sections.

Fig. 1 Block diagram of our research methodology

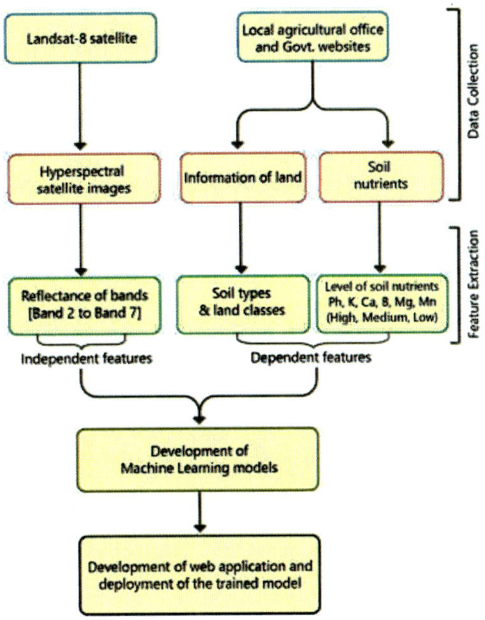

3 Data Collection

Satellite images are images of earth gathered by imaging satellites managed by governments and businesses around the world. There are numerous applications of satellite images and they are meteorology, agriculture, fishing, oceanography, biodiversity conservation, forestry, landscape, geology, etc. Satellite images are available from various sources. *Earth Explorer*,[1] *GloVis*,[2] and the Land Look Viewer all include data products built from Landsat-8 OLI (Operational Land Imager) or *TIRS* (Thermal Infrared Sensor) scenes that can be downloaded. *GloVis* provides images captured by numerous satellites for free to use for educational purposes. In this paper, we have collected imagery from Landsat-8 satellite for *Rangpur* division (Bodargonj, Fulbari, Gongachora, Kaunia, Kishorganj, Mithapukur, Nilfamari, Pathgram, Pirgacha, Pirgonj, Polashbari, Rajibpur, Rangpur Sadar, Soyodpur, and Taragonj). Table 1 presents the overall statistics of our dataset.

Landsat-8 contains two sensors. The first sensor collects 6 bands multi-spectral imagery at 30-meter resolution, as well as panchromatic imagery at 15 meter resolution. The second sensor collects thermal imagery at 100 meter resolution. In this paper, we use images collected from the first sensor in the time span of 2020 to 2021. In this study, most of the satellite data are downloaded from USGS (United States Geological Survey). These images are received (less than 10% cloud cover) in growing stage of the particular crop cycle. To download from Landsat-8 satellite, we fix the location from the earth explorer. After selecting date range and cloud coverage between 1 to 10, we download image dataset from Landsat-8 OLI/TIRS C1 level-1.

Randomly, we have taken values of soil nutrients in 1,957 points/places of Rangpur division as the ground truth data. We find out the reflectance index of the Rangpur division using the satellite imagery and make a correlation between the reflectance index and the number of soil nutrients. We work on the 6 different bands of satellite imagery. The *Landsat-8* satellite provides both multi-spectral and panchromatic images. We collect the relevant data from the 'DIVA GIS' website using the ArcMap software. For example, we can download 'BangladeshMap.shp' by selecting 'Bangladesh' as a nation and 'Administrative areas' as the subject. In the ArcGIS software, we create new layers and drag and drop each map separately. From the ArcToolbox shelf, we select and click on the 'Spatial Analyst Tools → Extraction → Extract by mask'.

The map will be sliced and saved to the 'Output raster' location after selecting the 'Input raster' and 'Input raster and feature mask data' from the 'Extract by mask' table. Figure 2 presents the image pre-processing steps where we show how hyperspectral image of a large area is pre-processed for a comparatively smaller target area for further analysis.

---

[1] https://earthexplorer.usgs.gov/.

[2] https://glovis.usgs.gov/.

Table 1 Statistics of our dataset

Category	Size
Number of districts	8
Number of upazilas (sub-unit of a district)	15
Number of collected images	1,560
Types of soil nutrients	6
Range of potassium (mg)	0.007–2.1
Range of phosphorus (mg)	0.000274–0.373
Range of boron (mg)	0.00016–0.0039
Range of calcium (mg)	0.1–14.6
Range of magnesium (mg)	0.007–3.91
Range of manganese (mg)	0.000121–0.268
Reflectance Range	0.03–0.40
Digital Number Range	4,988–12,366
Time span of data collection	January, 2020–July, 2021

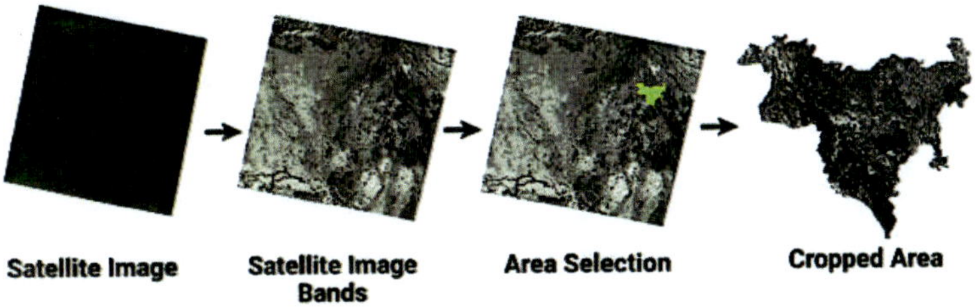

Fig. 2 Image pre-processing steps

4 Feature Engineering

In this section, we describe the process of generating different features and normalizing those features for the sake of uniformity and simplicity.

4.1 Reflectance Calculation

Reflectance is a kind of fraction of electromagnetic power which is reflected from a material or surface with their own reflectance spectra. In this paper, we calculate the reflectance to build up a relation between the index of reflectance and the imagery of the soil nutrients. To calculate reflectance, we first calculate the radiance. Finding the radiance values along with the digital numbers (DN) is mandatory. Here we use the Landsat-8 satellite to collect images, as the Landsat-8 OLI sensor is more sensitive, so these data are re-scaled into 16-bit DNs with a range from 0 and 65536. The formula to calculate radiance is as follows:

$$L_\lambda = ML * QCAL + AL$$

where, Ł$_\lambda$ is the Spectral Radiance, ML is the Radiance multiplicative scaling factor for the band, $QCAL$ is the Digital number and AL is the Radiance additive scaling factor for the band. Calculated radiance is then converted to the reflectance using the following formula:

$$\rho_\lambda = (\pi * L_\lambda * d^2)/(ESUN_\lambda * cos\theta_s)$$

where, ρ_λ = Unit-less planetary reflectance, L_λ = Spectral radiance, d = Earth–Sun distance in astronomical units, $ESUN_\lambda$ = Mean solar exoatmospheric irradiances, and θs = Solar zenith angle

Normalized reflectance: After calculating the reflectance and radiance, we find that there is quite a difference between the reflectance and the radiance. To convert all the values around 0 to 1, we have normalized the result by using the *min-max normalization* technique.

4.2 Band Extraction and Selection

Landsat-8 provides 11 bands of images for each area. From them, Band 1 is indigo and is utilized for the coastal applications, Band 8 is panchromatic, while Bands 10 and 11 are both thermal for technical reasons. Band 9 is similar to Band 8. Therefore, we cannot calculate the radiance and reflectance from those five bands. In this dataset, we are using the remaining 6 bands. Bands 2, 3, and 4 are visible as blue, green, and red, respectively. Near infrared (NIR) part of the spectrum is most commonly used in vegetation analysis because the water parts of the leaves of the healthy plants reflect this part and thus scatter the wavelengths back into the sky. Bands 6 and 7 cover various shortwave infrared (SWIR) slices. They are especially useful for distinguishing between the wet and dry areas of the earth and for geology.

4.3 Ground Data Extraction

Soil usually contains different types of materials. Among these materials, some are beneficial and some are harmful for the soil. In this work, we have worked with the main 6 nutrients of soil such as phosphorus, potassium, calcium, magnesium, boron, and manganese. These materials are important for increasing soil fertility and high yielding of crops. On the other hand, Rangpur[3] is Bangladesh's northernmost division, with a citizenry of 15,665,000 according to the census Of 2011. Rangpur division is made up of eight districts. These eight districts are divided into 58 upazilas. Among them, we work on 15 upazilas based on their population and cultivated land. In Rangpur division, there exist three types of soils: sandy, loam, and alluvium. In our study, we have found only loam soil most of our working places. Since all the soil types are same, we have deleted the feature, 'soil type' from the dataset.

5 Development of the ML Model

In this section, we describe the process of building the classification model. First, we find relevant features and we discretize those features into three class labels. Then, we apply different machine learning models to predict these class labels. Our hybrid model shows significantly better performance than the conventional and independent models.

5.1 Relevant Features

A raw dataset of 2083×15 samples was used to develop the prediction model of the soil nutrients. The reflectance of *six bands* (band 2 to 7), *soiltype* and *landclass* are the independent features that are used as an input to the classification models (see Table 2) and the soil nutrients are the target/dependent features to be predicted. Firstly, the numerical features of the dataset are normalized using the minimum and maximum values of the collected data for each feature. Secondly, we remove the outliers from the raw dataset by using the interquartile range (IQR) [32] method which leaves 1375×15 samples. Thirdly, we label the values of the soil nutrients into three categories: *high, medium*, and *low*. We create the range for each of the soil nutrients from Eqs. 1 and 2.

$$L_0 = Min + (Median - Mean)/2 \tag{1}$$

$$L_1 = Median + (Max - Min)/2 \tag{2}$$

where the *Low* class label ranges from Min to L_0, the *Medium* class label ranges from L_0 to L_1, and the *High* class label ranges from L_1 to Max.

[3] https://en.wikipedia.org/wiki/Rangpur/Division.

Table 2 Explored independent features

Name of the feature	Data type	Data range
Band2 to band7	Numerical	0.03–0.40
Soiltype	Categorical	Loam, ethelLoam, ethel
Landclass	Categorical	high land (HL), low land (LL), medium elevation land (MEL), medium low land (MLL), deep medium high land (DMHL), shallow medium high land (SMHL)

5.2 Performance of the Model

We evaluate the best performance for the soil nutrient prediction model by using 6 classification Models. We find that most of the cases low class label shows highly class imbalance problem (ranges from 6 to 9%). To solve the class imbalance problem, we use Synthetic Minority Over-sampling Technique (SMOTE) [9] over our imbalanced dataset and convert the dataset balanced. Then, we initially apply two conventional models (i.e., Gaussian Naïve Bayes [26], and K Nearest Neighbors [8]), Conventional ensembling model (Random Forest [6]) and two boosting models (Adaptive Boosting (ADB) [7] and Light Gradient Boosting (LGB) [3]) to predict the soil nutrients. We use 03 neighbors in case of KNN along with other default parameters. The parameters of developed RF, AdaBoost, and LGB model is selected through hyperparameter tuning using the Bayesian optimization technique [30] which provides the advantage of a better overall performance and non-requirement of certain values of parameters like Random Search [4] and Grid search [16]. The developed ADB model has a learning rate of 0.7 with 250 estimators. In case of RF, we use number of trees as 200 which helps to avoid over-confidence with a minimum leaf size of 75. For LGB model, we use the dart gradient boosting (Lgbm dart) as the boosting methods to avoid over specialization problem of gradient boosted decision tree (Lgbm gbdt). Additionally, the learning rate is taken 0.65 from the hyperparameter tuning along with 100 estimators, Number of leaves are taken 25 with minimum 05 data in each leaf and the bagging fraction is set 0.6, i.e., 60% of rows are used per tree building iteration. Additionally, the objective was set as multi-class with balanced weight for each class.Furthermore, we build a hybrid model (see Fig. 3) using the stacking of the developed KNN, Random Forest and LGB using a balanced weight.

We split the dataset into 70/30 ratio and train the model using a tenfold with a cross validation of 10 iterations. Being a multi-classification problem, the performance of the models are compared using the micro average of the F1-score and the area under the receiver-operating curve (AUC-ROC). The models are used to predict the labels of each nutrient separately. Table 3 shows the performance of the models measured using the micro average of F1-score. Gaussian NB has the weakest performance with a score ranging from 0.34 to 0.49. In contrast, KNN and RF have performed moderately well with a minimum score of 0.69 to 0.81 and 0.70 to 0.80, respectively.

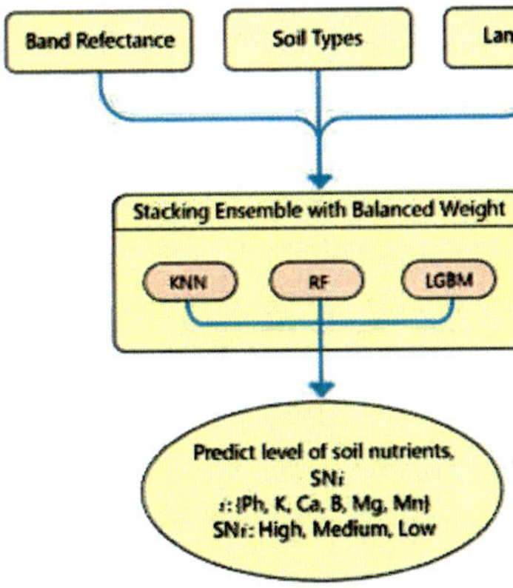

Fig. 3 Architecture of our developed hybrid model

Table 3 Micro-average of F1-score for different prediction models

Models	Phosphorus	Potassium	Boron	Calcium	Magnesium	Manganese
GNB	42	41	39	49	37	34
KNN	81	77	69	79	77	75
RF	79	76	70	80	78	76
AdaBoost	71	73	63	70	68	63
LGBM	82	79	75	82	80	78
Hybrid	88	85	80	87	82	83

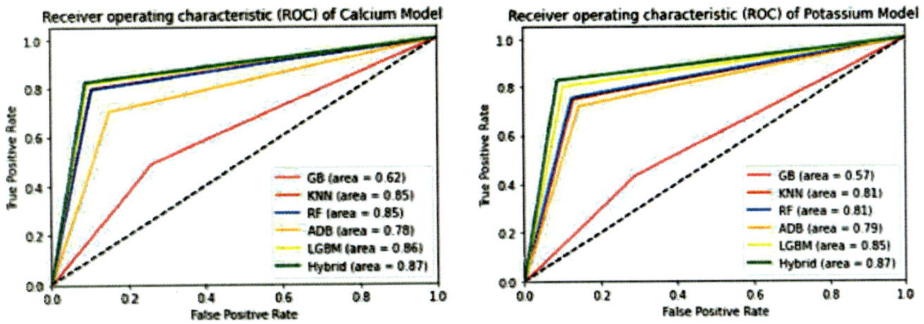

Fig. 4 ROC-AUC curves for predicting (**a**) calcium and (**b**) potassium

LGB has outperformed them with a score of 0.75 to 0.82. However, our developed hybrid model has outperformed others for all the nutrients with an outstanding score of 0.80 to 0.88.

On a similar note, Fig. 5 illustrates the performance of the models in terms of AUC-ROC for each of the soil nutrients. The hybrid model proves to be superior in terms of AUC-ROC for all types of soil nutrients. The AUC-ROC scores of the hybrid

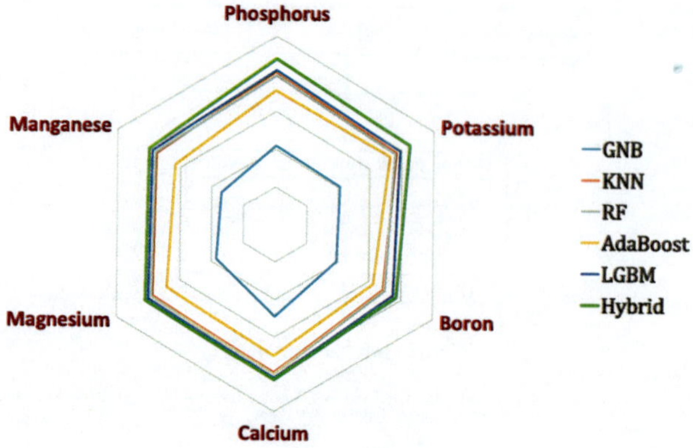

Fig. 5 Performance of the models in terms of AUC-ROC

model for phosphorus, potassium, boron, calcium, magnesium and manganese are 0.88, 0.85, 0.80, 0.87, 0.82 and 0.83, respectively, which outperform the AUC-ROC scores of the other models. The lowest score of 0.83 for boron depicts that in 83% of the cases, the model is predicting the level of boron in the soil correctly, whereas the highest score of 0.88 for potassium indicates 88% accurate prediction while predicting the level of potassium from the soil. For the sake of space, we only show the performance of different classifiers for predicting calcium and potassium (see Fig. 4). We make our codes[4] available for the researchers for further improvement.

5.3 *Web-Based Tool Implementation*

As a proof of concept, we implement a web-based tool[5] to inspect the interactive soil nutrients from the multi-spectral bands of satellite images of different regions of the northern parts of Bangladesh. The main goal of the tool is that the agriculturists, enthusiasts, and government service providers, in this field, can get an overall idea about the soil nutrients.

Figure 6 shows the interactive page of our web-based tool. We insert different values for the Bands of 2–7, soil type and label, then our hybrid model shows the prediction results with different soil nutrients.

[4] https://github.com/Sriad076/Soil_Nutrients_Prediction.

[5] https://soilnutrientprediction.sociocognitive.com/.

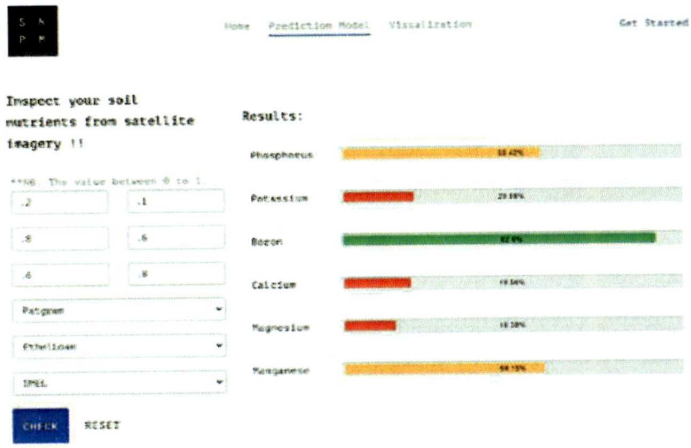

Fig. 6 Prediction of the soil nutrients from the reflectance of different bands

6 Discussion

Surveying cultivable lands physically by the agriculture department regularly is a costly and time intensive process. Since satellite imagery is available from different providers, government and policy makers can take this opportunity to analyze these imagery and receive an overall idea about the soil nutrients of those agricultural lands. These imagery give static information about this land. If we could capture real-time images, we could easily find the actual soil nutrient information.

From the literature, we find that a few studies [28] show the correlation between the soil nutrients and the satellite imagery. In a developing country where farmers have to spend exceedingly large amount to buy seeds, fertilizers, and pesticides, they are not likely to analyze the land condition to minimize the expense. However, if the farmers could analyze land information easily, they could take better decision by yielding higher production of crops.

In this study, we face several challenges such as the scarcity of dataset and relevant studies of satellite imagery in the region of Bangladesh. On the other hand, we have faced serious obstacle to collect images. The images that we have collected from the Landset-8 satellite covered a large area. The image of a large area also includes some areas which are not under analysis, and thus may create noise in our image dataset. Therefore, we divided the images into small area through the Arcmap website. Since we are only working with upazillas, we extract ground data from various upazillas in Bangladesh from local and head office of agriculture extensions due to the ground data crisis.

Our final classification model shows an outstanding performance. When we use two conventional models, i.e., Gaussian Naive Bayes and KNN classifiers, our models show weaker F1-score. We find that for all of the six soil nutrients at least one class label makes our dataset highly imbalance. Majority of the cases that we observe are significantly less in number when compared to two other class labels in the class distribution. Though, we use oversampling technique, a lazy classifier like KNN cannot improve the performance of a classifier much. Later, we find that our applied boosting algorithms perform better than the previous models, because they provide more importance on the instances which likely show wrong output. Finally, we can see that our hybrid classifier demonstrates the best performance because we integrate all other models and each model works when it has better strength on a set of instances.

Finally, the major problem of our current version of the web tool is that it can be benefited to the agriculturists and enthusiasts only. To make our system more usable for the root level users, i.e., the farmers, we plan to develop a recommender system which can take user's desired land area's latitude and longitude as input, and can generate suggestions of farmer's potential crops to be cultivated in the upcoming season based on the presence of the nutrients of that input area. We can also consider weather-aware [25] applications where changing of climate can be an important factor based on temporality [21, 22].

7 Conclusion

To the best of our knowledge, this is the first study that has conducted a satellite imagery-based soil nutrient prediction in the context of Bangladesh. We have also built a gold standard dataset by physically visiting and collecting agriculture extension offices for the actual soil nutrients' information. Then, we have found association between the reflectance of these imagery and the amount of soil nutrients. Later, we have developed a multi-label multi-class classification model to predict the soil nutrients in a single run of the model. We have observed that our model has shown outstanding performance. Soil nutrient measurement through soil testing is still a costly procedure for the farmers. Moreover, in remote places, soil testing facility is not readily available. The main focus of this paper is to deploy an evaluation system for forecasting the soil nutrients based on the training dataset.

References

1. Ahmadien, O., Ates, H.F., Baykas, T., Gunturk, B.K.: Predicting path loss distribution of an area from satellite images using deep learning. IEEE Access **8**, 64982–64991 (2020)
2. Alam, M.S., Kalpoma, K., Karim, M.S., Al Sefat, A., Kudoh, J.i.: Boro rice yield estimation model using modis ndvi data for bangladesh. In: IGARSS. pp. 7330–7333. IEEE (2019)
3. Alzamzami, F., Hoda, M., El Saddik, A.: Light gradient boosting machine for general sentiment classification on short texts: A comparative evaluation. IEEE Access **8**, 101840–101858 (2020)
4. Bergstra, J., Bengio, Y.: Random search for hyper-parameter optimization. Journal of machine learning research **13**(2) (2012)
5. Binte Mostafiz, R., Noguchi, R., Ahamed, T.: Agricultural land suitability assessment using satellite remote sensing-derived soil-vegetation indices. Land **10**(2), 223 (2021)

6. Chen, Y., Zheng, W., Li, W., Huang, Y.: Large group activity security risk assessment and risk early warning based on random forest algorithm. Pattern Recognition Letters **144**, 1–5 (2021)
7. Feng, D.C., Liu, Z.T., Wang, X.D., Chen, Y., Chang, J.Q., Wei, D.F., Jiang, Z.M.: Machine learning-based compressive strength prediction for concrete: An adaptive boosting approach. Construction and Building Materials **230**, 117000 (2020)
8. Gou, J., Ma, H., Ou, W., Zeng, S., Rao, Y., Yang, H.: A generalized mean distance-based k-nearest neighbor classifier. Expert Systems with App. **115**, 356–372 (2019)
9. Guo, S., Liu, Y., Chen, R., Sun, X., Wang, X.: Improved smote algorithm to deal with imbalanced activity classes in smart homes. Neural Processing Letters **50**(2), 1503–1526 (2019)
10. Jobbagy, E.G., Jackson, R.B.: The distribution of soil nutrients with depth: global patterns and the imprint of plants. Biogeochemistry **53**(1), 51–77 (2001)
11. Kalpoma, K.A., Haramoto, Y., Kudoh, J.I.: Real time fire monitoring system for russia and north asian region using noaa avhrr images. In: IGARSS. vol. 3, p. 1792 (2005)
12. Kalpoma, K.A., Kawano, K., Kudoh, J.i.: Ikonos image fusion process using steepest descent method with bi-linear interpolation. International journal of remote sensing **34**(2), 505–518 (2013)
13. Kalpoma, K.A., Kudoh, J.i.: Image fusion processing for ikonos 1-m color imagery. IEEE Transactions on Geoscience and Remote Sensing **45**(10), 3075–3086 (2007)
14. Kazi A, K., Nagatani, I., Kawano, K., Kudoh, J.I.: Development of a new dust index ndli for asian dust extraction system based on aqua modis data and monitoring of trans-boundary asian dust events in japan. IJRS **40**(3), 1030–1047 (2019)
15. Kudoh, J.I., Kalpoma, K., Kurita, Y.: An ikonos image fusion process using steepest decent method. In: IGARSS. pp. 3248–3250. IEEE (2006)
16. LaValle, S.M., Branicky, M.S., Lindemann, S.R.: On the relationship between classical grid search and probabilistic roadmaps. The International Journal of Robotics Research **23**(7–8), 673–692 (2004)
17. Lozano-Tello, e.a.: Crop identification by massive processing of multiannual satellite imagery for eu common agriculture policy subsidy control. European Journal of Remote Sensing **54**(1), 1–12 (2021)
18. Mahboob, M.G., Islam, A., Deshapriya, L.: Rice mapping and monitoring in sylhet region of bangladesh using modis ndvi. In: Proceedings of Asia Flux Mini-Workshop on RSEM. pp. 26–30 (2016)
19. Marage, D., Gégout, J.C.: Importance of soil nutrients in the distribution of forest communities on a large geographical scale. Global Ecology and Biogeography **18**(1), 88–97 (2009)
20. Mohammed, A.A., Khamees, H.T.: Categorizing and measurement satellite image processing of fire in the forest greece using remote sensing. IJEECS **21**(2), 843–853 (2021)
21. Mukta, M.S.H., Ali, M.E., Mahmud, J.: Identifying and predicting temporal change of basic human values from social network usage. In: ASONAM. pp. 619–620. ACM (2017)
22. Mukta, M.S.H., Ali, M.E., Mahmud, J.: Temporal modeling of basic human values from social network usage. Journal of the Association for Information Science and Technology **70**(2), 151–163 (2019)
23. Nakano, M., Kalpoma, K.A., Kudoh, J.I.: Automatic gcp creation for noaa avhrr image geometric correction. In: IGARSS 2004. vol. 6, pp. 3733–3735. IEEE (2004)
24. Nakano, M., Kalpoma, K.A., Nakamura, T., Kudoh, J.i.: A design of category classification system for high resolution satellite. In: IGARSS. vol. 7, pp. 4516–4518. IEEE (2004)
25. Nawshin, S., Mukta, M.S.H., Ali, M.E., Islam, A.N.: Modeling weather-aware prediction of user activities and future visits. IEEE Access **8**, 105127–105138 (2020)
26. Ontivero-Ortega, M., Lage-Castellanos, A., Valente, G., Goebel, R., Valdes-Sosa, M.: Fast gaussian naïve bayes for searchlight classification analysis. Neuroimage **163**, 471–479 (2017)
27. Priyanka, T., Soni, P., Malathy, C.: Agricultural crop yield prediction using artificial intelligence and satellite imagery. Eurasian Journal of Analytical Chemistry **13**(Engineering & Science SP), 6–12 (2018)
28. Saleh, S.M., Islam, S.R., Hossain, M.I., Pavel, K.A.A.: Study of bangladesh soil nutrient over satellite images. IJSR (2015)

29. Scheck, L., Weissmann, M., Bach, L.: Assimilating visible satellite images for convective-scale numerical weather prediction: A case-study. Quarterly Journal of the Royal Meteorological Society **146**(732), 3165–3186 (2020)
30. Snoek, J., Larochelle, H., Adams, R.P.: Practical bayesian optimization of machine learning algorithms. NIPS **25** (2012)
31. Turker, M., Rahimzadeganasl, A.: Agricultural field detection from satellite imagery using the combined otsu's thresholding algorithm and marker-controlled watershed-based transform. Journal of the Indian Society of Remote Sensing **49**(5), 1035–1050 (2021)
32. Vinutha, H., Poornima, B., Sagar, B.: Detection of outliers using interquartile range technique from intrusion dataset. In: Information and Decision Sciences, pp. 511–518. Springer (2018)
33. Yoshiaki, H., Muneto, I., Kalpoma, K.A., Kudoh, J.I.: Snow category extraction of noaa/avhrr images by using three dimensional histogram. In: IGARSS. vol. 6, pp. 3702–3704. IEEE (2004)

A Machine Learning Approach to Season Prediction Based on Agricultural Input Orders

Majwega Jackson

Abstract Agriculture is a source of income to many people around the globe. The evolution of modern agriculture led to an increase in the use of modern agricultural inputs, which has led to a sustainable growth in crop production. Based on their previous experiences, agricultural input suppliers stock a lot of products in their stores as they wait for right seasons. Unfortunately, due to the lack of precise knowledge about seasonal changes, their products overstay and some get tainted, contaminated, and expired leading to losses caused by dumping because of low sales made; hence, the closure of some of the businesses. In this fourth industrial revolution, which is smart agriculture based, agricultural input suppliers need the help of artificial intelligence to facilitate decision-making that can make their businesses more profitable and sustainable. In this research, we help suppliers predict the current season based on farmers' agricultural input orders in a given time period in order to make right inputs' stocking decisions. The method in this paper successfully models and analyzes various machine learning classifiers that illustrate features which facilitate supplier's decision before new inputs are stocked, which avoids losses hence leading to sustainable supplier business growth and development. We used classical metrics to compute the results of the different algorithms used and XGBoost, Random Forest, Decision tree, and K-NN all indicated a high percentage of overall dataset.

Keywords Artificial intelligence · Machine learning · Precision agriculture · Agricultural inputs · Supply chain

---

M. Jackson (✉)
Brac University, Dhaka, Bangladesh
e-mail: majwega.jackson@g.bracu.ac.bd

S. Hossain et al. (eds.), *Proceedings of International Conference on Fourth Industrial Revolution and Beyond 2021*, Lecture Notes in Networks and Systems 437,
https://doi.org/10.1007/978-981-19-2445-3_13

1 Introduction

1.1 Background and Motivation

Agriculture is playing an important role in global economic development, however, the rapid growth rate in the human population demands much from agriculture. To meet these high food demands, modern-day agriculture is faced with the big task of increasing the amount of food produced. Therefore, the most recent technologies are emerging in the field of agriculture to foster net productivity by collecting and having information processed [1]. The 17 sustainable development goals of 2015, which were embraced by all Member States of the UN, calls upon all Member States to immediately act upon ending poverty and hunger, improve health department and education, reduce inequality, and enhance economic growth, while handling the changes in the climate and working to save the planet's natural resources [2].

Bangladesh farmers are advised to apply agricultural chemicals to all crops to increase crop production so as to meet the demand of the rapidly growing population. An agricultural chemical refers to any element used in the growth or utilization of plants or animals for economic value to humans. These agricultural chemicals consist of fertilizers, pesticides, growth regulators, animal feed supplements, and raw materials divided into those that promote the growth of a plant or animal and that protect plants or animals.

Since fertilizers are chemical substances, they have an impact on the properties of soil (physical, chemical, and biological) as well as on crop yields [3]. Since nutrients are the key factors in the soil, they need appropriate management practices to ensure high crop yields. The application of both organic and inorganic sources will boost soil health and help to increase production as it involves the use of local sources, and hence, turned to be a rational, accurate, and economically feasible way of having nutrients supplied to crops [4]. Many benefits have been observed as a result of using agricultural chemicals, a sector upon which many low- and middle-income countries largely depend. This result has been attained due to the use of high-yield varieties of seeds, modern irrigation technologies, and agricultural chemicals [5]. Over the years, there has been an increase in crop output and productivity in many countries. Enhancements in productivity are a result of various factors such as the use of fertilizers, better seed varieties, and adoption of machinery. Pesticides have played an important role in the process by reducing losses caused by the weeds, diseases, and insect pests that would reduce the number of products harvested. Without the use of pesticides, "considerable economic losses" would be suffered and enumerated the substantial increases in yield and economic margin that come from the use of pesticides [6]. The increased use of agricultural chemicals has not all helped farmers to increase their productivity but also created employment opportunities for both the manufacturers and the suppliers (retail). However, the changes in climate lead to changes in the agricultural seasons. The inability to predict the current seasons has caused losses to agricultural input suppliers due to the fact that most of them use their previous experience and the available systems are only in the favor of farmers.

Therefore, there is a need for suppliers to predict the season before stocking inputs to reduce losses.

1.2 Our Contribution

In this paper, we model classifiers that illustrate the features that affect and predict the probability of farmers' chances to purchase before the supplier agricultural input suppliers stock new inputs. We test the classifiers on a real dataset gathered from farmers in MAAIF and ACDP with 21 features and about 16,392 samples. By analyzing the correlations in the heat map, we derive the features that have high correlations related to decision-making. Classical metrics are used to express the results of the algorithms used and XGBoost, Random Forest Classifier, K-NN, and Decision tree with high percentage in accuracy on the overall dataset. The outcomes of the data analysis prove that the implementation of machine learning systems that use the above four classification algorithms can support the supplier in season prediction before stocking to ensure business sustainability. Below is the contributions summary.

The proposed models predict the season before the supplier goes ahead to stock inputs.

(1) We illustrate the different features that affect supplier's stocking decision.
(2) We test different mathematical expressions to ensure the accuracy of the results.
(3) We test various classification models and recommend the best classifiers related to season prediction.
(4) We propose a model which helps agricultural input suppliers predict the season to make a strategic purchasing decision.
(5) We present an illustration of the top-level diagram, which diagrammatically explains the overall operation of the models which helps the supplier to understand the season and then go to stock new inputs.
(6) The proposed model, helps the supplier determine the amount of various inputs to stock based on the incoming order.

The rest of the paper is organized as follows: Sect. 2 is the Literature Review, Problem and Solution Formulation, Sect. 3 Data preparation, Data Processing, Training, and Sampling, Sect. 4 represents simulated results and performance evaluation of the classifiers. Section 5 is the Final Conclusion and recommendations for future work.

2 Existing Work

2.1 *Predictive Decision-Making*

Advanced analytical techniques are becoming popular in resolving complex classification-type decision problems in many fields. Similar works involved a wide range of fascinating systems with many attempting to perform climate figures. Many of the current determining innovations involve reenactments in light of material science and differential conditions, however, various new methodologies from computerized reasoning utilized essentially machine learning strategies, generally neural systems while others used models which had a probabilistic approach. Various researchers have embarked on season prediction both in agriculture, business, and sports arena [7, 8]. Most studies focus on analyzing the impact of season prediction on farmers' yields and productivity, not considering how agro-input suppliers' sales can be impacted [9]. Other researchers are focusing on having efficient techniques for predicting seasons and recommend farmers the best crops to cultivate, lending them a hand in their production and management but none of them advises agro-input suppliers of seasonal changes to stock accordingly [10]. In Marvin et al. [11], researchers are trying to establish an intelligent information prediction analysis on farming that will help farmers make decisions that can make their farming more efficient and profitable. In Datacamp [12], research focuses on identifying areas most affected by drought for certain crops, and 20 to 30 days in advance are provided to farmers for warning to facilitate planting decision is made which opportunity only available for farmers, not agro-input suppliers which may lead to losses on suppliers' side due to stock overstay and expiry. In general, most of the work aims at predicting farmer productivity, and all concentrate mainly on supporting farmers' side to avoid losses. No work is considering agro-input suppliers being affected by changes in seasons, and that is the research gap we are tackling.

2.2 *Machine Learning Classifiers*

There are three types of Machine Learning Algorithms, which include Supervised learning comprising a target variable dependent to be predicted by independent variables, for unsupervised learning, there is no target or outcome variable to estimate and for reinforcement, the machine is trained to make a specific decision.

3 Proposed Method

3.1 *Problem*

Agricultural suppliers' inability to predict the ongoing agricultural season.

Many agricultural input suppliers have been kicked out of the business mainly due to losses caused by a change in farming seasons. These losses are caused by a lack of knowledge regarding farmers' need for agricultural resources before purchasing a new stock for supply. Many suppliers stock new agricultural inputs based on their previous knowledge about seasons and farmers demand history. The unpredictable climatic conditions which change over time have impacted the agricultural sector due to changes in farming seasons. Due to a lack of systems in the support of these suppliers, currently, it's been so hard on their side to predict the right time to buy new stock, especially in low- and middle-income countries. They are always ready to supply but unfortunately only depend on farmers who only purchase during the right season. Because of this, they have stocked products and ended up making losses in terms of cash caused by pollution due to their overstay and poor storage facilities. Most research that has been done is focusing on the supply chain of these products with no focus on a system that favors these suppliers. Therefore, the need to address this problem is high hence the proposed model.

3.2 *Solution*

Machine learning and artificial intelligence are appealing to support sustainable agricultural investments through season prediction. Our proposed machine learning application is typically limited to the objective prediction of decisions and interpretation of the different factors that affect farmers' agricultural inputs purchase decisions and that allow for suppliers strategically planning. We propose ML classification of potential features that affect the stocking decisions of agricultural inputs based on predictive analysis of purchasing information. The proposed decision-making is performed through the application of machine learning classification algorithms to implement a probabilistic approach for predicting the current season with a clear interpretation of all the models that illustrate the features which affect supplier's decision-making. The figure below shows an interpretation of how the model works (Fig. 1).

3.3 *Dataset Description and Features*

The dataset used was collected from the MAAIF data repository from the data for orders that farmers make for different resources. But using models with the orders

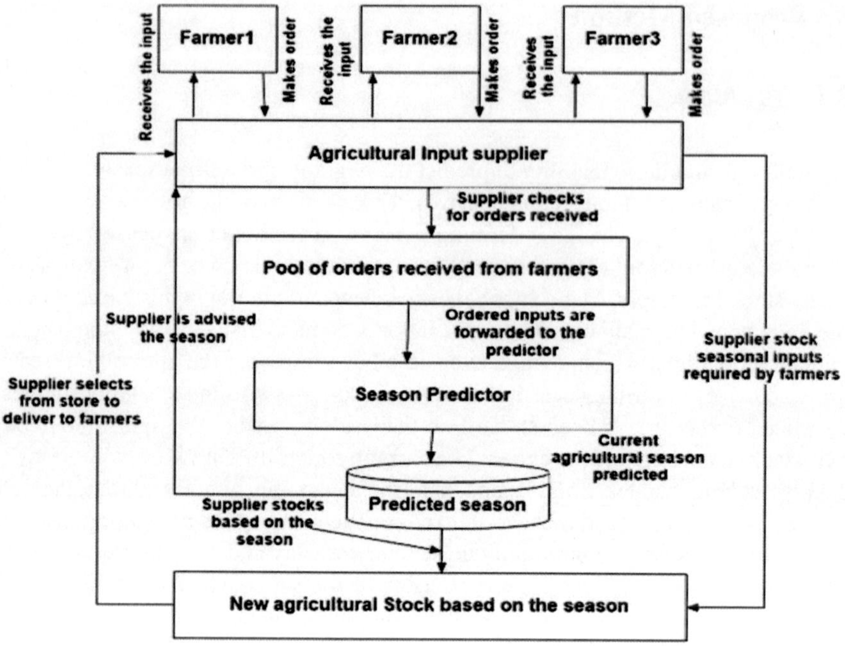

Fig. 1 A top-level diagram for the model

made, order date, and the number of items, we can predict the current agricultural season which will influence the supplier's new input stocking decision. The whole data is divided into train and test. Target is not included in the test but the test target values are available for the related tasks, and therefore, it is useful for us to demonstrate our idea. The dataset is imbalanced and most features (Table 1) are categorical, some with high cardinality and possibilities of missing imputation.

3.4 Data Exploration

We first explored the data types in our dataset, the frequency of each variable is visually illustrated below. In the figure, we plotted different types of distribution as shown below. The long left tail represents a negatively skewed distribution, a long right tail represents a positively skewed distribution and a bell curve represents a normal distribution whose skewness is zero (Fig. 2).

Table 1 Dataset description

OrderDate	Date a farmer received agricultural
Voucher	Unique ordering number
Farmer_name	Name of the farmer receiving resources
Sub_ccounty	Location of the farmer
Farmer_contact	Farmers' contact number
Panga	0—Farmer didn't order for pangs 1—Farmer ordered for pangs
Hermetic_bags	1—Farmer received hermetic bags 0—Farmer didn't receive hermetic bags
Tarpaulin	1—Farmer received tarpaulins 0—Farmer didn't receive tarpaulins
fDap_sacks	1—Farmer received dap fertilizer 0—Farmer didn't dap fertilizer
fUrea_sacks	1—Farmer received urea fertilizer 0—Farmer didn't receive urea fertilizer
fNpk25_5_5_sacks	1—Farmer received NPK25:5:5 fertilizer 0—Farmer didn't receive NPK25:5:5 fertilizer
fNpk17_17_17_sacks	1—Farmer received NPK17:17:17 fertilizer 0—Farmer didn't receive NPK17:17 fertilizer
Irish	1—Farmer needs the resource for the Irish crop 0—Farmer doesn't need the resource for Irish crop
Banana	1—Farmer needs the resource for the banana crop 0—Farmer doesn't need the resource for the banana crop
Coffee	1—Farmer needs the resource for the coffee crop 0—Farmer doesn't need the resource for the coffee crop
Beans	1—Farmer needs the resource for the bean crop 0—Farmer doesn't need the resource for the bean
Maize	1—Farmer needs the resource for the maize crop 0—Farmer doesn't need the resource the maize crop

(continued)

Table 1 (continued)

OrderDate	Date a farmer received agricultural
Number_of_items	Number of Items received by a farmer
Season_time	1—Planting time 2—Harvesting time 3—Pre-planting
Quantity_of_land	Number of acres of land possessed by a farmer

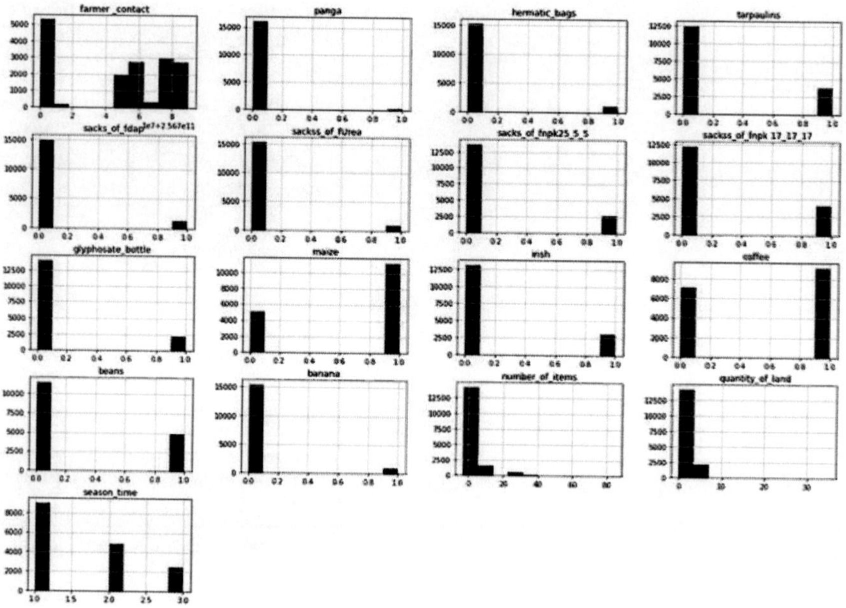

Fig. 2 Distribution of attributes

3.5 Analyzing Target Variable

Here, we are analyzing the agricultural input performance based on the orders made per season time. This helps the supplier to effectively locate resources when planning on what stock and how much to stock considering his budget. It is very good to understand when most orders are made. This enhances the supplier's profitability and growth (Figs. 3 and 4).

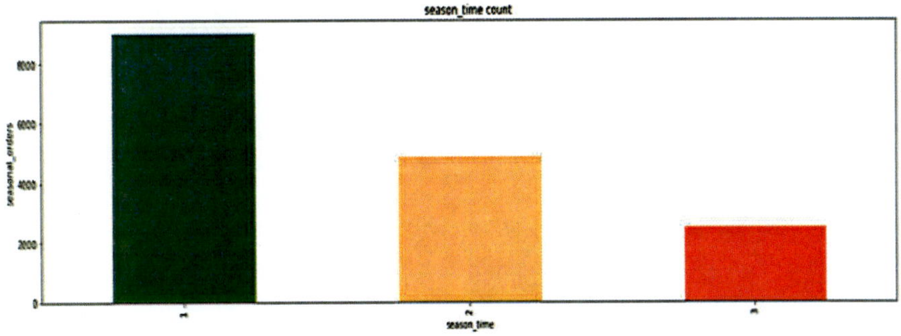

Fig. 3 A bar graph showing target variables

Fig. 4 A pie chart showing percentages for orders received in different seasons

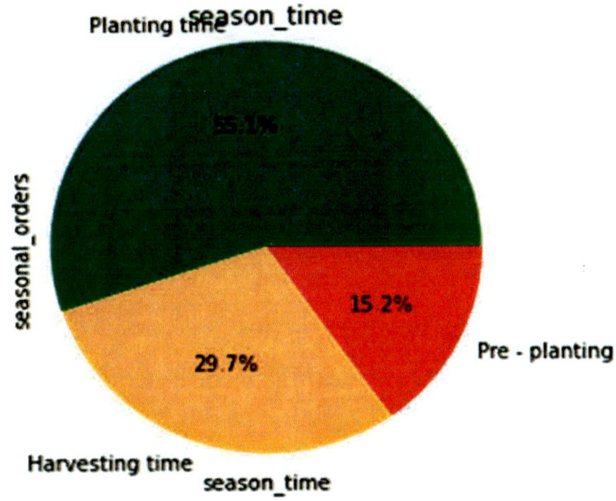

3.6 *Data Cleaning*

Correlation: Fig. 5 shows the correlations with all variables as Grey fields representing no correlation while pale white fields indicate very low correlation. Red implies a positive or direct correlation and blue reveals a negative or indirect correlation. Therefore, a high positive red represents a strong positive correlation and high negative blue field represents a high negative correlation. The figure shows that correlation decreases as we move from both high negative blue and high positive red toward zero. As per the matrix Heatmap below, there is a correlation between features in the dataset. The existence of a correlation between the variables in the data in (Fig. 5) indicates that good predictive results could be expected if a good train result is obtained.

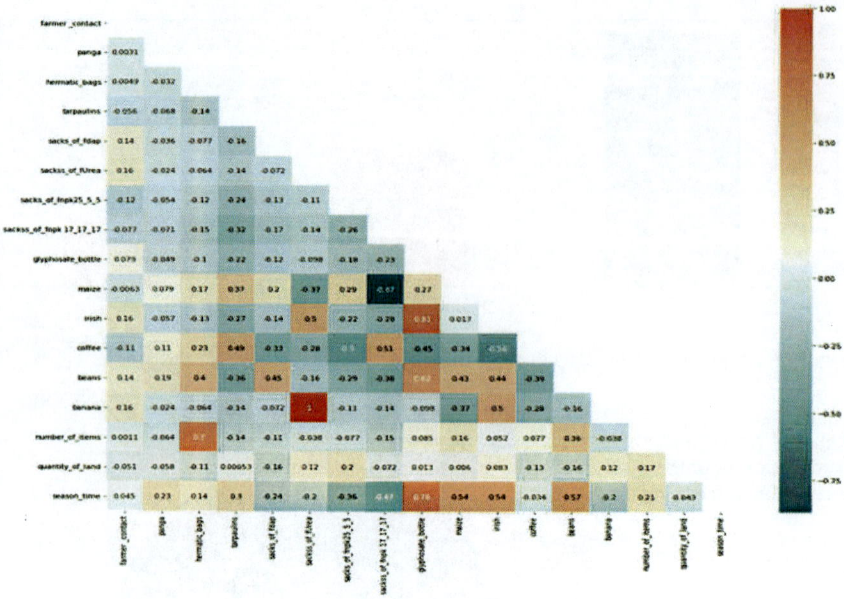

Fig. 5 Correlation matrix, Heatmap, of all features in the test dataset

Graphing Prominent Variables.

Discrete variables that are highly correlated to season time imply an increase in the number of items ordered and can influence suppliers to stock new agricultural inputs (Figs. 6 and 7).

Continuous variables are highly correlated to season time, and high results indicate that the product is needed in that season.

4 Results and Discussion

Since our dataset had clean data, we had a good fitting and after testing, the results showed that KNN, XGBoost, Random Forest, and Decision Tree models had the same percentage of accuracy implying they all can be used in the predictions. Results are indicated in the various confusion matrix diagrams for the used classification algorithms below.

kNN (k- Nearest Neighbors): KNN is the most commonly used model for classification as compared to regression, comprehensible, adaptable, modest, and among the most recommendable machine learning algorithms, however, it is computationally expensive. It uses variable normalization, which is required to prevent bias from high range variables and results if normalization between 0 and 1 is done for the same

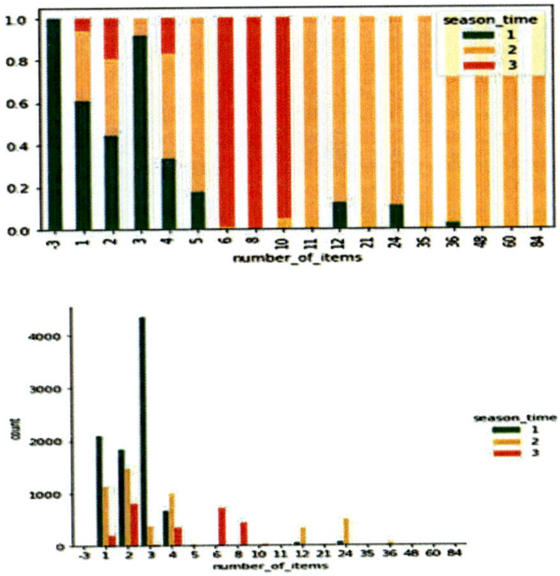

Fig. 6 Discrete variables that are highly correlated to season time

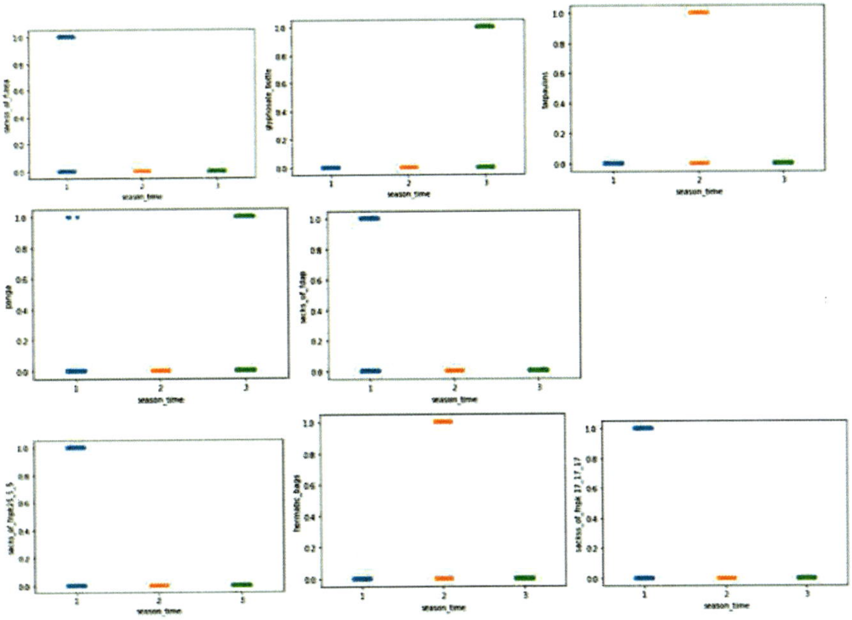

Fig. 7 Continuous variables that are highly correlated to season time

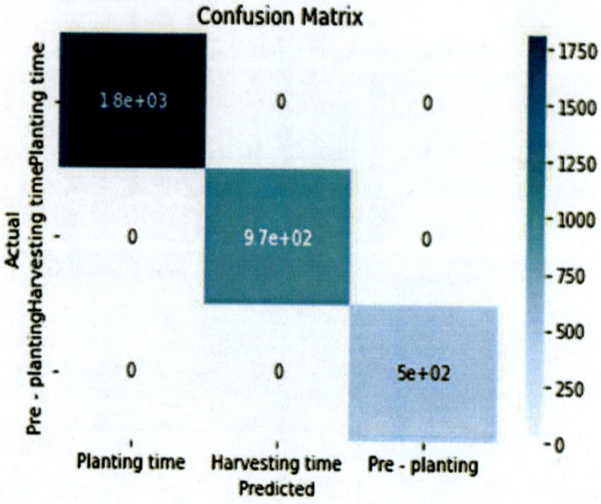

Fig. 8 A representation of the k-nearest neighbors' algorithm

scale data. kNN is not perfect for handling large dimensional data otherwise dimensions need to be reduced to improve its performance. However, handling missing values is another way for improving results. Using the below mathematical expression, we were able to generate a KNN confusion matrix and represented it on a diagram in Fig. 8.

$$sf(i', i'') = sf(n, l, m) = \frac{l+1}{2}\left(\frac{1}{l+n+2} + \frac{1}{l+m+2}\right)$$

We expressed the Random Forest confusion matrix with the help of the above mathematical expression and illustrated it in Fig. 10.

Decision Tree: A type of supervised ML algorithm learning algorithm that is mostly used for classification problems that work for both categorical and continuous dependent variables with a training time faster compared to that of neural network algorithms. It is a distribution-free or nonparametric technique that does not depend on probability dissemination assumptions and it can resolve high dimensional data with decent accuracy. We expressed the Decision Tree confusion matrix with the help of the above mathematical expression and illustrated the results in Fig. 9.

$$\text{Entropy}(t) = -\sum_{i=0}^{c-1} p(i|t) \log_2 p(i|t),$$

$$\text{Gini}(t) = 1 - \sum_{i=0}^{c-1} [p(i|t)]^2,$$

Random Forest: RF is a supervised machine learning algorithm. It ensembles learning, where different types of algorithms that are similar or dissimilar are merged many times in order to form a very strong prediction model, which can be used

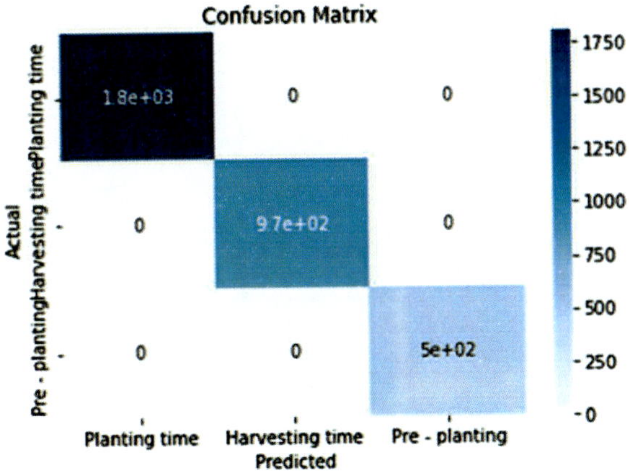

Fig. 9 A Representation of decision tree algorithm

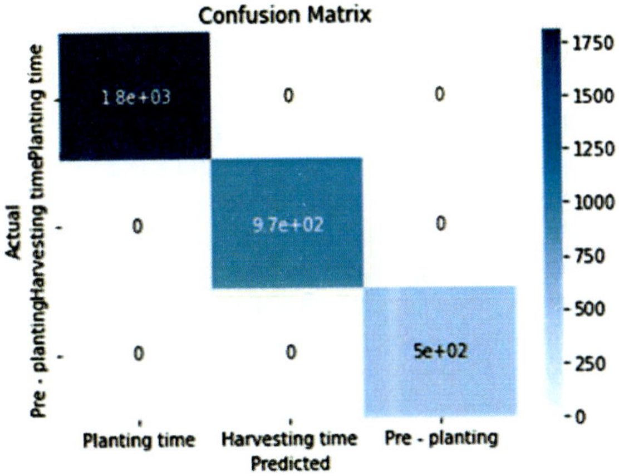

Fig. 10 Representation of random forest algorithm

for both regression and classification tasks. Although this algorithm is very stable, effective with a mixture of categorical and numerical features, efficient when data has missing values or it has not been scaled well, they require much more computational resources, due to a large number of decision trees joined together and their complexity makes them more time consuming to train than other comparable algorithms. We expressed the Random Forest confusion matrix with the help of the below mathematical expression and the illustrated it in Fig. 10.

$$fi_i = \sum_{j:nodes\ j\ splits\ on\ feature\ i} s_j C_j$$

$$norm\ fi_i = \frac{fi_i}{\sum_{j \in\ all\ features} fi_i}$$

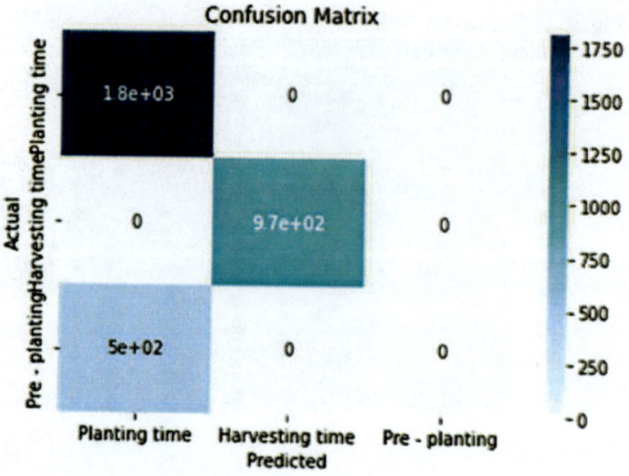

Fig. 11 A representation of the AdaBoost algorithm

$$RFfi_i = \frac{\sum_j norm\ fi_{ij}}{\sum_{j \in all\ features, k \in all\ trees} norm\ fi_{jk}}$$

AdaBoost: This is an iterative ensemble boosting classifier which combines various classifiers as a mechanism to increase the accuracy of classifiers. It builds a strong classifier by combining various low-performing classifiers in order to obtain highly accurate, strong classifier. The hidden concept in AdaBoost is setting the weights of classifiers and having the data sample in each iteration trained so as to ensure accurate predictions for the unusual observations. If an algorithm accepts weights on the training set, it can then be used as a base classifier. AdaBoost has two conditions to meet, namely: (a) It has to be trained interactively on various weighted training examples and (b) In each iteration, it provided an excellent fit for these examples by minimizing the training error. We expressed the AdaBoost confusion matrix with the help of the below mathematical expression (Fig. 11).

$$Z_t = \sum_{i=1}^{m} D_t(i) exp(-\alpha_t y_i h_t(x_i))$$

$$Z_t = \sum_{i=1}^{m} D_t(i) exp(-y_i h_t(x_i))$$

$$C_j = \frac{1}{2} ln \left(\frac{W^j_{\mathrm{merge}} + \in}{W^j_{\mathrm{split}} + \in} \right)$$

XGBoost: It is an algorithm with a high predictive power proving high levels of accuracy. The algorithm has both a linear model and a tree learning algorithm, which make it a faster algorithm as compared to already existing gradient booster techniques. It is the best machine learning algorithm for handling structured data

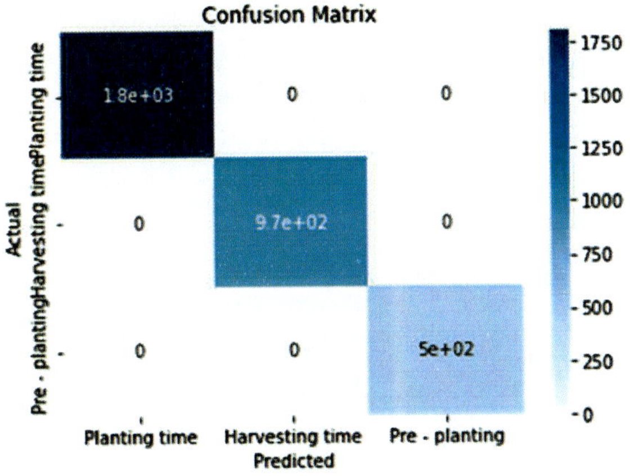

Fig. 12 A representation of the XGBoost algorithm

because of its ability to overcome several objective functions besides regression, classification, and ranking.

We expressed the XGBoost confusion matrix with the help of the below mathematical expression (Fig. 12).

$$F_0(x) = argmin_{\gamma} \sum_{i=1}^{n} L(y_i, \gamma)$$

$$\gamma_{\text{im}} = -\alpha \left[\frac{\partial (L(\nu_i,\ F(\mathrm{x}_i))}{\partial F(\mathrm{x}_i)} \right]_{F(\mathrm{x}) = F_{m-1}(\mathrm{x})}, \text{ where } \alpha \text{ is the learning rate}$$

$$F_m(\mathrm{x}) = F_{m-1}(\mathrm{x}) + \gamma_m h_m(\mathrm{x})$$

The overall results show that models can be used in prediction if the data is collected well and the dataset gets trained well. Therefore, due to clean data which was used any model used will give perfect results since they all gave the same predictive results.

5 Conclusion

In this work, we predicted the current agricultural season by considering the orders made by farmers to allow for effective seasonal stocking of agricultural inputs and interpreted ML models that illustrate the features that affect farmer's purchasing decisions and analyzed their impact on supplier's stocking decisions. Our proposed model can be modified by identifying more features that can influence supplier's stocking decisions. And for more accurate and effective results, we hope to integrate explainable Artificial intelligence in our model to allow for interpretability of the results. We are also looking forward to utilizing quantum computing in model training as a technique for improved model performance. Also, understanding the supplier's status in terms of finance and location will boost the accuracy of the prediction and increase the robustness of the ML models used.

References

1. Chakrabarty, A., Mansoor, N., Uddin, M.I., Al-adaileh, M.H., Alsharif, N., Alsaade, F.W.: Prediction approaches for smart cultivation: a comparative study. Complexity **2021**, 1–16 (2021)
2. Mostofa, B., Raihan, M.Z., Hossain, M.F., Farhana, T., Mia, M.M., Kader, M.A.: Effects of long-term mineral fertilization and manuring on rice-rice cropping pattern in sub-tropical floodplain soil. Asian J. Med. Biol. Res. **1**(2), 222–229 (2015)
3. Hossain, M.F., Farhana, T., Raihan, M.Z., Hasan, M.S., Mia, M.M., Rahman, M.M.: Effect of different fertilization practices on the growth and yield of cabbage. Asian J. Med. Biol. Res. **1**(2), 182–186 (2015)
4. Thornton, P., Dinesh, D., Cramer, L., Loboguerrero, A.M., Campbell, B.: Agriculture in a changing climate: Keeping our cool in the face of the hothouse. Outlook Agric. **47**(4), 283–290 (2018)
5. Aktar, M.W., Sengupta, D., Chowdhury, A.: Impact of pesticides use in agriculture: their benefits and hazards. Interdiscip. Toxicol. **2**(1), 1–12 (2009)
6. Chen, M., Chen, C.: Data mining computing of predicting NBA 2019–2020 regular season MVP winner. In: 2020 International Conference on Advances in Computing and Communication Engineering (ICACCE) (2020)
7. Medar, R., Rajpurohit, V.S., Shweta, S.: Crop yield prediction using machine learning techniques. In: 2019 IEEE 5th International Conference for Convergence in Technology (I2CT) (2019)
8. Shakoor, M.T., Rahman, K., Rayta, S.N., Chakrabarty, A.: Agricultural production output prediction using Supervised Machine Learning techniques. In: 2017 1st International Conference on Next Generation Computing Applications (NextComp) (2017)
9. G. Origel-Gutierrez and A. Perez-Flores, "Monitoring the 2019 agricultural drought in the state of San Luis Potosi, Mexico," in IGARSS 2020 - 2020 IEEE International Geoscience and Remote Sensing Symposium, 2020
10. Schmullius, C., Nithack, J.: Seasonal variations of agricultural fields in multifrequency SAR images. In: Proceedings of IGARSS '93-IEEE International Geoscience and Remote Sensing Symposium (2002)
11. Marvin, G., Jackson, M., Alam, M.G.R.: A machine learning approach for employee retention prediction. In: 2021 IEEE Region 10 Symposium (TENSYMP) (2021)
12. Datacamp.com. [Online]. https://www.datacamp.com/community/tutorials/adaboost-classifier-python. Accessed 17 Oct 2021

Assessment of Water Quality in Smart City Environment Leveraging ML-IoT

Sabbir Ahmed, Mayeesha Mahzabin, Shahrin Shahpar, Surraiya Islam Tonni, and Md. Sazzadur Rahman

Abstract Water quality measurement and potability for human residents are critical for health concerns in smart cities. Citizens are perplexed by data about the quality of their drinking water derived from chemical measurements. As a result, data on water potability derived from chemical sensor values must be interpreted prior to being made publicly available. This paper proposed an edge-cloud ubiquitous sensor network for low-cost water quality measurement to supplement existing IoT-based infrastructure. Machine learning algorithms are applied to a dataset containing eight fields related to water potability. Following that, a total of 16 machine learning algorithms for potability prediction were compiled, including 11 shallow learning algorithms and 5 deep learning algorithms. The performance of multiple machine learning algorithms for determining the potability of water based on chemical and laboratory measurements was compared. These results were then compared to those obtained using deep learning algorithms such as ANN, CNN-Resnet, and CNN-LSTM. CNN-Batch Normalization, the most accurate of these algorithms, achieved a maximum testing accuracy of 85.03%.

Keywords Potable water · Water quality · Machine learning · IoT

S. Ahmed (✉) · Md. S. Rahman
Institute of Information Technology, Jahangirnagar University, Savar, Dhaka 1342, Bangladesh
e-mail: sabbir.iit.ju@gmail.com

Md. S. Rahman
e-mail: sazzad@juniv.edu

M. Mahzabin · S. Shahpar · S. I. Tonni
Department of Information and Communication Technology, Bangladesh University of Professionals, Dhaka 1216, Bangladesh

S. Hossain et al. (eds.), *Proceedings of International Conference on Fourth Industrial Revolution and Beyond 2021*, Lecture Notes in Networks and Systems 437,
https://doi.org/10.1007/978-981-19-2445-3_14

1 Introduction

Clean and safe potable water [5] is the most crucial resource for humanity as human health, prosperity, and security are all dependent on it. The world's population is anticipated to reach nine billion people around 2050 [7]. Considering that the standard of living continues to expand, the volume of drinkable water required for human use will be equivalent to the assets of three planets earth [9]. Consequently, 785 million people, i.e., one in every nine, do not have access to safe drinking water [24]. Diarrhea kills an estimated 842,000 people per year throughout the world [24]. As a result, improved techniques for assessing water quality are required to ensure everyone's right to safe potable water.

The criteria used to evaluate water quality vary depending on their property and where it is applied. There are a variety of metrics that could be used to identify the quality of the water; however, monitoring each of them will increase effort, lowering the credibility of the analysis [21]. Conventional water quality procedure is inefficient because it has many disadvantages, including a long time intervention and the need to determine water samples from only a few places simultaneously. These strategies are also expensive and ineffective in densely populated countries such as Bangladesh, India, and China [14]. So, researchers are proposing smart monitoring systems to overcome these drawbacks. However, because numerous elements might affect water quality, traditional data processing approaches are no longer enough for resolving this problem and statistical approaches have a low level of accuracy [1]. Alongside the development of the Internet of Things (IoT), the water distribution system is a perfect pick for instrumentation with a wide range of sensors that can interact with one another and aggregate data for analytics, allowing for more effective water distribution and management in smart cities [22].

Machine Learning (ML) provides a fascinating approach for extracting relevant and effective data for forecasting water quality [13]. With IoT, water quality data of different quality parameters are gathered, analyzed, and can be later shown to users to be aware of their water condition. However, users can easily be intimidated by chemical values alone, so the data must be interpreted before being available to the public. This research aims to mitigate the health risks caused by unsafe water by incorporating IoT and ML to forecast water quality components. This paper proposed an edge cloud ubiquitous sensor network for low-cost water quality measurement to supplement existing IoT-based infrastructure. After that, we made a comparison with the performance of multiple ML algorithms to assess water potability from chemical and lab measurements.

The rest of the paper is arranged as follows: The previous works on IoT-based water quality monitoring systems are shown in Sect. 2. The overall flow diagram of the proposed method and its associated description are shown in Sect. 3. The experimental result is demonstrated in Sect. 4. The conclusion of our proposed framework is presented in Sect. 5.

2 Related Works

Auditing and evaluating the quality of the water is centered on the water's primary structural, chemical, and ecological attributes. Following this, Daigavane et al. [6] demonstrated the design and implementation of a low-cost, automatic, and adaptable device for water quality monitoring in real time with the IoT. Gowthamy et al. [2] employed water flow sensors, pH and turbidity sensors, and ultrasonic sensors to develop a comparable technique. However, Moparthi et al. [16] created a water quality monitoring system that may be used in municipal water tanks, such as those that store drinking water. Patil et al. [19] aimed to conserve water, improve quality, and lower the environmental effects of water use. AlMetwally et al. [3] developed a system that makes efficient use of the many filters used to optimize water quality. There will only be one time when the filters will be utilized. In the recent work of water management using IoT, Pincheira et al. [20] adopted heavily restricted wearable sensors as legitimate sources of data for a permissionless blockchain.

Developing accurate models for predicting water quality has been one of the major areas of research in water environmental science in recent years. One of the recent studies made by Liu et al. [15] suggested a water quality forecasting model based on Long Short-Term Memory (LSTM) and Deep Neural Networks(DNN). The anticipated and actual values were quite close, demonstrating that LSTM DNNs can reliably assess drinkable water condition. Koditala et al. [14] proposed a minimal water quality monitoring system that can monitor the temperature of water and change it to match the temperature of the atmosphere using multivariable linear regression. Ashwini et al. [4] presented a low-cost device for real-time water quality surveillance using IoT. Here, Random Forest (RF) along with K-Nearest Neighbor (KNN) algorithms are used to evaluate the water's parameters and classify them into various categories based on their intended usage. The accuracy was 96% and 97% accordingly. Pappu et al. [18] developed an intelligent water quality monitoring system based on IoT for water tanks used by housing areas. They acquired 100% accurate results. Wang et al. [23] presented a novel water quality prediction approach focusing on LSTM NN, with a view to predicting water quality. Haghiabi et al. [8] implemented an Artificial Neural Network (ANN) along with Group Method of Data Handling (GMDH) and Support Vector Machine (SVM) for determining water quality attributes in the river named Tireh. Wu et al. [26], on the other hand, used the ML algorithm attention neural network to categorize hygienic and unhygienic water using images. They made use of Wu et al. [25] dataset, which included 1000 pictures of water. Four standard metrics were used to assess classification: precision, recall, F1-Score, and accuracy. They got an accuracy of 66.4% for clean water and 71.2% for polluted water. In recent work, Jain et al. [10] sought to decipher use patterns, and the correlation of countless components might aid in further assessing and estimating the intake tendency of the highly crucial resource water using ARIMA model.

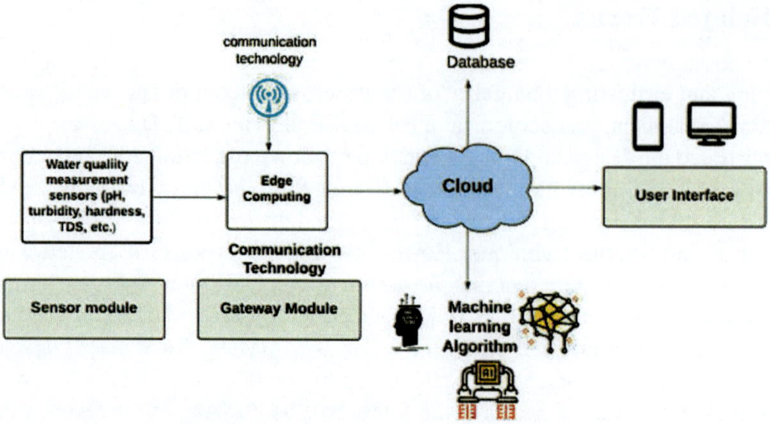

Fig. 1 Water quality management system

3 Proposed System

We have proposed a water quality management system which is cost-effective than the existing water quality systems. The water potability prediction system is divided into two phases in this study: sensor integration and machine learning application. Since the ML algorithm is probabilistic, the performance can vary a lot in terms of usability. However, if the ML architecture is appropriately designed, the rest of the system can be developed using existing sensing technology. In this study, we've provided a theoretical outline to demonstrate how the total system will function. We have examined roughly 16 ML algorithms to find the most accurate and suitable one for the system. Since there are many existing sensing technologies, but few ML implementations, our proposed system can be developed and perform well in the future if we have adequate resources.

3.1 IoT Infrastructure

From Fig. 1 we can see that our proposed water quality management system consists of four sections: Sensor module, Gateway module, Cloud server module, and User Interface Module.

Sensor module is a significant factor of IoT infrastructure. As our system is data-driven, therefore, all the data are taken from water quality sensors. These sensors are set in the water in the water distribution area. We employed sensors to measure water quality characteristics. The sensors are connected to an Arduino Uno micro-controller. The controller reads data from the water quality measurement sensors, analyzes it, and then sends it over the internet to a specified cloud server using the internet. Sensor module is in charge of transferring data to a gateway module, or local gateway. A gateway connects a device to the cloud. This communication can happen using Bluetooth, Zigbee, WiFi, Wireless Sensor Network (WSN), etc. Sensor data is instantly stored, displayed, and analyzed via a cloud server. In addition, water distribution system management might be more easily obtainable and maintained. Cloud servers may forecast the potability of the water using ML and deep learning (DL) algorithms. The end-user can see the classification model's output and decide whether the water is potable or not based on it. End-users may connect, control, and operate their own IoT things from anywhere using the user interface. Users can see the potability result on the website and in the mobile app also. The overall water quality management system follows edge computing, cloud computing, and ubiquitous computing.

3.1.1 Edge Computing

Edge computing is a decentralized network design in which data processing occurs at the network's edge, as close to the source as possible. The gateway module, where data from sensor modules is collected via various communication technologies such as WiFi, Zigbee, and WSN, will be used for edge computing in the water management system.

3.1.2 Cloud Computing

Cloud computing has the potential to be a viable replacement for data centers. The sensing data from edge devices must be saved and preprocessed in cloud systems in order to produce the dataset for ML architectures, as described in this article. In this scenario, the difference between cloud and edge computing is that cloud allows data to be concatenated, whereas edge computing will significantly reduce computation through decentralized approaches.

3.2 Dataset Description

We have used "water quality" dataset [12] for training our neural network models. The dataset contains 3276 different water bodies for 10 water quality metrics which are pH, hardness, total dissolved solids, chloramines, sulfate, conductivity, organic carbon, trihalomethanes, turbidity, and potability that indicates if water is safe for human consumption, where 1 means Potable and 0 means Not potable. These data are synthetically generated. In the dataset, there are 1998 non-potable and 1278 potable data. Table 1 contains brief description of dataset.

Table 1 Dataset description

	Count	Mean	Std	Min	50%	Max
pH	2785.00	7.08	1.59	0.00	7.04	14.00
Hardness	3276.00	196.36	32.88	47.43	196.97	323.12
Solids	3276.00	22014.09	8768.57	320.94	20927.83	61227.20
Chloramines	3276.00	7.12	1.58	0.35	7.13	13.13
Sulfate	2495.00	333.77	41.42	129.00	333.07	481.03
Conductivity	3276.00	426.21	80.82	181.48	421.88	753.34
Organic Carbon	3276.00	14.28	3.31	2.20	14.22	28.30
Trihalomethanes	3114.00	66.4	16.18	0.74	66.62	124.00
Turbidity	3276.00	3.97	0.78	1.45	3.96	6.74
Potability	3276.00	0.39	0.49	0.00	0.00	1.00

3.3 Data Preprocessing

We must first preprocess the data before applying it to the neural networks. There were several missing values for pH, sulfate, and trihalomethanes in the dataset. We balance the potability class once we have satisfied the datasets' null values. Missing values have been filled out by the mean value of the specific column. Standard Scaler subtracts the mean from every characteristic and scales that to unit variance. This process is carried out feature-by-feature. Variables on different scaling or minimum–maximum do not always yield optimum training of ML algorithms. Hence, standard scaling is applied to the input features of the dataset.

3.4 Machine Learning Classifiers

We have evaluated total 16 ML models where 5 of them are DL models. Then we have compared the performance of these models. The models are given below:

1. **Logistic Regression Classifier (LR)**: When a decision criterion is provided, logistic regression transforms into a classification method. The classification problem determines the threshold value, which is an important aspect of logistic regression. We have used class weight = balanced parameter in our model.
2. **Decision Tree Classifier (DT)**: By learning simple decision rules inferred from past data, the purpose of employing a decision tree is to develop a training model that can predict the class or value of the training data [17].
3. **K-Nearest Neighbor Classifier (KNN)**: K-Nearest Neighbor is a fundamental algorithm that keeps track of all available data and classifies data using a similarity metric. It is typically used to classify a data point based on the classification of its neighbors. In our model, we employed three neighbor parameters of 2, 3, and 4.

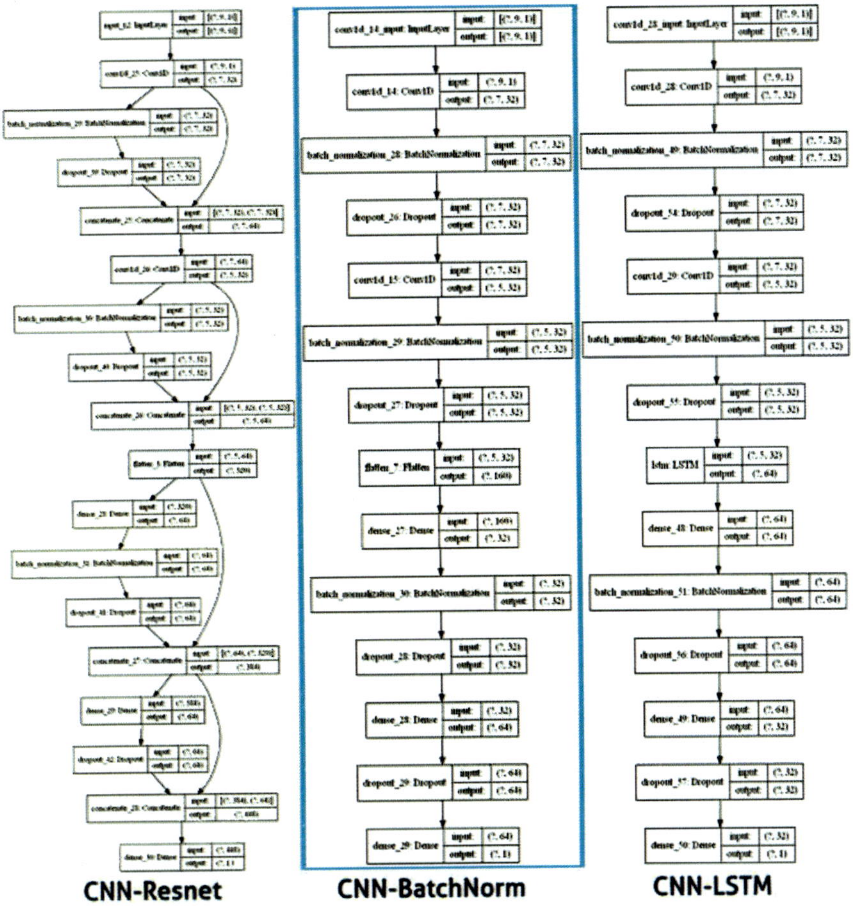

Fig. 2 Proposed model architecture

4. **Random Forest Classifier (RF)**: Random forest constructs many decision trees and integrates them, therefore, the accuracy and consistency of the predictions improve. In our model, we chose a maximum depth of 3 and a random state of 3.
5. **AdaBoost Classifier (ABoost)**: In an AdaBoost classifier, the weights of incorrectly classified instances are adjusted such that successive classifiers emphasize more on complex cases.
6. **Gradient Boosting Classifier (GBoost)**: Gradient boosting algorithm minimizes a loss function by iteratively choosing one that points in the direction of a negative gradient. It can rapidly overfit a training dataset.
7. **Extreme Gradient Boosting Classifier (XGBoost)**: Extreme Gradient Boosting Classifier is a fast and efficient gradient-boosted decision tree implementation. We employed a random state of 3, a learning rate of 0.05, a maximum depth of 10, and an estimator of 100 in our model.

8. **Support Vector Classifier (SVC)**: Support Vector Classifier is a supervised machine learning. It uses the kernel trick to convert data and then calculates an optimal boundary between the possible outputs.
9. **Stochastic Gradient Descent Classifier (SGD)**: Stochastic Gradient Descent is a simple but effective method for reducing a cost function. It's used to learn discriminative linear classifiers like SVM and logistic regression. We have used loss=hinge parameter in our model.
10. **Gaussian Naive Bayes Classifier (GNB)**: Naive Bayes is a classification technique for supervised machine learning that is based on the Bayes theorem. Gaussian Naive Bayes is a Naive Bayes variation that uses a Gaussian normal distribution and can handle continuous data.
11. **Bernoulli Naive Bayes Classifier (BNB)**: Bernoulli Naive Bayes is a member of the Naive Bayes family which only accepts binary values as input.
12. **Artificial Neural Network**: Non-linear problems are modeled using ANNs, which can anticipate output values for given input parameters based on their training data. The algorithm is trained on 100 epochs. Our model uses 1 input layer, 4 dense layers, 2 batch normalization layers, and 2 dropout layers [17].
13. **Convolutional Neural Network**: Convolutional neural networks (CNN) employ filtering to input to generate a feature map summarizing the existence of observed features. Thus, the features can be reduced by number without losing important details. This feature reduction with an appropriate activation function annihilates the local minima problem. Thus in this work, a one-dimensional convolution layer is used to improve classifier performance.
14. **CNN-Batch Normalization**: Batch Normalization (BN) is an algorithmic method for training Deep Neural Networks (DNN) quickly and more reliably. BN makes it possible for the network's various layers to learn on their own without interfering. It takes the output from the preceding layers and normalizes it.
15. **CNN-Resnet**: The challenge of deep network training is solved with ResNet or residual networks. When dealing with complex challenges, adding layers to a CNN enhances accuracy and performance. The objective of these layers is to learn about the features over time.
16. **CNN-Long Short-Term Memory Network**: CNN-LSTM is a LSTM architecture built for sequence prediction with spatial inputs. It combines the advantages of CNN, which can extract important characteristics from input, with LSTM, which can select the proper mode for relevant data automatically [11].

The architecture of CNN-resnet, CNN-BN, CNN-LSTM of the proposed classifiers is illustrated in Fig. 2.

4 Experimental Analysis

The experiment was conducted using a PC with an Intel Core i5 processor, 8 GB of RAM, and a 2 GB Nvidia 920mx GPU. Scikit-learn was used to create shallow

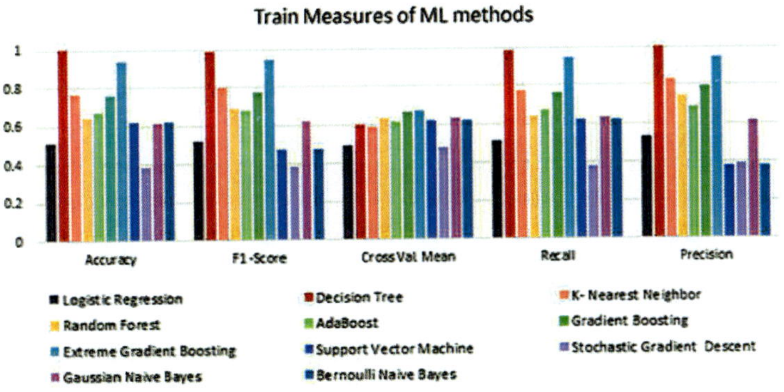

Fig. 3 Comparison of performance metrics in training data among ML classifiers

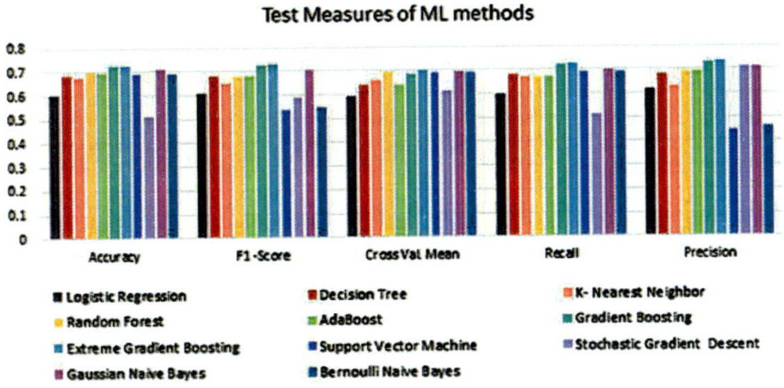

Fig. 4 Comparison of performance metrics in testing data among ML classifiers

machine learning algorithms, and TensorFlow was used to implement DNNs. The confusion matrix is computed initially and then the rest of the assessment measures are determined. The confusion matrix compares anticipated and actual output and produces true positive, false positive, true negative, and false negative values; these values are then used to compute various performance measures, including accuracy, recall, precision, F1-Score, Area Under Curve (AUC), cross-validation mean, and binary cross-entropy loss.

To determine the optimal machine model, 16 alternative models were developed and evaluated. As a consequence, a series of tests was conducted to fine-tune the parameters. For DL models, various factors, including the number of the convolutional layer, the LSTM layer, dense layers, and the density layer parameters, were considered throughout the selection process. Each model underwent a total of 100 epochs of training. The dataset has been divided into 20% for testing and the rest for training.

Eleven models of shallow machine learning were implemented and fine-tuned. Among these models are the LR, DT, KNN, RF, Aboost, Gboost, XGboost, SVM,

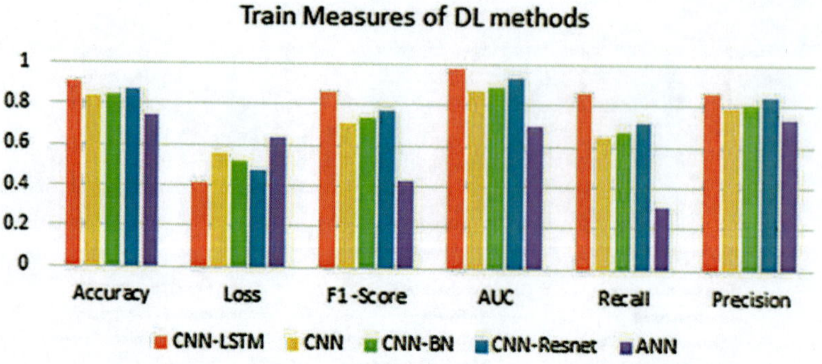

Fig. 5 Comparison of performance metrics in training data among DL classifiers

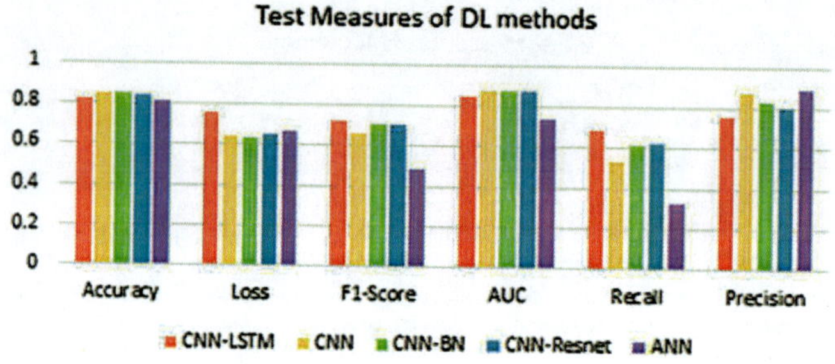

Fig. 6 Comparison of performance metrics in testing data among DL classifiers

SGD, BNB, and GNB classifiers, all of which are discussed in detail in the methodology section. After training these classifiers, they are each given trained data, and the resulting prediction is used to calculate the trained evolution measurements shown in Fig. 3. Among these classifiers, decision trees and extreme gradient boosting have demonstrated exceptional accuracy and F1-Scores. DT's training accuracy and F1-Score are both 100%, whereas XGBoost achieved a training accuracy of 93.89% and an F1-Score of 94%. XGBoost achieved a cross-validation mean score of 67.07% and Gboost achieved a cross-validation mean score of 66.33%. Again, these models were used to forecast the testing data, with XGboost achieving a maximum accuracy of 72.75% and F1-Score achieving a maximum accuracy of 73.13%. Figure 4 illustrates detailed comparisons of performance metrics.

Additionally, five DL models were fine-tuned and tested on the dataset. CNN, CNN with batch normalization, CNN with residual connection, and CNN-LSTM classifiers are among these models. Each of these classifiers has a training accuracy of 90.58%, 83.17%, 84.56%, 87.28%, and 74.79%, respectively. The results of the performance metrics are plotted against the training data in Fig. 5. CNN-BN achieved a maximum accuracy of 85.03%, an F1-Score of 70.30%, and an AUC of 87.63% during testing. Detailed comparisons of performance metrics of DL classifiers are depicted in Fig. 6.

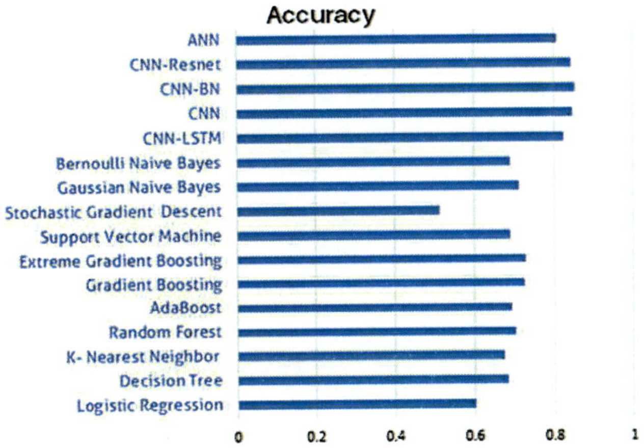

Fig. 7 Comparison of accuracy in testing data among classifiers

The findings of this demonstrate unequivocally that water potability can be determined using several of the previously mentioned water quality sensors. We evaluated 16 novel classifiers in order to determine the most effective one. Figure 7 illustrates a comparison of classifiers' test accuracy. According to the figure, one-dimensional convulsion-based classifiers (CNN) perform significantly better than other classifiers, with a maximum testing accuracy of 85.03% for CNN-BN.

5 Conclusion

Potable water is a core human right and an important part of any health protection program. This is a matter of national, regional, and local importance in terms of health and development. As a result of the decreases in unfavorable health impacts and health care expenses, several regions have found that expenditures on water supply and sanitation have a net economic benefit. Thus, water quality monitoring plays an important role in smart city to ensure the safety of people. In this paper, we have proposed an edge-cloud ubiquitous sensor network for low-cost water quality measurement to supplement existing IoT-based infrastructure. We have also performed different ML and DL models to assess water potability from chemical and lab measurements and compared the performance of these models to ensure user-safe drinkable water.

References

1. Ahmed, A.N., Othman, F.B., Afan, H.A., Ibrahim, R.K., Fai, C.M., Hossain, M.S., Ehteram, M., Elshafie, A.: Machine learning methods for better water quality prediction. J. Hydrol. **578**, 124084 (2019)
2. AlMetwally, S.A.H., Hassan, M.K., Mourad, M.H.: Smart water monitoring system using IoT. Int. Res. J. Eng. Technol. **5**(10), 1170–1173 (2018)
3. AlMetwally, S.A.H., Hassan, M.K., Mourad, M.H.: Real time internet of things (IoT) based water quality management system. Proc. CIRP **91**, 478–485 (2020)
4. Ashwini, K., Vedha, J.J., Priya, M.: Intelligent model for predicting water quality. Int. J. Adv. Res. Ideas Innov. Technol. ISSN **5**(2), 70–75 (2019)
5. Brenniman, G.R.: Potable water. In: Environmental Geology, pp. 492–492. Springer (1999)
6. Daigavane, V.V., Gaikwad, M.: Water quality monitoring system based on IoT. Adv. Wwirel. Mobile Commun. **10**(5), 1107–1116 (2017)
7. Daigger, G.T.: Tools for future success. Water Environ. Technol. **15**(12), 38–45 (2003)
8. Haghiabi, A.H., Nasrolahi, A.H., Parsaie, A.: Water quality prediction using machine learning methods. Water Quality Res. J. **53**(1), 3–13 (2018)
9. Hussam, A.: Potable water: nature and purification. In: Monitoring Water Quality, pp. 261–283 (2013)
10. Jain, H., Buch, M., Babu, P.: Water management system using machine learning. In: Data Engineering and Intelligent Computing, pp. 481–492. Springer (2021)
11. John, A., Cardiff, B., John, D.: A 1d-CNN based deep learning technique for sleep apnea detection in IoT sensors. In: 2021 IEEE International Symposium on Circuits and Systems (ISCAS), pp. 1–5. IEEE (2021)
12. kadiwal, A.: Water quality. https://www.kaggle.com/adityakadiwal/water-potability. Accessed 30 Sept 2021
13. Khalaf, M., Alaskar, H., Hussain, A.J., Baker, T., Maamar, Z., Buyya, R., Liatsis, P., Khan, W., Tawfik, H., Al-Jumeily, D.: IoT-enabled flood severity prediction via ensemble machine learning models. IEEE Access **8**, 70375–70386 (2020)
14. Koditala, N.K., Pandey, P.S.: Water quality monitoring system using IoT and machine learning. In: 2018 International Conference on Research in Intelligent and Computing in Engineering (RICE), pp. 1–5. IEEE (2018)
15. Liu, P., Wang, J., Sangaiah, A.K., Xie, Y., Yin, X.: Analysis and prediction of water quality using LSTM deep neural networks in IoT environment. Sustainability **11**(7), 2058 (2019)
16. Moparthi, N.R., Mukesh, C., Sagar, P.V.: Water quality monitoring system using IoT. In: 2018 Fourth International Conference on Advances in Electrical, Electronics, Information, Communication and Bio-Informatics (AEEICB), pp. 1–5. IEEE (2018)
17. Mosavi, A., Salimi, M., Faizollahzadeh Ardabili, S., Rabczuk, T., Shamshirband, S., Varkonyi-Koczy, A.R.: State of the art of machine learning models in energy systems, a systematic review. Energies **12**(7), 1301 (2019)
18. Pappu, S., Vudatha, P., Niharika, A., Karthick, T., Sankaranarayanan, S.: Intelligent IoT based water quality monitoring system. Int. J. Appl. Eng. Res. **12**(16), 5447–5454 (2017)
19. Patil, J., Grampurohit, A., Yadav, V., Nair, A., Selvan, S.: Iot based water monitoring and alert system. IRJET **7**, 3396–3401 (2020)
20. Pincheira, M., Vecchio, M., Giaffreda, R., Kanhere, S.S.: Cost-effective IoT devices as trustworthy data sources for a blockchain-based water management system in precision agriculture. Comput. Electron. Agric. **180**, 105889 (2021)
21. Pule, M., Yahya, A., Chuma, J.: Wireless sensor networks: a survey on monitoring water quality. J. Appl. Res. Technol. **15**(6), 562–570 (2017)
22. Verma, P., Kumar, A., Rathod, N., Jain, P., Mallikarjun, S., Subramanian, R., Amrutur, B., Kumar, M.M., Sundaresan, R.: Towards an iot based water management system for a campus. In: 2015 IEEE First International Smart Cities Conference (ISC2), pp. 1–6. IEEE (2015)

23. Wang, Y., Zhou, J., Chen, K., Wang, Y., Liu, L.: Water quality prediction method based on LSTM neural network. In: 2017 12th International Conference on Intelligent Systems and Knowledge Engineering (ISKE), pp. 1–5. IEEE (2017)
24. WHO: Drinking-water. https://www.who.int/news-room/fact-sheets/detail/drinking-water (June 2019). Accessed 30 Sept 2021
25. Wu, X., Shivakumara, P., Zhu, L., Zhang, H., Shi, J., Lu, T., Pal, U., Blumenstein, M.: Fourier transform based features for clean and polluted water image classification. In: 2018 24th International Conference on Pattern Recognition (ICPR), pp. 1707–1712. IEEE (2018)
26. Wu, Y., Zhang, X., Xiao, Y., Feng, J.: Attention neural network for water image classification under IoT environment. Appl. Sci. **10**(3), 909 (2020)

Smart IoT Biofloc Water Management System Using Decision Regression Tree

Samsil Arefin Mozumder and A. S. M. Sharifuzzaman Sagar

Abstract The conventional fishing industry has several difficulties: water contamination, temperature instability, nutrition, area, expense, etc. In fish farming, Biofloc technology turns traditional farming into a sophisticated infrastructure that enables the utilization of leftover food by turning it into bacterial biomass. The purpose of our study is to propose an intelligent IoT Biofloc system that improves efficiency and production. This article introduced a system that gathers data from sensors, stores data in the cloud, analyzes it using a machine learning model such as a decision regression tree model to predict the water condition, and provides real-time monitoring through an Android app. The proposed system achieved a satisfactory accuracy of 79% during the experiment.

Keywords Biofloc · IoT · Machine learning · Decision regression tree · Aquaculture

1 Introduction

Aquaculture is proliferating across the globe these days, and an indoor fish farm known as Biofloc technology (BFT) is becoming more widespread in fish farming. Aquatic creatures are cultivated in a large tank using Biofloc technology by altering the water quality. Biofloc technology can help us save money on foodstuffs, while an intelligent aquaculture infrastructure can help us save money on the workforce. It is a cost-effective method that is helpful to the fish's health [1, 2]. Moreover, it is regarded as a water-reuse system that is effective.

In this method, the tank's water supply is restricted, and the fish are allowed to grow over an extended period without receiving any water changes. Water samples

S. A. Mozumder
East Delta University, Chattogram, Bangladesh

A. S. M. Sharifuzzaman Sagar (✉)
Sejong University, Seoul, South Korea
e-mail: sharifsagar80@sju.ac.kr

S. Hossain et al. (eds.), *Proceedings of International Conference on Fourth Industrial Revolution and Beyond 2021*, Lecture Notes in Networks and Systems 437,
https://doi.org/10.1007/978-981-19-2445-3_15

are traditionally taken and analyzed, and the results take time to get back. It's challenging to keep track of water quality indicators and change water according to impurity levels regularly. Furthermore, Bangladesh is prone to various natural disasters, such as floods and cyclones, which significantly impact aquaculture in both reservoirs and marine areas. For such catastrophes, fish producers must suffer a significant loss due to contaminated water and rising salinity in coastal waters. Conventional fish farming contributes to a number of additional difficulties, including carbon dioxide, ammonia, and nitrogen pollution in the water supply. Detoxification is a costly and time-consuming process that involves additional treatment. Biofloc is a great alternative to the conventional farming approach since it is more cost-efficient. Biofloc aids in the natural purification of water, reducing the need for additional equipment or substances. Maintaining water quality may result in increased output and decreased fish mortality. The most critical elements affecting the operation of a fish farm employing Bioflocs are the water quality characteristics.

IoT is now playing a significant role in the intelligent aquaculture industry, allowing for remote monitoring of many conditions. The Internet of Things based fish farming is an excellent option for resolving the significant issue that occurs in Biofloc. Additionally, there is a significant economic disparity between intelligent farming and conventional farming. The Internet of Things simplifies fish farming by using a range of sensors and actuators in combination with a robust communication infrastructure.

The article proposes an Internet of Things based intelligent water monitoring system for Biofloc technology in Bangladesh. The proposed system monitors and measures water quality using sensors capable of sensing the quality of water components such as dissolved oxygen, nitrogen, pH, water temperature, nitrate, ammonium, and carbon dioxide, and transmitting the measured data to a mobile device through a mobile application. Moreover, the system automatically controls the actuators to resolve the water quality issues if the water quality level exceeds a certain threshold. Fish producers can also resolve this issue by examining the measured values of water quality parameters from their smartphones.

2 Related Works

Researchers have suggested different water quality management systems and control methods for aquaculture; however, they concentrated on a few types of sensors such as pH, temperature, and water level, while excluding water regulating actuators. Zougmorc et al. [3] developed low-cost IoT solutions for African agricultural fish producers; however, the system lacks an automated actuator that regulates water parameters. Encinas et al. [4] presented a ZigBee-based wireless sensor network for aquaculture systems that included a temperature sensor, a pH sensor, and a dissolved oxygen sensor. However, water parameter management is not performed here as well. Liu et al. used the Recirculating Aquaculture System (RAS) to conduct an experiment on Ras Carpio [5]. In 2011, RAS was a better solution to pond-based aquaculture. The water parameters were continuously checked, and if the value of a parameter exceeded specific parameters, the water automatically recirculated. WATT TriO Matic, WATT Sensolyt, and WATT Trioxy Therm type sensors were used to measure dissolved oxygen, pH, and temperature. While the method had many advantages and was quickly superseded by a conventional aquaculture system, it did have certain drawbacks. It requires water exchange, which is a time-consuming and expensive procedure. Dzulqornain et al. developed a novel IoT system based on the IFTTT architecture [6]. They measured dissolved oxygen levels, the temperature of the water, and the PH level. Sensors were used to determine the water level, and an aerator system was used to regulate the system. The aerator system was integrated with a microcontroller NodeMCU, a relay, a power source, and a propeller. Sensor data was transferred to the web, so the customer could see it from any location. Teja et al. [7] proposed a real-time management system for fish ponds using the ESP 32 development board, AWS cloud, and sensor technologies. This system is made up of a pH sensor, ultrasonic sensor, and DHT 11 sensor for determining the quality of water. This system monitored sensors; however, the suggested system's shortcoming is that actuator control of water quality parameters was not performed.

To optimize production, individuals must examine the following water quality parameters and adhere to the ethical requirements for fish farming. Water temperature is a critical factor that has a significant impact on fish farming. Fish larvae can live at varying temperatures. Fish have a poor tolerance for temperature changes that occur rapidly. The temperature has other detrimental effects since algae and zooplankton are also temperature sensitive. It controls oxygen levels, pH, salinity, and other temperature-dependent water characteristics both directly and indirectly. As the temperature rises, dissolved oxygen levels drop, and more carbon dioxide is generated in the water due to the lack of oxygen in hot water [8, 9]. For better fish health and development, the water in the tank must be between 24 and 30 °C [10]. However, the water temperature has an effect on the fish, as does the pH (Potential of Hydrogen) value indicating the acidity or alkalinity of a solution. The water in the tank may be acidic (pH 7.0)—alkaline (pH greater than 7.0). As the pH of the water decreases, the ammonia ions (NH_3) chemically react to form ammonium ions (NH_4+) and hydroxyl ions (OH−) [11]. And carbon dioxide ($C0_2$) as pH levels rise, they combine chemically with water to produce toxic ammonia ions (NH_3) that harm and kill fish [10].

3 Proposed System Architecture

The architecture of our intelligent Biofloc water quality monitoring system is shown in Fig. 1. The system constantly monitors and transmits alerts to a smartphone through cloud infrastructure using an ESP32 development kit module, allowing users to access real-time data from anywhere. The project's primary goal is to examine the water's pH, temperature, TDS, and electrical conductivity. We also utilized machine learning techniques such as a decision regression tree on the Biofloc dataset to train and predict

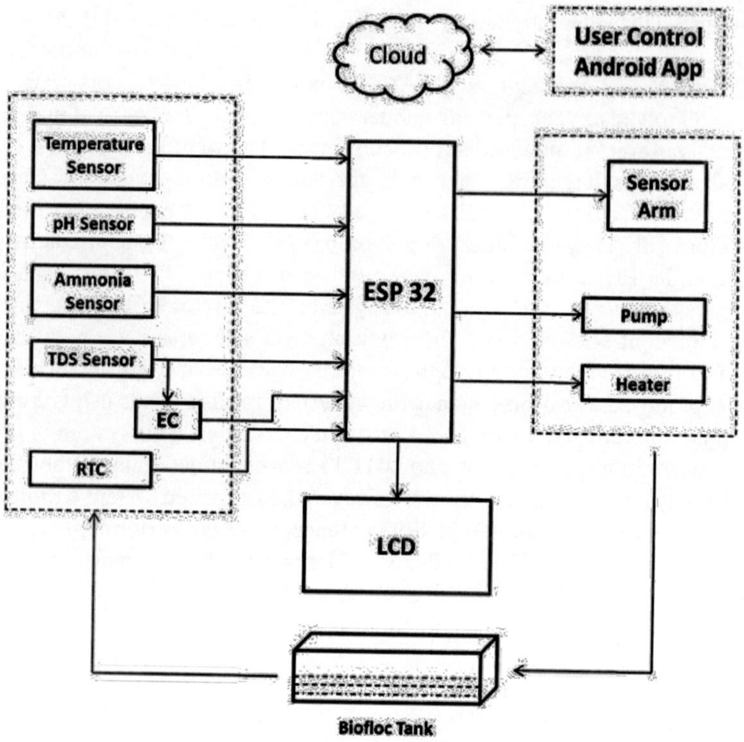

Fig. 1 The proposed system architecture of the smart IoT Biofloc water management system

the water quality of the Biofloc container. The system's sensors give a real-time measurement of the water's qualities. The system produces output by analyzing the training data and the sensor readings. It forecasts water quality, assesses the situation, and automatically controls actuators to maintain a specific level of water quality. It notifies the user of the result and option through an app, displaying real-time water metrics.

3.1 *Hardware Architecture*

The hardware architecture mainly consists of three parts such as the perception, IoT cloud, and graphical user interface (GUI). The components used in the perception parts are pH sensor, temperature sensor, TDS sensor, ammonia sensor, RTC, LCD, ESP32 development Kit, sensor arm, pump, and heater. We used Google Firebase cloud service to store the real-time data. Moreover, we developed an Android application for the user to check and control the water quality remotely.

PH Sensor: The PH sensor detects the flow of hydrogen ions in liquid systems, showing causticity or alkalinity as PH. For ESP32, we utilize a PH sensor, and model the PH sensor module probe. This sensor is used in our research to measure soil acidity and water quality.

Water Temperature Sensor: A water temperature sensor, model DS18B20, is used in this system. It can accurately detect temperatures ranging from 55 to 125 °C.

Ammonia Sensor: One of the most serious threats to fish and shrimp farming viability is ammonia. We used an ammonia sensor to monitor ammonia in the Biofloc system. We used the MQ-135 Gas sensor to collect ammonia data.

TDS Sensor: The TDS value shows the number of soluble solids dissolved in one liter of water. Water with a high total dissolved solids (TDS) value is generally considered to be less clean than water with a lower TDS value. As a result, the TDS value may be utilized as one of the indicators of water quality. We have used the TDS meter to measure the turbidity of the water.

Electrical Conductivity: Biofloc's primary objective is to provide and maintain an appropriate and stable habitat for fish. Electrical conductivity (EC) is a property of water that indicates its capacity to "transport" an electrical current, and therefore, an approximation for the concentration of dissolved particles or ions in the water. Pure water has a very low conductivity; therefore, the more dissolved particles and ions in the water, the more electrical current it can conduct. EC is critical in checking the water's quality.

ESP32 Development Board: The ESP32 development board is the successor of the ESP8266 kit. The ESP32 has several additional features over its predecessor. It has dual-core processing and wireless connectivity through Wi-Fi and Bluetooth.

IoT Cloud: We can store all sensor data in the Biofloc water monitoring system with the help of the IoT cloud service. It also provides the system's visual analysis. In our research, we used the Google Firebase cloud servers to store and transfer data to the user end.

GUI: GUI is essential for real-time IoT monitoring systems. Users can monitor sensor data and control actuators using the GUI remotely. We developed an Android app using Android Studio for the proposed system. The app is connected with the Google Cloud and acquires the stored data from the cloud. Users can check real-time data using the app. We have also made it available for the user to control the actuators through the app.

3.2 Control System Architecture

Algorithm 1 describes the control procedure of the proposed Biofloc water monitoring system. The proposed system uses pH, water temperature, ammonia, TDS, and electrical conductivity to process and analyze the data. The system then predicts the water condition and control actuator based on the results of the water condition prediction. Firstly, the system saves the sensor data to the relevant variables; then it sends the data to the IoT cloud for storing and real-time monitoring. Secondly, the

IoT cloud sends the data to the user's Android app, where users can check real-time sensor data and control the actuators manually. Thirdly, sensor data are also fed into the classifier to predict the water condition of the Biofloc system. If the water condition is not good, the system automatically checks every sensor data to control the actuators.

Algorithm 1: Smart IoT Biofloc water monitoring system

Input : PH sensor, Water temperature sensor, Ammonia sensor, TDS sensor, Electrical conductivity

Output: Water condition prediction, actuator control

```
var ← PHsensor, Watertemperaturesensor, Ammoniasensor, TDSsensor, Electricalconductivity
var → IoTcloud
IoTcloud → Androidapp
var → classifier
condition ← classifier
if condition ≠ good then
    if temperature > 30 then
        Turn on Oxygen Pump and cooling fan
    if temperature < 24 then
        Turn on heater
    if PH > 9 then
        Turn on oxygen pump and add phosphoric acid
    if PH < 6.5 then
        Turn on water filter and add baking soda
    if turbidity > 1800||turbidity < 1100 then
        Turn on water filter
else
    Turn off all actuators
```

3.3 System Prototype

Figure 2 shows the proposed system prototype. We have used a water bulb to demonstrate the water heater. A small pump is used to demonstrate the oxygen pump, and a servo motor is used to demonstrate the cooling function of the proposed system. We have also attached a sensor arm equipped with a TDS and temperature sensor probe to clean the probe from time to time. The sensor arm is controlled through a servo motor, where an Android application can control the servo motor.

Figure 3 shows the central control and sensing board of the prototype. The control board consists of a microcontroller, relay, and an LCD panel to show the real-time sensor data. Sensors and actuators are attached to the ESP32 microcontroller through the PCB board.

Figure 4 shows the raw sensor data stored in the Google Firebase cloud from the prototype. Real-time sensor data with actuator conditions are stored and sent to the user's Android application through the IoT cloud. Users can access this data from anywhere in the world.

Figure 5 shows the raw sensor data collected from the prototype in our developed Android application. The app is connected to the IoT cloud, and it updates the sensor

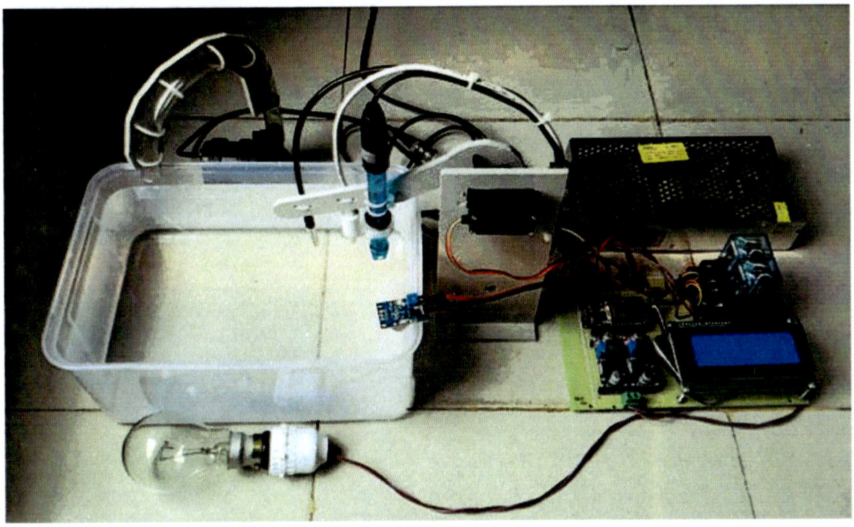

Fig. 2 Proposed smart IoT Biofloc water management system prototype

Fig. 3 The proposed central control unit for sensors and actuators

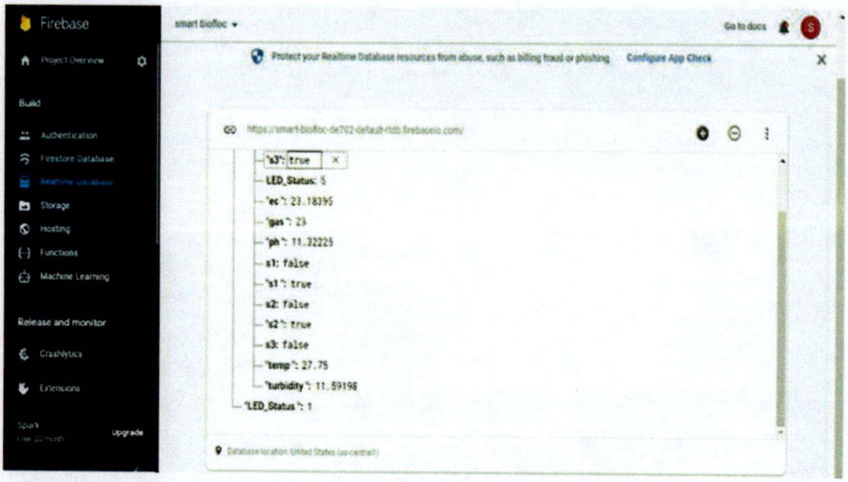

Fig. 4 Sensor data stored in the Google Firebase IoT cloud

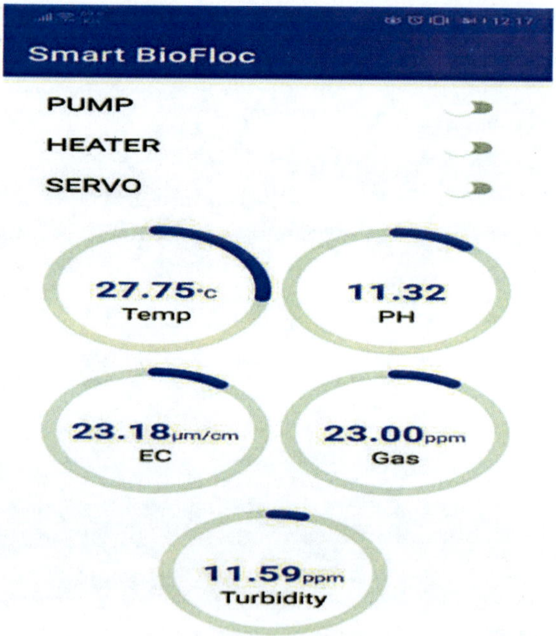

Fig. 5 The GUI of the developed smart IoT Biofloc water management system Android application

data in real-time. Temp represents the temperature, pH represents the pH level, EC represents the electrical conductivity, Gas represents the ammonia gas level, and turbidity represents the water's turbidity level. We have also added three switches to control the actuators remotely in an emergency if the automatic system fails to work.

3.4 Machine Learning Model

The decision regression tree machine learning model is used in the proposed system to predict the water condition of the Biofloc container. The decision tree model performs very well when dealing with tabular data with numerical characteristics or reference variables with less than hundreds of categories. Decision trees, as opposed to linear models, may capture nonlinear interactions between characteristics and targets.

We have taken 4000 samples using the temperature, pH, and TDS sensor to train our classifier. The data was processed and stored in a specific string format to extract the features from the sensor data. Many Python libraries provide decision regression trees to train datasets. We used the Microml Python library to train the model because it also provides the porting facility to the microcontroller. Then, we used the EloquentArduino library in the Arduino IDE for ESP32 to make use of the ported classifier to predict real-time water conditions.

4 Result Analysis

The proposed system is successfully implemented in the Biofloc system. We observed simultaneous data transmission from the system to the cloud and Android app. Figure 6 shows the real-time sensor data in the LCD panel during the experiment. It is easily observable that our system successfully collects the sensor data. As water condition is very important for healthy fish production, actual sensing data will keep the water condition at a suitable level to accelerate the fish production without any need for humans.

We prepared the dataset of 4000 samples for regression tree classification training. We used temperature, pH, and turbidity as key parameters to determine the water condition of the Biofloc system. We have also set a certain threshold for the key parameters for the good condition, such as temperature (24–30), PH (6–9), and turbidity (1200–1800). If any of the corresponding values of the parameters go

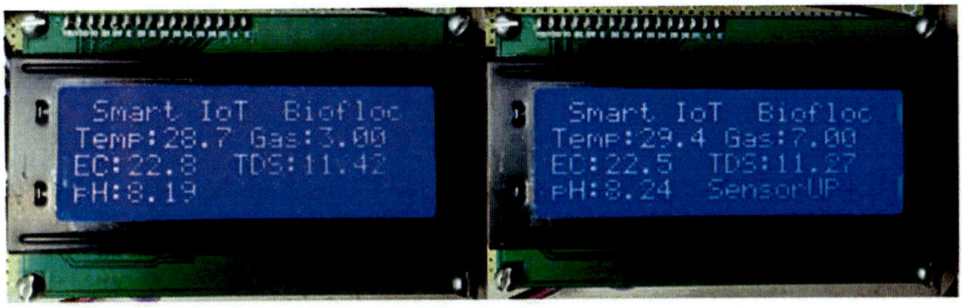

Fig. 6 Sensor value displayed in the LCD panel during the experiments

Table 1 Dataset samples used in the machine learning model

Condition	Temperature (°C)	PH	TDS
Good	24.98	7.81	1350
Good	25.59	7.81	1760
Good	26.03	7.98	1750
Good	25.95	7.95	1740
Good	26.31	7.66	1740
Bad	23.00	9.90	1850
Bad	31.00	7.00	1400
Bad	23.00	9.00	1600
Bad	33.00	5.70	1050
Bad	27.90	7.16	1100

beyond the threshold, we label the data as bad water conditions. Table 1 shows the 10 samples from the prepared dataset used to train the classifier. These data were fed into the regression tree machine learning model for training. We have used the Micromlgen Python library to train the data. Micromlgen provides different machine learning models to train and port the results in C++ files for use in the microcontrollers. The generated C++ file can be used as a classifier library in the microcontroller for the water condition prediction.

Table 2 shows the different sensor values collected from the Biofloc system for a day. We can see that the early morning and the evening temperature of the water was very low. During the mid-day, the temperature was higher than in the morning, which also affects the other parameters of the Biofloc system, such as pH, EC, and turbidity. The gas level was stable throughout the day.

Figure 7 shows the predicted water condition in the Arduino IDE. The classifier takes 500 ms to predict the water condition from the acquired sensor data of the Biofloc system. We can see that the water condition of the Biofloc was good during the experiment.

Previous research works on smart Biofloc systems are based on Long Short-Term Memory (LSTM) [13], Least-Squares Support Vector Regression (LSSVR) [14], Group Data Handling Method (GMDH) [15], Support Vector Machine (SVM) [16], and Artificial Neural Network (ANN) [17, 18], which measure pH, DO, BOD, COD, TDS, EC, PO_{43}-,NO_3-N, and NH_3-N, and continuously compare with the acceptable range.

Some research proposed manual control over the app; most of them could take the reading from the sensors by a microcontroller and display them on LCD. Those systems always had to have a person to monitor and manually manipulate parameters accordingly. Our proposed system can predict water conditions and maintain the whole process in a precise way. It needs only a small portion of space as compared to the pond and wide swampland. A container is enough to produce a large number of fish in the house.

Table 2 Sensor output during the experiment in the Biofloc system

Water quality Parameter					
Time	Temperature (°C)	pH	TDS (Mg/L)	EC μm/cm	NH_3 ppm
5:00 AM	25.56	8.1	1752	45.85	5.95
5:30 AM	25.59	7.98	1760	45.65	5.96
6:00 AM	26.03	7.98	1750	45.65	5.98
6:30 AM	25.95	7.95	1740	45.62	5.98
7:00 AM	26.31	7.66	1740	45.63	5.65
7:30 AM	27.45	7.54	1740	45.64	5.89
8:00 AM	27.32	7.36	1741	45.58	5.91
8:30 AM	27.35	7.20	1739	45.59	5.91
9:00 AM	27.86	7.19	1734	46.02	5.97
9:30 AM	27.90	7.16	1734	45.99	6.07
10:00 AM	28.21	7.18	1732	46.00	6.09
10:30 AM	28.25	7.17	1730	46.10	6.08
11:00 AM	28.43	7.16	1728	46.11	6.11
11:30 AM	29.06	7.16	1735	46.15	6.13
12:00 PM	29.12	7.16	1729	46.16	6.11
12:30 PM	29.45	7.17	1728	46.17	6.13
1:00 PM	29.33	7.14	1726	46.22	6.26

Fig. 7 Predicted water condition level in the ESP32 serial monitor

Table 3 Performance comparison between different proposed water monitoring systems

Method	Accuracy (%)	Parameters	Solution	Smart decision
LSTM [13]	82	pH, temperature, DO	Manual control	Yes
LSSVR [14]	76	Temperature, DO, TDS, pH, EC, PO43-, NO3-N	Manual control	Yes
GMDH [15]	74	Temperature, DO, TDS, pH	Manual control	Yes
SVM [16]	70	Temperature, DO, TDS, pH	Manual control	Yes
ANN [17]	72	Temperature, DO, TDS, pH, BOD, COD, TSS	Manual control	Yes
ANN [18]	77	Temperature, DO, TDS, pH, Floc	Automatic control	Yes
Proposed method	79	Temperature, DO, TDS, EC, pH, NH_3	Automatic control	Yes

We conducted a comparison of our implemented system to pre-existing systems. We highlighted the method's precision, parameters, prediction, and decision-making capabilities. The comparison is shown in Table 3. We can see that our approach achieved a good accuracy of 79%, which outperforms some existing methods. Moreover, we have also monitored different parameters that were not included in the existing system, such as ammonia gas monitoring. However, the LSTM method [13] yields good accuracy, but they did not provide any automatic control system for water monitoring, which is very important for IoT-based systems. Therefore, our proposed system can be considered a robust Biofloc water condition monitoring system.

5 Conclusion

Fish farming has been thriving for decades despite economic limitations, increasing prices, or even nonavailability of personnel, regular water management, and rapid increases in pollution. The smart system analyzes water conditions in real-time, lowering production costs, increasing productivity, reducing human dependence, and ensuring socioeconomic sustainability. The suggested system checks water quality in real time and provides an immediate warning to the user. A machine learning method such as a decision regression tree was used to maintain a sustainable water condition. Experiments on the implemented features were conducted to validate the proposed IoT system. According to the tests, 79% accuracy was achieved. In the future, we want to enhance the model's accuracy and assess how well it performs in terms of the fish growth in the Biofloc.

References

1. Ghose, B.: Fisheries and aquaculture in Bangladesh: challenges and opportunities. Aquaculture Research. In: Bashar A. (ed.) Biofloc aquaculture: prospects and challenges in Bangladesh (2014). https://doi.org/10.13140/RG.2.2.13233.94560
2. Das, S.K., Mandal, A.: Biofloc Technology (BFT): an effective tool for remediation of environmental issues and cost effective novel technology in aquaculture. Int. J. Oceanogr. Aquacul. **2**(2) (2018). https://doi.org/10.23880/ijoac-16000135
3. Zougmore, T.-W., Malo, S., Kagembega, F., Toguevini, A.: Low cost IoT solutions for agricultures fish fanners in Afirca: a case study from Burkina Faso. In: 2018 1st International Conference on Smart Cities and Communities (SCCIC). Ouagadougou, pp. 1–7 (2018)
4. Encinas, C., Ruiz, E., Cortez, J., Espinoza, A.: Design and implementation of a distributed IoT system for the monitoring of water quality in aquaculture. In: 2017 Wireless Telecommunications Symposium (WTS). Chicago, IL, USA, pp. 1–7 (2017)
5. Liu, S., et al.: Prediction of dissolved oxygen content in river crab culture based on least squares support vector regression optimized by improved particle swarm optimization. Comput. Electr. Agricul. **95**, 82–91 (2013). https://doi.org/10.1016/j.compag.2013.03.009
6. Dzulqornain, M.I., Rasyid, M.U., Sukaridhoto, S.: Design and development of smart aquaculture system based on IFTTT model and cloud integration (2018)
7. Teja, K.B.R., Monika, M., Chandravathi, C., Kodali, P.: Smart monitoring system for pond management and automation in aquaculture. In: 2020 International Conference on Communication and Signal Processing (ICCSP). Chennai, India, pp. 204–208 (2020)
8. Saaid, M.F., Fadhil, N.S.M., Ali, M.S.A.M., Noor, M.Z.H.: Automated indoor Aquaponic cultivation technique. In: 2013 IEEE 3rd International Conference oil System Engineering and Technology, pp. 285–289 (2013)
9. Boyd, C.E.: Water temperature in aquaculture « Global Aquaculture Advocate. Global Aquaculture Alliance. https://www.aquaculturcalhance.org/advocate/water-temperature-in/aquaculture
10. Bhatnagar, A., Devi, P.: Water quality guidelines for the management of [Kind fish culture 3, 30 (2013)
11. Tucker, C.S., D'Abramo, L.R.: Managing high pi I in freshwater ponds. Stoncvillc: Southern Regional Aquaculture Center (2008)
12. Kubitza, F.: The oft-overlooked water quality parameter:pH « Global Aquaculture Adv cate. Global Aquaculture Alliance. https://www.aquaculturealliance.oig/advocaterihe-of-overlookedwater-quality-parameter-ph
13. Hu, Z., Zhang, Y., Zhao, Y., Xie, M., Zhong, J., Tu, Z., Liu, J.: A water quality prediction method based on the DEEP LSTM network considering correlation in Smart Mariculture. Sensors **19**(6), 1420 (2019). https://doi.org/10.3390/s19061420
14. Tan, G., Yan, J., Gao, C., Yang, S.: Prediction of water quality time series data based on least squares support vector machine. Proc. Eng. **31**, 1194–1199 (2012). https://doi.org/10.1016/j.proeng.2012.01.1162
15. Varis, O.: Water quality models: tools for the analysis of data, knowledge, and decisions. Water Sci. Technol. 30(2), 13–19 (1994). https://doi.org/10.2166/wst.1994.0024
16. Haghiabi, A.H., Nasrolahi, A.H., Parsaie, A.: Water quality prediction using machine learning methods. Water Qual. Res. J. **53**(1), 3–13 (2018). https://doi.org/10.2166/wqrj.2018.025
17. Kadam, A.K., Wagh, V.M., Muley, A.A., Umrikar, B.N., Sankhua, R.N.: Prediction of water quality index using artificial neural network and multiple linear regression modelling approach in Shivganga River Basin India. Model. Earth Syst. Env. **5**(3), 951–962 (2019). https://doi.org/10.1007/s40808-019-00581-3
18. Rashid, M.M., Nayan, A., Simi, S.A., Saha, J., Rahman, M.O., Kibria, M.G.: IoT based smart water quality prediction for biofloc aquaculture (2021)

Temperature and Current Density Analysis of Thermoelectric Generator for Regenerative Breaking of the Hybrid Electric Vehicle

Md. Zilan Uddin Saif, Farhan Tasnim, Md. Abu Hanif, Md. Rasel Ahmed, Kawshik Shikder, and Chowdhury Akram Hossain

Abstract The advancement in the area of Hybrid Electric Vehicles (HEVs) is one of the most dynamic in the modern world with the concern of global warming issues. The incorporation of a Regenerative Braking System (RBS) into this technology is also a primary element in keeping this field up to date fiercely. It is shown in this article that the suggested Thermo-Electric Generator (TEG) model can be compared to other current RBS technologies using the comparative research method described above. Aside from that, a representation of the model is provided together with the results of the temperature and current density study performed using the TEG. The suggested design was simulated using the Ansys mechanical model and the Ansys 2021 Workbench software, which was used to generate a test simulation of the concept. This device captures the wasted heat of the vehicle energy from the brake pad of the wheels and turns it into useful electrical energy. In accordance with the principles of RBS, this design will be one of the most important sources of extending the driving range of HEVs while simultaneously reducing the net cost of recharging for the end-user.

Keywords TEG · HEV · Ansys · Heat transfer · Regenerative braking · Temperature

Md. Z. U. Saif (✉) · F. Tasnim · Md. A. Hanif · Md. R. Ahmed · K. Shikder · C. A. Hossain
American International University-Bangladesh, Dhaka, Bangladesh
e-mail: zilanuddin69@gmail.com

F. Tasnim
e-mail: farhantasnim1649@gmail.com

Md. A. Hanif
e-mail: ahrobinkhan123@gmail.com

Md. R. Ahmed
e-mail: mdraselahmed779@gmail.com

K. Shikder
e-mail: kawshik.shikder@aiub.edu

C. A. Hossain
e-mail: chowdhury.akram@aiub.edu

S. Hossain et al. (eds.), *Proceedings of International Conference on Fourth Industrial Revolution and Beyond 2021*, Lecture Notes in Networks and Systems 437,
https://doi.org/10.1007/978-981-19-2445-3_16

1 Introduction

The automobile industry has shown interest in incorporating new technology into vehicle design throughout the previous few decades. These technologies are employed in the emerging electric vehicles to enhance road safety and reduce the environmental impact. In terms of environmental impact, the transportation sector has been responsible for Greenhouse Gas emissions for the past 30 years or more, and attempts have been undertaken to mitigate the environmental effects of such emissions [1]. Electric Vehicles (EVs) are pushed as a clean alternative to traditional vehicles that use fossil fuels as a swift way to enhance environmental quality parameters in the transportation sector [2]. Hybrid electric vehicles (HEVs), which combine an engine with one or more electric motors, are widely regarded as the most promising clean vehicle option in the short to medium term. Various manufacturers are currently developing hybrid electric vehicles. These automobiles are currently being marketed as a way to improve the efficiency of our transportation system while also lowering our dependency on and use of fossil fuels. A recent study has been on how energy-efficient and cost-effective HEVs are [3]. In urban road conditions with frequent start–stop or acceleration–deceleration drive patterns, regenerative braking plays a significant part in improving energy efficiency and extending the driving range of electric vehicles. When an electric vehicle (EV) is braking, the regenerative braking system (RBS) can convert the vehicle's kinetic or potential energy into electrical energy, which can be used to charge the power battery and give motor braking torque to aid braking and prevent brake temperature rise [4]. In recent years, the global has focused on thermoelectric electric generators for waste heat recovery. Thermoelectric electric generation (TEG) is one of the most prominent technologies for producing electrical current directly from a temperature differential. With respect to the performance of thermoelectric materials, thermoelectric technology is currently facing issues in efficiency and physical property promotion, and a complete evaluation of thermoelectric research can address this issue [5]. In this manuscript, the new model of TEG has been displayed which took the waste heat from the disk pad of the vehicle and converted it to useful electrical energy. The converted energy is stored in the battery of a hybrid electric vehicle (HEV) to further enhance the driving efficiency. And also shows the current density of the TEG module and how much current flows through the chosen cross-sectional area of this module.

2 Literature Review

To enhance the TEG's net output, the authors [6] propose a resilient design optimization (RDO) that optimizes the mean of the performance function while eliminating its variance, resulting in an optimum design that is less susceptible to the TEG system uncertainties. Compared to the deterministic optimal design derived from previous TEG optimization research, the compromise design ensures a consistent and high

TEG net power output. The authors [5] concentrated on the growing interest in thermoelectric technology and applications, thermoelectricity principles, and a description of current and future materials. It also explains which models were created and how nonequilibrium and finite-time thermodynamics were used to improve various thermoelectric applications. The usage of nonlinear loads produces harmonics, and it is desirable to eliminate harmonics while maximizing TEG voltage utilization. The performance of a thermoelectric generator fed LVDC system in nonlinear loads was described in [7]. TEG system can be turned into a hybrid system in combination with solar cells in the future to boost vehicle economy by lowering the usage of nonrenewable fossil fuels and thereby reducing CO2 emissions, making the car eco-friendly. Here, a TEG module generates electrical energy from heat dissipated in the vehicle's engines and silencer surface [8]. The importance of using the self-consistency technique to compute current density in a resonant tunneling device is explained. Because of peaks at specific system compositions, this result is critical for RTD applications with minimal bias [9]. The microscopic processes of ions production, space migration, space recombination, and diffusion are all taken into account in a novel mathematical model of SF 6 gas DC current density and conductivity. Furthermore, authors [10] indicate that the new model enhances the ohm conductance and saturation current transition processes. The ground electrode's geographical current density distribution in a horizontal wide-area multilayer earth, and construct the current density calculation formula and the proposed method addresses the issue of determining the geographical current density distribution of the DC grounding electrode [11]. In our previous research the net output is 41.89 W also the output current is 97.001A [5, 12, 13]

3 Modeling

TEG was investigated and modeled according to a specific method with identical external dimensions, as illustrated in Fig. 1. SOLIDWORKS 2020 was used to model

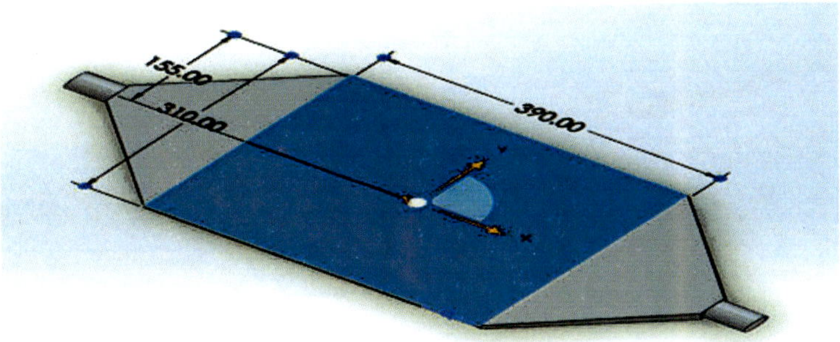

Fig. 1 Modell of TEG in SolidWorks

the heat exchangers' internal area. Each of these models is divided into three sections: front, center, and rear. The central part is fashioned like a rectangular cuboid and is 390 mm × 310 mm × 22 mm in size. According to the calculations, Fig. 1 depicts the input and outflow sections of the heat exchanger, which are 36 mm in diameter. Additionally, Fig. 3 illustrates the back diameter of the matching heat exchanger, while Fig. 4 illustrates the arrangement of thermoelectric modules, which illustrates the kind of heat that will be absorbed. The heat exchangers incorporated in each method are unique in terms of size and angle at the intake. According to the TEG model results, it is clear that the model was evaluated carefully, gradually, and incrementally. A key element influencing the generator's efficiency is its ability to distribute energy evenly across the different TEM modules. A high ZT value is needed in this design, as indicated by the equation since it requires a material with a high α (Seebeck coefficient), σ (high electrical conductivity), and λ (thermal conductivity) all at the same time (1). In particular, Eq. 1 contains the dimensionless coefficient of thermoelectric efficiency, ZT, which specifies the following key device characteristics: (Seebeck coefficient), (electric conductivity), and (thermal conductivity), all of which are temperature dependent [14] (Fig. 2).

Fig. 2 Diameter of the inlet and outlet sections of the heat exchanger

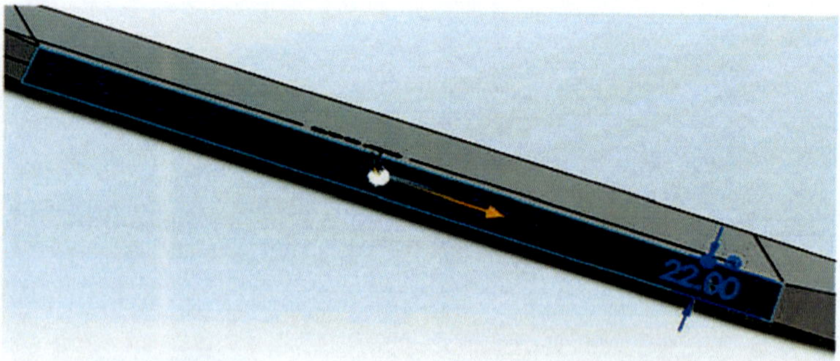

Fig. 3 Rear diameter of the inlet and outlet sections of the heat exchanger

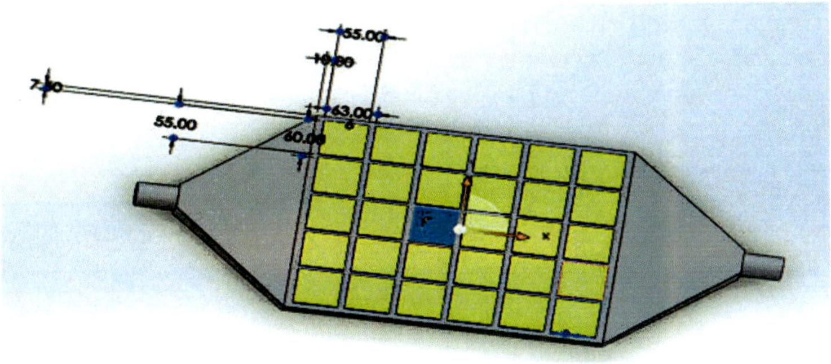

Fig. 4 Arrangement of thermoelectric modules

$$ZT = \frac{\alpha 2.\sigma}{\lambda}.T \quad (1)$$

4 Simulation

A simulation of the proposed design was created using the Ansys mechanical model and the Ansys 2021 Workbench software, which was utilized to build a test simulation of the proposed design. Several parameters are given in the following Table 1 to help you start the simulation. The required material, such as copper alloy, was supplied as input to get things started.

It has been chosen to include the geometry model in the Ansys 2021 R1 Workbench in order to simplify the execution of new simulation processes. This will save time and money. This portion of the mesh analysis process utilized body sizing calculated from the default model size of 0.015 mm in the previous section to determine the mesh size. For the purpose of achieving our desired outcome during the analysis phase, we have selected various sections of the TEG to make our drawings. The temperature range of the disc pad on the surface is 326 to 346 °C, with the anticipated peak temperature, in this case, being 326 °C. In order to get more realistic findings, a variety of temperature settings were simulated in this model, including 290 and 260 °C, among other temperatures [17, 18]. The cold junction temperature is assumed to be 25 °C in this computation, which should be taken into consideration.

Table 1 Parameters of PN materials [15, 16]

Parameter	P-type	N-type
Seebeck coefficient/μV K^{-1})	250	$^{-}$250
Electrical resistivity/Ω m	1.64×10^{-5}	1.64×10^{-5}
Thermal conductivity/W m^{-1} K^{-1}	1.46	1.46

The temperature difference T between the hot and cold sides was determined in this case using Eq. (2). The thermoelectric field, Ethermf, is derived by multiplying the Seebeck coefficient, of TEG, by the Seebeck coefficient of TEG, as shown in Eq. (3) [14, 19–21].

$$\Delta T = T_h - T_c \tag{2}$$

$$E_{thermf} = \alpha.\Delta T \tag{3}$$

R is the internal resistance of the generator; and the computed current is denoted by I, where we get

$$I = \alpha \Delta T R + R_L \tag{4}$$

The electric output power delivered by TEG to the load is

$$P = n.\left(\alpha_{PN}.I.\Delta T - R.I^2\right) \tag{5}$$

When the two voltage ends are linked together, the current density will flow through them from a high potential to a low potential as a result of the voltage difference. Following the setting of all of the parameters for the research, the whole simulation process was carried out, resulting in the results described below in the result analysis section of this report. In the initial stage of the TEG, as shown in Figs. 5, 6, and 7, it was anticipated that the temperature would rise to 326 °C, 290 °C, and 250 °C, respectively, before settling at 326 °C in the most effective way possible (Fig. 8).

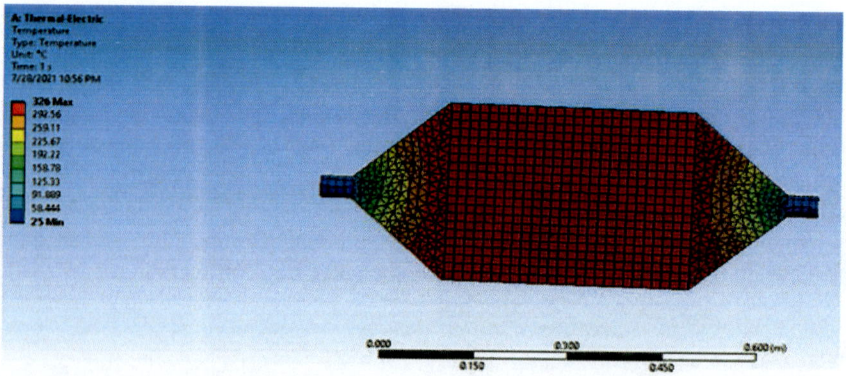

Fig. 5 Result of the temperature distribution at 326 °C over the surfaces with different heat exchangers

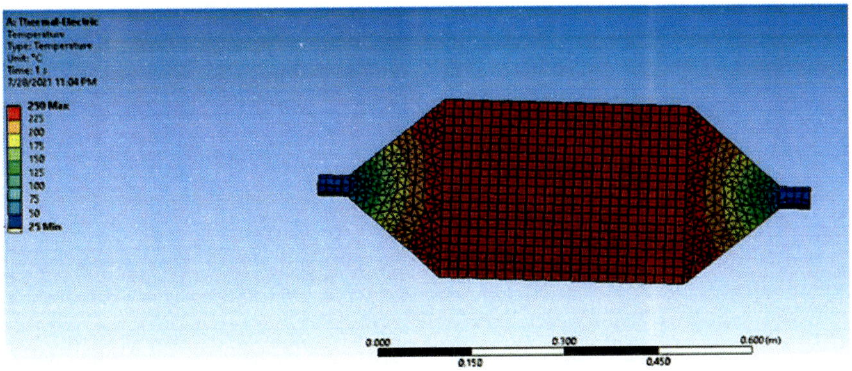

Fig. 6 Result of the temperature distribution at 290 °C over the surfaces with different the heat exchangers

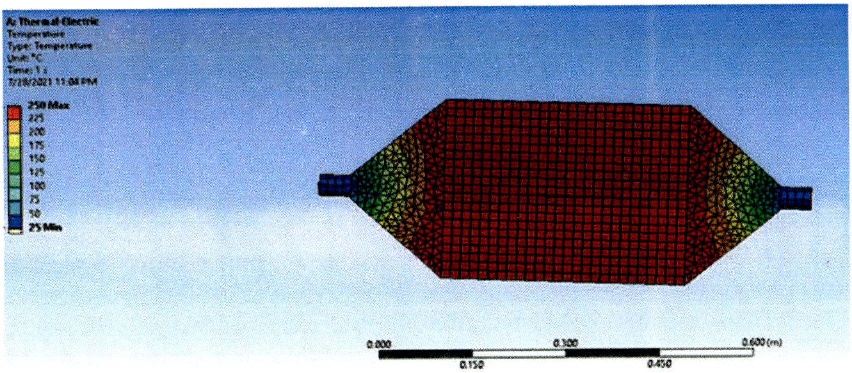

Fig. 7 Result of the temperature distribution at 250 °C over the surfaces with different heat exchangers

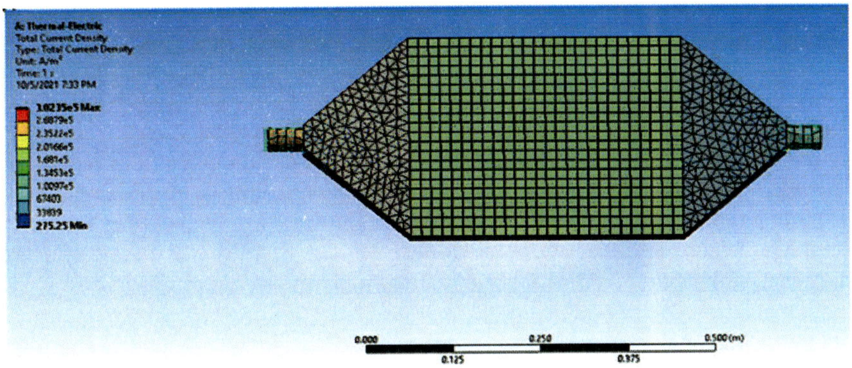

Fig. 8 Current density at 326 °C

Table 2 Output current for current density

Temperature (°C)	Current density (Minimum) [A/m^2]	Current density (Maximum) [A/m^2]	Current density (Average) [A/m^2]
326	275.25	3.0235e + 005	21,228
320	269.38	2.9598e + 005	20,782
310	259.61	2.8537e + 005	20,037
300	249.84	2.7476e + 005	19,292
290	240.09	2.6414e + 005	18,547
280	230.35	2.5353e + 005	17,802
270	220.62	2.4292e + 005	17,058
260	210.89	2.3231e + 005	16,313
250	201.18	2.217e + 005	15,568

5 Result Analysis

Table 2 shows different temperature stages concerning current density. At 326 °C, the current density was 3.0235e+005 A/m^2 at that stage, the average current density is 21228 A/m^2. Also the minimum current density is 275.25 A/m^2; on the other hand, the minimum applied temperature was 250 °C. The current density is 2.217e+005 A/m^2. At that temperature point, the average current density is 15568 A/m^2; the minimum current density is 201.18 A/m^2 as it can be seen that the ratio between temperature and current is better when the temperature is low compared to the current density at a higher temperature according to the table. Therefore, a better current density has been developed with a moderate temperature value than the higher one in terms of efficiency. The wasted heat of the brake pad can be converted into electrical energy.

6 Graphical Analysis

The values of current density and temperature have been plotted in the graph for the minimum, average, and maximum flow of electrons, respectively, in Figs. 9, 10, and11. In Fig. 9, due to the minimum current density, the amount of electric current flowing through a unit cross-sectional area is very less resulting in the lower amplitude of current density in the graph. Here, for the minimum taken value of temperature (250 K), the current density is 201.18 A/m^2; whereas, the value is 15568 A/m^2 for the average and 221700 A/m^2 for the maximum density. Also, in a similar way, for the maximum value of temperature from the simulation, i.e., for 326 K, the values are 275.25 A/m^2 for minimum, 21228 A/m^2 for average, and 302350 A/m^2 for the maximum density. Though all the graphs are linearly increasing with the increase of temperature, the values are differing to a great extent due to the motion of the flow of electrons.

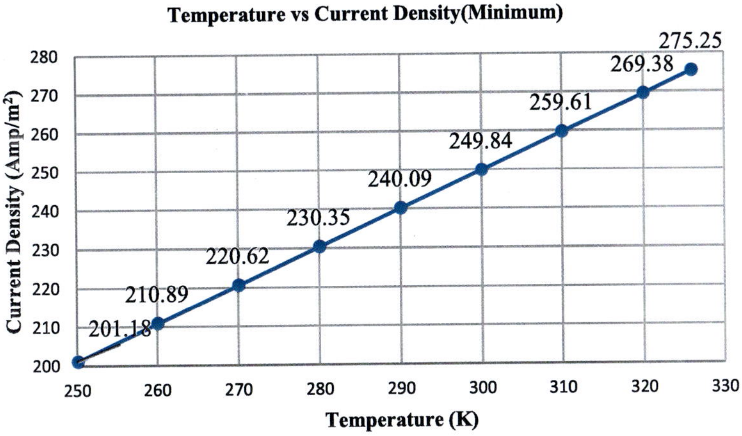

Fig. 9 Temperature versus current density graph for minimum

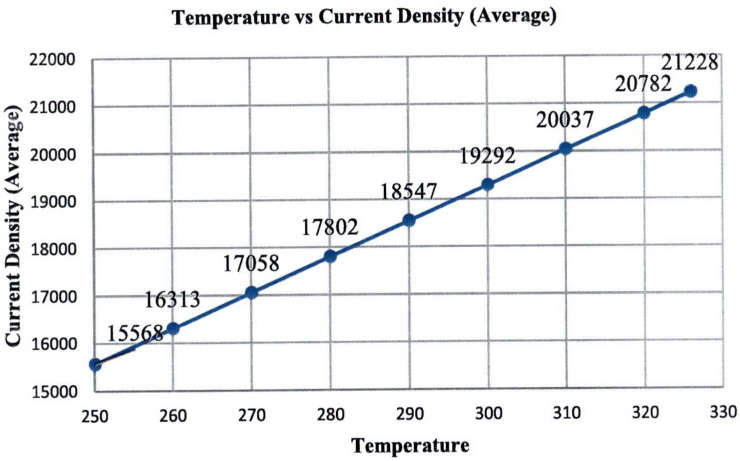

Fig. 10 Temperature versus current density graph for average

7 Comparative Study

TEGs are used to convert the thermal energy lost from the brake pad of an HEV to electrical energy, while kinetic energy is needed to complete the conversion process in the basic regenerative braking system. Rather than using the brake pad's excess kinetic energy, the TEG utilizes the brake pad's thermal energy. This element of the proposed model lends it a particular identity since it has not been seen before in the assessment of this technology in the literature.

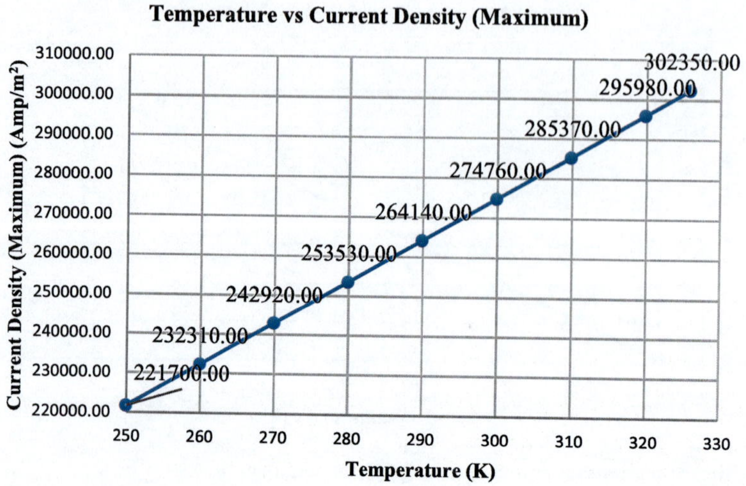

Fig. 11 Temperature versus current density graph for maximum

8 Future Works

The design of a current controller, which would regulate the flow of current from the output of the TEG and maintain the necessary characteristics before reaching the battery of the HEV, would be part of the system's extended development. As the controller's input, the output of TEG will serve as the controller's input, and the electrical output must be maintained at the necessary level to ensure that the energy stored in the battery is preserved. This controller will be built in such a manner that it will be operational even in the event of unexpected uncertainties, taking into account the safety problem and the efficient design for the collected energy of the vehicle that has been proposed. At some point in the future, this energy will be used to expand vehicle range by supplying the necessary power to run the vehicle in its most energy-efficient mode, allowing the vehicle to travel longer distances.

9 Conclusion

TEG is improving as a result of advancements in thermoelectric materials, as well as the ongoing development of heat exchangers for this system. Also taken into consideration are innovations that seek to reduce thermal resistance while simultaneously improving heat transfer. Rather than being utilized less often in this area, TEG is now being used more frequently and for a broader range of applications. Many papers have been written on the geometry of thermoelectric modules, and these publications may be helpful in the development of future designs for these devices. According to this paper, a TEG model was created, and the relationship between output power and temperature was studied in terms of heat production. Preparation of the TEG model was carried out in SOLIDWORKS 2020, with simulation carried out in Ansys R1 2021, which was used as the simulation program. When compared to this model, hot brake pads from hybrid cars were collected and converted to electrical energy according to the idea stated.

References

1. Kopelias, P., Demiridi, E., Vogiatzis, K., Skabardonis, A., Zafiropoulou, V.: Connected & autonomous vehicles–environmental impacts–a review. Sci. Total Environ. **712**, 135237 (2020)
2. Filote, C., Felseghi, R.A., Raboaca, M.S., Aşchilean, I.: Environmental impact assessment of green energy systems for power supply of electric vehicle charging station. Int. J. Energy Res. **44**(13), 10471–10494 (2020)
3. Saleki, A., Rezazade, S., Changizian, M.: Analysis and simulation of hybrid electric vehicles for sedan vehicle. In: 2017 Iranian Conference on Electrical Engineering (ICEE). IEEE, pp. 1412–1416 (2017)
4. Ma, Z., Sun, D.: Energy recovery strategy based on ideal braking force distribution for regenerative braking system of a four-wheel drive electric vehicle. IEEE Access **8**, 136234–136242 (2020)
5. Pourkiaei, S.M., Ahmadi, M.H., Sadeghzadeh, M., Moosavi, S., Pourfayaz, F., Chen, L., Yazdi, M.A.P., Kumar, R.: Thermoelectric cooler and thermoelectric generator devices: a review of present and potential applications. Model. Mater. Energy **186**, 115249 (2019)
6. Lee, U., Park, S., Lee, I.: Robust design optimization (RDO) of thermoelectric generator system using non-dominated sorting genetic algorithm II (NSGA-II). Energy **196**, 117090 (2020)
7. Sreekala, P., Ramkumar, A., Rajesh, K.: Thermo electric generator fed LVDC system analyzed with non linear loads. In: 2019 IEEE International Conference on Clean Energy and Energy Efficient Electronics Circuit for Sustainable Development (INCCES). IEEE, pp. 1–4 (2019)
8. Praveena, B., Sowmiya, A., Logeshwari, P., Kowshalya, G.A., Sonia, B.M., Kavipriya, R.K.: Thermo electric generator module in driving the vehicle and monitoring using IoT. In: 2018 2nd International Conference on Inventive Systems and Control (ICISC). IEEE, pp. 822–825 (2018)
9. Deyasi, A., Lodh, R., Karmakar, B., Biswas, P.: Effect of self-consistency technique on current density profile of resonant tunneling diode. In: 2018 2nd International Conference on Electronics, Materials Engineering & Nano-Technology (IEMENTech). IEEE, pp. 1–4 (2018)
10. Zhou-Sheng, Z.H.A.N.G., Na, D.O.N.G.: Study on current density and conductivity mathematical model of SF 6 under DC uniform field strength. In: 2018 IEEE International Conference on High Voltage Engineering and Application (ICHVE). IEEE, pp. 1–4 (2018)
11. Fu, H., Zhao, J., Li, J., Liu, Y., Li, Z., Pan, Z.: High-precision calculation of geographical current density of DC grounding electrodes. In: 2019 IEEE 3rd Conference on Energy Internet and Energy System Integration (EI2). IEEE, pp. 1242–1245 (2019)
12. Saif, M.Z.U., Tasnim, F., Hanif, M.A., Ahmed, M.R., Shikder, K., Hossain, C.A.: Temperature and Power Analysis of the Thermoelectric Generator in Hybrid Electric Vehicles. In: 2021 International Conference on Computational Performance Evaluation (ComPE), pp. 128-132 (2021). https://doi.org/10.1109/ComPE53109.2021.9752285
13. Saif, M.Z.U., Mahmud, D.M., Tasnim, F., Hanif, M.A., Ahmed, M.R., Hossain, C.A.: Current and Temperature Analysis of Thermoelectric Generator for Regenerative Breaking of the Hybrid Electric Vehicle. In: 2021 IEEE 6th International Conference on Computing, Communication and Automation (ICCCA), pp. 504-508 (2021). https://doi.org/10.1109/ICCCA52192.2021.9666431

14. Merkisz, J., Siedlecki, M.: Specific emissions analysis for a combustion engine in dynamometer operation in relation to the thermal state of the exhaust gas af-tertreatment systems in a modified NRSC test. In: MATEC Web of Conferences. EDP Sciences, vol. 118, p. 00027 (2017)
15. Min, G., Rowe, D.M.: Recent concepts in thermoelectric power generation. In: Twenty-First International Conference on Thermoelectrics, 2002. Proceedings ICT'02. IEEE, pp. 365–374 (2002)
16. Liu, X., Li, C., Deng, Y.D., Su, C.Q.: An energy-harvesting system using thermoelectric power generation for automotive application. Int. J. Electr. Power Energy Syst. **1**(67), 510–516 (2015)
17. Singh, H., Shergill, H.: Thermal Analysis of Disc Brake Using Comsol. Int. J. Emer. Technol. **3**(1), 84–88 (2012)
18. Belhocine, A.: FE prediction of thermal performance and stresses in an automotive disc brake system. Int. J. Adv. Manuf. Technol. **89**(9–12), 3563–3578 (2017)
19. Korotkov, A.S., Loboda, V.V., Makarov, S.B., Feldhoff, A.: Modeling thermoelectric generators using the ANSYS software platform: Methodology, practical applications, and prospects. Russ. Microlectron **46**(2), 131–138 (2017)
20. Snyder, G.J., Toberer, E.S.: Complex thermoelectric materials. Materials for sustainable energy: a collection of peerreviewed research and review articles from Nature Publishing Group, pp. 101–110 (2011)
21. Enescu, D.: Thermoelectric energy harvesting: basic principles and applications. Green Energy Adv. **1** (2019)

Automatic pH Controlling System of Dyeing Machine

Md. Mynul Islam, Md. Rafiqul Islam, Md. Harunur Rashid Bhuiyan, Tarik Reza Toha, Anupom Mondol, and Shaikh Md. Mominul Alam

Abstract Dyeing is the most important and prime section of the textile manufacturing process. During dyeing, different types of faults occur which may cause huge loss of production cost, energy, and time. Batch-to-Batch shade variation, Patchy dyeing effect, Dye spots, Uneven dyeing, and Lighter shades are the most common dyeing fault in our textile dyeing manufacturing process. From a thorough investigation and a significant number of research works, we identified some common causes of those dyeing faults. Among them, improper chemical mixing in the mixing tank is one of them which changes the pH condition of the Dye-Bath and causes huge dyeing faults. In most cases, chemicals or dyes are applied to the dyeing machine manually and create these problems. During every production after a while pH of the Dye-Bath is checked manually. To avoid this manual pH checking system we propose a device that detects pH, monitor, and control it automatically. The device is designed with an Analog pH meter as pH sensor and 5v DC micro submersible

Md. M. Islam (✉) · Md. R. Islam · Md. H. R. Bhuiyan
Bangladesh University of Textiles, Dhaka 1208, Bangladesh
e-mail: 201718017@tmdm.butex.edu.bd

Md. R. Islam
e-mail: 201718001@tmdm.butex.edu.bd

Md. H. R. Bhuiyan
e-mail: 201718004@tmdm.butex.edu.bd

T. R. Toha
Lecturer, Department of Textile Machinery Design and Maintenance, Bangladesh University of Textiles, Dhaka 1208, Bangladesh
e-mail: toha@tmdm.butex.edu.bd

A. Mondol
Assistant Professor, Department of Textile Machinery Design and Maintenance, Bangladesh University of Textiles, Dhaka 1208, Bangladesh
e-mail: anupommondol@tmdm.butex.edu.bd

S. Md. M. Alam
Professor, Department of Textile Machinery Design and Maintenance, Bangladesh University of Textiles, Dhaka 1208, Bangladesh
e-mail: mominulalam@tmdm.butex.edu.bd

S. Hossain et al. (eds.), *Proceedings of International Conference on Fourth Industrial Revolution and Beyond 2021*, Lecture Notes in Networks and Systems 437,
https://doi.org/10.1007/978-981-19-2445-3_17

motor pump, with an Arduino UNO R3 connected with a 4-channel Relay module, HC-05 Bluetooth module, and other parts. The sensor senses the pH of the Dye-Bath and the pumps will help to maintain the required pH in the vessel by pumping the proper amount of chemicals or dyes in the dye bath. Dyeing faults can be reduced significantly by controlling the pH of Dye bath and our developed system offers to do it automatically.

Keywords Dye-Bath · Dyeing fault · pH monitoring · pH controlling · IoT-based automatic controlling system

1 Introduction

Bangladesh has become home to the second-largest garment industry in the world. This is the highest foreign currency earning sector. After China Bangladesh is the second-largest exporter of clothing in the world [13]. The RMG sector generated USD 28.14 billion in the financial year 2016–2017. This was 12.36% of the GDP and 80.7% of the total export earnings [19, 20]. Bangladesh currently has 424 spinning industries, 794 textile weaving industries, 241 dyeing and finishing industries, and a total of over 6,500 registered and over 500 un-registered textile factories [21]

Among all parts of the textile manufacturing process, Dyeing is the most complex and critical process. In the textile industry, Knit dyeing, Woven dyeing, Yarn dyeing, and finishing factories are available in Bangladesh. All of these factories have great importance in the readymade apparel export-import business. There is over 220 dyeing industry in Bangladesh The dyeing industry has a huge production per day [7].

Dyeing is the coloration process of textile fiber, yarn, or fabric by using dyes and chemicals. According to the required shade, color, or order, the specific amount of dyes and chemicals are mixed [8]. This is the most complex part of the fabric manufacturing process. Desizing, Scouring, Bleaching, Dyeing, and Finishing processes are done in the dyeing industry and maximum fabric faults are created in these processes. It's not an easy job to match buyers approved sample shade with preproduction sample and production sample shade. In a few cases, it's needed to re-dye the whole batch for not matching shade, and it causes huge production loss.

During this whole dyeing process, observe different types of faults which are called dyeing faults. Various kinds of dyeing faults are available. Uneven Dyeing, Batch-to-Batch Shade Variation, Running Shade, Pin Hole, Soda Spots, Fly Dye Stains, Softener Marks, Hand Feel problems, Crease Marks, Dye Spots, and Patchy Dyeing Effects are very common in our dyeing industry [2, 12]. Many research works are done to identify the causes of these dyeing faults. Lack of controlling dyeing parameters, very rapid addition of dyes and chemicals, due to improper pre-treatment, higher machine loading, running at lower nozzle pressure, High bath draining temperature, excessive loading of fabric during dyeing, sudden change in temperature during cooling, due to lack of synchronization of winch speed and pump pressure, improper mixing and dissolving of dyestuff, dye bath hardness, soaping start before wash, $CaCO_3$ or $MgCO_3$ in soaping bath, quick soda dosing, mixing the dye in the dyeing floor and then flying dust particles may come in contact with the dissolved dyestuffs, too dirty dyeing floor, not maintaining proper pH, poor emulsification of softener applied, inferior quality of softener is responsible for those dyeing faults. Among them uneven scouring and bleaching, improper scouring wash and acid, low liquor ratio, improper dosing of dyes and chemicals, improper pH and temperature are commonly responsible for those dyeing faults and cause huge production loss [8].

In our dying industry, most of the dyeing faults happen to improper mixing of chemicals in the dye bath. There are some dyes used in our industry that are hard to dissolve or mix in the dye bath. Sometimes they remain unmixed and create a fault in the fabric. The manual operating system is another important fact for the dyeing faults. Due to the manual operating system, sometimes proper timing of chemical mixing is disrupted, and controlling some parameters including pH becomes difficult. Improper pH condition of the dye bath is a common cause for many dyeing faults. These faults make the production complex and are caused for a huge loss of time and money. It is needed to take another step to make faultless fabric.

Many research works are done for preventing dyeing faults. But there is no rectify process for improper chemical mixing in the dye bath automatically in real-time. So, we focused on an automatic dosing system so that no improper chemical mixing happened and at the same time pH of the dye bath will be controlled automatically. It reduces dyeing faults significantly. We proposed an automatic pH controlling device that controls the pH of the dye bath.

Based on our work, we make the following contributions to this paper:

- We propose a low-cost real-time wireless pH monitoring and controlling system.
- We develop a real-time pH controlling system by using three different pumps which actuate based on pH data.
- Our automated system will decrease the production cost.
- By using this device authority can monitor the real condition of dye bath and can take any necessary step.
- This device reduces dyeing fault significantly and authority can save time and money.
- We test our developed system on different textile industries that shows the feasibility of our system in the real environment.

2 Background and Related Work

To avoid dyeing faults and make the dyeing system automated, many works have already been done and many research works are also done for automatic pH control systems. In this section, we discussed the background of these research works.

2.1 Related Work on pH Controlling of Dyeing Machine

Jahmeerbacus et al. [14], Hossain et al. [9], and Koh et al. [15] proposed a fuzzy control of dye baths. They minimized pH tracking error using pH error derivation. But for the acidic solution when excessive acid needs to reduce, base pumping may cause the dyeing fault. It may change the whole dye bath environment and damage the product. We proposed such a device where water will be pumped to reduce excessive acid solution. The water tank will be mixed with other chemicals for maintaining a proper dye bath environment without hampering pH conditions.

2.2 Related Work on Automation in Dyeing Machine

Dai et al. [5] proposed an automatic temperature controlling system in a dyeing machine. For maintaining temperature in the dyeing section of the textile industry PLC-based automation system is used in this project. When the temperature will exceed the required temperature cooling valve will open and minimize temperature. Xie et al. [22] proposed a pressure, heating or cooling, and frequency conversion control system on an overflow dyeing machine. Automation can improve productivity and the efficiency of textile production is explained in [4]. Application of automation in the dyeing section can precisely control various parameters which are most important for dyeing like temperature, water level, pressure, time of treatment. But in this paper pH controlling was missing where PH is one compulsory parameter of dyeing. Without proper pH maintaining, faultless fabric produced is quite difficult. Ananya et al. [3] parameters of dyeing are predicted by an artificial neural network. For proper maintaining desired shade salt concentration, pressure, run time, etc. parameters are the most important part. In this project, monitoring has been shown but there is no controlling system for achieving desired shade. Hudedmani [10], Rahul et al. [17] show temperature controlling in the dyeing machine automatically by using a PLC controller. Temperature controlling is already an existing system in dyeing machines. But we need to control pH for avoiding dyeing faults. We designed an automatic pH controlling device to reduce dyeing fault.

2.3 Related Work on Dyeing Faults

Majharul et al. [11] identify some causes and remedies of dyeing faults. Rico et al. [12] worked on dyeing faults of knit fabric and also identify the causes and remedies. Ashaduzzamanet al. [2] worked on especially particular faults, Batch to batch shade variation. From all of these research works we identify that among all the causes of pH controlling, temperature, improper dyes, soda, or another chemical dosing, manual operating system are commonly responsible for these faults. And pH controlling is one of the main common causes of these dyeing faults. But this research works don't show any solution for controlling pH in Dye-Bath. So, instead of manual pH monitoring and controlling system, we proposed an automatic pH controlling system for the main vessel of dyeing machines.

2.4 *Related Work on Automatic pH Controlling System*

Photsathian et al. [18] proposed a cloud-based crayfish breeding system by controlling water temperature and pH monitoring. For keeping optimum temperature and pH in crayfish this system is developed. In this project, all data will be stored in cloud storage. Temperature and pH sensors are used for sensing the condition of water and WiFi module are used for processing and sending data to cloud storage. No pH controlling system is added here. Rico et al. [1], Defa et al. [6] proposed a pH controlling system for Hydroponics and aquaponics systems respectively. A pH sensor, a microcontroller, and peristaltic pumps are used in this device. Defa et al. [6] designed their device with a WiFi system which is hardly applicable in our industry, we use Bluetooth modules to make our device easily applicable in the dyeing industry. Rico et al. [1], Defa et al. [6] designed their system with two pumping systems included acid and base. We use three pumping systems including acid, base, and water for controlling pH. Using only acid and the base pumps may cause our fabric or yarn. And it will also decrease the chemical cost by using a water pump.

So, after analyzing all of these research works about dyeing faults and pH controlling, we propose a sensor-based automatic operating system using a pH sensor, pump, relay module, Bluetooth module, and Arduino UNO. When the pH value of the main vessel will be changed because of improper dosing or other reasons, automatically our sensor detects the wrong pH value and at the same time, our pump will start mixing proper chemicals to maintain the required pH value in the main vessel.

3 Methodology

In this section, we describe the system design, device implementation, and the working process of our proposed system.

3.1 *System Design*

Our proposed system is designed with a sensor [shown in Fig. 1b], microcontroller, Bluetooth module [shown in Fig. 1a], relay module [shown in Fig. 1c], and pump. Sensors sense the data from the main vessel. The microcontroller collects the data and sends it to another microcontroller of the receiver part via Bluetooth module and

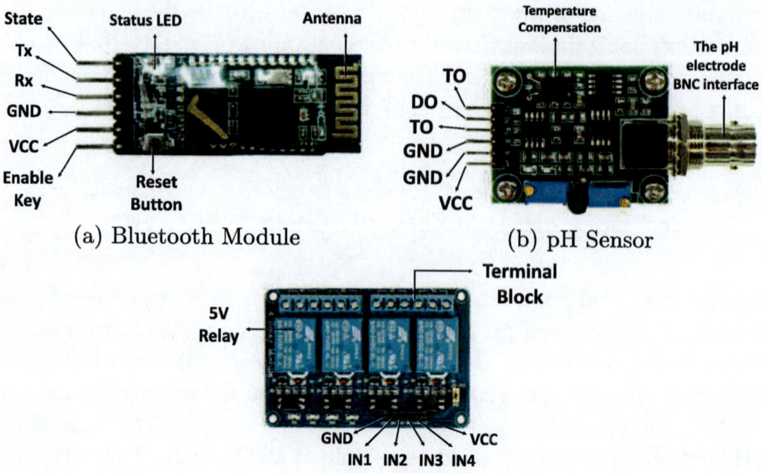

(a) Bluetooth Module (b) pH Sensor (c) Relay Module

Fig. 1 Major instruments of our device

the microcontroller controls the pumping system by relay module. Relay module act as an actuator in our proposed device.

3.1.1 Sender Part

The sender part is shown in Fig. 2a. This part is designed with an analog pH meter/ sensor kit, Arduino UNO R3. The pH sensor senses the pH of the main vessel and Arduino R3 collects the data from the sensor. Arduino UNO sends the data to the receiver part through the Bluetooth module. We use the HC-05 Bluetooth module in our system.

3.1.2 Receiver Part

The receiver part is shown in Fig. 2b. This part is designed with a microcontroller Arduino UNO R3, 5V 10A 4-Channel relay module, and 5V DC mini water pump. The relay module gets signals from the microcontroller and operates the pump according to the system. We use 3 pumps for water, acid, and base. These are the micro submersible motor pump that can pump 120 liters per hour.

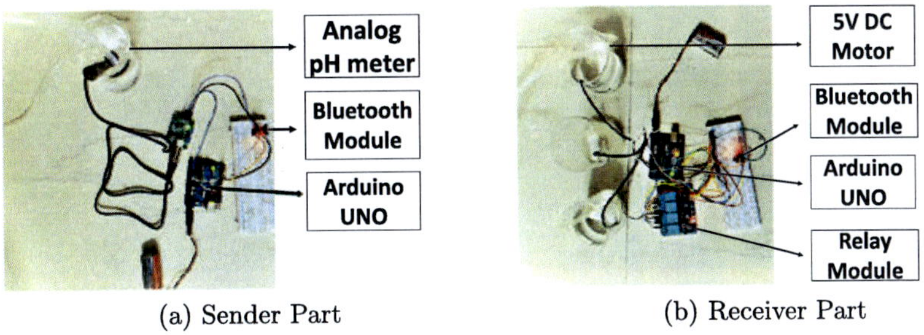

Fig. 2 Design of our proposed system

Fig. 3 Flowchart of the system

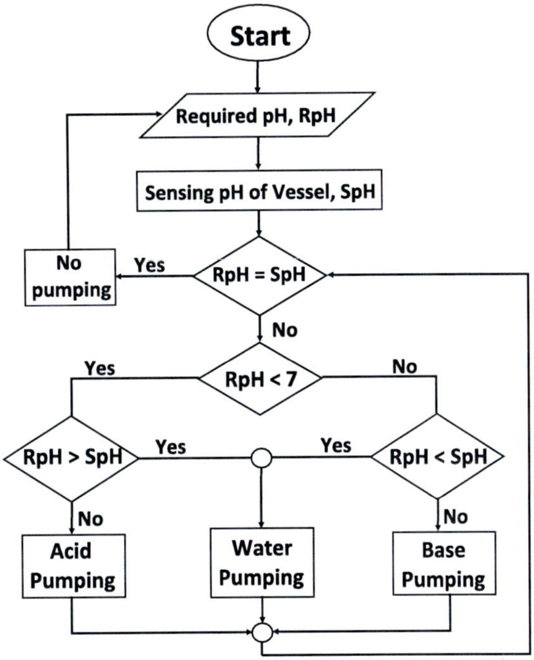

3.2 *System Flowchart*

Figure 3 is the flowchart of our automatic pH controlling system. When the required pH value is 7 and the pH value decreases or increases at the main vessel, the sensor senses the wrong value and sends a signal to the relay module and it starts pumping water until the desired pH value is achieved.

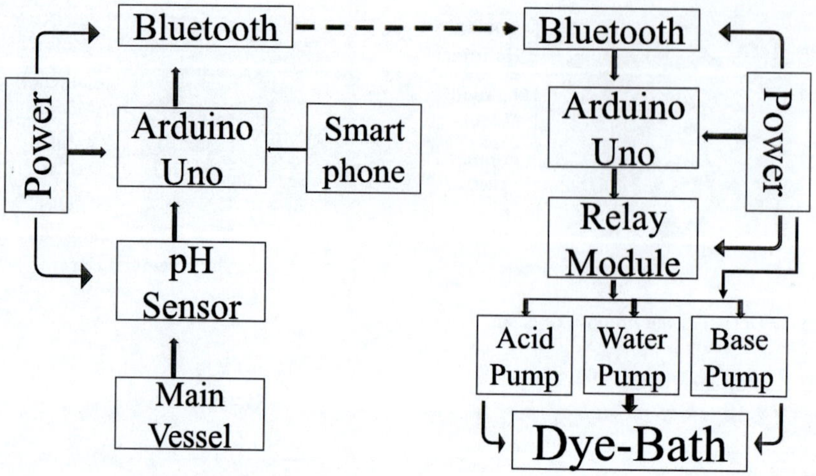

Fig. 4 Block diagram of our proposed system

3.3 *Block Diagram*

Figure 4 shows the block diagram of our proposed system. The first analog pH sensor measures the pH conditions of the main vessel and the microcontroller collects it and sends the data to the receiver part via HC-05 Bluetooth module. Another microcontroller receives the data from the Bluetooth module and sends signals to the 5V 10A 4-channel relay module. After getting the signal from the microcontroller (Arduino UNO R3) and according to the signal, the relay module pumps water, acid, or base.

3.4 *Cost Analysis*

We proposed a low-cost Auto dosing and pH controlling system for the Dyeing industry. Our proposed system costing is shown in Table 1a. Our total cost is about 70.83 USD, which is very cost-effective compare with McLean et al. [16] which is shown in Table 1b.

4 Experimental Evaluation

We test our device and collect data from Gangeetal Thread Factory at Tejgaon, Dhaka and Epyllion Group at Rajendrapur, Dhaka. From Gangeetal Thread Factory we collect data for the polyester yarn dyeing process and from Epyllion Group we

Table 1 Comparison of costing with existing system

(a) Our proposed System

Component name	Quantity	Unit price (USD)
5V Mini Water Pump Motor	3	5.28
9V Battery	4	3.35
Breadboard	3	3.52
4-Channel Relay	1	3.24
Bluetooth Module Breakout (HC-05)	2	7.71
Arduino UNO R3	2	17.17
Analog pH meter/Sensor Kit	1	30.56
Total cost		70.83

(b) Others proposed System

Component name	Quantity	Unit price (USD)
Arduino Nano	1	19.80
USB to microUSB cable adapter	1	5.09
Solderless breadboard	1	9.88
2G microSD flash memory card	1	9.75
4-channel relay module	1	9.99
Breadboard jumper wires	4	6.98
Push-to-connect tube fitting	1	6.99
Push-to-connect unions/joints	1	6.99
Electronic solenoid valve -$\frac{1}{4}$" 12 V DC	1	10.99
12 V AC-DC power converter	1	10.00
CO2gas regulator: CGA-320,Tubing$\frac{1}{4}$" OD	1	31.14
Check valve	1	9.06
Total cost		136.76

collect data for the cotton dyeing process. Figure 5 shows the data collection process from the Dyeing machine during cotton dyeing.

4.1 Experiment on Polyester Yarn Dyeing Process and Findings

In Gangeetal Thread Factory we collected data on polyester dyeing. Formic acid, dispersing agent, reduction cleaner, and softener are used as a chemical for this dyeing. This dyeing is done in an acidic medium. The pH of the initial solution we checked was 4.9. When it starts to dye, the machine remains fully closed. It's impossible to check the pH of the dye bath during dyeing. Here no system of run

(a) Dye-Bath pH Measuring (b) Controlling pH by Using Our Device

Fig. 5 Experiment on dyeing floor

Table 2 Chemical cost analysis of polyester dyeing

Formic acid			Buffer acid		
Chemicals	Quantity	Cost (USD)	Chemicals	Quantity	Cost (USD)
Formic acid	100 kg	129.08	Buffer	100 kg	369.63
Dispersing agent	10	53.98	Dispersing agent	10	53.98
Reduction cleaner	10	28.82	Reduction cleaner	10	28.82
Softener	5	17.60	Softener	5	17.60
Total chemical cost		229.48	Total chemical cost		470.03

back, that we can measure the pH of the dye bath during dyeing. That's why after finishing the dyeing process we collect the solution before draining and measure pH. it was 6.38, which means the acidic solution becomes neutral gradually. The main reason for this problem is formic acid which volatilizes above 100C, the acid starts to volatilize, and pH value increases. It harms the quality of the yarn. Sometimes for ensuring good quality they use buffer acid instead of formic acid. But this buffer acid is four-time higher in cost. Which increases the production cost. Table 2 shows that chemical cost is much higher for buffer acid. Here no system to add acid during dyeing or pH value increases. They can't monitor that.

We proposed such a system that pH will be monitored automatically and also pH will be controlled in the dye bath by adding acid when it is required. By using this device they can use formic acid instead of buffer acid to ensure good quality and it will reduce the cost of the production.

We set our device in the solution of 6.38 pH value, that we collected after the dyeing process. Our device controlled the pH value of that solution according to the required pH value. Our device pumping the proper amount of acid to reach near the required pH value of 4.9 [shown in Fig. 6]

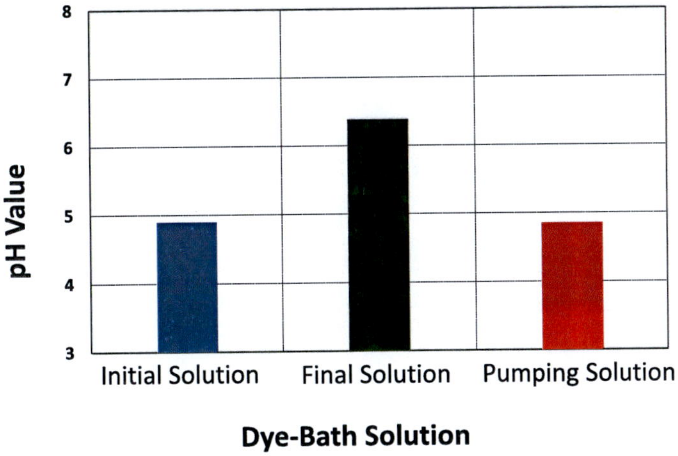

Fig. 6 pH controlling for polyester yarn dyeing

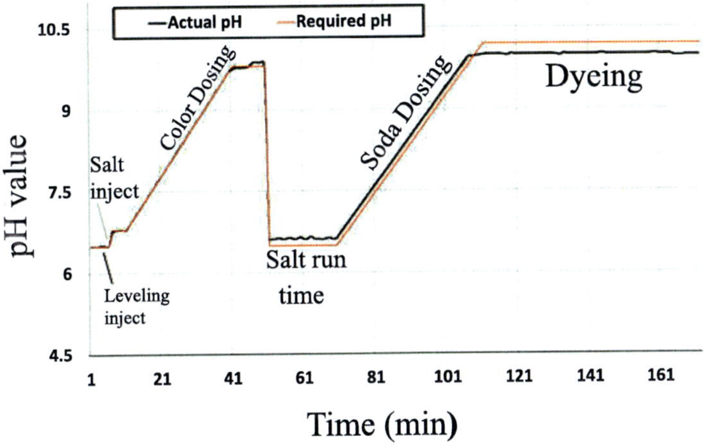

Fig. 7 pH monitoring on cotton dyeing process

4.2 Experiment on Cotton Dyeing Process and Findings

In the Epyllion Group, we collect data on cotton dyeing. Throughout the whole dyeing process, we collected data. They gave us the dye bath sample by running back system. We collect the pH value of the dye bath solution at every step of the dyeing process.

pH checking is the first step of dyeing, in this stage pH must keep at 6 to 6.5. Our observation by our device reading was 6.52. Then Neutral Enzyme and leveling agent were injected into the vessel. After injecting the enzyme and leveling agent, salt is injected. The working time for these injections is 10 min for 55°c temperature.

The average pH was 6.8. Color Dosing for dyeing is 30 min at 55°C temperature. In this stage pH was 9.7. Then the temperature is raised to 60°C. Then working time for salt is 20 min. In this stage, the average pH value was 6.63 Soda dosing is the most important part of dyeing. After soda dosing pH was 10. At 60°C temperature soda dosing are happened for 40 min. After dyeing, we measured the pH value of solutions which was 9.8. Figure 7 shows the graphical view of pH condition through the whole Dyeing process.

5 Commercialization

This device can be used in wastewater treatment, aquaculture, food and beverage, pool and spa water, aquarium and fish tanks, research works, hydroponics, and other industries where water quality should be controlled.

During the commercialization of this project, a few points should be maintained.

- We need to calibrate pH sensor regularly.
- At the end of every production, the Pipes of every pump should be washed properly. Otherwise previous chemicals, remaining inside the pipe, will mix with new products and create a fault in the fabric.

6 Conclusion

Dyeing faults are very common in our dyeing industry. Different dyeing faults have been highlighted in this paper where manual pH checking system in dye bath is very common. To avoid a manual pH checking system for dye bath this paper presents a low-cost automated device that can measure the pH of dye bath and can control the pH according to the required pH. So, the authority of the textile industry especially the dyeing section can use this device to maintain the pH of dye bath automatically to prevent faults that occurred for lacking proper pH maintenance in dye bath. These faults are caused for the wrong operating system and increased production costs. By using this device industry will decrease faults and increase profit. The time needed to neutralize the pH level is not measured in this proposed study. We will take steps to get the time needed to neutralize the pH level and better sensor reading by using a better quality sensor in the future. We proposed an automatic pH controlling system which is more developed system compare with existing system. In future we will directly apply our device in dye bath to compare our result with existing system.

Acknowledgements We would like to acknowledge the Gangeetal Thread Factory and Epyllion Group for providing us the opportunity of data collection from running machines during dyeing production. We also give our gratitude to the Bangladesh University of Textiles for providing financial support to this research work.

References

1. Alam, R.L., Nasuha, A.: Sistem pengendali ph air dan pemantauan lingkungan tanaman hidroponik menggunakan fuzzy logic berbasis iot. Electron. Inf. Vocat. Educ. (ELINVO) **5**, 11–20 (2020)
2. Ashaduzzaman, H.K., Rahman, M., Taher, A., Mia, S., et al.: Causes and remedies of batch to batch shade variation in dyeing textile floor. J. Text. Sci. Eng. **6**, 2 (2016)
3. Chakraborty, A., Kaur, P.D., Chakraborty, J., et al.: Automation in colouration technology to predict dyeing parameters for desired shade and fastness. Indian J. Fibre Text. Res. (IJFTR) **44**, 450–458 (2019)
4. Choubey, N.S., Agrawal, M.: Automation in textile industry. Methods **2** (2016)
5. Dai, J.: The design of automatic temperature control system for the dyeing machine. Mod. Appl. Sci. (2008)
6. Defa, R., Ramdhani, M., Priramadhi, R., Aprillia, B.: Automatic controlling system and iot based monitoring for ph rate on the aquaponics system. J. Phys. Conf. Series **1367**, 012072. IOP Publishing (2019)
7. dhakatribune: Bangladesh textile sector saw investment of tk6,900 crore in last five years|dhaka tribune. https://www.dhakatribune.com/business/2019/01/07(2019). Accessed 31 Aug 2021
8. Hasan, K.F., Sultana, M.Z., Taher, M.A., Deb, H., ur Rahman, M.M.: Coloration process & parameters for knit fabric dyeing along with different forms of dyeing faults & correlated remedies in textile dyeing industries. IOSR J. Polymer Text. Eng. (IOSR-JPTE) 2348–0181 (2016)
9. Hossain, I., Hossain, A., Choudhury, I.A.: Color strength modeling of viscose/lycra blended fabrics using a fuzzy logic approach. J. Eng. Fibers Fabr. **10**, 155892501501000130 (2015)
10. Hudedmani, M.G., Umayal, R., Kabberalli, S.K., Hittalamani, R.: Programmable logic controller (plc) in automation. Adv. J. Graduate Res. **2**, 37–45 (2017)
11. Islam, M.M.: Identification of different types of dyeing fault, causes and controlling method for quality and productivity improvement. Int. Res. J. Modern. Eng. Technol. Sci. (2021)
12. Islam, M.M., Mahmud, S.: Study on different types of dyeing faults, causes and remedies of knit fabrics. Int. J. Eng. Technol. **15**, 6–15 (2015)
13. Islam, M.M., Khan, A.M., Islam, M.M.: Textile industries in Bangladesh and challenges of growth. Res. J. Eng. Sci. **2278**, 9472 (2013)
14. Jahmeerbacus, M., Kistamah, N., Ramgulam, R.: Fuzzy control of dyebath ph in exhaust dyeing. Color. Technol. **120**, 51–55 (2004)
15. Koh, J., Shim, G., Kim, J.: ph control for dyeing polyamide in a reused acid dyebath. Color. Technol. **117**, 337–341 (2001)
16. McLean, K.M., Pasulka, A.L., Bockmon, E.E.: A low-cost, accessible, and high-performing arduino-based seawater ph control system for biological applications. HardwareX, e00247 (2021)
17. Pawar, R., Bhasme, N.: Application of plc's for automation of processes in industries. Int. J. Eng. Res. Appl.**6**, 53–59 (2016)
18. Photsathian, T., Suttikul, T., Tangsrirat, W.: Design and improvement of wireless crayfish breeding system by controlling water temperature and monitoring ph via cloud system services. In: 2018 3rd International Conference on Control and Robotics Engineering (ICCRE), pp. 205–208. IEEE (2018)
19. Sarkar, P., Anjum, A., Khan, E.A., et al.: Overview of major industries in Bangladesh. J. Chem. Eng. **30**, 51–58 (2017)
20. Sikder, M.A.R.: A review of textile industry in Bangladesh. Int. J. Adv. Multidiscip. Res. **6**, 9–14 (2019)
21. Textile network: Bangladesh: Growing textile industry|textile network. https://textile-network.com/en/Technical-Textiles/Fasern-Garne/Bangladesh-Growing-textile-industry(2018). Accessed 31 Aug 2021
22. Xie, S.: The design of auto-control system on overflow dyeing machine. Phys. Proc. **24**, 2094–2098 (2012)

IoT-Based Automated Crack and Object Identifier Vehicle for Railway System

Tareq Anwar Shikdar, Fahad Bin Ayub, Sekh Faisal, and Md. Moontasir Rashid

Abstract This paper is based on the crack and object detection of rail tracks for safe railway transportation in Bangladesh and to reduce the rail accidents due to derailment, which is caused by cracks of the railway tracks. An IoT-based solar-powered automatic crack and object detection vehicle called "Crack and object Identifier Vehicle" is proposed to detect the cracks of the rail line and send the coordinates to the station master. As the vehicle examines the rail tracks by image processing method with ESP32 CAM as well as an ultrasonic sensor, IR sensor, and NPN proximity sensors sense to detect the cracks and other components which can be the cause of derailments of the rail line. A dedicated application is built for monitoring the rail tracks and getting the coordinates of the crack location and object detected location to the authority. Also, the coordinates of the location are consigned to the station master by SMS using GSM and GPS modules. In this paper, an automated crack and object identifier vehicle is proposed for detecting cracks and obstacles in the railway track that vehicle can help to reduce the railway accidents from the rail track, especially in those countries whose railway infrastructure is poor.

Keywords Railway · Crack detection · IoT · Image processing · ESP32 CAM · Lora Network

1 Introduction

Railway transportation is one of the major transportations in Bangladesh for traveling from one corner to another. In this case, there are so many dangerous rail accidents that happen in Bangladesh [1]. Since 1972, Bangladesh Railway has been experiencing a large number of accidents [2], especially from 2001 to 2010 and it is about 5938 [3] due to derailments approximately 60%, which are caused by the crack of the rail line [4, 5]. Although, Bangladesh Railways is the safest transportation its management conditions and accident rates do not satisfy the statement [6]. Also, Bangladesh

T. A. Shikdar (✉) · F. B. Ayub · S. Faisal · Md. M. Rashid
Department of Electrical and Electronic Engineering, Leading University, Sylhet, Bangladesh
e-mail: tareqanwar82737@yahoo.com

S. Hossain et al. (eds.), *Proceedings of International Conference on Fourth Industrial Revolution and Beyond 2021*, Lecture Notes in Networks and Systems 437,
https://doi.org/10.1007/978-981-19-2445-3_18

Railway is not a fully automated maintenance system as a bunch of people called track men walk on the rail line using a manual vehicle with maintenance components to detect the crack, deformation, and obstacle on the rail line as well as this present process does not achieve a good result and acceptable for detecting the derailments because it is a very slow process and track men needs to know the method of detecting cracks [2]. Focusing on this problem, this paper is proposing an automated crack detection vehicle that can scan the railway path for detecting the crack and send the coordinates of the cracked location to the station master by dedicated apps and SMS system. For minimizing the accidents due to derailments, Bangladesh Railway System needs an automated system to detect the cracks and obstacles on the rail line. The aim of this paper is to reduce the railway accidents of the Bangladesh Railway System which are caused by derailments.

2 Literature Review

Poor railway infrastructure countries have not automated crack and object detection devices. As a result, there is so much research that has been done before to detect railway cracks and objects for reducing accident rates. A research team proposed a model with an IR sensor that could detect the cracks in the railway track and be monitored by Bluetooth technology [7]. Besides, for long-distance data transfer, so many teams have modeled several crack detection systems using GSM and GPS Module [8–10]. By using ultrasonic sensors for detecting cracks and obstacles, the researchers proposed separate modules. In 2016, a research team proposed a system that uses ultrasonic sensors to detect the crack in the track [11], and a PIR sensor is also used in this field to detect the crack [12]. In the preceding years, another system was introduced that used image processing for detecting cracks [13]. Moreover, some teams were investigating to find out faults in the track by using recurrent neural networks [14]. To detect the inconvenience obstacles on the rail line, image processing methods can be unique and faultless. There are so many surveys that have been recognized for detecting objects. In 2019, Shatabdi and his team partners proposed a cost-effective object detection robot that could detect objects in real time [15]. Also, in [16, 17], object detection by image processing methods was discussed. For data-transferring, Bluetooth technology [7], and IoT technology [18] were used, but that was not more effective for long-range transmission. However, all the research contributed to making a new automated system. All things considered, this paper introduced a unique, sustainable, cost-effective, and faultless automated crack detection vehicle called the "Crack Identifier Vehicle" for reducing the death ratio, which is caused by the crack on the railway path.

3 Methodology

3.1 System Process

Figure 1 shows the entire system process of the proposed system. Before implementation of the proposed system for the experiment, state-of-the-art and related works is observed for identifying the problems. After analyzing the problems, planning, accuracy, and architecture of the proposed system are designed with requirements and specifications. Then, the proposed system was implemented for the experiment and the result is collected for observation.

Fig. 1 System process of the proposed system

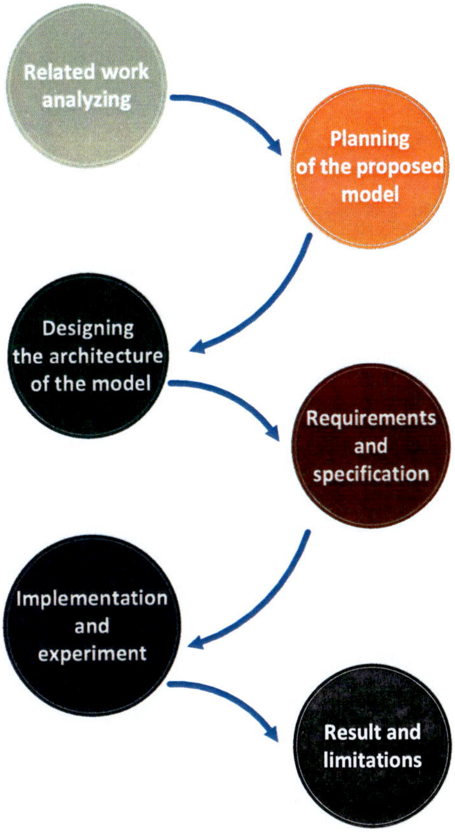

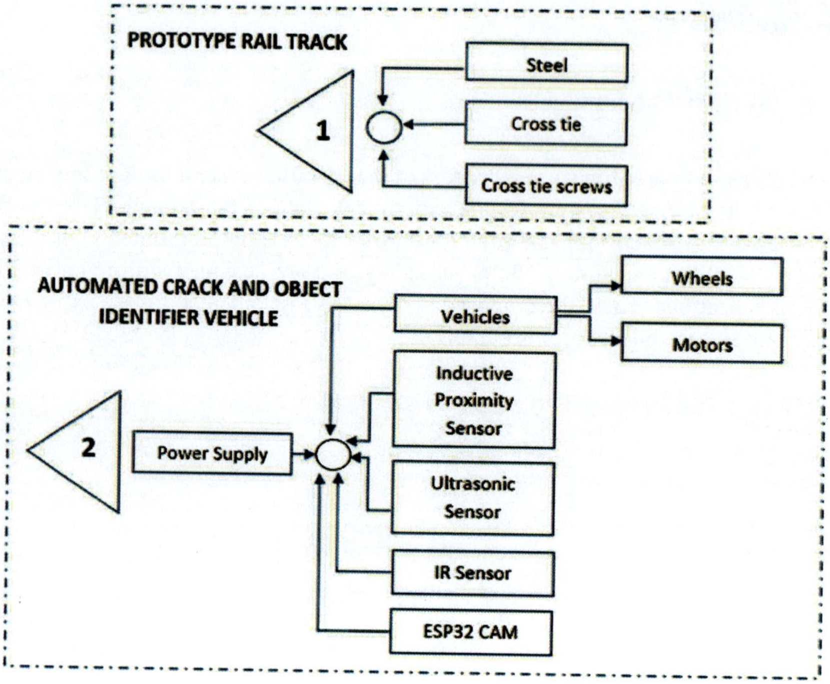

Fig. 2 System design of the proposed system

3.2 *System Design*

Figure 2 shows the system design of the proposed system. The equipment, sensors, vehicle constructions, and camera were installed in the crack identifier vehicle as shown in Fig. 2 described. Firstly, a rail track was constructed to experiment with the vehicle on the track as a real-life scenario. Then the vehicle is designed with the specification of the proposed system need [19]. For testing the vehicle, one lithium battery was connected and the vehicle was started to see the vehicle's performance on the rail track. Lastly, the sensors were added to the vehicle for analyzing the experimental results of the proposed system.

3.3 *Schematic Diagram*

Figure 3 shows the general scheme of the crack and object identifier vehicle, where it describes the working process at a glance.

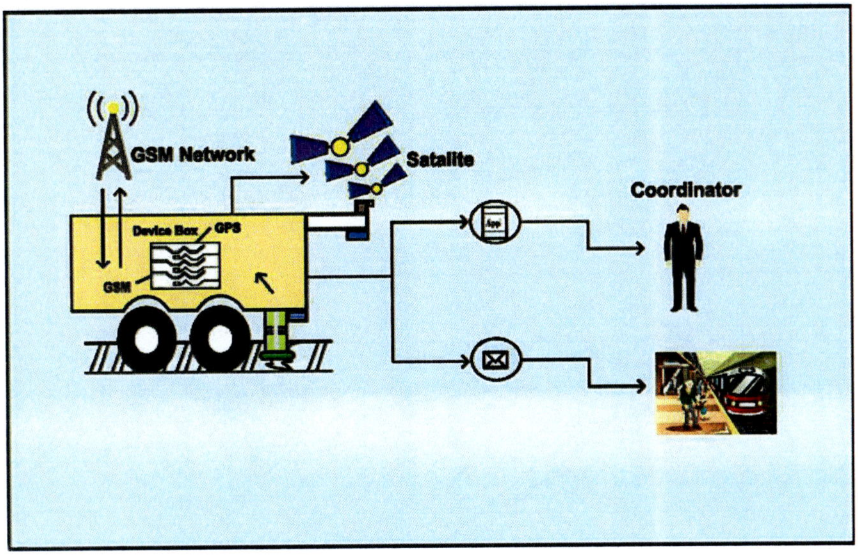

Fig. 3 General scheme of the proposed system

4 Crack and Object Detected Vehicle and Experimental Track Infrastructure Construction

Rail tracks and vehicles have some basic geometrical configuration, dynamics, and congratulations [20, 21]. Considering this, in this paper, a rail track is built for experimenting with the crack and object identifying vehicles in real-life scenarios and take the observation.

4.1 Prototype of Experimental Rail Track

A 1m long experimental rail track was built for observing the vehicle's behavior. The distances between rail widths and cross-ties are 11.8 cm and 22.5 cm, respectively, as shown in Fig. 4. The track is built of pure steel and cross-ties are metallic as well as other components that are used in the track.

4.2 Prototype of Vehicle Construction

A 20×12.2 cm vehicle was built with 4 wheels shown in Fig. 5 and a 3D view of the vehicle is shown in Fig. 6.

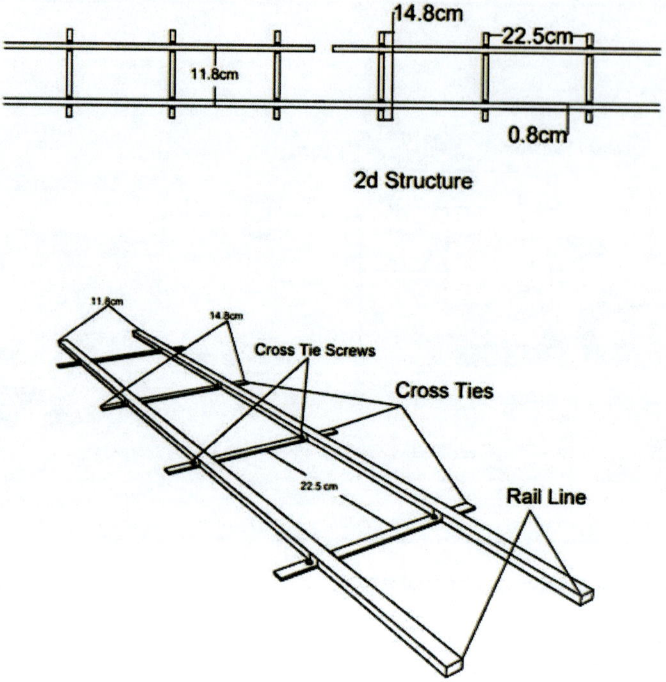

Fig. 4 Structure and measurement of the experimental track

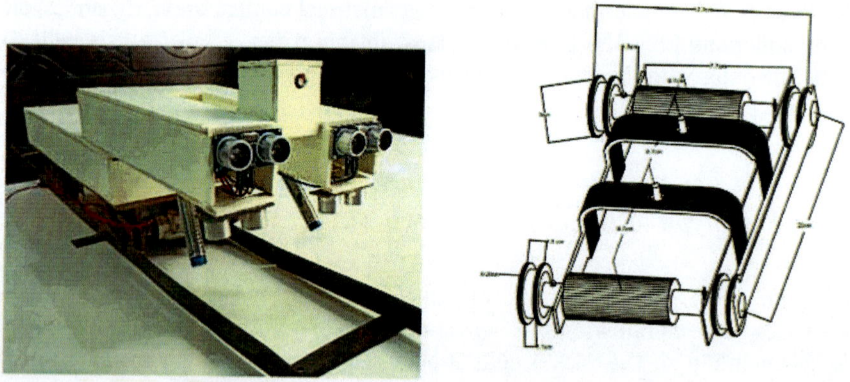

(a) (b)

Fig. 5 **a** Prototype of crack and object identifier vehicle, **b** Structure of the crack and object identifier vehicle

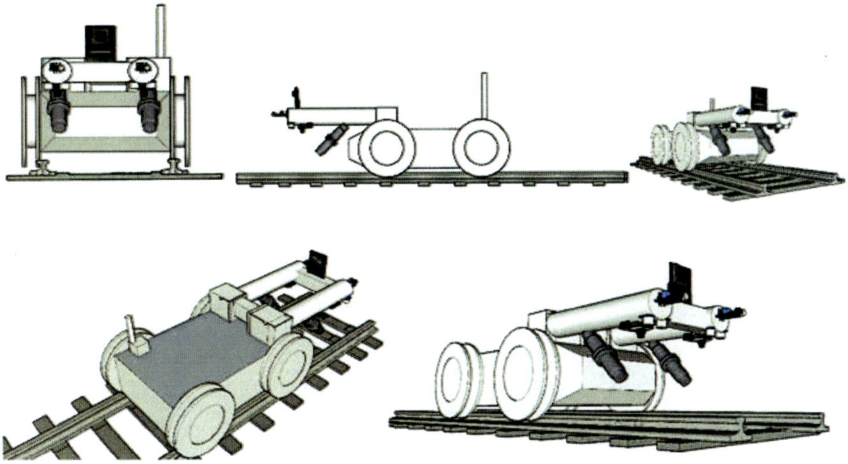

Fig. 6 3D view of the crack and object identifier vehicle

1. Wheels: In order to run the vehicle on the track, two pair's metallic wheels were built. The diameters of each wheel are 4 cm.
2. Gear Motors: Two Gear DC motors were fixed with wheels that can carry the weight of the vehicle and internal devices. Each gear DC motor is rated at 12v and 200 rpm.
3. Motor Driver: One L298N motor driver is used in this system for controlling the motor and circuit.

5 IoT-Based Crack and Object Identifier Vehicle

The automated crack and object detection system is built for detecting the cracks and objects on a rail path called "Crack and Object Identifier Vehicle". It is a vehicle designed to scan the path in order to find out the faults, cracks, disconnection of the track, and obstacles on the rail line and sends the GPS coordinates to the station master with monitoring apps and SMS method.

5.1 *Block Diagram*

This proposed system has two parts such as the vehicle side and the data receiving side as shown in Fig. 1. On the vehicle side, an Arduino Uno is used as the brain of this device. GSM and GPS modules were used in this system to get the coordinates for sending messages and the Tx LoRa module is used for transferring the data to the data receiving side. On the data receiving side, the Rx LoRa module received the

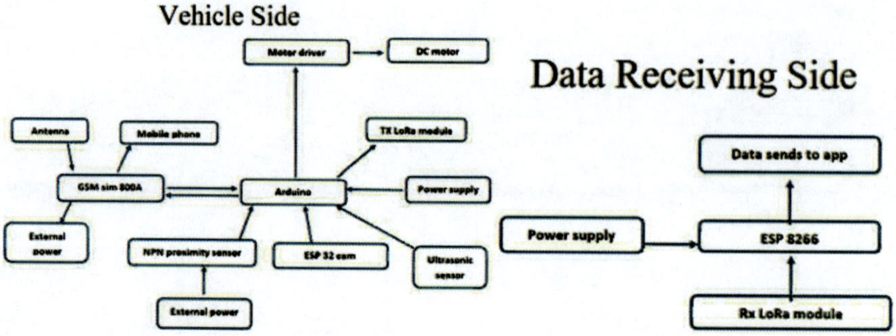

Fig. 7 Block diagram of the proposed system

data with ESP8266 and send it to the dedicated apps for monitoring. Figure 7 shows the block diagram of this proposed system

5.2 *Flowchart*

The proposed system is experimented as shown in Fig. 8. Figure 8 shows the algorithm of the proposed system.

5.3 *Circuit Diagram*

On the transmitting side, the ultrasonic sensor, GSM module, GPS module, and NPN proximity sensor have two digital pin outputs, which are connected to Arduino's digital pin. In the circuit board, Optocoupler is used to trigger the NPN proximity sensor.

LoRa module and ESP32 CAM are also connected to Arduino. In addition to that, two motors are connected to the Arduino by motor drivers.

All the components get power supply from the solar panels through Arduino; additionally, a 12 volts battery is used for powering up GSM and NPN proximity sensors, and a relay module is connected to the Arduino for controlling as a switch.

On receiving side, the LoRa module, ESP8266 is connected to the Arduino and all the components get the power from the 12V adaptor. Figure 9 shows the circuit diagram of this device (Fig. 10).

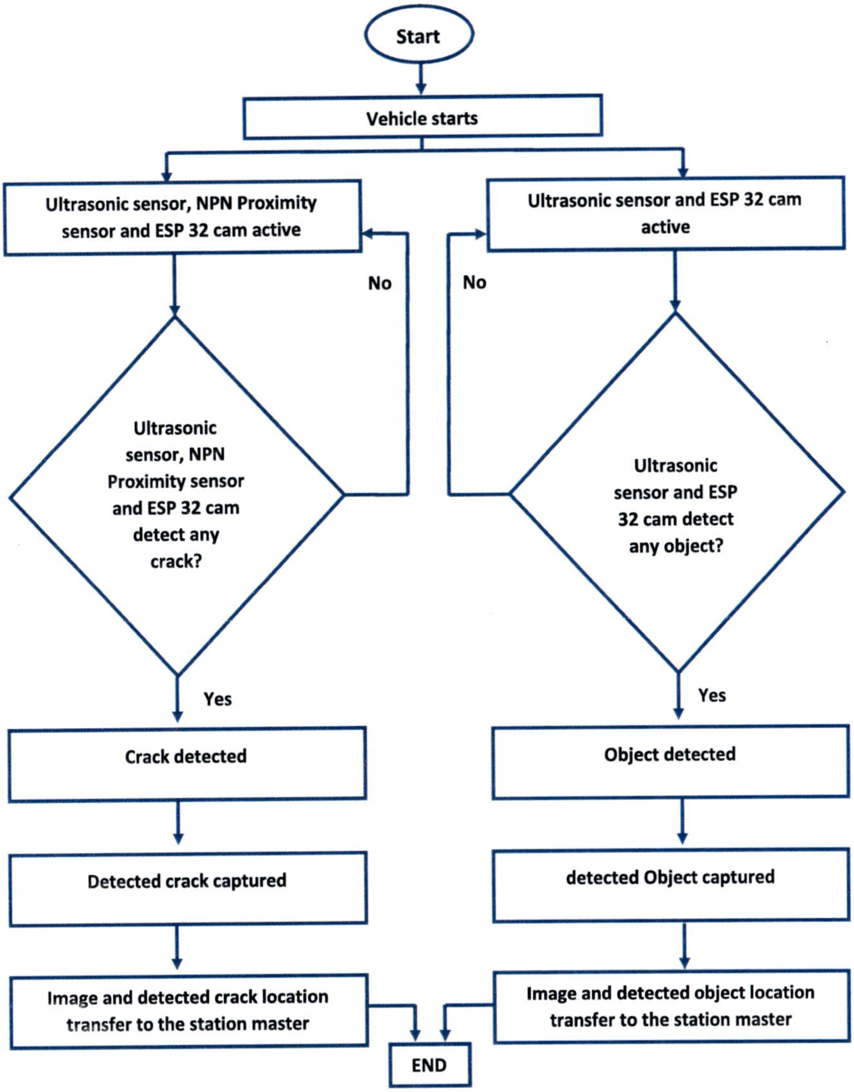

Fig. 8 Flowchart of the proposed system

5.4 Power Supply

1. Solar panel: Four solar panels were used in this proposed system for the power source of the vehicle. Solar panels are rated at 9 and 6 V
2. Buck Module: One buck module is used in this proposed system for converting the electrical power and distributing the steady from the solar panel

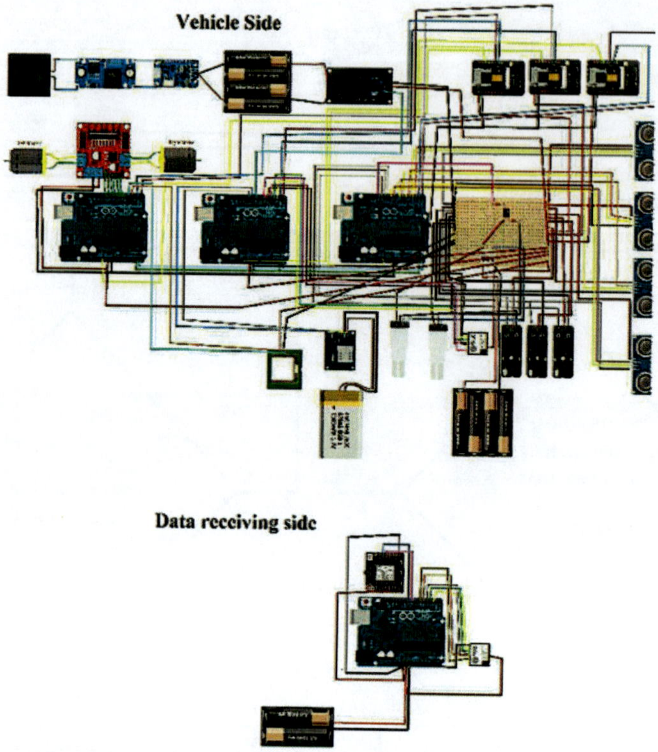

Fig. 9 Circuit diagram of the proposed system

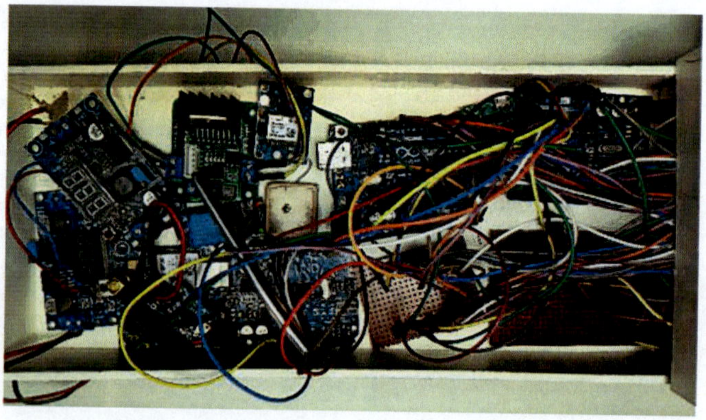

Fig.10 Internal connections of the proposed system

3. Battery: One 4 s Li-ion battery is used in this proposed system for storing the power from the solar cell to distribute to the Arduino. Each cell battery is rated at 3 V and also 12 V battery is used for the NPN proximity sensor and GSM module and two additional 12 V batteries are used for GSM and at the data receiving side.
4. Adaptor: One adaptor is used on receiving side to convert the AC power supply to DC at 12-2A.

5.5 Hardware Development for Detecting the Crack on the Railway Track

There are mainly two kinds of sensors used in this system for detecting the crack of the railway track and an ESP32 CAM is used for capturing the detected crack. The operational description is as follows.

1. Ultrasonic sensor: Two HCSR-04 ultrasonic sensors are used in this system for detecting the crack. The sensors are fitted in front of the vehicle. Two sensors are sensing the upper rail line as shown in Fig. 5. When there is a crack or disconnection in the rail line, the sensor detects the crack and sends the data to further operation.
2. NPN proximity sensor: Two NPN inductive proximity sensors were used in this system to find out the crack in the track. The sensor was fixed on the front side of the vehicle shown in Fig. 5. An inductive proximity sensor can detect metal objects. The sensor has a range of 4 mm or 0.4 cm or 0.156 inches. This sensor is more difficult to hook up so it is triggered by a circuit as shown in Fig. 9. For triggering, a PC- 817 Optocoupler is used. This transistor can transfer electrical signals between two isolated circuits. When there is a gap between the railway path, the sensor sends data as no metal to the Arduino for further operation.
3. ESP32 CAM: One ESP32 CAM is used in this system for capturing the detected crack, disconnection, and deformation of the rail line. When the ultrasonic sensor detects the crack of the rail line, ESP32 CAM captures it and sends it to the vehicle's apps for monitoring. Samples of the cracks, disconnections, and deformation are shown in Fig. 11. Also, ESP32 CAM is used as a monitoring device using OpenCV python. A dataset was considered for the image processing to detect the crack, disconnection, and deformation of the rail line. Some samples of the datasets are shown in Fig. 11.

5.6 Hardware Development for Object Detection

1. Ultrasonic and IR sensor: Two ultrasonic sensors and three IR are used in this system for detecting the objects on the rail line as shown in Fig. 5. When an

Fig. 11 Samples of the cracks, disconnections, and deformation

object is detected by the sensor, the signal is sent to the Arduino for further operation.

2. ESP32CAM: One ESP32 CAM is used for capturing the objects, which is sensed by the ultrasonic sensor using OpenCV Python. Some samples of the objects are shown in Fig. 12. Overall operation is similar to the crack detection module so that the operation's description is not described separately. Table 1 shows the

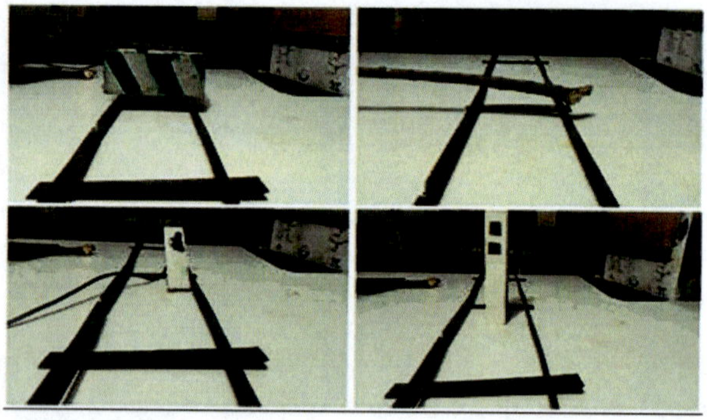

Fig. 12 Some samples of the objects

Table 1 Average accurate rate of image processing of crack and object detection

Type	Image resolution	Accuracy rate (%)	Real-time speed
Cracks, disconnections	360 * 240	94.00	YES
Cracks, disconnections	800 * 600	91.00	YES
Cracks, disconnections	1024 * 768	94.83	YES
Objects	800 * 600	93.00	YES
Objects	1024 * 768	95.00	YES

accuracy rate of the image processing for detecting cracks and objects on the rail track in real time.

5.7 Hardware Development for Data Transmitting to the Receiving Side

For data transferring to the receiver side, one SX1278 (Ra-01) LoRa module is used in the vehicle. LoRa module is used for long-range data transfer. This LoRa module can transfer data about 10 km. All the collected data is transferred to the receiver side Lora Module

5.8 Hardware Development for Data Receiving from Transmitting Side

1. LoRa module: On the receiver side, an SX1278 (Ra-01) LoRa module was installed. The transmitting side sends the data to the receiving side for further operation such as checking the experimental results, monitoring, and observation of the experimental results.
2. ESP8266: One ESP8266 microcontroller is used on the receiver side. LoRa module sends the collected data to the ESP8266 for real-time monitoring by dedicated apps.
3. "Crack Identifier Vehicle" apps: A dedicated app is built with Android studio for controlling and monitoring the experimental results. All the real-time experimented results, crack location, live monitoring, and controlling of the vehicle can be observed by the apps.

5.9 Hardware Development for GSM and GPS Module

Besides the "Crack Identifier Vehicle" apps, the SMS method is used in the system for collecting detected crack locations. In our system, SIM 800A and GY-NEO-6M as GSM and GPS modules were used, respectively. The modules were connected to the Arduino. When sensors detect the cracks, disconnections, and deformations of the rail path, it sends the signal to the Arduino. The Arduino sends an alert message to the coordinator and station master with the GPS coordinates of the cracked track location. A sim card was inserted into the module, and AT+CMGF and AT+CMCS commands were used in the code which instructed the SIM card to send a message with GPS coordinates to a particular number.

Table 2 Required components with rating and quantity

Serial no	Components name	Rating	Quantity
1	DC Gear Motor	12 v, 300 RPM	2
2	Micro-controller	Arduino Uno	3
3	GSM Module	SIM800A controls with AT Command	1
4	Antenna	Quad-banned antenna, operates four frequencies	1
5	SIM Network	Local mobile network	1
6	Wheels	Metal	4
7	Chassis	Metal	1
8	Rail Track	Metal	As required
9	Cross-ties and screws	Metal	As required
10	Inductive Proximity Sensor	NPN type, range 0.4 cm	2
11	Ultrasonic Sensor	Range 1 feet	4
12	Buck Module	DC–DC,12 v to 5 v	1
13	Relay Module	1 Channel, 12 v	1
14	Battery	4 cells at 3 v	1
15	Power Source	Lipo battery, 12 v, and 1800 mAh and 2200mAh	2
16	Wire	Silicon wire (max current up to 70A)	As required
17	Solar Panel	6 v–9 v	4
18	PVC Board	2 square feet	1

6 Components and Specification

Table 2 shows the components and apparatuses used in this proposed system.

7 Experimental Result

The proposed systems experimental is observed by two methods; one is real-time experimental results with “Crack Identifier Vehicle” and the second is crack location and object detected location by SMS.

a) App opening page *b) Home page* *c) Sensor status page*

Fig. 13 Application interference of this proposed system

7.1 "Crack and Object Identifier Vehicle" App

Figure 13 shows the application interference of this proposed system.

Figure 14 shows the experimental results of this model by "Crack and Object Identifier Vehicle"

Crack location and object detected location are shown in "Cracked Location" and "Object Detected Location". The cracked image and object detected image are shown on the "Cracked Image and Object Detected Image" page. Also, real-time monitoring of the vehicle is shown on the "Live Monitoring" page.

7.2 "SMS" System

Figure 15 shows the experimental results of GPS Coordinates. A message notification with GPS coordinates was sent to the selected number when the vehicle finds a crack or disconnection.

Table 3 shows the observation taken during experimentation

8 Limitations and Future Work

The proposed system has some limitations such as recorded video cannot be transferred to the receiving side and live monitoring did not show accurate results all the time, as well as camera, did not perform at the night so that powerful sensors and camera can be implemented in the future work and industrialization can proceed.

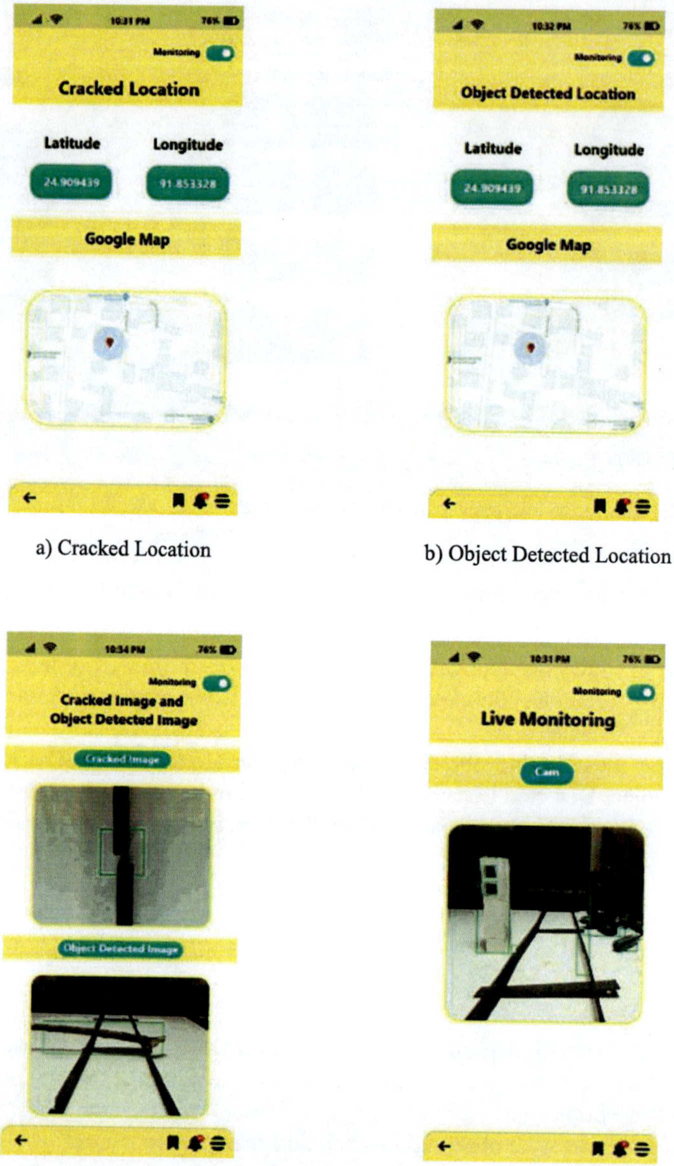

a) Cracked Location

b) Object Detected Location

a) Cracked image and object detected image

d) Live monitoring

Fig. 14 Experimental result of this proposed system

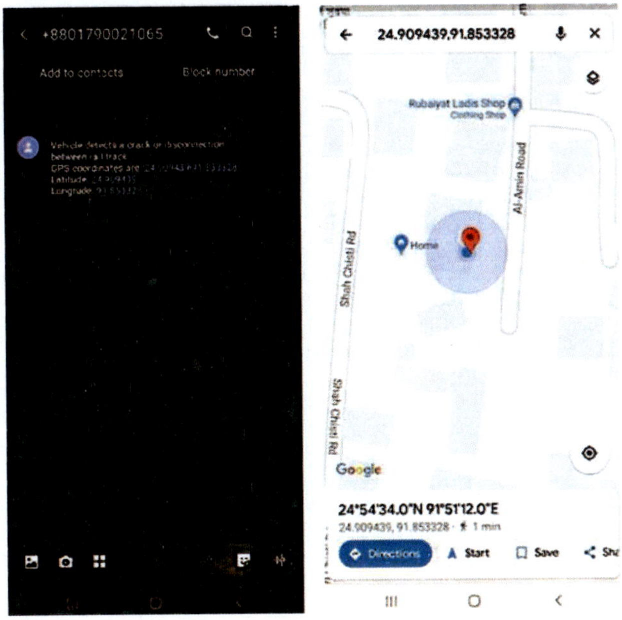

Fig. 15 Experimental results of GPS Coordinates

Table 3 Experimental observation of the SMS system

Ex. No	Crack detected	Disconnection detected	Alert sent (seconds) for cracked detection	Alert sent (seconds) for disconnection
1	YES	YES	5.82	4.92
2	YES	YES	5.34	5.22
3	YES	YES	6.12	5.67

9 Conclusion

An automated crack detection vehicle is designed to reduce the accident rate in Bangladesh due to the derailment by considering the real-life scenario. As a focus on sustainability, the vehicle is being experimented with solar power. But the proper power could not be obtained from the solar panel; so additional power was also used in this system. The main aim of this paper is to provide an automated system for detecting cracks and objects on the rail line. However, the system is getting a 100% performance rate. "Bangladesh Railway Ministry" invited us to appraise the idea as well as instructed us to experiment with this model on a real railway path. This research paper can be an ideal model for detecting the cracks, disconnections, and objects on the railway track as a low-cost and effective automated crack detection vehicle.

References

1. Pwint, H.N.Y., Tun, Z.M., Tun, H.M.: Automatic railway gate control system using microcontroller. Int. J. Sci. Eng. Technol. Res. (IJSETR) **3**(5) (2014)
2. Samia, N.K. et al.: Automated railway maintenance system in context of Bangladesh. In: 2019 1st International Conference on Advances in Science, Engineering and Robotics Technology (ICASERT), pp. 1–5 (2019). https://doi.org/10.1109/ICASERT.2019.8934529
3. Barkan, C., Tyler Dick, C., Anderson, R.: Railroad derailment factors affecting hazardous materials transportation risk. Transp. Res. Record J. Transp. Res. Board **1825**, 64–74 (2003)
4. Mahfuz, N., Dhali, O.A., Ahmed, S., Nigar, M.: Autonomous railway crack detector robot for Bangladesh: SCANOBOT. In: 2017 IEEE Region 10 Humanitarian Technology Conference (R10-HTC), pp. 524–527 (2017). https://doi.org/10.1109/R10-HTC.2017.8289014
5. Rao, C.M., BalaJaswanth, B.R.B.: Crack sensing scheme in rail tracking system. Int. J. Eng. Res. Appl. **4**(1), 13–18 (2014)
6. Bangladesh railway (br), people's republic of Bangladesh. Accessed July 21, 2018, at 08:18:00PM. http://www.railway.gov.bd/
7. Krishna, B.R., Seshendra, D., Raja, G., Sudharshan, T., Srikanth, K.: Railway track fault detection system by using ir sensors and bluetooth technology. Asian J. Appl. Sci. Technol. (AJAST), 82–84 (2017)
8. Kumar, S.: Study of rail breaks: associated risks and maintenance strategies (2006)
9. Cannon, D.F., Edel, K., Grassie, S.L., Sawley, K.J.: Rail defects: an overview **26**(10), 865–886 (2003)
10. Reddy, V.: Development of an integrated model for assessment of operational risks in rail track. Master Thesis, Queensland University of Technology School of Engineering Systems (2007)
11. Kumar, S.S.J., Titus, T.J., Ganesh, V., Sanjana Devi, V.S.: Automotive crack detection for railway track using ultrasonic sensorz. Int. J. Eng. Technol. Comput. Res. (IJETCR) **4**(6), 34–37 (2016). ISSN: 2348–2117
12. Navaraja, P.: Crack detection system for railway track by using ultrasonic and PIR sensor. IJAICT **1**(1) (2014)
13. Rizvi, A.R., Khan, P.R., Ahmad, S.: Crack detection in railway track using image processing. Int. J. Adv. Res. Ideas Innovations Technol. (2017) (Corpus ID: 64931603)
14. De Bruin, T., Verbert, K., Babuska, R.: Railway track circuit fault diagnosis using recurrent neural networks. IEEE Trans. Neural Netw. Learn. Syst. **28**(3), 523–533 (2017)
15. Roy, S., Hazera, C.T., Das, D., et al.: A computer vision and artificial intelligence based cost-effective object sensing robot. Int. J. Intell. Robot Appl. **3**, 457–470 (2019). https://doi.org/10.1007/s41315-019-00107-1
16. Premachandra, C., Ueda, S., Suzuki, Y.: Detection and tracking of moving objects at road intersections using a 360-degree camera for driver assistance and automated driving. IEEE Access **8**, 135652–135660 (2020)
17. Han, Y., Ma, S., Xu, Y., He, L., Li, S., Zhu, M.: Effective complex airport object detection in remote sensing images based on improved end-to-end convolutional neural network. IEEE Access **8**, 172652–172663 (2020)
18. Salvi, S., Shetty, S.: AI based solar powered railway track crack detection and notification system with chatbot support. In: 2019 Third International conference on I-SMAC (IoT in Social, Mobile, Analytics and Cloud) (I-SMAC), pp. 565–571 (2019). https://doi.org/10.1109/I-SMAC47947.2019.9032670
19. Shikdar, T.A., Dey, S., Mumtahina, S., Rashid, M.M., Chowdhury, G.M.: Design and simulation of single phase and three phase wireless power transfer in electric vehicle using MATLAB/Simulink. In: Mekhilef, S., Shaw, R.N., Siano, P. (eds.) Innovations in Electrical and Electronic Engineering. ICEEE 2022. Lecture Notes in Electrical Engineering, vol. 894. Springer, Singapore (2022). https://doi.org/10.1007/978-981-19-1677-9_8
20. Jovanovic, S.: Railway track quality assessment and related decision making. In: The American Railway Engineeing and Maintenance of Way Association (AREMA) 2006 Annual Conference, Lousville, USA (2006)

21. Shikdar, T.A., Rashid, M.M., Ayub, F.B., Faisal, S.: Development of automated protection and monitoring system for poor railway infrastructure. In: Bindhu, V., Tavares, J.M.R.S., Du, KL. (eds.) Proceedings of Third International Conference on Communication, Computing and Electronics Systems. Lecture Notes in Electrical Engineering, vol. 844. Springer, Singapore (2022). https://doi.org/10.1007/978-981-16-8862-1_35

Data and Network Security

Towards Developing a Smart Air Quality Monitoring and Security System to Ensure Workplace Health and Safety

Faria Tashnim Mazumder, Pritom Goswami, Tarik Reza Toha, Anupom Mondol, and Shaikh Md. Mominul Alam

Abstract Air pollution has caused a vicious impact on the climate and human health. As a consequence, it has drawn the attention of researchers, policymakers, and citizens. Traditionally, air pollution is measured using heavy and expensive instruments at fixed monitoring stations. But the number of these monitoring stations is very few. Also, the scale of fire accidents due to flammable gas leakage is increasing rapidly. To control air pollution and to reduce fire accidents, a vastly installed device is a dire need. In the following research, we have proposed a low-cost sensor-based monitoring system with multiple network options, along with some preliminary data analysis results. Our system is capable of sharing quantitative data of air pollution globally using Global System for Mobile Communications (GSM), Wi-Fi, and Bluetooth. It also has a smart alarm system to alert authorized personals. The foreground of this research is implementing a feasible system that can benefit the 3rd world countries as well as the developed ones.

Keywords Air pollution · Mobile sensing · Gas sensors · Wireless communication · Arduino · IoT · Security System

1 Introduction

Due to recent events, air pollution has become a matter of concern to the government, scientific community, and general people. Pollution has grown a tremendous drawback, especially in the developing countries, because of the rapid urbanization.In the underdeveloped or developing nations, almost 90 percent of the population are exposed to the precarious levels of contaminated air [7]. As per global estimates,

F. T. Mazumder (✉) · P. Goswami · T. R. Toha · A. Mondol · S. Md. M. Alam
Bangladesh University of Textiles, Tejgaon, Dhaka 1208, Bangladesh
e-mail: mazumder@gmail.com

S. Hossain et al. (eds.), *Proceedings of International Conference on Fourth Industrial Revolution and Beyond 2021*, Lecture Notes in Networks and Systems 437,
https://doi.org/10.1007/978-981-19-2445-3_19

air pollution, both ambient air pollution (AAP) and household air pollution (HAP), accounts for 7 million premature deaths worldwide [22]. People are more at risk of occupational exposure to ambient air pollutants, particularly vehicle drivers, and manufacturing-based industrial workers [8, 9]. Vehicle drivers are openly inhaling burned fuel pollutants like CO, CO_2, SO_2, etc. that cause permanent damage. Industrial workers are also uncovered to chemically produced pollutants like TVOCs, H_2, NO_x, etc. and they are ignorant of the side effects. With the recent advancements, fire safety has also become one of the fundamental problems. Fire hazards are diminishing in case of business and home security, and devastating regarding human life. The apparent way to reduce this kind of loss is to respond to these emergencies as fast as possible [19]. Therefore, an air quality monitoring and security system is mandatory for monitoring air pollution and fire accident efficiently before the situation converts critical.

In 2013, air contamination took lives of around 5.5 million people, and it ranks fifth worldwide among all risks. This caused human suffering and decreased financial improvement.Pollution-related deaths strike mainly young children and the elderly. Furthermore, premature deaths also cause lost labor income for working-age people. A report finds that due to air pollution, annual labor income losses cost the equivalent of almost 0.83–1% of Gross Domestic Product (GDP) in South Asia [4].

Improvement of environment and minimization of human exposure to pollutant gases can be achieved by constant monitoring of pollutants and green-house gases. Traditional AQI monitoring systems use complex instruments to achieve data. As a result, these types of equipment are generally extremely costly [3]. Furthermore, these monitoring stations are generally located in a few selected areas. This can significantly affect the accuracy of the pollutants' distribution estimation as air pollution is location-dependent [11]. The purpose of AQI monitoring stations is not only to collect data but also to provide valuable information. This will help on managing and improving the environment through the process of informed decision making by the policy-makers, scientists, and planners [21].

In this situation, low-cost sensor-based air pollutant monitoring technologies promise a revolutionary advance in air quality monitoring. In this paper, we have introduced a smart air quality monitoring and security system. The main contributions of this paper:

- We have proposed a modern approach in developing low-cost air pollution monitoring and security system, which consists of multiple sensors that can measure the quantity of 11 environmental parameters and also detect the the presence of smoke, LPG, and flame at the same time.
- We have built a system with multiple networking options for communication, i.e., Bluetooth, nRF, Wi-Fi, and GPRS. The implementation of IoT can provide real-time air quality data to the users. We have also developed a web application for data visualization.
- We have evaluated our developed system in real life scenarios, which demonstrates the applicability of our proposed approach. We have also provided a small-scale comparison of our device with existing systems.

Our proposed method provides both safety and environmental protection. Portable sensors programmed by microcontroller and multiple network options have made the device user-friendly and cost-effective. The alarm system is designed to get the quickest local response.

2 Literature Review

Air quality monitoring is a subject of many research projects and community-based initiatives. To make any step ahead of controlling the pollution rate, it is mandatory to monitor the quality of air. With analytical information on air pollution, scientists and policymakers can make a smart move on improving the environment. Currently, there is no device available on the market that could automatically monitor multiple types of air quality parameters and offer security measures for fire accidents at the same time [13].

Husain et al. [10] have proposed an air quality monitoring system that can sense air quality values, CO, and dust, and uses Bluetooth but this device is not compatible with the remote users. Setiawan et al. [20] have proposed an IoT-based system that can detect smoke, CO, and particulate matter, and uses the Wi-Fi network to send data to the cloud. Postolache et al. [18], have proposed an air pollution monitoring architecture that is based on Wi-Fi. In this paper, sensor nodes consist of multiple sensors i.e. temperature, humidity, CO, methane, solvent vapors sensors etc. Lungu et al. [16] have described a product which is a hand-held mobile device, that covers different locations, but the data-set is bounded as the evaluation was done manually. Tsow et al. [23] have developed a sensor node that consists of multiple sensors which can detect VOCs, temperature, and also relative humidity, with a Bluetooth module that can communicate with the smartphone. An IoT based system is proposed by Dhas et al. [6]. The system consists of a Raspberry PI module and sensors which are connected to a computer system for monitoring the fluctuation of parameters like Carbon Monoxide (CO) and other gases from. User can access real-time air pollution data via a web page from anywhere within the network's range.

Most of the existing systems are not cost-efficient and widely accessible. These devices can not offer any user-friendly solutions because of their data and network limitations. Moreover, they monitor very few pollutants that are not enough to understand the whole picture of pollution.

Taking this scenario into account, we then decided to build a device with multiple network options that can monitor many air quality parameters and has a security system that can alert the proper authority to reduce accidents. With multiple network options, the system has become more robust because the appliance control or monitoring becomes cheaper and possible regardless of the distance.

3 Methodology

3.1 Overall System Architecture

Our system is developed by integrating many hardware and software components. It has multiple sensor nodes, a base node, and a service station. Each sensor node consists of multiple sensors. The sensor node is primarily responsible for the information or sensor data collection and distribution. The sensor data can be monitored by using an android phone via Bluetooth, GSM, and Wi-Fi. Data can also be monitored through a laptop or computer. Data collected by the sensor nodes are then transmitted back to the server station using the accessing points. Fig. 1 shows the overall system architecture of our monitoring system.

As the data are being uploaded to the server using a Wi-Fi module, we considered two scenarios: (i) Scenario-A, where the sensor node has direct access to Wi-Fi or within the Wi-Fi range. (ii) Scenario-B, where the sensor node is out of the Wi-Fi zone. We designed two different methodologies for both scenarios.

For Scenario-A, the sensor node can send data via Bluetooth, GSM and Wi-Fi. Then data are directly uploaded to the website via Wi-Fi. For Scenario-B, the sensor node consists of the nRF24L01 module instead of the Wi-Fi module. The nRF24L01 module sends data as a packet wirelessly to a base node which consists of a NodeMCU, an nRF, and has Wi-Fi accessibility. The distance between the sensor node and the base node can be 60 meter. It also has Bluetooth and GSM modules that can transmit data.

We included cellular IoT in the project to transmit data to the server using the GSM module. It is also programmed to text the authorized users if there is a leakage of LPG gas, an outbreak of fire, and smoke. If there is no Wi-Fi available in the area, then the data can be uploaded to the website via the GSM module.

3.2 Cost of System Components

The proposed system in Scenario-A costs around 180.67 USD and Scenario-B costs around 215.48 USD. In total, it costs 227.2 USD. All the unit prices are collected and calculated from TechShopBD [2] and AMAZON [1].

3.3 Sensor Node Development

We have developed two types of sensor nodes, for scenario-A and scenario-B. The sensor node consists of an Arduino Mega, Wi-Fi module ESP-8266 (Scenario-A) or nRF24L01 (Scenario-B), MQ135, MQ2, MQ8, MQ136, CJMCU4541, CCS811 gas sensors, Flame sensor, DHT11 temperature and humidity sensor, Sharp GP2Y1010

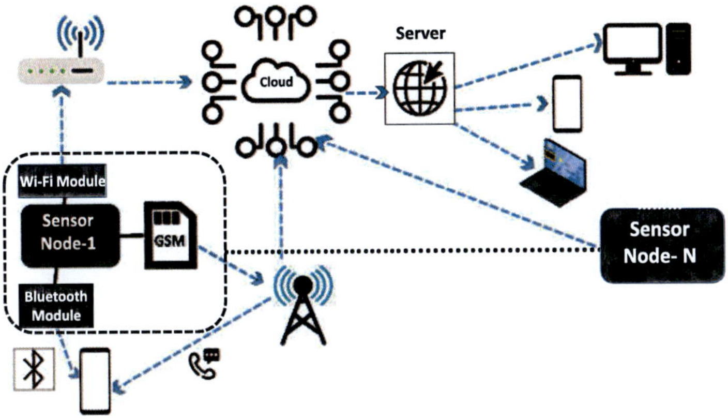

(a) System architecture for Scenario-A.

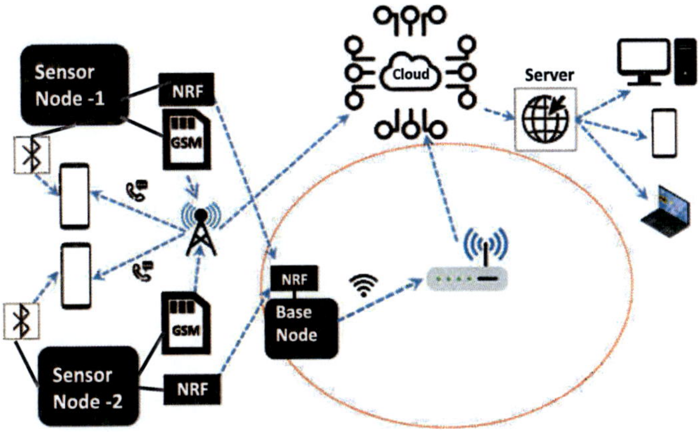

(b) System architecture for Scenario-B.

Fig. 1 Overall system architecture

dust sensor, GSM SIM 900 module, RTC module, SD card module, and HC-5 Bluetooth module. The following environmental parameters can measure air pollution levels: H_2, CO, CO_2, NO_x, Total Volatile Organic Compounds (TVOCs), SO_2, AQI, and Particulate Matters ($PM_{2.5}$ and PM_{10}). Other parameters like temperature, humidity, heat index, smoke, LPG, flame, etc. can also be measured. We can monitor the data on a smartphone via Bluetooth or in our customized web platform through Wi-Fi or a 3G network (GSM Module). The data can be stored on an SD card. Figure 2 shows diagram of our proposed sensor node for both scenarios.

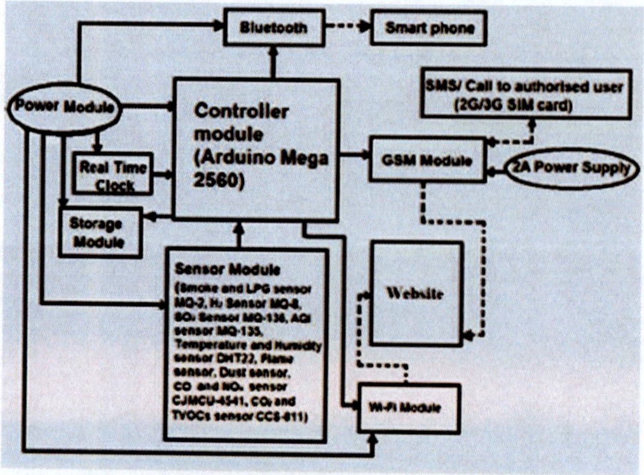

(a) Sensor node for Scenario-A

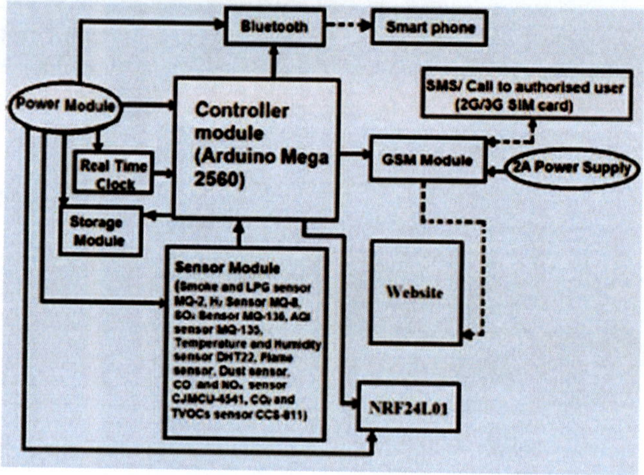

(b) Sensor node for Scenario-B

Fig. 2 Block diagram of the sensor node

3.4 Base Node Development (For Scenario-B Only)

The base node has a ESP8266 NodeMCU module, an nRF24L01 module, and a power supply module. This node receives data from the sensor node via the nRF24L01 module and then uploads the data to the server station.

3.5 Server Station

The Laravel framework version 8 and PHP (version 8.0.5) are used to build the server station. HTTP connection protocol does the task of the communication between the base node and server station. The server station saves all data from the sensor nodes for a long period and displays it in graphical and tabular form.

4 Experimental Evaluation

Based on the proposed system portrayed in Figs. 1 and 2, the overall evaluations are illustrated with experimental results in this section.

4.1 System Implementation

The real-life system implementation is shown in Fig. 7. Portable battery was used for the power supply as we intended to use this for outdoor and indoor air quality measurement.

4.2 Testing and Data Collection

After developing our proposed prototype, we proceeded in testing the compatibility and reliability of our device. We deployed the device in the environment and collected real-time data of industrial air pollution from the dyeing section and ETP of Sublime Greentex ltd. We also collected data from vehicular emissions and the indoor air quality of a residential area. After that, we analyzed the data. The measured results for several air pollution parameters can be seen in Figs. 3, 4 and 5. Data were successfully uploaded to our website as integer value, and also visualized as a graphical view (Fig. 6).

4.3 Performance Analysis

In this section, we have reevaluated the overall performance of our proposed system (Fig. 7).

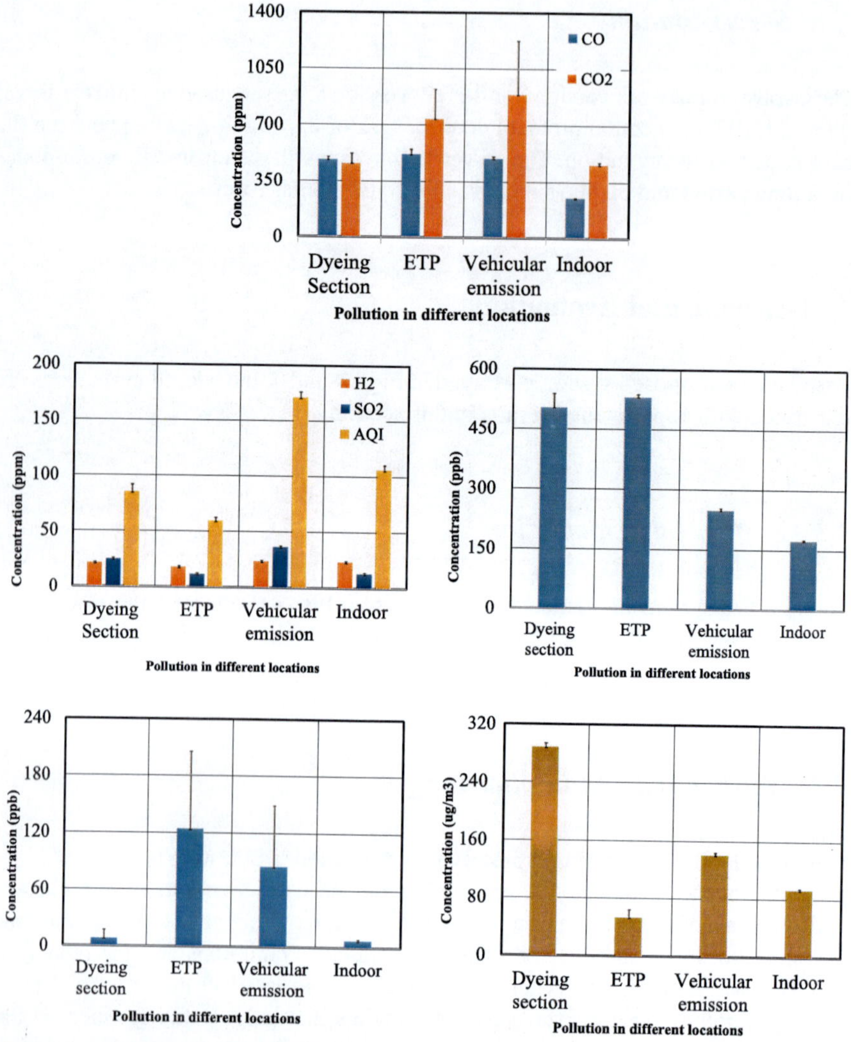

Fig. 3 Experimental data of the major pollutants

4.3.1 Network Reliability

We tested the performance of different networking modules. We sent an ID with each data packet and counted how many packets were received. By doing this, we could determine the packet drop rate in the time duration of 200 seconds. From what we observed, Bluetooth and GSM module are more reliable for continuous data collection than the other modules. The following table contains the test results (Table 1).

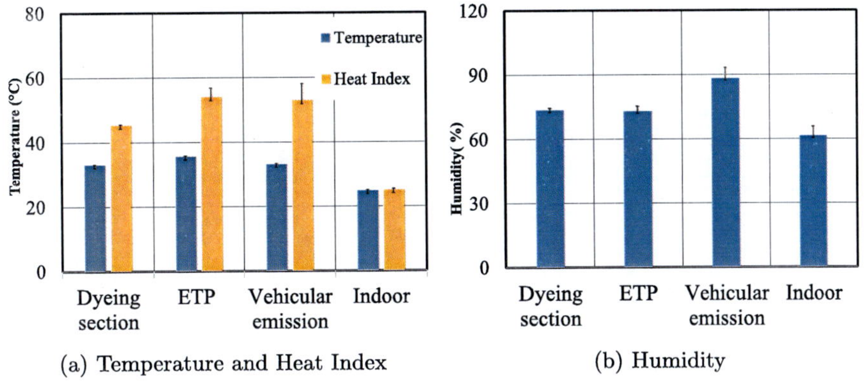

(a) Temperature and Heat Index (b) Humidity

Fig. 4 Experimental data of the environmental parameters

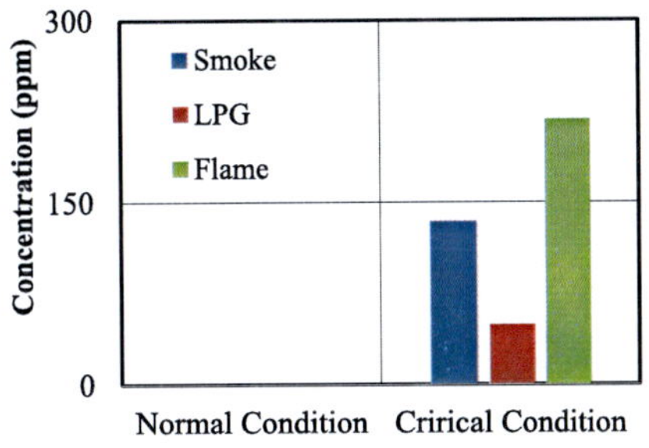

Fig. 5 Experimental data of the security parameters in artificial test condition

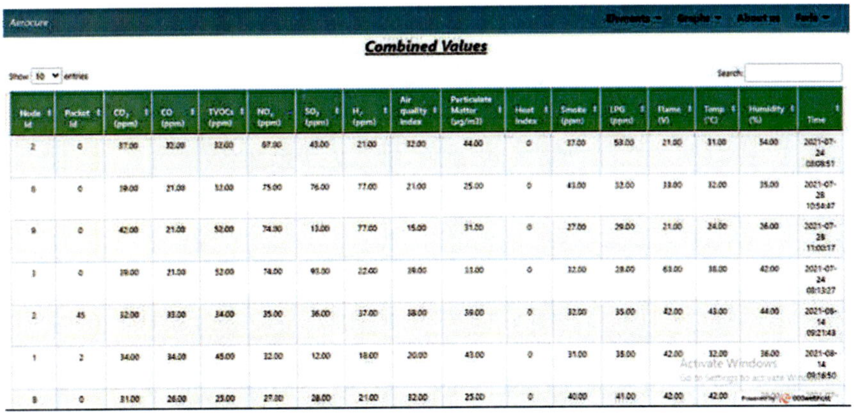

Combined Values

Node Id	Packet Id	CO_2 (ppm)	CO (ppm)	TVOCs (ppm)	NO_x (ppm)	SO_2 (ppm)	H_2 (ppm)	Air quality index	Particulate Matter (µg/m3)	Heat Index	Smoke (ppm)	LPG (ppm)	Flame (V)	Temp (°C)	Humidity (%)	Time
2	0	37.00	32.00	32.00	67.00	43.00	21.00	32.00	44.00	0	37.00	53.00	21.00	31.00	54.00	2021-07-24 08:08:51
8	0	19.00	21.00	32.00	75.00	76.00	77.00	21.00	25.00	0	43.00	32.00	33.00	32.00	35.00	2021-07-28 10:54:47
9	0	42.00	21.00	52.00	74.00	13.00	77.00	15.00	31.00	0	27.00	29.00	21.00	24.00	36.00	2021-07-28 11:00:17
3	0	39.00	21.00	52.00	74.00	93.00	22.00	39.00	33.00	0	32.00	28.00	63.00	38.00	42.00	2021-07-24 08:13:27
2	45	32.00	33.00	34.00	35.00	36.00	37.00	38.00	39.00	0	32.00	35.00	42.00	43.00	44.00	2021-08-14 09:21:43
1	2	34.00	34.20	45.00	32.00	12.00	18.00	20.00	43.00	0	31.00	35.00	42.00	32.00	36.00	2021-08-14 09:16:50
8	0	31.00	26.00	25.00	27.00	28.00	21.00	32.00	25.00	0	40.00	41.00	42.00	42.00	[illegible]	[illegible]

Fig. 6 Display of sensor data on web page

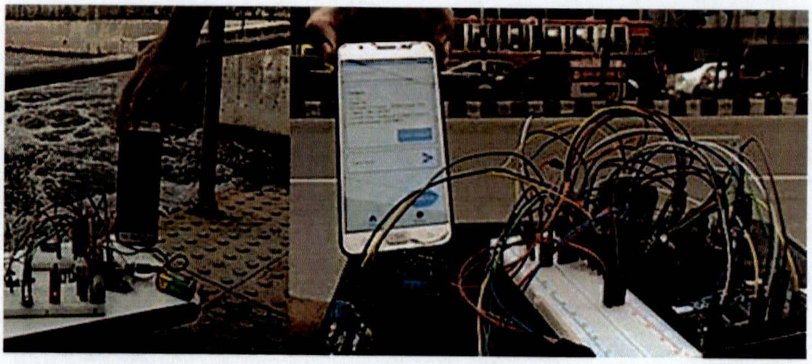

Fig. 7 Real life implementation

Table 1 Network Reliability Test

Networking Module	Time (s)	Packet sent	Packet received	Packet loss
Bluetooth	200	100	100	0
GSM	200	30	30	0
Wi-fi Module (ESP-8266): Scenario-A	200	100	97	3
NodeMCU+ nRF24L01: Scenario-B	200	100	79	21

Table 2 Comparison of Networking Options with Other Research

Reference	Bluetooth	Wi-Fi	Cellular network (GSM)	nRF
This Work	Yes	Yes	Yes	Yes
Marques et al. [17]	–	Yes	–	–
Lohani et al. [15]	Yes	–	–	–
Liu et al. [14]	–	–	Yes	–
Zheng et al. [25]	–	–	Yes	Yes

4.3.2 Comparison with Other Works

We have compared our proposed method with existing research work to analyze the quality. We have compared our work from two different aspects. In Table 2, a comparison of networking options provided by the system is shown. Our proposed method has multiple networking options that other research work lacks. In Table 3, we have compared our system cost with other existing systems.

From analyzing the information found from Tables 1, 2, and 3, we can see that our developed system offers maximum facilities in a cheaper cost.

Table 3 Comparison of Cost with Other Research

Reference	Pollutants	Cost (in USD)
This Work	CO, CO_2, SO_2, NO_x, TVOCs, AQI, Temperature, Heat Index, Humidity, H_2, $PM_{2.5}$, PM_{10}, Smoke, LPG, Flame	218.96
Cavaliere et al. [5]	PM_{10}, $PM_{2.5}$	1000
Williams et al. [24]	PM_{10}	2000
Laquai et al. [12]	NO_x, O_3, PM_{10}, $PM_{2.5}$	3700

5 Findings

Observing Figs. 3, 4, and 5 we can summarize the situation in these following points:

1. The dyeing section carries the highest value of particulate matter and high value of NO_x as most of the chemicals used in the process contain macro, micro molecules, and nitrogen compounds. Apart from CO_2 the remaining parameters are almost the same as indoor.
2. ETP section shows the highest value of temperature and heat index, CO, CO_2, TVOCs, and NO_x emission as it is a treatment process to recycle chemical wastage. This chemical process creates the least amount of dust pollutants.
3. Vehicular emission shows the highest value of CO_2, CO, and SO_2 pollution. It also carries a higher value in particulate pollutants. Vast uses of fossil fuel and poor fitness quality of the engine are the main reasons.
4. The indoor pollution value was the least of all as expected. But the comparatively high value in particulate pollutants indicates the mismanagement of trash and wastage.
5. During an artificial test condition, LPG, smoke, and flame sensor worked perfectly (Fig. 5).
6. According to Sect. 4.3.1, for security purposes, Bluetooth and GSM modules should be used to get the quickest response. On the other hand, for pollution data collection purposes, all the networking modules performed quite well.

6 Conclusion

In this research, we have developed a device capable of monitoring major air pollutants, and to make it effective in maximum industries, we have included a security system. Our pilot device has performed remarkably in measuring the mentioned factors. In the future, we will generate and apply a machine learning algorithm on the data sets that can predict pollution incrimination. Our focus will be on making the device more robust and improving the sensor quality for more data accuracy.

Acknowledgements Our research work is funded by Bangladesh University of Textiles. We would like to acknowledge Sublime Greentex Ltd. for real data collection purposes.

References

1. Amazon.com. Spend less. Smile more. https://www.amazon.com/. Accessed 10 May 2022
2. Techshopbd.com-largest robotics-electronics shop in bangladesh. https://www.techshopbd.com/. Accessed 10 May 2022
3. Amorim, L.C., Carneiro, J.P., Cardeal, Z.L.: An optimized method for determination of benzene in exhaled air by gas chromatography-mass spectrometry using solid phase microextraction as a sampling technique. J. Chromatogr. B **865**(1–2), 141–146 (2008)
4. Bank, W.: World Development Indicators 2014. The World Bank (2014)
5. Cavaliere, A., Carotenuto, F., Di Gennaro, F., Gioli, B., Gualtieri, G., Martelli, F., Matese, A., Toscano, P., Vagnoli, C., Zaldei, A.: Development of low-cost air quality stations for next generation monitoring networks: Calibration and validation of pm2. 5 and pm10 sensors. Sensors **18**(9), 2843 (2018)
6. Dhas, Y.J., Jeyanthi, P.: Environmental pollution monitoring system using internet of things (iot). J. Chem. Pharm. Sci. (2017)
7. Dhingra, S., Madda, R.B., Gandomi, A.H., Patan, R., Daneshmand, M.: Internet of things mobile-air pollution monitoring system (iot-mobair). IEEE Internet Things J. **6**(3), 5577–5584 (2019)
8. Ekpenyong, C.E., Ettebong, E., Akpan, E., Samson, T., Daniel, N.E.: Urban city transportation mode and respiratory health effect of air pollution: a cross-sectional study among transit and non-transit workers in nigeria. BMJ Open **2**(5), e001253 (2012)
9. Fund, U.N.C., UNICEF., Organization, W.H.: Children in the New Millennium: Environmental Impact on Health. UNEP/Earthprint (2002)
10. Husain, A.M., Rini, T.H., Haque, M.I., Alam, M.R.: Air quality monitoring: The use of arduino and android. J. Mod. Sci. Technol. **4**(1), 86–96 (2016)
11. Kracht, O., Gerboles, M.: Spatial representativeness evaluation of air quality monitoring sites by point-centred variography. Int. J. Environ. Pollut. **65**(1–3), 229–245 (2019)
12. Laquai, B.: Particle distribution dependent inaccuracy of the plantower pms5003 low-cost pm-sensor (2017)
13. Li, Z.: Arduino based environmental air monitoring system. Ph.D. thesis, Case Western Reserve University (2017)
14. Liu, J.H., Chen, Y.F., Lin, T.S., Lai, D.W., Wen, T.H., Sun, C.H., Juang, J.Y., Jiang, J.A.: Developed urban air quality monitoring system based on wireless sensor networks. In: 2011 Fifth International Conference on Sensing Technology, pp. 549–554. IEEE (2011)
15. Lohani, D., Acharya, D.: Smartvent: A context aware iot system to measure indoor air quality and ventilation rate. In: 2016 17th IEEE International Conference on Mobile Data Management (MDM), vol. 2, pp. 64–69. IEEE (2016)
16. Lungu, M., Stefu, N.: Study on particulate matter dispersion by correlating direct measurements with numerical simulations: Case study-timisoara urban area. Int. J. Environ. Sci. Technol. **15**(7), 1441–1452 (2018)
17. Marques, G., Pitarma, R.: Monitoring health factors in indoor living environments using internet of things. In: World Conference on Information Systems and Technologies, pp. 785–794. Springer (2017)

18. Postolache, O.A., Pereira, J.D., Girao, P.S.: Smart sensors network for air quality monitoring applications. IEEE Trans. Instrum. Meas. **58**(9), 3253–3262 (2009)
19. Reddy, M.S., Rao, K.R.: Fire accident detection and prevention monitoring system using wireless sensor network enabled android application. Indian J. Sci. Technol. **9**(17), 1–5 (2016)
20. Setiawan, F., Kustiawan, I.: Iot based air quality monitoring. In: IOP Conference Series: Materials Science and Engineering, vol. 384, p. 012008 (2018)
21. Shaban, K.B., Kadri, A., Rezk, E.: Urban air pollution monitoring system with forecasting models. IEEE Sens. J. **16**(8), 2598–2606 (2016)
22. Shaddick, G., Thomas, M., Mudu, P., Ruggeri, G., Gumy, S.: Half the world's population are exposed to increasing air pollution. NPJ Clim. Atmos. Sci. **3**(1), 1–5 (2020)
23. Tsow, F., Forzani, E., Rai, A., Wang, R., Tsui, R., Mastroianni, S., Knobbe, C., Gandolfi, A.J., Tao, N.: A wearable and wireless sensor system for real-time monitoring of toxic environmental volatile organic compounds. IEEE Sens. J. **9**(12), 1734–1740 (2009)
24. Williams, R., Kilaru, V., Snyder, E., Kaufman, A., Dye, T., Rutter, A., Russel, A., Hafner, H.: Air sensor guidebook, US Environmental Protection Agency, Washington, DC. Tech. rep., EPA/600/R-14/159 (NTIS PB2015-100610) (2014)
25. Zheng, Z.B.: Design of distributed indoor air quality remote monitoring network. In: Advanced Materials Research, vol. 850, pp. 500–503. Trans Tech Publ (2014)

Traffic Analysis-Based Android SMS Malware Detection Using Machine Learning

Kanis R. Mim, Md. Sakir Hossain, Sumona A. Tisha, Kallul R. Kalpo, Mehedi H. Bakul, and Md. Shakhawat Hossain

Abstract Devices with the Android operating system (OS) are more vulnerable to cyberattack due to its popularity. Among the attacks on the Android mobile devices, attacks due to SMS malware are growing and estimated to be one of the major cyber attacks in near future as messaging is more frequently used compared to voice communication and the SMS malware is spread through text message. Existing effort to detect the SMS malware attack is not sufficient to curb its spread. There exists no work to date to investigate the detection accuracy of each type of SMS malware. In this paper, based on the traffic analysis we carry out a comprehensive investigation on detecting different kinds of SMS malware. We employ state-of-the-art machine learning algorithms along with feature selection in detecting the SMS malware. We consider two detection scenarios: (1) binary classification between SMS malware and benign traffic, and (2) multiclass classification between all classes (11 classes) of SMS malware and benign traffic. Furthermore, we figure out the types of SMS malware which are most difficult to detect. Such kind of investigation was never done before. In both scenarios, the decision tree is found to be able to detect the SMS malware with the highest accuracy. The accuracies are 99.96 and 91.12% in the scenarios 1 and 2, respectively. The FakeNotify and Beanbot types of SMS malware are most difficult to detect.

Keywords Malware · Android · SMS malware · Machine learning · Detection · Traffic

1 Introduction

Information technology (IT) has become an unavoidable part of our daily lives. It has transformed our lives and is now a staple part of our everyday lives. Due to the corona virus pandemic, we witness a sudden soar of the use of the Internet in all

K. R. Mim · Md. S. Hossain (✉) · S. A. Tisha · K. R. Kalpo · M. H. Bakul · Md. S. Hossain
American International University-Bangladesh, Dhaka 1229, Bangladesh
e-mail: sakir.hossain@aiub.edu

S. Hossain et al. (eds.), *Proceedings of International Conference on Fourth Industrial Revolution and Beyond 2021*, Lecture Notes in Networks and Systems 437,
https://doi.org/10.1007/978-981-19-2445-3_20

spheres of our lives. Nowadays, we are much more dependent on computers, mobile devices, and the Internet than ever before.

People enjoy a greater amount of mobility and applications in mobile devices compared to computers. With the increase of the dependency on mobile devices and the increasing popularity of mobile devices, hackers are becoming more interested in the devices to know the confidential information of the users. Mobile applications work as a rider to penetrate the security of the mobile devices. Most people do not consider mobile applications as a potential source of their smartphone security breach because of their inadequate knowledge about IT. In the public mindset, mobile application installations are not commonly correlated with malware. Malicious software, also known as malware, are programs that interfere with a computer's functionality, steal sensitive data, or take control of the device without the user's permission [6]. By exploiting a device flaw, the attacker deceives the user into running the malicious program or receives unwanted remote access. Computer viruses, trojan horses, worms, and spyware are examples of such malware.

To identify malware families and discover new malware, an incremental malware detection scheme is proposed in [12]. There are three key elements of this system: feature extraction and selection, decision-making, and new malware detection. Gray-scale files, Opcode n-grams, and import functions all are part of the data processing module. Improved information gain is used to reduce the number of high-dimensional features. The features are used by the decision-making module to classify malware and recognize the suspicious malware. Finally, the detection module searches for new malware families using the shared nearest neighbor (SNN) clustering algorithm. The detection accuracy against the test data set is 86.7%. In this case, a binary classification problem is solved. The classes include malware and benign.

Another machine learning assisted binary classification of malware is proposed in [15]. However, instead of selecting multiple features [12], this is a classification based on a single feature: opcode frequency. It uses the Malica [14] dataset which consists of data for the Windows operating system. The deep neural network consisting of 2, 4, and 7 hidden layers is used in detecting malware. The random forest with variance threshold-based feature selection achieves the highest accuracy of 99.78%.

While there are the highest number of malware for Windows among the operating systems used in PCs, Android is similarly the worst hit operating system of mobile devices by malware. One of the common features of most applications for mobile devices is that they require some permission to access which part of the mobile devices they can access. These permissions are highly related with whether an application is a malware or benign. In [13], a framework for extracting a collection of 123 complex permissions from 11000 Android applications is presented. On the newly developed dataset, for detecting malicious Android applications, a range of machine learning classification algorithms are considered. Of them, the simple logistic provides the highest accuracy.

A hardware-based approach is proposed in [18] which exploits the behaviors of an application in accessing memory. It uses F-score feature selection. For classification, it uses random forest, support vector machine (SVM), and logistic regression. Similar to the previous works, a binary classification is carried out in [18]. Up to 99% accuracy with a false positive rate of 5% can be achieved by employing this technique.

Those sophisticated malware which can identify whether they are being analyzed by an emulator can effectively stop their activities in a bid to avoid being identified. Thus, the emulator-based solutions (e.g., [17]) cannot detect sophisticated malware with a considerable accuracy. To overcome such limitations, a machine learning-based malware detection technique is proposed in [2] which can dynamically extract the features of an Android package (APK). The dataset is extracted from the Android Malware Genome Project [20]. Information gain-based feature selection is employed. Machine learning techniques such as SVM, Naïve Bayes, simple logistic, partial decision tree (PART), random forest, and multilayer perceptron (MLP) are used for classification. For early identification of Android malware, a parallel machine learning-based classification method is proposed in [19]. A hybrid classification model is built from a parallel mixture of heterogeneous classifiers using actual malware and benign applications. The effectiveness and ability of the model to increase the detection performance is demonstrated by an empirical testing under various mixture schemes. The capabilities of many classifiers with different characteristics can be harnessed not only for improved Android malware identification but also for faster white box review using the more interpretable constituent classifiers.

In [11], the CICAndMal2017 dataset is created which consists of samples of four categories of Android malware, namely, adware, scareware, ransomware, and SMS malware. Three classifiers, such as random forest, decision tree, and K-nearest neighbor (KNN), are used to classify malware in three different scenarios: (i) if a sample is a malware (ii) detection of the categories of the malware (iii) detection of the families of the categories where the term family indicates the types of the malware which belong to any category of the malware. All four classes of the malware are considered collectively. The detection of each type of malware is not investigated.

In [1], a little more comprehensive study is carried out on the CICAndMal2017 Android malware dataset. Two improved models of the random forest and decision tree are proposed. In all three scenarios, the proposed models outperform the models proposed in [11]. Binary classification is carried out between benign and malware traffic. However, the SMS malware detection accuracy is not investigated.

A more detailed analysis of the Android malware detection is carried out in [5]. To improve the performance of the machine learning model, a total of 25 features are selected experimentally. The detection performance of difference machine learning models for the types of each category of the Android malware is investigated. However, no investigation is carried out to reveal the detection of each family of each class. For example, the family classification is carried out. However, the accuracy of detecting a particular family of a specific category is not known. Furthermore, the number of samples used to classify the SMS malware from other malware is very limited (only few hundreds). Training a machine learning model against such extremely limited number of samples cannot be effective in practice in a production network.

From the literature review, it is evident that there is a limited effort to date in the detection of the Android malware. The traffic flow-based Android malware detection is found to be more effective. In [1, 5, 11], the Android malware detection is analyzed. However, there is no investigation to know the detection accuracy for each family of the each category. In literature, the SMS malware is largely overlooked and gained comparatively less attention. Especially, to the best of our knowledge, the detection accuracy of each type of the SMS malware is never investigated. In this paper, we look into this matter and fill the research gap. We carry out a thorough investigation about the detection of the SMS malware and the families of the SMS malware. Furthermore, we also investigate the accuracies of the detection of each family of the SMS malware. Through experiments, we find that the decision tree provides the highest detection accuracy. It is also revealed that the Beanbot and FakeNotify SMS malware are the most difficult to detect.

2 SMS Malware Attacks

The word malware is a malicious software contraction. It usually consists of cyber attacker-developed code, designed to inflict significant data and device damage or to obtain unauthorized access to a network [16]. There are different kinds of malware such as viruses, worms, spyware, adware, scareware, trojan horse, and so on. Of them, the trojan horse is considered as the most severe malware as it attaches itself to the useful non-executable files such as images, audio, and video files. As people usually consider the executable files as suspicious, the Trojan horse malware is largely ignored.

SMS malware is defined as a malware that allows unauthorized calls and/or sends text messages without user permission. It belongs to the trojan horse type of malware. Usually, it consists of a URL-containing text message. By making the message appear to come from a valid source, such as their bank, the idea is to get the subscriber to open the connection [8]. Often these calls and texts are then directed to chargeable SMS text providers or cyber criminal-operated premium-charge numbers, creating large sources of revenue for cyber criminal networks. SMS malware also often plunders the victim's contacts list, sends links to more people hoping people will click the links sent from the people they know and trust.

Next, we shortly describe different types (i.e., families) of SMS malware:

1. **Beanbot**: Beanbot is a type of malware that transmits information from the user to a remote server and sends premium-rate SMS from the infected device. It is remotely operated. Not only will it transport personally identifiable data to its command and control server, but it will also send text messages in the background stealthily, resulting in unnecessary charges on the phone bill of the customer.
2. **Fakeinst**: Fakeinst malware tends to be an installer for other programs, but the malware sends text messages to premium-rate numbers or services when executed [7].

3. **Fakemart**: A malware that includes a trojan aimed at mobile devices running Android OS. When hiding itself as a black market application, it can execute the following actions: clear the contents of XMBPSP.xml, where the XMBPSP.xml is a shared preference file where settings are written, in the mutual choice and reconfigure it to send premium rated SMS messages to 81211 or 81308, set the system to silent mode, delete SMS received from 81211, etc.
4. **FakeNotify**: FakeNotify is a Trojan horse for Android devices that sends premium-rate phone numbers with SMS messages, captures and sends information and shows web pages periodically. It sends premium rated SMS messages to get through AV software by using obfuscation and identification techniques. SMSs are silently sent to premium-rate numbers via FakeNotify, then the recipient is routed to an unsolicited page.
5. **Biige**: By tracking SMS messages, outgoing calls, location information, etc., the Biige malware is used to spy on a victim's phone. It also offers an 'Uninstallation Security' feature that sends an SMS to a user-defined number from the phone any time when an attempt is made to uninstall the application.
6. **Jifake**: Jifake is a malicious trojan infection that uses a targeted user account to link and distributes premium SMS messages [4]. It runs in the background, and to camouflage itself, it produces extra unexpected files on the device. Other malware can be downloaded and installed without user knowledge. It would also block the installation of applications that will delete malware from the device.
7. **Mazarbot**: Mazarbot takes the administrative control of a device. To become infected by the Mazarbot, a user has to click on a link that arrives in an unsolicited text from an anonymous number [3]. It has the power to take a wide range of actions on the devices, including sending messages, receiving texts, collecting accurate device details, and even erasing the phone memory.
8. **Nandrobox**: Nandrobox steals data from a compromised device. It also deletes SMS texts from the device. It sends out a particular number of SMS messages, and then intercepts incoming messages from that number to protect its track.
9. **Plankton**: Plankton can come as part of Android applications that are repackaged and downloaded from third-party Android app markets. It switches the configuration of the device and steals information that is saved on the device. It secretly transmits information about the device to a remote location.
10. **SMS sniffer**: The SMS sniffer is somewhat similar to the adware category of malware. It automatically shows commercial advertisements. Sometimes, it is not malicious, but unwanted.
11. **Zsone**: Zsone delivers SMS to premium-rate numbers connected with SMS-based subscription services.

3 Experimental Results and Analysis

In this section, we will carry out extensive experiments to detect Android SMS malware. The block diagram of the system model is shown in Fig. 1. In this paper, we use the CICAndMal2017 dataset [11]. The part of the dataset containing different types of SMS malware is used in this paper. The types are Beanbot, Biige, Fakeinst, FakeNotify, Mazarbot, Nandrobox, Plankton, Jifake, SMSsniffer, and Zsone. This dataset contains 473411 samples and 83 features. We perform preprocessing on the dataset to make it suitable for training a machine learning model. The following preprocessing is carried out on the dataset:

- While generating the dataset, there was a duplicate feature. Not only the name but also the values were duplicated. Here, the dataset does not need the same name feature with the same value. To solve this problem, one of them is removed from the dataset.
- To use a classifier in Python, we needed to have integers only as feature values. However, in the dataset, there were some values which were of string type. To solve this problem, we use a label encoder. The label encoder assigns a unique integer value to every possible value of a categorical variable [9].

Next, we employ ANOVA-F feature selection algorithm to select the most important features to reduce the computational overhead and improve detection accuracy. Out of the 83 features of the dataset, we select the most important 18 features (see Table 1). Bwd and Fwd represent the backward and forward direction. The PSH flag is used to indicate that the sender will send data even if its buffer is not filled. The URG flag is used to mark a specific byte of segment so that the byte gets priority over other bytes. The RST flag is used to reset the connection between the sender and receiver. The CWE and ECE flags stand for congestion window extrapolator and echo of

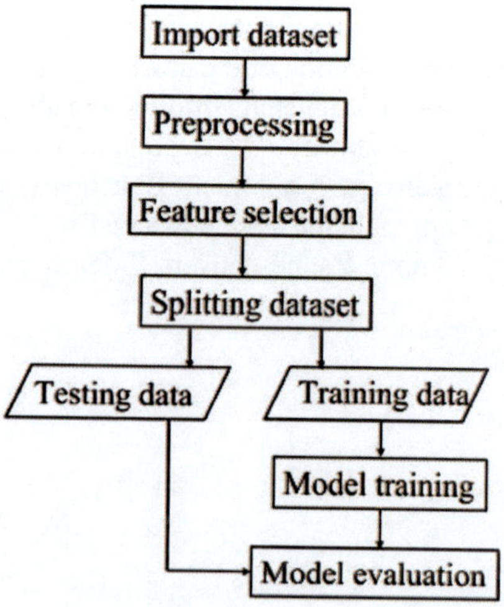

Fig. 1 Block diagram of the system model

Table 1 Selected features

SL.	Features	SL	Features
1	Bwd PSH Flags	10	Bwd Avg Bytes/Bulk
2	Fwd URG Flags	11	Bwd Avg Packets/Bulk
3	Bwd URG Flags	12	Bwd Avg Bulk Rate
4	RST Flag Count	13	Timestamp
5	CWE Flag Count	14	min_seg_size_forward
6	ECE Flag Count	15	Min Packet Length
7	Fwd Avg Bytes/Bulk	16	Protocol
8	Fwd Avg Packets/Bulk	17	Fwd IAT Mean
9	Fwd Avg Bulk Rate	18	Fwd IAT Total

congestion encountered, respectively. These two flags along with two IP flags help control network congestion. The Avg Bulk Rate represents the average bulk rate in a direction. The features Avg Bytes/Bulk and Avg Packets/Bulk represent the average number of bytes and packets per Bulk, respectively. The timestamp feature indicates the data sending time. The min_seg_size_forward feature indicates the minimum size of a segment in the forward direction. The minimum packet length is the minimum size of a packet. The mean and total inter-arrival time between segments/packets is represented by IAT Mean and IAT Total, respectively. Finally, the Protocol feature indicates if the used protocol is TCP or UDP. As shown in Fig. 1, once the preprocessing is done, the dataset is split into the train and test parts. The ratio of the number of train and test samples is 0.7: 0.3. The training samples are used to train a machine learning model, while we evaluate the prediction accuracy of the model using the testing samples.

3.1 Performance Metrics

To evaluate the performance of the trained classifiers, some evaluation metrics are used. These metrics measure and summarize the quality of a trained classifier when it is tested against the test data. There are some important metrics for classification evaluation [10]. These are mentioned below:

1. Accuracy: It measures the ratio of the number of correct predictions over the total number of instances evaluated.

$$Accuracy = \frac{TP + TN}{TP + FP + TN + FN} \tag{1}$$

   where TP, FP, TN, and FN represent true positive, false positive, true negative, and false negative, respectively.

2. Precision: It measures the positive patterns that are correctly predicted from the total predicted patterns in a positive class.

$$Precision = \frac{TP}{TP + FP} \tag{2}$$

3. Recall: This measures the fraction of positive patterns that are correctly classified.

$$Recall = \frac{TP}{TP + TN} \tag{3}$$

4. F1-score: This metric represents the mean between recall and precision values, and it must be harmonic mean.

$$F1 - score = \frac{2 * Pricision * Recall}{Precision + Recall} \tag{4}$$

3.2 *Performance Evaluation*

We use various classifiers in detecting the SMS malware. The classifiers are built and tested using Python. First, we investigate the detection accuracy of various machine learning algorithms in detecting SMS malware traffic in the presence of benign traffic (see Fig. 2). The notations used in Fig. 2 include: feature selection (FS), decision tree (DT), K-nearest neighbor (KNN), multilayer perceptron (MLP) of artificial neural network, random forest (RF), and Naïve Bayes (NB). We consider the classification with and without feature selection algorithm. While the decision tree provides the highest detection accuracy, the Naïve Bayes classifier underperforms all other classifiers. The decision tree provides almost 100% accuracy. The accuracies are 99.96% and 99. 64% in the scenarios without and with ANOVA-F. Although the accuracy falls slightly due to the feature selection, the feature selection is still important as it reduces the number of features by 78.31%. The reduction of the number of features decreases computational complexity as in most cases the computational complexity is proportional to the number of features if other parameters are considered constant. The random forest achieves 99.24% accuracy which is 0.72% lower than the highest achiever decision tree. Another important aspect is that the feature selection improves the accuracy of all classifiers except decision tree. The highest improvement (42.93%) is obtained for MLP. The detection accuracies due to the KNN and Naïve Bayes classifiers are 98.25% and 59.6%, respectively.

Next, we investigate the performance of the different machine learning algorithms in detecting different families of SMS malware. Figure 3 shows the detection accuracies for both without and with ANOVA-F feature selection algorithms. The highest detection accuracy of 91.12% is found due to the decision tree classifier and the second highest accuracy of 89.3% is achieved by the random forest. Similar to the binary classification shown in Fig. 2, we see that the feature selection leads to an

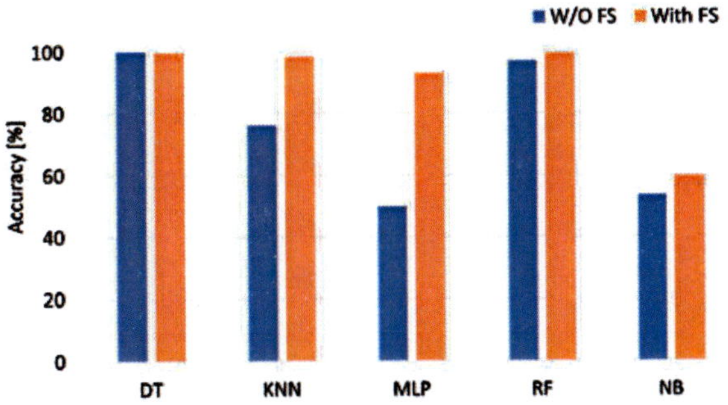

Fig. 2 Accuracy of binary classification between the benign and SMS malware traffic

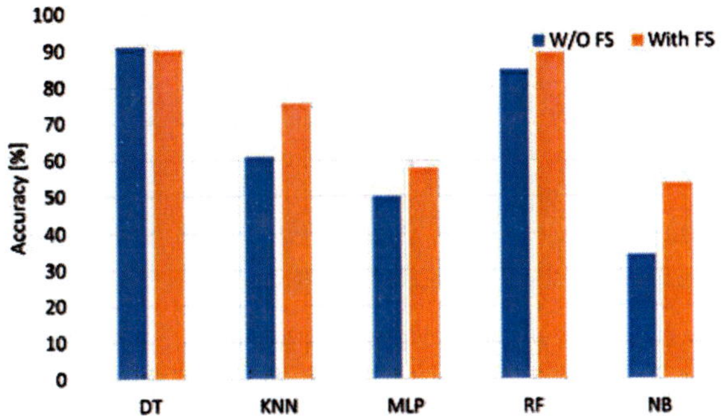

Fig. 3 Detection accuracy of the classes of SMS malware traffic in the presence of benign traffic

improvement of the detection accuracy of all classifiers except the decision tree in which case the detection accuracy degrades by less than 1%. The difference between the best (decision tree) and worst (Naïve Bayes) performers is 37.62%. To clearly illustrate the impact of the multiclass family classification and the binary SMS malware classification, we consider Fig. 4 which shows the detection accuracies achieved by different machine learning algorithms with the feature selection. The reason for choosing the feature selection case for the comparison can be explained as: the feature selection significantly improves the computational complexity. Furthermore, it improves the detection accuracy significantly. The detection accuracy improves significantly in all classifiers except the decision tree where a little degradation of the throughput is observed. Thus, the benefits far outweigh the limitations. In Fig. 4, the detection accuracy of each type of the SMS malware is significantly lower compared to the binary classification. The reduction of the detection accuracy varies depending on the classifiers. While the most degradation of 35% is found for artificial neural network, the least difference is obtained for Naïve Bayes classifier. About 10% less accuracy in family classification is obtained in both decision tree and random forest.

Next, we investigate which types of SMS malware are difficult to detect. Instead of considering all classifiers, we consider the best classifier only, decision tree, for the problem. The precision, recall, F1-score of the detection of various types of SMS malware are shown in Fig. 5. We consider the benign traffic and 11 classes of SMS malware. We see that the benign, Biige, Mazarbot, Nandorbox, and Zsone traffic

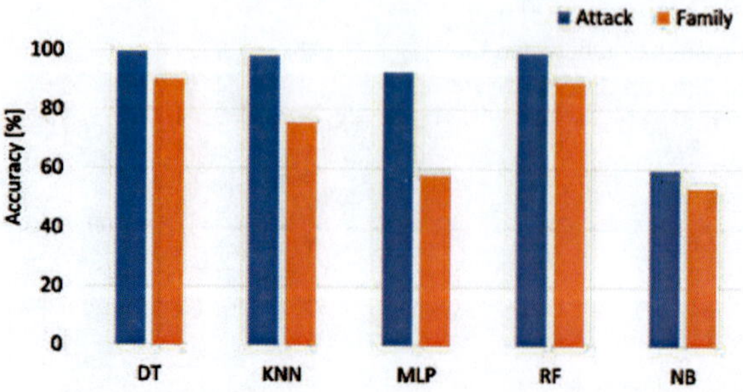

Fig. 4 Comparison of the detection accuracy of the SMS malware and various types of SMS malware

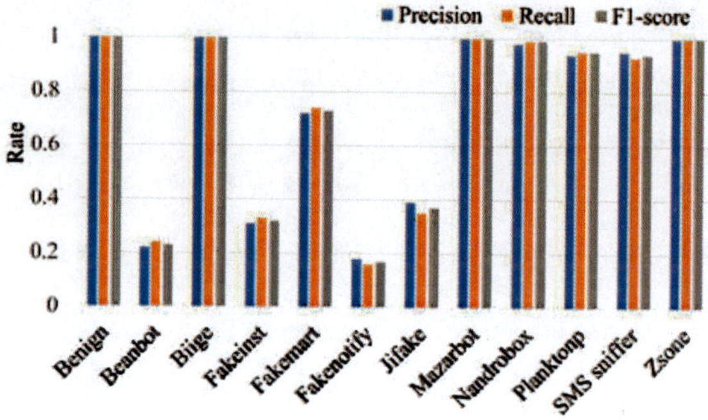

Fig. 5 Comparative precision, recall, F1-score of various SMS malware

can be detected with almost 100% precision, recall, and F1-score. This indicates that the amount of FP and FN are insignificant. On the other hand, the Beanbot, FakeNotify, Jifake, and Fakeinst are detected with less than 40% accuracy with the Beanbot and FakeNotify have around 20% precision, recall, and F1-score. We also find that there is no significant difference between the number of FP and FN cases as the precision and recall values are very close to each other. To find the reason behind such a low detection rate for the four types of SMS malware, we consider the confusion matrix. As before, the confusion matrix due to the decision tree is considered (see Table 2). It is revealed from the confusion matrix that there is a high degree of resemblance among the traffic of the Beanbot, Fakeinst, and FakeNotify SMS malware. For the Beanbot, some 55% and 18% of traffic are wrongly classified as the FakeNotify and Jifake SMS malware. Almost two-third of the Fakeinst SMS malware is wrongly classified as the FakeNotify. The TP of the FakeNotify is the minimum among all SMS malware. However, 48% and 34% of the FakeNotify traffic are wrongly classified as the Fakeinst and Beanbot traffic, respectively. Regarding Jifake traffic, more than 40% traffic is wrongly classified as the traffic of the Beanbot malware.

Table 2 Confusion matrix of the decision tree

	Ben ign	Bean bot	Biige	Fake Inst	Fake mart	Fake notify	Jif ake	Maza rbot	Nand robox	Plan kton	SMS sniffer	Zsone
Benign	100%	0	0	0	0	0	0	0	0	0	0	0
Beanbot	1%	24%	0	2%	0	55%	18%	0	0	0	0	0
Biige	0	0	100%	0	0	0	0	0	0	0	0	0
FakeInst	1%	2%	0	33%	0	63%	0	0	0	0	0	0
Fakemart	0	0	0	0	74%	0	15%	0	0	10%	1%	0
FakeNotify	1%	34%	0	48%	0	16%	0	0	0	0	0	0
Jifake	0	41%	0	3%	20%	1%	35%	0	0	0	0	0
Mazarbot	0	0	0	0	0	0	0	100%	0	0	0	0
Nandrobox	0	0	0	0	0	0	0	0	99%	0	1%	0
Plankton	0	0	0	0	1%	0	0	0	0	95%	3%	0
SMS sniffer	0	0	0	0	0	0	0	0	2%	5%	93%	0
Zsone	0	0	0	0	0	0	0	0	0	0	0	100%

To protect our devices from the SMS malware, we should refrain from downloading any unknown app, clicking an untrusted URL, downloading an Email attachment sent from an untrusted sender, and so on. Furthermore, a good anti-virus and anti-spyware should be installed in our devices. Furthermore, restricting access to our devices from others is also a precautionary step.

4 Conclusions

In this paper, we built different machine learning models for detecting SMS malware of Android mobile devices. We considered 11 types of SMS malware such as Beanbot, Biige, Fakeinst, Fakemart, FakeNotify, Jifake, Mazarbot, Nandrobox, Plankton, SMS sniffer, and Zsone. We employed the ANOVA-F value feature selection algorithm, which selected 18 features from the available 83 features. Thereafter, we built models using five popular classification algorithms: decision tree, random forest, ANN, KNN, and Naïve Bayes. The decision tree was found to outperform all other algorithms. SMS malware detection accuracy was more than 99% in the detection of the SMS malware, while it was about 90% in the detection of the specific type of the SMS malware. We also found that the Biige, Mazarbot, Zsone, and Nandrobox could be detected with almost 100% accuracy, while the detection accuracy of the Beanbot and FakeNotify went down to around 20%. However, still there are scopes to further investigate to further improve the accuracy in the SMS malware type classification. The use of ensemble classifiers can be investigated to further improve the accuracy. We are leaving it as a future research.

References

1. Abuthawabeh, M., Mahmoud, K.: Enhanced android malware detection and family classification, using conversation-level network traffic features. Int. Arab J. Inf. Technol. **17**(4A), 607–614 (2020)
2. Alzaylaee, M.K., Yerima, S.Y., Sezer, S.: Emulator vs real phone: Android malware detection using machine learning. In: Proceedings of the 3rd ACM on International Workshop on Security and Privacy Analytics, pp. 65–72 (2017)
3. Belous, A., Saladukha, V.: Viruses, Hardware and Software Trojans: Attacks and Countermeasures. Springer Nature (2020)
4. Dunham, K. et al.: Android Malware and Analysis. CRC Press (2014)
5. Fallah, S., Bidgoly, A.J.: Benchmarking machine learning algorithms for android malware detection. Jordanian J. Comput. Inf. Technol. (JJCIT) **5**(03) (2019)
6. Feldman, S., Stadther, D., Wang, B.: Manilyzer: automated android malware detection through manifest analysis. In: Proceedings of the International Conference on Mobile Ad Hoc and Sensor Systems, pp. 767–772. IEEE (2014)
7. Haag, C.: Security and privacy in communication networks (2012)
8. Hamandi, K., Chehab, A., Elhajj, I.H., Kayssi, A.: Android sms malware: Vulnerability and mitigation. In: Proceedings of the 27th International Conference on Advanced Information Networking and Applications Workshops, pp. 1004–1009. IEEE (2013)

9. Hancock, J.T., Khoshgoftaar, T.M.: Survey on categorical data for neural networks. J. Big Data 7(1), 1–41 (2020)
10. Hossin, M., Sulaiman, M.N.: A review on evaluation metrics for data classification evaluations. Int. J. Data Mining Knowl. Manage. Process **5**(2), 1 (2015)
11. Lashkari, A.H., et al.: Toward developing a systematic approach to generate benchmark android malware datasets and classification. In: Proceedings of the International Carnahan Conference on Security Technology (ICCST), pp. 1–7. IEEE (2018)
12. Liu, L., Wang, B.S., Yu, B., Zhong, Q.X.: Automatic malware classification and new malware detection using machine learning. Front. Inf. Technol. Electr. Eng. **18**(9), 1336–1347 (2017)
13. Mahindru, A., Singh, P.: Dynamic permissions based android malware detection using machine learning techniques. In: Proceedings of the Innovations in Software Engineering Conference, pp. 202–210 (2017)
14. Nappa, A., Rafique, M.Z., Caballero, J.: Driving in the cloud: An analysis of drive-by download operations and abuse reporting. In: Proceedings of the International Conference on Detection of Intrusions and Malware, and Vulnerability Assessment, pp. 1–20. Springer (2013)
15. Rathore, H., Agarwal, S., Sahay, S.K., Sewak, M.: Malware detection using machine learning and deep learning. In: Proceedings of the International Conference on Big Data Analytics, pp. 402–411. Springer (2018)
16. Rieck, K., Holz, T., Willems, C., Düssel, P., Laskov, P.: Learning and classification of malware behavior. In: Proceedings of the International Conference on Detection of Intrusions and Malware, and Vulnerability Assessment, pp. 108–125. Springer (2008)
17. Wang, C.Y., You, C.Y., Hsu, F.H., Lee, C.H., Liu, C.H., Zhuang, Y.: SMS observer: a dynamic mechanism to analyze the behavior of sms-based malware. J. Parallel Distrib. Comput. **156** (2021)
18. Xu, Z., Ray, S., Subramanyan, P., Malik, S.: Malware detection using machine learning based analysis of virtual memory access patterns. In: Proceedings of the Design, Automation & Test in Europe Conference & Exhibition (DATE), 2017, pp. 169–174. IEEE (2017)
19. Yerima, S.Y., Sezer, S., Muttik, I.: Android malware detection using parallel machine learning classifiers. In: Proceedings of the International Conference on Next Generation Mobile Apps, Services and Technologies, pp. 37–42. IEEE (2014)
20. Zhou, Y., Jiang, X.: Android malware genome project. http://www.malgenomeproject.org/ (2020)

Toward Tolerating Soft Errors for Network Data Packets

Rumana Afrin, Muhammad Sheikh Sadi, and Md. Liakot Ali

Abstract Nowadays, network data packets use checksum methodology to detect the soft error. But the drawback of this methodology is that it cannot detect exactly which data bit is erroneous. To overcome and minimize soft errors for network data packets, we have proposed a new method, which can detect and correct single-bit and multiple bits' soft errors without using any backup, i.e., Error Correction Code (ECC). It also uses the compressed packed down data bits' technique, which helps to minimize the data packet size. The experimental study shows that the error detection and correction capability of the proposed method are much higher than the checksum method by incurring a considerable bit overhead and the time requirement is lessened as well in the proposed method.

Keywords Soft error tolerance · Check-sum · Multiple bit upset · Single bit upset · Data packets

1 Introduction

The modern communication system of information technology is at high risk due to soft errors [1]. This problem has become an obstacle for network communication. The previous studies show that network device is often unable to communicate with other network devices due to failure in network communication, which occurred by soft errors [2]. The soft error rate in network communication is quickly growing

R. Afrin (✉) · M. S. Sadi (✉)
Department of Computer Science and Engineering, Khulna University of Engineering and Technology, Khulna, Bangladesh
e-mail: r.afrin07@gmail.com

M. S. Sadi
e-mail: sadi@cse.kuet.ac.bd

M. S. Sadi · Md. L. Ali (✉)
Institute of Information and Communication Technology (IICT), University of Engineering Technology (BUET), Dhaka, Bangladesh
e-mail: liakot@iict.buet.ac.bd

S. Hossain et al. (eds.), *Proceedings of International Conference on Fourth Industrial Revolution and Beyond 2021*, Lecture Notes in Networks and Systems 437,
https://doi.org/10.1007/978-981-19-2445-3_21

because of various natural reasons such as alpha-particle radiation, cosmic radiation, ionizing effects, high-voltage particles, protons, heavy electrons, neutrons, and so on. The increasing unpredictability of soft errors is responsible for ungraceful network communication [3], particularly for networking equipment whose requirements are obligatory to filter, encode and decode, forward and receive data to the endpoint equipment such as servers, personal computers, and smart devices [4]. For example, in network communication, the network devices must afford a high level of accessibility and reliability [5, 6]. However, these types of devices suffer from an efficient soft error tolerance methodology. Besides this, the network communication protocols (e.g., IPv4) are also at high risk due to soft errors.

Conventional error tolerance methodologies need modification in the architecture of network data packets. Nevertheless, modifying architectures is too costly to implement [7]. Usually, the checksum method is used to detect a soft error in the IPv4 network. But this method can only detect (but cannot correct) soft error; moreover, this method cannot detect exactly which bit or bits are erroneous. In this method, after detecting the soft error, the receiver sends an acknowledgment to the sender to send the data packet again. Thus, the whole process becomes costlier and consumes much time.

This paper offers a new method to increase the reliability and efficiency of network data packets. The proposed technique can provide more efficient and trustworthy network communication among senders and receivers. It also provides better performance by minimizing data packet size.

This paper is organized as follows. In Sect. 2, related work about soft error tolerance methodologies for the network data packet is mentioned. The proposed methodology is introduced in Sect. 3; experimental results with case studies are described in Sect. 4. In the end, the conclusion is entailed in Sect. 5.

2 Related Work

Several approaches have been proposed to tolerate soft errors for IPv4 network data packets. These approaches are outlined shortly as follows.

New parity computations and checksum methodology was proposed in [8], where input and output data were compared to detect the soft error. This method can detect only single-bit errors.

A user datagram protocol (UDP) checksum was proposed in [9], where UDP checksum information was used to detect and correct up to two bits of errors. The performance efficiency of this method was 79%.

A numerical packing-based method was proposed in [10] to detect single data corruption. However, this methodology was proposed to detect only a single bit error.

A sparse recovery and compressive sensing-based method were proposed in [11]. It protected a partial packet from soft errors in an arbitrary linear coding system but it had a higher computational and time complexity.

A guided probabilistic checksum technique was proposed in [12]. This method used checksum code to detect and correct soft errors. This method can detect and correct up to 75% of soft errors.

In [13], a checksum method was used for an embedded control network. The error detection rate was great but this method cannot detect specific erroneous data bits and had no error correction capability.

A Compound Poisson (CP) methodology was proposed in [14], which can detect and correct single and multiple erroneous bits by calculating clustered random events. Nevertheless, the probability of failing is higher due to independent supposition in input data.

Almost none of the existing techniques (so far we have reviewed) can tolerate soft errors for IPv4 network data packets. Moreover, most of the methods can detect and correct single-bit errors. Hence, there is a great need for a new methodology that can detect and correct soft errors for IPv4 network data packets with low time consumption and cost.

3 The Proposed Methodology

The proposed methodology to provide soft error tolerance for the network data packet is outlined shortly as follows.

3.1 Error Tolerance for Data Packet of IPv4 Network

The total encoding procedure is completed in two steps. In the first step, the data packet is minimized and in the second step, check bits are generated and the data packet is encrypted. In the IPv4 protocol, the data packet header is divided into 14 different portions. In Fig. 1, an IPv4 data packet is shown. There are some specific values to define some fields. For example, in the version field, "0100" is used to indicate the IPv4 version and "0110" is used to define the IPv6 version. By shortening, "0" is used instead of "0100" and "1" is used instead of "0110" in this paper. Similarly, flag "10" is used instead of "010" to indicate no fragment, and "00" is used instead

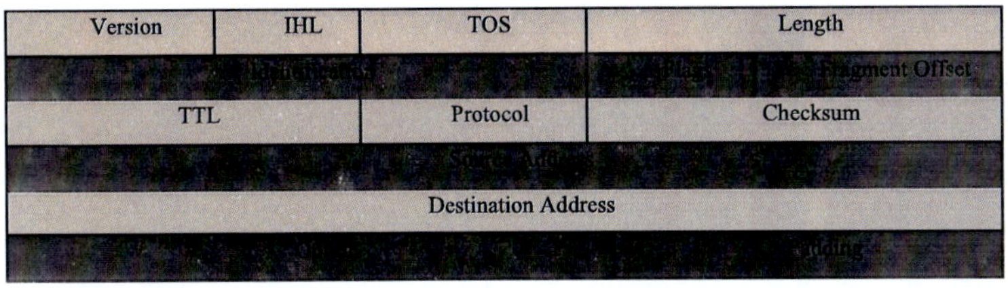

Fig. 1 Packet header of IPv4 data

Table 1 Compressed data value for fixed data bits

Field name	Value	Compressed value
IPv4 version	0100	0
IPv6 version	0110	1
Flag (no fragment)	000	00
Flag (last fragment)	001	01
Flag (more fragment)	101	10
Checksum	x	x
Protocol (ICMP)	00,000,001	000
Protocol (IGMP)	00,000,011	001
Protocol (TCP)	00,000,110	010
Protocol (UDP)	00,010,001	011
Protocol (EncaAP)	00,101,001	100
Protocol (OSPF)	01,011,001	101
Protocol (SCTP)	10,000,100	110

of "000" to indicate the last fragment and "01" is used instead of "001" to indicate more fragments.

There are several protocols used for network communication. But all protocols are not used for IPv4 data packets. As shown in Table 1, listed protocols are generally used for IPv4 network data packets. The proposed methodology is used for IPv4 network data packets, where 192 packet header bits are divided into 6 rows. Each row has or multiple columns. First row has 4 columns, second and third rows have 3 columns, fourth and fifth rows have 1 column each and sixth row has 2 columns. In this paper, fixed data bits are compressed as shown in Table 1.

The proposed methodology to provide soft error tolerance for the network data packet is outlined shortly as follows.

After compressing data bits, check bits are generated by using the flowing methodology.

- First check bit C_0 is generated by using (1).

$$D_{r.i} \oplus D_{r.n} = C_{r.i} \tag{1}$$

where D represents data bits, r represents row number, and i represents column number within a row and C represents check bits.

- Second check bit C_1 is generated by using (2).

$$D_{r.i} \oplus D_{r.(i+1)} = C_{r.(i+1)} \tag{2}$$

- Finally, the rest of the check bits $C_2 \ldots C_n$ are generated by using (3).

$$D_{r.(i+1)} \oplus C_{r.(i+1)} = C_{r.(i+1)} \quad (3)$$

where I = 1.....n.

3.2 Decoding Procedure of Data Packet for IPv4

For decoding data packets and detecting errors in data, the procedure is executed in several steps. These are outlined shortly as follows.

First, regenerating data bits from check bits is done by using (4).

$$C_{r.i} \oplus C_{r.(i-1)} = D'_n \quad (4)$$

where I = n...0.1. For generating data bits D_0 and D_1, (5) and (6) are used.

$$D'_n \oplus C_0 = D'_0 \quad (5)$$

$$D'_0 \oplus C_1 = D'_1 \quad (6)$$

For detecting errors in network packet data, (7) is used.

$$D_n \oplus D'_n = H_n \quad (7)$$

If H_n equals to zero, then the data bit is correct which means no error has occurred. If H_n equals to nonzero, then an error has occurred and the erroneous data bit is corrected by using (8).

$$D_{correct} = D_n \oplus H_n \quad (8)$$

Now data packet is decompressed by using Table 1 and original data bits are stored in the receiver memory.

In Fig. 2a, the check bits generation methodology using data bits is shown. Figure 2b shows how data bits are generated from check bits by using XOR operations for each row.

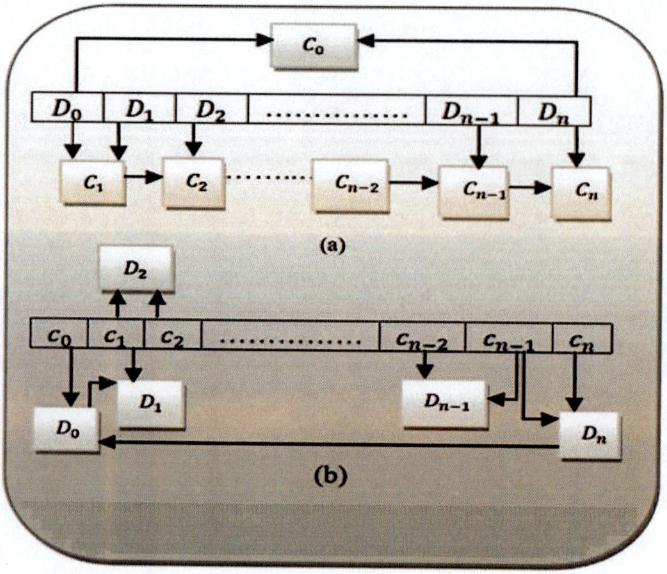

Fig. 2 Encryption and decryption procedure of the proposed method

3.3 An Example of Encoding and Decoding Procedure for IPv4 Data Packet

In Fig. 3, an IPv4 data packet is shown with data values. Now data bits are compressed by using Table 1. Here the version is IPv4, where data bits are "0100" and it is replaced by "0".

Version	IHL	TOS	Length
0100	0101	01110010	1111000011110000
Identification		**Flags**	**Fragment Offset**
0000111100110010		000	0000111100001
TTL	**Protocol**	**Checksum**	
00011000	00000110	B861	
Source Address			
10101100000010000111111110000000001			
Destination Address			
01010101011111110000001111010110			
Option		**Padding**	
000000000000000000000000		000000000	

Fig. 3 An IPv4 packet header with a data value

Table 2 IPv4 compressed data values

<table>
<tr><th colspan="6">Compressed data value</th></tr>
<tr><td>Version
0</td><td>IHL
0101</td><td>TOS
01110010</td><td colspan="3">Length
1111000011110000</td></tr>
<tr><td colspan="3">Identification
0000111100110010</td><td colspan="2">Flags
00</td><td>Fragment offset
0000111100001</td></tr>
<tr><td colspan="2">TTL
00011000</td><td>Protocol
010</td><td colspan="3">Checksum</td></tr>
<tr><td colspan="6">Source address
10101100000100001111111000000001</td></tr>
<tr><td colspan="6">Destination address
01010101011111110000001111010110</td></tr>
<tr><td colspan="4">Option
000000000000000000000000</td><td colspan="2">Padding
000000000</td></tr>
</table>

Similarly, Flag, Checksum, and Protocol field's data bits are also replaced by using Table 1 and then compressed data bits are placed in Table 2.

Now check bits for D_{28}, D_{59}, D_{75}, D_{107}, D_{139}, and D_{171} are generated by using (1).

$$D_0 \oplus D_{28} = 0 \oplus 0 = 0 = C_0$$

Similarly, the rest of the check bits C_{29},C_{60}, C_{76}, C_{108}, and C_{140} are generated by using (1).

Second check bits of each row are generated by using (2).

$$D_0 \oplus D_{28} = 0 \oplus 0 = 0 = C_0$$

And by using (3), rest of the check bits are generated.

$$D_2 \oplus C_1 = 0 \oplus 0 = 0 = C_2$$

Now some erroneous bits are injected into the data packet. To detect and correct erroneous data bits, the first data bits are regenerated by using (4).

$$C_{28} \oplus C_{27} = 0 \oplus 0 = 0 = D'_{28}$$

Similarly, the rest of the data bits of each row are generated by using (4).

The first data bit of each row is generated by using (5).

$$D'_{28} \oplus C_0 = 0 \oplus 0 = 0 = D'_0$$

Similarly, the rest of the first data bits of each row are also generated by using (5).

And second data bits of each row are generated by using (6).

$$D'_0 \oplus C_1 = 0 \oplus 0 = 0 = D'_1$$

Now erroneous data bits are checked by using (7).

$$D'_0 \oplus D_0 = 0 \oplus 0 = 0 = H_0$$

Here H_2, H_{78}, H_{79}, H_{109}, H_{139}, and H_{171} are nonzero. Therefore, D_2, D_{79}, D_{80}, D_{109}, D_{139}, and D_{170} are erroneous data bits. These erroneous data bits are corrected by using (8).

$$D_2 \oplus H_2 = 0 \oplus 1 = 1 = D_{2_{Correct}}$$

And the rest of the correct data bits for D_{79}, D_{80}, D_{109}, D_{139}, and D_{170} are 1, 1, 1, 0, and 0, respectively.

By using this methodology, any single and multiple erroneous data bits can be detected and corrected anywhere in the data word.

4 Experimental Analysis

The experimental setup, fault injection procedure, and results of the proposed methodology are shown as follows.

4.1 Experimental Setup

For the implementation of the proposed methodology, Intel Processor with 4 GB RAM is used and as a programming language, Turbo C++ is used.

4.2 Fault Injection

For experimental analysis of the proposed method, faults are injected in data bits and the result is compared with the existing checksum method [13]. Both arbitrary and sequential faults are injected and then the proposed method is applied to the data bits.

Table 3 The comparison of error detection rate between the checksum method and the proposed method

Corrupt bits in words	Checksum method error detection rate (%)	Proposed method error detection rate (%)	Checksum method error correction rate (%)	Proposed method error correction rate (%)
1	100	100	0	100
2	100	100	0	100
3	100	100	0	100
4	80	100	0	100
5	65	100	0	100
6	45	100	0	100
7	12	100	0	100

4.3 *Experimental Results Analysis for IPv4 Network Data Packets*

To simulate the proposed method, 192 data bits are divided into 6*32 matrixes. After that, data bits are compressed by using Table 1. Then for encryption and generating check bits, the proposed method is used. For exploring the proficiency of the conventional checksum method [13] and the proposed method, error detection and correction rate, execution time complexity, code overhead, codeword, and required memory size are analyzed.

From Table 3, it can be observed that the proposed method has better error detection and correction rate than the conventional checksum method. The checksum method can detect 100% error up to 4 bits then its working efficiency gradually decreases and for 8 bits, the chance of error detection is only 4%. Furthermore, its error correction capability is zero. On the other hand, the proposed method can detect 100% erroneous data bits.

The comparison of error detection and correction rate between the proposed method and the checksum method is shown in Fig. 4. After analyzing Table 3 and Fig. 4, it can be observed that the proposed method is more efficient than the checksum method.

Table 4 shows that the checksum method needs 16 check bits for 192 data bits to detect soft errors (it cannot correct errors), whereas the proposed method needs 172 check bits to detect and correct soft errors. Code overhead, codeword, and coding efficiency of the checksum method are 8.34%, 208%, and 7.692%, respectively, whereas for the proposed method those are 89.58%, 364%, and 47.252%, respectively. Code overhead and codeword of the proposed method are higher than the checksum method but the coding efficiency of the proposed method is better than the checksum method because the checksum method can only detect the occurrence of the soft error; it does not have any capability to detect the region of erroneous data bits and it cannot correct erroneous data as well whereas the proposed method can detect the region

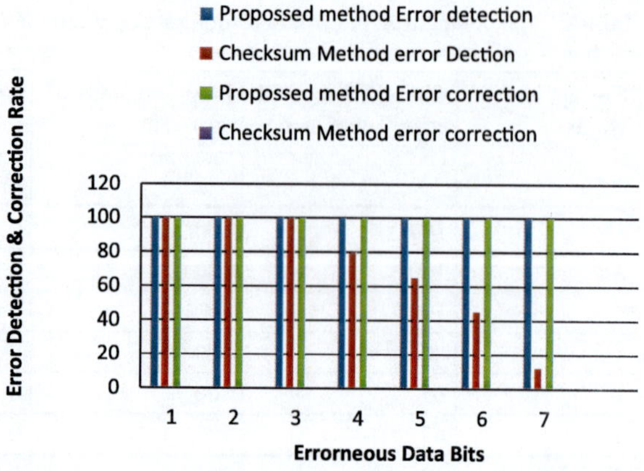

Fig. 4 An evaluation of the error detection correction rate between the proposed method and the checksum method

Table 4 The comparison of coding efficiency between the proposed method and the checksum method

Methods	Data bits	Check bits	Error correction rate	Code overhead (%)	Code word	Code efficiency
Checksum method	192	16	0	8.34	208	7.692
Proposed method	192	172	192	89.58	364	47.25

of erroneous data bits and can also correct those data bits. Therefore, the proposed method's working ability is better than the conventional checksum method.

Code overhead, codeword, and coding efficiency of the proposed method and checksum method are shown in Fig. 5. The proposed method requires larger code overhead, codeword, and coding efficiency than the checksum method.

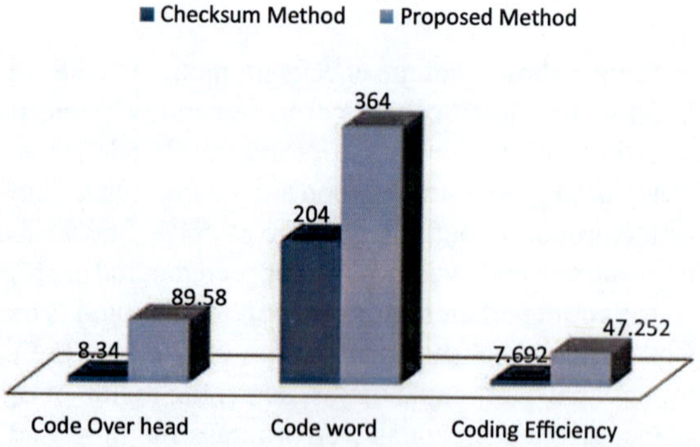

Fig. 5 Evaluation of code overhead codeword and coding efficiency between proposed method and checksum method

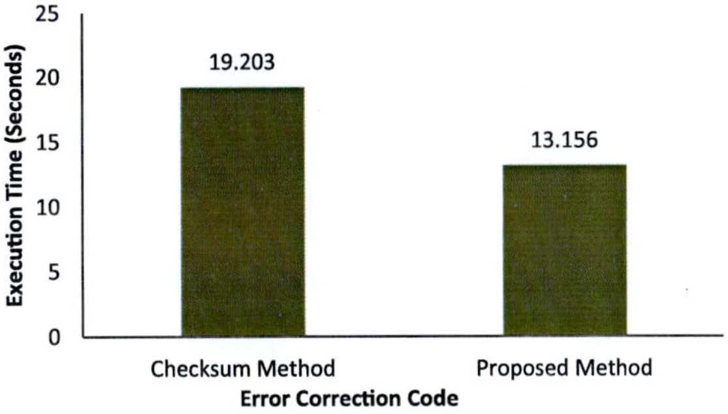

Fig. 6 A graphical presentation of the required execution time between the proposed method and the checksum method

Figure 6 shows the time required for executing the proposed method and checksum method. The proposed method needs 13.156 s to detect erroneous data bits, whereas the checksum method takes 19.203 s to identify that an error has occurred. Then it sends an acknowledgment to the sender for resending the data packet. Therefore, it needs more time than the proposed method because the proposed method can correct erroneous data bits without the help of the sender.

5 Conclusions

This paper proposes a new error correction technique to tolerate soft errors in IPv4 network data packets. It does not need any checksum schema to detect and correct soft errors in IPv4 network data packets. The effectiveness of the error detection and correction method proposed in this paper is illustrated using software simulation. The experimental study shows that the error detection and correction capability of the proposed method are much higher than the existing checksum method. The future enhancement of this paper could work with IPV6 network data packets and could work on lowering bit overhead by adopting new methodologies.

Acknowledgement This paper is a part of MSc Engineering thesis of the first author. We do acknowledge the support and cooperation of the Department of Computer Science and Engineering, Khulna University of Engineering & Technology for performing this research.

References

1. Sadi, M.S., Ahmed, W., Jürjens, J.: Towards tolerating soft errors for embedded systems. Springer Nature Comput. Sci. **2**(2), 1–10 (2021). https://doi.org/10.1007/s42979-021-00497-9
2. Gao, Y., Xiao, F., Liu, J., Wang, R.: Distributed soft fault detection for interval type-2 fuzzy-model-based stochastic systems with wireless sensor networks. IEEE Trans. Ind. Inf. **15**(1), 334–347 (2019)

3. Boraten, T., Kodi, A.: Runtime techniques to mitigate soft errors in network-on-chip (NoC) architectures. IEEE Trans. Comput. Aided Design Integr. Circuits Syst. **37**(3), 682–695 (2018)
4. Syafalni, I., Sasao, T., Wen, X.: A method to detect bit flips in a soft-error resilient TCAM. IEEE Trans. Comput. Aided Design Integr. Circuits Syst. **37**(6), 1185–1196 (2018)
5. Saggese, G.P., Wang, N.J., Kalbarczyk, Z.T., Patel, S.J., Iyer, R.K.: An experimental study of soft errors in microprocessors. IEEE Trans. Very Large Scale Integr. (VLSI) Syst. **27**(2) 486–490 (2019)
6. Silburt, A.L., Evans, A., Perryman, I., Wen, S.J., Alexandrescu, D.: Design for soft error resiliency in Internet core routers. IEEE Trans. Nucl. Sci. **56**(6), 3551–3555 (2009)
7. Eftaxiopoulos, N., Axelos, N., Pekmestzi, K.: Low leakage radiation-tolerant CAM/TCAM cell. In: 21st International On-Line Testing Symposium (IOLTS), pp.206–211 (2015)
8. Redinbo, G.R.: Designing checksums for detecting errors in fast unitary transforms. IEEE Trans. Comput. **67**(4) 566–572 (2018)
9. Golaghazadeh, F., Coulombe, S., Coudoux, F., Corlay, P.: Checksum-filtered list decoding applied to H.264 and H.265 video error correction. IEEE Trans. Circuits Syst. Video Technol. **28**(8), 1993–2006 (2017)
10. IjeomaAnarado, M.A., Anam, F.V., Andreopoulos, Y.: Mitigating silent data corruptions in integer matrix products: toward reliable multimedia computing on unreliable hardware. IEEE Trans. Circuits Syst. Video Technol. **27**(11), 192476–192489 (2016)
11. Mohammadi, M.S., Zhang, Q., Dutkiewicz, E.: Reading damaged scripts: partial packet recovery based on compressive sensing for efficient random linear coded transmission. IEEE Trans. Commun. **64**(8), 3296–3310 (2016)
12. Nisar, M.M., Chatterjee, A.: Guided probabilistic checksums for error control in low-power digital filters. IEEE Trans. Comput. **60**(3), 1313–1326 (2011)
13. Maxinoand, T.C., Koopman, P.J.: The effectiveness of checksums for embedded control networks. IEEE Trans. Dependable Secure Comput. **6**(1), 1313–1326 (2009)
14. Baeg, S., Wen, S., Wong, R.: Minimizing soft errors in TCAM devices: a probabilistic approach to determining scrubbing intervals. IEEE Trans. Circuits Syst. **57**(4), 814–822 (2010)

An Ensemble Machine Learning Approach for Detecting Fraudulent Banking Transactions

Sifat Bin Zaman, Noshin Tasnim, Nusrat Jahan, Md. Mahbubur Rahman, and Samrat Kumar Dey

Abstract With the increased number of banking institutions and their clients, the number of fraudulent transaction has increased dramatically nowadays. Banks and financial institutions are passing a difficult time in safeguarding their online transaction channels. One of the main reasons is the availability of online access channels to the customers and those are increasingly exploited by the attacker. In this paper, a fraud detection approach using an ensemble of machine learning classifiers has been proposed. In the model, an ensemble of K-Nearest Neighbor, Random Forest, and XGBoost algorithms are combined to detect frauds in online transactions. To archive better accuracy in detection, an exploratory data analysis is conducted. Feature engineering is used to add useful features to the data to improve detection accuracy and oversampling is done to solve the imbalances in data. The proposed classifier shows better results compared to some well-known standalone and ensemble classifiers in terms of accuracy, f1 score and roc-auc score.

Keywords Random forest · K-Nearest Neighbors · XGBoost · Ensemble model · Feature engineering · F1 score · Roc-auc score

1 Introduction

A survey by BBC shows that individuals increasingly handle their financial affairs on the internet and so losses from online banking fraud increased by 48% in 2014 compared to 2013. In 2017, this figure is expected to grow to 46.4%. Due to the abundance of cash resources nowadays, losses are constantly rising. The banking

S. B. Zaman · N. Tasnim (✉) · N. Jahan · Md. M. Rahman
Department of Computer Science and Engineering, Military Institute of Science and Technology, Mirpur Cantonment, Dhaka 1216, Bangladesh
e-mail: nosh.tas.cse@gmail.com

S. K. Dey
School of Science and Technology, Bangladesh Open University, Gazipur 1705, Bangladesh

S. Hossain et al. (eds.), *Proceedings of International Conference on Fourth Industrial Revolution and Beyond 2021*, Lecture Notes in Networks and Systems 437,
https://doi.org/10.1007/978-981-19-2445-3_22

industry has always been an enticing target for fraudsters. All sorts of banking fraud refer to the act of obtaining money or other assets held by a financial organization through illicit means. With the rising automation of banking services, bank account and transaction fraud have skyrocketed. Fraud detection [1] includes tracking user behavior in order to predict, detect, or avoid adverse behavior. Understanding the technologies involved [2] in detecting banking frauds is crucial to effectively combat banking fraud. There are multiple algorithms for banking fraud detection, for example, artificial neural-network models which are based upon artificial intelligence and machine learning approach, data mining systems, intelligent decision engines which is based on artificial intelligence, meta learning agents, and fuzzy systems. In general, there are two types of machine learning algorithms to detect the fraud in banking transactions: supervised and unsupervised learning. The machine learning algorithms enable to run of automated transaction processing on a large complex dataset and detect the fraudulent activity and flag that fraudulent activity.

This paper represents the ensemble of K-Nearest Neighbor, Random Forest and XGBoost algorithms to build the classifier to prevent frauds in banking transactions and for the evaluation of performance measurement techniques, accuracy, f1 score, and roc-auc score are used.

The rest of this paper is organized as follows. Section 2 reviews a brief discussion and loopholes of the related research works, Sect. 3 presents algorithms discussion implemented for fraud detection, Sect. 4 contains the detail description of methodology, Sect. 5 represents the description of evaluation metrics related to the performance measurement of our techniques. The result analysis is discussed in Sect. 6. Finally, we conclude the paper in Sect. 7.

2 Related Works

There are different supervised and unsupervised machine learning algorithms used for fraud detection in banking transactions. This section briefly discusses the research works related to fraud detection in banking system using machine learning algorithms which have a number of advantages as well as some shortcomings.

An article [3] by Shirgave et al. represents fraud detection techniques using machine learning and shows a comparison among them using performance measure like accuracy, precision, and specificity. The article also proposes a fraud detection system which uses supervised Random Forest algorithm and thus increases the accuracy of detecting fraud in credit card. The proposed system uses learning to rank approach and to rank the alert. Also it addresses the problem of concept drift in fraud detection. But in the proposed system, semi-supervised learning methods for classification of alert have not been applied. Maniraj et al. [4] shows an analysis and pre-processing data sets of German Bank, 2006 as well as the use of several anomaly detection algorithms on credit card transaction data, such as the Local Outlier Factor and the Isolation Forest algorithm. While the method achieves a precision of over

99.6%, when only a tenth of the data set is considered, it only achieves a precision of 28%. The precision increases to 33% when the complete dataset is given into the algorithm. However, there is no result analysis for combining both algorithms to detect the frauds in transactions.

A research [5] by Abhay et al. presents an analysis and pre-processing data sets (kaggle) and the implementations of a number of anomaly detection algorithms on the transformed credit card transaction data by using Local Outlier Factor and Isolation Forest algorithm and shows that f1 score is better for the Isolation Forest then Local Outlier factor. A paper Koushika et al. [6] proposes an algorithm based on machine learning credit card fraud detection. It solves the issue of fraudulent transactions. This framework increases the probability of card fraud by exponential activity nominally. The results show that the accuracy of Random Forest, Support Vector Machine, and KNN classifiers achieves, respectively, 94.84%, 89.4%. The paper determines that Random Forest can predict new fraud cases very quickly.

Varun Kumar et al. [7] present a model which predict the fraudulent and non-fraudulent transactions using machine learning algorithms and neural networks. This paper uses a dataset from kaggle. The result shows that ANN model has the best accuracy.

Dhiman Sarma et al. [8] represents a web-based application, a fraud detection system using a community detection algorithm. Here Neo4j, a graph database system, was implemented for searching and filtering fraud cases. Another work [9] by Eweoya et al. uses decision tree to formulate a model to detect and predict fraud transactions and describes an analysis of the model and calculated the accuracy, TPR, NPR of the built model. An article [10] by Ishu Trivedi et al. shows genetic algorithm for finding optimal solution for the problem and implicitly generating the result of the fraudulent transaction. The main aim is to detect the fraudulent transaction and to develop a method of generating test data. Accuracy level is not mentioned here in the article.

Raj et al. [11] shows fuzzy darwinian, dempster, and Bayesian theory have very high accuracy in detecting credit card fraud. Anusorn Charleonnan et al. [12] proposes RUSMRN ensemble for classifying default of the data sets for client credit cards. It is based on RUS data sampling technique and MRN algorithm. The proposed RUSMRN model gives the highest accuracy. A paper [13] by Sahil Dhankhad et al. presents comparison between quite a few ML models based on their accuracy, TPR, FPR, g-mean, recall, precision, specificity, and F1-Score. Applied on a dataset they determine stacking classifier which is used LR as meta classifier is most promising for predicting fraud transaction.

3 Algorithms

This research works with a few machine learning algorithms for detecting fraudulent banking transactions. These are discussed briefly in this section.

3.1 *Random Forest*

Random Forest developed by Leo Breiman, supervised learning algorithm that can be used for classification as well as regression related problems. It creates a "forest" out of an ensemble of decision trees, which are commonly trained using the "bagging" method. Bagging predictors is a method of producing many versions of a predictor and integrating them into a single aggregated prediction. The risk of over-fitting is reduced in a forest with a large number of trees.

3.2 *K-Nearest Neighbors Algorithm*

The k-nearest neighbors (KNN), a supervised machine learning technique. It can be used to solve the regression problems. The KNN [18] algorithm stores all available data and classifies a new data point based on its similarity to the existing data. However, in order to use KNN, we must choose a proper value for k, and the classification success is highly dependent on this number. There are other methods for determining the k value, but one of the most straightforward is to run the algorithm multiple times with various k values and select the one that performs the best.

3.3 *XGBoost Algorithm*

XGBoost [19] is a common implementation of gradient boosting. It has proven its determination in terms of speed and performance. Gradient boosting, gradient boosting machine, and other terms are used to describe ensemble strategies in which earlier model errors are rectified in new models. The AdaBoost algorithm is one of the best instances of such an algorithm. Complex models can be penalized using both L1 and L2 regularization in XGBoost. Data is sparse due to missing values or data processing operations such as one-hot encoding.

3.4 *Logistic Regression*

The supervised learning [20] classification algorithm logistic regression is used to predict a target variable. Because the nature of the target or dependent variable is dichotomous, there are only two viable classes. In simple terms, the dependent variable is binary in nature, with data represented as 1 (representing success/yes) or 0 (representing failure/no).

4 Methodology

In this research, we try to detect the fraudulent transactions in a banking transaction data set [21] named as PS_20174392719_1491204439457_log. This data set contains payments from a variety of customers made during a variety of time periods and in a variety of quantities. The highly uneven distribution between positive and negative classes in around 6 million rows of data is the key technical problem it provides for predicting fraud. Another stumbling hurdle to this data's utility is the potential for missing or null values to cause disparities. We attempt to address these difficulties by performing a thorough data exploration and cleaning process, followed by oversampling the data set. Finally, we selected some appropriate machine learning methods for training and analyzing our classification models.

The following block diagram highlights the methodology of our proposed work (Fig. 1).

4.1 *Data Analysis*

In this section, we explore various insights within the data set using Data frame methods. This is the most efficient way to get insights of the data set.

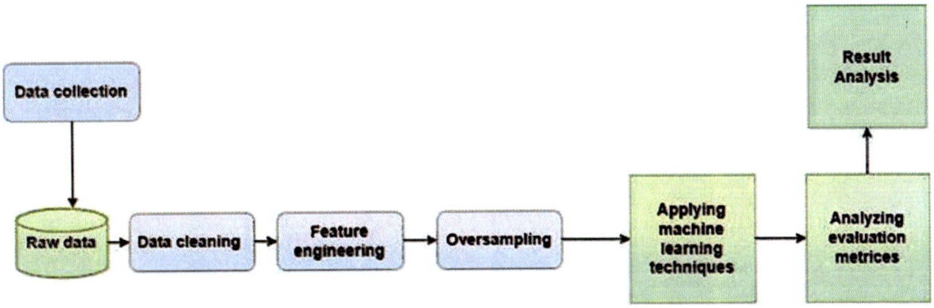

Fig. 1 Block diagram of the methodology of the proposed work

	step	type	amount	nameOrig	oldBalanceOrig	newBalanceOrig
0	1	PAYMENT	9839.64	C1231006815	170136.0	160296.36
1	1	PAYMENT	1864.28	C1666544295	21249.0	19384.72
2	1	TRANSFER	181.00	C1305486145	181.0	0.00
3	1	CASH_OUT	181.00	C840083671	181.0	0.00
4	1	PAYMENT	11668.14	C2048537720	41554.0	29885.86

	nameDest	oldBalanceDest	newBalanceDest	isFraud	isFlaggedFraud
0	M1979787155	0.0	0.0	0	0
1	M2044282225	0.0	0.0	0	0
2	C553264065	0.0	0.0	1	0
3	C38997010	21182.0	0.0	1	0
4	M1230701703	0.0	0.0	0	0

Fig. 2 First 5 rows of the data set

Figure 2 shows a snapshot of the features and the first 5 rows of the data set. We found out in the data set that there are five types of transactions, and out of those five, fraud occurs only in two of them: 'TRANSFER' and 'CASH_OUT'. Remarkably, the number of fraudulent transactions having type 'TRANSFER' almost equals the number of fraudulent transactions having type 'CASH_OUT'. The research [21] on the same data set showed that three features: nameOrig, nameDest, and isFlagged-Fraud are the least meaningful and hence can be dropped from the data set.

4.2 *Data Cleaning*

From the data analysis of the previous subsection, we know that fraud only occurs in 'TRANSFER's and 'CASH_OUT's. So we drop the data rows having other types for our analysis in order to get better performance. We performed binary encoding of the feature 'type'. We also dropped the previously mentioned features from the data set.

The data has multiple transactions with zero balances in the destination account before and after a non-zero amount is transacted. Because zero destination account balances are a prominent indicator of fraud, we do not impute the account balance (before the transaction is made) with a statistic or from a distribution with a subsequent adjustment for the amount transacted. This would obscure this fraud indicator and make fraudulent transactions appear legitimate. Instead, we substituted 0 with -1 for these cases, which will be more beneficial to a machine learning (ML) system for detecting fraud.

There are also multiple transactions in the data that have zero balances in the originating account before and after a non-zero amount is transacted. Instead of imputing a numerical value, we substitute the value of 0 with a null value, following the same logic as before.

4.3 Feature Engineering

We introduced two new features (columns) to record mistakes in the originating and destination accounts for each transaction, motivated by the prospect of zero-balances assisting to distinguish between fraudulent and real transactions. These new features have proven to be crucial in achieving the greatest results from the ML algorithms (as shown in Fig. 3).

X['errorBalanceOrig'] = X.newBalanceOrig + X.amount − X.oldBalanceOrig
X['errorBalanceDest'] = X.oldBalanceDest + X.amount − X.newBalanceDest

4.4 Oversampling

Because the percentage of fraudulent transactions is such a small percentage of the entire data, fraud data sets have a class imbalance problem. We can use oversampling or undersampling strategies to balance the data set. By producing instances from the minority class, oversampling increases the number of members of the minority class. Undersampling is the process of lowering the number of instances in the majority class by randomly selecting points from it until it equalizes with the minority class. Both operations are fraught with dangers: oversampling creates duplicates or similar data points, which aren't always useful for fraud detection because illicit transactions vary. We might data points and thus information due to undersampling. As a result, [22] we used an oversampling technique known as SMOTE (Synthetic Minority Oversampling Technique). SMOTE will use the neighbor instances to generate new data points from the minority class, thus the created samples are not exact replicas, but they are comparable to the instances we have.

4.5 Machine Learning for Fraud Detection

We did a train test split to measure the performance after appropriately cleaning and altering the data set. We divided the data into two categories: training data (70%) and testing data (30%). The models were then trained with the training data set and tested on the testing data set using the following machine learning classifiers: K-Nearest Neighbors (KNN), Random Forest (RF), Extreme Gradient Boosting (XGBoost), Logistic Regression (LR), Ensemble of KNN, RF and XGBoost, Ensemble of KNN, LR and XGBoost, Ensemble of KNN, RF, LR, and XGBoost.

Testing the models on the testing data set, we attained the confusion matrix for each model and then we evaluated the performance of those models. We also generated feature importance histogram showing the most relevant features for the model (as shown in Fig. 3).

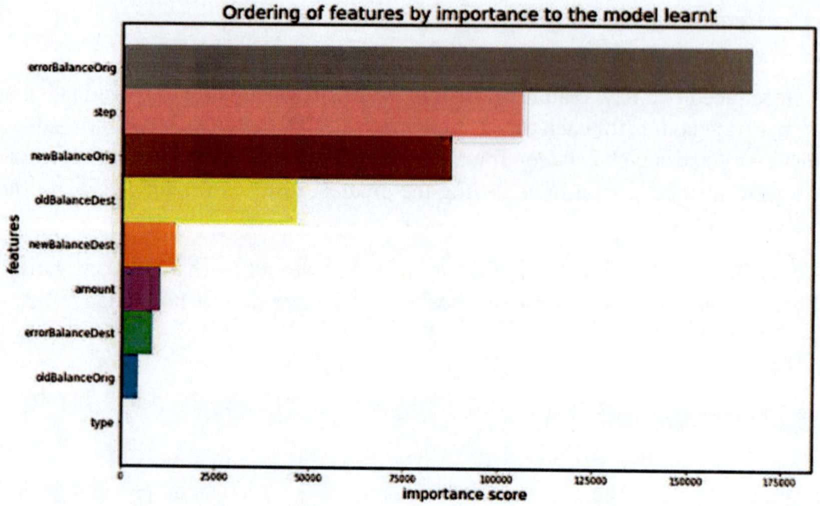

Fig. 3 Feature importance histogram

5 Evaluation Metrics

In this section, some terminologies that are relevant to our evaluation are presented. This chapter will be convenient for the readers to understand the Sect. 6 of this paper.

5.1 *False Positive and False Negative*

When you obtain a positive test result when you should get a negative result, it's called a false positive. When we get a negative result when we should get a positive result, we call it a false negative. The true definition of a positive result is when we get a positive result where we should have gotten a positive result. We receive a true negative result when we should get a true negative result.

5.2 *Confusion Matrix*

A confusion matrix [23] is a table that summarizes the performance of a classification model (or "classifier") on a set of test data with known truth values. Each column in the matrix denotes an actual class instance, while each row in the matrix denotes a predicted class instance (and vice versa).

5.3 *Precision and Recall*

Precision and recall are used to evaluate the performance of categorization or information retrieval systems when they are combined. The percentage of relevant occurrences among all retrieved instances is defined as precision. Recall, or sensitivity, is the percentage of retrieved occurrences among all relevant cases. A perfect classifier having a recall and accuracy of one. Accuracy and recall should be reported jointly at all times. When a single digital measurement of a system's performance is required, accuracy and recall are sometimes combined in F1-Score.

$$Precision = \frac{(TP)}{(TP) + (FP)} \tag{1}$$

$$Recall = \frac{(TP)}{(TP) + (FN)} \tag{2}$$

5.4 *F1-Score*

Generally, the accuracy rate and the recovery rate are given together and are not listed separately. This is because it is easy to change the sensitivity of the model to improve accuracy at the expense of recovery and vice versa. If you need a single number to describe the performance of the model, the most convenient number is F-score, which is the harmonic average of accuracy and recovery rate:

$$FScore = \frac{2 * Precision * Recall}{Precision + Recall} \tag{3}$$

5.5 *ROC-AUC Score*

The tp (true positive) rate is plotted on the Y pivot and the fp (false positive) rate is plotted on the X hub in ROC charts, which are two-dimensional diagrams. The relative compromises between benefits (true positives) and expenses are depicted in a ROC diagram (bogus positives). A discrete classifier produces only one class label. Each discrete classifier generates a pair of (fp rate, tp rate) values that are compared to a single point in ROC space. There are a few key points to remember in the ROC space.

The strategy of never giving a positive order is addressed at the lower left point (0, 0); such a classifier submits no false positive mistakes while also acquiring no clear positives. The upper right point addresses the opposing method of unequivocally giving positive directives (1, 1). The point (0, 1) is about incredible characterization. As you can see, D's execution is incredible. In general, one point in ROC space is

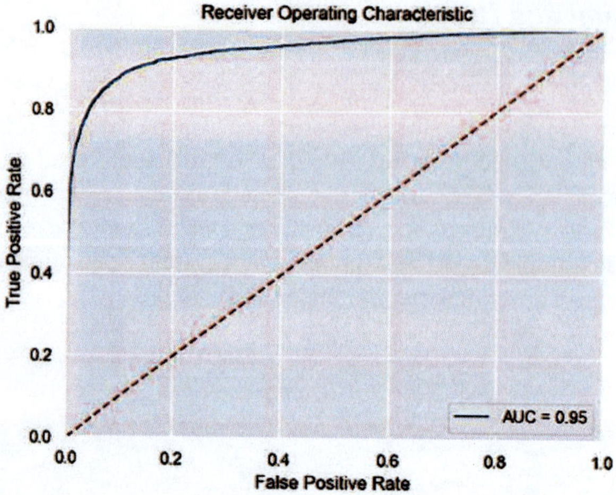

Fig. 4 ROC-AUC curve

better than another if it is to the northwest of the first (tp rate is greater, fp rate is lower, or both). Classifiers that appear near the X pivot on the left side of a ROC chart may be termed "moderate": they create positive arrangements only with sufficient proof, so they don't make many false positives, but they also have low evident positive rates. Classifiers on the top right-hand side of a ROC chart could be termed "liberal," in that they create positive characterizations with minimal proof, which allows them to appropriately order almost all positives, but they have large fake positive rates.

The AUC measures how successfully a model can distinguish between positive and negative classifications. The higher the AUC number, the better. Figure 4 shows an ROC-AUC curve with a score of 0.95.

6 Result and Analysis

After using four of the most well-known ML classifiers and three ensemble models of these ML classifiers (as mentioned in Sect. 4.5), we have come forth to the result which helps us to analyze the best model for predicting fraudulent banking transactions based on the evaluation metrics mentioned in the Sect. 5.

After implementing the ML classifiers on the training data set, the outcome turned out with the following measures.

From Table 1, we can say that XGBoost ML classifier provides best results in terms of accuracy, f1 score and roc-auc score for detecting fraudulent and non-fraudulent transactions. In this research, we will mainly focus on the f1 score and the roc-auc score since accuracy is not the perfect measure for imbalanced datasets. These are

Table 1 Results for ML classifiers

ML classifiers	(TP)	(FP)	(FN)	(TN)	Accuracy (%)	Precision (%)	Recall (%)	F1-Score (%)	ROC-AUC score (%)
KNN	175871	4234	362	171999	98.696	97.6491	99.7946	98.7102	98.9860
RF	175154	6127	1079	170106	97.9555	96.6202	99.3877	97.9844	97.9555
XGBoost	175527	2186	706	174047	99.1795	98.7699	99.5994	99.1829	99.1794
LR	157168	8695	19065	167538	92.1241	94.7577	89.1819	91.8853	92.1240

Table 2 Results for ML classifiers (ensemble)

ML classifiers	(TP)	(FP)	(FN)	(TN)	Accuracy (%)	Precision (%)	Recall (%)	F1-Score (%)	ROC-AUC score (%)
KNN, RF, XGB	175402	1019	831	175214	99.4751	99.4224	99.4224	99.4754	99.4322
KNN, LR, XGB	175575	2743	658	173490	99.0351	98.4617	99.6266	99.0408	99.0350
KNN, RF, LR, XGB	175734	2955	499	173278	99.02	98.3463	99.7169	99.0268	99.0200

the best measures for imbalanced datasets with binary classification. So, we can consider XGBoost classifier as more efficient standalone ML model to detect fraud transaction.

From Table 2, we can see that the performance measures do not always improve for any ensemble ML classifier. Rather, only the ensemble classifier of (KNN, RF, XGB) gives the better results based on the F1 score and Roc-auc score. surpassing the XGBoost classifier. Here, the results for the proposed ensemble of KNN, RF, and XGB are the best among all other standalone and ensemble classifiers.

7 Conclusion

Analyzing the data set and cleaning it according to the requirements and then applying oversampling on it, we could apply the ML classifiers on the testing data set. For the four ML classifiers, Logistic Regression, K-Nearest Neighbors, Random Forest, and XGBoost, we can estimate that they produce different result considering different error percentages. We have also applied 3 ensemble ML classifiers: (KNN, RF, XGB), (KNN, LR, XGB), and (KNN, RF, LR, XGB) and observed the accuracy, precision, f1 score, and roc-auc score for these classifiers. However, it is almost impossible to gain 100% accurate result all the time from machine learning classifiers, so if we can consider the minimum error percentage, then the results for the proposed ensemble classifier of KNN, RF, and XGB are very promising for detecting fraudulent and non-fraudulent banking transactions and is much better than the other compared classifiers.

References

1. Butaru, F., Chen, Q., Clark, B., Das, S., Lo, A.W., Siddique, A.: Risk and risk management in the credit card industry. J. Banking Finan. **72**, 218–239 (2016)
2. Praveen Kumar Donepudi: Machine learning and artificial intelligence in banking. Eng. Int. **5**(2), 83–86 (2017)
3. Shirgave, S., Awati, C., More, R., Patil, S.: A review on credit card fraud detection using machine learning. Int. J. Sci. Technol. Res **8**, 1217–1220 (2019)
4. Maniraj, S.P., Saini, A., Ahmed, S., Sarkar, S.: Credit card fraud detection using machine learning and data science. Int. J. Eng. Res. **8**(09) (2019)
5. Goel, A., Mathur, A., Tripathi, A., Agrawal, KK.: Credit card fraud detection using machine learning. J. Xi'an Univ. Arch. Technol. **XII**(IV), 50–54 (2020)
6. Kousika, N., Vishali, G., Sunandhana, S., Arvind Vijay, M.: Machine learning based fraud analysis and detection system. In: Journal of Physics: Conference Series, vol. 1916, pp. 120–125. IOP Publishing (2021)
7. Varun Kumar, K.S., Vijaya Kumar, V.G., Vijay Shankar, A., Pratibha, K.: Credit card fraud detection using machine learning algorithms. Int. J. Eng. Res. Technol. (IJERT), **9**, 26–30 (2020)
8. Sarma, D., Alam, W., Saha, I., Alam, M.N., Alam, M.J., Hossain, S.: Bank fraud detection using community detection algorithm. In: 2020 Second International Conference on Inventive Research in Computing Applications (ICIRCA), pp. 642–646. IEEE (2020)
9. Eweoya, IO., Adebiyi, A.A., Azeta, A.A., Angela, E.A.: Fraud prediction in bank loan administration using decision tree. In: Journal of Physics: Conference Series, vol. 1299, pp. 012037. IOP Publishing (2019)
10. Trivedi, I., Mridushi Monika, M.: Credit card fraud detection. Int. J. Adv. Res. Comput. Commun. Eng. **5**(1) (2016)
11. Benson Edwin Raj, S., Annie Portia, A.: Analysis on credit card fraud detection methods. In: 2011 International Conference on Computer, Communication and Electrical Technology (ICCCET), pp. 152–156. IEEE (2011)
12. Charleonnan, A.: Credit card fraud detection using RUS and MRN algorithms. In: 2016 Management and Innovation Technology International Conference (MITicon), pp. MIT–73. IEEE (2016)
13. Dhankhad, S., Mohammed, E., Far, B.: Supervised machine learning algorithms for credit card fraudulent transaction detection: a comparative study. In: 2018 IEEE International Conference on information Reuse and Integration (IRI), pp. 122–125. IEEE (2018)
14. Liu, J.-M., Tian, J., Cai, Z.-X., Zhou, Y., Luo, R.-H., Wang, R.-R.: A hybrid semi-supervised approach for financial fraud detection. In: 2017 International Conference on Machine Learning and Cybernetics (ICMLC), vol. 1, pp. 217–222. IEEE (2017)
15. John, S.N., Anele, C., Kennedy, O.O., Olajide, F., Kennedy, C.G.: Realtime fraud detection in the banking sector using data mining techniques/algorithm. In: 2016 International Conference on Computational Science and Computational Intelligence (CSCI), pp. 1186–1191. IEEE (2016)
16. Huang, J., Chai, J., Cho, S.: Deep learning in finance and banking: a literature review and classification. Front. Bus. Res. China **14**, 1–24 (2020)
17. Uma Pavan Kumar, K., Gandhi, O., Venkata Reddy, M., Srinivasu, S.V.N.: Usage of KNN, decision tree and random forest algorithms in machine learning and performance analysis with a comparative measure. Mach. Intell. Soft Comput. 473–479. Springer (2021)
18. Wang, L.: Research and implementation of machine learning classifier based on KNN. In: IOP Conference Series: Materials Science and Engineering, vol. 677, pp. 052038. IOP Publishing (2019)
19. Bentéjac, C., Csörgő, A., Martínez-Muñoz, G.: A comparative analysis of gradient boosting algorithms. Artif. Intell. Rev. **54**(3), 1937–1967 (2021)
20. Joanne Peng, C.-Y., Lida Lee, K., Ingersoll, G.M.: An introduction to logistic regression analysis and reporting. J. Educ. Res. **96**(1), 3–14 (2002)

21. Llabhishekll. Fraud transaction detection, Aug 2018
22. Brownlee, J.: Smote for imbalanced classification with python, March 2021
23. Lewis, H.G., Brown, M.: A generalized confusion matrix for assessing area estimates from remotely sensed data. Int. J. Remote Sens. **22**(16), 3223–3235 (2001)

LANDCHAIN: A Blockchain-Based Lightweight Land Administration System for Bangladesh

Fardin Ahmed Niloy, Nozib Ud Dowla, Md. Samiul Alam, Jobair Hossain, Fahim Muntasir, Shahadat Hossain, and Md. Manzurul Hasan

Abstract In a developing country like Bangladesh, the land is seen as a valuable asset. Therefore, investing in land assets appears to be the best way to put people's money to utilize. However, purchasing land is time-consuming due to the need for a sequence of communications among buyers, sellers, and many government agencies to obtain the required documentations for approval. In addition, most land administration operations in Bangladesh are paper-based, making them more difficult to manage and secure. This paper presents *LANDCHAIN*, a blockchain-based land administration system that automates the land administration process while maintaining security and transparency. Instead of a typical database, the system is built on the blockchain, ensuring that data is tamper-proof. Our proposed approach is lightweight, making it simple to deploy and to scale automatically. This proposed model is tested on the Ethereum Virtual Machine (EVM) and is shown to be more reliable, secure, efficient, and easier to scale than that of a typical database management system. Thus, the *LANDCHAIN* solves the existing issues of land administration while also assisting the government in optimizing the processes.

Keywords Land administration · Digital mutation · Ethereum virtual machine

F. A. Niloy (✉) · N. U. Dowla · Md. S. Alam · J. Hossain · F. Muntasir · Md. M. Hasan
American International University-Bangladesh (AIUB), Dhaka, Bangladesh
e-mail: niloyfardinahmed@gmail.com

Md. M. Hasan
e-mail: manzurul@aiub.edu

S. Hossain
City University, Dhaka, Bangladesh

S. Hossain et al. (eds.), *Proceedings of International Conference on Fourth Industrial Revolution and Beyond 2021*, Lecture Notes in Networks and Systems 437,
https://doi.org/10.1007/978-981-19-2445-3_23

1 Introduction

Land registration is the procedure by which government agencies record land titles and ownership transfers. Land information are used to verify land ownerships, prevent frauds, and provide financial services such as loan applications [1]. In addition, such information is utilized by governments to collect taxes, deliver services, and exercise regional authority. However, the traditional paper-based method of keeping land-related information takes long time to process and requires manual labor to maintain. Furthermore, paper-based records can be harmed in a variety of ways, including natural disasters. Hence, buyers must rely on third parties to purchase or to verify land documents, producing additional fees and time [2]. A paper-based system cannot solve this problem because it is quite difficult to maintain and verify. As a result, many governments, particularly in developing nations, look to store land information digitally. Blockchain is an immutable and transparent method. Therefore, it may be a potential solution to the present challenges of storing and verifying land-related information. This paper discusses the shortcomings of previous schemes and proposes a more effective one. Its contributions are as follows:

- Identifying the deficiency of existing traditional methods and proposing a lightweight blockchain model for land administration.
- Reducing the complexity of the current process and minimizing document forgery to protect the citizens according to Bangladesh's land rights.
- To easily obtain information about public properties and to collect taxes from landowners.
- Eliminating corruption from Bangladesh's land management system, i.e., illegally acquiring land through forged deeds, records, and so on.

The rest of the paper is organized as follows. In Sect. 2, related works are discussed. A discussion on Bangladesh's existing land administration system is given in Sect. 3. Section 4 illustrates blockchain-based land administration of Bangladesh. In Sect. 5, we present performance and cost analyses. Finally, in Sect. 6, we conclude our work with possible future research directions.

2 Related Work

Satoshi Nakamoto first officially introduced blockchain technology in the form of a digital currency called Bitcoin in 2008 [3]; but, due to the versatility of blockchains, they are now being used in a variety of financial and non-financial areas. The blockchain's underlying technology provides a cryptographic way to ensure that the data stored in blocks remain tamper-proof. Because of its immutability and transparency, blockchain is commonly employed in applications requiring data security. According to BRAC and the Copenhagen Consensus Center, digitizing land records is Bangladesh's fourth most important development intervention, with a 619 dollar return on investment [4, 5]. As a result, Bangladesh's government has made digitizing land records a top priority. Our research creates LANDCHAIN, a lightweight framework for securely storing land record data using blockchain technology, to overcome flow issues with conventional paper-based systems. The lightweight framework takes advantage of blockchain's unique ability to store data securely. For example, a customer can use LANDCHAIN to request that the sub-registry update land ownership and validate transactions before adding them to the blockchain. As a result of using a digital platform, land document forgeries can be avoided, and owners can obtain documents in an expedient and hassle-free way. Land is a valuable asset in developing countries such as Bangladesh, and most people save their lives to own it. However, due to the paper-based trading system of land management, buying can be a time-consuming procedure, and in many circumstances, people are forced to face the prospect of buying or selling land. These procedures are streamlined with the help of LANDCHAIN in a way that is hassle-free, secure, and transparent.

According to Spielman [6], Blockchain technology would have a transformative impact on the land registration procedure, dramatically altering the course of the process. The author proposed in the paper that by incorporating blockchain technology into the land administration process, the efficiency of ownership transfer can be enhanced, document forgery can be avoided, and transaction management can be handled with better security. Javed et al. [7] presented a caching approach to boost computational processing power and optimize the system. Furthermore, the suggested system would benefit from the Proof of Authority (PoA) consensus process and the implementation of a symmetric key cryptography technique.

Georgia is one of the first countries to integrate blockchain technology into its land management system successfully [8, 9]. Georgia's government altered the traditional registration process to accomplish this achievement, placing Georgia third in the World Bank's Doing Business report for ease of property registration [10, 11].

3 Existing Land Administration System of Bangladesh

Bangladesh is a country of more than 163 million people but with only 147,570 sq. km. So, land scarcity is abundant here. Outdated laws and infrastructure regarding land administration resulted in high tenure insecurity here [12]. It causes an increasing number of conflicts regarding land ownership and incidents of land-grabbing. Many wealthy and corrupt individuals misuse their power to grab public lands using false documents and are winning court decrees. Bangladesh has been planning to use blockchain technology in different sectors of e-governance in order to enter the fourth industrial revolution as a significant power [13] and to create Digital Bangladesh. The finance minister expressed his desire to implement blockchain technology on an experimental basis in his budget speech last year [14]. One of the sectors they want to use is blockchain technology to manage land resources and make them immutable and more transparent.

3.1 Problem with Current Paper-Based System

- **Lengthy survey process**: The process of land surveying requires ten phases to be completed in Bangladesh. These are field surveying, review processes, editing, and final record printing.
- **Too many intermediaries**: At present, completing a land transfer task requires so many mediators that could be done with fewer people involvement.
- **Problems with manual record keeping**: The current system maintains records and manages information for the whole land sector manually, including all land surveys (CS, SA, and RS) dating back more than a century.
- **Illegal sale of land**: Counterfeit deeds, forged signatures, and the cause of illegal land sales can be quickly produced in an ongoing paper-based system. Unfortunately, these scams are hard to identify.
- **False information when leasing khas land**: People or organizations lease Khas land from the government by providing duplicated documents or false information.
- **Degradable data storage system**: Anyone can easily steal paper-based systems and non-digitized data, and it can also be destroyed or lost.
- **Dual ownership**: Dual ownership is frequently generated due to the National Land Revenue Board's failure to perform necessary mutation work in existing traditional system.

3.2 Current Process of Land Management

In Fig. 1, we illustrate the current process of land management. A buyer and a seller have to go through many government ministries to buy/sell their land. As a result, the process takes much time, which also opens up the possibilities for various problems such as bribery for faster processing, document forgery, and many more problems.

In Bangladesh, managing blockchain-based solutions has not been implemented yet. However, researchers have suggested their models for a blockchain-based land information management system.

3.3 Problems with Existing Blockchain Models

In Table 1, we give a broad discussion for each of those proposed solutions. Upon going through each proposed solution properly, some pros and cons of each system have been evaluated. That information are then analyzed and further developed to improve our proposed solution.

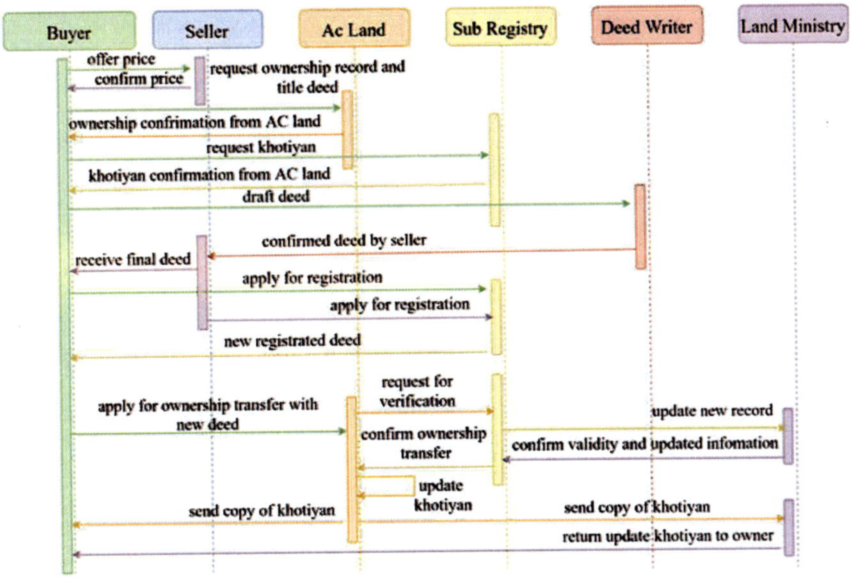

Fig. 1 Sequence diagram of traditional paper-based system of land administration of Bangladesh

Table 1 Analyses on existing proposed systems in Bangladesh

Authors	Problem identified and solution	Technique/results	Limitation
Islam et al. [15]	Ethereum based blockchain solution to solve the problem with the existing traditional system	Test data was not provided	Efficiency of the system was not discussed
Alam et al. [16]	The authors have proposed a hybrid blockchain model to solve the problem with the current paper-based system in Bangladesh	Efficiency testing for multi-ownership transfer was done for a single test. In addition, data were provided for each transaction cost	In phase 1, the authors proposed public blockchain, which raises privacy concerns
Majumdar et al. [17]	They proposed a system with Delegated Proof of Stake (DPoS) algorithm	Hypothetical system	No actual results were discussed
Ashiquzzaman et al. [18]	They proposed a hypothetical system	Hypothetical system	Public blockchain had been proposed thus, raises a privacy concern of users' data

3.4 Situation in Other Countries

According to the World Bank, about 70% of the people on Earth is lack authorized proof of ownership to their land [19]. Storing land information data in a database is not enough because it might enable corrupt government officials to manipulate land title data easily. According to Transparency International, 20% of land service users worldwide had to use some forms of bribe in order to register their properties or to verify land ownerships information [20]. In order to tackle this situation many countries are adopting blockchain technology [21].

- **India**: In terms of land administration, India and Bangladesh have similar challenges. Different departments keep additional documentations to establish ownership [22]. Although certain land records in India have been digitized, they remain vulnerable to alteration [23]. Furthermore, a statistics shows that nearly $700 million in bribes are paid among land registrars in India [24, 25].
- **Ghana**: Ghana has various laws governing land ownership [26]. There are primarily two systems—customary law and statute law—that operate concurrently to resolve property conflicts [27]. The Land Administration Project (LAP) was established to implement institutional norms and increase service delivery efficiencies [26]. They have, however, encountered numerous obstacles in implementing proper land administration since the program's inception in 1999 [28].

In Table 2, a detailed discussion on the various system from other countries has been provided. Upon analyzing those, their limitations are also included in the table.

4 Blockchain-Based Land Administration System

A blockchain-based solution is necessary in order to provide transparency, efficient land management, and security. Various countries have already started using the blockchain-based solution for managing information related to blockchain. In our proposed system, we try to keep our system simple and easy to implement so that our system can be implemented and adopted by various ministries of Government. As of now, most of the land-related works are done manually, and data related to land are stored in the paper. Thus, to modernize the land registry and verification of land ownership, we are proposing our model.

4.1 Proposed System

The LANDCHAIN platform, which uses blockchain technology, may be used on the web, mobile, and desktop to enable quick, fraud-resistant, secure, and private land management activities. The user can verify ownership, inspect the land's history,

Table 2 Analyses on existing systems in different countries

	Problem Identification and solution	Technique/results	Limitation
Sladić et al. [29]	Analyzed possible blockchain solution for LAS	Public blockchain on Ethereum network	Transaction feeis much higher on Ethereum network
Shang et al. [9]	Discussed about importance of land information management in Georgia	Proposed theoretical two phase blockchain-based solution	Theoretical solution was proposed
Balkiz et al. [30]	Discussion on the importance of blockchain in land administration for Cyprus	Explains the necessity of blockchain based solution for Cyprus	Does not provide any direction on implementation
Aquib et al. [31]	A blockchain based land information management for Pakistan.	Blockchain based platform for seculary store land-related data.	Implementation of the system is provided but no data was given for performance the system
Nguyen et al. [32]	Managing land valuation certificate	Using blockchain to manage land valuation certificate	Performance test was not provided

execute land transactions, get legal help with land administration, and lease land from other owners or the government. Furthermore, the government can protect its Khas land and keep Khatiyan in a secure storage facility. The data can be safely stored on the blockchain in a distributed fashion across government-owned computing machines.

To mitigate the issues mentioned above with the traditional land registration and verification process in Fig. 2, we propose a modern solution based on blockchain technology. By implementing and using our solution, the buyer/seller can transfer land ownership instantly. Also, they have access to various information related to the land they own or the land they are purchasing. As those data are securely stored in blockchain, those data are prone to any forgery or tampering. To open an account, first, it is checked whether there is an existing account. If there is no account as mentioned in Fig. 3, then the user can create a new account. Our system gathers users' necessary information from the existing national id database. In addition, we are using digital KYC (Know Your Customer or, Know Your Client) so that users can open their accounts quickly.

In Fig. 4, flowchart is described for buying and selling any land. First, a buyer sends the request to the seller. After the positive response of the seller, an e-Deed is generated and finalizes the land prices through buyer and seller negotiations. After

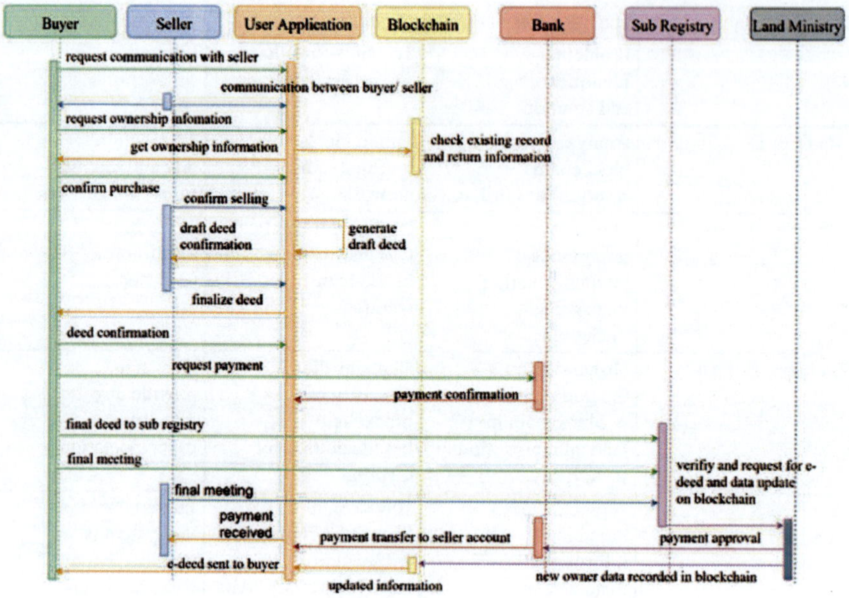

Fig. 2 Proposed blockchain-based model for land administration of Bangladesh

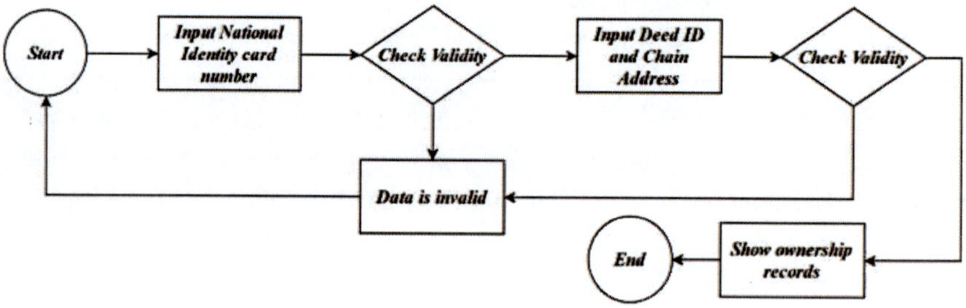

Fig. 3 Workflow diagram of verification of ownership

sending the e-Deeds to the both buyer and seller, a new blockchain is being generated. Data is being added to the block and distributed to the other nodes if the buyer clears the payments.

4.2 Blockchain Architecture for LANDCHAIN

In Fig. 5, our system architecture for the proposed system is given. Our model will be on the cloud in Linux Virtual Machine. For easy maintenance and distribution, docker has been used for packaging the system. By using docker, it will be easy to

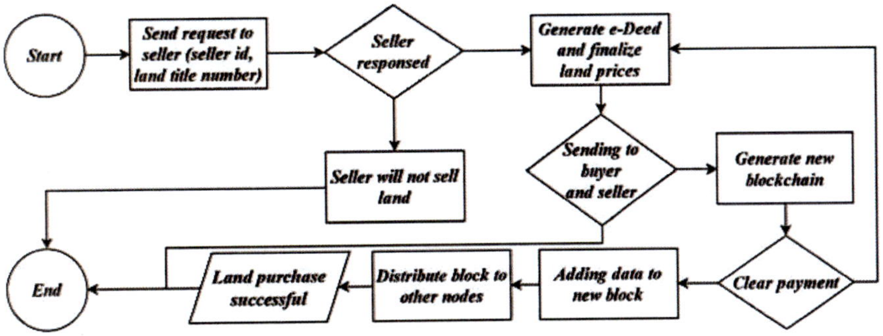

Fig. 4 Workflow diagram of buying or selling process of lands

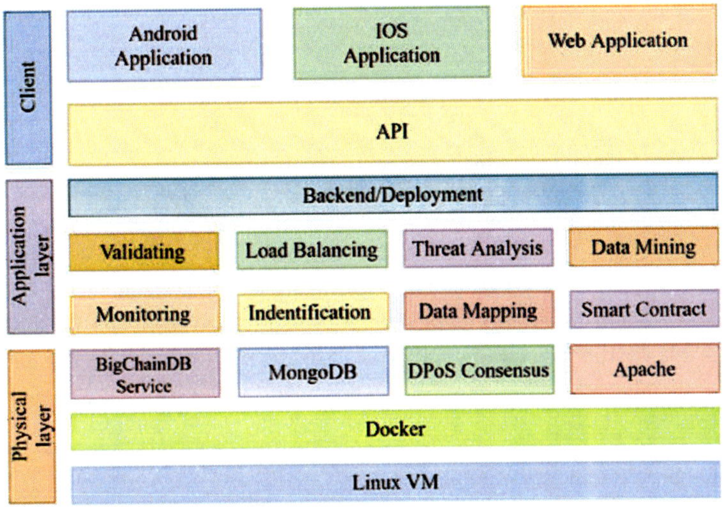

Fig. 5 Blockchain architecture for LANDCHAIN

deploy the system and also scale when necessary. In our application layer, core parts of system logic will be running. System users will be able to interact with the system by using their smartphone application or the web application.

5 Performance and Cost Analyses

Our proposed model has simulated and tested in Ethereum Virtual Machine (EVM). For each operation we have analyzed the cost and measured performance as well. In current paper-based process, it takes time to process land administration-related

Table 3 Performance analyses between traditional paper-based system and LANDCHAIN

Operation	Paper-based system	LANDCHAIN
Document verification	10–15 days	Instant
Ownership transfer	1–2 week	10 min and 1 visit to registry office
Avg. visit to ministries	4–5 visits	1 visit
Avg. cost	1500–2000 BDT	5 BDT per blockchain operation

Table 4 Cost analyses between Alam et al. model and LANDCHAIN

	Average cost of operation (in BDT)	
Operation	Kazi Masudul Alam et al. model	LANDCHAIN
Adding user	8.16	4.35
Document verification	Not applicable	1.86
Khatiyan mutation	18.85	6.12
Khatiyan entry	16.70	7.43
Adding plot	6.06	1.37

works. However, in our system all the works are instant and easily verifiable. In terms of cost analyses LANDCHAIN outperforms Alam et al.'s model and paper-based system.

In Table 3, a comparison between our model and current paper-based system is given. In Table 4, we have compared pricing against a proposed system [16] and found that our system cost significant lower than the proposed system.

6 Future Research Direction and Conclusion

In this paper, we have proposed a blockchain-based land administration system for Bangladesh. We have analyzed existing blockchain-based solutions from Bangladesh and other countries and provide improved models than other models. We have also discussed the proposed model's system architecture. Each step of the land administration process has been shown using a flowchart diagram. Finally, we have tested our model and provided the necessary verdict in the discussion section to validate our system. In the future, remote sensor data can be added to e-Deed. Sensor data may help recognize the land area, and machine learning approaches can utilize to determine land prices. Real-life implementations, prototypes along with sufficient discussions and analyses with them may be nice future works.

References

1. Dobhal, A., Regan, M.: Immutability & auditability: the critical elements in property rights registries. In: Annual World Bank Conference on Land and Property (2016)
2. Domeher, D., Abdulai, R.: Access to credit in the developing world: does land registration matter? Third World Quart. **33**(1), 161–175 (2012)
3. Nakamoto, S.: Bitcoin: a peer-to-peer electronic cash system. In: Decentralized Business Review, p. 21260 (2008)
4. How to spend it. https://www.economist.com/finance-and-economics/2016/05/05/how-to-spend-it, May 2016. Accessed 29 August 2021
5. Rahman, S. H., Talukder, S.K.: Service process simplification of land administration in Bangladesh: a case study of AC (land) office, Savar Uapazilla, Dhaka. BRAC Institute of Governance and Development
6. Spielman, A.: Blockchain: digitally rebuilding the real estate industry. Ph.D. thesis, Massachusetts Institute of Technology (2016)
7. Javed, M.U., Rehman, M., Javaid, N., Aldegheishem, A., Alrajeh, N., Tahir, M.: Blockchain-based secure data storage for distributed vehicular networks. Appl. Sci. **10**(6), 2011 (2020)
8. Moné, L.: Which governments are using blockchain right now? https://consensys.net/blog/enterprise-blockchain/which-governments-are-using-blockchain-right-now/, April 2021
9. Shang, Q., Price, A.: A blockchain-based land titling project in the republic of Georgia: rebuilding public trust and lessons for future pilot projects. Innov. Technol. Govern. Global. **12**(3-4), 72–78 (2019)
10. Shin, L.: Republic of Georgia to pilot land titling on blockchain with economist hernando de soto, bitfury. https://www.forbes.com/sites/laurashin/2016/04/21/republic-of-georgia-to-pilot-land-titling-on-blockchain-with-economist-hernando-de-soto-bitfury/?sh=6776059b44da, July 2017. Accessed 29 Aug 2021
11. Benbunan-Fich, R., Castellanos, A.: Digitization of land records: from paper to blockchain (2018)
12. Hilhorst, T., Porche, N.: Food security and land governance factsheet. https://www.landgovernance.org/assets/20160608-Factsheet-Benin.pdf (2016). Accessed 29 Aug 2021
13. Ferdous, M.S., Sultana, J., Reza, M.S., Ahmed, S.: National blockchain strategy, Bangladesh (2020)
14. Bhowmik, R.K., Bala, S.K.: National budget 2020–2021. *The Institute of Cost and Management Accountants of Bangladesh*
15. Islam, M.S., Iqbal, F.S., Islam, M.: A novel framework for implementation of land registration and ownership management via blockchain in Bangladesh. In: 2020 IEEE Region 10 Symposium (TENSYMP), pp. 859–862. IEEE (2020)
16. Masudul Alam, K., Ashfiqur Rahman, J.M., Tasnim, A., Akther, A.: A blockchain-based land title management system for Bangladesh. J. King Saud Univ.-Comput. Inf. Sci. (2020)
17. Majumdar, M.A., Monim, M., Shahriyer, M.M.: Blockchain based land registry with delegated proof of stake (DPoS) consensus in Bangladesh. In: 2020 IEEE Region 10 Symposium (TENSYMP), pp. 1756–1759. IEEE (2020)
18. Ashiquzzaman, R.I.S., Pipash, H.K., Elme, K.M., Mahmud, D.M.: Ethereum based land registry system for Bangladesh
19. Deininger, K., Augustinus, C., Enemark, S., Munro-Faure, P.: Innovations in land rights recognition, administration, and governance. World Bank Publications (2010)
20. Aitken, R.: Bitland's African blockchain initiative putting land on the ledger. Forbes Magazine (2016)
21. Müller, H., Seifert, M.: Blockchain, a feasible technology for land administration. In: FIG Working Week: Geospatial Information for a Smarter Life and Environmental Resilience, pp. 22–26 (2019)
22. Sekhari, A., Chatterjee, R., Dwivedi, R., Negi, R., Shukla, S.: Entangled blockchains in land registry management. In: BW3, India (2019)

23. Bhattacharya, A.: Blockchain is helping build a new Indian city, but it's no cure for corruption (2017)
24. Gupta, M., Haris Minal, M.: An empirical analysis of forecast performance of the GDP growth in India. Glob. Bus. Rev. **20**(2), 368–386 (2019)
25. Thakur, V., Doja, M.N., Dwivedi, Y.K., Ahmad, T., Khadanga, G.: Land records on blockchain for implementation of land titling in India. Int. J. Inf. Manag. **52**, 101940 (2020)
26. Cobbinah, P.B., Asibey, M.O., Gyedu-Pensang, Y.A.: Urban land use planning in Ghana: navigating complex coalescence of land ownership and administration. Land Use Policy **99**, 105054 (2020)
27. Mwanza, K., Wilkins, H.: African startups bet on blockchain to tackle land fraud. https://www.reuters.com/article/us-africa-landrights-blockchain-idUSKCN1G00YK, February 2018. Accessed 29 Aug 2021
28. Mintah, K., Baako, K.T., Kavaarpuo, G., Otchere, G.K.: Skin lands in Ghana and application of blockchain technology for acquisition and title registration. J. Prop. Plann. Environ. Law (2020)
29. Sladić, G., Milosavljević, B., Nikolić, S., Sladić, D., Radulović, A.: A blockchain solution for securing real property transactions: a case study for Serbia. ISPRS Int. J. Geo Inf. **10**(1), 35 (2021)
30. Yapicioglu, B., Leshinsky, R.: Blockchain as a tool for land rights: ownership of land in Cyprus. J. Prop. Plann. Environ. Law (2020)
31. Aquib, M., Dhomeja, L.D., Dahri, K., Malkani, Y.A.: Blockchain-based land record management in Pakistan. In: 2020 3rd International Conference on Computing, Mathematics and Engineering Technologies (iCoMET), pp. 1–5. IEEE (2020)
32. Nguyen, N.H., Nguyen, B.M., Dao, T.C., Do, B.L.: Towards blockchainizing land valuation certificate management procedures in Vietnam. In: 2020 RIVF International Conference on Computing and Communication Technologies (RIVF), pp. 1–6. IEEE (2020)

A Machine Learning Approach for Phishing Websites Prediction with Novel Feature Selection Framework

Pankaj Bhowmik and Pulak Chandra Bhowmik

Abstract Considering the recent events, Bangladesh stands highly vulnerable to cyberattacks. During this COVID-19 pandemic, since people have become more active on the Internet, the frequency of phishing attacks is escalating concurrently. However, this study used the two latest phishing datasets and proposed an efficient machine learning approach to predict the potential phishing and legitimate websites by analyzing the features of URLs. The proposed study introduced a two-phase empirical hybrid framework for feature selection, which includes the filter, embedded, and wrapper methods. In the primary phase—four filter methods are applied, and to obtain the cumulative feature subset the function perturbation ensemble is performed using set union operation. In the final stage, the filtered feature subset is fed in the wrapper method—where a statistical model is employed to select features with a p-value test, and in the embedded method—which used an algorithmic technique (random forest). The best features are derived from the final feature subsets of the wrapper and embedded methods applying set intersection operation. Using this hybrid ensemble approach, this case study obtained only 37.84% of the features from the datasets for experimentation. Afterward, the study performed cross-validation using three optimized machine learning models (Random Forest Classifier, XGBoost Classifier, Artificial Neural Network) on Data-I, and the XGBoost model outperformed with a 96.10% accuracy score. Besides, the model secured a significant accuracy of 97.40% on Data-II using <38% of the original features.

Keywords Phishing · Ensemble · Hybrid feature selection · Machine learning

P. Bhowmik (✉)
Department of Computer Science and Engineering, Hajee Mohammad Danesh Science and Technology University (HSTU), Dinajpur 5200, Bangladesh
e-mail: pankaj.cshstu@gmail.com

P. C. Bhowmik
Department of Computer Science and Engineering, Stamford University Bangladesh, Dhaka 1217, Bangladesh

S. Hossain et al. (eds.), *Proceedings of International Conference on Fourth Industrial Revolution and Beyond 2021*, Lecture Notes in Networks and Systems 437,
https://doi.org/10.1007/978-981-19-2445-3_24

1 Introduction

In this age of the Internet—diverse service deliveries, communication, networking, and several other facilities are getting digitalized. Especially, banking transactions, online marketing, and money transfer are on top-notch. But cyber-criminals are continuously trying to steal sensitive information either by breaking down the security protocols or by seducing the users [1]. From Bangladesh's perspective, the cyberspace vulnerability level is booming consistently. According to the BGD e-GOV CIRT statistics, the number of registered incidents is currently at its peak. In this COVID-19 crisis, phishing and spam attacks—closely associated with each other, are reported as the top cyber threats in Bangladesh. These cynical attacks target specific industries such as banking and finance, local government officials, defense, and law enforcement agencies. For instance, the Bangladesh Bank cyber heist incident (2016) is one of the historical cyberattacks, worth US$81 M [2]. Phishing attacks are well-known cybercrimes which are functioned through forged websites, often lead to financial scams. Users follow the spam links unknowingly, which redirects them to phishing websites. And, thus, innocent people get trapped and reveal personal data such as credit card info, passwords, and scammers take access to users' confidential data. Phishing attacks have become a common scenario and are much more sophisticated these days [3].

To address this cynical issue, researchers and cybersecurity experts are constantly working hard, besides, they developed different possible solutions [4]. Phishing attack detection based on "Blacklist" is one of the popular preventive approaches. The "Blacklist" contains the universal resource locator (URL) of all known phishing websites. If a surfed URL is listed on the Blacklist, then it's a phishing website and legitimates otherwise, the web browsers provide warnings accordingly. But the drawback is, as tons of phishing websites are developing in a daily manner, and if the "Blacklist" is not updated, browsers will not be able to detect the newly generated phishing URLs [5]. Besides, in the heuristic-based technique, which is an extended variant of the listing-based method, URLs and other features are extracted from websites and compared with ground-list to decide websites' legitimacy [6, 7]. Machine learning (ML)-based phishing detection is becoming more dominating lately. In ML approaches, lists of features are extracted from URLs to predict phishing sites, as a result, these methods can effectively combat the dynamic changes of phishing attacks [1, 8].

However, in URL-based phishing website detection, many features are extracted, but not all of them are equally important. Consequently, researchers have introduced several techniques to rank and derive the best attributes from the feature space [9]. To cite an example, Ali [10] experimented with two feature selection approaches for phishing website detection. The performance of his proposed ML models revealed that wrapper-based feature selection outperformed the filter method. Embedded methods, on the other hand, use learning algorithms for variable selection. Decision tree-based algorithms, for instance, CART, random forest, ID3, C4.5 are common embedded approaches for feature selection. Liu et al. proposed an embedded feature selection approach, where CART algorithm is used as the key component. To deal with imbalanced data, they added index-weighting and developed a weighted Gini index method. In experimental results, the method showed excellent performance using around 20% of the total features [11, 12]. Pes et al. [13] established an ensemble-based substantial approach for feature selection. They experimented with the data perturbation strategy using a set of feature selection methods. In general, they found that the ensemble feature subsets provide good outcomes, particularly in terms of stability.

In this article, a novel hybrid feature selection framework is proposed to derive the best features from phishing datasets and thus detect phishing websites effectively. This framework has two phases—at the beginning phase, a set of filter methods are applied to select the primary feature subsets from the dataset and then obtained the secondary (cumulative) feature subset with the function perturbation ensemble. In the final phase, the embedded and wrapper method is applied. Afterward, from the feature subsets of the wrapper and embedded method, the best feature set is obtained using a set intersection operation. In this study, the two latest phishing datasets of Vrbančič [14] are used. The study performed cross-validation using three optimized ML models, namely, Random Forest Classifier (RFC), XGBoost Classifier (XGBC), and Artificial Neural Network (ANN) on Data-I. These ML models are optimized by applying Grid Search hyperparameter tuning. Besides, the outperformed model on Data-I is used to perform cross-validation (CV) on Data-II. Finally, a comparative performance analysis of the proposed and existing research works is established.

2 Related Works

Over the past years, researchers and cybersecurity experts have proposed several solutions and methods to combat phishing attacks. Statistically, the performance of the ML-based approaches reflects their superiority over conventional methods. In this section, the authors of this proposed study shed some light on the contemporary ML-based solutions introduced by the researchers to fight against phishing attacks.

A deep learning approach is conducted by Somesha et al. [1] to predict the legitimacy of websites based on URLs. They applied the Information Gain feature selection method for ranking the attributes, then selected the ten best features for their experiment. Among the three deep learning models, the LSTM outperformed securing an accuracy of 99.57%. Apart from neural networks, the tree-based ensemble algorithms performed significantly well in phishing website prediction. In particular, the Random Forest, Extra Trees, and Gradient Tree Boosting showed promising outcomes [8, 15]. Based on the datasets used in this article, Vrbančič et al. presented a method to address the parameter setting issue of deep neural networks (DNN). They applied swarm intelligence meta-heuristic algorithms (bat algorithm, firefly algorithm) to optimize DNNs parameters. They used four phishing datasets for their experiment and achieved promising outcomes in classifying phishing websites—the proposed firefly method outperformed. Considering Vrbančič's small and full dataset, the firefly method showed an accuracy rate of 90.17% and 94.39%, respectively, in CV [16].

Al-Sarem et al. [3] proposed an optimized ensemble method considering the stacking approach. They experimented with six ML classifier models and the parameters of those models were optimized using a genetic algorithm (GA). However, the ensemble model with SVM as a meta-learner showed an accuracy of 97.39% on Vrbančič full_data. In another study, Alsariera et al. [8] developed meta-learner models in which the Extra-tree classifier is considered the base classifier. The LogitBoost-Extra Tree model achieved the highest accuracy of 97.58% and false positive rate of 0.018 in CV. To resolve the dynamism of phishing attacks, Adeyemo et al. [17] proposed an ensemble-based approach that combines tree induction and logistic regression techniques. They integrated the bagging and boosting methods with the base algorithm "Logistic Model Tree" to build more effective models. However, they used two phishing datasets and achieved a minimum accuracy of 97.18%.

In a study, Chiew et al. [18] established a hybrid ensemble feature selection, in which a set of filter methods are applied to derive feature subsets. It had two perturbation phases—the data perturbation cycle derived the primary and secondary feature subsets gradually and the final features are obtained from the function perturbation ensemble cycle. The combination of both these cycles results in a hybrid ensemble feature selection. They determined the cut-off ranks for selecting features with considering the gradient changes. Besides, their proposed framework selected only 20.8% of features from the data but achieved promising outcomes from the ML models.

Since phishing attacks are dynamic in nature, after exploring the previous researches—in general, we can reach a consensus that ML-based methods are much reliable in addressing these cynical attacks. In this article, the authors introduced a hybrid framework of feature selection to boost the proposed models' performance empirically. It's a two-phase hybrid approach in which filter, embedded, and wrapper methods are applied accordingly.

3 Research Methodology

In this article, an empirical feature selection approach is introduced, which is more like a hybrid method to aggregate the best features from the dataset. In this section, the overall methodology employed in this study for phishing website detection is discussed and illustrated in Fig. 1. It has two distinct layers, namely Feature Engineering and Cross-Validation. In this study, two phishing website datasets of Vrbančič are used, publicly available in Mendeley. For reading convenience, the Vrbančič small_dataset is tagged as "Data-I", where Vrbančič full_dataset is tagged as "Data-II".

In the early stage, the process of how the datasets are collected and converted into CSV files is depicted. Vrbančič collected 58,645 and 88,647 website URLs for Data-I

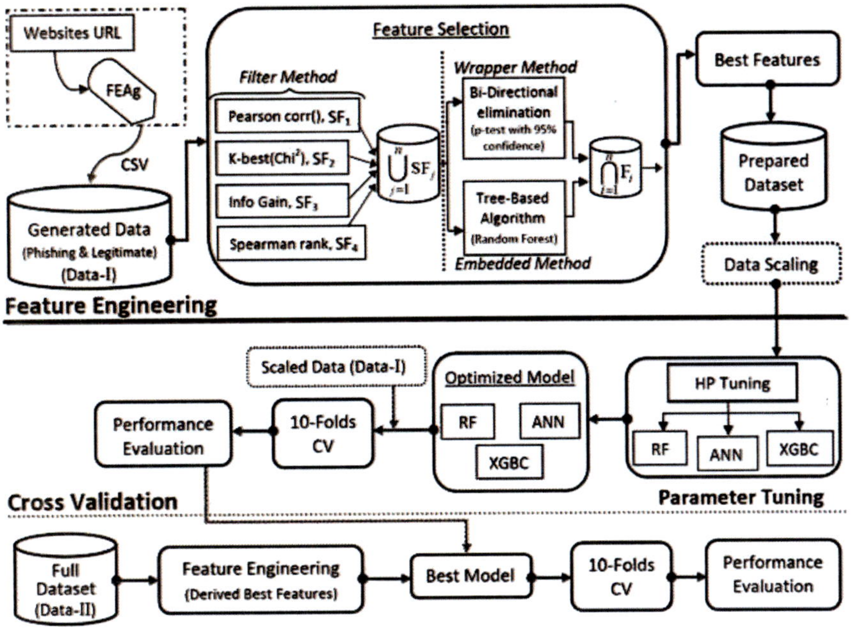

Fig. 1 The structure of proposed methodology

and Data-II, respectively, and designed a Feature Extraction Algorithm (FEAg). The FEAg extracts 111 unique features from each URL and saves them in a CSV file [19].

The "Feature Engineering" layer depicts that the feature selection approach, in this study, used a hybrid framework. In this process to obtain the most important attributes from Data-I, the raw features are refined through three methods, namely, the filter method, embedded method, and wrapper method. In the filter method, the study applied four distinct techniques, e.g., Pearson correlation, Information Gain, Chi-square (Chi^2), and Spearman correlation rank, and combined the filtered features with a set union operation ($\cup$). Then the feature subset from the filter method is fed to a wrapper method known as bi-directional (or stepwise) elimination and an embedded method to select the best features. In the wrapper method, the features are selected with the statistical model which runs a p-value test with conforming 95% confidence intervals, i.e., the probability to accept the null hypothesis is only 5%. On the other hand, in the embedded method, the features are derived using "random forest" algorithm. Following that, the best features are obtained from the final stage with a set intersection operation ($\cap$). Afterward, the prepared dataset containing the best features is scaled and fed to the ML models (RFC, XGBC, and ANN) for hyperparameter (HP) tuning. Grid Search is applied for tuning the HP of the models, which are passed to the next layer.

In the "Cross-Validation" phase, the study performed a ten-fold CV on the three optimized models and evaluated their performance. Besides, the study made a comparative assessment to find out the outperformed model which is used to perform CV on Data-II in the succeeding step. The study estimated the model's evaluation metrics to assess its performance in detecting phishing and legitimate websites.

The experiment was conducted on Jupyter notebook environment using Python programming language. The proposed feature selection algorithm (in Algorithm 1) and ML models are implemented with different python machine learning libraries.

4 Experimentation, Results Analysis, and Discussion

The experiment is carried out according to the proposed methodology. In this section, the outcomes, i.e., statistical modeling and numerical simulation of this experiment are discussed and highlighted.

4.1 Dataset Descriptions

The phishing websites datasets used in this experiment are gathered from Mendeley [19]. The dataset was found by Grega Vrbančič, and it has two variants. He used a feature extraction algorithm (FEAg) to get a list of features from the input URLs. The features of the datasets can be grouped into six classes. To estimate the value of features, the website URL strings are divided into four sub-strings (domain, directory, file, parameter), besides other external services are considered. The number of observations in Data-I and Data-II is 58,645 (phishing-30,647, legitimate-27,998) and 88,647 (phishing-30,647, legitimate-58,000), respectively, but both the datasets have identical and equal number of features (112). The target variable defined as "phishing", concludes if an observation of URL falls into legitimate or phishing class [14].

4.2 Feature Engineering

Data-I and Data-II contain 111 features (excluding the target variable) each. However, all these features are not reasonably significant, and as the number of features increases, the dimension of the dataset also increases. Consequently, this study proposed an empirical feature selection approach to select the best features. It's a hybrid approach that applies three methods, namely, the filter, wrapper, and embedded method. The proposed hybrid feature selection algorithm is noted in Algorithm 1.

Algorithm 1 Hybrid Feature Selection

Input: Dataset (O,F) // O= observation, F= feature
Output: Best Features
Initialization:
N← FILTERMETHODS(Pearson,Chi^2,Info Gain,Spearman)
for each N **do** // Filter method
  FS_i = CALCULATEFILTERSCORE(Dataset)
end for
T_N← INITIALIZECUTOFF(T_k) // T_k= cut-off range of filter methods (N)
for each F **do** // Primary feature subset
  if $FS_i \in T_N$ **then**
    $SF_j = \bigcup_{i=1}^{F} FS_i$ // Set union operation
  end if
end for // BFF = Best filtered feature
  $BFF = \bigcup_{j=1}^{N} SF_j$ // Function perturbation
Bi_feature = [] // Empty list
STEPWISEELIMINATION(BFF,SL_{in}=0.05,SL_{out}=0.05) // Wrapper method
for each feature in BFF **do**
  p_value = CALCULATEPVALUE(BFF)
  if $p_{min} < SL_{in}$ **then** // Forward Selection
    add to Bi_feature
    for each Bi_feature **do**
      if $p_{max} >= SL_{out}$ **then** // Backward Selection
        remove from Bi_feature
      else break
    end for
  else break
end for
return Bi_feature
for each feature in BFF **do**
  FI_k = CALCULATEIMPORTANCESCORE(BFF) // Apply Embedded Method
  if $FI_k \in T_{k\text{-}em}$ **then** // $T_{k\text{-}em}$ = cut-off range of embedded method
    $Em\_feature = \bigcup_{k=1}^{F} FI_k$
  end if
end for
best_features = $\bigcap$(Bi_feature, Em_feature) // Set intersection operation

Feature Selection. The filter method, in the study, used four feature selection techniques. The techniques estimate the importance score of individual features considering their frequency or correlation with other features. Each technique provides a feature subset (SF) that has the most important features in it. At the end of this method, all the feature subsets are combined with a set union operation—this process is known as "function perturbation" [18]. Besides, from a technique to select the most

Table 1 Cut-off range of feature selection methods and number of features selected

Feature selection method		Cut-off range (Tk)	Selected feature
Filter	Pearson correlation (SF1)	0.087–0.627	68
	Chi-square (SF2)	0.00015–0.688	61
	Information Gain (SF3)	0.0001–0.336	59
	Spearman correlation rank (SF4)	0.084–0.686	67
	BFF = (SF1 ∪ SF2 ∪ SF3 ∪ SF4)	—	89
Embedded	Random Forest	0.001–0.10	65
Wrapper	Stepwise elimination	p-test (95% confidence)	51
best_features = (Embedded ∩ Wrapper)		—	**42**

correlated features, a cut-off range (Tk) is set. Each technique has an identical cut-off, determined by analyzing and ranking the feature importance scores. A cut-off range aka, cut-off rank [13, 18] defines an optimal extent to select features regarding their importance score. For instance, in Table 1, the cut-off range of Pearson correlation 0.087–0.627 demonstrates, only the features having an importance score in this threshold point will be selected. In this study, an ad-hoc cut-off of distinct techniques is gauged between the high and low feature importance scores, i.e., poorly correlated features are excluded. The output of the filter method is a cumulative feature subset (BFF), which is fed to the wrapper and embedded method. Subsequently, the study obtained 89 most important filtered features (excluding target feature) from Data-I.

In the wrapper method, the bi-directional elimination (aka, stepwise elimination) technique is applied to get the optimal feature subset. Bi-directional elimination is a combination of the forward and backward elimination techniques, which aims to find out the features' best correlation. In this technique, a statistical model is built where a p-value test is carried out to select features with peak confidence intervals. Unlike the filter method, the computational time required to perform the wrapper method is considerably high, but it selects features with maximum prediction accuracy [20]. Hence, from the filtered features, 51 features were selected with the wrapper method conforming 95% confidence intervals. Then, in the embedded method, a random forest algorithm is employed which derived 65 important features from the cumulative feature subset. Finally, applying a set intersection operation between the wrapper and embedded feature subset, the study obtained 42 best features from the dataset.

The output feature set of the hybrid method is used as the final features of Data-I and Data-II to perform CV. The shape of Data-I and Data-II before feature selection was (58,645,112) and (88,647,112), respectively. But after applying the proposed feature selection method, the shape reduced significantly to (58,645,43) and (88,647,43).

Afterward, the datasets are scaled with the Min–Max feature scaling method, since data in the real world does not available in a fixed range. Thus, scaling will control the bias of the features having higher weight. Besides, the scaled data allow

Table 2 Tuned hyperparameters of the proposed ML models

Model	Best parameter selected with Grid search CV
ANN	Optimizer = Adam, learning_rate = 0.001, epochs = 95, batch_size = 64
RFC	max_depth = 75, n_estimators = 900, criterion = 'entropy', min_samples_split = 2
XGBC	max_depth = 6, n_estimators = 800, learning_rate = 0.1, min_child_weight = 4

each feature to pay a uniform contribution in optimizing the target function. In the Min–Max feature scaling method, the data are scaled in a range of "0–1".

Hyperparameter Optimization. This study used three machine learning models, i.e., RFC, XGBC, and ANN to detect phishing websites from the datasets. ML algorithms with default parameters are less likely to perform well than algorithms with tuned parameters [21]. Therefore, the Grid Search HP tuning technique is applied to tune the parameters of the three models, in Table 2. This technique used Data-I (split: train-80%, test-20%) to perform parameter tuning and provides optimized learning models as output. These optimized models are employed to perform CV in layer-2.

4.3 Cross-Validation

The study performed ten-fold CV using three optimized models (RFC, XGBC, ANN) on Data-I and estimated their performance evaluation metrics. The study also calculated the Mean Square Error (MSE) rate and area under curve (AUC) score of the models. Besides, the mean Receiver Operating Curve (ROC) and Precision-Recall (PR) Curve of each model is estimated. A comparative assessment is performed considering the mean scores in the validation of the models, shown in Table 3.

In cross-validation, the XGBC model surpassed ANN and RFC in all possible performance evaluation metrics considered in this experiment—accuracy (ACC), for example, is 96.10%, followed by an F1-score of 96.09% and an MSE of 0.039. Besides, the RFC model showed a notable performance with an accuracy of 95.88%

Table 3 Cross-validation performance comparison of proposed models on Data-I

Evaluation Metrics	ANN	RFC	XGBC
Accuracy (%)	93.95	95.88	96.10
Precision (%)	93.87	95.89	96.08
Recall (%)	94.64	95.85	96.09
F1-Score (%)	94.23	95.87	96.09
False Positive Rate (FPR)	0.0735	0.0429	0.0375
False Negative Rate (FNR)	0.0607	0.0357	0.0372
AUC Score	0.9723	0.9893	0.9910
MSE	0.0455	0.0412	0.0390

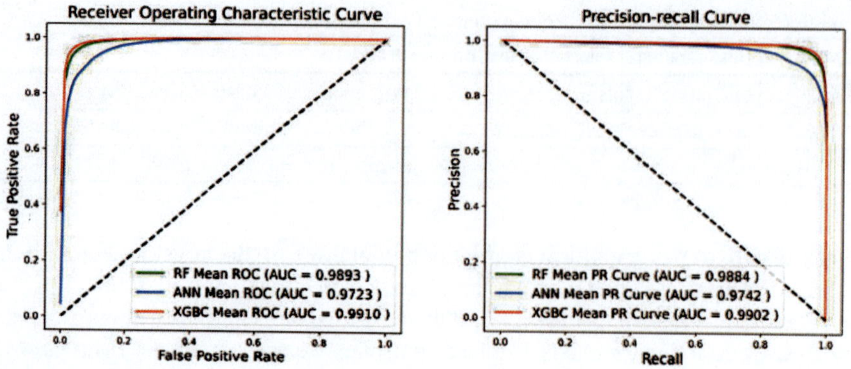

Fig. 2 Mean ROC and PR curve of the ML models (ANN, RFC, XGBC) in cross-validation

which is close to XGBC. ANN model, on the other hand, provided a decent outcome securing a 93.95% accuracy rate and 94.23% F1-score. The average AUC score of XGBC, RFC, and ANN is 0.9723, 0.9893, and 0.9910, respectively. Besides, the XGBC model achieved the minimum FPR of 0.0375 and RFC showed an FNR of 0.0357. The ROC curves and PR curves of the classifier models are illustrated in Fig. 2. The ROC curves are based on the true positive and false positive rates, where the precision-recall curves are derived from precision and recall rates.

The study also performed CV on Data-II with the best model XGBC. Data-I (small_data) and Data-II (full_data) are very much similar—they have precisely the same features (name and numbers). In this circumstance, the study used the selected top 42 features for Data-II and performed ten-fold CV. Based on that, the study evaluated the model's performance by calculating the evaluation metrics, MSE of each fold, and estimated the AUC score, ROC curve, and precision-recall curve.

The XGBC model performed the validation prominently good on Data-II. The model secured significant accuracy of 97.40%, F1-score of 97.12%, and an error rate of only 2.60%. Besides, the model showed more than 97% of precision and recall score and a high AUC score of 0.9942. It also gained a favorable MSE rate of 0.0260. In Table 4, the performance assessment metrics of the model XGBC are noted considering each fold in CV. Besides, the confusion matrix of the model gives the information of the FPR and FNR of about 0.0372 and 0.0381, respectively. The ROC curve and PR curve of the XGBC model is illustrated in Fig. 3.

With this proposed study, a considerable performance in detecting phishing and legitimate websites is achieved. As the study used the latest phishing datasets for this experiment, only a few pieces of research have been conducted previously on these data. The highlighted outcome of this study is that it used only about 38% of the original features of the data for the experiment. Furthermore, Data-II is an imbalanced dataset, but this study did not consider resampling or other techniques to tackle the class imbalance issues—because the study used neural networks and tree-based learning algorithms which are highly competent to handle them. Besides, resampling can impose overfitting on ML models or lose invaluable information.

Table 4 Cross-validation performance of XGBC model on Data-II

Fold	XGBC model				
	Accuracy	Precision	Recall	F1-Score	MSE
1	97.28	96.92	97.07	97.01	0.0272
2	97.55	97.32	97.20	97.26	0.0245
3	96.98	96.71	96.68	96.70	0.0302
4	97.43	97.16	97.13	97.15	0.0257
5	97.23	96.93	96.91	96.94	0.0277
6	97.47	97.28	97.12	97.20	0.0253
7	97.30	96.94	97.05	96.99	0.0269
8	97.61	97.32	97.36	97.34	0.0239
9	97.58	97.35	97.30	97.32	0.0242
10	97.53	97.38	97.21	97.30	0.0247
Avg	97.40	97.13	97.10	97.12	0.0260

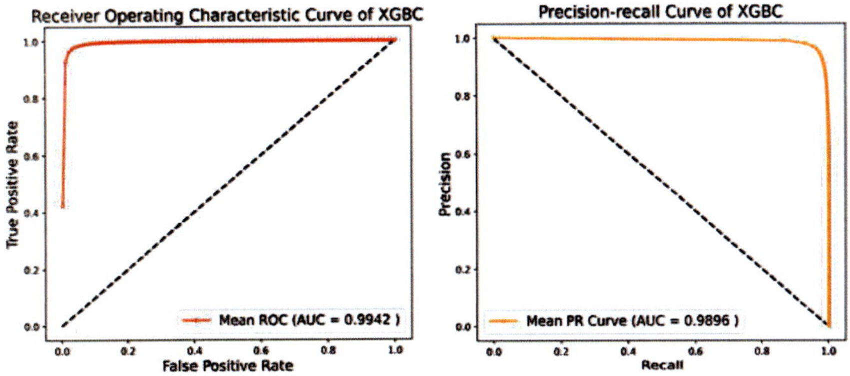

Fig. 3 ROC and PR curve of the model XGBC for Data-II

Now, let's compare the performance metrics of this proposed study with the existing research works.

Overall, from Table 5, it is noticeable that the proposed model XGBC outruns the existing research works using only 42 features. Besides, compared to the traditional individual feature selection methods, the proposed hybrid approach obtained about 16% lesser features (Table 1) which reduced the models' computational complexity.

Table 5 Outcomes comparison of the previous and proposed research studies

Dataset	Research Ref	Best method	Features	Performance
Data-I	Grega Vrbančič et al. [16]	DNN, optimized with firefly algorithm	111 FSA (×)	Accuracy 90.17% F1-score 90.11%
	Proposed method	XGBoost, with hybrid feature selection approach	**42** FSA (√)	Accuracy **96.10%** F1-score **96.09%**
Data-II	Grega Vrbančič et al. [16]	DNN, optimized with firefly algorithm	111 FSA (×)	Accuracy 94.39% F1-score 93.83%
	Mohammed Al-Sarem et al. [3]	Stacking ensemble, GA-based optimization	111 FSA (×)	Accuracy 97.39%
	Proposed method	XGBoost, with hybrid feature selection approach	**42** FSA (√)	Accuracy **97.40%** F1-score **97.12%**

[a]FSA = Feature Selection Approach, if applied (√) or not applied (×)
The asterisk denotes the elaboration of FSA along with relevant information

5 Conclusion

Bangladesh ranks as one of the leading vulnerable regions in South Asia, susceptible to cyberattacks [2]. Consequently, the country needs to strengthen its reliance and cybersecurity for combating prospective threats such as phishing attacks. In this article, the study introduced an empirical hybrid feature selection approach to establish an optimized machine learning model for phishing website detection. The study used the two latest URL-based phishing datasets. Applying this hybrid feature selection method, only 42 features (out of 111) were derived from the dataset. In this method, the raw features are refined through the filter method, embedded method, and wrapper method accordingly. This novel hybrid approach used six different (four filters, one wrapper, and one embedded) feature selection techniques in total. Besides, the proposed models (ANN, RFC, and XGBC) are optimized with Grid Search-based hyperparameter tuning. Ten-fold CV is performed using the optimized models on Data-I, and the result showed XGBC came up with the maximum validation accuracy of 96.10%, with F1-score of 96.09%. Following that, the study performed CV on Data-II using the best model XGBC. The model showed a significant performance securing an accuracy of 97.40%, an F1-score of 97.12%, and an MSE rate of 0.0260. However, the study chose a hybrid method to ensemble the strength of individual feature selection techniques. Every feature selection techniques have its advantages, but previous studies showed that hybrid ensemble methods improve the overall outcomes [18].

From the top of the authors' knowledge, in most of the existing research studies on URL-based phishing website detection, researchers did not consider feature selection methods since they might be wondering that this would cost massive information loss. However, in feature selection, information might be lost, but only the least significant ones. In addition, if you are working with high dimensional data, it's hard to achieve greater accuracy because the model needs to consider tons of attributes for decision-making. As a consequence, if the dimension of data is reduced, the complexity and degree of computations of models will also be minimized. Considering this case study, the number of features of Data-I and Data-II is reduced to about 38%, yet the proposed model outperformed the existing studies [3, 16]. Consequently, in ML studies to deal with high-dimensional and imbalanced data, the feature selection methods could be a substantial paradigm. However, if the size of features in a dataset is small, reducing the dimension of data would not be a good practice.

In the future, the study will attempt to apply the proposed methodology on other available phishing datasets and experiment with state-of-the-art DNNs to improve the efficiency of phishing website detection. Although the manual setting of the cut-off range (Table 1) in feature selection methods showed a favorable reflection on the model's performance, the study will continue to develop an automated assignment method for the best cut-off ranges.

References

1. Somesha, M., et al.: Efficient deep learning techniques for the detection of phishing websites. Sādhanā **45**, 165 (2020). https://doi.org/10.1007/s12046-020-01392-4
2. BGD e-GOV CIRT Homepage. https://www.cirt.gov.bd/. Accessed 21 Sept 2021
3. Al-Sarem, M., et al.: An optimized stacking ensemble model for phishing websites detection. Electronics **10**(11), 1285 (2021). https://doi.org/10.3390/electronics10111285
4. Kalaharsha, P., Mehtre, B.M.: Detecting phishing sites—an overview. (2021). https://arxiv.org/abs/2103.12739
5. Sarma, D., et al.: Comparative analysis of machine learning algorithms for phishing website detection. In: Smys S., et al. (eds.) Inventive Computation and Information Tec-hnologies, vol. 173. LNNS, Springer, Singapore (2021). https://doi.org/10.1007/978-981-33-4305-4_64
6. da Silva, C.M.R., Feitosa, E.L., Garcia, V.C.: Heuristic-based strategy for phishing prediction: a survey of URL-based approach. Comput. Secur. **88**, 101613 (2020)
7. Zuraiq, A.A., Alkasassbeh, M.: Review: phishing detection approaches. In: 2nd International Conference on new Trends in Computing Sciences, pp. 1–6 (2019)
8. Alsariera, Y.A., et al.: AI meta-learners and extra-trees algorithm for the detection of phishing websites. IEEE Access. **8**, 142532–142542 (2020). https://doi.org/10.1109/ACCESS.2020.3013699
9. Hannousse, A., Yahiouche, S.: Towards benchmark datasets for machine learning based website phishing detection: an experimental study. Eng. Appl. Artif. Intell. **104**, 104347 (2021). https://doi.org/10.1016/j.engappai.2021.104347
10. Ali, W.: Phishing website detection based on supervised machine learning with wrapper features selection. Int. J. Adv. Comput. Sci. Appl. (IJACSA) **8**(9) (2017). https://doi.org/10.14569/IJACSA.2017.080910
11. Guyon, I., Elisseeff, A.: An introduction to variable and feature selection. J. Mach. Learn. Res. **3**, 1157–1182 (2003)
12. Liu, H.Y., Zhou, M.C., Liu, Q.G.: An embedded feature selection method for imbalanced data classification. IEEE/CAA J. Autom. Sinica. **6**(3) (2019)
13. Pes, B., Dessì, N., Angioni, M.: Exploiting the ensemble paradigm for stable feature selection: a case study on high-dimensional genomic data. Inf. Fusion **35**, 132–147 (2017). https://doi.org/10.1016/j.inffus.2016.10.001
14. Vrbančič, G., Fister, I., Jr., Podgorelec, V.: Datasets for phishing websites detection. Data Brief **33**, 106438 (2020). https://doi.org/10.1016/j.dib.2020.106438

15. Vaitkevicius, P., Marcinkevicius, V.: Comparison of classification algorithms for detection of phishing websites. Informatica **31**(1), 143–160 (2020)
16. Vrbančič, G., Fister, I., Jr., Podgorelec, V.: Parameter setting for deep neural networks using swarm intelligence on phishing websites classification. Int. J. Artif. Intell. Tools **28**(06), 1960008 (2019)
17. Adeyemo, V.E., et al.: Ensemble-based logistic model trees for website phishing detection. In: Anbar M., et al. (eds.) ACeS 2020. CCIS, vol. 1347. Springer, Singapore (2021)
18. Chiew, K.L., et al.: A new hybrid ensemble feature selection framework for machine learning-based phishing detection system. Inf. Sci. **484**, 153–166 (2019). https://doi.org/10.1016/j.ins.2019.01.064
19. Vrbančič, G.: Phishing Websites Dataset. Mendeley Data. V1 (2020). https://doi.org/10.17632/72ptz43s9v.1
20. Mochammad, S., et al.: Stable hybrid feature selection method for compressor fault diagnosis. IEEE Access. **9**, 97415–97429 (2021). https://doi.org/10.1109/ACCESS.2021.3092884
21. Bhowmik, P., et al.: Analysis of social media data to classify and detect frequent issues using machine learning approach. In: 2020 2nd International Conference on Advanced Information and Communication Technology (ICAICT), pp. 394–399 (2020). https://doi.org/10.1109/ICAICT51780.2020.9333452

Activity Classification from First-Person Office Videos with Visual Privacy Protection

Partho Ghosh, Md. Abrar Istiak, Nayeeb Rashid, Ahsan Habib Akash, Ridwan Abrar, Ankan Ghosh Dastider, Asif Shahriyar Sushmit, and Taufiq Hasan

Abstract With the advent of wearable body cameras, human activity classification from First-Person Videos (FPV) has become a topic of increasing importance for various applications, including life-logging, law enforcement, sports, workplace, and health care. One of the challenging aspects of FPV is its exposure to potentially sensitive objects within the user's field of view. In this work, we developed a visual privacy-aware activity classification system focusing on office videos. We utilized a Mask-RCNN with an Inception–ResNet hybrid as a feature extractor for detecting and later blurring out sensitive objects (e.g., digital screens, human face, paper) from the videos. We incorporate an ensemble of Recurrent Neural Networks (RNNs) with ResNet, ResNext, and DenseNet-based feature extractors for activity classification. The proposed system was trained and evaluated on the FPV office video dataset which is a subset of the BON (Tadesse et al. Bon—egocentric vision dataset for office activity recognition [1]) dataset and includes 18 classes made available through the IEEE Video and Image Processing (VIP) Cup 2019 competition. On the original unprotected FPVs, the proposed activity classifier ensemble reached an accuracy of 85.078% with precision, recall, and F1-scores of 0.88, 0.85, and 0.86, respectively. The performances were slightly degraded on privacy-protected videos, with accuracy, precision, recall, and F1-scores at 73.68%, 0.79, 0.75, and 0.74, respectively.

Keywords Activity classification · Visual privacy protection · First-person video

P. Ghosh · Md. A. Istiak · N. Rashid · A. H. Akash · R. Abrar · A. G. Dastider
Department of Electrical and Electronic Engineering (EEE), Bangladesh University of Engineering and Technology (BUET), Dhaka 1205, Bangladesh

A. S. Sushmit · T. Hasan (✉)
mHealth Research Group, Department of Biomedical Engineering (BME), Bangladesh University of Engineering and Technology (BUET), Dhaka 1205, Bangladesh
e-mail: taufiq@bme.buet.ac.bd

S. Hossain et al. (eds.), *Proceedings of International Conference on Fourth Industrial Revolution and Beyond 2021*, Lecture Notes in Networks and Systems 437,
https://doi.org/10.1007/978-981-19-2445-3_25

1 Introduction

As body-worn cameras are becoming more ubiquitous, automatic logging of human activity from FPVs has become a topic of increasing interest. At present, FPV or PoV (point-of-view) cameras are mostly utilized by athletes, motor drivers, and police officers [2, 3]. Generally, it is not feasible to manually analyze large numbers of such videos. Automatically processing FPVs can thus be beneficial in several major applications, including activity logging, law enforcement, search and rescue missions, inspections, home-based rehabilitation, and wildlife observation [4]. As the Augmented Reality (AR) glasses become mainstream, the availability of FPV data is expected to increase while also raising associated concerns for security and privacy of the users [5].

Activity classification can be done in two different modalities: one using sensor data and the other using video data. Recent developments in the deep learning field have made tremendous progress in classifying activity in sensor data. In [6] they have described such a state-of-the-art method, and all of the other methods follow a similar path. One of the major problems with working with sensor data is that the amount of activity becomes very limited. On the other hand, automatic activity classification from videos has been undertaken using two major approaches in the past. Traditional methods generally extract hand-engineered features from the videos to train a machine learning algorithm for classification. Frequently employed features include average pooling (AP), robust motion features (RMF), and pooled appearance features (PAF) [7, 8]. In recent times, deep neural networks are being used to learn the features and subsequently classify the activities. One of the effective methods involves using a Convolutional Neural Network (ConvNet) with Long Short-term Memory (LSTM) [9] units as back-ends [10, 11]. In this approach, the ConvNet learns the front-end image features, whereas the LSTMs recognize the temporal features within the videos that are relevant for human activities. Another well-known method found in literature consists of using a 3D ConvNet to recognize the actions [12]. 3D ConvNets are functionally similar to 2D ConvNets, except that they incorporate spatio-temporal filters [13], and thus do not specifically require the LSTM layers. However, none of the previous work in the area of activity classification using FPVs has addressed the issue of privacy. For visual privacy protection, recently developed iPrivacy [14] approach protects the most significant privacy-sensitive object via blurring. Serdar et al. proposed a reliable and reversible privacy protection scheme based on false color [15] which affects the whole frame's content. Some of the most popular ways to protect visual privacy are silhouette, scrambling, in-painting [16], and filtering [17]. Latest research on action recognition with visual privacy protection proposed a compression-based coding scheme named VSCS [18] method for privacy protection. It employs a coding scheme to compress images but VSCS frames lose all content of the frame except the main content which yields visually distracting results.

In this work, we develop a privacy-aware activity classification framework for office videos (Fig. 1). In most cases, the personal information contained within FPV videos does not carry useful features for detecting the user's activity. Accordingly, we use a deep learning model to identify the sensitive regions of the video and make them unintelligible visually (e.g., blurred). Next, these privacy-protected videos are used to train deep learning models for activity recognition using the FPV dataset of office activities made available for the IEEE VIP Cup 2019 [7]. Performance comparisons between activity classification from original versus privacy-protected videos are performed to demonstrate the effectiveness of the proposed framework.

2 Dataset

The dataset we used in this study is a part of a bigger dataset named BON—Egocentric Vision Dataset for Office Activity Recognition [1] and this subset was provided through the IEEE VIP Cup 2019 competition. The dataset was collected using a chest-mounted GoPro Hero3+ Camera with 1280×760 pixels resolution, and a frame rate of 30 fps [7]. Human activities can be broadly categorized as (i) human-to-human activity, (ii) human-to-object activity, and (iii) ambulatory activity. The BON dataset contains a total of 18 activity in classes that include all of these broad categories. The dataset and distribution of the 18 classes are summarized in Fig. 2. The figure shows that the class imbalance problem is a major issue in the dataset that we need to address.

During the competition, the dataset was released in two phases, 400 and 932 videos, respectively. We found label noise in 32 videos through manual annotation,

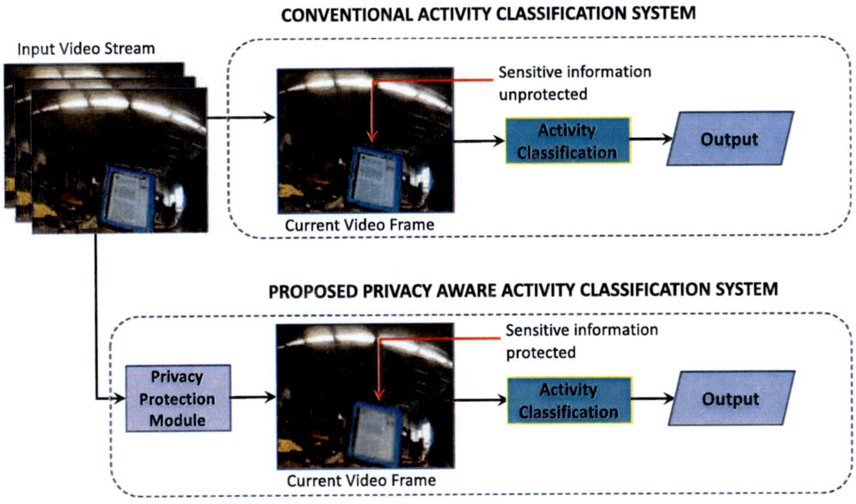

Fig. 1 Conventional versus proposed privacy-aware activity classification systems

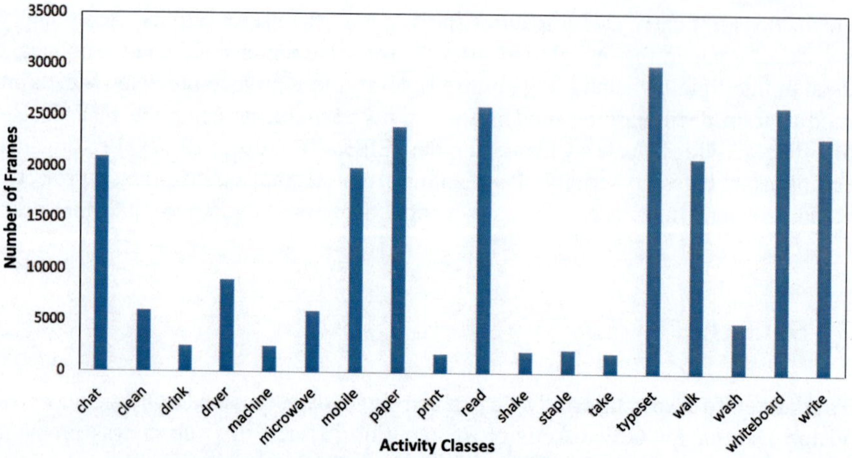

Fig. 2 Bar plot showing the number of video files available for each activity class. The uneven distribution of data in different classes illustrates the issue of class imbalance

Table 1 Training/test splits in the sub-datasets used

Sub-datasets	# Videos in training	# Videos in test
Original	873 original	364 original
Blurred	873 blurred	364 blurred
Mixed	873 org. + 873 blurred (1746 total)	364 org. + 364 blurred (728 total)

and these are excluded. The remaining videos from the two phases are used to prepare our training and test set containing 873 and 364 videos, respectively.

To train the privacy-aware system, we prepared three different sub-datasets for training and testing our deep learning models (Table 1). We refer them to as the (i) *original*, (ii) *blurred*, and (iii) *mixed* sub-datasets. The *original* set contains the original videos from the BON dataset, *blurred* set contains visual privacy protected videos, and the *mixed* subset contains videos from both subsets (i) and (ii) in an equal amount. The process of generating visual privacy-protected videos in the *blurred*-set are described in the following sections.

3 Proposed Architecture

3.1 Pre-processing

Analyzing the histogram of the average brightness of the video frames (mean value of each image pixel), we observed that some videos have a significantly higher

Table 2 Performance of different feature extractor-based Mask R-CNN models on the COCO dataset for object segmentation

Model name	mAP
Mask R-CNN ResNet101 Atrous [30]	33
Mask R-CNN Inception V2	25
Mask R-CNN ResNet50 Atrous	29
Mask R-CNN Inception ResNet V2 Atrous [30]	**36**

brightness compared to others (e.g., data provided in the "Oxford" folder). Thus, we applied gamma correction [19] to the images to normalize this effect. Next, we resize the video frames and normalize them to fit them in the particular deep learning pre-trained networks, Wide ResNet [20], ResNext [21], and DenseNet [22], as described later in Sect. 3.3.

3.2 *Visual Privacy Protection Module*

In this section, we describe the proposed visual privacy protection and security enhancement module that addresses the visual privacy concerns in the BON dataset. The framework consists of two steps: (i) Identifying the sensitive regions from a video frame and (ii) Making these parts unintelligible by blurring [23, 24].

3.2.1 Identification of Sensitive Regions

First, we manually screened the dataset for objects that are prone to privacy violations of the users or may contain visually sensitive information that can lead to a security breach. In the data, we found a total of seven objects that may contain sensitive information: (i) digital screen, (ii) laptop, (iii) mobile, (iv) book, (v) person, (vi) keyboard, and (vii) toilet/urinal. We used the Mask R-CNN [25] network to identify these objects in our video frames [26].

We utilize the Mask R-CNN approach as it performs instance segmentation [27] to produce a mask for sensitive objects that may not be of uniform shape (e.g., a person). Traditional methods that generate a bounding box [28, 29] are not suitable for our application since it will result in the blurring of a larger box-shaped region as compared to the actual sensitive object. This may degrade the activity classifier performance. Examples of object localization (bounding box) and instance segmentation for an image frame of the activity class "Chat" are shown in Fig. 3a and b, respectively.

Different feature extractor networks can be utilized with a Mask R-CNN model, including InceptionV2 [24], Resnet50 [31], Resnet101 [31], and InceptionResNetV2

[24] with Atrous convolution. The performance of these models for object segmentation, have been compared on the Common Objects in Context (COCO) dataset [32] with respect to the Mean Average Precision (mAP) metric [33, 34]. These results are available online in the TensorFlow GitHub repository.[1] According to these mAP scores reproduced in Table 2, InceptionResNetV2 with Atrous convolution model provides the best performance on the COCO dataset. Thus, we select this model for our sensitive object segmentation. Since the BON dataset does not contain any object mask labels, it was not possible to objectively evaluate the performance of the object segmentation module on this data. However, we validated the performance of the mask detection model based on visual observation of a subset of BON frames.

The final architecture of the sensitive region segmentation is shown in Fig. 4. We used a pre-trained Inception–Resnet Hybrid-based Mask R-CNN model trained on the COCO dataset. First, we classify between 80 objects available in the COCO dataset. Next, we selected seven potentially sensitive objects, as previously described. Finally, the frames are converted to RGB from BGR and are resized according to the required input dimension of the activity classifier network.

3.2.2 Protecting the Sensitive Objects

After identifying the visually sensitive regions from the video frames, a Gaussian filter is used to blur the regions where the sensitive objects are detected. An example video frame from the "chat" class is shown in Fig. 5a along with the detected mask using the Mask R-CNN network, while Fig. 5b shows the same frame with visually sensitive objects blurred (hands, screens, and keyboard).

Object localization Instance segmentation

Fig. 3 Example image frames illustrating object localization and instance segmentation for a video obtained from the class "Chat" of the BON dataset

[1] https://github.com/tensorflow.

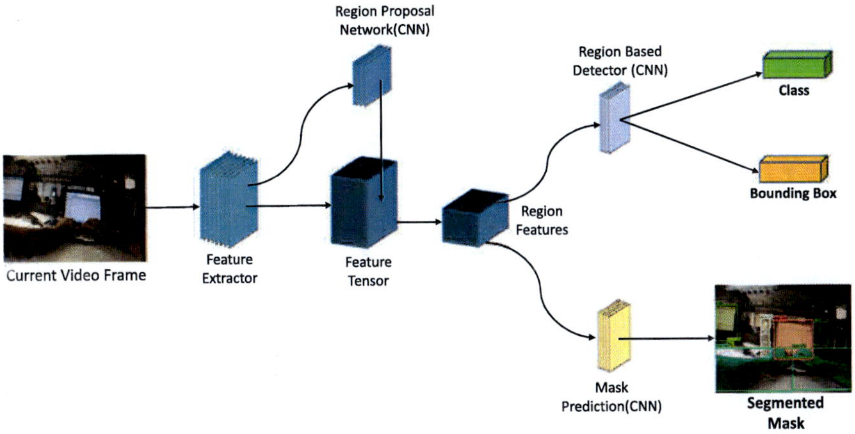

Fig. 4 Model architecture for identification of visually sensitive regions from the video frames. A Mask R-CNN model with an Inception Resnet V2 Atrous feature extractor is used

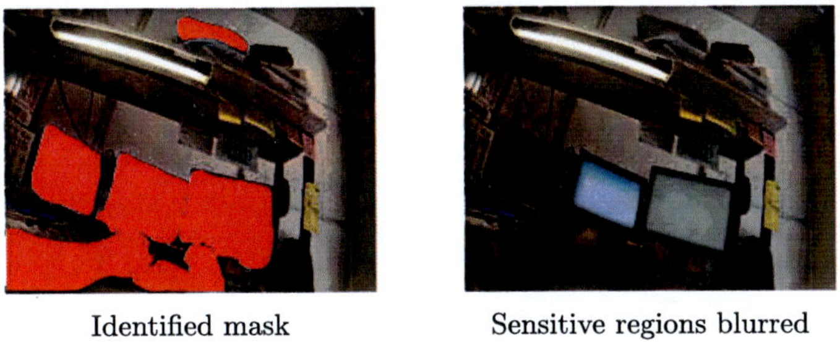

Fig. 5 Example of generated masks using the Mask R-CNN network using the Inception ResNetV2 with Atrous convolution model as a feature extractor. The video frame was labeled as the "Typeset" class

3.3 Activity Classification Model

The proposed activity classification module consists of a three-channel single stream network inspired by [35]. The network is illustrated in Fig. 6. The authors of [35] used AlexNet [36] as the feature extractor with two uni-directional LSTM layers for temporal sequence modeling. In contrast, we propose to utilize an ensemble of Wide ResNet [20], ResNext [21], and DenseNet [22] feature extractors and a single layer bi-directional LSTM with framewise attention. The proposed framework is detailed in the following subsections.

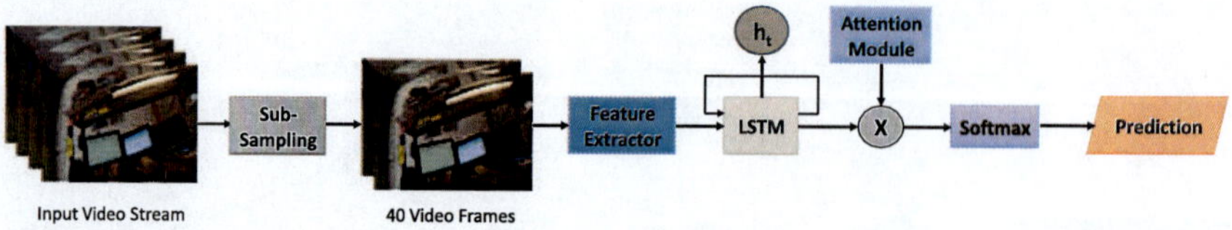

Fig. 6 Proposed activity classification framework excluding the privacy protection module

3.3.1 Video Frame Selection

We sub-sampled video frames to reduce the computational load. Regardless of video duration, we extracted 40 frames. To select these 40 frames, we first determine a sampling period by dividing the total number of frames in the video by 40 and taking the floor of the result. Then we randomly took the first frame in the range of 1 to the sampling period. After selecting the first frame, we jump the sampling period and take the next frame, and so on until all 40 frames are taken. In this way, we were able to fit almost all the video information into these 40 frames. We experimented with different values except 40, but 40 frames per video produced the best result. These image frames are then gamma-corrected and normalized prior to loading into the model.

3.3.2 Feature Extraction Layer

We experimented with different feature extractor models including, DenseNet121 [22], DenseNet161 [22], ResNet50 [31], ResNet101 [31], ResNet152 [31], InceptionV3 [37], InceptionResNetV2 [24], ResNext [21], and Wide ResNet 101 [20]. In these experiments, we used pre-trained models on ImageNet [38]. We froze the feature extractor and trained all the remaining layers of our model.

3.3.3 LSTM Layer

We pass the video frames to a bi-directional single-layer LSTM network [39]. Our LSTM layer had an input dimension of 512, returned by the feature extractor, and a hidden state size of 1024.

3.3.4 Frame Wise Attention

We hypothesize that, for activity classification, all the sampled video frames are not of equal temporal importance. Some frames are temporally more important to activity recognition than others. Accordingly, we use a frame-wise attention module with a softmax activation function after the LSTM layer to give more weightage to the feature vectors corresponding to the frames that are more important to recognize the activity happening in the video.

3.3.5 Fully Connected Layers

The output of the frame-wise attention module is passed to the fully connected layer, followed by batch normalization and a softmax decision layer.

3.3.6 Training Scheme

The proposed activity classification network is trained for 20 epochs with the Adam optimizer [40] using cross-entropy as the loss function. To address the class-imbalance issue in the data, we implemented a balanced training scheme where every mini-batch contained an equal number of samples from each class. We oversampled the underrepresented classes while creating these balanced mini-batches. We begin training with an initial learning rate of 0.001 and implemented a learning rate scheduler with a patience value of 5 and with a decay factor of $1/10$ to facilitate model convergence.

4 Experimental Evaluation

4.1 Performance of Activity Classification Models

In our activity classification network, we implemented a variety of feature extractors and compared their results in different settings. During our experiments, we observed that each model tested performs best at a specific frame size. The top-performing models and their performance on the *original* test data are shown in Table 3.

Table 3 Performance comparison of different networks on the *original* test dataset

Model	Resolution	Accuracy (%)
1. ResNext + Attention	248 × 248	75.92
2. DenseNet + Attention	512 × 512	74.87
3. Wide ResNet 101 + Attention	324 × 324	79.84
4. Wide ResNet 101	324 × 324	75.39

Table 4 Accuracy of the proposed model ensemble on different test sets while using different training sets

Ensemble training data	Accuracy (%)	
	Original-test	*Blurred*-test
Original-train dataset	85.07	68.8
Mixed-train dataset	82.72	75.8

4.2 Performance of Model Ensemble

We propose to use an ensemble of the four classifiers described in the previous section. Ensemble weights are calculated based on the class-wise F1-score of each model. We first compare the performance of the ensemble using models trained and tested on different data subsets. The results provided in Table 4 show that the ensemble model trained only on the *original* training set performs poorly on the *blurred* test set. However, training on *mixed* dataset improves the *blurred* test performance with a drop of accuracy in the *original* test condition. We address this trade-off by adding fine-tuned models to our final ensemble.

4.3 Model Tuning and Final Ensemble

To address the issue of performance degradation on the *blurred* test, we select our models trained on the *original* sub-dataset and use transfer learning to fine-tune them using the *blurred* (privacy-protected) set. We prepare the final ensemble using the four original models mentioned in Table 3 and the proposed fine-tuned versions. Prediction time of ensemble classifier is extensively discussed in the IEEE VIP Cup final evaluation [41]. On 5-s samples of each task, this ensemble classifier took 94.42 s on average.

4.4 Results and Discussion

The results of the final ensemble and class-wise F1-scores are shown in Table 5 and Fig. 7, respectively. From the results, we first observe that the class imbalance did not affect our system performance as we used the balanced-batch training technique. As an illustrative example, from Fig. 2 we observe that the class "typeset" included a significantly larger amount of data compared to the "take" class. The final class-wise F1-scores depicted in Fig. 7 show that the system performed better in detecting the minority class. Secondly, we observe that due to the visual privacy protection the performance of the initial network dropped quite significantly. The performance gap

Table 5 Overall performance of the final ensemble classifier

Sub-dataset	Precision	Recall	F1-Score	Accuracy (%)
Original	0.88(±0.1)	0.85(±0.1)	0.86(±0.1)	85.08(±11.4)
Blurred	0.79(±0.2)	0.75(±0.2)	0.74(±0.2)	73.68(±19.97)

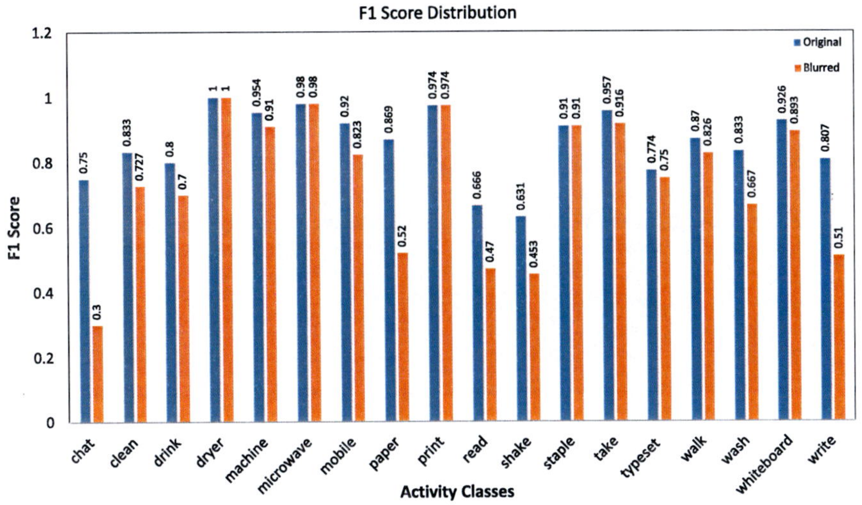

Fig. 7 F1-score distribution on *original* test and *blurred* test dataset videos across the class labels

between original videos and protected videos was about 20%. To address it, we train our network using visual protected videos. That improved the result on protected videos but degraded the results on original videos and again created a huge performance gap between them. We achieve the best results when we trained the model on original videos and fine-tuned those models using the protected videos. In this way, the performance gap between the original videos and visual privacy-protected videos was significantly reduced and we reached a more generalized approach. On average, the performance metrics in Table 5 degraded about 10% from their original values due to visual privacy protection, demonstrating the effectiveness of the proposed privacy-aware activity classification system.

5 Conclusions

In this work, we have developed a privacy-aware activity classification from FPV videos using an ensemble of deep learning models. The visual privacy protection module utilized pre-trained networks to identify sensitive regions within the video frames and performed Gaussian blurring on those pixels. The system has been trained using balanced mini-batches to effectively address the issue of class imbalance in the training data. The ensemble of models trained on unprotected videos and later fine-tuned on privacy-protected videos provided the best overall performance in both conditions.

Acknowledgements We want to thank the department of BME and Brain Station 23 (Dhaka, Bangladesh) for supporting this research. The NVIDIA Corporation donated the TITAN Xp GPU used for this work.

References

1. Tadesse, G.A., Bent, O., Weldemariam, K., Istiak, M.A., Hasan, T.: Bon—egocentric vision dataset for office activity recognition. https://dx.doi.org/10.21227/ehzr-w794(2021)
2. Ryoo, M.S., Matthies, L.: First-person activity recognition: what are they doing to me? In: Proceedings of IEEE CVPR, pp. 2730–2737 (2013)
3. Chen, H., Li, H., Song, A., Haberland, M., Akar, O., Dhillon, A., Zhou, T., Bertozzi, A.L., Brantingham, P.J.: Semi-supervised first-person activity recognition in body-worn video. *arXiv preprint* arXiv:1904.09062 (2019)
4. Tadesse, G.A., Cavallaro, A.: Visual features for ego-centric activity recognition: a survey. In: Proceedings of 4th ACM WearSys. ACM, pp. 48–53 (2018)
5. Roesner, F., Kohno, T., Molnar, D.: Security and privacy for augmented reality systems. Comm. ACM **57**(4), 88–96 (2014)
6. Sikder, N., Ahad, M.A.R., Nahid, A.-A.: Human action recognition based on a sequential deep learning model. In: 2021 Joint 10th International Conference on Informatics, Electronics & Vision (ICIEV) and 2021 5th International Conference on Imaging, Vision & Pattern Recognition (icIVPR). IEEE (2021), pp. 1–7
7. Abebe, G., Catala, A., Cavallaro, A.: A first-person vision dataset of office activities. In: Proceedings IAPR MPRSS-HCI, pp. 27–37. Springer (2018)
8. Sun, S., Kuang, Z., Sheng, L., Ouyang, W., Zhang, W.: Optical flow guided feature: a fast and robust motion representation for video action recognition. In: Proceeding IEEE CVPR, pp. 1390–1399 (2018)
9. Hochreiter, S., Schmidhuber, J.: Long short-term memory. Neural Comput. **9**(8), 1735–1780 (1997)
10. Donahue, J., Anne Hendricks, L., Guadarrama, S., Rohrbach, M., Venugopalan, S., Saenko, K., Darrell, T.: Long-term recurrent convolutional networks for visual recognition and description. In: Proceedings of IEEE CVPR, pp. 2625–2634 (2015)
11. Kahani, R., Talebpour, A., Mahmoudi-Aznaveh, A.: A correlation based feature representation for first-person activity recognition. Multimedia Tools Appl. **78**(15), 21 673–21 694 (2019)
12. Ji, S., Xu, W., Yang, M., Yu, K.: 3D convolutional neural networks for human action recognition. IEEE Trans. Pattern Anal. **35**(1), 221–231 (2012)
13. Carreira, J., Zisserman, A.: Quo vadis, action recognition? A new model and the kinetics dataset. In: Proceedings of IEEE CVPR, pp. 6299–6308 (2017)
14. Yu, J., Zhang, B., Kuang, Z., Lin, D., Fan, J.: iprivacy: image privacy protection by identifying sensitive objects via deep multi-task learning. IEEE Trans. Inf. Forensics Secur. **12**(5), 1005–1016 (2016)
15. Çiftçi, S., Akyüz, A.O., Ebrahimi, T.: A reliable and reversible image privacy protection based on false colors. IEEE Ttrans. Multimedia **20**(1), 68–81 (2017)

16. Paruchuri, J., Cheung, S.-C., Hail, M.: Video data hiding for managing privacy information in surveillance systems. EURASIP J. Inf. Secur. **2009**, 1–18 (2009)
17. Frome, A., Cheung, G., Abdulkader, A., Zennaro, M., Wu, B., Bissacco, A., Adam, H., Neven, H., Vincent, L.: Large-scale privacy protection in google street view. In: IEEE 12th International Conference on Computer Vision. IEEE, pp. 2373–2380 (2009)
18. Liu, J., Zhang, R., Han, G., Sun, N., Kwong, S.: Video action recognition with visual privacy protection based on compressed sensing. J. Syst. Archit. **113**, 101882 (2021)
19. Marcu, G., Zhong, J., Swen, S.: Dynamic gamma correction, December 21 2006, US Patent App. 11/153,959
20. Zagoruyko, S., Komodakis, N.: Wide residual networks, *arXiv preprint* arXiv:1605.07146 (2016)
21. Xie, S., Girshick, R., DollÁir, P., Tu, Z., He, K.: Aggregated residual transformations for deep neural networks. In: Proceedings of IEEE CVPR, pp. 1492–1500 (2017)
22. Huang, G., Liu, Z., Van Der Maaten, L., Weinberger, K.Q.: Densely connected convolutional networks. In: Proceedings of IEEE CVPR, pp. 4700–4708 (2017)
23. Xu, W., Cheung, S.-C.S., Soares, N.: Affect-preserving privacy protection of video. In: Proceedings of ICIP. IEEE, pp. 158–162 (2015)
24. Szegedy, C., Ioffe, S., Vanhoucke, V., Alemi, A.A.: Inception-v4, inception-resnet and the impact of residual connections on learning. In: Proceeding of 31st AAAI (2017)
25. He, K., Gkioxari, G., Dollár, P., Girshick, R.: Mask r-cnn. In: Proceedings of IEEE ICCV, pp. 2961–2969 (2017)
26. Lim, J.J., Salakhutdinov, R., Torralba, A.: Transfer learning by borrowing examples for multiclass object detection. In: Proceedings of NIPS, pp. 118–126 (2011)
27. Romera-Paredes, B., Torr, P.H.S.: Recurrent instance segmentation. In: Proceedings of ECCV, pp. 312–329. Springer (2016)
28. Long, Y., Gong, Y., Xiao, Z., Liu, Q.: Accurate object localization in remote sensing images based on convolutional neural networks. IEEE Trans. Geosci. Remote **55**(5), 2486–2498 (2017)
29. Cinbis, R.G., Verbeek, J., Schmid, C.: Weakly supervised object localization with multi-fold multiple instance learning. IEEE Trans. Pattern Anal. **39**(1), 189–203 (2016)
30. Rosenfeld, A., Zemel, R., Tsotsos, J.K.: The elephant in the room. *arXiv preprint* arXiv:1808.03305 (2018)
31. He, K., Zhang, X., Ren, S., Sun, J.: Deep residual learning for image recognition. In: Proceedings of IEEE CVPR, pp. 770–778 (2016)
32. Lin, T.-Y., Maire, M., Belongie, S., Hays, J., Perona, P., Ramanan, D., Dollár, P., Zitnick, C.L.: Microsoft coco: common objects in context. In: Proceedings of ECCV, pp. 740–755. Springer (2014)
33. Kuznetsova, A., Rom, H., Alldrin, N., Uijlings, J., Krasin, I., Pont-Tuset, J., Kamali, S., Popov, S., Malloci, M., Kolesnikov, A., Duerig, T., et al.: The open images dataset v4: unified image classification, object detection, and visual relationship detection at scale, *arXiv preprint* arXiv:1811.00982 (2018)
34. Huang, J., Rathod, V., Sun, C., Zhu, M., Korattikara, A., Fathi, A., Fischer, I., Wojna, Z., Song, Y., Guadarrama, S., Murphy, K., et al.: Speed/accuracy trade-offs for modern convolutional object detectors. In: Proceedings of IEEE CVPR, pp. 7310–7311 (2017)
35. Ullah, A., Ahmad, J., Muhammad, K., Sajjad, M., Baik, S.W.: Action recognition in video sequences using deep bi-directional LSTM with CNN features. IEEE Access **6**, 1155–1166 (2017)
36. Krizhevsky, A., Sutskever, I., Hinton, G.E.: Imagenet classification with deep convolutional neural networks. In: Proceedings of NIPS, pp. 1097–1105 (2012)
37. Szegedy, C., Vanhoucke, V., Ioffe, S., Shlens, J., Wojna, Z.: Rethinking the inception architecture for computer vision. In: Proceedings of IEEE CVPR, pp. 2818–2826 (2016)
38. Deng, J., Dong, W., Socher, R., Li, L.J., Li, K., Fei-Fei, L.: Imagenet: a large-scale hierarchical image database. In: Proceedings of IEEE CVPR. IEEE, pp. 248–255 (2009)
39. Huang, Z., Xu, W., Yu, K.: Bidirectional LSTM-CRF models for sequence tagging, *arXiv preprint* arXiv:1508.01991 (2015)

40. Kingma, D.P., Ba, J.: Adam: a method for stochastic optimization, *arXiv preprint* arXiv:1412.6980 (2014)
41. Tadesse, G.A., Bent, O., Marcenaro, L., Weldemariam, K., Cavallaro, A.: Privacy-aware human activity recognition from a wearable camera: highlights from the IEEE video and image processing cup 2019 student competition [sp competitions]. IEEE Signal Process. Mag. **37**(3), 168–172 (2020)

IOTA-Based Efficient and Reliable Scheme for Internet of Vehicles

Tamara Islam Meghla, Md Whaiduzzaman, Alistair Barros, Md. Julkar Nayeen Mahi, Mehdi Sookhak, Colin Fidge, and Rajkumar Buyya

Abstract Distributed Ledger Technology (DLT) has become popular with the creation of the Bitcoin Blockchain. However, it does not fully address issues such as scalability, transaction fees, and quantum security in application to the Internet of Vehicles (IoV). The IOTA Foundation has claimed to handle these issues: The Tangle. Directed Acyclic Graph (DAG)-based DLT named IoTA, encapsulating Tangles on the IoV environment. In this research, we demonstrate tip selection strategies to affect the IOTA tangle's structure and behavior in IoV. Our proposed scheme describes Tangle's implementation within the IoV use case scenario, providing a highly scalable and fee-less DLT mechanism. We use bundle constructing algorithms for making secured transactions among vehicles across the consensus process. This strategic platform strikes a good balance between the punishment of lazy or busy tips and managed to tackle the left-behind tips, reducing computational latency and approving higher inter-arrival data rates. Our findings show that the random data execution

T. I. Meghla
Faculty of Information Technology and Communication Sciences, Tampere University, Tampere, Finland

Md Whaiduzzaman · A. Barros
School of Information Systems, Queensland University of Technology, Brisbane, QLD 4000, Australia

Md Whaiduzzaman (✉) · Md. J. Nayeen Mahi
Institute of Information Technology, Jahangirnagar University, Savar, Bangladesh
e-mail: wzaman@juniv.edu

M. Sookhak
Department of Computer Science, Texas A&M University, Corpus Christ, TX 78412, USA

C. Fidge
School of Computer Science, Queensland University of Technology, Brisbane, QLD 4000, Australia

R. Buyya
School of Computing and Information Systems, University of Melbourne, Parkville, VIC 3010, Australia

S. Hossain et al. (eds.), *Proceedings of International Conference on Fourth Industrial Revolution and Beyond 2021*, Lecture Notes in Networks and Systems 437,
https://doi.org/10.1007/978-981-19-2445-3_26

rate (α) and the transaction data arrival rate (λ) are effective. Finally, our scheme shows an efficiency of 80% node execution and low latency-based reliable fee-less data delivery over the IoV network ends.

Keywords Internet of vehicles · Directed acyclic graph · Tangle · IOTA

1 Introduction

The rapid rise of the Interconnected IoT devices allows the flow of information through the Intelligent Transportation System (ITS), which has mobilized the Internet of Vehicle (IoV) ecosystem. The IoV ecosystem is a network of IoT-enabled vehicles integrating information into the network such as real-time traffic problems, vehicles' locations, distance-based incentives, increasing traffic, and even route guidance in terms of traffic efficiency. These data are gathered through the use of sensors and devices within vehicles. The increasing number of vehicles linked to the IoT has turned the IoV ecosystem into an interactive network of significance with the continued evolution of communication and computing technologies [1, 2].

The IoV leads to a massive increase in vehicle data collected through the IoV ecosystem integrating cloud computing services (see Fig. 1). Consequently, the overwhelming amount of data collected through IoV [3] can be outsourced to a centralized cloud. Over the next five years, global IP traffic is expected to increase fivefold, with monthly IP traffic reaching 31 gigabytes per capital in 2021. Therefore, the emerging technological pattern of the next wave of IoV is expected to be featured decentralization, distributed management, and distributed storage. However, as decentralized infrastructure is adopted, data and communication must have high-security and privacy-preserving requirements [4].

Blockchain platforms have gained much attention in IoV solutions recently for lifetime maintenance and control of ITS apps [5]. Blockchains are secure; however, their degree of security depends on the network's amount of hash power. The people rely on mining, and more massive mining machines impose a challenge that affects the network [6].

IOTA is a permissionless, open-source, distributed ledger, which uses "Tangle" as its underlying technology, and the DAG-based data structure is used to store the ledger state as Tangle has no chain, no blocks, and no miners either. In addition to the data structure, the IOTA also takes a different approach to consensus. A DAG consensus mechanism could be a way to overcome inefficiencies in the Blockchain. Since all parties decide on the longest chain and discard forking and branching, Blockchain has a limited throughput Transaction Per Second (TPS). In comparison, the Tangle allows the ultimate merger of different branches of DAG, leading to a rapid overall performance.

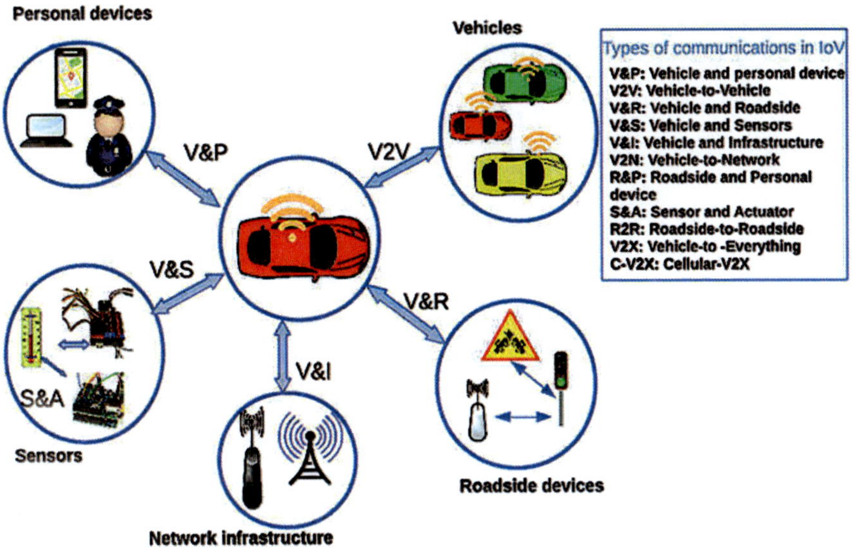

Fig. 1 Generic IoV ecosystem scenario

This research aims to introduce the IOTA concept in the IoV platform and explain its exciting features, incorporating the algorithmic approach to explain its step-by-step functionality and simulations that established the viability of IoTA in the IOV ecosystem. Therefore, we address these research objectives:

RO1: To propose an IOTA-based IoV scheme incorporating an algorithmic approach for fee-less transactions.

RO2: To establish efficient data transfer for IOTA in heterogeneous active vehicles within IoV network.

In this paper, we use the IOTA [7] to overcome the shortcomings of existing public blockchain technologies, focusing on tackling critical limitations such as transaction fees verification rates and the lack of scalability and sustainability. This research presents a fast, fee-less, secure, lightweight, and distributed ledger platform for vehicular networks. The IOTA decentralized ledger is structured as a Direct Acyclic Graph (DAG) called the Tangle—a novel strategy to distributed ledgers due to fee-less and quantum secure transactions and without theoretical scaling limitations. A miner is not required, and a zero transaction fee is possible in the IoV ecosystem. Nevertheless, IOTA's structure is promising, and several experimental studies have been done on the IOTA scenario. More precisely, the key contributions to this work are summarized as follows:

1. We present the IOTA concept for IoV to allow vehicles to connect and execute microtransactions in real-time without any fees.
2. We use DAG to store and verify network transactions for the IOTA consensus mechanism without any fees.

Table 1 Essential abbreviations

Short form	Full form or definition
AWRS	Anti-Withholding Reward System
BWH	Block Withholding
DAG	Direct Acyclic Graph
DLT	Distributed Ledger Technology
FAW	Fork-After-Withholding
IOTA	Open-source distributed ledger and cryptocurrency for Internet of Things
IoV	Internet of Vehicles
ITS	Intelligent Transport System
MWM	Minimum Weight Magnitude
PIR	Partial Initiative Release
PoS	Proof of Stake
PoW	Proof of Work

3. We present two algorithms to create transactions to approve different tip selection strategies.
4. We perform a transaction simulation scenario to analyze a tangled structure and confirm the viability of the new scheme.

The remainder of this paper is organized as follows: Sect. 2 presents recent works related to Distributed Ledger Technology for the IoV. Section 3 describes the essential information about the IOTA Tangle. Section 4 presents the research methodology. Section 5 describes our new scheme and algorithm. Section 6 provides simulation and results. Section 7 concludes the paper. Essential abbreviations are shown in Table 1.

2 Background and Related Work

In this section, we discuss the relevant background and related works to the state of the art of IOTA in IoV.

Christopher Natoli et al. proposed an evaluation framework [8] with insight into a system model, desired properties, and analyzed criteria. They explored the membership selection consensus and an integrated structure that influenced each other in existing systems. Roland Schmid and Roger Wattenhofer showed a Byzantine Fault Tolerance (BFT) [9] protocol to achieve an optimistic latency of just two message delays despite tolerating byzantine failures throughout the “fast track”. A generator block governed the transactions included in a new block; however, the block was not included in the block graph.

Hanqing et al. proposed an attacking model in a proof of work Blockchain that an internal and external attacker [10] consist of two phases. They demonstrated the original system became the two attackers' multi-attacker scheme: the internal and external attackers [11]. A semi-honest mining strategy in a multi-attacker system called the Partial Initiative Release (PIR) was used. Muoi Tranand and Inho Choi presented a stealthier Bitcoin attack—EREBUS attack. It allowed a network adversary to control a targeted Bitcoin node's peer connections for route manipulation without control-plane evidence of the attack. They used real bitcoin address messages from their live node to obtain highly accurate results.

Federico Matteo Bencic and Ivana Podnar Zarko analyzed distributed ledger technology to emphasize the features relevant to distributed systems. It showed the DAG [12] paradigm usage in the context of distributed ledgers and compared them with Blockchain-based solutions. They differentiate between the Blockchain and DAG consistently to optimize scalability, performance, and speed. Yixin Li and Bin Cao identified a DAG-based ledger and the corresponding consensus algorithm as a promising technology for IoT [13]. DAG framework solved certain limitations such as long confirmation delay, high transaction fee, high resource consumption, and low transaction throughput. They showed the trade-off between delay [14] in confirmation and security, which can act as a reference for the practical implementation of DAG [15].

Research in IOTA is emerging. This includes research article [16, 17]. The IOTA foundation provided the IOTA blogs, and developers have published several articles on features, specifications, and in technology [18–21] of IOTA.

3 IOTA: A Permission-Less Distributed Ledger

In this paper, we present the Tip selection Algorithm (TSA), which is biased toward the ones with larger cumulative weight. In the IOTA network, each node's weight is used to achieve a consensus ultimately. In the Tangle, the node's weight is proportional to how much effort the issuing node invests [22]. The acceptance of a new transaction indicates the validity of its history [23] and implies that every account has positive balances and ensures no double-spending or new illegal token making (see Fig. 2).

In a Tangle graph, the first transaction is called the genesis. In the beginning, all IOTA [24] tokens have been created, and no new ones will ever be made. All of the tangle transactions directly or indirectly approve of the genesis. When a transaction is accepted, it becomes part of a consensus and cannot be changed fundamentally. This is achieved by allowing each transaction to perform a small Proof of Work (PoW) calculation, and this PoW is a minor effort for each transaction, making it expensive to spam or fork the main Tangle after consensus has been established.

In Fig. 3, orange node depicts "unconfirmed transaction", green nodes depict "confirmed transaction', gray nodes depict the newest transaction referred to as "tip", and red nodes depict transaction deadlocks. Tips contain the newest transaction across

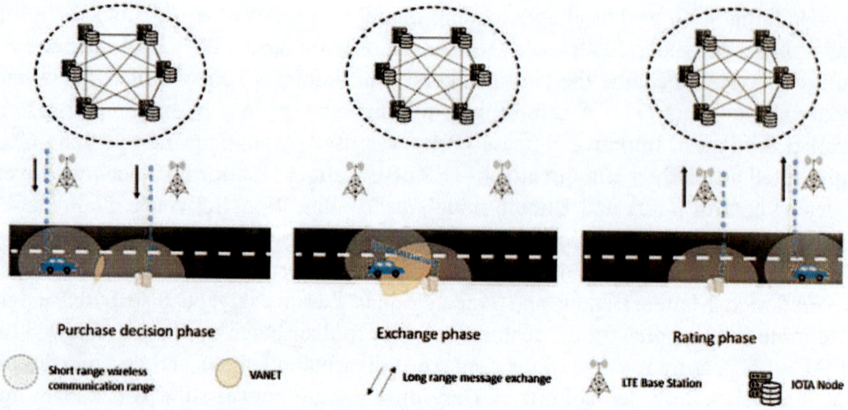

Fig. 2 IoTA cluster working pattern in IoV ecosystem [33]

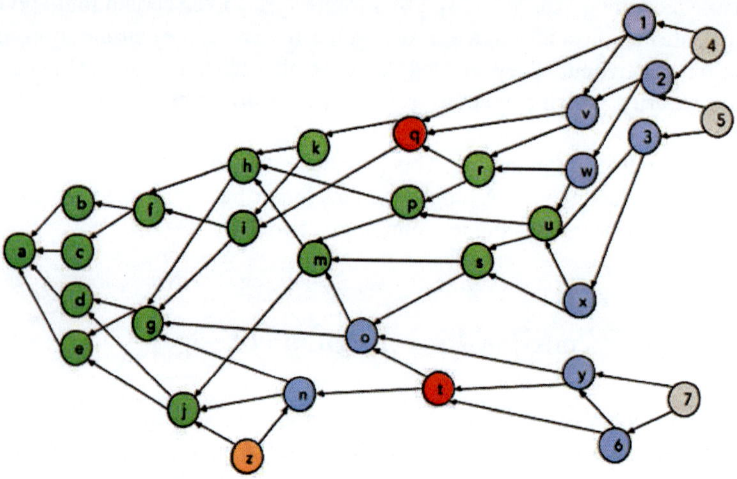

Fig. 3 A typical IoTA tangle communication

the network and have not been indicated by other transactions. Tips “4”,“5” and “7” contain the newest transaction just been added to the network. The IOTA architecture uses Trinary logic based on trits and trytes instead of binary logic [25]. The distributed IOTA ledger platform improves throughput and scalability, offers zero transaction fees, and scales with usability in the IoV ecosystem.

4 Research Methodology

Our solution uses IOTA technology to transfer a message from one vehicle to another vehicle to make a secure transaction over the insecure Internet. Transactions are broadcast via bundle which consists of multiple transactions. In this approach, vehicles must perform several steps for transaction processing.

Address generation and address validation are essential to realize why seeds, addresses, and address validation are required [26] when two vehicles want to share a data seed should be generated as randomly as possible because this acts like a password to a user's account [27]. The address is only the final result that comes from the Seed, Private Key, through digests. A node first creates a bundle to add a transaction to the Tangle. IOTA uses an Unspent Transaction Output (i.e., Utransaction) scheme, which means that inputs (i.e., addresses) are needed to transfer tokens from them to the outputs (i.e., addresses). Once all transactions have taken place, the list of the approved transactions should be attached to the bundle. After bundle construction, we attach a transaction to the Tangle. However, every new transaction must approve two previous non-confirmed transactions (i.e., tips), a process called validation. The node performs the Tip Selection Algorithm.

After tip selection, validation is completed. Verifying a site on the Tangle is a recursive process. A node must ensure that the transaction only references verified transactions and does not set a negative account balance for transaction verification. A valid transaction must include a hash [28] to fulfill the Hashcash requirements, which means that the PoW was done on this transaction. The last step before a node issues a transaction is performing a Proof of Work (PoW). Once the bundle is constructed and signed and the tips are added to the bundle, the PoW must be done for each transaction in the bundle. In each bundle transaction, we must fill the trunk and the branch and find a nonce (see Fig. 4). Every transaction in a bundle requires a nonce to be accepted by the Tangle network (see Fig. 3). The IOTA team designed

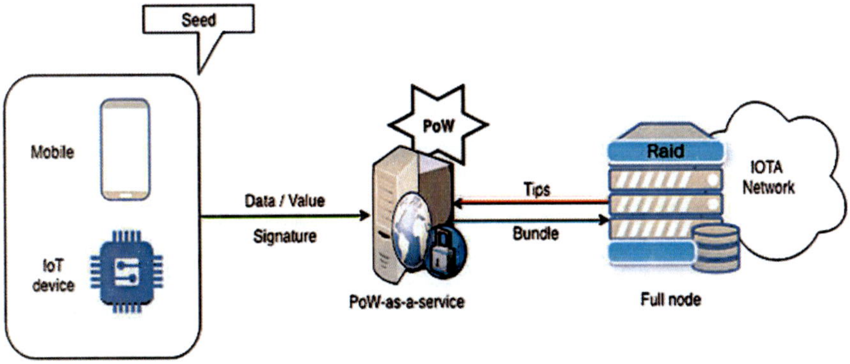

Fig. 4 Generic proof of work activity in IoTA

their cryptographic hash function called Curl to perform the PoW. After calculating a nonce, the modified bundles are broadcast all over the Tangle network.

5 New Proposed Scheme and Algorithm

IOTA tangle builds on the DAG framework concept as a substitute for Blockchain. For creating a transaction, a Bundle Construction Algorithm 1. is used, and the Tip Selection Algorithm 2. is used by the new transaction to select two-parent transactions to approve.

We have developed an IOTA-based Ledger Technology model. In Fig. 2, vehicles are connected and share their data [29] from vehicle to vehicle. Vehicles use a wide range of technology [30] to perceive their surroundings, including radar, sonar, GPS, lidar, and odometry. Our new model uses IOTA Tangle technology to secure and manage vehicular data [31], [32], microtransactions, and ride-sharing transactions in the IoV scenario.

We present a bundle construction algorithm (see Algorithm 1 and Table 2) to make a transaction among vehicles to vehicle-based on IOTA, and Tangle will be available for a bundle when a PoW operation has successfully verified all transactions. A seed is needed for address generation when a vehicle wants to communicate with another vehicle, and Tangle will validate this address. If the address is validated, they can create a transaction for further processing. The algorithm signs all transactions using a private key and bundle hash to gain a more secure transportation system.

Table 2 The notations are used in the algorithms

Symbol	Description
α	Random Data Execution Rate
λ	Data Arrival Rate
PB	Proposed Bundle
PT	Proposed Transaction
RNDEX	Random Data Execution
RgeApp	Return get Approvers
RNDEXM	Random Data Execution Module
RNDEXMS	Random Data Execution Module Scheduler
S	Key Balance Address Value (Seed)
TSA	Tip Selection Algorithm

Algorithm 1: Bundle Construction Algorithm

Input: Seed, Tag, Address value, Key, Proposed Transaction (PT), Proposed Bundle (PB)
Output: Bundle Construction from Seed S, Tag T, Address value A, PT and PB

```
foreach Transactions T append in Process Bundle sequence PB do
    Step 1: Actor: IoTA System Execution IoTASE
        Get input ← Seed S
        Balance Key Generation BG to appear in T ← Address A AND Tag t
        if S is Verified from valid A and t
        Balance Key input B ← index I AND Security Key K
        Proposed Transaction PT is Verified then
            PT ← true
            Security Range SR ← PB Compress in T AND
            PT Verification ← A, T, BG
            Crypt Signing Key is Verified ← Finalize PB AND PT
        else
            PT Not Verified ← Stop PT AND
            Check Verified S
        else
            Proposed Bundle Generator PBG ← true
            Compile Address Key of PBG AND Security Level SL of PBG
            Crypt Signing Key ← import From Signature Fragment
            Generator SFG Private Key Address PKA AND Security Level
            Transaction Hash SLTH import To ← SFG
            Compiled SFG ← PBG AND SLTH
    if PBG is Verified then
        Step 2: Actor: PBG Execution Layer PBGEL
            Execution Range ER ← PKA AND SL
            Newly Created PBG PBGNC ← SFG, PBG AND SL
            PBGNC ← Compiled Hash Trunk Generation HTG
            From tryte String St AND Tip tt
            Transaction Hash String HStr AND St import To ← HTG
    else
        Step 4: Actor: PBG tag Compilation PBGtC
            HTG Not Verified From ← T, PT, St
            PBGtC ← PBG Obsolete tag PBGOt
            Continue to PBGtC Verification
            Get Verified PBG From PBGEL AND Finalize PB, PT
```

After signing all transactions, we can consider them as a bundle. This bundle is validated by selecting tips from Tangle using a Tip Selection Algorithm (see Algorithm 2 and Table 2). The Proof of Work (PoW) validates this bundle again and broadcasts it to the Tangle environment.

Algorithm 2: Tip Selection Algorithm

```
Input: Number of Node Execution, Arrival Rate (λ), Random Data Execution Rate (α)
Output: Find out the tips from Tangle

foreach Tangle Generation GENTan AND Call Function CF From Tangle Tan do
    Step 1: Actor: Tip Selection Primary
        Generate Tangle GENTan ← genesis node declaration GENnodec
        GENnodec ← Tip Selection Call TSeC AND
        Return Links Node RLN
        if GENTan is Verified From valid RLN
        GENTan ← GENnodec, TSeC AND return Candidates rC
        then
            import Random Data Exec RNDEX To ← Return
            get Approvers RgeApp
            Generated RgeApp From ← RNDEX AND rC
            Generate Tangle Call GenTaC ← Finalize
            RNDEX Module RNDEXM
        else
            Step 2: Actor: Tip Selection Secondary
                RgeApp AND RNDEX Not Verified From ← RNDEXM
                Compile Tip Select Link Node TSLN
                Stop RNDEXM To Verify TSLN
                Reallocate Scheduler RNDEXM ← RNDEXMS
        else
            Step 3: Actor: Tip Selection Tertiary
                TSLN ← true
                Compile RNDEXM AND RNDEXMS
                Compress RNDEXMS ← Get Return Ancestor Set
                GetRAS AND RNDEX
                Complete RNDEXM AND RNDEXMS From ← Verified
                GENTan AND Not Tip Select NTSel
                Get Verified GENTan From ← Verified RgeApp,
                RNDEX, RNDEXM AND RNDEXMS
```

6 Simulation and Results

The core of the simulation is in ReactJs, a JavaScript library for building the Tangle for the tip selection while constructing the bundles. Furthermore, we used IoTA visualizer to visualize the comparison of different Tip Selection Algorithms and also compared the confirmation probability and confirmation delay with respect to the arrival rate (λ), random data execution rate (α), and vehicle node execution.

In the simulation, we designed three scenarios to track the algorithm's behavior and observe their performances concerning efficient data executions. Each scenario reveals necessary information for the analysis considering arrival rate, random data execution rate, and vehicle node execution. We define and identify three main considerable parameters throughout experiments that match our defined research objectives.

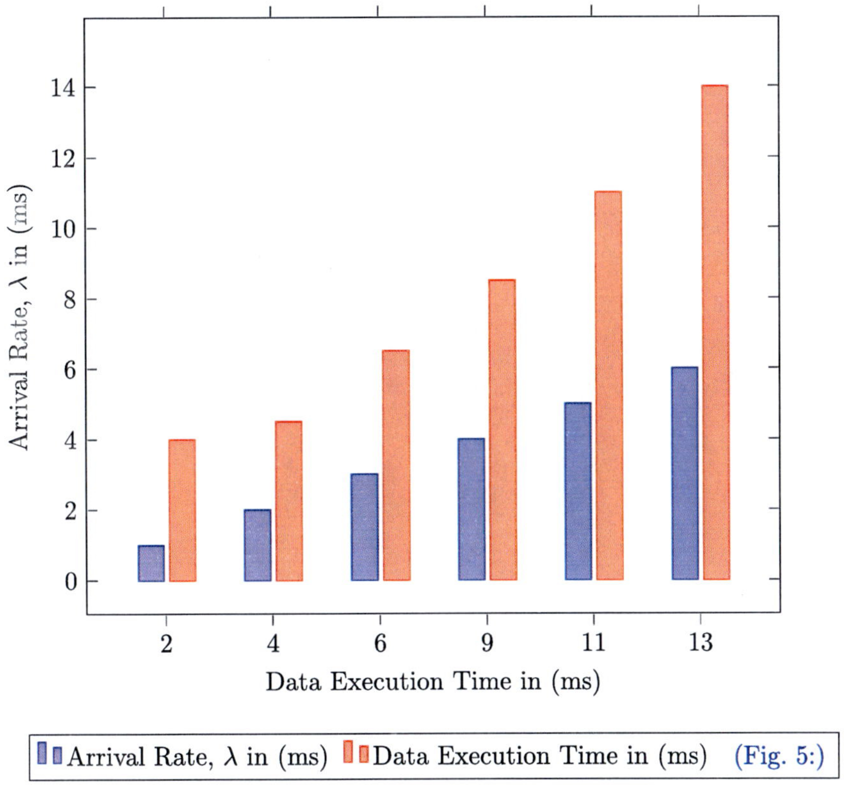

Fig. 5 Arrival rate (λ) vs. data execution time (Dt) in our proposed IoTA mechanism

In Fig. 5, we can see that the data execution time relies on the arrival rate. When the arrival rate increases, the data execution time goes double in accumulation. This means our procedure delivers the existing process (i.e., task) blocks while maintaining and considering newly created process stacks within algorithmic recursion. In 2 ms of process arrival rate, we get 4 ms of data execution time, and in 6 ms arrival rate, we get the data execution time of 14 ms. This result considerably supports our primary research objective of an efficient data delivery scheme where the existing processes are compiled with the new processes leaving no leftovers of process stacks or data offloads.

In Fig. 6, we can see that the random data execution rate relies on the data execution time. When the random data execution rate increases, the data execution time also increases in cumulative order. This means our procedure delivers the existing data handoff process (i.e., task offloads) blocks, inter-arrival process blocks, and existing processed data blocks. The random data execution rate considers and verifies the newly created or incoming process stack throughout the algorithmic recursive strategy. This recursive policy of the algorithm approach supports the reliability of our proposed scheme. In 50 ms of random data execution rate, we get 60 ms of data

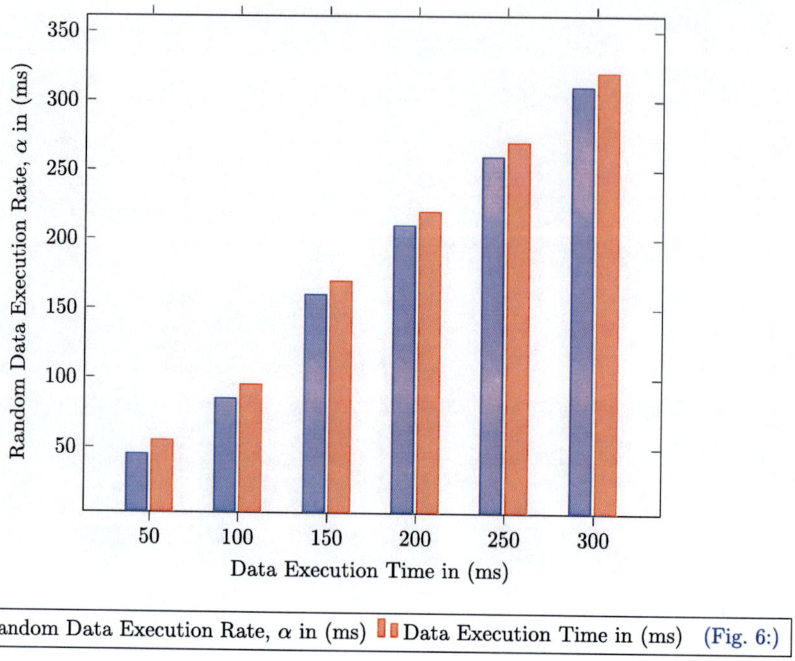

Fig. 6 Random data execution rate (α) vs. data execution time (Dt) in our proposed IoTA mechanism

execution time, and in 300 ms random data execution rate, we get the data execution time of 330 ms. This result considerably supports our primary research objective of an efficient and reliable data delivery scheme where the existing processes are compiled with the new processes providing fewer data offloads toward IoTA network occupied (i.e., task deliverable) nodes.

In Fig. 7, we can see that the vehicle node execution relies on the packet delivery time. When the vehicle node execution increases, the packet delivery time also increases in an augmenting order. This means our scheme delivers the existing and newly generated data packets (i.e., task-dependent stacks) toward the occupied vehicular nodes within the IoTA network. The inter-arrival task process blocks and existing task processed data blocks are accumulated while delivering to end-to-end IoTA network hops. The vehicle node execution considers and verifies newly created or incoming packet data throughout our developed scheme. This algorithmic approach supports the efficiency of 80% node execution rate within the IoTA scenario. In 20 ms of vehicle node execution, we get 50 ms of packet delivery time, and in 125 ms node execution rate, we get the packet delivery time of 150 ms. This result considerably supports our primary research objective of a reliable and low latency-based data delivery scheme. Moreover, the existing processes are compiled with the new processes delivering less delay over hop-to-hop data distribution toward the network occupied (i.e., task deliverable) nodes of the IoTA scenario.

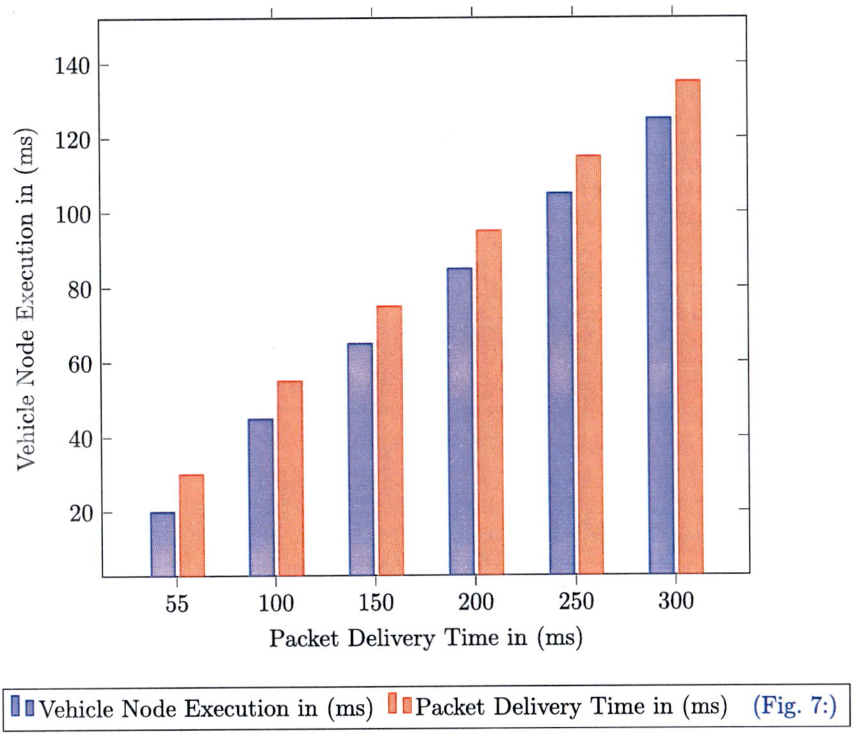

Fig. 7 Vehicle node execution (Vn) vs. packet delivery time (PDt) in our proposed IoTA mechanism

6.1 *Research Objectives*

In this section, we justify and establish our research objectives stated in this paper.

RO1: To propose an IOTA-based IoV scheme incorporating an algorithmic approach for fee-less transactions.

We answer the first objective by discussing the implementation overview algorithmic design and implementation based on practical scenarios. The fee-less distributed transaction property enhances the secured policies through a fast execution pattern. However, blockchain needs bidding and allocated transaction fees to every bit of block generation and distributed toward the intended users. The PoW allocated bidding process and transaction fee generation for each block, minimizing this extra latency; blockchain needs a congestion control strategy over moving vehicles, which is challenging. The IoTA successfully eliminates the network node congestion control problem (see Fig. 5). Since IoTA needs no transaction fee for each block generation, it distributes data via a secured channel toward the intended users in no time. The IoTA mechanism itself is aware of node congestion analogy and useful for high-speed moving entities (see Fig. 3). The vehicles within an IoV (i.e., Internet of vehicles) control area usually are not free from congestion awareness because of maintaining large network data execution. Partially or fully, data handoff occurs now and then in an IoV area since moving vehicles employ location awareness and RSU or BST allocation attribute problems. An additional delay mechanism (i.e., PoW-based sequential bidding for the transaction) may hamper the configured delay (i.e., optimized scheduled process to avoid extra delay) allocation toward the intended nodes as well as users within an IoV equipped area (see Figs. 6 and 2). For this reason, IoTA-based secured data transaction and distribution is effective in an IoV accumulated area.

RO2: To establish efficient data transfer for IOTA in heterogeneous active vehicles within IoV network.

We answer our second objective by experiment results and further discussion. In IoTA, the decentralized fashion of data distribution increases ends user activity, authenticity, and availability in a secured motion (see Fig. 3). Blockchain works as a sequential data distribution hierarchy, distribution, and transaction formulation across the network in a one-to-one pattern as a sequential order block. This causes extra network traffic and data execution delays which are not feasible for the frequent moving vehicle within the IoV network (see Fig. 2). For heterogeneous data delivery and execution toward the independent vehicle, we need a decentralized architecture of data execution and accession to minimize network congestion and data handoff delays (see Fig. 6). The IoTA property successfully alleviates the data accession (see Fig. 7) and verification and validation latencies (see Fig. 5) by deploying demand-based data transferring across IoV controlled area. Considering the data execution (i.e., cluster node data deliveries) around IoV (see Fig. 7), the IoTA architecture employs promising results. Thus, our proposed IoTA-based scheme is beneficial for efficient data transfer in the IoV network.

7 Conclusion

In this research, we investigated the possibility of IoTA technology, including Tangles in the IoV environment. We demonstrated the tip selection strategy to modify the data distribution structure module and node accession behavior across an IOTA tangle within IoV. Our newly configured algorithmic approach uses a modified tip selection and bundles construction of data block that deploys scheduled packet stacks efficiently toward the intended IoTA nodes. The implemented tip strategy scales the tangle module in an efficient data deliverable property toward the occupied IoTA nodes. The experimental result showed that our proposed scheme supports maximum data delivery, providing a low latency within higher inter-arrival rates. In the 6 ms arrival rate and 50 ms of random data execution rate (α), we get the data execution time of 14 and 60 ms, respectively. However, in the 125 ms node execution rate, we get the packet delivery time of 150 ms while supporting the efficiency of 80% node execution within the IoTA-occupied network. The overall result predominantly expressed our primary research objective of building a reliable and efficient low latency-based data delivery scheme within the IoV use case scenario.

Acknowledgements This research is partly supported through the Australian Research Council Discovery Project: DP190100314, "Re-Engineering Enterprise Systems for Microservices in the Cloud".

References

1. Whaiduzzaman, M., Sookhak, M., Gani, A., Buyya, R.: A survey on vehicular cloud computing. J. Netw. Comput. Appl. **40**, 325–344 (2014)
2. Whaiduzzaman, M., Barros, A., Shovon, A. R., Hossain, M.R., Fidge, C.: A Resilient Fog-IoT framework for seamless microservice execution. In: IEEE International Conference on Services Computing (SCC), pp. 213–221 (2021)
3. Zichichi, M., Ferretti, S., D'Angelo, G.: Are distributed ledger technologies ready for smart transportation systems? Preprint at arXiv:2001.09018 (2020)
4. Satu, M.S., Roy, S., Akhter, F., Whaiduzzaman, M.: IoLT: an IOT based collaborative blended learning platform in higher education. In: IEEE International Conference on Innovation in Engineering and technology (ICIET), pp. 1–6 (2018)
5. Whaiduzzaman, M., Mahi, M.J.N., Barros, A., Khalil, M.I., Fidge, C., Buyya, R.: BFIM: performance measurement of a blockchain based hierarchical tree layered Fog-IoT microservice architecture. IEEE Access (2021)
6. Cardellini, V., Presti, F.L., Nardelli, M., Russo, G.R.: Decentralized self-adaptation for elastic data stream processing. Futur. Gener. Comput. Syst. **87**, 171–185 (2018)
7. Sønstebø, D.: The transparency compendium. IOTA Blog (2017). June, 16
8. Natoli, C., Yu, J., Gramoli, V., Esteves-Verissimo, P.: Deconstructing blockchains: A comprehensive survey on consensus, membership and structure. arXiv preprint arXiv:1908.08316 (2019)
9. Schmid, R., Wattenhofer, R.: PermitBFT: Exploring the Byzantine Fast-Path. Preprint at arXiv:1906.10368 (2019)
10. Liu, H., Ruan, N., Liu, J.K.: Catfish effect between internal and external attackers: being semi-honest is Helpful. Preprint at arXiv:1907.03720 (2019)
11. Tran, M., Choi, I., Moon, G.J., Vu, A.V., Kang, M.S.: A stealthier partitioning attack against bitcoin peer-to-peer network. In: 2020 IEEE Symposium on Security and Privacy (SP), pp. 894–909. IEEE (2020)
12. Benić, F.M., Žarko, I.P.: Distributed ledger technology: blockchain compared to directed acyclic graph. In: 2018 IEEE 38th International Conference on Distributed Computing Systems (ICDCS), pp. 1569–1570. IEEE (2018)
13. Li, Y., Cao, B., Peng, M., Zhang, L., Zhang, L., Feng, D., Yu, J.: Direct acyclic graph-based ledger for Internet of Things: performance and security analysis. IEEE/ACM Trans. Netw. **28**(4), 1643–1656 (2020)
14. Cao, B., Li, Y., Zhang, L., Zhang, L., Mumtaz, S., Zhou, Z., Peng, M.: When Internet of Things meets blockchain: challenges in distributed consensus. IEEE Network **33**(6), 133–139 (2019)
15. Popov, S.: The Tangle, October 1, 2017 Version 1.3 (2017)
16. Popov, S., Saa, O., Finardi, P.: Equilibria in the Tangle. Comput. Ind. Eng. **136**, 160–172 (2019)
17. Fiedler, I.: Properties of a token-based machine economy (2017). Accessed 04 Feb 2018
18. Schiener, D.: A primer on IOTA (with presentation). IOTA Blog. May, 21 (2017)
19. Silvano, W.F., Marcelino, R.: Iota Tangle: A cryptocurrency to communicate Internet-of-Things data. Futur. Gener. Comput. Syst. **112**, 307–319 (2020)
20. Nakamoto, S.: Bitcoin: A peer-to-peer electronic cash system. Manubot. Tech, Rep (2019)
21. Elhabob, R., Zhao, Y., Sella, I., Xiong, H.: Efficient certificateless public key cryptography with equality test for internet of vehicles. IEEE Access **7**, 68957–68969 (2019)
22. Divya, M., Biradar, N.B.: IOTA-next generation block chain. Int. J. Eng. Comput. Sci **7**(04), 23823–23826 (2018)

23. Janeko, T., Zelinka, I.: Impact of security aspects at the IOTA protocol. In: International Conference on Intelligent Information Technologies for Industry, pp. 41–48. Springer, Cham (2018)
24. Park, J., Chitchyan, R., Angelopoulou, A., Murkin, J.: A block-free distributed ledger for p2p energy trading: Case with iota?. In: International Conference on Advanced Information Systems Engineering, pp. 111–125. Springer, Cham (2019)
25. Elsts, A., Mitskas, E., Oikonomou, G.: Distributed ledger technology and the internet of things: a feasibility study. In: Proceedings of the 1st Workshop on Blockchain-enabled Networked Sensor Systems, pp. 7–12 (2018)
26. Tuyisenge, L., Ayaida, M., Tohme, S., Afilal, L.E.: Network architectures in internet of vehicles (IoV): review, protocols analysis, challenges and issues. In: International Conference on Internet of Vehicles, pp. 3–13. Springer, Cham (2018)
27. Nasir, M.K., Whaiduzzaman, M.: Use of cell phone density for Intelligent Transportation System (ITS) in Bangladesh. Jahangirnagar Univ. J. Inf. Technol. **1**, 49–54 (2012)
28. Hossain, M.R., Whaiduzzaman, M., Barros, A., Tuly, S.R., Mahi, M.J.N., Roy, S., Fidge, C., Buyya, R.: A scheduling-based dynamic fog computing framework for augmenting resource utilization. Simul. Model. Pract. Theory **111**, 102336 (2021)
29. Whaiduzzaman, M., Tuly, S.R., Haque, N., Hossain, M., Barros, A.: Credit based task scheduling process management in fog computing (2020)
30. Brogan, J., Baskaran, I., Ramachandran, N.: Authenticating health activity data using distributed ledger technologies. Comput. Struct. Biotechnol. J. **16**, 257–266 (2018)
31. Whaiduzzaman, M., Oliullah, K., Mahi, M.J.N., Barros, A.: AUASF: an anonymous users authentication scheme for Fog-IoT environment. In: 2020 11th International Conference on Computing, Communication and Networking Technologies (ICCCNT), pp. 1–7. IEEE (2020)
32. Whaiduzzaman, M., Ismail Sumi, A., Barros, A., Satu, S., Hossain, M.: Towards latency aware emerging technology for internet of vehicles (2021)
33. Lücking, M., Kretzer, F., Kannengießer, N., Beigl, M., Sunyaev, A., Stork, W.: When data fly: an open data trading system in vehicular ad hoc networks. Electronics **10**(6), 654 (2021)

Emerging Applications

Pandemic Effect on Education System Among University Students

Farzana Yesmin, Md. Mosfikur Rahman, Mohd. Saifuzzaman, and Nazmun Nessa Moon

Abstract Online exam systems are pieces of software used in online education to evaluate students' performance. However, both the security of the system they respond to and a broad understanding of the risks they may face must be addressed. In this article, we will look into the most recent issue, which is online exams and their security systems. Because most students depended on Google, their friends, and other resources for help, we focused on software-based security and online support for cheating like IP address tracing, block local file accessing, etc. We did this project step by step like collecting data, data pre-processing, data cleaning, training data, and trying to make a prediction of how the student cheated the most. Then we can start our future work as we want to increase the security of online exams. According to the findings of a few study papers we read, individuals that use e-testing do so for all main kinds of assessments, including quizzes, midterms, and final exams. The most common format for these tests is multiple choice, which is certainly the easiest to cheat on, but the faculty admits they aren't routinely employing anti-cheating tactics. So, this research paper will help the new world in the field of online exam systems.

Keywords E-learning · Online exam · Machine learning · KNN · Naïve Bayes · Logistic regression

1 Introduction

Education has not been spared as technology has rapidly digitized all services and offerings, bringing them to the web platform. The pervasiveness of technology has

F. Yesmin (✉) · Md. M. Rahman · Mohd. Saifuzzaman · N. N. Moon
Daffodil International University, Dhaka, Bangladesh
e-mail: yesminfarzana.cse@gmail.com

Mohd. Saifuzzaman
e-mail: saifuzzaman.cse@diu.edu.bd

N. N. Moon
e-mail: moon@daffodilvarsity.edu.bd

S. Hossain et al. (eds.), *Proceedings of International Conference on Fourth Industrial Revolution and Beyond 2021*, Lecture Notes in Networks and Systems 437,
https://doi.org/10.1007/978-981-19-2445-3_27

facilitated a seamless shift to the online environment, thanks to ubiquitous high-speed internet and computers. Learning management systems, which are software applications that track, report, administer, and document materials shared with students, have advanced and adapted in universities, schools, and other educational institutions. This teaching and evaluation technique allows for a balanced approach to knowledge transmission as well as easy candidate grading. The goal of online exams is for assessment givers to be able to make the transition from offline to online processes. The online market is constantly expanding, and factors including convenience, scalability, expanded reach, and customization are influencing its adoption. Established evaluation methodologies are rapidly approaching saturation and will become obsolete in the near future. Exams are designed to fairly assess a student's ability. The exam results would be erroneous in representing their true competence if the exam's impartiality was compromised. Cheating raises various difficulties that we as an educational institution must be aware of, in addition to the incorrect assessment of students' learning progress. It's no longer necessary to scribble math formulas on your palms. Nowadays, students have devised ingenious techniques to cheat, most of which take advantage of the technology available to them. Sharing student's screen with another computer, utilizing cutting edge technological devices, using mobile apps and taking notes on a smartphone, creating false identities in order to obtain third-party assistance, etc. Exam proctoring is the most prevalent solution. It is specifically developed to allow students to take online tests safely, with features like automated monitoring, ID verification, and machine lockdown, as well as IP address tracing and blocking local file access. This option, however, may be fraught with technical difficulties and significant expenditures. But, in order to discover a solution, we must first determine which elements are appropriate for pupils to cheat on. In this study work, my team and I used machine learning to forecast how students cheated the most and how they could improve their grades. We use three different sorts of algorithms in our project. To accurately train our data, we used logistic regression, KNN, and Naive Bayes. We test 20% of the data first, then train the remaining data our project's core theme is fairly straightforward. The title encapsulates the entire viewpoint. Though it is a recent issue, we need to enhance the security mechanism of online tests. Students employ a variety of strategies to pass. Students, on the other hand, constitute a country's backbone. As a result, we must give it considerable consideration. As a result, our initiative first identifies a problem before attempting to address it. As previously stated, it is a new trend, so just a few people are beginning to work with it. As a result, we were unable to locate any datasets suitable for our investigation. We manually collect data using a Google form. We put certain qualities there, and individuals entered data using these attribute questions. We gather approximately 1000 data points, transform them into numerical values, train them, and then write our research using the trained data. Better and safer online exams would be provided by a secure online examination system. The whole process and quality of the exam process can be hampered by online examination systems that lack sufficient information security protocols. Our future work has a high level of ambition. We create a low-cost, practical solution that will benefit all types of students and educational institutions.

2 Review Works

Giustolisi et al. (2013) aim to provide a research roadmap in the field. Safe, verifiable, and accountable e-exam systems have identified three separate study goals, each of which is led by particular research issues. They've just suggested a safe online examination method there. [1] Shraim (2019), despite the fact that higher education institutions in Palestine are increasingly using online exams, the perspectives of students on their use remain unexplored. As a result, the major focus of this study is to look into the online testing processes used by higher education institutions from the perspective of students who have taken online two exams at Palestine Technical University-Kadoorie (PTUK). Its findings will assist the institution in identifying essential and strategic components of effective e-exam design and administration to support students in higher education learning. [2] Muzaffar et al. (2021) develop a review procedure that entails six phases according to the standard SLR standards. The first two phases are already addressed in the introduction to the article, and the next four phases are detailed in the accompanying segments. [3] Sarrayrih (2016) creates a completely automated Online Exam System that can gather, combine, and analyze data while also evaluating the program's impact. The system will only be accessible to authorized users with certain duties (Administrator: maintains the website, Institutes: register to administer examinations, Students: give exams online). [4] Ramu and Arivoli (2013) state because of the high stakes involved in online exams, reliable student authentication is highly important. Authentication verifies that the users are who they claim to be. Authentication is used to verify the identity of online students and is a critical component of security in an online test setting. [5]. Alotaibi (2010) to protect templates against fraudulent use, recommended utilizing a distorted signal or feature vector. Using data hiding and watermarking techniques, fingerprint images can be made more secure by detecting changes and hiding one biometric within another. To protect against intruders, the scanned fingerprints will be stored in encrypted form. However, the time between fingerprint scans may be so short that it creates storage issues for such a large amount of data. As a result, the time interval can be increased to at least three seconds, allowing for a reasonable reduction in data volume. [6] Feinman (2018) according to this research article, a quasi-experimental study was conducted with the primary objective of determining the effectiveness of nonbiometric security measures. Using a taxonomy of cheating reduction approaches based on the fraud triangle hypothesis, the security measures were selected for use. According to this study, the fact that the number of students who replied to all test items was higher on the proctored tests than on the unproctored exams may imply that students tend to skip more questions in unproctored contexts than they do in proctored environments Despite this, the sample was varied in terms of majors and included typical community college students. In addition, the sample of students in the study was not drawn at random from the population of all community college students who take introductory statistics in web-assisted environments. Whatever the case, the application of the security methods employed in the study may make it possible to assess student knowledge in a credible, low-cost, and convenient manner [7]. Apampa et al. (2010) investigation into the current user security paradigm and the vulnerabilities that it contains is the subject of this study. According to the authors, a paradigm for user security in summative electronic evaluations has been developed specifically for this situation. Data security, location security, software security, and the network security of summative e-assessments are some of the issues to be concerned about. Instead of categorizing all impersonation threats as Type A, Type B, or Type C, the authors of this research categorize them into three types: A, B, and C. The tracking module computes the difference between two video frames in order to determine the location and position of the learner on the track. Student risk levels are determined at various points during a test using information from the tracker, which is fed into a classifier module. The current model's identification and authentication security goals are insufficient in terms of protecting student information from impersonation attacks [8]. Lin et al. (2004) study analyses how tool creation can work in conjunction with security to provide a learning environment that is safe, efficient, and effective. They detailed different e-learning apps and categorized them according to their functions, which they then listed. They stated the security and privacy needs in terms of the situations and the types of data shared, based on this categorization. In addition, they planned to upgrade our e-learning systems in order to make them comply with the new distance education standard, known as the Sharable Content Object Reference Model (SCORM). They worked on student performance evaluation, Course Authoring and Student Assessment Application, Distributed Multimedia Presentation Application with Floor Control, Security and Privacy: Requirements and Solutions, Copyright Protection, and Floor Control Security, while at this location. [9] Weiner and Hurtz (2017) aim of the research seems to be to compare online remote proctoring of high-stakes licensure examinations to traditional test center proctoring of the same examinations in order to determine whether or not these alternative test administration conditions are equivalent or whether or not they have potential differences. The findings of the study suggested that examinee scores achieved under online kiosk-based proctoring conditions were psychometrically sound and equivalent to the identical exams taken under traditional test center proctoring circumstances. Furthermore, examinees assessed online kiosk proctored examinations well, and ratings of testing conditions were essentially uncorrelated with exam performance, implying that examinees' experiences under the various testing conditions had no impact on their performance on the exams. Globally speaking, the study's findings endorsed the use of a kiosk-based remote proctoring approach for high-stakes test delivery that was equivalent to using an onsite test center proctoring method. Despite this, there was no indication that remote proctoring had any effect on test results. [10] Al-Saleem and Ullah (2014); in this study, the authors present a variety of security concerns in computer-based and online testing. In addition, they investigated a variety of authentication techniques for examinee authentication. Furthermore, they suggested a novel authentication system that combines the username/password with palm-based biometrics. The proposed scheme will enhance the authentication process to an optimum level during computer-based or online-based testing. The video capturing technique should also be merged with the existing system, in order to make it more secure in the future. Also, the CATSIM simulation tool should be used to simulate the entire authentication and verification process. The suggested system should improve the authentication procedure during computer-based or online testing to an optimal level. The video capture technology should also be integrated into the existing system in the future to make it more secure. In addition, the CATSIM simulation tool should be utilized to model the whole authentication and verification process [11].

Jung and Yeom (2009)—This author shows how the SeCOnE system provides both secure online exam management and an e-monitoring method for the prevention and detection of cheating. The strategies described in this article for preventing and detecting cheating encompass cheating methods known for the online test procedure through computer or Internet, although they may not address all conceivable cheating methods. This paper is intended for tests conducted over the Internet at a defined time with a single problem set, but with no restrictions on testing location. The fact that the second can be applied to exams given at various periods is a strong feature. To prevent cheating during the test, the examiner should prepare as many problem sets as there are exam hours. One overhead expense for this system is the preparation of the equipment that should be used to monitor and authenticate the entities, such as webcams and microphones. Another overhead to consider is the net burden caused by monitoring data transport and storage, but this is not a big issue when data compression is utilized and more monitor servers are provided. They should look at the consequences of hostile entities in the system, as well as the procedures involved in dealing with failures to improve this paper. The author should do some more work for improving this paper for enhancing security for online exams using group cryptography [12]. Mathapatiat et al. (2017) This proposed technique increases the level of security for conducting online tests. The primary concept of this approach is the incorporation of live video and real tokens held by the user. This study demonstrates the viability of the results by testing a fully functional model. Through an evaluation of time, porn rate, and effort, it is demonstrated that the novel technique performance corresponds to the achievement level in traditional. The new approach is strong since it can survive any unfavorable circumstance of severe testing. The new method's performance is compatible with the standards and delivers steady, consistent results, making the new graphical password scheme reliable. When compared to previous approaches, the data show that the new methodology considerably enhances confrontation to attacks. Thus, they should improve their paper as well as system to meet the criteria of online examination systems. Finally, the new system secures and ensures the highest level of security during examinations. It can be improved by using compound character and image size [13]. Urosevic (2019)—The aims and research questions were justified and discussed throughout the literature review, and the findings were afterward compared and contrasted with the findings from the semi-structured interviews. The research topic is: How can institutions assure proper student identity when examinations are taken outside of school by using various authentication methods? was examined using a variety of authentication techniques Institutions must keep up with what is going on across the world, which is justified by the rapid advancement of technology and potential applications of machine learning and artificial intelligence in e-learning. As observed throughout the literature study, potential outcomes and online tests have previously been introduced; however, security safeguards and fully supervised services may be lacking. Students need test time to focus on what they're doing, not to switch off and deal with any issues with third-party monitoring providers, for example. Unethical behavior is a problem that has been handled at institutions where tests are held on-site and hence does not pose a greater risk while working online. They conducted an empirical study on a group of online students from local and overseas Universities. The result shows the impact of question type on usability, in particular the amount of time taken by the introduction of the proposed approach. They should conduct a post-experiment survey to collect student feedback on the proposed technique [14]. Lin et al. (2016)—A plethora of methods has been developed in this paper to anonymize relational data. However, preventing the leakage of sensitive information in transactional data is also a key concern. Algorithms suggested for anonymizing transactional data often result in significant information loss. In this work, they introduce PTA, a unique anonymization system composed of three modules. It anonymized transactional data with little information loss and is also highly fast when compared to state-of-the-art transactional data anonymization techniques. Significant tests should be conducted to evaluate the performance of the developed system against state-of-the-art techniques for anonymizing transactional data in terms of runtime and information loss. The outcome of this work is quite impressive and also it outperforms the existing handcrafted feature extraction methods. It is obvious that integration of different approaches will improve the performance of the whole system [15] within the open-source software development Moodle.

3 Research Methodology

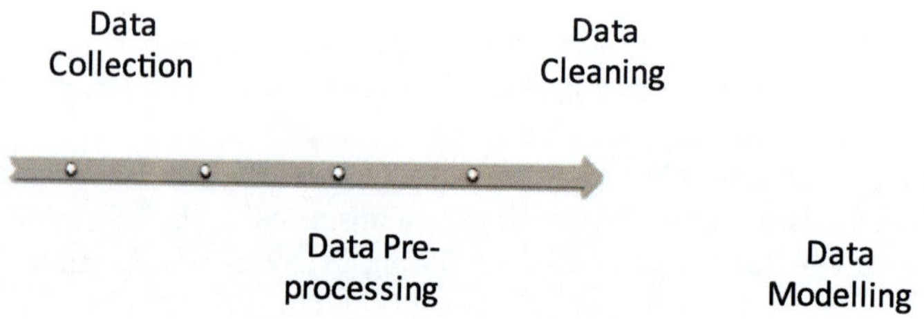

3.1 Data Collection

The authors are making use of the real-time dataset that was acquired through a Google form on their website. The form was distributed on social media, and from there, all of the necessary information was gathered. The dataset contains information about students' online exam experiences, including whether or not they received any online assistance. There were some questions about the dataset. Questions are:

- Your Grade in Online Examination System.
- Get help in exams from your classmates.
- Get help in exams from Google and YouTube.
- Friend circle grouping exam.
- Get help from any other source?

Answer to the first question was two; the first one was increased, and the second one was decreased, resulting in a total of two answers. The responses to the second question were given in percentages. The answer to the third question is the same as the answer to the second question. The fourth and fifth questions have two possible answers: yes and no. The final question had four responses, which were: electricity problems, network problems, device management problems, and others.

3.2 Data Pre-processing

Following the collection of the data, the dataset was discovered to be in Microsoft Excel format. For the purposes of this project, data was first converted into a CSV format. Many nominal values were discovered after the conversion process was complete. Those nominal values are manually processed and converted to numerical values by a computer program. These values were then translated into integer numbers using a string conversion algorithm. Take, for example, the following:

- In Result
  - Increase: 1
  - Decrease: 0
- Online Help and Friend Help
  - Percentage deviation from typical numerical values.
- Group Exam and other source
  - Yes: 1,
  - No: 0
- Problem During exam
  - Electricity Problem: 1
  - Net Problem: 2

- Device Management Problem: 3
- Others: 4

3.3 Data Cleaning

Following the cleaning of the data, we proceed to the identification of the dependent and independent variables. In this particular instance, we have chosen X and Y. Y is the dependent variable, whereas X is the independent variable in this situation. Following that, we selected 20% of the data to train on and 80% of the data to give accuracy to the model.

3.4 Data Modeling

(1) K-Nearest Neighbor: KNN is the abbreviation for K-Nearest Neighbors, which is a type of proximity search. A learning algorithm is used, and it is under the supervision of a trained individual. It establishes the distance between each row of training data and each row of test data using the distance function. By their respective values, arrange the distances determined in ascending order. The first k rows of a sorted array are returned by this function. In this row, find out which class is the most frequently seen. Should be returned as the projected class.

$$Pr(Y = j|X = x0) = 1k \sum i \in N0I(yi = j) \tag{1}$$

(2) Logistic Regression: Machine learning algorithm logistic regression can be used to predict if an observation belongs to one of two classes based on its characteristics. In financial analysis, linear regression is a machine learning technique based on supervised learning. It's in charge of carrying out a regression operation. These models use independent variables to predict the value of a prediction. I linear regression is used to predict the value of a dependent (y) variable given the value of an independent I (x).

$$Log(p1 - p/) \tag{2}$$

(3) Gaussian Naïve Bayes: A simple learning algorithm, Naive Bayes makes use of the Bayes rule in conjunction with a strong assumption that the attributes are conditionally independent, given the class in which they are used. It is possible to build a Bayesian classifier using a probabilistic model in which the classification is represented by a latent variable that is probabilistically related to the observed variables.

$$P(A|B) = P(B|A) \cdot P(A)P(B) \tag{3}$$

4 Result and Discussion

(1) Confusion Matrix: The values of the KNN, Logistic, and Naive Bayes Confusion Matrix are listed in Table 1. Here, the formula of this Confusion Matrix is True positive 3 (TP), True negative (TN), False positive (FP), and False negative (FN).

(2) Classification Report: A Classification report must be created to evaluate the accuracy of the predictions provided by a classification algorithm. Classification metrics precision, recall, and F1-score are displayed in a single report row for each class. Metrics like true positives and false negatives are factored into the formula for calculating performance. A comparison table has been created with both the decreased and raised values. The algorithm's precision is also taken into account.

(3) Classification Report: The machine learning model measures the accuracy with which a model finds correlations and patterns between variables in a dataset based on the input data, or training data. This accuracy chart illustrates the relative importance of KNN, Logistic, and Naive Bayes methods. And KNN had the highest Accuracy rating, earning a score of 95%.

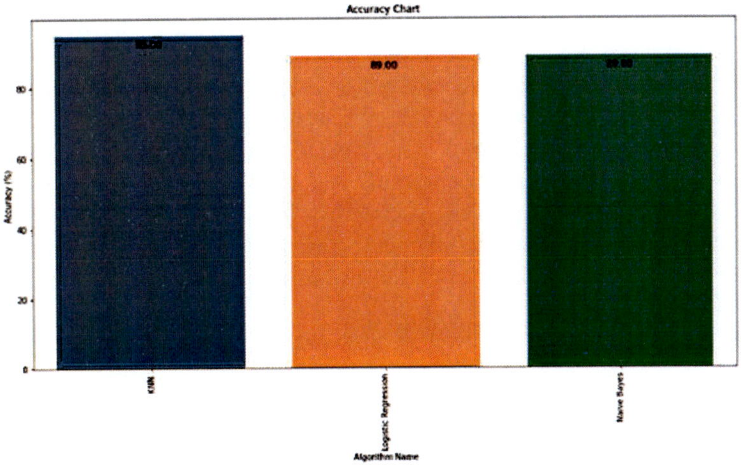

Table 1 Confusion matrix

Algorithm		Increased	Decreased
KNN	Increased	174	14
	Decreased	6	4
Logistic regression	Increased	178	0
	Decreased	22	0
Naïve Bayes	Increased	178	0
	Decreased	22	0

5 Comparative Analysis

An online examination system makes it possible to prepare and deliver exams entirely online. Because of Internet-based configuration options, this is a real possibility. This paper's goal is to make the online test system more secure. KNN, Logistic Regression, and Gaussian Naive Bayes were a few of the confusion matrices we used in our research. The classification report is summarized in Table 2.

On 95% of all occasions, the KNN matrix performed flawlessly.

There were two methods used to achieve the same level of accuracy: logistic regression and Gaussian Naive Bayes. The precision of the results used by different authors and researchers, on the other hand, has led to conflicting conclusions. Online exams can be made more secure by utilizing biometric authentication methods such as fingerprint recognition. Through the use of online testing methods, the objectivity of evaluations can be improved, while the effort required to make corrections is reduced. For new student orientation and registration, this is very useful when utilized in large classrooms (STEOP). Modern assessment appears to be essential for modern education at this time. The data for this report was gathered using a google form. Oftentimes, authors may gather their information from a variety of various websites all over the Internet. This leads me to believe that updating exam security using real data is preferable in this instance.

Students who use e-testing for quizzes, midterms, and exams of all kinds, according to research after research.

However, instructors report they do not regularly use procedures to avoid academic dishonesty despite the most common exam format being multiple choice, which makes cheating the easiest. The WebCT courseware platform offers numerous preventative measures for educators that are as easy to implement as using good judgment and selecting a single directive. If academic dishonesty is an issue, more should be done to prevent it for obvious reasons. There isn't a solution in adding more work to the faculty; rather, swift decisions about basic options and responsibilities must be made.

Table 2 Classification report

Algorithm	Class	Precision	Recall	F1-Score	Accuracy (%)
KNN	Decreased	0.80	0.73	0.76	95
	Increased	0.97	0.98	0.97	
Logistic regression	Decreased	0	0	0	89
	Increased	0.89	1	0.94	
Naïve Bayes	Decreased	0	0	0	89
	Increased	0.89	1	0.94	

6 Conclusion and Future Work

This paper explains how our online test system will be safer and more secure, as well as how students might improve their grades by cheating online. The measures presented in this work for preventing and detecting cheating cover cheating methods identified for online exam processes through computer or Internet, however, they may not cover all possible 95%, 89%, 89% Accuracy Chart KNN, Logistic Regression, Naïve Bayes four cheating methods. Our primary goal is to improve the examination atmosphere. We will take things one step at a time. To begin, this study will investigate how students cheated, as well as the difficulties they encountered during the online exam. Following that, we will address their concerns as well as the subject of cheating. As a result, the online testing method improved day by day. We concentrated on software-based security and their Internet assistance for cheating because most students relied on Google, their friends, and other resources for aid. That is why we concentrate mostly on that aspect. To classify our paper's intention, we employ KNN, Logistic Regression, and Naive Bayes algorithms in our coding section. In computer-based and online-based testing, we have looked into many security issues. We have also looked into a number of other authentication mechanisms for examinees. Finally, the lack of in-person supervision owing to geographical limits, along with accessible technologies, has made it easier for students to cheat on online tests. Knowing how students cheat in novel ways can help you find the proper answer and alleviate security concerns.

In the future, we intend to improve our projection by incorporating more and deeper records from our own database, which we are continuously building with increasingly complex data. We would like to incorporate more security solutions, as well as how to create them from a Bangladeshi standpoint. We currently do not have enough datasets; but, in the future, we will have a large dataset with many variables that will indicate how we should improve security in the online examination system. The consequences of hostile entities in the system, as well as the mechanisms involved in handling errors, will be the focus of future research.

Acknowledgements First and foremost, we would like to express our profound gratitude to Allah for His tremendous grace, which has enabled us to complete this work successfully. We also express our gratitude to my respectable course instructor. Because he is with us throughout our whole journey. Finally, we'd want to thank everyone who helped us, including well-wishers, friends, family, and elders. In addition to concepts. This study is the result of a great deal of hard work.

References

1. Giustolisi, R., Lenzini, G., Bella, G.: What security for electronic exams?. In: 2013 International Conference on Risks and Security of Internet and Systems (CRiSIS), pp. 1–5. IEEE
2. Shraim, K.: Online examination practices in higher education institutions: learners' perspectives. Turk. Online J. Distance Educ. **20**(4), 185–196 (2019)

3. Muzaffar, A.W., Tahir, M., Anwar, M.W., Chaudry, Q., Mir, S.R., Rasheed, Y.: A systematic review of online exams solutions in e-learning: techniques, tools, and global adoption. IEEE Access **9**, 32689–32712 (2021)
4. Sarrayrih, M.A.: Implementation and security development online exam, performances and problems for online university exam. Int. J. Comput. Sci. Inform. Secur. **14**(1), 24 (2016)
5. Ramu, T., Arivoli, T.: A framework of secure biometric based online exam authentication: an alternative to traditional exam. Int. J. Sci. Eng. Res. **4**(11), 52–60 (2013)
6. Alotaibi, S.: Using biometrics authentication via fingerprint recognition in e-exams in an e-learning environment (2010)
7. Feinman, Y.: Security mechanisms on web-based exams in introductory statistics community college courses. J. Soc. Behav. Health Sci. **12**(1), 11 (2018)
8. Apampa, K.M., Wills, G., Argles, D.: User security issues in summative e-assessment security. Int. J. Digit. Soc. (IJDS) **1**(2), 1–13 (2010)
9. Lin, N.H., Korba, L., Yee, G., Shih, T.K., Lin, H.W.: Security and privacy technologies for distance education applications. In: 18th International Conference on Advanced Information Networking and Applications, 2004. AINA 2004, vol. 1, pp. 580–585. IEEE (2004)
10. Weiner, J.A., Hurtz, G.M.: A comparative study of online remote proctored versus onsite proctored high-stakes exams. J. Appl. Test. Technol. **18**(1), 13–20 (2017)
11. Al-Saleem, S.M., Ullah, H.: Security Considerations and Recommendations in Computer-Based Testing, Hindawi Publishing Corporation (2014)
12. Jung, I.Y., Yeom, H.Y.: Enhanced security for online exams using group cryptography. In: IEEE Transactions on Education, pp. 340–349. IEEE (2009)
13. Mathapati, M., Senthil Kumaran, T., Krishna Kumara, A., Vinoth Kumar, S.: International Conference on Smart Technologies and Management for Computing, Communication, Controls, Energy, and Materials. IEEE (2017)
14. Urosevic, A.K.: Student authentication framework for online exams outside of school, Urosevic2019 (2019)
15. Lin, J.C.-W., Liu, Q., Fournier-Viger, P., Hong, T.-P.: PTA: An Efficient System for Transaction Database Anonymization. Institute of Electrical and Electronics Engineers Inc (2016)

Designing a Virtual Machine Placement Algorithm to Enhance Energy Efficiency in a Cloud Environment

Litan Islam, Md. Nasim Akhtar, Mostafijur Rahman, MD. Shakil, Rathindranath Baromon, and R. B. Ahmed

Abstract One of the most popular technologies of the present generation is Cloud Computing. Cloud data center operating costs are very high because they consume a large amount of electrical energy. Many kinds of research have been worked on Virtual Machine (VM) consolidations to optimize energy consumption in cloud environments. Energy efficiency is one of the major issues of a big data center, because the data center consumes a lot of energy to run and serve its clients. VM consolidation techniques are the best utilization of cloud data center resources and improvements to energy efficiency in the cloud environment. The time complexity VM placement of the existing algorithm is O(n*m), where n denotes the number of active physical hosts and m denotes the number of the selected VMs. Now, the time complexity VM placement of the proposed algorithm is O(n), where n denotes the number of active physical hosts. Our proposed system is saving on average 12% of energy from the existing systems. This energy efficiency is extremely useful in cloud data centers.

Keywords Cloud computing · Big data center · Virtual machine · Energy efficiency

L. Islam (✉) · Md. N. Akhtar · MD. Shakil · R. Baromon
Dhaka University of Engineering and Technology, Gazipur, Bangladesh
e-mail: litandpi2@gmail.com

Md. N. Akhtar
Chandpur Science and Technology University, Chandpur, Bangladesh

M. Rahman
Green University of Bangladesh, Dhaka, Bangladesh

R. B. Ahmed
University Malaysia Perlis, Perlis, Malaysia

S. Hossain et al. (eds.), *Proceedings of International Conference on Fourth Industrial Revolution and Beyond 2021*, Lecture Notes in Networks and Systems 437,
https://doi.org/10.1007/978-981-19-2445-3_28

1 Introduction

Cloud computing is now one accepting impressive consideration from both academic and industrial communities. The reason behind ever-growing applications and the economy of cloud computing. Nowadays, cloud computing has been gaining significant momentum. In the wide range, data centers are set up in different areas of the world [1]. Extensive utilization of energy by information centers is getting more unique providing unique facilities. From the research, it is found that worldwide around 1.5% of power is used only for information centers [2]. It is also notable that the control utilization depends on the physical host, which runs on the information center. In any case, based on the investigation [3], it was evaluated that around 15–20% of power is used for the normal utilization of host resources of the information centers. Information collected for the survey of green computing has shown that the vitality utilization by the physical host contributed around 60% to the generally fetched an information center [4]. Subsequently, it is basic to center on making strides in the utilization of information center assets.

One strategy to improve resource utilization and spare vitality is dynamic Virtual Machine (VM) consolidation. It broadly utilizes the resources in data centers. This approach intermittently reallocates VMs by utilizing live VM movement. For a data center, the VMs dynamically decrease the number of hosts which reduce the use of energy and make a profitable investment [5]. The VM technique ensures Quality of Service (QoS) and improves the energy efficiency in cloud data centers.

2 Related Work

A dynamic VM consolidation strategy was first introduced [6] for minimizing the energy cost of big data centers. By using VM live migrations, they developed an energy-saving strategy that consolidated the data center. Paper [7] classifies the dynamic VM consolidation into four parts: (1) host overload detection, (2) host underload detection, (3) VM selection, and (4) VM placement. Beloglazov et al. [8] proposed the energy-aware resource allocation heuristics method for efficient management of data centers for cloud computing. Hanini et al. [9] proposed the strategy of dynamic VM consolidation, an upper static CPU utilization threshold to decide when a host has overloaded. Farahnakian et al. [10] explored a CPU utilization expectation strategy based on the linear regression strategy to distinguish overloaded hosts. The authors [11, 12] improved energy efficiency and Quality of Service (QoS) to decrease the number of hosts and power, where a small number of virtual machine (VM) migrations happened.

Most of the researchers are focusing on dynamic VM consolidation, but still there are limitations, such as the number of virtual machine (VM) migrations. Generally, the VM consolidation process is performed using four steps. The host overload detection algorithm [11] is used to determine the overload condition of the host, and then migrate one or more VMs to the host. In the VM consolidation approach, one of the primary steps is to recognize the overloaded host. The goal of this algorithm is to decrease the possibility of host overload and improve the quality of service [13, 14]. The algorithm can be found in the reference [11].

The host underload detection algorithm [11] functions when a host is considered underloaded. In this instant, the VMs from the host need to migrate and cause the host to shut down. After finding an underloaded host, the host must be shut down for improving resource usage and save energy. This type of method is extremely energy efficient in the big data centers in the cloud environment. The algorithm can be found in the reference [11].

The VM selection algorithm [11] selects the VM that is migrated from the overloaded host. After discovering a set of overloaded hosts, some VMs must be migrated to those hosts. This migration ensures the QoS for the client. In research [11], we analyze the decrease of CPU capacity in an overloaded host. The VM selection algorithm selects the VM for the overloaded hosts. This algorithm decreases invalid VM migrations. For this reason, this algorithm decreases the loss of CPU capacity and increases energy efficiency. This approach ensures the guarantee of the QoS. The algorithm can be found in the reference [11]. The VM placement algorithm [11] finds the new hosts from the overloaded and underloaded hosts.

3 Methodology

Cloud computing services are based on the Infrastructure as a Service (IaaS) service model [25]. The amount of RAM, the network bandwidth, and millions of instructions per second are used to measure the performance of each host in the cloud environment.

3.1 VM Placement

The VM placement algorithm distributes the new VMs to the hosts. The VM selection algorithm selects the new VM to idle hosts. This algorithm increases the utilization of the resources and decreases the energy consumption of the big data center. The time complexity of VM placement of the existing algorithm [11] is O(n*m), where n is denoted by the number of active physical hosts and m is denoted by the number of the selected VMs. Our proposed VM placement algorithm is shown below.

Algorithm: VM Placement

```
Input: HostList Output: Migration_Map
foreach Host in HostList do
    Check if Host is in Overloaded then
        VMsToMigrate.add(getVMsToMigrateFromOverloadedHost(Host))
Migration_Map.add(getNewVMPlacement(VMsToMigrate))
```

```
VMsToMigrate.clear()
foreach Host in HostList do
    Check if Host is in underloaded then
        VMsToMigrate.add(Host.getVM_List())
            Migration_Map.add(getNewVMPlacement(VMsToMigrate))
return Migration_Map
```

In this algorithm, 1st for loop checks whether a physical host is overloaded or not. If the host is overloaded, then the VM selection algorithm selects the VMs that need to be migrated from the host. In 2nd for loop checks whether a host is underloaded or not. If the host is underloaded, then find a placement of the VMs from these hosts. Finally, the algorithm returns a Migration_Map. The Migration_Map contains detailed information on the new VM placement. Thus, the run time complexity of our proposed algorithm is O(n), where n is denoted by the number of hosts. VM placement process [15–17] is shown in Fig. 1.

For managing data centers, live VM migration [18–21] is one of the leading technologies, where selected VMs to migrate overloaded hosts. VM migration allows transferring a VM from one physical machine to another without disrupting the

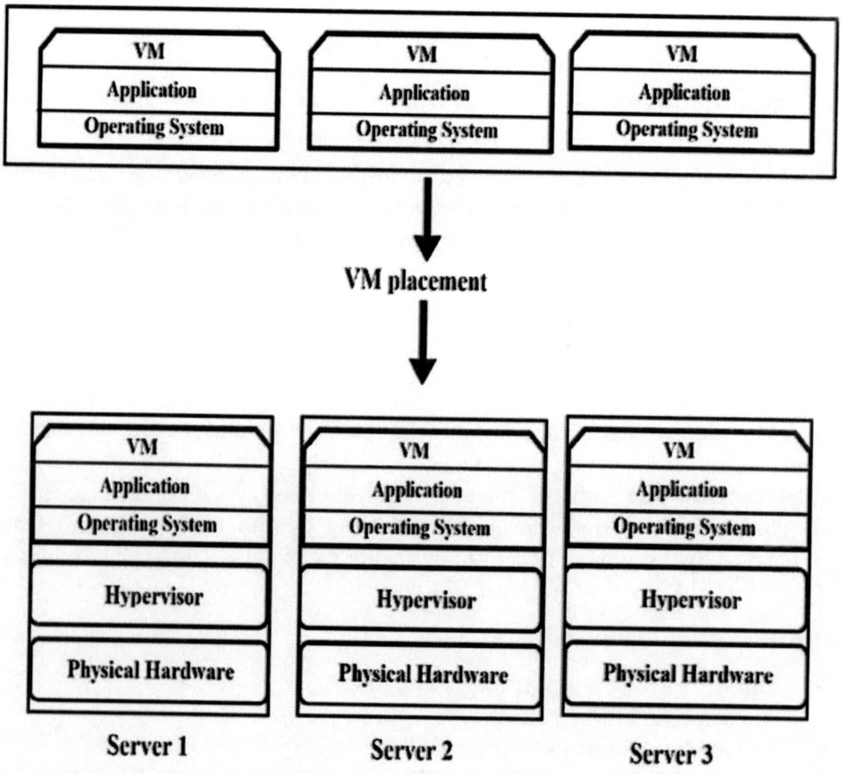

Fig. 1 VM placement

services. VM migration also achieves load balancing by alleviating loads from overloaded servers. VM live migration is additionally capable of managing data centers and optimizing procedures to improve the live migration process [22]. However, it minimizes the time of migration, which means achieving the best utilization of the network bandwidth.

3.2 VM Consolidation Process

VM consolidation process transfers the VM among active physical servers.VM consolidation is one of the key techniques in cloud resource management systems. This approach increases the energy efficiency of cloud data centers. The VM consolidation process can be divided into two steps [11]. In the first step, it identifies overloaded physical hosts using the host overload detection algorithm among the active physical hosts and migrates the VMs from the hosts. The VM selection algorithm selects VMs that need to be migrated to the overloaded hosts. In the second step, it recognizes underloaded physical hosts using the host underload detection algorithm. Then the VMs are allocated to the other hosts using the VM placement algorithm.

4 Results and Discussion

Simulation Results and Analysis: The workload used for this setup is 20110303 in which there are 1052 virtual machines deployed on 800 host machines. Simulation is compared with all the existing VM placement algorithms and proposed algorithms to show the comparative analysis.

Compare with Service Level Agreement Time per Active Host (SLATAH): Figure 2 shows the comparison of the Service Level Agreement Time per Active Host (SLATAH). THR-MMT-0.8 existing system SLATAH was shown 12.99% in Fig. 2 and our proposed system shown as 22.86% in Fig. 3, which improved SLATAH with our system was 9.87%. THR-MC-0.8 existing system SLATAH was shown 13.79% in Fig. 2 and our proposed system shown as 23.45% in Fig. 3, which improved SLATAH with our system was 9.66%. THR-RS-0.8 existing system SLATAH was shown 14.01% in Fig. 2 and our proposed system shown as 24.12% in Fig. 3, which improved SLATAH with our system was 10.11%. Figure 3 shows the proposed system SLATAH (%) to be better than the existing system.

Compare with Performance Degradation due to Migrations (PDM): THR-MMT-0.8 existing system PDM shows 23% in Fig. 4 and our proposed system shows 25% in Fig. 5, which improved PDM with our system as 2%. THR-MC-0.8 existing system PDM shows 27% in Fig. 4 and our proposed system shows 30% in Fig. 5, which improved PDM with our system as 3%. THR-RS-0.8 existing system PDM shows

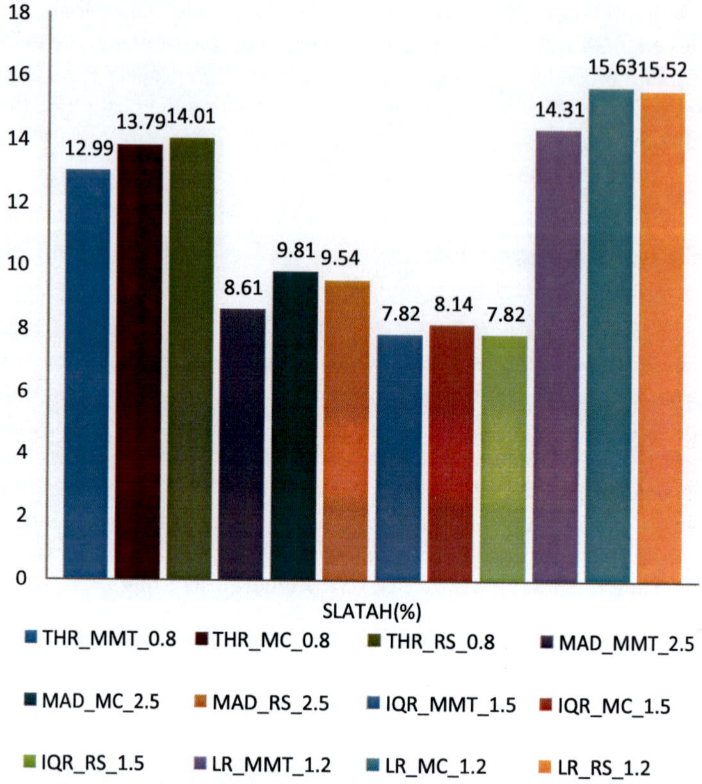

Fig. 2 SLATAH (%) in existing system

27% in Fig. 4 and our proposed system shown 30% in Fig. 5, which improved PDM with our system as 3%. Figure 5 shows the proposed system PDM (%) better most of the system than the existing system.

Compare with Mean Execution Time: THR-MMT-0.8 existing system's mean execution time shows 3.41 ms in Fig. 6 and our proposed system shows 2.93 ms in Fig. 7, which reduce the mean execution time with our system as 0.5 ms. THR-MC-0.8 existing system shows 5.38 ms in Fig. 6 and our proposed system shows 4.15 ms in Fig. 7, which reduce mean execution time with our system as 1.23 ms. THR-RS-0.8 existing system shows 3.99 ms Fig. 6 and our proposed system shows 2.33 millisecond Fig. 3, which reduce mean execution time with our system as 1.66 ms. Figure 7 shows our proposed system mean execution time is better for most of the system than the existing system.

Compare with Energy consumption: THR-MMT-0.8 existing system energy consumption shows 41.81 Kwh in Fig. 8 and our proposed system shows 34.37 Kwh in Fig. 9, which reduce the energy consumption with our system as 7.44 Kwh. THR-MC-0.8 existing system energy consumption shows 40.85 Kwh in Fig. 8 and our

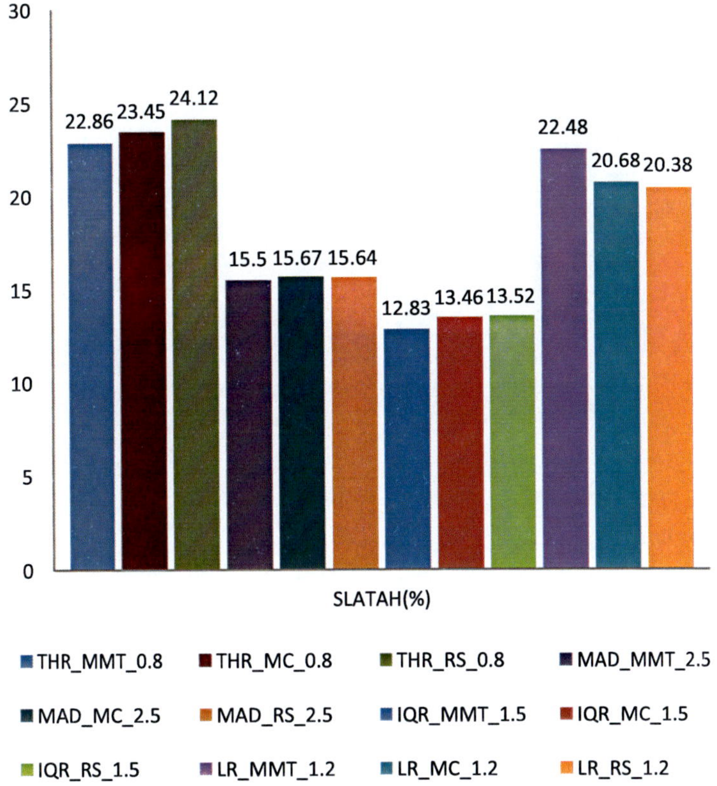

Fig. 3 SLATAH (%) in the proposed system

proposed system shows 33.94 Kwh in Fig. 9, which reduce the energy consumption with our system was 6.91 Kwh. THR-RS-0.8 existing system energy consumption shows 41.2 Kwh in Fig. 8 and our proposed system shows 34.34 Kwh in Fig. 9, which reduce mean execution time with our system as 6.86 Kwh. Figure 9 shows our proposed system is more energy-efficient than the existing system.

Our proposed system is saving on average 12% of energy from the existing different systems.

5 Conclusion and Future Works

A large amount of electrical energy is consumed by cloud data centers. The energy efficiency of the proposed system is compared with the existing system [11]. It is found that around 12% more energy efficiency can be obtained from the proposed system. The achievement source is to reduce the time complexity of the VM placement algorithm. The time complexity of VM placement is O(n), where n is the

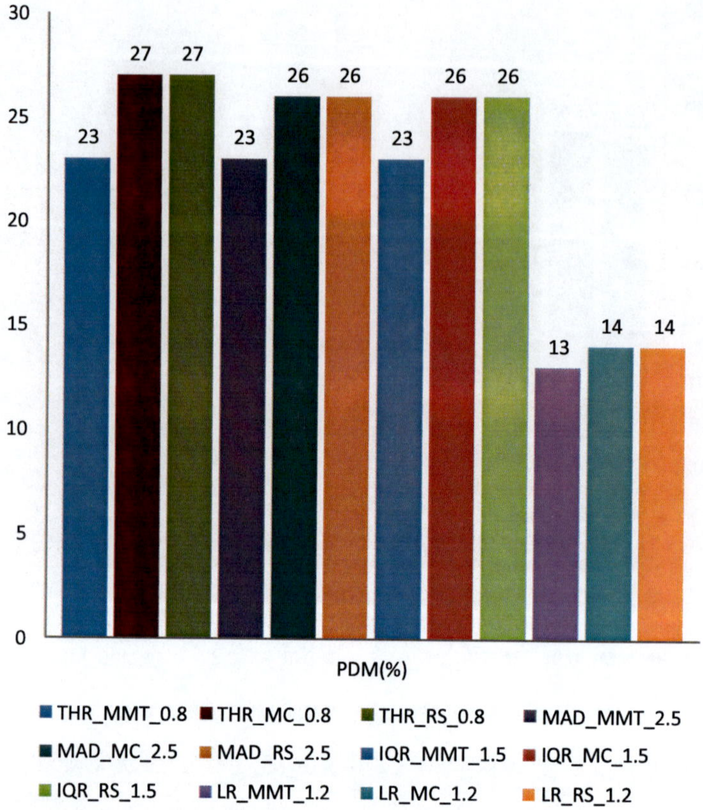

Fig. 4 PDM (%) in existing system

number of active physical hosts. This research can be extended by deploying the VM placement algorithm within the OpenStack cloud platform to evaluate the algorithm in a real system as a part of an energy-efficient dynamic VM consolidation.

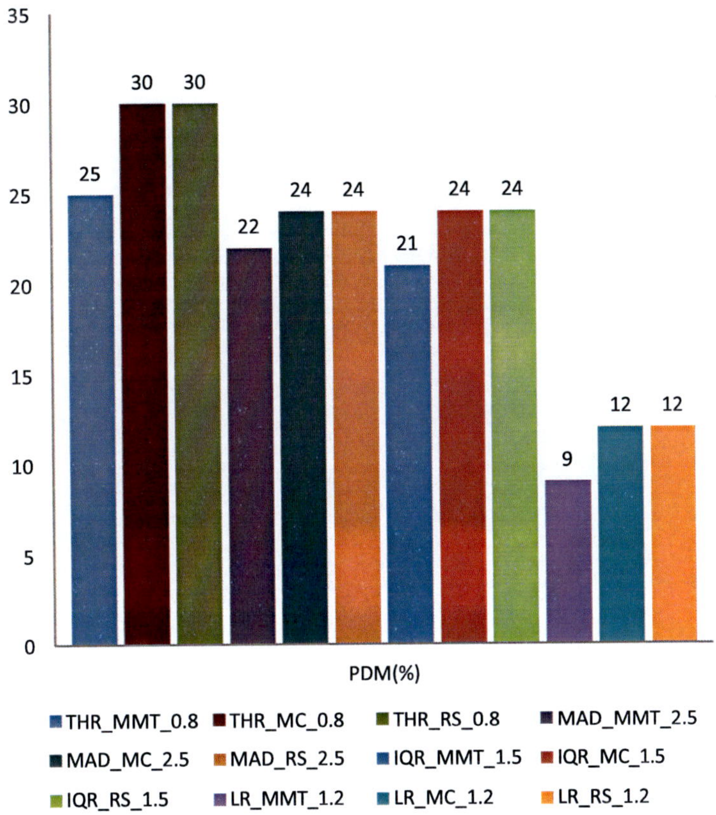

Fig. 5 PDM (%) in proposed system

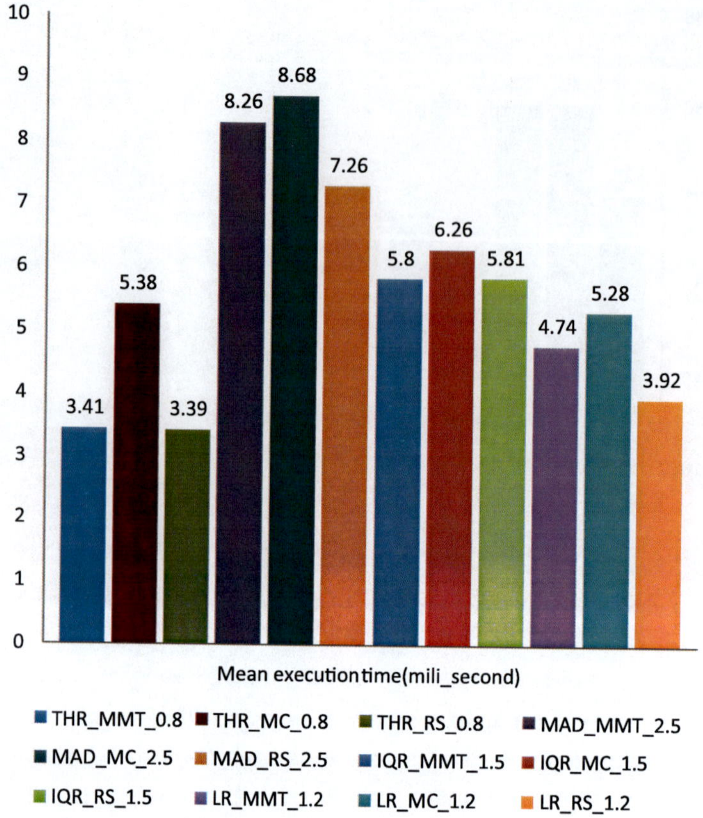

Fig. 6 Mean execution time (ms) in existing system

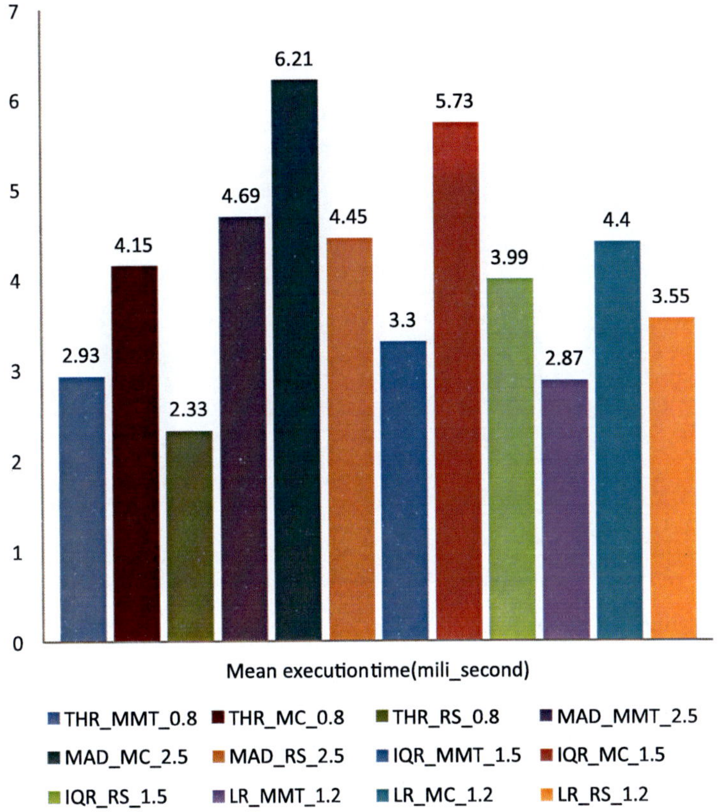

Fig. 7 Mean execution time (ms) in proposed system

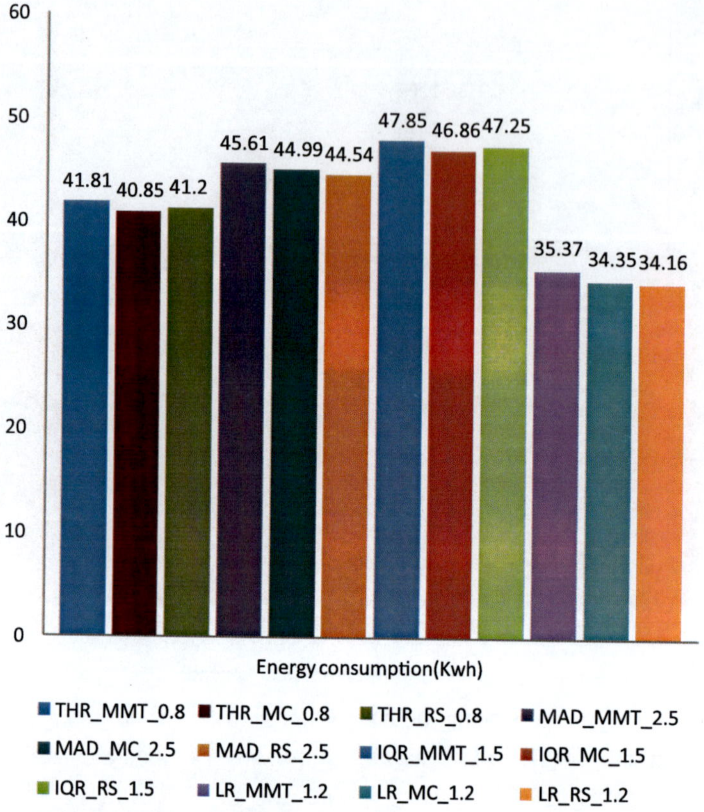

Fig. 8 Energy consumption in existing system

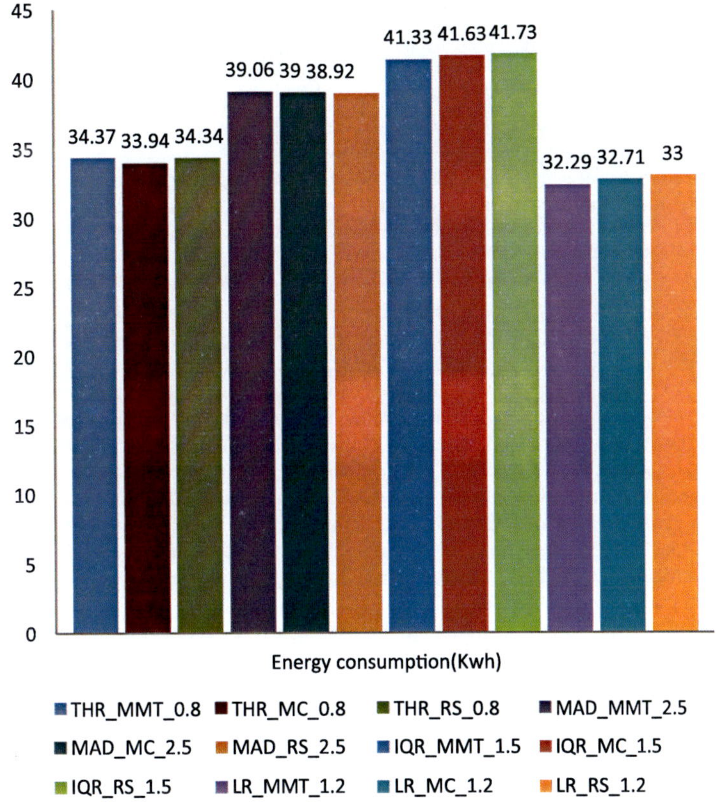

Fig. 9 Energy consumption in proposed system

References

1. Sadiku, M.N.O., Musa, S.M., Momoh, O.D.: Cloud computing: Opportunities and challenges. IEEE Potentials **33**(1), 34–36 (2014)
2. Koomey, J.: Growth in Data Center Electricity Use 2005 to 2010. Analytics Press, Oakland, CA, USA (2011). https://www.analyticspress.com/datacenters.html
3. De-Assunção, M.D., Gelas, J.-P., Lefèvre, L., Orgerie, A.-C.: The green grid'5000: Instrumenting and using a grid with energy sensors. Remote Instrum. eSci. Rel. Aspects **5931**, 25–42 (2011)
4. Sharma, S., Sharma, G.: A review on secure and energy efficient approaches for green computing. Int. J. Comput. Appl. **138**(11), 25–32 (2016)
5. Voorsluys, W., Broberg, J., Venugopal, S., Buyya, R.: Cost of virtual machine live migration in clouds: a performance evaluation. In: Proceedings of 1st International Conference on Cloud Computing, Beijing, China, pp. 254–265 (2009)
6. Nathuji, R., Schwan, K.: VirtualPower: coordinated power management in virtualized enterprise systems. ACM SIGOPS Oper. Syst. Rev. **41**(6), 265–278 (2007)
7. Beloglazov, A., Buyya, R.: Optimal online deterministic algorithms and adaptive heuristics for energy and performance efficient dynamic consolidation of virtual machines in cloud data centers. Concurr. Comput. Pract. Exp. **24**(13), 1397–1420 (2012)

8. Beloglazov, A., Abawajy, J., Buyya, R.: Energy-aware resource allocation heuristics for efficient management of data centers for cloud computing. Future Generat. Comput. Syst. **28**(5), 755–768 (2012)
9. Hanini, M., El Kafhali, S., Salah, K.: Dynamic VM allocation and traffic control to manage QoS and energy consumption in cloud computing environment. Int. J. Comput. Appl. Technol. **60**(4), 307–316 (2019)
10. Farahnakian, F., Liljeberg, P., Plosila, J.: LiRCUP: Linear regression-based CPU usage prediction algorithm for live migration of virtual machines in data centers. In: Proceedings of 39th IEEE EUROMICRO Conference on Software Engineering and Advanced Applications (SEAA), pp. 357–364 (2013)
11. Liu, Y., Sun, X., Wei, W., Jing, W.: Enhancing energy-efficient and QOS dynamic virtual machine consolidation method in cloud environment. IEEE Access 6, 31224–31235 (2018)
12. Yadav, R., Zhang, W., Li, K., Liu, C., Laghari, A.A.: Managing overloaded hosts for energy-efficiency in cloud data centers. Cluster Computing, pp. 1–15 (2021)
13. Zhang, Q., Zhani, M.F., Zhu, Q., Zhang, S., Boutaba, R., Hellerstein, J.L.: Dynamic energy-aware capacity provisioning for cloud computing environments. In: Proceedings of ACM International Conference on Autonomic Computing (ICAC), pp. 145–154 (2012)
14. Zhang, Q., Zhani, M.F., Boutaba, R., Hellerstein, J.L.: Dynamic heterogeneity-aware resource provisioning in the cloud. IEEE Trans. Cloud Comput. **2**(1), 14–28 (2014)
15. Gao, Y., Guan, H., Qi, Z., Hou, Y., Liu, L.: A multi-objective ant colony system algorithm for virtual machine placement in cloud computing. J. Comput. Syst. Sci. **79**(8), 1230–1242 (2013)
16. Fu, X., Zhou, C.: Predicted affinity based virtual machine placement in cloud computing environments. IEEE Trans. Cloud Comput. **8**(1), 246–255 (2017)
17. Nabavi, S.S., Gill, S.S., Xu, M., Masdari, M., Garraghan, P.: Tractor: Traffic-aware and power-efficient virtual machine placement in edge-cloud data centers using artificial bee colony optimization. Int. J. Commun. Syst. e4747 (2021)
18. Masdari, M., Khezri, H.: Efficient VM migrations using forecasting techniques in cloud computing: a comprehensive review. Cluster Comput. 1–30 (2020)
19. Li, C., Feng, D., Hua, Y., Qin, L.: Efficient live virtual machine migration for memory write-intensive workloads. Future Gener. Comput. Syst. **95**, 126–139 (2019)
20. Ibrahim, A., Noshy, M., Ali, H.A., Badawy, M.: Papso: A power-aware VM placement technique based on particle swarm optimization. IEEE Access 8, 81747–81764 (2020)
21. Kumar, A.V., Krishnakumar, V., Kumar, A.N.: Efficient performance upsurge in live migration with a downturn in the migration time and down-time. Cluster Comput. **22**(S5), 12737–12747 (2019)
22. Noshy, M., Ibrahim, A., Ali, H.A.: Optimization of live virtual machine migration in cloud computing: a survey and future directions. J. Netw. Comput. Appl. 110, 1–10 (2018)

A Comparative Analysis of AWS Cloud-Native Application Deployment Model

Khandakar Razoan Ahmed and Motaharul Islam

Abstract Application deployment in cloud computing has become a widely used system where we follow some specific models. For a long time, we have deployed our application in virtual machines. Nevertheless, nowadays the Docker container model is gaining popularity due to its important features. Therefore, we need to know which model is better in terms of request–response, and which modelis error-prone and gives better throughput. In our paper, we have done a comparative study between these two native cloud application deployment models. The main parameters of this study are HTTP response in a fixed time, error percentage, and throughput. We have used the Amazon EC2 service for the Virtual Machine model and Amazon ECS service for the Docker Container model in the implementation part and the Apache JMeter tool in the test part. In our study, we have found that the docker container has performed better in terms of total and average responses to the HTTP request and throughput.

Keywords Virtual machine · Docker container · Amazon EC2 · Amazon ECS · Cloud · AWS · Apache JMeter

1 Introduction

Cloud computing depends on shared computing resources through the internet instead of having local machines or servers. Cloud computing is the most ideal choice for storing, handling, and maintaining millions of data [1]. Now we can store and retrieve our data from anywhere at any time [2]. A cloud can be private or public. There are some service providers available who give us the public cloud service, for example,

K. R. Ahmed (✉)
Associate System Administrator Therap BD Limited, Dhaka, Bangladesh
e-mail: razoanahmed421@gmail.com

M. Islam
Department of Computer Science and Engineering, United International University, Dhaka, Bangladesh
e-mail: motaharul@cse.uiu.ac.bd

S. Hossain et al. (eds.), *Proceedings of International Conference on Fourth Industrial Revolution and Beyond 2021*, Lecture Notes in Networks and Systems 437,
https://doi.org/10.1007/978-981-19-2445-3_29

Amazon Web Services (AWS), Google Cloud Platform, and Microsoft Azure [3]. Among them, AWS is the better public cloud service provider in light of four advantages such as fault-tolerant, high scalability, resilience, and availability zone. Fault tolerance is the property that empowers a framework to keep on working appropriately in case of the disappointment of a portion of its segments. Scalability is the attribute of a system that depicts its capacity to adapt and perform well under an expanding workload; likewise will have the option to increment or lessen its level of performance according to the demands. Resilience is the ability to recoup from troublesome and harder conditions. AWS gives us the resources we need, as well as gives us the highest fault-tolerant advantage, high scalability, and best resilience facilities. AWS also has the most availability zone 48 with the 20 regions all over the world. So, we can choose any zone from any region where we want to deploy our application. Also, our data will be stored in another two data centers for backup, so under any misfortune or pandemic, our information will be protected.

AWS provides us with more than 90+ services like Amazon Elastic Compute Cloud (Amazon EC2), Amazon Elastic Container Service (Amazon ECS), AWS Lambda, Amazon Simple Storage Service (Amazon S3), AWS IoT Greengrass, AWS CloudFormation, virtual private cloud (VPC), AWS Identity and Access Management (IAM), etc. [4, 5]. And among these services, we have implemented some of them in our project. Amazon EC2 is a web service that gives us a secure and resizable process limit in the cloud [5]. EC2 is intended to make web-scale [6] distributed computing simpler for developers. It does not just give us the unlimited authority of computing resources yet, it lets us run on Amazon's proven computing environment. Amazon VPC gives us serious security features, for example, security groups and network access control lists, to empower inbound and outbound filtering at the instance and subnet level [5]. Amazon S3 is a scalable, rapid, web-based cloud storage service intended for online backup and archiving of information and applications on AWS. Additionally, it is an object storage administration that offers industry-driving scalability, data accessibility, security, and performance. ECS is an Amazon container administration service. AWS ECS is an exceptionally adaptable, super-quick, and fault-tolerant service help that makes it easy to create, launch, stop, and maintain our containers on the cluster [7]. We can run our tasks on a serverless infrastructure (managed by AWS Fargate) for full command over our infrastructure. We additionally can run our services on a cluster of EC2 instances that we can oversee.

To deploy applications, the cloud is the briskest, reliable, and useful approach. From the very beginning, people deploy applications on-premises basis. For quite a while, this was the main method of deployment. On-premises always require a greater expense in getting off the ground as hardware and installation are necessary for the housing of on-site servers. In addition, maintaining the infrastructure and wasting resources are some big disadvantages of on-premises [8]. To diminish these issues, cloud computing is very much helpful. Every year, many applications are deployed, so it is necessary to look for smooth computation for large data and the performance of those applications. To deploy any kind of application or software, we have four kinds of deployment models such as Public Cloud, Private Cloud, Community Cloud, and Hybrid Cloud. Among them, the public cloud is the best model [8]. Some of the advantages are unsophisticated installation and use, easy access to data, cost-effectiveness, continuous operation time, eliminated need for software, and simple admittance to data, cost-effectiveness, and scalability [8].

A virtual machine (VM) is an image file that is overseen by the hypervisor [1]. Hypervisor permits the computing resources (RAM, CPU, Memory, Storage, etc.) of the server to every single Virtual Machine [9], and keeps them isolated to avoid interference. VM represents the behavior of a separate machine and the capacity to perform tasks as a separate machine. In our proposed project, we have built the infrastructure for deploying our web application in the VM. AWS provides us the VM's services by the Amazon EC2 service and we have utilized the EC2 instance for the execution part.

Docker containers are primarily used to deploy our applications in microservice-based architecture [10]. We divide the whole service into small parts that we call microservices [11]. The benefits of Docker containers are fast speed, lightweight, compactness, quick conveyance, and scalability [3, 12]. Because of these advantages, it becomes attractive to develop containerized services nowadays. Docker container is less costly than virtual machine [10]. In a container environment, we virtualized the applications and executed them [13]. Containers are running in an isolated way on top of the operating system's kernel [12]. Docker container guarantees that our application is going to work consistently in every environment. In our proposed project, we are going to build the infrastructure for deploying our web application by using the Docker container model. AWS provides us with the Amazon ECS services and we have used Cluster and Task Definition of the Amazon ECS services for the implementation part [12]. The healthcare application we have deployed has been customized by us. This website is a healthcare website that we have named WeCare Hospital. We deployed the website using both models and analyzed their performances.

In our paper, we have proposed a comparative study between two native cloud application deployment models. Cloud-native application deployment implies where we build the whole infrastructure in the cloud with the help of services of cloud service providers like AWS. And afterward, we deploy our application in that infrastructure for quick and easy deployment, superfast performance, and simplicity to oversee. In this deployment, we follow some models like Virtual Machine, Docker Container, etc.

This paper aims to solve two challenges of the native cloud application deployment. First, our paper helps to present the accurate way of deploying applications in the Virtual Machine and Docker Container models which will also guide about deploying applications using Amazon EC2 and Amazon ECS services. Consequently, it will give proper knowledge about which model performs better in which performance test. There is a lot of work we have done to build up the infrastructure and deploy our application, but some major contributions are:

- We have used AWS services to implement this comparative study such as Amazon EC2, Amazon ECS, CloudFormation, Amazon S3, VPC, and IAM.

- To build the infrastructure in AWS, we have utilized the CloudFormation service by using the automation concept.
- The healthcare website template that we have deployed in the models is customized by us.
- We have used the Apache JMeter tool in the performance comparison test.
- We have composed a realistic performance test plan and divided it into three parts: Linear Load, Step Load, and Peak Load.

The rest of the paper is organized into different sections. Section 2 discusses the previous research works. In Sects. 3.1 and 3.2, we describe our system architecture. Sections 4.1 and 4.2, we implement and describe our realistic test plan using Apache JMeter. Section 5 is about experimental evaluations. Finally, the last section puts a conclusion to our research work.

2 Related Works

In [1], the authors have analyzed the performance of VM and containers with the same hardware and software configuration based on performance efficiency, security, storage, isolation, and networking.

In [3], a performance comparison has been shown between Linux containers and VM where the author gave the reasons why the container is more useful and attractive to software developers and infrastructure administrators. The later conducted a comparison on the basis of performance and scalability. In [10], the authors have described Docker System, Docker's applications, and advantages in cloud Computing through an application instance.

In [14], the authors proposed a comparative study between VM and Docker Containers. They discussed some basics of VM, Docker containers, virtualization techniques, and their advantages and disadvantage. Lastly, they compared them based on booting time, standardization, portability, security, low redundancy, resource distribution, etc.

In [13], the authors have operated a comparative study between Xen and Docker. They have used one extensive micro-benchmarks to compare virtualization upward of distinct elements on a host system between these two virtualization technologies. They have also added another two macro-benchmarks called HPL and YCSB, where HPL is for a case of high-performance computing applications and YCSB is for a case of online transaction processing applications, to understand the virtualization impact on real-world application.

In [12], although authors have pointed out that services deployed by AWS EC2 VMs performed better than services deployed by AWS ECS. They have conducted the performance test based on throughput, response time, and CPU utilization. However, in our paper, we have tried to conduct the performance test with a proper testing plan and strategy where we have conducted real-life scenario-based performance tests.

In [4], the authors presented a performance study to examine the proper way to utilize containers from different viewpoints. They have led some experiments to measure performance variations between application containers and system containers. Furthermore, they have inspected the service quality of ECS and Google Container Engine.

3 System Architecture

3.1 EC2-Based System Architecture

The EC2-based cloud-native application system architecture is introduced in Fig. 1. We have deployed a sample healthcare website application and analyzed this web application's performance by Apache JMeter from our local machine. We need to do this performance test through the Internet. So we have to set up Apache JMeter in our local machine and then we can send an HTTP request through the internet to our web address. Here we need Amazon Route 53 service, which is a highly accessible and scalable cloud DNS web service. Amazon Route 53 connects user requests to the infrastructure successfully. And these infrastructure run in AWS using services like Amazon EC2, Elastic load balancer, Amazon S3 etc. Amazon Route 53 service connects our HTTP request to our infrastructure so that we can access the hospital website. When our HTTP request tries to access the hospital application hosted in EC2 Webserver, HTTP requests are served through the Internet gateway attached to the VPC Network. Internet gateway sends the HTTP request to the EC2 instance or machine which is working as a web server.

In Fig. 1. the peripheral box signifies the AWS cloud. The immediate inner box signifies a region which implies that we are working in a specific region of the AWS cloud. We have deployed the EC2 instance in the closest location which is in Mumbai (India) and the short name of the region is ap-south-1, Mumbai. Here two availability zones mean that there are two data centers available for this region. And AWS has more than 48 availability zones to deploy our application. As Fig. 1 shows we have

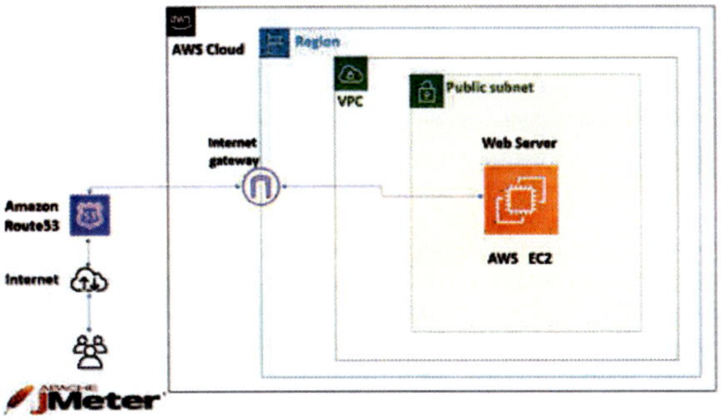

Fig. 1 EC2 system architecture

built a VPC network for the network requirements of the application. There is a public subnet in the VPC to isolate network traffics, and the EC2 box is placed in the public subnet. Our hospital website application runs on the Apache HTTPD service on our web server.

3.2 ECS-Based System Architecture

The ECS-based cloud-native application system architecture is introduced in Fig. 2. We have deployed our web application and analyzed the performance of the Apache JMeter from our local machine. The main difference between Figs. 1 and 2 is the ECS service in the EC2 infrastructure. Here, we have used the ECS service in the EC2-based infrastructure to implement the Docker container model.

To deploy our web application in the ECS service, first, we need an Amazon EC2 instance where we can create an ECS Cluster. Amazon ECS cluster is a logical gathering of tasks or services [7]. We can use one or more EC2 instances as container instances with the cluster to run the task definition. As we know cluster manages a group of services, so to prepare our application to run on ECS, we need to create a task definition which we can call a blueprint for our application. Task definition is a text file in JSON format. It creates one or more containers up to a certain limit, that scales the application. We can determine various parameters in the Task definitions for our application. The parameters that we need to determine are which containers to utilize, how much data volumes ought to be utilized, and which ports ought to be opened for the application. The particular parameters for the task definition rely upon the requirements of our application. So this cluster will manage the task definition that we have created and register the EC2 instance with the container where our webserver is running.

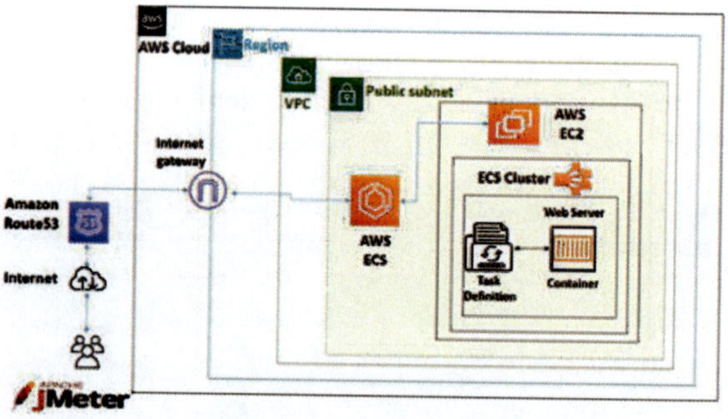

Fig. 2 ECS system architecture

Table 1 Environment and tools

AWS Services	AWS IAM
	Amazon ECS
	Amazon VPC
	Amazon EC2
	Cloudformation
Tools	Apache JMeter
	Visual Studio Code
Network performance	EC2: Low to Moderate
	ECS: Moderate
Instance type	EC2: T2.Nano: 1 vCPU and 500 MB
	ECS: T2.2xlarge: 8 vCPU and 32 GB

4 Implementation Details

4.1 Infrastructure Build

For the implementation part, we have utilized some AWS services. First of all, to construct the network part, we have used the VPC service where we have set the subnets, route tables, internet gateway, and elastic IPs. To store our data and files, we have utilized the S3 bucket where we need to create our bucket and store the data. We have used AWS IAM which is a web service. By IAM, we obtain secure control access to our AWS resources. The complete infrastructure is built using the automation concept and the automation is done by using an AWS service called CloudFormation. This service is considered an Infrastructure as a Code (IAAC) service. We have made a YML layout document for building the EC2 and ECS administration. In the format document, we have referenced and selected everything that we have to do manually for making these services. In table 1, we can see that we have chosen the T2.Nano type instance for EC2 service which consists of 1 vCPU and 500 MB Memory. For our project, this type of instance is a decent choice. But for the ECS service, we have chosen a T2.2xlarge type instance whichconsist of 8 vCPUs and 32 GB of Memory [7]. To support the microservices, we chose this instance and we restricted it to create one container consisting of 1 vCPU and 500 MB memory for a similar condition.

4.2 Performance Test

We have stored the healthcare website template in the S3 bucket and run it on the webserver. The website has been customized by us for the test to create a real-life scenario-based comparison. For the comparison test part, we have utilized Apache

JMeter [15]. Apache JMeter is an adaptable and smart tool where we can test the performance of both static and dynamic web applications. We expect to lead a wise and realistic test plan. Our customized website ran and tested on the webserver using both models. In the deployment part, we have selected the apsouth-1(Mumbai) region which is the closest to our destination. However, during the test period, we have selected the us-east-1(North Virginia) region in the JMeter with the goal that latency can be there and make this test realistic. To lead this test, we have to pick and select some particular ideas and plans [15]. We have set the test in three steps to take the test more realistic and real-life scenario-based. The three steps are Linear Load, Step Load, and Peak Load. Native JMeter does not support three complex load tests like this, so we have used a third-party plugin called Ultimate Thread Group. This plugin helps to set the thread group parameters and examine the system behavior after every new group of users is added. JMeter does not offer a native listener to identify the number of active virtual users for a thread group at every given point of time, so we have utilized another third-party plugin called Active Threads Over Time. It presents this number on some simple yet concise graphs.

In our test plan, the initial step is Linear Load where both of the models give their basic performance because this is one of the typical performance testings. Here we have chosen 500 threads which created 500 virtual users for the test. We have set the parameters that this test begins with 10 s of initial delay and then it increases from 1 to 500 in a few seconds. After that 500 threads hit consistently, we need to see how our web server performs. The load had been generated for 600 s. In Fig. 3, you can see the active threads over time where Jmeter is visualizing how the load is generating. Jmeter also collects the data that during the test time how many requests are being responded from our website in total, how many requests are being responded in every second,, how many seconds this website is taking to respond to the request, and how many errors are being created by the webserver. This Linear Load test runs in both models EC2 and ECS.

Secondly, we need to conduct a step load test. Here the initial stage is similar to the linear load test, but after a particular time, one more thread group is added. In this way, a total of 6 threads are added through a loop. For example, we have set the initial parameters as the initial stage is 500 threads send the HTTP request to our webservers in every second in 70 s afterward one more step is added where more

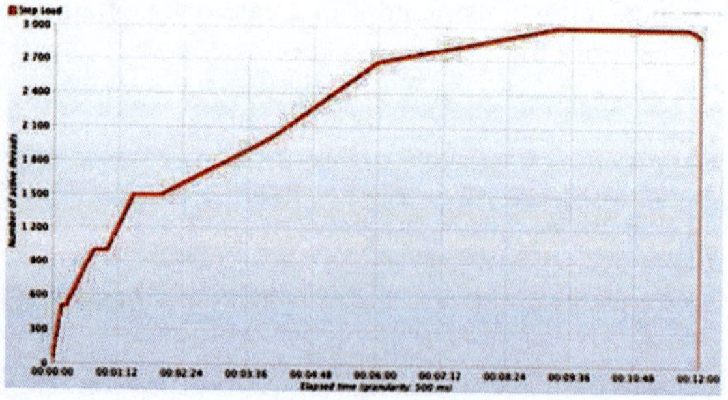

Fig. 3 Active threads over time EC2 step load

500 threads start to hit with 15 s initial delay and hit for 220 s. Another 500 threads start to hit after 30 s with the initial delay of 60 s and hit for 180 s. In this way, a total of 6 thread groups send the request to our website, so that in this Step load test, our web server has to respond to almost 3000 threads, and this increase from 500 step by step. We need to check our website for how it keeps pace and increases with the requirement. The ultimate Thread Group plugin helps to set the thread group parameters and examine the system behavior after every new group of users is added.

Lastly, we have utilized a short time intensive load test called Peak load. It creates an unusual environment to evaluate the reaction of the server. The server needs to respond to such extreme load jumps and we can see that in a normal load test when an extreme situation takes place, how our webserver faces it, and also how well it recovers from the traffic burst. Both models react differently, so we need to monitor their reactions. In the Peak load test, JMeter starts with the usual 500 threads with the 0 s initial delay. 500 users hit our website for 600 s and then suddenly the threads number jump from 500 to 3000 at a glance, then we perceive how the server responds in this situation, and we assemble the data in the meantime.

5 Experimental Evaluations

This section describes the way we have collected data. We have applied the Apache JMeter tool for this performance test. In this section, we also examine our data, and based on the data, we decide which models perform better.

In Fig. 3. we can see what it looks like when the step load is running. On the y-axis, the number of active threads is from 0 to 3000, and on the x-axis, the elapsed time is from 0 to 12 min. This graph shows how many threads are hitting per second to the server and how 500 threads group joins after a specific time. In this way, a total of 3000 threads hit the server every second and it continuously hits for 12 min. The whole step load test has worked according to our plan just like other load tests.

In Fig. 4, we can see the response time over time of the EC2 model in the Step Load test. In the graph, the x-axis presents elapsed time from 0 to 12 min and the y-axis presents response time in milliseconds. As the graph shows that EC2 performs

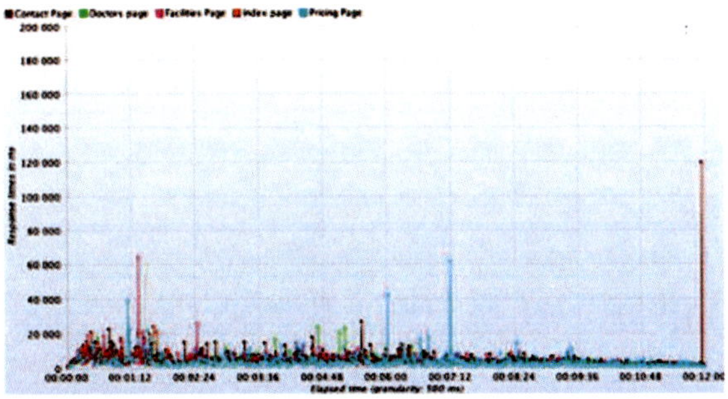

Fig. 4 Response time over time EC2 step load

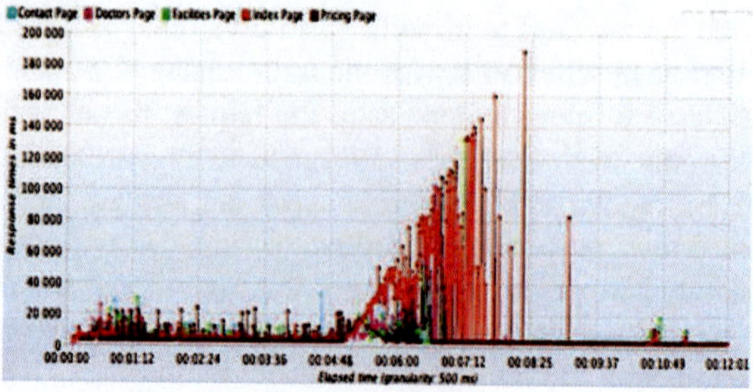

Fig. 5 Response time over time ECS step load

well in the step load test and it has taken an average of 20,000 ms, it means EC2 has taken an average of 2 s to respond to all the HTTP requests. Only in some moments when new chunks of threads have been added, the server then takes 4–6 s. But at one moment, it takes 10 s to respond, though it has happened only one time, so we can overlook it.

In Fig. 5, the ECS model surprisingly performed better than the EC2 model because, as we can see ECS-based webserver has taken almost 0 s and sometimes 2–4 s to respond to the HTTP request, whereas EC2 has taken an average of 2 s sometimes 4 to 6. The threads increase on a stepped basis but still, it never reacts but rather performs constantly.

In Fig. 6, EC2 performs well from the start but when the threads become 3000 from just 500, it doesn't react well. We can after the peak load average response increase much higher which means the server has become slow now it has responded at an average of 12–14 s.

ECS performs significantly well in Fig. 7. We can see the server was responding at an average time of almost 0s, but when the peak started, it gets slower. Though it was a tough situation still they responded at an average of 8-10s which is much lesser than EC2. Therefore, it took less time than the EC2 model during most of the times.

In Table 1, we can see ECS performs superior to EC2 as far as the total response in the entire test time. In the Linear Load test, EC2 responded to 33,946 requests, whereas ECS responded to 78,163 requests which are practically twofold in contrast with EC2. In the Step Load, EC2 responded to 29,043 requests, though ECS reacted to

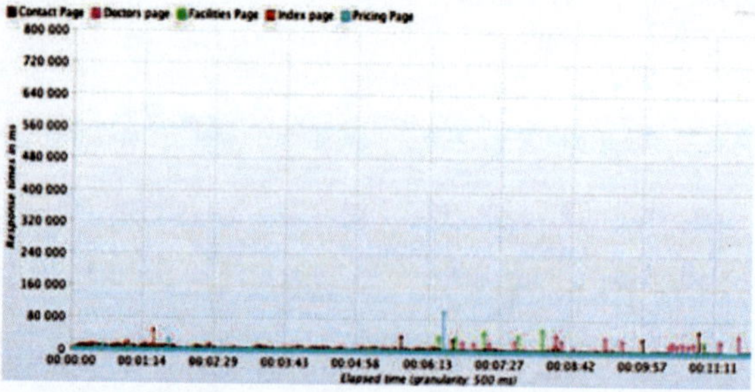

Fig. 6 Response time over time EC2 peak load

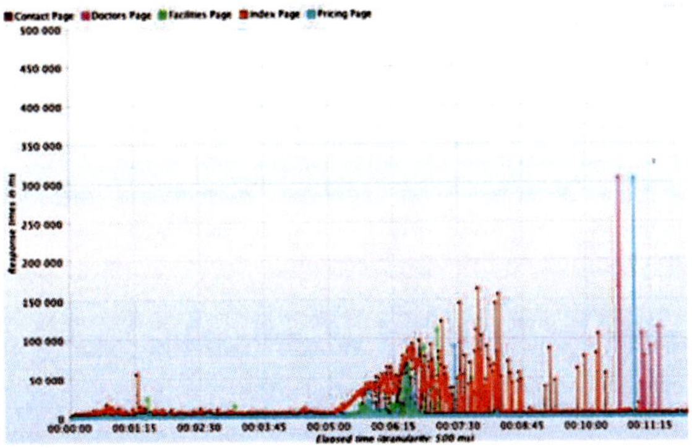

Fig. 7 Response time over time ECS peak load

64,094 requests which are a similar practically double in contrast with EC2. In Peak, Load EC2 responded to 18,337 requests, while ECS responded to 34,003 requests which are again almost double in comparison to EC2. In error percentage, EC2 made 0% blunder, while ECS made a 1.05% mistake in the Linear Load. In the Step Load, EC2 made a 0.01% error which is right around 0 while ECS made a 2.22% mistake. In the Peak Load, EC2 made a 2.93% blunder which is nearly 3% though whereas ECS created an 8.86% error which is almost three times then EC2. In throughput, EC2 responded to 47.77 requests per second, while ECS responded to 92.10 requests per second which are twofold that of EC2 in the Linear Load. In the Step Load, EC2 responded to 35.89437.

Table 2 shows the data that we get from the Performance test. In the data, we have presented Samples, Average, Error, and Throughput requests per second, whereas ECS responded to 80.95 requests per second which are practically triple thanthat of EC2. In the Peak, Load EC2 responded to 25.41 requests every second, whereas ECS responded to 34.46 requests every second which is way more than EC2.

If we analyze the performance table, the number of requests responded by ECS is almost double that by EC2. Additionally in the throughput part, ECS performs much more than double. Be that as it may, in error percentage, ECS is creating more than EC2, particularly in Peak Load, when an abrupt traffic burst happens, it begins to make far more blunders than EC2. However, when we run our application, the first

Table 2 Performance analysis table

Model	Label	Samples	Avg	Error (%)	Throughput
EC2	Linear load	33,946	1893	0.00	47.77
EC2	Step load	29,043	3365	0.01	35.89
EC2	Peak load	18,337	6747	2.93	25.41
ECS	Linear load	78,163	4945	1.05	92.10
ECS	Step load	64,094	1816	2.22	80.95
ECS	Peak load	34,003	44,776	8.86	34.45

thing we investigate is what number of requests it can respond, and in that manner, ECS is far better than EC2. Moreover, traffic burst is certainly not a customary thing, so we need to see which method gives us better performance most of the time. In terms of network performance, ECS did better than EC2. But EC2 gets low to moderate network performances. However, to support Docker containers in the EC2 instance, we need a better network performance. And it is not much more than ECS. Even after getting slightly better network performances, ECS is taking much fewer resources than EC2. Furthermore, suppose if one container in one instance can give a better response than the entire EC2 instance, then if we use 10 containers in the same instance, it can give multiple times better performance which is the fundamental advantage of ECS. Therefore, we can conclude that the Docker Container model is performing superior to Virtual Machines as far as HTTP reactions and throughput. But, Virtual Machine performs better than Docker in terms of error generalization.

6 Conclusions

In this paper, we have presented a comparative study between two native cloud application deployment models. We have developed the infrastructure for those models and deployed our healthcare application there. We have to lead a comparison test that how both models perform by using the Apache JMeter tool. Additionally, to make the performance test more realistic, we have utilized a three-step based test plan. The parameters of our comparison test were HTTP response, error percentage, and throughput. Docker container performed essentially better than Virtual Machine in the HTTP response and throughput part, although Virtual Machine did better than Docker Container in the error percentage part.

References

1. Barik, R. K., Lenka, R. K., Rao, K. R., Ghose, D.: Performance analysis of virtual machines and containers in cloud computing. In: 2016 International Conference on Computing, Communication and Automation (ICCCA), Noida, pp. 1204–1210 (2016)
2. Zeng, H., Wang, B., Deng, W., Zhang, W.: Measurement and evaluation for Docker container networking. In: 2017 International Conference on Cyber-Enabled Distributed Computing and Knowledge Discovery (CyberC), Nanjing, pp. 105–108 (2017)
3. Joy, A.M.: Performance comparison between Linux containers and virtual machines. In: 2015 International Conference on Advances in Computer Engineering and Applications, Ghaziabad, pp. 342–346 (2015)
4. Sutton, R.S., Barto, A.: Reinforcement Learning: An Introduction. MIT Press (1998)
5. Ruan, B., Huang, H., Wu, S., Jin, H.: A performance study of containers in cloud environment. In: 2016 Asia-Pacific Services Computing Conference (APSCC), Zhangjiajie, pp. 343–356 (2016)
6. Islam, M.M., Khan, Z., Alsaawy, Y.: A framework for harmonizing internet of things (IoT) in cloud: analyses and implementation. Wirel. Netw. (2019)
7. AWS Homepage. www.aws.amazon.com. Accessed 30 Aug 2020

8. Medium Homepage. www.medium.com. Accessed 30 Aug 2020
9. Qu, C., Calheiros, R.N., Buyya, R.: Auto-scaling web applications in clouds: a taxonomy and survey. ACM Comput. Surv. 51(4) (2018)
10. Islam, M.M., Khan, Z., Alsaawy, Y.: An Implementation of Harmonizing Internet of Things (IoT) in Cloud. Lecture Notes of the Institute for Computer Sciences, Social Informatics and Telecommunications Engineering, pp. 3–12. Springer, Cham (2018)
11. Yadav, A.K., Garg, M.L., Ritika: Docker containers versus virtual machine-based virtualization. In: 2018 International Conference on Emerging Technologies in Data Mining and Information Security (IEMIS), Kolkata, pp. 141–150 (2018)
12. Liu, D., Zhao, L.: The research and implementation of cloud computing platform based on docker. In: 2014 11th International Computer Conference on Wavelet Actiev Media Technology and Information Processing (ICCWAMTIP), Chengdu, pp. 475–478 (2014)
13. Salah, T., Zemerly, M.J., Yeun, C.Y., Al-Qutayri, M., Al-Hammadi, Y.: Performance comparison between container-based and VM-based services. In: 2017 20th Conference on Innovations in Clouds, Internet and Networks (ICIN), Paris, pp. 185–190 (2017)
14. Wang, B., Song, Y., Cui, X., Cao, J.: Performance comparison between hypervisorand container-based virtualizations for cloud users. In: 2017 4th International Conference on Systems and Informatics (ICSAI), Hangzhou, pp. 684–689 (2017)
15. DZone Homepage. www.dzone.com. Accessed 30 August 2020

Global Undergraduate Research Landscape Analysis and Its Effect on 4th Industrial Revolution

Javed Mostafa, Shamim Ashrafi, Kazi Sadman Ahmed, Md. Imranul Islam, Shadman Islam Chowdhury, and Muntasir Khaleque

Abstract Undergraduate level research engagements can aid the students in learning about the current ongoing innovations and shape them to become an efficient workforce for the future. However, it is poorly managed globally, including in Bangladesh, because it is difficult to track due to the scattered nature of this information. InNeed Intelligent Cloud, along with the University of North Carolina Chapel Hill and National Science Foundation(NSF) of USA, aims to tackle this issue through the BLUMER research project. Based on the researchers' research study, we created a searchable database with a web-based search interface that searches through the awarded projects data of the National Science Foundation (NSF) website, which is stored in the AWS cloud database. It also provides an analytical overview of the undergraduate research data through an analytical dashboard that provides numerical insights into the current research landscape around the USA. The aim of the system is to guide the students, researchers, and supervisors in their desired research interests. By establishing an evaluation metric for understanding undergraduate level research production, in the future, we can build upon this and focus on the commercialization of this project which would enable global outreach and connection. This will help create an iterative process that would promote undergraduate research among students at an early stage and build a technologically enabled and industry-focused workforce. So, it can aid in students' theoretical and applied skills development and create an efficient workforce that can lead the 4th industrial revolution in various fields such as Artificial Intelligence, Robotics and Automation, IoT and Smart Agriculture, Data Analytics, Cloud Computing, etc.

Keywords Education · Research · Cloud computing · Data analysis · Behavioral analysis · 4th industrial revolution · Workforce development

---

J. Mostafa (✉)
University of North Carolina at Chapel Hill, Chapel Hill, North Carolina, United States
e-mail: info@inneed.cloud

S. Ashrafi · K. S. Ahmed · Md. I. Islam · S. I. Chowdhury · M. Khaleque
InNeed Intelligent Cloud, Dallas, TX, United States

S. Hossain et al. (eds.), *Proceedings of International Conference on Fourth Industrial Revolution and Beyond 2021*, Lecture Notes in Networks and Systems 437,
https://doi.org/10.1007/978-981-19-2445-3_30

1 Introduction

Every year, millions of research products are produced throughout the world. Research outputs including publications, conference presentations, and patents do not commonly contain information that would help undergraduate students to get affiliated with them, which can later be used as a launching pad to help them build a career in that research field. In the context of the USA, in recent researches, it was shown that the undergraduate research opportunities are limited to a single site [2], participants of NSF REU that are specially selected [6], lastly structured research opportunities in the first year of study are very rare [1]. In a country like Bangladesh, where the socio-economical situation is very dynamic, it is very important that we give a platform to the various categories of students that exist, to help them go beyond typical academic work and actual research work that would aid them in gaining higher studies opportunities, and also cater to specific student's needs. This iterative process will create a sustainable and scalable workforce for the various industries within the country. By understanding the undergraduate research outcomes, especially the goal of developing the knowledge mine that would aid in developing "Digital Bangladesh", we can create a cycle that would help future students in the decision-making process.

1.1 Problem Statement

There is a lack of a proper system that can track, analyze, and consolidate scattered research project data that can enable students to search for research projects and create an international network with other fellow researchers across the world. Students need to search in a very scattered and time-consuming way through google or other platforms. Institutions also don't have a tool to track their undergraduates' research output, the current comparative studies are comparatively labor, resource, and time-intensive due to the use of longitudinal survey data [4]. All of these lead to a workforce that is not well adapted to the current technologies that would accelerate the 4th industrial revolution in our country. The universities can benefit by having a portal where they can post project opportunities for the student thus creating a collaboration platform for everyone to use. The administrators can track student engagements and determine the strategy to influence more students' interest in research participation and understand their career trajectories. The students themselves can find a single source of information that aggregates research projects from different sources and find opportunities for research work based on their own interests. A combination of all of these can provide a national platform that can generate a technologically efficient workforce capable of bringing changes to various industries.

1.2 Literature Review

We tried to find research ideas that dealt with how research can bring innovation to the industry and if there are any tools that empower students to get affiliated with their desired research projects.

After reading through the Institute of Entrepreneurship and Innovation's multi-year study on research participation of graduate and undergraduate students, we found that these respective students had contributed greatly to the industry. This further enforced our original idea, as reading further, we found that students who participated in research grants that enabled work between a third-party organization already in the industry had further pushed the industry path and contribution of said students [3].

We came across another study conducted by The University of Tennessee regarding the extent to which perceived learning gains from undergraduate research affected degree (and consequently industry) aspirations. The study found it statistically significant that a student involved in undergraduate research would be more inclined to continue education in a graduate degree, and consequently, the contributions to the industry he/she specialized in, in an industrial application [5].

Another article we followed was conducted by the Milwaukee School of Engineering, which also focused on tracking the influence of undergraduate research programs on a student's future career path, either graduate school pursuit or industry-related. Though mainly focused on higher education, there was a particularly interesting statistic that suggested almost 71% of the sourced students would continue education and further contribute to the industry standard. This makes it essential that providing these open opportunities to participate in research grants is great in furthering industry development [7].

2 The BLUMER Project

To tackle the aforementioned need for such a tool, InNeed Intelligent Cloud partnered with the researchers of the University of North Carolina at Chapel Hill and the National Science Foundation of USA to create a web application that facilitates the BLUMER vision. It is a web application that promotes the following:

- Establishing an evaluation metric for understanding undergraduate research production.
- Creating an initial database of undergraduate research products.
- Developing, refining, and validating the tool to be scalable to understand the impact of undergraduate research experiences on the nation's scientific workforce.
- Using the evaluation metrics to contribute to the science of science and innovation policy within the country.

- Allowing institutions that do not currently track undergraduate research work a repository which can, in turn, be a collaborative workspace for the students and other professors.
- Creating an iterative process that ensures the creation of a research-minded and technologically advanced workforce that sustains industrial advancements within the country.

By incorporating institutional awarded projects as our data source, we provide a single source of information for the users.

3 The Model Of BLUMER System

Our work on BLUMER started from the EAGER proposal that was done by the team in the Carolina Health-Informatics Program(CHIP) at the University of North Carolina at Chapel hill. Our team in InNeed Intelligent Cloud worked with the researchers at UNC to define the scope of our technical foundation. Based on the requirements we got, we decided that the architecture of the BLUMER system will contain the following modules:

- A web-based search engine that can search through thousands of awarded research projects based on topics chosen by the user
- Email search capabilities through which users can also query the database.
- Admin capability to add newly awarded projects inside the database that can later be searched
- An analytical dashboard that allows the admins of the web application to see the state of the awarded projects throughout the country through various graphs, that would allow them to make informed decisions.

3.1 Process of Creating The Architecture

Based on the proposal we had from the University of North Carolina at Chapel Hill, we based our architecture on the following action items:

1. Data Modeling and DB design: We analyzed and designed data models for the undergrad research database. This database stores structured data on undergraduate research which will be accessed by the undergraduate students through an Omni-channel search interface. This will be established as the central datastore of the system.
2. Development of Basic Web Search Front-End/UI: Our next set of action plans involved creating a search application that allowed searching the database and retrieving project information. The application features both simple and advanced

search features and was designed with cloud-native technology. The technology is serverless, thus it is flexible and scalable to the user number.

3. Basic Email Front-End Query and Retrieval System: To establish the omnichannel system, we implemented the search to be done through basic email clients like Gmail or yahoo, which give users access to the database through an email interface.
4. Basic Admin Panel: The universities that adopt our application would require an admin panel that would give them the ability to put newly awarded projects and other related information in our database. So, We developed a database data entry, update, and maintenance system for BLUMER system administrators for the universities.
5. Data Loading: We added National Science Foundation or NSF's awarded project data to our database.
6. Testing: To test the usability and scalability of the system, we did extensive testing which involved black and white box testing to understand the unit capabilities of the application. We also had a sample group from the University of North Carolina at Chapel Hill, which acted as the user to test out the web application from a user perspective.

The whole architecture of the system is given below (Fig. 1).

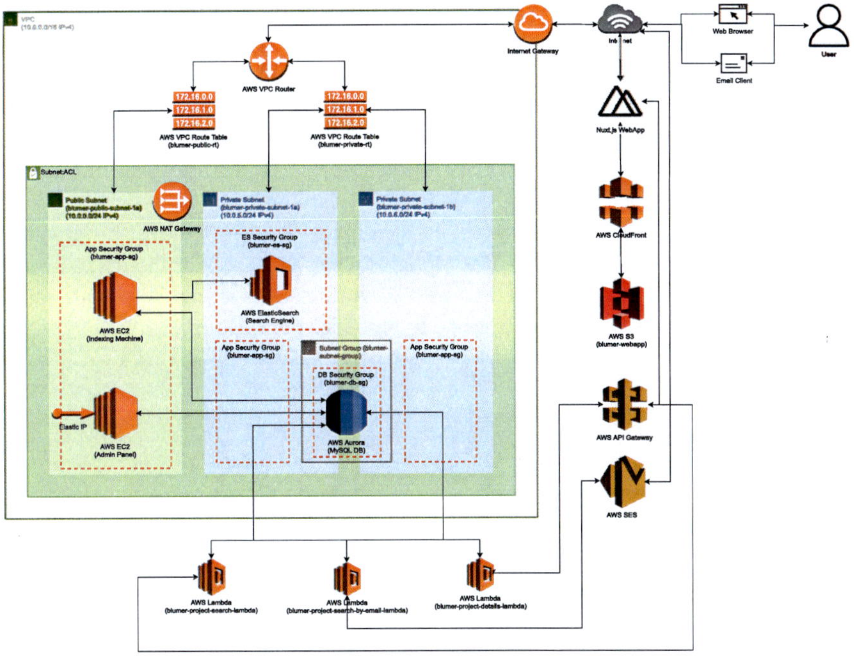

Fig. 1 The AWS architecture of the BLUMER application

We had built the whole architecture of the system in the Amazon cloud environment. For computing, we had used AWS lambdas, which would interact with the RDS Database we had used to store the data. We used API gateway to capture the user request from the front-end web application that also interacted with AWS Simple Email service or SES to provide the search capability through the email client. The admin web app that we have built is created through a Django framework and it is run on an EC2 machine. The whole infrastructure is secure through AWS VPC with the databases in private subnets and user-facing components on public subnets.

3.2 Data Collection

From the initial proposal of the University of North Carolina Chapel Hill, we ingested the database with National Science Foundation's awarded projects. The whole data flow pipeline diagram is given below.

The award was given out to various universities in the United States of America. The individual Data consisted of various fields such as start time, principal Investigators, University, University states, abstract, etc. We scrapped the data from their website and ingested it within our own database. We extracted the data from 2021 to 1990 to give the database a considerable amount of data to start on. The application performed considerably well by retrieving the data in 3 s from the database we had. By having the database in the cloud, we were able to be this efficient, also the serverless compute tool we used, AWS lambda, scaled up based on the request.

As seen from Fig. 2, we can see that after we had ingested the data into our database, we also indexed the data to Elasticsearch to provide the admins a dashboard that would allow them to see the state of awarded projects from a high level throughout the nation.

As previously mentioned, we modified the data we got from the NSF website in such a way that it adheres to the data model we had created. For the database, we had used the aurora database which is also serverless, thus it is able to scale really well. Below is the ERD Diagram we had to build for the system (Fig. 3).

4 Results and Output

4.1 The Web Application

The main portion of the system is the web application that will allow users to search through the awarded projects and get details of an individual project, that one might get by searching the web application. We can see the workings of the web application through Figs. 4, 5, and 6.

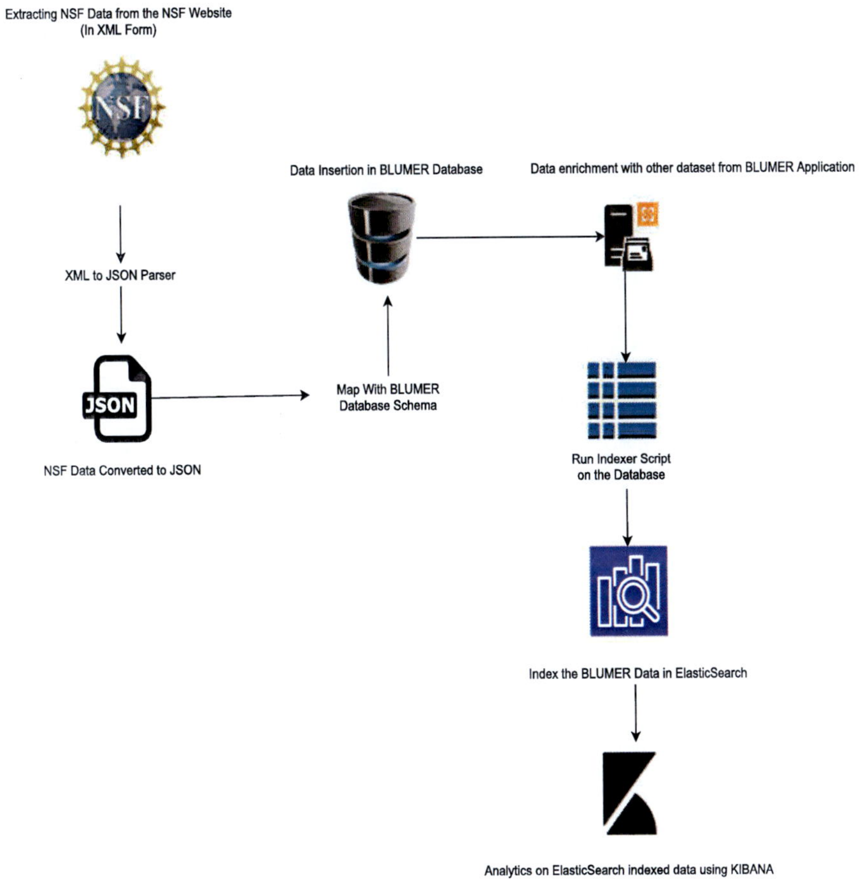

Fig. 2 The data pipeline of BLUMER

The working process of navigating the web application is given below:

- The user uses the search bar to write down desired keywords and choose related fields such as computer science, chemical science, etc.
- The resulting page gives the results based on the search criteria a user provides.
- The results can be further filtered using university name, university states, supervisor, and date.
- The user can click on a particular project to get the details regarding the project such as project description, start date, related researchers, project duration, the type of project, etc.

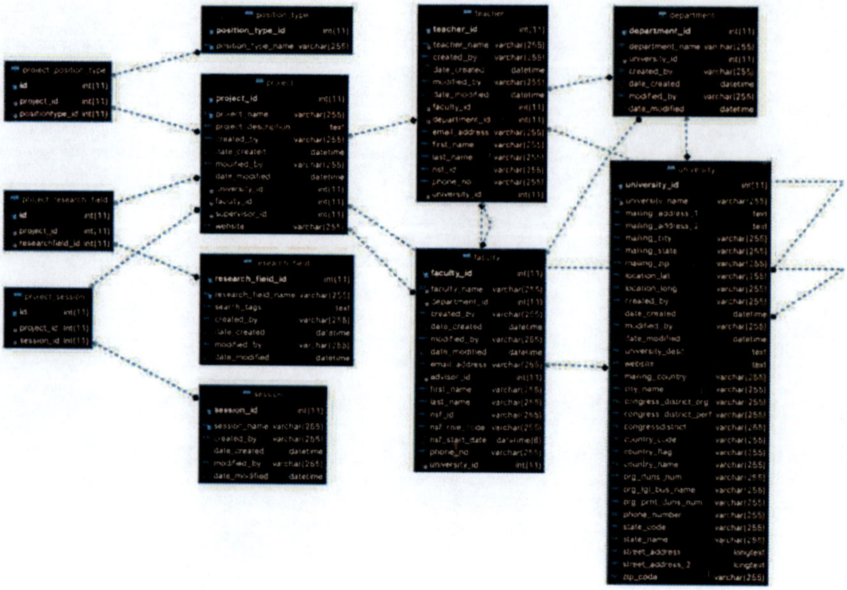

Fig. 3 The ERD of BLUMER

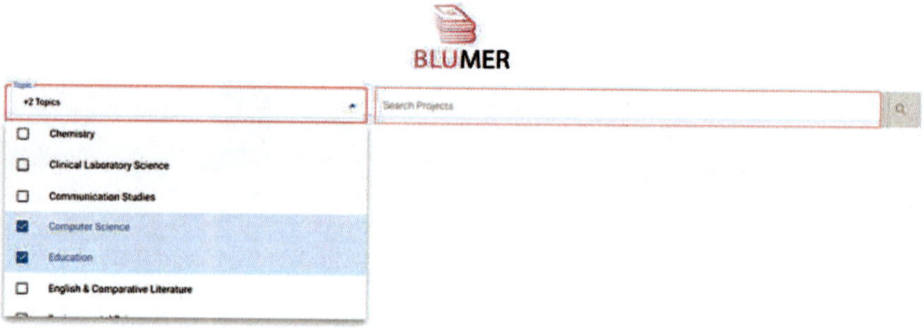

Fig. 4 The frontpage of the web application

4.2 The Admin Panel

The admin panel allows the proper authority within a university to add their own data to the database by adding newly awarded projects. They can also add information such as Faculty, Research type/position type, Research Fields, and The sessions needed for the research (Fall/spring/summer). Under those topics, there is an extensive array of information that can be entered to further enhance this knowledge mine. The topics that an admin user can change or add inside the BLUMER admin dashboards are—Faculty, Teacher, Research Type/Position Type, Projects, Research Fields, Sessions

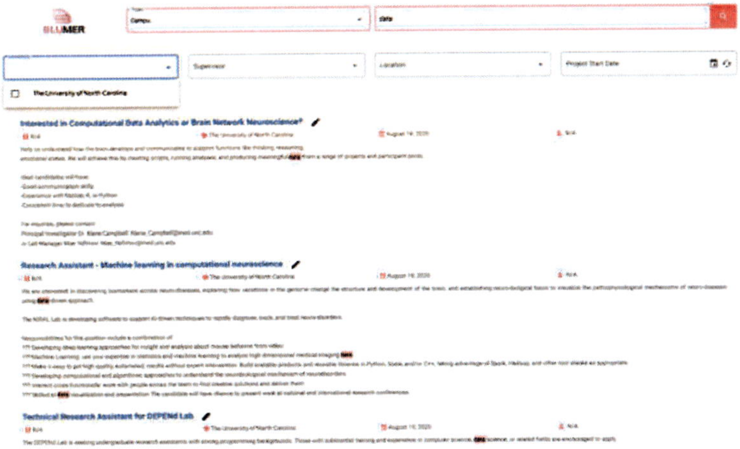

Fig. 5 The result page of the web application

(Fall/Spring/Summer), University User (A root admin user can enter other users), and Group (Similar users can be placed inside a group).

The following Fig. 7 shows the admin pages of the web application.

4.3 The BLUMER Dashboard

The BLUMER dashboard consists of the analytical dashboard that gives the administrators of the universities a look into the state of the awarded research projects throughout the country. It gives the admins a view into how the projects are distributed throughout different areas, what are the fields where most awards are given, and the type of projects that are typically available. Our goal is that the admins can use this dashboard to provide useful information to the students as advisors such that students can get informed advice regarding their future intentions. Below we can see the view of the dashboard (Fig. 8).

5 Conclusion and Future goals

In conclusion, Bangladesh is a country with a lot of potential to make the youth into a research-oriented outcome-based workforce. With the advent of technology and with the vision of "Digital Bangladesh" looming, our focus on creating and making BLUMER accessible to Bangladesh will fulfill that vision. The outcome of

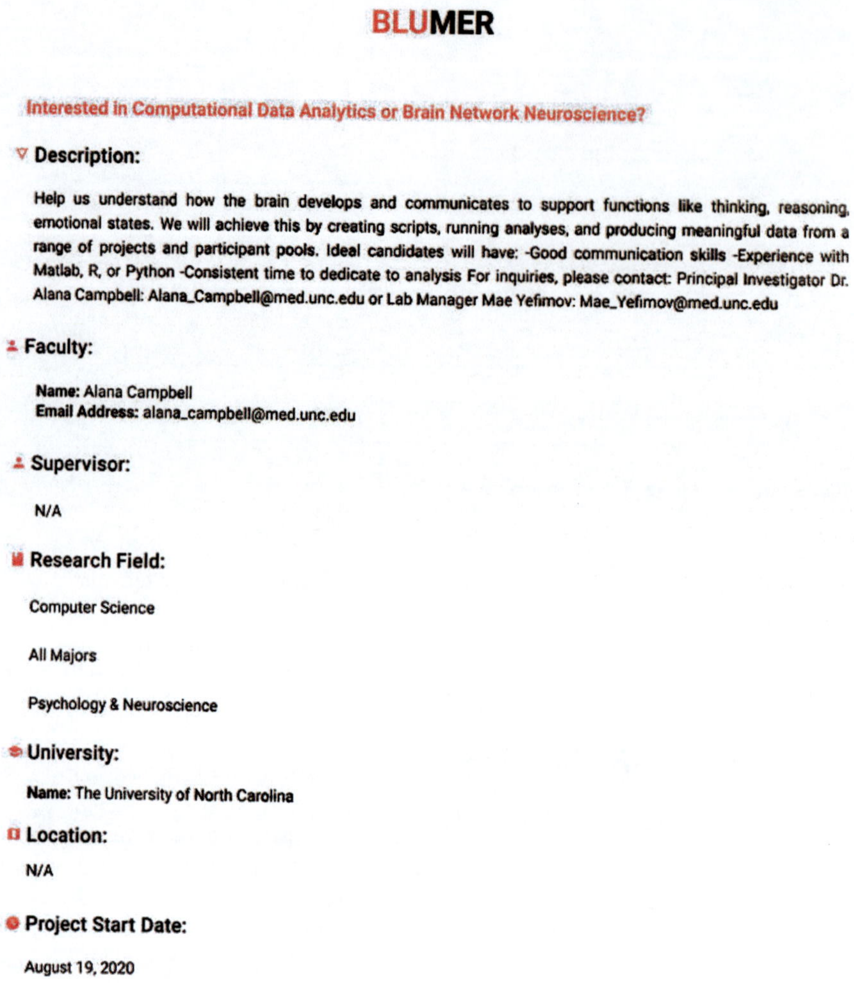
BLUMER

Interested in Computational Data Analytics or Brain Network Neuroscience?

Description:

Help us understand how the brain develops and communicates to support functions like thinking, reasoning, emotional states. We will achieve this by creating scripts, running analyses, and producing meaningful data from a range of projects and participant pools. Ideal candidates will have: -Good communication skills -Experience with Matlab, R, or Python -Consistent time to dedicate to analysis For inquiries, please contact: Principal Investigator Dr. Alana Campbell: Alana_Campbell@med.unc.edu or Lab Manager Mae Yefimov: Mae_Yefimov@med.unc.edu

Faculty:

Name: Alana Campbell
Email Address: alana_campbell@med.unc.edu

Supervisor:

N/A

Research Field:

Computer Science

All Majors

Psychology & Neuroscience

University:

Name: The University of North Carolina

Location:

N/A

Project Start Date:

August 19, 2020

Fig. 6 The description page of a project

our current system facilitates this by enabling the students to start research work at a very early stage. In the future, we plan to use machine learning that will also show the administrator and industry leader's future career projections of the students using our application. It will help us to collect data and do trend analysis and behavioral analysis through machine learning that would allow us to observe student behavior on a mass scale. We are also open to collaborating with universities in Bangladesh and their researchers to run localized BLUMER pilot research on undergraduate research data, outcomes, and future careers in industry and academia. Through the commercialization of BLUMER, we can connect researchers, students, and industry experts on a national and international level. It can provide a collaborative workspace

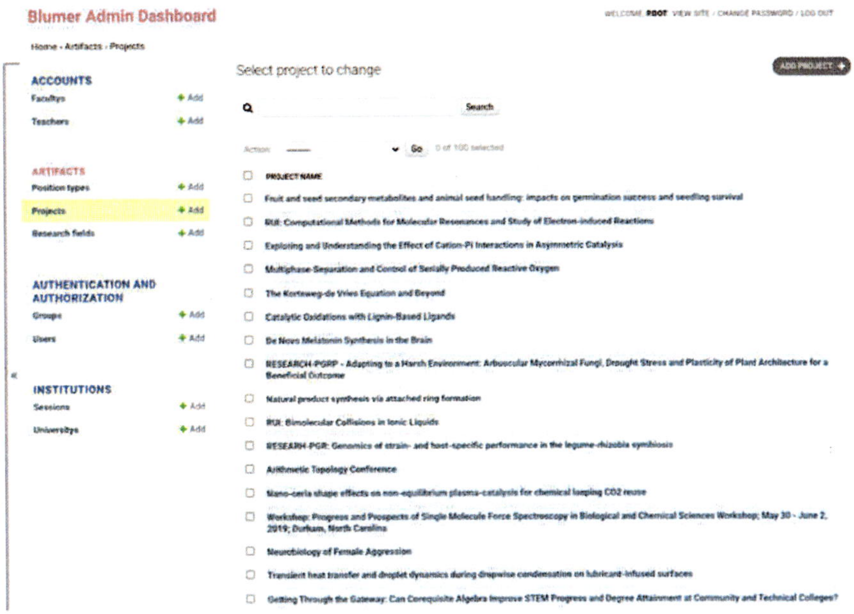

Fig. 7 The admin panel result

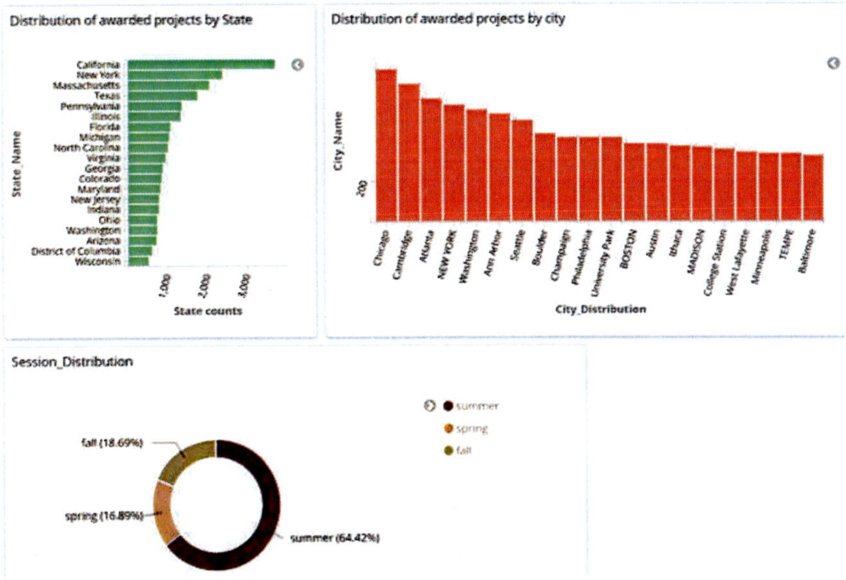

Fig. 8 The BLUMER dashboard

for the professors and students of various universities in Bangladesh and outside of Bangladesh. This can be a revolutionary platform that would help students be a better workforce for the 4th Industrial Revolution and beyond.

References

1. Bowman, N.A., Holmes, J.M.: Getting off to a good start? First-year undergraduate research experiences and student outcomes. Higher Educ. **76**(1), 17–33 (2018)
2. Grineski, S., Daniels, H., Collins, T., Morales, D.X., Frederick, A., Garcia, M.: The conundrum of social class: disparities in publishing among stem students in undergraduate research programs at a hispanic majority institution. Sci. Educ. **102**(2), 283–303 (2018)
3. Libaers, D.P.: Time allocation decisions of academic scientists and their impact on technology commercialization. IEEE Trans. Eng. Manag. **59**(4), 705–716 (2012)
4. Rahman, F., Hu, H., Brylow, D., Kussmaul, C.: Bringing undergraduate research experience in non-r1 institutions. In: Proceedings of the 2017 ACM SIGCSE Technical Symposium on Computer Science Education, pp. 671–672 (2017)
5. Strayhorn, T.L.: Work in progress-does learning about research affect the graduate degree aspirations of stem undergraduates? A path analysis. In: 2010 IEEE Frontiers in Education Conference (FIE), pp. T3D–1. IEEE (2010)
6. Wilson, A.E., Pollock, J.L., Billick, I., Domingo, C., Fernandez-Figueroa, E.G., Nagy, E.S., Steury, T.D., Summers, A.: Assessing science training programs: structured undergraduate research programs make a difference. BioScience **68**(7), 529–534 (2018)
7. Yauch, C.A.: The impact of undergraduate research experiences on graduate school pursuit by industrial engineers. In: 2007 37th Annual Frontiers In Education Conference-Global Engineering: Knowledge Without Borders, Opportunities Without Passports, pp. S4B–17. IEEE (2007)

Exploring the Capabilities Toward Embracing 4IR for Sustainable Business in Bangladesh

Mafi Rahman and Abu Taher

Abstract The prime objective of the study is to explore the capabilities of the organizations and universities toward embracing 4IR for sustainable business. To achieve this objective, the researchers used both primary and secondary data. The capabilities of the organizations and universities have been measured using the Five-Point Likert Scale ranging from very unlikely (1) to very likely (5). For primary data collection, 200 samples were selected from organizations and universities based on convenient and purposive sampling. The collected data were analyzed with the help of statistical tools and techniques. Study findings show that there is a lacking in the capabilities of the organizations and universities to incorporate the 4IR components. The implication of the study was that the policy makers of government and different stakeholders should work together to sensitize and create awareness for the proper application of 4IR-based digital technologies in business. The study faced some limitations and finally pinpointed the future directions for further research in this area.

Keywords Fourth Industrial Revolution (4IR) · Digital technologies · Innovative education system · University–Industry Collaboration (UIC) · Dynamic capabilities · Sustainable business

1 Introduction

When the whole world is advancing technologically, Bangladesh being a developing country shouldn't also lag behind. It is because the technologies will be able to bring a revolutionary change in many aspects of the country such as its economy, supply chain, globalization, international trade system, and so on. The technology

M. Rahman (✉)
Department of Business Administration-General, Faculty of Business Studies, Bangladesh University of Professionals (BUP), Dhaka, Bangladesh
e-mail: mafirahman@bup.edu.bd

A. Taher
University Grants Commission of Bangladesh (UGC), Dhaka, Bangladesh
e-mail: taher_member@ugc.gov.bd

S. Hossain et al. (eds.), *Proceedings of International Conference on Fourth Industrial Revolution and Beyond 2021*, Lecture Notes in Networks and Systems 437,
https://doi.org/10.1007/978-981-19-2445-3_31

will also have an impact on the human capital transformation. If the organizations of Bangladesh can adapt the digital technologies like Artificial Intelligence, IoT (Internet of Things), Communication and Networking, Robotics and Automation, Data Analytics and Cloud Computing, the businesses will be able to thrive in the 4th Industrial Revolution [3]. Making this possible will help us reshape the businesses of the country and bring about sustainable development in Bangladesh.

Though Bangladesh is already termed as "Digital Bangladesh", there are many areas where technological evolution is still substandard. The reflection of Digital Bangladesh can be seen in few areas only such as ICT industry, internet availability, and transition of many companies from labor to machine. But the areas which need development are Artificial Intelligence, Robotics and Automation, and many others. It is high time we should think about fostering the emerging industrial revolution for the development of our Digital Bangladesh.

As mentioned earlier, by thriving in the Fourth Industrial Revolution (4IR), we will be able to reshape the businesses of the country. It is because 4IR will bring diverse changes in the businesses. Some of these changes are already visible such as transition from labor to machine. Many garment factories and private corporations are minimizing their labor number by replacing them with machines. The reason is that a machine can do tasks of many humans in lesser time with minimal waste. This is raising the demand for more skilled human resources who are well accustomed to digital technologies. This will also lead to a change from labor-management relations to machine-management relations.

1.1 Statement of the Problem

Embracing this "4IR" or "Industry 4.0" will not be an easy task for a developing country like Bangladesh. The first challenge that will come in the way is the shortage of skilled human resources. It is because the university graduates lack the required skills that are demanded by today's advanced organizations. Chidera [5] mentioned that there are top 10 skills that are required among the graduates to face the challenges of the Fourth Industrial Revolution. These are Active Listening, Complex Problem Solving, Coordinating with Others, Creativity, Critical Thinking, Judgment and Decision-Making, Negotiation, People Management, Quality Control, and Service Orientation. Unfortunately, most of the university graduates don't possess these skills. This gap between graduates' knowledge and the skills demanded by employers can only be solved if the education system is totally reformed. The universities need to focus on the Outcome-Based Education (OBE) curriculum in light of the challenges of 4IR. Necessary measures need to be taken to transform the Traditional Education System into Outcome-Based Education (OBE) System. Both universities and industries of Bangladesh can go for University–Industry Collaboration for enhancing the education quality, and skilled professionals can be involved in the university curriculum committees to formulate the need-based curriculum. The

University–Industry Collaboration will pave the way for fulfilling the employers' demand for the diverse range of competencies in university graduates.

Even though many universities are changing their curriculum, the application of this change in the classroom is still imperceptible. This is happening because the faculties became comfortable with the traditional teaching method and thus don't want to leave their comfort zone. Resistance to change is a common scenario almost everywhere, be it in our professional life or personal life.

If we talk about the business world, the digital revolution that is taking place all over the world seems like a double-edged sword. The reason is that it comes with a package of both benefits and vulnerabilities to business and society. It is seen that new technologies bring varied complexities for the businesses to cope with such transforming business models or ways of working [4]. But this change needs to be accepted for the future benefit of the country. When all other countries are becoming technologically advanced and increasingly competitive, Bangladesh needs to keep up with this global change and competition. And for this, the first change needs to be made in the education system of the country. A need-based, action-oriented education system will lead to competent and capable graduates with the required skills and knowledge who will be able to easily cope with any kind of technological complexities. This will help the organizations to embrace the 4th Industrial Revolution more smoothly. If organizations can succeed, it will lead to the sustainable development of Bangladesh that would eventually help to build digital Bangladesh in a true sense.

1.2 Literature Review

If we look around, we can see that our environment is gradually becoming digitalized, thus making our economy a global one. Brenner [4] states that digitalization is radically reordering the traditional industry boundaries and reshaping our businesses. With the help of diverse new technologies, a digital toolbox of solutions has been created which is challenging the status quo [7]. Now, the businesses are not only using these technologies in a smart way, but also in a more sustainable way which will aid the countries to tackle the change in climate [4]. This entrance of the businesses into the 4IR revolution by embracing digital technologies such as Robotics, IoT, and Predictive Analytics, is making them smarter. Industry 4.0 or i4.0 is playing a vital role in integrating networks, organizations, and people, thus building potential for artificial intelligence, IoT, and advanced automation. Digitalization is bringing many benefits to the organizations such as speeding up the production process, improving productivity, increasing profitability, and reducing errors. Among all these benefits, the most notable benefit of digitalization is sustainability. It is because, digitalization is making the economy grow by minimizing the environmental impacts, ensuring proper utilization of resources, and confirming efficient waste management [2].

Today's tertiary level education system is regarded as a role player in modern society. As the institutions are operating in a complicated era which is, Industry

4.0, they need to prepare their graduates accordingly [12]. As mentioned earlier, the 4IR demands some skills from the future graduates. Enhancing the students' skills is only possible if tertiary education is reformed. For this, the universities need to incorporate Outcome-Based Education (OBE) curriculum aligned with SDGs. OBE differs from the traditional teaching method, in that, the emphasis is given to the learning outcomes of the students. To make this happen, the teachers need to focus on students' learnings rather than their teachings. A huge responsibility lies on the tertiary education level regarding making the graduates prepared for the twenty-first century society's demands. If all the universities of Bangladesh can incorporate OBE in their curriculum, they will certainly be able to create a huge number of 4th Industrial Revolution capable graduates [10]. To make this happen, the teachers require some skills too. Kemp [9] states that the teachers need to possess 5 skills in order to flourish in the 4IR. These skills are Embrace Change with Positivity, Collaborate, Be Creative and Take Risks, Have a Sense of Humor, and Teach Holistically.

The universities and businesses should collaborate with each other as they got divergent strengths that will likely benefit each other. Through University–Industry Collaboration (UIC), the universities will be able to understand which skills the companies of the twenty-first century are demanding. For this reason, UIC is now regarded as a vital innovation resource for the sustainable development of the country. This collaboration will strengthen the knowledge and skills flow. Thus, it will become easier to prepare the students for Industry 4.0. If UIC occurs, it will be possible to easily transmit the technical information where the industry will play the role of technology demanders and universities will be the technology suppliers [14]. Guimón [8] states that UIC will aid in developing the students' skills and generating, acquiring, and adopting their knowledge (technology and innovation transfer). This will help the students learn about the efficient and effective use of resources, thus leading to a sustainable business.

Technological changes are occurring very rapidly all over the world due to the Fourth Industrial Revolution. If the organizations want to survive in this race and take part in the transformational process, they need to understand and build the right capabilities. This is also necessary for organizational growth and development. It is because, through the application of the 4IR technologies, the organizations will be able to solve many industrial problems. Organizations need to have an understanding regarding the dynamic capabilities that are needed to effectively utilize their internal and external competencies in order to achieve sustained competitive advantage. If this can be done, the dynamic capabilities will be able to have an impact on the organizations' resources which will lead to a competitive advantage for the firm. Effective management of the dynamic capabilities which are rare, valuable, non-substitutable, and inimitable resources will pave the way for sustainable business [1]. Teece [13] identifies that the dynamic capabilities comprise of Resources (employees, equipment, buildings, and intangible asset), VRIN Resources (valuable, rare, imperfectly imitable, and non-substitutable), Strategies, and Capabilities.

1.3 Research Objectives

The main objective of this study is to explore the capabilities of the organizations and universities toward embracing 4IR for sustainable business. To attain this main objective, the following specific objectives are spelled out:

1. To assess the capability of the organizations toward embracing the digital technologies for sustainable business.
2. To examine the ability of the universities to embrace the innovative education system for sustainable business.
3. To find out the status of University–Industry collaboration for sustainable business.
4. To spell out the capability of the organizations to embrace the dynamic capabilities for sustainable business.
5. To address some policy implications for sustainable business with a view to mitigate the challenges of 4IR in Bangladesh.

1.4 Research Model

See Fig. 1.

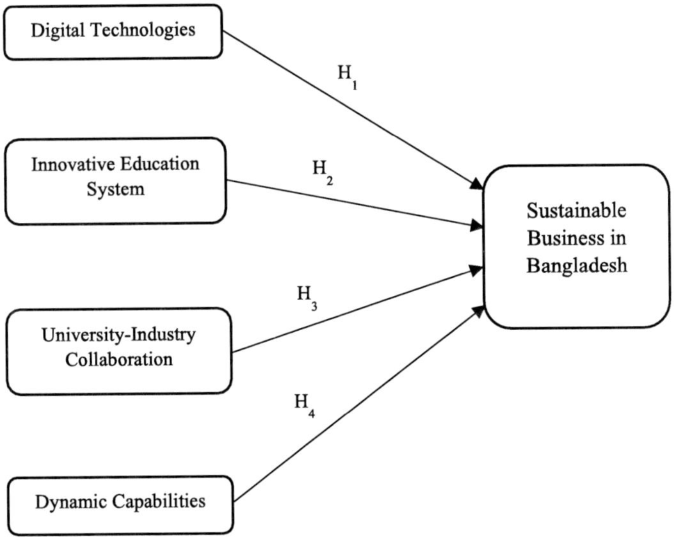

Fig. 1 Conceptual model

1.5 Research Hypothesis

The hypotheses were set up after an intensive review of the literature and research studies done earlier:

H_1: Organizations show a positive approach toward embracing Digital Technologies for Sustainable Business.
H_2: Universities show a positive approach toward embracing an Innovative Education System for Sustainable Business.
H_3: Organizations and Universities show a positive approach to collaborating and having an impact on Sustainable Business.
H_4: Organizations possess Dynamic Capabilities for Sustainable Business.

2 Methods

2.1 Study Design

This study was empirical in nature. A survey method was used for collecting this study's data.

2.2 Sampling Procedure and Sample Size

To do the study, convenient and purposive sampling techniques have been used. A total of 250 structured questionnaires were distributed among different universities' faculties and organizations' officials. From them, 200 valid questionnaires were returned for the study's analysis of which 100 came from faculties and 100 from the officials.

2.3 Period of the Study

The survey was conducted between August to September 2021.

2.4 Survey Instrument

As no standard scale associated with the study was not found in existence, the authors prepared the survey instrument by doing a comprehensive literature review. Based on pilot testing, the irrelevant variables were omitted from the questionnaire. While developing the survey instrument, the authors kept in mind that the instrument is short and practical to administer in terms of the time needed to complete the study. The scale was adapted from Parasuraman et al. [11] so that consistency is ensured. Five-Point Likert Scale has been used ranging from very unlikely (1) to very likely (5).

2.5 Data Collection Procedure

For conducting the study, both primary and secondary data have been used. While collecting data, the researchers were cautious of the study's objectives and hypotheses. For collecting primary data, direct contact was made with the respondents via emails and phone calls in which the survey purpose and general outlines were introduced. Then, the online survey link was emailed to the respondents which contained a cover page introducing the study, and assurance of confidentiality. Secondary data were collected from relevant articles in Google Scholar.

2.6 Reliability and Validity

The reliability was to be considered with the help of Cronbach's Alpha. Validation procedures were involved by consulting with expert researchers.

2.7 Data Analysis

All collected data were tabulated, interpreted, and simplified in order to make them suitable for research purposes. The collected raw data were inputted into SPSS (Version 20) for statistical analysis. Then, the analyzed data were arranged into a Word file for summarization.

3 Results

The findings of this study have been discussed under the subsequent headings.

3.1 Sample Profile

Table 1 exhibits the respondents' profiles by using the information gathered from the questionnaire.

The total number of respondents in this study is 200, out of which the number of officials is n = 100 and faculties is n = 100. In case of *officials*, most of the respondents are female (55%). The majority of the respondents (80%) are found to be under the age of 30 and most of them hold an under-graduation degree (53%).

Table 1 Respondents' profile

List of variables attributes		Frequency		Percent		Valid percent		Cumulative percent	
		O	F	O	F	O	F	O	F
Gender	Male	45	46	45	46	45	46	45	46
	Female	55	54	55	54	55	54	100	100
Age	18–30 years	80	78	80	78	80	78	80	78
	31–40 years	14	17	14	17	14	17	94	95
	41–50 years	4	5	4	5	4	5	98	100
	51–60 years	2	0	2	0	2	0	100	0
	Above 60 years	0	0	0	0	0	0	0	0
Educational level	Under-Graduation	53	0	53	0	53	0	53	0
	Post-Graduation	47	95	47	95	47	95	100	95
	M.Phil/Ph.D.	0	5	0	5	0	5	0	100
	Postdoc	0	0	0	0	0	0	0	0
Org./Institute type	Public	20	95	20	95	20	95	20	95
	Private	80	5	80	5	80	5	100	100
Designation	Employee	61	N/A	61	N/A	61	N/A	61	N/A
	Employer	39	N/A	39	N/A	39	N/A	100	N/A
	Lecturer	N/A	81	N/A	81	N/A	81	N/A	81
	Assistant professor	N/A	14	N/A	14	N/A	14	N/A	95
	Associate professor	N/A	3	N/A	3	N/A	3	N/A	98
	Professor	N/A	2	N/A	2	N/A	2	N/A	100
Total years of working experience	Less than a year	20	39	20	39	20	39	20	39
	1–2 years	19	21	19	21	19	21	39	60
	3–5 years	37	27	37	27	37	27	76	87
	6–10 years	10	2	10	2	10	2	86	89
	10+ years	14	11	14	11	14	11	100	100

Note O = Officials; F = Faculties
Source Field Survey

Most of the respondents are officials of a private organization (80%). The number of employees (61%) is more than that of the employers (39%). The total years of working experience of most of the respondents (37%) is between 3 and 5 years. In case of *faculties*, the majority of the respondents are female (54%) and aged under 30 (78%). Most of them hold postgraduation degrees (95%) and are teachers at public universities (95%). The majority of the respondents are lecturers (81%). The total years of working experience of most of the respondents (39%) is less than a year.

3.2 Reliability Test

Cronbach's Alpha has been used to conduct the reliability test for measuring the internal consistency of the study instrument, i.e., the questionnaire, as well as its stability. The reliability test indicates the extent to which the study instrument (questionnaire) is without error or bias, and thus ensures consistent measurement across various items in the study instrument. For ensuring reliability, the value of the coefficient must range between 0 and 1 and must cross 60% (0.60). Then the result of the value will be regarded as having satisfactory internal consistency reliability [6]. In our study, the Cronbach's Alpha of each item is found to be above 0.60 which is more than satisfactory level of internal consistency.

3.3 Regression Analysis

From the above regression analysis table, it is visible that the R-value of all the hypotheses is below the accepted rate. This shouldn't come as a surprise to us as we all are well aware of the fact that Bangladesh's technological and educational quality are still not up to the mark (Table 2).

3.4 Outcome of the Hypothesis Testing

See Table 3.

Table 2 Regression analysis

H1				
Model	**R**	**R square**	**Adjusted R square**	**Std. error of the estimate**
1	0.255[a]	0.065	0.015	0.75128

[a]Predictors: (Constant), Artificial Intelligence, Internet of Things, Communication and Networking, Robotics and Automation, Data Analytics and Cloud Computing

H_2				
Model	**R**	**R square**	**Adjusted R square**	**Std. error of the estimate**
1	0.040[a]	0.002	−0.009	0.675

[a]Predictors: (Constant), Innovative Education System

H_3								
Model	**R**		**R square**		**Adjusted R square**		**Std. error of the estimate**	
	O	**U**	**O**	**U**	**O**	**U**	**O**	**U**
1	0.022[a]	0.339[a]	0.000	0.115	−0.010	0.106	0.761	0.636

[a]Predictors: (Constant), University-Industry Collaboration

Note O = Organizations; U = Universities

H_4				
Model	**R**	**R square**	**Adjusted R square**	**Std. error of the estimate**
1	0.406[a]	0.165	0.130	0.706

[a]Predictors: (Constant), Resources, VRIN Resources, Strategies, Capabilities

Source Field Survey

Table 3 Empirically tested conceptual model

Hypothesis no.	Hypotheses	Accepted/Rejected
1.	Organizations show a positive approach towards embracing digital technologies for sustainable business	Rejected
2.	Universities show a positive approach towards embracing innovative education system for sustainable business	Rejected
3.	Organizations and universities show a positive approach for collaborating and having an impact on sustainable business	Rejected
4.	Organizations possess dynamic capabilities for sustainable business	Rejected

4 Discussion

4.1 Recapitulation and Policy Implication

The study has identified the capabilities of the organizations and universities toward embracing 4IR for sustainable business. Through thorough analysis and statistical tests, it has been observed that the capabilities related to 4IR are still absent in the organizations and universities of Bangladesh. As a result, all the hypotheses of this study were rejected. This result creates concern among the researchers regarding the implication of Digital Bangladesh in a true sense. If instant planning and resource allocation for incorporating the 4IR components are not done, Bangladesh will fall far behind in this technologically competitive world. It is because, in digital transformation, Bangladesh is still in a nascent stage.

In order to make ourselves capable enough to embrace the 4IR and pave the way for sustainable business, the following guidelines can be taken into consideration:

- Policy makers of government and different stakeholders should work together for sensitization and awareness about the proper application of 4IR-based digital technologies such as Artificial Intelligence, IoT (Internet of Things), Communication and Networking, Robotics and Automation, and Data Analytics and Cloud Computing, in the organizations of Bangladesh. The organizations should also arrange training programs for all stakeholders to make them equipped with digital technologies.
- The university authorities should fully reform the curriculum with respect to industry-required skills related to 4IR. They need to provide needed training to the faculties so that they are properly motivated enough to apply innovative teaching strategies. Also, training should be given to the faculties for assisting them to possess Kemp's [9] mentioned 5 skills that they require to flourish in the 4IR. A similar focus needs to be given to help the faculties in imparting the top 10 skills of 4IR, as mentioned by Chidera [5], to their students.
- The academia and industry people have to jointly organize informal meetings on different issues of the corporate world.
- The organizations should ensure the availability and usability of all the dynamic capabilities such as Resources (employees, equipment, buildings, and intangible asset), VRIN Resources (valuable, rare, imperfectly imitable, and nonsubstitutable), Strategies, and Capabilities, that would eventually help for sustainable business in Bangladesh.

4.2 Limitations

This study got a number of limitations. *Firstly*, the study was based on the data collected from a small sample size which has been taken from a small geographical area. This makes the implications of the study restricted. *Secondly*, the survey instrument was mainly constructed with a Likert scale. So, there exists a chance of central tendency bias and social desirability bias. *Thirdly*, the questionnaire can be developed by discussion with relevant field experts. *Finally,* this research study was confined to officials and faculties of public and private organizations and universities located in Dhaka and Mymensingh of Bangladesh. Hence, this research study on Bangladesh may not be the representative one in other developing countries.

4.3 Directions for Future Research

Based on the literature review, data produced, findings, and references of the existing studies, the following further research may be considered:

- Impact of 4th Industrial Revolution (4IR) on Sustainable Business in Bangladesh.
- Empirical Analysis on Fourth Industrial Revolution 2030 and beyond.
- The Fourth Industrial Revolution: Reshaping the Business for Sustainable Development in Bangladesh.
- Strengthening Technology-Based Teaching and Learning Process in 4IR Era.

5 Conclusions

The findings of this study have successfully attained the research objectives and may contribute to the literature on the Fourth Industrial Revolution in Bangladesh and other developing countries. The findings also provided insights into the limitations of Bangladesh regarding incorporating 4IR and the areas where improvements are needed. Detailed suggestions have been given by the researchers to mitigate the existing challenges of incorporating 4IR and enhance its effectiveness in Bangladesh.

Embracing this 4IR will pave the way for the country's development and will also aid to sustain in this competitive world. Though this transition from traditional methods of working and teaching to a more technology-based one won't be easy, if it can be done, it will help Bangladesh to achieve the Sustainable Development Goals (SDGs).

References

1. Aghimien, D., Aigbavboa, C., Matabane, K.: Dynamic capabilities for construction organizations in the fourth industrial revolution era. Int. J. Constr. Manage. 1–10 (2021).https://doi.org/10.1080/15623599.2021.1940745
2. Agrawal, R., Wankhede, V.A., Kumar, A., Upadhyay, A., GarzaReyes, J.A.: Nexus of circular economy and sustainable performance in the era of digitalization. Int. J. Prod. Perform. Manage., pp. 1–27 (2021). https://doi.org/10.1108/IJPPM-12-2020-0676

3. Bhuiyan, A.B., Ali, M.J., Zulkifli, N., Kumarasamy, M.M.: Industry 4.0: challenges, opportunities, and strategic solutions for Bangladesh. Int. J. Bus. Manage. Future **4**(2), 41–56 (2020). https://doi.org/10.46281/ijbmf.v4i2.832
4. Brenner, B.: Transformative sustainable business models in the light of the digital imperative—a global business economics perspective. Sustainability **10**(12), 4428 (2018). https://doi.org/10.3390/su10124428
5. Chidera, U.: Top 10 Skills You Need to Thrive in the 4th Industrial Revolution. After School Africa (2020). Retrieved Sept. 18, 2021, from https://www.afterschoolafrica.com/51364/top-10-skills-you-need-to-thrive-in-the-4th-industrial-revolution/
6. Cronbach, L.J.: Essentials of Psychological Testing. Harper and Row, New York, NY (1970)
7. George, G., Merrill, R.K., Schillebeeckx, S.J.: Digital sustainability and entrepreneurship: how digital innovations are helping tackle climate change and sustainable development. Entrep. Theory Pract. **45**(5), 999–1027 (2021). https://doi.org/10.1177/1042258719899425
8. Guimón, J.: Promoting university-industry collaboration in developing countries. World Bank **3**, 12–48 (2013)
9. Kemp, C.: Top 5 Skills Teachers Need to Flourish in the Fourth Industrial Revolution. Ed Tech, Global Connections, mrkempnz, Personal, Professional Development (2018). Retrieved Sept. 19, 2021, from http://mrkempnz.com/2018/05/top-5-skills-teachers-need-to-flourish-in-the-fourth-industrial-revolution.html
10. Nakkeeran, R., Babu, R., Manimaran, R., Gnanasivam, P.: Importance of outcome based education (OBE) to advance educational quality and enhance global mobility. Int. J. Pure Appl. Math. **119**(17), 1483–1492 (2018)
11. Parasuraman, A., Zeithaml, V., Berry, L.: A conceptual model of service quality and its implication for future research (SERVQUAL). J. Market. Autumn **49**(4), 41–50 (1985)
12. Ramraj, U., Marimuthu, F.: Preparing undergraduate learners with skills required by a transformative work environment. Int. J. High. Educ. **10**(1), 287–294 (2021). https://doi.org/10.5430/ijhe.v10n1p287
13. Teece, D.: Dynamic capabilities as (workable) management systems theory. J. Manag. Organ. **24**(3), 359–368 (2018). https://doi.org/10.1017/jmo.2017.75
14. Yu, S., Zhang, S., Yuizono, T.: Exploring the influences of innovation climate and resource endowments through two types of university-industry collaborative activities on regional sustainable development. Sustainability **13**(14), 7559 (2021). https://doi.org/10.3390/su13147559

Fourth Industrial Revolution: Shaping the New Bangladesh

Vincent Chang, Charalabos C. Doumanidis, Arshad M. Chowdhury, and Dave Dowland

Abstract A broader framework for preparing a national strategy for Bangladesh to participate in the global fourth Industrial Revolution (4IR) has been presented. Specific challenges and unique opportunities for Bangladesh considering the global nature of 4IR are introduced. A conceptual functional model, a set of action items, and a sustainable financial model including a tentative roadmap have been proposed.

Keyword 4IR

1 Introduction

The first three industrial revolutions, enabled by steam power, electricity and hard automation, and robotics and computer software, respectively, disrupted industry at a *localized* scale. By contrast, the fourth industrial revolution (4IR), empowered by the internet, cloud, big data, artificial intelligence and machine learning, blockchain, industrial analytics, nano/bio-technology, Internet of Things (IoT), robotics and unmanned aerial vehicle (UAVs), augmented reality and virtual reality (AR/VR), etc., is by its nature a *globalized* transformation enterprise. This is because it relies on industrial, academic, and government synergy at an international scale, and depends on know-how, skillsets, data, and human and material resources all over the world, with no country being self-sufficient or independent; While at the same time, it opens pathways to global markets and socioeconomic impacts. For developing nations such as Bangladesh, this upcoming reality entails several special ***challenges*** such as:

i. Need for better and functional connectivity with global markets and networks, from the academic, scholarly and commercial all the way to linguistic, cultural and geopolitical connectivity systems;

V. Chang · C. C. Doumanidis · A. M. Chowdhury · D. Dowland (✉)
Brac University, 66 Mohakhali, Dhaka, Bangladesh
e-mail: dave.dowland@bracu.ac.bd

S. Hossain et al. (eds.), *Proceedings of International Conference on Fourth Industrial Revolution and Beyond 2021*, Lecture Notes in Networks and Systems 437,
https://doi.org/10.1007/978-981-19-2445-3_32

ii. Weakness in the knowledge supply chain of research, development, education, innovation, and technology transfer, affecting adversely their industrial productivity;
iii. Struggling with human and physical *infrastructure* challenges such as training and motivation opportunities, aged or semi-functional facilities, threats by pandemics, climate change, etc.;
iv. Requirement of modern, efficient, and effective administrative structures, policies, procedures, and regulations supportive of requisite fast-paced cooperation with the rest of the world.

Simultaneously, the global character of 4IR also presents unique ***opportunities*** for Bangladesh, such as:

i. Taking advantage of its unique critical mass and strong professional drive of its *human resources* to direct personnel to new-generation, high-tech professional choices of the future;
ii. Capitalizing on its exceptional *development* rate and natural *wealth* to direct generated resources to modern economic development threads aligning with 4IR initiatives;
iii. Further development of its worldwide *connectivity* with scientifically leading diaspora community, such as in the USA, UK, Australia, and across the world;
iv. Benefiting from the newer expatriate community of prominent 4IR *technocrat* residents of Bangladesh, and investing in the current emphasis trend toward technological research

2 Objectives

This concept note is intended to provide a framework outlining modalities of Bangladesh's participation in the international 4IR renaissance for the best benefit of the country, its people, and the entire world. The ***objectives*** of such a 4IR initiative in Bangladesh are:

i. Reskilling and training of its *workforce* toward 4IR technologies, such as IoT, VR/AR, Three-Dimensional Printing (3DP), imaging/visualization, robots and UAVs; all promising exciting, well-paid, sustainable careers;
ii. Upgrading its physical *infrastructure* for scientific and technological education, research, development, innovation and implementation, to conform with modern international quality standards;
iii. Establishing and/or improving its *management* mechanisms for a seamless connection and synergy of the *public* with academe, industry, and government toward participation in 4IR;
iv. Modernizing of its *industry* in all active sectors such as textile, garment, pharmaceutics, agriculture, etc., via effective *technology transfer* from university and other research labs of the country and world;

v. Broadening and deepening its network of *international* partnerships and cooperation with leading science and technology actors of the 4IR universal rebirth, with many socioeconomic gains.

3 Success Examples

Successful accomplishment of similar objectives has been recently or is presently witnessed in many developing countries, serving as ***models*** for 4IR in Bangladesh, namely, those in East Asia (such as Singapore, https://smart.mit.edu/), Middle East (e.g., Israel, https://innovationisrael.org.il/en/), Central Asia (such as Kazakhstan, https://astanatimes.com/2020/01/aifc-to-establish-fourth-industrial-revolution-affiliation-centre/), etc.

In Bangladesh, the office of the Prime Minister has also emphasized skilled manpower creation for the 4IR (https://www.thedailystar.net/country/news/pm-emphasises-skilled-manpower-creation-4th-industrial-revolution-2001013, https://www.albd.org/articles/news/35302/Govt-creating-skilled-manpower-for-%E2%80%984th-Industrial-Revolution%E2%80%99:-HPM-Sheikh-Hasina). It appears that such fruitful efforts toward national development "*from the third to the first world in just one generation*" (*Lee Kuan Yew, Singapore*) via an industrial revolution share certain common ***attributes***:

A. A strong and visionary executive *leadership*, coordinating all aspects of the industrial revolution and directing them toward development in critical industry sectors and needs of each country;
B. A strategic *investment* by the government to its higher education system, often with external support such as by the World Bank HEAT project (https://projects.worldbank.org/en/projects-operations/project-detail/P168961)
C. A focused national *partnership* with pre-eminent international technological institutions to accelerate and transform its industry and socioeconomic impact (e.g., Singapore–MIT Alliance).

4 Functional Model

A conceptual *functional model* for a "**4IR engine**" is illustrated in Fig. 1. Similar to an actual engine consuming fuel, air, and supplies and producing work, heat, and exhaust products, 4IR relies on industry, higher education, and the broader society to yield career and economic benefits, know-how property, and social-cultural impacts. Like a real engine, 4IR needs a starter initiative with government leadership, stimulus investment, and national and international partnerships to enable feedback acceleration and lead to its productive and transformative operation.

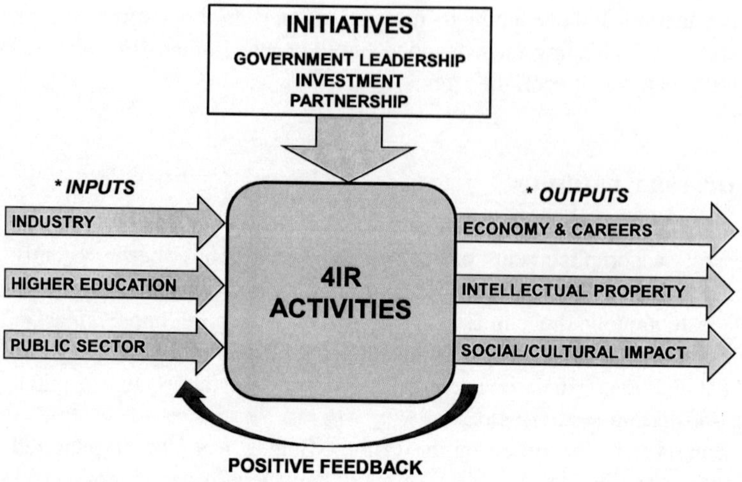

Fig. 1 A conceptual functional model of 4IR engine

5 Action Plan

This analogy clearly calls for a close public–private partnership (PPP) of the Bangladesh government with both local and international industry and higher education in order to implement an ***action plan*** addressing the above stated objective. Such action items should include, but are not limited to the following activities:

1. Developing 4IR focused academic programs and trained personnel:
   a. Establishment of new 4IR training, reskilling, and retraining programs across *higher education*, including degree courses and minor programs offered also for open associate certification, in topics such as artificial intelligence, big data, industrial analytics, internet of things, nano/biotechnology, robotics and UAVs, virtual and augmented reality, etc.
   b. Establishment of dedicated 4IR Master of Science *graduate programs*, to be combined with undergraduate curricula in (4 + 1) year joint BS–MS programs for selected high-performing students, so that they could acquire the requisite interdisciplinary skills in more than one field, e.g., BS in Engineering plus MS in AI, in 5 years taking advantage of summer brakes for research electives.
   c. Enabling of research projects of industrial interest in higher education through advanced graduate programs for the *doctoral degree*, including in private universities, and relaxing limitations for academic program development in private universities in order to synchronize with the 4IR aggressive time scales.

   d. Aligning academic curriculum to produce an effective workforce for a 'knowledge economy', where hard skills are complemented by high-level soft skills and aptitudes, critical thinking skills, resourcefulness, imagination and creativity, ability to improvise in fast-changing situations, teamworking, communication, leadership, entrepreneurship, life-long learning to operate in a global and borderless world.

2. Building 4IR focused R&D capacity:

   a. Establishment of at least one modern, fully equipped, and manned research and innovation *laboratory, center or institute* in each university with industrial sponsorship, such as named facilities, endowed chairs, and student scholarships. This should be implemented in a complementary fashion, so as to promote research and development interests of the Bangladesh industry.
   b. Foundation of a *4IR Innovation Park*, with physical and administrative resources to facilitate innovation via the establishment of startup joint ventures between industry, academe, and government, in a *Special Economic Zone* with incentives for international corporate participation and investment, along with the legal framework to support its operation and impact to the Bangladeshi industry.

3. Outreaching:

   a. Setting up one or more state-of-the-art *4IR Museums of Science and Technology*, open to the public with interactive exhibits of industrial technologies including robotics, holograms, simulators, etc., (e.g., https://mitmuseum.mit.edu/) to promote, along with online educational materials and webinars, public awareness and literacy on 4IR and solicit the interest and professional direction of the next generation of young students, their parents, and teachers.
   b. Setting up strong and interconnected *Alumni Associations* of universities, bringing together their graduates working in the industry back to their alma mater, along with parents and other stakeholders, in order to support 4IR projects, internships, and career fairs of industrial interest in academe, and therefore, facilitating graduation and transition of students and trainees to professional careers.

4. Technology Transfer:

   a. Data analysis and selection of a model *technology transfer* mechanism of intellectual property from universities and research institutes to Bangladesh industry, such as the MIT joint ventures model, with academic co-ownership of the startup/JV and with national industry securing the rights of the first negotiation for commercializing patents, copyrights, etc., resulting from the joint effort.

b. Establishment of cross-sector *networks* such as the NSF Industry-University Cooperative Research Centers (IUCRC, https://iucrc.nsf.gov/), Grant Opportunities for Academic Liaisons with Industry (GOALI, https://www.nsf.gov/eng/iip/goali.jsp), i-Corps™, etc., to bring together the previous university labs and centers with international institutions and industry partners.
c. Establishment of an independent *Foundation for International Research in Science and Technology* (FIRST, or *Fourth IR Synergy for Transformation*) as a top-end public–private partnership to coordinate hierarchically all previous activities, provide support and report outcomes, and align the 4IR initiatives to the global trends and impacts to Bangladesh industry, academe, and society.

5. Partnership:

   a. Alignment and synergy of new programs to partnerships with eminent *international institutions*; e.g., co-design of Master of Science curricula in coordination with the MIT MicroMasters program, allowing students to train themselves on 4IR subjects with quality online materials (EdX); and then travel briefly to the USA to take an examination granting them an MIT degree. This is also expedient for certification and accreditation by international societies such as ABET, Washington Accord, etc.
   b. Cooperation of Bangladeshi research and innovation with *international funding* and support agencies, such as the European Commission Horizon Europe 4th cluster (Global Challenges in Digital Industry and Space), its Marie S. Curie Mobility Program, and its European Innovation Council and Ecosystem (https://ec.europa.eu/info/research-and-innovation/); the United States NSF Partnerships for International Research and Education (PIRE, https://www.nsf.gov/pire) and Fulbright Program; the UK Newton and Australian Research Council (ARC), etc.

6 Functional Model

A sustainable *financial model* for supporting such a 4IR initiative in Bangladesh could follow the paradigm of the *US* Small Business Innovation Research (SBIR) and Small Business Technology Transfer (STTR) programs (https://www.sbir.gov/about), illustrated in Fig. 2. In this model, Phase I in the low-low-*technology readiness levels* (*TRL*) research usually in academe and research institutions is stimulated by government funding to prove the feasibility of a 4IR technology. This is followed by Phase II toward mid-TRL development of a working prototype, co-funded by industry with government matching support. The final Phase III toward commercialization of the product and production line is funded by industry and the private investment sector, yielding taxable revenue for the government to recover its stimulus investment and continue 4IR support. The US Government invests 2.5% of the annual budget

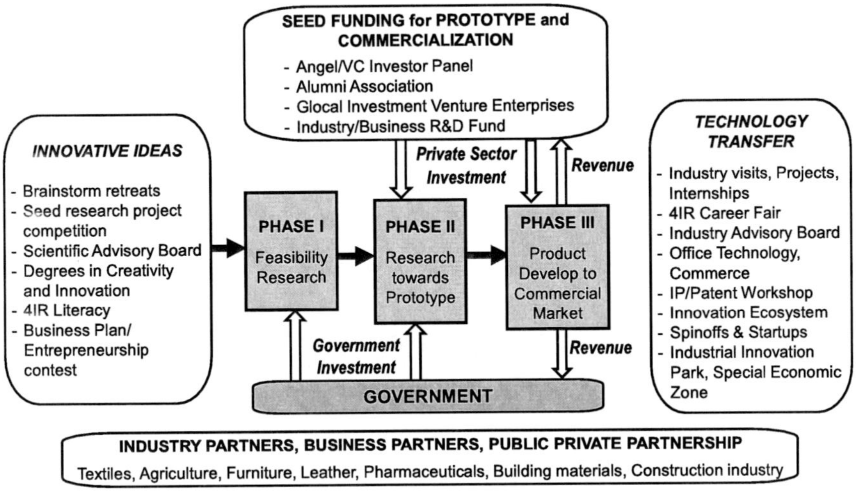

Fig. 2 A sustainable financial model of 4IR initiatives

of all participating agencies toward this SBIR/STTR investment to the 4IR. This is implemented with solicitation of competitive proposals by small startup firms and joint industry–university ventures with innovative ideas; support by private seed funding toward technology transfer; and investment by industry and business partners into product development to commercial markets.

This model was also recently and successfully adopted by the Government of *Kazakhstan* in its "Subsoil Order" to support the national mining and hydrocarbon industry toward 4IR directions; The government taxes the beneficiary industry by 1% of production costs, invested into the national university system (Nazarbayev Fund) toward spin-off of academic–industry joint ventures and small businesses to develop 4IR science and technology of interest to and for modernization of this industry (to a total of $250 M annually). In the case of Bangladesh, which is a designated international cooperation country to the EU Horizon Europe program, this could be co-funded toward its participation in the European Innovation Ecosystem and other 4IR-oriented initiatives. Several global NGOs with activity in Bangladesh, such as the B&M Gates Foundation, OSUN, and the World Bank could also contribute to research and outreach aspects of 4IR projects, e.g., through the international BRAC network, etc., contributing to their worldwide impact.

7 Performance Metrics

Some useful ***metrics*** for the outcomes in various areas of a 4IR initiative in Bangladesh could include:

4IR activity	Performance metrics
1. Developing 4IR focused academic programs and trained personnel	i. The number of undergraduate minor/vocational 4IR new retraining/reskilling programs in higher education ii. The number of students and graduates taking advantage of 4IR education and training iii. The number of graduate doctoral, masters and fast-track 5-year graduate programs established
2. Building 4IR focused R&D capacity	i. The number and research capacity of new laboratories/centers/institutes in universities addressing 4IR technology development, and the monetary/in-kind industry investment in them ii. The number of joint ventures and other small businesses incubated in the 4IR Innovation Park, and the total industrial, government, and foreign investment attracted to it
3. Outreaching	i. The number of annual visitors of the 4IR Museum(s) of Science and Technology, and the number of online visits to the web materials of the 4IR portal ii. The number and total value of new professional opportunities, i.e., careers, traineeships and internships generated by the 4IR initiative in Bangladesh industry
4. Technology Transfer	i. The total invested value, return, and ROI of the Bangladesh government, academe, and industry from 4IR programs, at all levels and cumulatively as a contribution to the GDP of the country ii. The total number of industry–university networks and their cumulative financial activity
5. Partnership	i. The number of MoUs and MoAs with international universities and research institutions toward 4IR projects, and the global standing of these foreign partners ii. The total value of research and other support by international foundations, programs, and other activities related to 4IR science and technology, that were carried out in Bangladesh

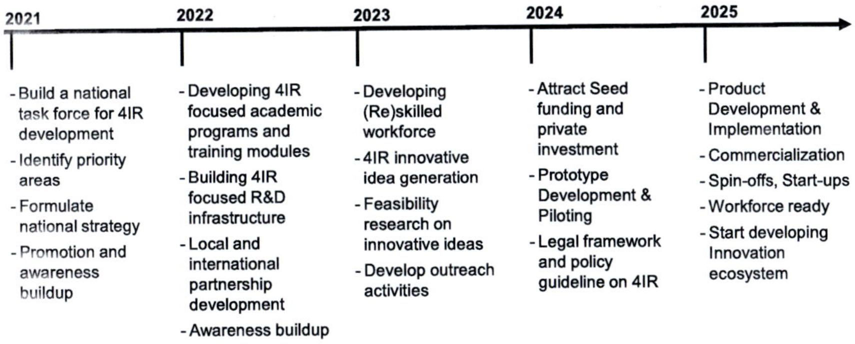

Fig. 3 Tentative roadmap of 4IR initiatives in Bangladesh

8 Tentative Roadmap

A tentative ***roadmap*** of the 4IR initiative of Bangladesh is shown in Fig. 3. The rigorousness and consideration of the proposed timeline are developed in accordance with the similar roadmap of the National Strategy for Artificial Intelligence Bangladesh (2019–2024).

9 Conclusion

The fourth industrial revolution (4IR), a globalized transformation enterprise, opens pathways to global markets and socioeconomic impacts for all participating countries. However, preparing a nation for participating in 4IR requires strategic planning and initiatives. We have proposed a broader framework outlining various modalities to develop a national strategy for Bangladesh to participate in global 4IR. The specific challenges and unique opportunities for Bangladesh have been discussed followed by the objectives and goals of the 4IR initiative and a few recent success stories on similar 4IR initiatives by other developing countries. We have also presented a conceptual functional model, a set of action items, and a sustainable financial model for the proposed 4IR initiatives. Finally, the performance metrics for the outcomes and a tentative roadmap indicating the timeline of 4IR initiatives are presented. We strongly believe that the proposed model will be helpful to formalize a national strategy for Bangladesh in its preparation for participating in the global 4IR renaissance of the twenty-first century.

Bibliography

1. UGC International Conference on Fourth Industrial Revolution and Beyond, 9–11/12/2021, Dhaka, Bangladesh—Proceedings to be published by Springer & Elsevier
2. Mannan, A.: Challenges and Opportunities of 4IR in Bangladesh, a2i Cabinet Division/ICT Division, 1/4/2021 (presentation)
3. The World Bank, Higher Education Acceleration and Transformation Project (P168961), Project Information Document and ESRS Appraisal Stage, 23–24/12/2020
4. Information and Communication Technology Division of GPRB, National Strategy for Artificial Intelligence Bangladesh, March 2020
5. https://ictd.portal.gov.bd/sites/default/files/files/ictd.portal.gov.bd/page/6c9773a2_7556_4395_bbec_f132b9d819f0/Draft%20-%20Mastering%20National%20Strategy%20for%20Artificial%20Intellgence%20-%20Bangladesh.pdf
6. LICT Project, Education and Skill for Leveraging Fourth Industrial Revolution, June 2019
7. Rumi, M.H., Rashid, M.H., Makhdum, N., Nahid, N.U.: Fourth industrial resolution in bangladesh: prospects and challenges. Asian J. Soc. Sci. Legal Stud. **2**(5), 104–114 (2020)

Development of an IoT-Enabled Digital Noticeboard

Sihabul Islam, Md. Liakot Ali, and Muhammad Sheikh Sadi

Abstract A noticeboard is an essential part of any educational institution because it is a common medium of communicating with every entity of that institution. In educational institutes, permission and signatures are generally taken from the academicians and institution authorities on the notice paper. However, this process creates different complexities in most cases. In this paper, we have developed a digital noticeboard for the academicians of educational institutions that can reduce a huge load of paper and pencils while displaying the notices on the digital board in a comfortable way. Moreover, a web application has been developed through which users can send data through the Internet of Things (IoT). In addition to these, this system provides an easy, fast, and reliable way of showing the notices attractively on an LCD monitor for an educational institution where faculty members, students, and university administration would be able to communicate conveniently.

Keywords Digital noticeboard · Internet of things (IoT) · Academicians · Raspberry Pi · Cloud computing

1 Introduction

Digital Noticeboard is a demand in this era of Information and Communication Technology (ICT). Conventional noticeboards are wooden where information is disseminated in the papers stick to it usually seen in various places like educational institutions, railway stations, bus terminals, shopping malls, offices, airports, etc. [1–3].

S. Islam · Md. L. Ali · M. S. Sadi (✉)
Institute of Information and Communication Technology (IICT), Bangladesh University of Engineering Technology (BUET), Dhaka, Bangladesh
e-mail: mdsadi@ieee.org

S. Islam
e-mail: 1018311004@iict.buet.ac.bd

Md. L. Ali
e-mail: liakot@iict.buet.ac.bd

S. Hossain et al. (eds.), *Proceedings of International Conference on Fourth Industrial Revolution and Beyond 2021*, Lecture Notes in Networks and Systems 437,
https://doi.org/10.1007/978-981-19-2445-3_33

The problem here is that a huge volume of papers and accessories are needed for displaying day-to-day information. Moreover, a dedicated person is required to place the notice on the wooden board. It uses the workforce as well as kills huge time.

Previously, a digital noticeboard has been proposed to address this problem. Some of the research in this area is published in the literature [4–7]. A digital noticeboard using GSM technology is proposed by Gurav and Jagtap [4] where only text messages can be transferred. Kadam et al. [5] have proposed Bluetooth technology and Bhoyar et al. [6] have proposed Zigbee technology for the implementation of the digital noticeboard. However, these methods are suffering from several limitations, such as working in a smaller area, slow processing speed, and failure to support video or PDF files in the system.

This paper facilitates communications among the university administration, teachers, and students through an IoT-enabled digital noticeboard. It is user-friendly, cost-effective, highly visual, and easy to update. Whenever the university authorities pass a notice, it can be shown directly to the LCD monitor. There is no need to print the paper and attach it manually to the noticeboard. Many students do not have the habit of reading noticeboards. But by using the proposed system, we can get the attention of the students. As a result, many students will be able to know the important messages very easily. A web application has been developed through which the administration or the faculty members can individually assign text, image, audio, or video to the cloud according to their role. The system loads the data from the cloud and shows it on the LCD monitor in the form of a web page. Students can view this notice from anywhere by browsing this URL. Although this system has been developed for the university environment, it can also be used for various purposes such as railway stations, bus terminals, shopping malls, offices, and airports. In that case, the web application needs to be customized based on its use case diagram for different environments.

The rest of this paper is arranged into several sections as follows. Section 2 presents the existing methods for implementing the digital noticeboard system. The proposed methodology is shown in Sect. 3. The methodology is divided into four subsections. Section 4 shows the results of the proposed system. Finally, the paper has ended with a short conclusion.

2 Related Work

In this section, we have explained several existing methods in the proposed area of research.

Some researchers proposed a digital noticeboard system using GSM technology [7, 8]. Here, a 16 × 2 LCD screen and a GSM module connected with an Arduino board are used. After receiving the text information from the authorized user via the GSM module, it is displayed on the LCD screen. However, the system deals with the text information only.

Table 1 The summary of the findings obtained by the existing methods

Existing methods	Interfaces	Necessary components	Drawbacks
Lavanya et al. [7]	Microcontroller	ATMEGA 328P, LCD, GSM module	Can only send a text within 100 m
Basarkod et al. [8]	Microcontroller	Arduino UNO R3, GSM modem SIM 900A, LCD monitor	Can send only text
Sakthieswaran et al. [9]	Wi-Fi	Microcontroller, display panel	Cannot send image, pdf
Sunitha et al. [10]	Bluetooth	Arduino board, LCD	Only messages can show in text format. Can only send a text within 10 m
Jadhav et al. [11]	PC, Wi-Fi module	Raspberry Pi, ARM	Access for smaller distance
Lavanya et al. [12]	Cloud	Arduino, LCD monitor	Only one message can be sent

Bluetooth technology has been introduced by implementing the digital noticeboard [10]. Here, the text is sent via Android phone. The receiver receives the text and displays it on an LCD. Bluetooth technology cannot transfer data over a distance of more than 10 m. Nonetheless, working within 10 m is not practical in most of the scenarios.

The Wi-Fi-based digital noticeboard was proposed by the researchers [9, 11, 12]. In that system, the data is first sent to the webserver. The receiver then takes the data from the server and shows it on the LCD monitor. But there are still some imperfections in this system such as images, PDF files, audio, and video type information that cannot be transmitted here.

A common method has been proposed by several authors in [13–18]. Here, a website has been created through which users provide the necessary information in the cloud. Then, Raspberry Pi shows the given information from the cloud to an LCD monitor. Notice can be among any one of the formats: text, image, PDF, audio, and video. However, this method does not support all the formats above.

The summary of the findings obtained by the existing methods is shown in Table 1.

3 Methodology

This section starts with a description (as shown in the first sub-section) of the overview of the proposed system. The second subsection of the section explains a web application that has been developed for displaying university notices. The process to pass and show the notice on a digital display is described in the third subsection. The fourth subsection illustrates the interface to store files in the developed system. The section is ended with a discussion of system configuration (as shown in the fifth subsection).

3.1 The Overview of the Proposed System

The architecture of this proposed system is shown in Fig. 1. The main purpose of this system is to display university notices in text, image, pdf, audio, and video format through a digital noticeboard. It displays notices based on a web application. A web application has been developed for this implementation. There are two types of authority for uploading notices to the cloud. One is the role of academic authority and the other is the role of administrative authority. When the authority uploads,

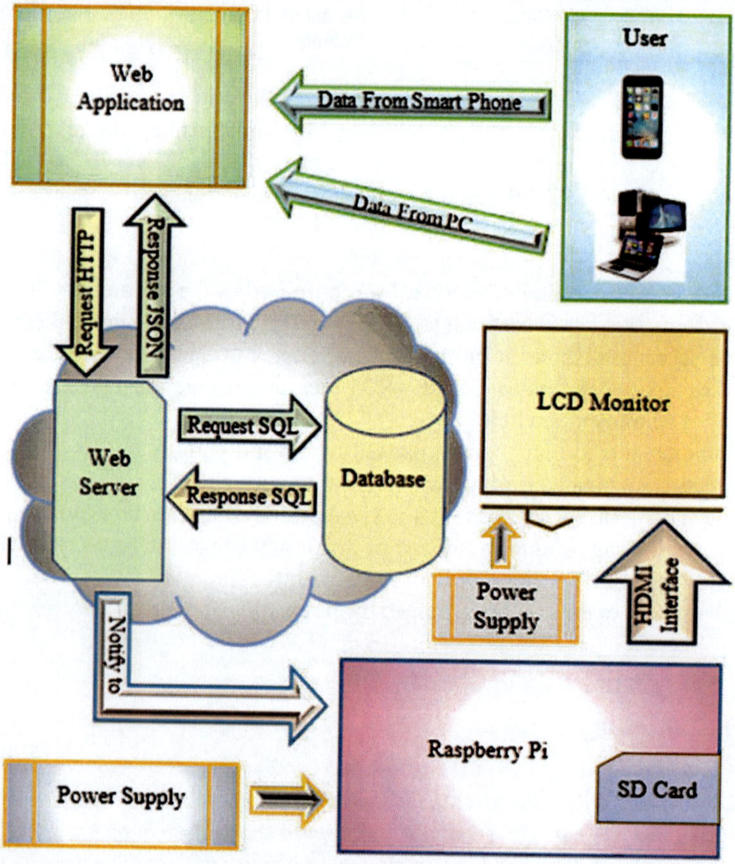

Fig. 1 The architecture of the proposed system

edits, or removes a notice, the web application communicates with the web server via HTTP request and JSON response. Then the webserver and the database meet with each other via SQL Request and SQL Response.

This system uses Raspberry Pi as a processor plugged into a high-definition LCD monitor. The Raspbian Pi OS is installed on an SD card and the card is attached to the Raspberry Pi. Raspberry Pi visits a specific web URL of the notice page from the cloud and displays notices on the LCD monitor via an HDMI interface.

System Requirements. A web application has been developed to implement this system. Different languages like HTML, CSS, Bootstrap, jQuery, MySQL, and PHP have been used to develop the web application. Raspberry Pi 3 Model B, 4 GB SD card, Raspbian OS, HDMI to VGA connector, and LCD Monitor have also been used for hardware implementation.

3.2 *Development of the Web Application*

The web application functionalities of the system are described by using the use case diagram as shown in Fig. 2. There are two kinds of registered users. One is the administrative admin and the other is the faculty member. The diagram presents

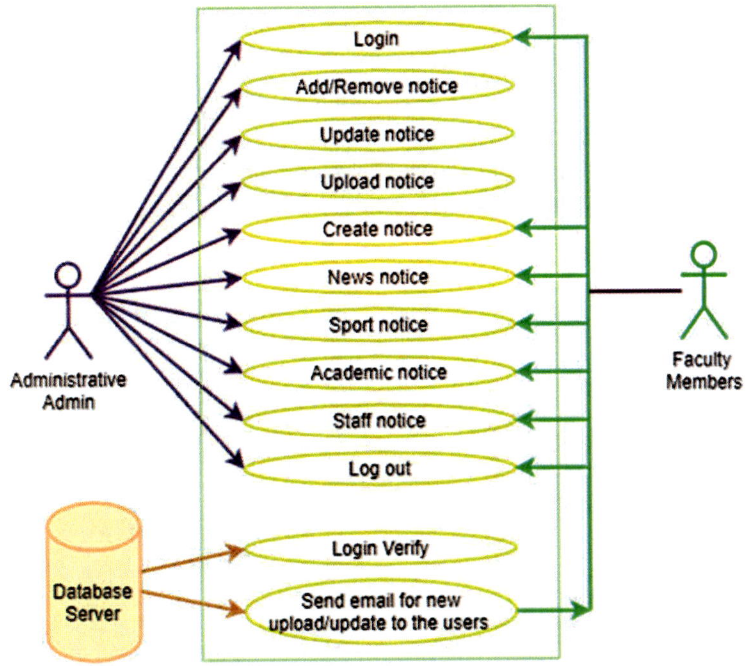

Fig. 2 The use case diagram of the system

the interactions between the system and administrative admin as well as the faculty members. The registered users can log in to the system and create notices according to their roles. The administrative admin maintains the flow of the noticeboard's message. He/She can add, edit, upload, or remove any notice on the noticeboard. He/She can log out after completing the required work. The server authorizes an user through email communication.

Users' Authentication Process. A user can be registered or unregistered. The authentication process of the system is given in Fig. 3.

Registered users will be able to login into the system with the correct username and password. An unregistered user can become a registered user by completing an online form. The information on the online form must be valid. Thus, the system processes the authentication step.

Users' Activity Analysis. A valid user has to complete a few steps to send a notice to a digital LCD monitor. After logging in, the user will get a web page to upload any notice according to the role. The flow diagram of the user's activity is illustrated in Fig. 4. Admin can create, upload, edit or remove all types of notices. And faculty members can only create, upload or edit notices, but cannot delete them. The user

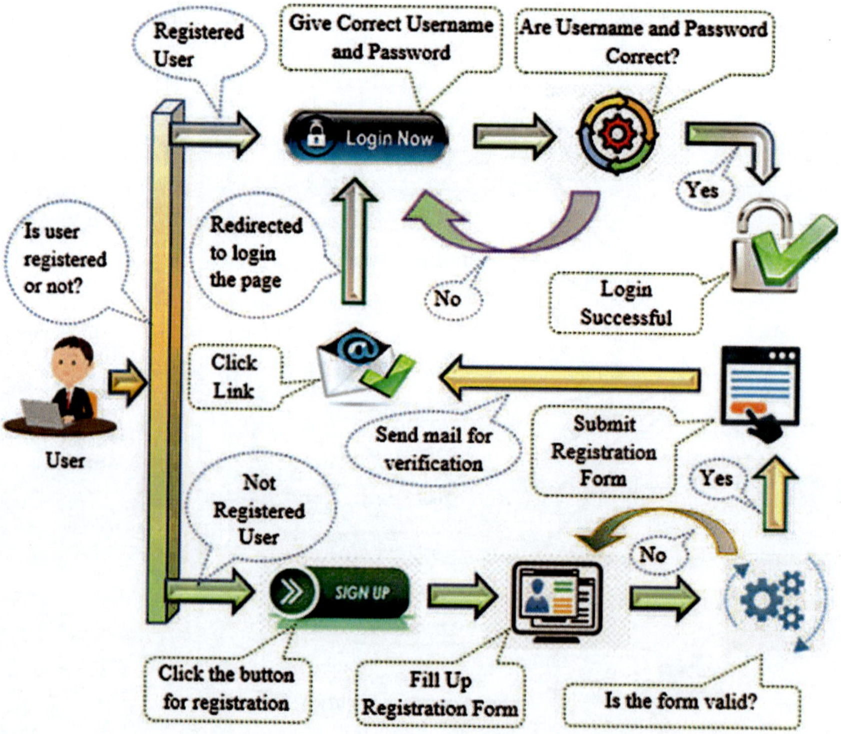

Fig. 3 The authentication process of the system

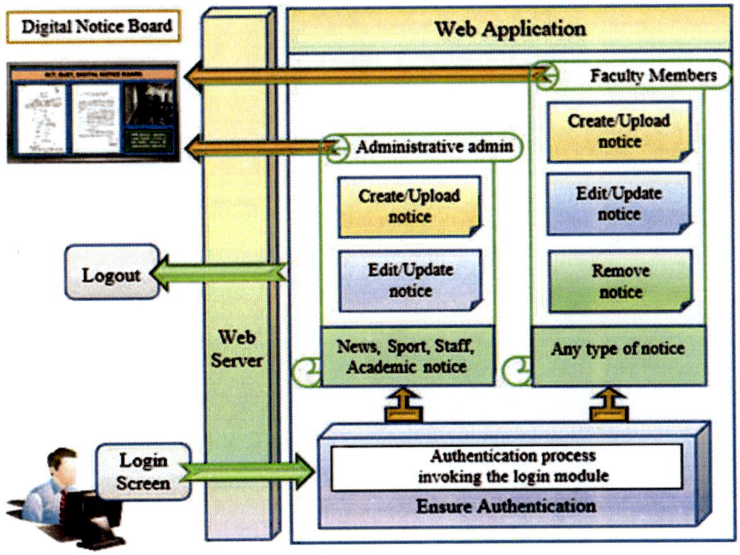

Fig. 4 Users' activity flow diagram for sending a notice to the digital board

will then log out manually. If he/she does not log out manually, it will be logged out automatically after 5 min.

3.3 The Process to Pass and Show the Notices on LCD Monitor

Figure 5 shows the notice upload page interface. This page has two mode options. Three pages of image or PDF type are shown on the monitor by selecting mode-1. Mode-2 can be used to display two images or PDF pages, a video, and text information. If one mode is selected, the other mode will be disabled automatically. The mode that will be selected in the system is displayed on the monitor. This page has a text area for writing text-type information. If any text information is typed and submitted here, it will be stored in the cloud. There is an option to select a media file for uploading audio/video files. An audio or video file is uploaded via file select or drag and drop.

Then if we click on the submit button, it will be submitted to the cloud server. In addition, this page has the option of uploading another file of images or PDF files. In this case, the file size can also be set manually. If the size of the image is smaller than the set size, it will be uploaded directly. Otherwise, the image will be resized by the web application and uploaded. When a PDF file is needed to show as notice, it will be converted to an image file and will be uploaded. Then the users will submit the files.

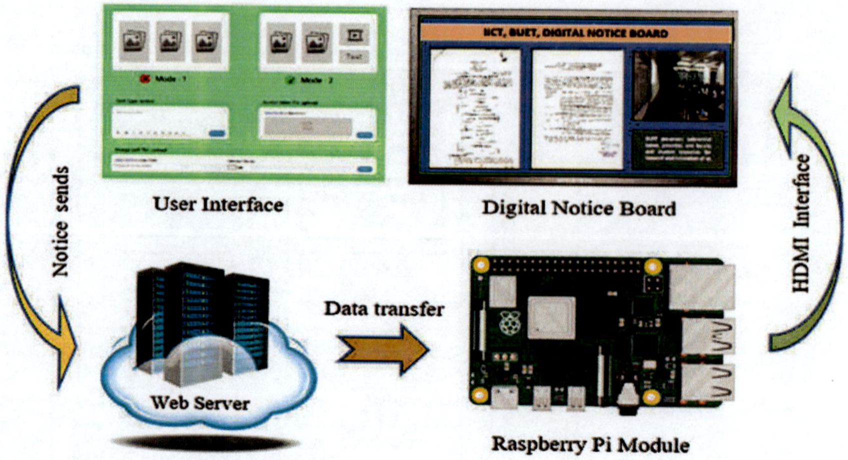

Fig. 5 Notice upload interface of the system

The overall process of the system is discussed step by step as below:

Step 1. Ensuring secure access to the System through the Login process.
Step 2. Choosing notice display mode.
Step 3. Setting the size of an image to display as a notice. Setting length for a text message.
Step 4. Selecting notice files in the format of image, pdf, audio, or video. Writing for the text message.
Step 5. Clicking the button to upload the notice.
Step 6. Repeating Step 1 for uploading another notice.
Step 7. Displaying notice on the noticeboard.

3.4 *Notice file Store Interface*

In this system, notices are sent from a smartphone or PC. Then, Raspberry Pi captures the notices. The Sender and receiver of the notices are interfaced through a wireless network. So, the users' devices must be connected to the Wi-Fi network to send data to the cloud. After establishing the internet connection, the user has to enter the login page. To access the login page, a user needs a valid username and password that must be assigned by the admin. After entering the correct username and password, he/she is redirected to the notice upload page. From this page, authorized users can submit various notices such as text, images, pdf, audio, and video files to the cloud.

These are stored in the database when a file or text is submitted from the notice uploading page. A separate interface is generated for each type of submitted file. Four different interfaces are shown in Fig. 6 for four different categories. Each notice contains information such as file name, notice type, and submission time.

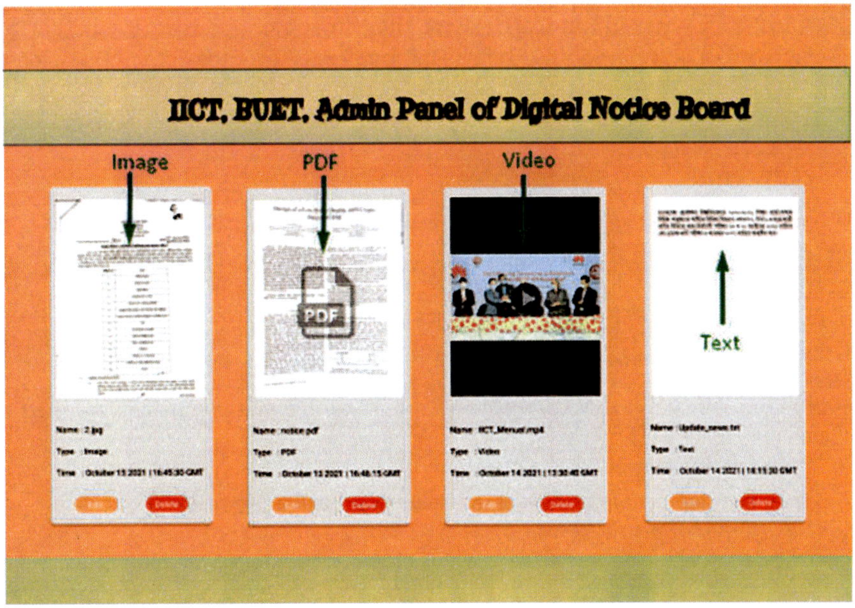

Fig. 6 Notice store interface of the system

Each interface has an edit and a delete button. Information can be edited through the edit button. Notice can be deleted from the database by clicking the Delete button.

3.5 System Configurations

A large LCD monitor is attached to the Raspberry Pi. It launches several functions according to commands automatically after connecting to the power supply. Some of the commands are discussed as follows.

Raspberry Pi's Auto Start file. This command is used to access Raspberry Pi's Auto Start file. Few necessary commands are set here through which the PI module automatically runs the command. The command is:

"*sudo nano/etc./xdg/lxsession/LXDEpi/autostart*".

Show the Window of the Web Page. This command is used to show the window of a web page in full screen automatically through the chromium browser when Raspberry Pi is open. Here, the page of http://demoone.xyz/IICT_project/board.html URL has been set which shows on the monitor. The command is:

"*@chromium-browser --start-fullscreen*
https://demoone.xyz/IICT_project/board.html".

Raspberry Pi's Screen Control System. These three commands are used to control Raspberry Pi's screen control system. The commands force the current lxsession to stop the screensaver and prevent it from blanking out. The last command turns off the dpms that prevent the desktop interface from blanking out. The commands are:

"*@xset s noblank*",
"*@xset s off*",
"*@xset -dpms*".

Sleep Mode Control. The system's sleep mode is turned off by these two commands. The commands are:

"*sudo nano /etc./lidhtdm/lightdm.conf*",
"*xserver-command = X -s 0 dpms*".

Installing Apache 2, PHP 5, and SQLite. The following commands are used to install the Apache HTTP server:

"*sudo apt-get install apache2*",
"*sudo apt-get install php5 php5-gd sqlite php5-sqlite php5-curl*",
"*sudo service apache2 restart*".

Installing ownCloud. The following commands are used to install ownCloud on the Raspberry Pi:

"*wget* https://download.owncloud.org/community/owncloud-9.1.0.zip",
"*sudo mv owncloud-9.0.4.zip /var/www/html*",
"*cd /var/www/html*",
"*sudo unzip -q owncloud-9.0.4.zip*",
"*sudo mkdir /var/www/html/owncloud/data*",
"*sudo chown www-data:www-data /var/www/html/owncloud/data*",
"*sudo chmod 750 /var/www/html/owncloud/data*",
"*sudo reboot*".

4 Results and Discussion

Extensive evaluation processes are accomplished to test how successful and effective this system will be. Figure 7 shows the implementation of the digital noticeboard. It shows two images and a PDF page on a large LCD monitor. When a new notice is uploaded to the server, it is automatically shown on the LCD monitor in the order of date and time.

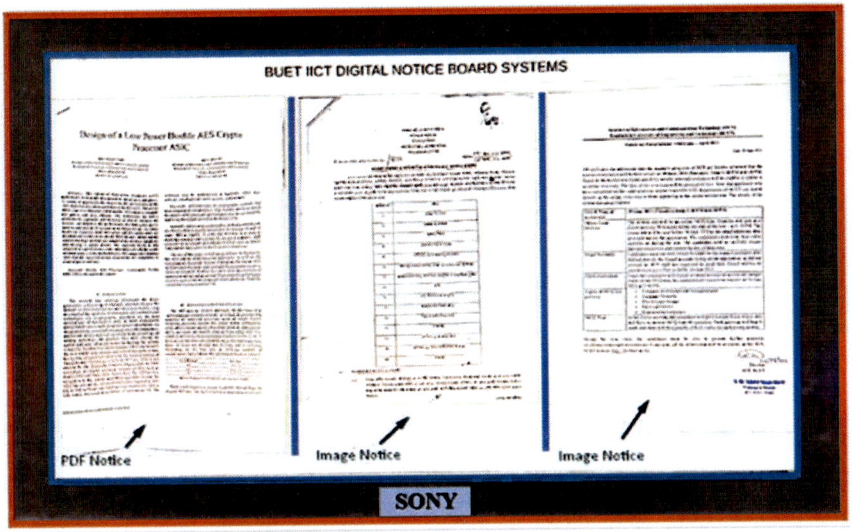

Fig. 7 Displaying notice on the monitor

4.1 *Time Requirement Analysis for Uploading Different Notices*

To show a notice on a digital monitor, it has to be uploaded to the server. Notices can be in different formats such as image, pdf, audio, and video.

These different formats require different times to upload to the server as shown in Table 2 and Fig. 8.

From Table 2, we can observe that the required time for the different files may vary for different types and formats. Here, the required time for three types of video files' sizes (e.g., 1.5 MB, 3 MB, and 10 MB) are 4.63 s, 8.49 s, and 9.18 s, respectively. Thus, we have got the various uploading time requirements for audio, pdf, and image type files to the system which are shown in Table 2. However, Fig. 8 is represented based on the data in Table 2, which shows the uploading time for different types of notices to the system.

Table 2 Time requirements of uploading Image, PDF, Audio, and video files

Video size (mb)	Required times (s)	Audio size (mb)	Required times (s)	PDF size	Required times (s)	Image size	Required times (s)
1.5	5.63	1	4.48	140 kb	4.42	220 kb	4.24
3.0	6.49	2	5.90	3.79 mb	7.05	2.5mb	5.79
10	9.17	5	8.25	10 mb	10.23	5 mb	8.13
–	–	10	10.23	–	–	9 mb	9.53

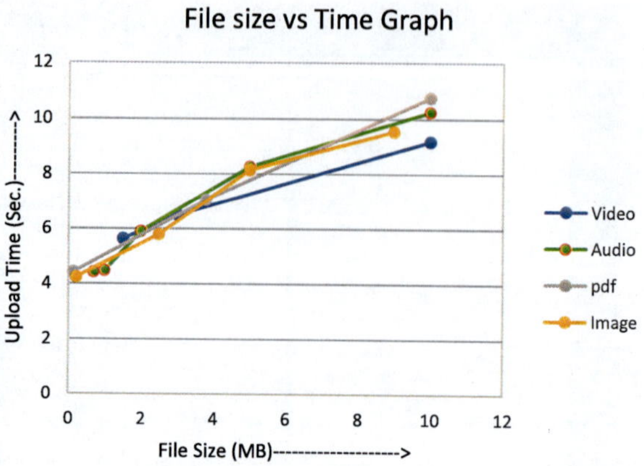

Fig. 8 Time requirements for uploading different types of files

4.2 Cost Estimation for the Proposed System

The cost of the Raspberry Pi 3 Model B complete set (including cooling fan, heat sink, power adapter switch, and 16 GB memory card pre-install Operating system) is approximately 5000 Taka. The cost for domain registration and hosting purposes per year is around 2000 taka. Note that the LCD monitor is an important part of this system. But it can be priced differently according to the system (to be applied) configuration.

4.3 The Comparison between the Proposed Method and Existing Methods

The comparison of several features between the proposed method and the existing methods is shown in Table 3.

Table 3 Comparison of several features between the proposed method and existing methods

Methods	Text	Image	PDF	Audio	Video	Time measurement	Cost estimation
Different methods in the paper [7–12]	✓	✓	×	×	×	×	×
Turkane et al. [13], Siripuri et al. [14] and Gupta et al. [15]	✓	✓	×	×	×	×	×
Ganesh [16]	✓	✓	✓	×	×	×	×
Darekar et al. [18]	✓	✓	×	×	×	×	×
Proposed method	✓	✓	✓	✓	✓	✓	✓

It can be observed from this table that the existing methods support at most three file types, whereas the proposed method supports all five categories as mentioned here. Moreover, the cost analysis of this proposed system and the required time to upload the notice files onto the server are not discussed in the existing papers, whereas this paper has considered this issue as a matter of concern.

5 Conclusions

This paper proposes a new method to develop an IoT-enabled digital noticeboard. Here, the system can upload any notice from the five categories, e.g., image, text, pdf, audio, and video files to the server and display it on the LCD monitor within a minimal time requirement. As a result, notice viewers can have soft real-time updates. And, the users can reduce their valuable time and paper wastage, as well as save huge costs, by using the proposed method. Although the system shows static notices, it would help the university administration, faculty members, and students with rapid and efficient communication. In the future, there are several scopes to enhance this research by adding additional features like showing dynamic notices, real-time editing options, supporting more categories of files, etc.

Acknowledgements The authors would like to thank IICT, BUET for allowing this research to be conducted in the Advanced Embedded System Laboratory of the Institute.

References

1. Sattiwale, G., Tongase, R., Kamble, V., Saut, R., Shriwas, M.K.: A past present and new features of digital noticeboard. Int. J. Sci. Res. Publ. **7**(5), 194–601 (2017)
2. Digital Noticeboard. https://dir.indiamart.com/impcat/digital-notice-board.html. Accessed 7 Oct 2021
3. How (And Why) to Make a Digital Bulletin Board. https://screen.cloud/blog/digital-bulletin-board. Accessed 7 Oct 2021
4. Gurav, R.K., Jagtap, R.: Wireless digital noticeboard using GSM technology. Int. Res. J. Eng. Technol. **2**(09), 2395–2456 (2015)
5. Kadam, S., Saxena, A., Gaurav, T.: Android based wireless noticeboard and printer. Int. J. Innov. Res. Comput. Commun. Eng. **3**, 12 (2015)
6. Bhoyar, C.N., Khobragade, S., Neware, S.: Zigbee based electronic noticeboard. Int. J. Eng. Sci. Comput. 5422–5424 (2017)
7. Priya, C., Ramya, C.: GSM based wireless noticeboard using graphical LCD. Ann. Rom. Soc. Cell Biol. 3601–3607 (2021)
8. Basarkod, P.I.: GSM based SMS driven automatic display noticeboard. Turk. J. Comput. Math. Edu. (TURCOMAT) **12**(11), 6919–6924 (2021)
9. Sakthieswaran, P., Sanothkumar, P., Rajesh, R., Sathish, P., Mohananthini, D.N., Saravanan, D.S.: Voice based digital noticeboard using WIFI. Int. J. Innov. Res. Technol. **6**(11), 229–235 (2020)

10. Sunitha, D., Patil, V.C., Manjula, H.N., Jebakani, S.: Digital noticeboard using smart phones-speech recognition voice command. In: International Conference on Current Trends towards Converging Technologies (ICCTCT), pp. 1–4. IEEE (2018)
11. Jadhav, V.B., Nagwanshi, T.S., Patil, Y.P., Patil, D.R.: Digital noticeboard using Raspberry Pi. Int. J. Res. Eng. Tech. (IJRET) 3 (2016)
12. Lavanya, G., Deepika, N., Sangeetha, T., Maheshwari, R., Josephine, R.: Internet of things based notifications using smart noticeboard. Int. J. Pure Appl. Math. **119**(10), 1915–1920 (2018)
13. Turkane, S.M., Taware, S.S., Kashid, P.S., Landge, C.D.: ARM-11 Based Wireless Noticeboard (2015)
14. Siripuri, K., Ailoni, P., Gudiyella, S.K., Nimmala, B.C., Emmadi, A.K.: Web Controlled Smart Noticeboard. EasyChair (2021)
15. Gupta, P., Parida, A., Naidu, A.: Digital noticeboard. Int. J. Trend Sci. Res. Dev. (2019)
16. Ganesh, E.N.: Implementation of digital noticeboard using raspberry Pi and IoT. Orient. J. Comput. Sci. Technol. **12**(1), 14–20 (2019)
17. Chen, P.W., Chen, Y.H., Wu, Y.H.: Pushing the digital noticeboard toward ubiquitous based on the concept of the internet of everything. In: Twelfth International Conference on Ubi-Media Computing (Ubi-Media) 2019, pp. 230–235. IEEE (2019)
18. Darekar, S., Davane, B., Khose, S., Panigrahi, A.: Real Time Notice Display System using Cloud (2019)

Augmented Reality-Based Lifelogging System for Reminiscence

Md. Nazmus Sakib, Mehrab Mustafy Rahman, Hasan Mahmud, and Md. Kamrul Hasan

Abstract In recent times, Human–Computer Interaction(HCI)-related research has shown an increasing interest in systems designed for supporting human memory. As regards, the extent to which Augmented Reality (AR) being used to experience the reality is blooming and being applied to each and every sector of HCI thoroughly. In this paper, we are presenting a mobile-based system and its design, build-out, and procedure to reminisce past memories using lifelogging system where users store images with related details in cloud storage based on location. When the user comes back to the location, using the concept of geo-location-based AR, the stored images are augmented into the pre-existing reality. As a result of tapping onto the objects, the pre-stored images with provided details are overlaid on the view screen. We propose that such approaches can encourage more mindful engagement with lifelogging system and AR promoting positive effects linked to reminiscing with a little to no cognitive work.

Keywords Augmented reality · Lifelogging · Reminiscence · Crowdsourcing · Memory retrieval

1 Introduction

Developments in technology have numerous applications and one of them is that people can record and store vast quantities of personal information which is necessary to recall past events. That being the case, retrieving memories has always been a substantial area in cognitive science. Apart from sentimental longing for a moment in the past, recollecting previous memories have other demands. In our aging popu-

Both of the authors have equal contribution in this research.

Md. N. Sakib · M. M. Rahman · H. Mahmud (✉) · Md. K. Hasan
Systems and Software Lab (SSL), Department of Computer Science and Engineering (CSE), Islamic University of Technology (IUT), Dhaka, Bangladesh
e-mail: hasan@iut-dhaka.edu

S. Hossain et al. (eds.), *Proceedings of International Conference on Fourth Industrial Revolution and Beyond 2021*, Lecture Notes in Networks and Systems 437,
https://doi.org/10.1007/978-981-19-2445-3_34

lation, the number of persons suffering from chronic diseases is increasing. Persons suffering from memory impairment may lose the ability to recall names, past activities, and objects. Not only for the aged population, if we think about the younger generation, but there are also many minute details of past events that are not being fetched properly, sometimes the important information is being overlooked, other times it is taking a long while to get the facts necessary at the moment. For this reason, lifelogging has become a necessity and with the increase in mobile devices available to the population, it now has an easy outlet or practice as mentioned in [1, 2].

In this article, [3] lifelogging is described as the practice of individuals recording their quotidian lives in different degrees of detail for a number of reasons, and it is a highly strong tool for reminiscence. The record comprises a database of a person's life and actions that is more or less complete. The information may be used to learn more about how individuals spend their lives. A lifelogger is a system that can attain this capacity. Reminiscence may be defined as the process of recalling prior events or experiences, as described in this study [4]. For example, if a person visits a place of interest today and after several years he returns to the same spot again, his brain tracks down that particular memory engraved deep inside years back and tries to incorporate it by relating to that location. This whole act of memory recall or recognition falls under reminiscence [5].

As per the studies, predominantly recognition works better than recall. One common explanation is that recognition can exhibit psychologically simple results of enhanced cognitive fluency induced by priming [6]. Priming is a process in which exposure to one stimulus impacts a reaction to a second stimulus without conscious instruction or purpose. Priming effects are quite strong; subjects will display priming on a cue they viewed or experienced once, a year ago, for less than a second [7]. Cues or memory cues are extensively necessary to trigger the brain responses to fetch the memories and the simplicity of the retrieval process is directly dependent on the comprehensiveness of the cue. In the case of memory recognition, all the conventional ways to fetch memories stand in need of real-world experience. AR, an interactive experience of a real-world environment where the objects that reside in the real world are enhanced by computer-generated perceptual information, widens engagement and interaction while providing an enhanced user experience [8]. Different types of AR methods include Projection-based, Recognition based, and Location-based AR.

The objective of this study is to establish a system, where subjects can experience an enhanced real-time experience while recognizing memories using their lifelog data. An amplified user experience to view the memory cues through a system centered on geo-location-based AR. We intend to make the system comprehensive yet simple and aim to retrieve memories with apparently little to no cognitive effort as our proposed system will be providing certain stimuli which are significantly easier to process than others.

Some gaps are prevalent in the existing methodologies and it is important to address those. The existing image-based lifelogging systems did not consider important inputs, e.g., time, date, location, weather, information associated with images, and the impact of the aforementioned inputs in lifelogging for reminiscence. Furthermore, some works have used multimodal data in lifelogging system but there is still considerable room to utilize those data to its zenith to reduce the cognitive effort for memory recall. In some approaches, data attaining procedure requires a lot of effort which reduces the usability of the system. In view of the issues presented above, this article aims to make the following contributions:

- Proposes an unconventional geo-location AR-based lifelogging approach.
- Recommends a quantitative metric of evaluation of memory recall (An event-specific questionnaire considering the important factors like visual, coherence, time, and perspective details).
- Introduces an approach of using relevant crowdsourced data as a tool for reminiscence.

In this study, we discuss the Related Works of Reminiscence and Lifelogging, followed by the Challenges and Problem Discussion. After that, we provide our Proposed Methodology and Architecture that tackle the problems followed by Experimental Results and Conclude this study.

2 Related Works

The review work by Ksibi et al. (2021) in [9] mentions that lifelogging is a popular research area because of increasing demand from a variety of applications but getting insights from egocentric experiences, utilizing a massive deluge of unlabeled and unstructured data, remains a serious difficulty. Lifelogging has proved itself to be a useful method in state-of-the-art technology in reminiscence. Caprani et al. (2011) in [10] proposed a technique through which a device named Sensecam will be used to record day-long activities and the user will be able to view those lifelogged records and use them as cues to recall memories. Josef Hallberg et al. (2009) in their research mentioned in [11] named their system Review Client where lifelog data include past activities and events with other people with a varying range of cues. This system proved to be helpful for people with dementia. Another lifelogging system was proposed by Matthew L. Lee (2008) in [12]. In their article, they proposed a system consisting of three parts. The first part is *Input Section* where the records or cues will be stored. The second part is *Cue Chooser* where subjects will be provided memory cues by their caregivers. This will be determined by the specific needs, since persons with memory impairment may require more than a variety of numbers and types of cues. And in the final part called *Viewer*, subjects will view those cues along with their caregivers so that they can guide them through if they struggle to remember the events. Paul McCullagh, H.Z. Chris Nugent (2014) in this article [13] developed a design for people with impaired memory and evaluates its efficacy for reminiscence using lifelogged data. The main focus was to make the system simple to use while attaining significant success in reminiscence.

In the research [14], Svetlana Nikitina et al. (2018) have proposed a smart conversational agent that can help older adults in the reminiscence process. They talk about conversational agents and cognitive services as the base for building upcoming reminiscence applications and establish the idea of a smart reminiscence agent. Interference from different sources is a part of our daily lives and some of these often become so severe that sometimes they can directly affect our memory. Benton (1957) states that when the source of interference is removed, forgetting drops from around 75% to about 25% over the course of 24 h. Other methodological factors can lower this amount by at least 10%, leaving 15% as an estimate of the forgetting during a 24-hour period in [15]. This makes it evident why there is a strong necessity for lifelogging. Furthermore, van Teijlingen et al. (2021) mentioned in [16] that memory loss is one of the primary causes of chronic impairment globally, affecting over 50 million people and resulting in over 10 million new cases each year and mild memory impairments might occur as a result of normal aging. Thus, in recent decades, there has been a significant surge in interest in rehabilitation strategies for dealing with memory loss.

Crowdsourcing has proven to be effective with reminiscence. Korovina (2016) mentions in their study [17] that by living active social lives, older people can reduce loneliness. Elder people-friendly tags need to be used during photo labeling. Publishing photos of those social contexts help refresh the memory of the people and this concludes that Reminiscence positively affects people. Taking the previously mentioned work to a further level, Svetlana et al. (2018) in [18] discussed the identification of effective and scalable crowd-based strategies for content preparation, conversation logic, and meaningful benchmark for a reminiscence conversational agent(chatbot) aimed at older adults. Yung-Chin Tsao et al. (2019) in [19], discussed the potential usefulness of incorporating AR and VR in reminiscence which engenders an enhancing effect in terms of user experience. The entire system is meant to apply as a therapy to the target subjects by alleviating the memory setbacks.

AR has a multitude of uses ranging from entertainment to educational purposes. The use of AR to increase memory recall and learning experiences in human anatomy and physiology is investigated in [20] by Chen et al. (2019). They have used Microsoft HoloLens for this purpose.

3 Problem Definition

Humans have always used visual-based methods to help them remember information; which started from cave drawings, clay tablets, and continued to text and images, or video. Lifelogging and other personal informatics systems help users collect data for self-monitoring and reflection. Human memory is unquestionably a vital cognitive ability but one that can often be unreliable due to memory fog, mentioned in the research of [21]. External memory aids such as diaries, photos, alarms, and calendars are often employed to assist in remembering important events in our past and future[22].

The existing methodologies of reminiscence are insufficient concerning robustness and comprehensiveness for memory recognition and in turn, fail to build real-time experiences in terms of visualization while retrieving memories. Consequently, they demand a great deal of cognitive work to meticulously remember past events. Therefore, for proper memory retrieval, a detailed and entrancing experience is essential. This study aims to explore the immersive experience offered by AR in reminiscence which intends to make the entire retrieving process more efficient while reducing the cognitive effort to the minimum.

4 Proposed Methodologies

We have divided our proposed system into three subsystems. Since smartphones have become the de facto lifelogging device, and smartphone-based lifelogging (SBL) research has become a popular pastime as a result of technical improvements mentioned in [23] by Ali et al. (2020), we have used a smartphone-based approach. The input subsystem takes input using the smartphone and sends it to the cloud subsystem which handles storage and processing capabilities. Lastly, the output subsystem provides an output that is meaningful for the user. As for input, the user will have to select some images that he would like to store for reminiscence and some additional information associated with it. The additional input parameters will be described later on in the article. Then comes the information stored in the cloud subsystem which will not only be responsible for storing the data sequentially and systematically but also have to do some processing there. Finally, our output subsystem will consist of several features. In this subsystem, the points known as Point of Interest or POI will be protruded whenever the user means to use the system where that POI was previously saved. An interesting aspect of the output subsystem is the crowdsourcing feature where the user can tag other users into his images, therefore, building a connected grid. So even if the user is not the creator of that POI, he can view that POI thus helping him to recognize the memories in a secondary way.

4.1 Input Subsystem

The input subsystem consists of the following properties: Image, Date, Time, Coordinates, Weather (Temperature, Humidity), Event Name, Person behind the camera, People associated with the image, and other descriptions.

4.2 Cloud Subsystem

The cloud subsystem will have two parts:

Storage: the cloud subsystem will incorporate data and store them respective to the POI. The information the user has stored will be saved accordingly there.
Crowdsourcing: The tagged people will be crowdsourced and connected via this feature. The user will get tagged if someone decides to mention him and he can do so as well to others. He will be able to view the POI of other people as well. This processing of crowdsourced data will be done on the cloud subsystem.

4.3 Output Subsystem

The output subsystem will be responsible to give the users comprehensive information about the ideal environment. When the user will revisit a place and turn on the camera through the app and view the area around him, all the POIs were created by him or the POIs where he was tagged by another user will be popped up. Upon accessing a POI, data will show the image(s) associated along with some of the information that he had previously saved. These cues will be containing all the major details which the user might require to recognize the past events without putting any cognitive effort. At the same time, the interface of the system will provide a real-world environment thus assisting the users to relive the memories without any interference.

4.4 System Flow

The system architecture in Fig. 1 will show how the data flow is occurring here:

From the input subsystem, after obtaining the required inputs the data will flow into the cloud storage and be saved there. During this phase, the processing part of the crowdsourcing and POI creation will occur virtually at that location. After that when the user will come to that location, and view through the application using his camera, he will be able to view those POI and while clicking the POI he will be able to view the data stored there, e.g., the images and the associated information.

4.5 System Implementation

For our system to be implemented, we picked the most available form of smartphone which was android and thus we worked on android studio for building our application. Also, we decided to use the WIKITUDE SDK[7], because it contains some built-in

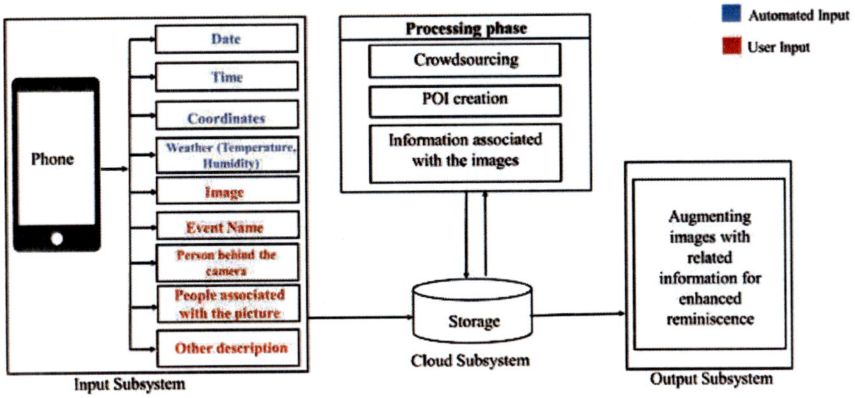

Fig. 1 The overall architecture of the entire proposed system

functions and API endpoints that ensure the easy build-up of geo-location-based AR implementations.

The overall algorithmic flow of the system is explained below:

1. A POI needs to be created using the mobile app which will enable the users to store their lifelog data. This POI creation is handled by a method 'createPOI()' which calls an API endpoint of WIKITUDE SDK.

2. The input subsystem will be responsible to upload the image the user has taken and its associated information to the cloud storage. The image and its associated information will be taken as an input in JSON format. The 'imageDetails' object shows an example of an input object.

```
imageDetails = {
     "id": "0"
     "image":"IUT.png"
     "date": "15-10-2019"
     "time": "15:40:34",
     "coordinates": "N-52.58.110, E-45.04.173",
     "weather":{"temperature":"30","humidity": "12%"},
     "eventName": "Farewell",
     "personBehindTheCamera": "Ifaz Ahmed Aflan",
     "personTagged": [{"id": "0","name": "Fahim Shahriar Reed"},
             {"id": "1","name": "Junaid Mahmud"}],
     "description":"it is an example of JSON data"
}
```

3. The cloud subsystem will be responsible to store and provide necessary data. The crowdsourcing feature mainly depends on the stored data.

Here, 'saveImageInformation()' demonstrates the post method of the cloud subsystem.

```
/* takes one parameter, 'imageDetails'.*/

saveImageInformation(imageDetails){
    /* posts the input data into cloud storage by calling
    an API endpoint of WIKITUDE SDK. */
}
```

Then, the previously created POI will be mapped with the stored image and its information for future purpose in 'mapImageInformationWithPOI()' method.

```
/* mapImageWithPOI is a map object where the key will be the POI
and the value will be corresponding image and its details.*/

mapImageInformationWithPOI(){
    mapImageWithPOI.set(POI, imageDetails)
}
```

Also the 'crowdSource()' method will be responsible to fetch POI accordance with the images where the user was tagged into.

```
/*using the get method of an API of WIKITUDE SDK, we will receive
'fetchedData' object and check whether the user was tagged in any
of the images.*/

crowdSource(){
    for each data d in fetchedData{
        if user == d.taggedUser {
            /* show corresponding POI where the user was tagged */
        }
    }
}
```

4. The output subsystem is accountable to exhibit the corresponding information to the user. After opening the app camera, the system using the geo-location-based AR shows the user all the POIs he had saved before or was tagged at. Since this is a geo-location-based concept, the user will be able to see his previously set POI only when he is nearby the location of his saved POI(s). Under this condition, the user is provided with a notification to open his camera from the application to view his previously saved POI. The view will be an interactive and augmented object containing all the necessary stored information. A snap of the user interface is shown in Fig. 2. The image shows that the user has some POIs and after tapping one of the POIs, related information appeared with associated details.

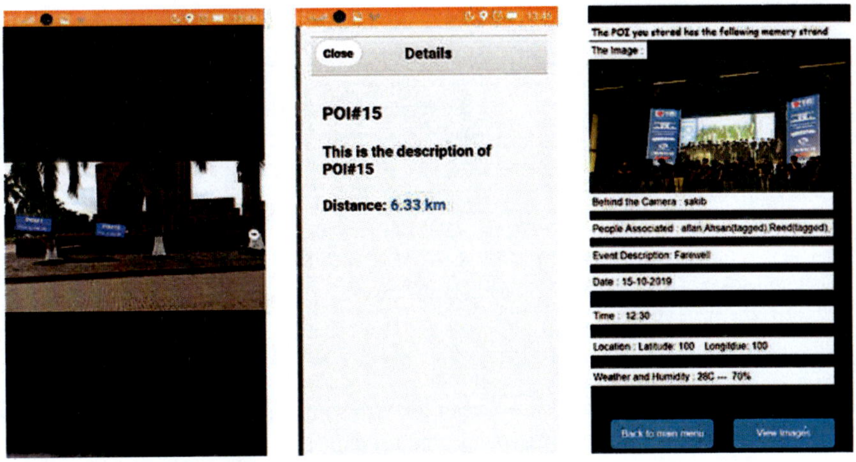

Fig. 2 A user interface with some POIs (left) and Some details after tapping one of the POIs (middle) and A user interface after selecting an image saved in a POI (right) (The interface is built using Wikitude SDK) [7]

5 Experimental Results

To see whether our system really boosts reminiscence as it is supposed to, we developed our own experiment [24]. We recruited 12 participants and after taking full consent from them we conducted the evaluation. The participants were all university students from the age group of 21–25 without any prior history of memory impairment issues. We considered the event of the farewell program of a university in Bangladesh and based on that event we prepared a set of questionnaires consisting of six questions which were answered before and after using the system. This questionnaire session took place 1 year after the event which gave us ample time to check for memory recognition. This set of questionnaires was not randomized rather we followed standard metrics. For recalling memories of any particular event, we need to consider several properties while setting up a questionnaire [4]. According to that, we have considered visual, coherence, time, and sensory details for the six questions we used. Instead of using any predefined and generalized questionnaire, we prepared a custom questionnaire, taking into consideration that self-generated cues provide a strong development for episodic memory [25].

5.1 Result Analysis and Discussion

After getting the answers from all 12 participants, we evaluated the answers, calculated the average of their scores, and used ANOVA (Analysis of Variance) test to evaluate the result. After that, we plotted the scores to a bar chart (shown in Fig. 3)

Questions	Using the System	Without Using the system
Q1	4.9	3.5
Q2	4	3.9
Q3	4.3	2.2
Q4	4.9	4.2
Q5	4.7	4.3
Q6	3	2.8
	Mean = 4.3	Mean = 3.5

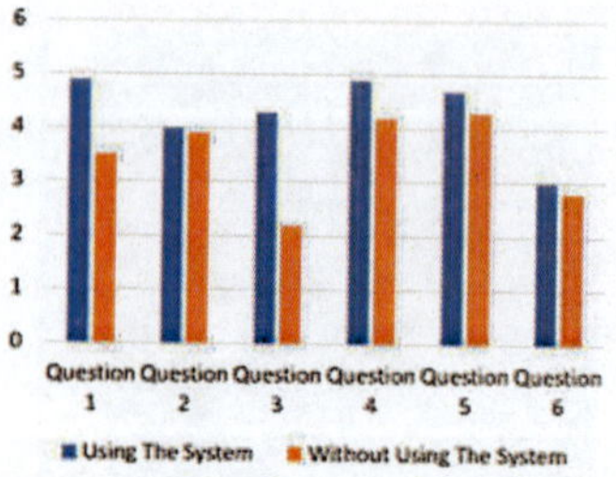

Fig. 3 Table of Mean scores of the 12 participants answering the questions (left) and Bar diagram of the average values (right)

which shows the dissimilarities between the two samples. The first sample consists of the mean scores of those participants who used the system. And the second sample consists of the mean scores of those participants who did not use the system.

To evaluate the F-statistics value, we need to evaluate the sum of squares within (SSW) and sum of squares between (SSB) and we have to denote the degree of freedom (DOF) for SSB, DOF for SSW, and DOF for both SSW and SSB. From this data gathered from Fig. 3, we evaluated all the values, and from this data value, we evaluated F-statistics value. As we aim to use null hypothesis testing, so we have to determine the F-value to compare whether our system withstands.

Null Hypothesis: There are no changes in term of reminiscence after using the system.

Alternate Hypothesis: There are changes in term of reminiscence after using the system.

Now, using the formula, we have got SSW = 5.03, DOF of SSW = 10, SSB = 2.03, and DOF of SSB = 1.
If we consider F(1,10) value from F-statistics table for level of significance = 0.1, then our evaluated value has to be greater than the F(1,10) value so that we can reject our null hypothesis.

After the calculation, we get the F-value = 4.04 and from the F-statistics table [26] for alpha = 0.1 gives us 3.28 which is smaller than our evaluated value. So, we can reject the null hypothesis. This interprets that our system helps in reminiscence. Further, as we asked our participants how the system helped them to recognize past events, it was discerned that the recall rate was good.

In our system, we used image data with text information, varying a wide range of input, in order to avail adequate cues for users to reminisce with little to no cognitive effort. In existing systems, the utilization of lifelog data for reminiscence is insufficient which we attempted to mitigate using more assistive cues. Associated information with the images, e.g., time, date, weather, photo taker name, tagged people, and event names have been used as context cues which will be beneficial for memory retrieval. Further, geo-location-based AR provides location-based cues by augmented POI which creates an extensive user experience. In the existing lifelogging systems, data collection demands a substantial amount of effort and in some cases, human intervention is required which contracts the usability of the system. Our proposed system consists of a simple mobile-based interface that is easily operable. Moreover, we have integrated the concepts of relevant crowdsourced data into our system. This feature enables the user to get access to POI which he had not initially created but contains information related to him, i.e., he has access to POIs where he was tagged in the images. This provides more data and more cues which will conspicuously help him recognize past events with less cognitive effort.

6 Conclusion

In this study, we have used only images with some related information as lifelog data. But for reminiscence, other data types are also important because images are not solely responsible for the recognition of memories. So, in the future, we intend to work with multimodal data like audio, video, noise, etc., alongside images and text data. Also, we wish to add a feature that will notify the user periodically by sending images of memories. These images will be blurred and the blurriness will be reduced over time which may help in memory recall. We also intend to work with patients with memory impairment since we were unable to do so in this study, given our limited time and resources. We have tried to utilize human memory as much as possible by implementing a system where the pre-stored images will be used to recognize memories using AR. Amassing lifelog data has a strong impact. Helping reminiscence through lifelogging will definitely create strong narratives for people of all ages, especially the elderly and thus assist in not leaving behind crucial knowledge, experience, and legacy.

References

1. Chen, Y., Jones, G.J.F.: Augmenting human memory using personal lifelogs. In: 1st Augmented Human International Conference (2010)
2. Lee, M.L., Dey, A.K.: Lifelogging memory appliance for people with episodic memory impairment. In: UbiComp '08 Proceedings of the 10th International Conference on Ubiquitous Computing (2018)
3. Arvola, M., Blomkvist, J., et al: Lifelogging in user experience research: supporting recall and improving data richness. Int. J. Aspects Des. (2017)
4. Munawar, K., Kuhn, S.K., Haque, S.: Understanding the reminiscence bump: a systematic review. NCBI 2008 (2008)
5. Webster, J.D., Haight, B.K.: The Art and Science of Reminiscing Theory, Research, Methods, and Applications. Taylor & Francis (2015)
6. Alter, A.L., Oppenheimer, D.M.: Uniting the tribes of fluency to form a metacognitive nation. Pers. Soc. Psychol. Rev. 219–235 (2009)
7. Wikitude sdk. (2018). https://www.wikitude.com/. Accessed 2021
8. Brady, T.F., Konkle, T., et al.: A review of visual memory capacity: beyond individual items and toward structured representations. J. Vis. (2011)

9. Ksibi, A., Alluhaidan, A.S.D., Salhi, A., El-Rahman, S.A.: Overview of lifelogging: Current challenges and advances. IEEE Access **9**, 62630–62641 (2021)
10. Caprani, O.N., et al.: Motivating lifelogging practices through shared family reminiscence. In: CHI 2011 Workshop: Bridging Practices, Theories, and Technologies to Support Reminiscence, vol. 60 (2011)
11. Hallberg, J., Kikhia, B., Bengtsson, J., et al. : Reminiscence processes using life-log entities for persons with mild dementia
12. Lee, M.L., Dey, A.K.: Using lifelogging to support recollection for people with episodic memory impairment and their caregivers. In: Proceedings of the 2nd International Workshop on Systems and Networking Support for Health Care and Assisted Living Environments (2008). https://doi.org/10.1145/1515747.1515765
13. McCullagh, P., Chris Nugent, H.Z.: Design and evaluation of a tool for reminiscence of life-logged data. In: IEEE International Conference on Bioinformatics and Biomedicine (BIBM) (2014)
14. Nikitina, S., Callaioli, S., Baez, M: Smart conversational agents for reminiscence. In: IEEE/ACM 1st International Workshop on Software Engineering for Cognitive Services (SE4COG), vol. 6 (2018)
15. Underwood, B.J.: Interference and forgetting. Psychol. Rev., 49–60 (1957). https://doi.org/10.1037/h0044616
16. van Teijlingen, T., Oudman, E., Postma, A.: Lifelogging as a rehabilitation tool in patients with amnesia: a narrative literature review on the effect of lifelogging on memory loss. Neuropsychol. Rehabil. 1–27 (2021)
17. Olga Evgenjevna, O.G.: Crowdsourcing for tagging photos. TPU (2016)
18. Svetlana, N., Daniel, F., et al: Crowdsourcing for reminiscence chatbot design arXiv:1805.12346v1
19. Tsao, Y.C., Shu, C.C., et al: Development of a reminiscence therapy system for the elderly using the integration of virtual reality and augmented reality. In: IEEE Eurasia Conference on Biomedical Engineering, Healthcare and Sustainability (IEEE ECBIOS 2019) (2019). https://doi.org/10.3390/su11174792
20. Zhang, L., Luczak, T., Smith, E., Burch, R.F., et al.: Using microsoft hololens to improve memory recall in anatomy and physiology: A pilot study to examine the efficacy of using augmented reality in education. J. Educ. Technol. Develop. Exch. (JETDE) **12**(1), 2 (2019)
21. Shimojima, Y., Koyazu, T.: Feeling of time gap (fog) in event memory. Shinrigaku Kenkyu Japan. J. Psychol. 136–142 (1999). https://doi.org/10.4992/jjpsy.70.136
22. Sellen, A.J., Fogg, A., et al: Do life-logging technologies support memory for the past?: an experimental study using SenseCam. In: CHI '07: Proceedings of the SIGCHI Conference on Human Factors in Computing Systems, 81–90 (2007). https://doi.org/10.1145/1240624.1240636
23. Ali, S., Khusro, S., Khan, A., Khan, I., Solheria, S.F.: An insight of smartphone-based lifelogging research: issues, challenges, and research opportunities: insight of smartphone-based lifelogging research. Proc. Pak. Acad. Sci. A Phys. Comput. Sci. **56**(3), 1–16 (2019)
24. Saidin, N.F., Halim, N.D.A., et al: A review of research on augmented reality in education: Advantages and applications. Int. Educ. Stud. (2015). https://doi.org/10.5539/ies.v8n13p1
25. Wheeler, R.L., Gabbert, F.: Using self-generated cues to facilitate recall: A narrative review (2017). https://doi.org/10.3389/fpsyg.2017.01830
26. F distribution tables. http://www.socr.ucla.edu/Applets.dir/F_Table.html (Accessed 2021)

The Bangabandhu Tariff—Improving Access to Sustainable Electricity and Co-Benefits of Climate Change Through Community Power Purchase Agreements

Sebastian Groh, Raluca Dumitrescu, and Daniel Philipp

Abstract The COVID-19 pandemic has long moved beyond a global health crisis and morphing into a social and economic crisis, which combined with climate change threats, can seriously jeopardize the advancements that have been made to date towards the Sustainable Development Goals (SDGs), globally and in particular in Bangladesh. Our collective actions towards economic recovery and building resilience against future pandemics represent a chance to pave the way for a more sustainable future. At its center lies sustainable energy access as defined in SDG 7 and a key enabler for the majority of the remaining SDGs. SDG 7 is essential to build future resilience. Public energy utilities need to recover from a dramatic shortfall in revenues, households, and businesses have to overcome financing challenges to be able to tap into significant and inclusive growth opportunities. At the same time, it is important that central grid expansion, or the threat thereof, does not negatively affect or prevent decentralized and decarbonized solutions. C-19 has also unearthed several shortcomings in our existing systems. The future of energy is fueled by the 5Ds: decarbonization, decentralization, democratization, digitization, and disruption. The 4th Industrial Revolution has provided us with the right tools to start turning this future vision of the power sector into a reality today. In this paper, we are proposing an innovation policy, the Bangabandhu Tariff, for both a greener and a more inclusive recovery based on smart IoT devices enabling the interconnection of remote solar home systems and their collective feed-in to the national grid. This policy is based on Community Power Purchase Agreements an effective tool towards a resilient and sustainable recovery, the development of national utilities

S. Groh (✉)
Brac University, Dhaka, Bangladesh
e-mail: sebastian.groh@bracu.ac.bd

R. Dumitrescu
TU Berlin, Berlin, Germany

D. Philipp
MicroEnergy International, Berlin, Germany

S. Hossain et al. (eds.), *Proceedings of International Conference on Fourth Industrial Revolution and Beyond 2021*, Lecture Notes in Networks and Systems 437,
https://doi.org/10.1007/978-981-19-2445-3_35

while empowering decentralized renewable energy prosumers with a financial incentive for electricity generation and storage usage to help power the nation starting at the base of the pyramid.

Keywords SDG 7 · Sustainable energy · Smart grids · Base of the pyramid · Power purchasing agreement · Prosumerism

1 Introduction

The COVID-19 (C-19) pandemic has transformed from a mere health crisis into a social and economic crisis, which combined with climate change threats, can seriously jeopardize the advancements that have been made to date towards the Sustainable Development Goals (SDGs), globally and in particular in Bangladesh. However, at the same time, C-19 has also managed to unearth a range of systematic challenges that need to be addressed urgently. As the saying goes "*every crisis contains the seeds of opportunity*", this crisis also offers us the chance with the help of the tools that the 4th Industrial Revolution (4IR) has put into our hands, to embed inclusive green finance into our recovery path. SDG 7, access to sustainable energy, is often referred to as the enabler for many if not all other SDGs. In the last couple of years, the Government of Bangladesh has made remarkable progress toward universal electrification, by now reaching close to 100% of the population, in a time span very few would have expected. However, for access to affordable, reliable, and clean energy in a manner that satisfies requirements based on all three dimensions, economic, environmental, and socially equitable, a strong push for innovation is required, not only technical but also business model and policy-wise. The age-old dogma of centralized electricity generation, transmission, and distribution via a rigid grid infrastructure is going to change very quickly. The future of energy is fueled by the 5 Ds: decarbonization, decentralization, democratization, digitization, and disruption. In this paper, we will showcase how Bangladesh with the help of smart grids, intelligently interconnected between groups of SHS and national grid infrastructure can be a lighthouse example of where this future of energy can begin today. In fond memory of Bangabandhu Sheikh Mujibur Rahman, the Father of the Nation, and his vision for a "*Golden Bangladesh*", we are proposing an innovation policy, the Bangabandhu Tariff, for both a greener and a more inclusive recovery based on smart IoT devices enabling the interconnection of remote solar home systems and their collective feed-in to the national grid. This policy is based on Community Power Purchase Agreements an effective tool towards a resilient and sustainable recovery, the development of national utilities while empowering decentralized renewable energy prosumers[1] with a financial incentive for electricity generation and storage usage to help power the nation starting at the base of the pyramid.

[1] A prosumer is a hybrid form of an energy consumer and an energy producer, depending on her respective consumption and production pattern. Prosumerism provides flexibility to the user.

1.1 COVID-19 and Climate Change Threaten to Cause a Humanitarian Crisis

With a GDP per capita that just crossed the USD 2000 mark, supported by a decade of continuous and strong GDP growth, always north of 6% [1], and a population of approximately 164 million, ranking Bangladesh the eighth largest country by population [2]. Bangladesh has emerged as one of the fastest-growing economies in the world. It is now scheduled to graduate to middle-income status from the group of least developed countries according to the recommendation of the United Nations Committee for Development Policy [3]. And yet, the C-19 crisis has also had its effects on Bangladesh's economy. The economy is reported to still have grown by around 3.8% despite the crisis and is projected to go up to 5.04% in 2021 [1] however, based on research conducted in the early phases of the pandemic by the BRAC Institute of Governance and Development (BIGD), the poor have experienced an 80% drop in income due to the pandemic, as well as the emergence of a group labeled as the new poor who used to be 40% above the poverty line but now have gone below the poverty line [4]. In mid-May 2020, Bangladesh has been hit by the strongest cyclone to form over the Bay of Bengal since 1999—super cyclone Amphan. The UN office in Bangladesh estimated that 10 million people were affected and some 500,000 people may have lost their homes, causing losses of more than USD $ 1bn [5]. The country's energy infrastructure was hit badly; around 20 million electricity connections out of a total of 36.4 million were affected due to uprooted trees which fell on electricity transmission and distribution lines [5]. Bangladesh is among the world's ten most vulnerable countries [6] to climate change. Bangladesh experiences frequent flooding and salinity problems which displace a growing number of people from the country's Ganges River Delta and force them to move to urban areas. While this could benefit energy distribution planning by focusing the grid extensions on consolidated urban regions, this trend also hinders the country's economic growth by limiting large-scale economic activity from growing beyond the two major urban cities: (a) Dhaka (the capital) and (b) Chittagong (the main port city). Furthermore, by concentrating a majority of the economic growth within these two cities, Bangladesh experiences further (negative) knock-on effects on the environment in these two areas.

1.2 Bangladesh's Progress Towards Universal Access to Electricity

Bangladesh is among the countries making the most progress with respect to bringing electricity to its people. The country grew at an annual rate of over 3% since 2010 [7], by way of decentralized and centralized electrification measures in its fight against energy poverty [8, 9]. On-grid electrification has reached close to 100% of the population according to statements by the Ministry of Power, Energy and Mineral Resources [10]. In the same time frame, generation capacity has been ramped up

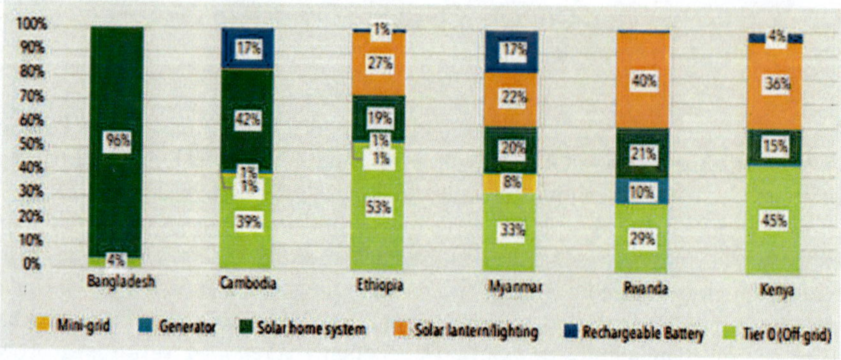

Fig. 1 Share in off-grid electrification (in percentage). *Source* [7]

from 5493 MW of installed capacity to 20,383 MW of installed capacity in 2020. This represents an increase of about 271% in its installed capacity.[2] In terms of per capita electricity consumption Bangladesh is relatively low at 470 kWh, compared to 1100 kWh for India, 4300 kWh for China, and 12,000 kWh for the US. The on-grid energy mix is dominated by fossil fuels, amounting to close to 98%. Off-grid penetration has reached approximately 5% of Bangladesh's population or close to 9.7 million households in remote rural communities, with solar home system (SHS) as the main type of solution [11]. Bangladesh has implemented one of the most successful off-grid access programs in the world. Managed by the governmental agency Infrastructure Development Company Limited (IDCOL) in partnership with non-governmental organizations which enabled financial inclusion, the official SHS program [12] has reached over 4.3 million households and businesses in remote rural areas of Bangladesh (Figs. 1 and 2).

1.3 *Resilience and Reliability of Bangladesh's National Grid Infrastructure*

Grid resilience is the ability to avoid or withstand grid stress events without suffering operational compromise or to adapt to and compensate for the resultant strains so as to minimize compromise via graceful degradation [13]. Bangladesh has enough generation capacity at approximately 22,000 MW against a demand of 9000 MW [14]. Yet brown and blackouts continue to occur, with power cuts longer than 1 hour and frequent voltage fluctuations [15], due to age-old infrastructure incapable of carrying the load needed, due to the overloading of transformers to a point of non-functionality or burning, and due to hampering with the distribution lines. While the peak demand occurs during the evening hours due to lighting load, there is

[2] https://www.bpdb.gov.bd/bpdb_new/index.php/site/annual_reports. Last accessed Sep 2, 2021.

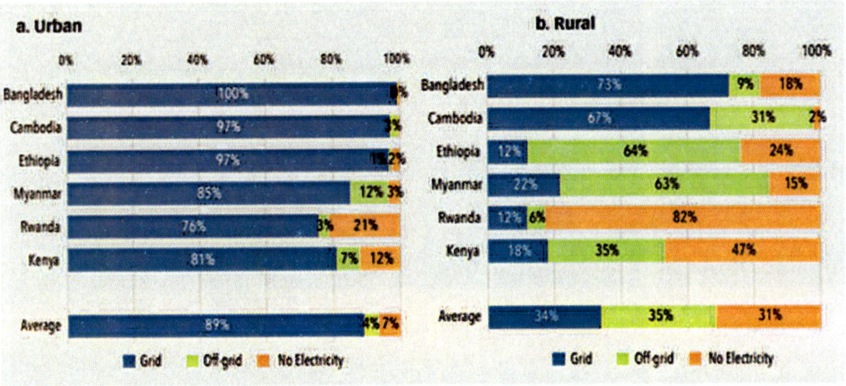

Fig. 2 Type of electricity connectivity by share of total and along the urban/rural divide, 2017. *Source* [7]

a significant use (500–2500 MW) of expensive liquid fuel-run generation during daytime, which can be cost-effectively replaced by renewable energy. Bangladesh is yet to achieve its potential for renewable energy for electricity generation. The total installed renewable energy generation capacity is currently only 628 MW and the share of energy in grid supply is merely 1.4%.

Due to unreliable electricity supplies from government utilities, captive power plants (usually powered by natural gas or diesel) continue to play a significant role in the country's electricity sector [16]. The share of captive power has come down from almost 18% in 2010, but still stands at close to 12% today. While in urban Bangladesh, it is feasible for households, apartment buildings, and industrial consumers to cater for their own temporary power generation system, the investment in such systems is often a high barrier for rural households [11]. Hence, for a sustainable energy sector to develop, the country needs to upgrade its infrastructure, integrated distributed energy generation, and storage solutions. Bangladesh's future energy system has to become more flexible and smarter—through digitization. Digitization is helping improve the safety, productivity, accessibility, and sustainability of energy systems around the world. If an electricity grid is top-down and rigid, the result we see today: we have a vast overcapacity in electricity generation but yet load shedding at the same time. An increase in digitization will help here, coupled with more decentralized generation, and especially storage capacity by creating better incentives and making it easier for prosumers to store and sell surplus electricity to the grid. This requires smart meters powered by the internet of thing (IoT) technology, as part of the 4IR.

1.4 *The Future of Energy Begins with Bangladesh or Bangladesh's 5 Ds: Decarbonization, Decentralization, Digitization, Democratization, and Disruption*

Bangladesh has proven to be an innovator in the off-grid sector [17–20]; it is the world's first country in which SHS are interconnected into peer-to-peer networks or microgrids [21], allowing for the sharing of surplus electricity among peers (transforming consumer households into prosumers), enabling a peer economy based on *prosumerism* and local value creation. This modular infrastructural approach allows microgrids to grow dynamically while the ownership and assets are distributed.

The challenge faced by the majority of utility companies in the developing world today is three-fold: (1) to increase the number of customers while (2) keeping the service quality high and (3) using as much green energy as possible. These goals are already achieved for the rural beneficiaries under the SHS program in Bangladesh; providing 100% electricity availability in the evening hours using solar energy. This is an impressive effort and a manifestation of the 5 Ds as it is a fully based on solar PV and storage (decarbonized), based on rooftop installations (decentralized), in part equipped with smart meters and facilitated via mobile money (digitized), off-grid populations have been turned from passive consumers into proactive prosumers in peer-to-peer grids (democratization) and those grids represent a disruption of the current electricity market [22]. Bangladesh needs to capitalize on this momentum in order to create a nationwide smart grid if it is to provide better and more inclusive electricity services for all 165 million people in Bangladesh as projected under SDG 7 [23]. This will require smartly integrated electrification pathways, instead of subsidized central grid extension posing a barrier toward off-grid solutions [24, 25]. The key strategic move required is the differentiation between the regional grids and microgrids. The national grid is required for power generation and power transfer to large urban and industrial loads. But to reach a large number of dispersed rural customers things need to stay simple and low cost for the utility. A single access point, a single meter for connecting a community microgrid to the national grid can make this a reality. This "Point of Common Coupling" (PCC) is the key to complete electrification across Bangladesh and set an example for an energy future that is decarbonized, decentralized, digitized, democratized, and disruptive.

2 The Bangabandhu Tariff

2.1 *Bangabandhu Sheikh Mujibur Rahman, The Father of the Nation*

Sheikh Mujibur Rahman, affectionately known as Bangabandhu (friend of Bangalis), was a Bangladeshi politician, statesman, and "Founding Father of Bangladesh"—hence the title Father of the Nation—who provided the political leadership for the independence of Bangladesh in 1971 and later served as the first president and then prime minister until his assassination in 1975 [26].

Energy plays a critical role in the development process of a community and as a nation as a whole, especially in the earlier stages of development where otherwise an energy poverty trap can hinder human and economic development [18]. Hence, powering up the nation remains one of the major policy agendas of the Bangladesh government since the country's independence [16]. Bangabandhu Sheikh Mujibur Rahman realized this very early on. Hence, he undertook several steps towards the development of the young country's energy sector by harnessing its own mineral resources and initiating the extraction, distillation, and marketing of mineral oil and natural gas. He strengthened institution-building by establishing both the Bangladesh Petroleum Exploration and Production Company Limited (BAPEX) and the Bangladesh Power Development Board (BPDB) which both exist today. He further purchased five gas fields (Bakhrabad, Titas, Rashidpur, Kailashtila, and Habiganj) from Shell Oil Company at, looking back at it today, an almost symbolic cost.

At the heart of Bangabandhu policy was the desired direction of inclusive development as the only way to true prosperity. It is in this spirit that we introduce the Bangabandhu tariff in this paper for both a greener and a more inclusive recovery enabled by the tools the 4IR has provided us with.

2.2 *The Point of Common Coupling and An Energy Policy Innovation*

The PCC is the integration of the country's two major electrification efforts: grid extension and SHS dissemination (Fig. 3). The Bangladesh Rural Electrification Board (BREB) grid connects to one side of the PCC. On the other side is a local microgrid where SHS are connected to each other and new users. This microgrid can not only sustain itself but also provide last mile grid infrastructure, which can also be used for energy balancing with the national grid, reducing load shedding, and increasing service stability. With this approach, the existing infrastructure of 4.5 million+ SHS (representing approximately 200 MAh of storage capacity and

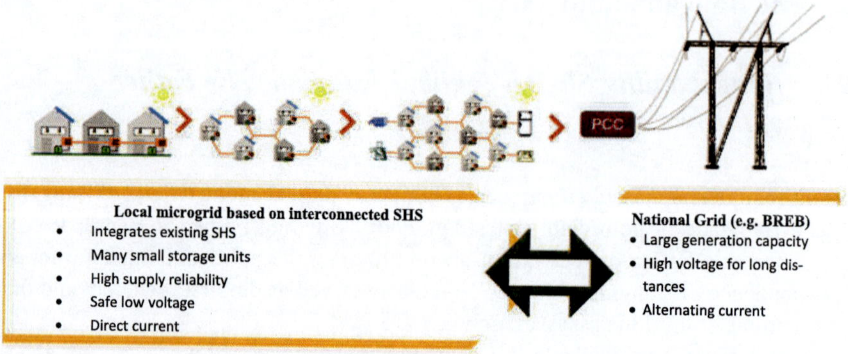

Fig. 3 Architectural structure of the PCC. *Source* [11]

approximately 175 MW of generation) can be leveraged to address Bangladesh's national grid resilience and reliability by:

- Smartening the grid—enabling advanced metering at the customer level to allow for automatic outage detection and service restoration.
- Distributing generation—decentralized microgrids can integrate with the grid to be part of its operational and resource mix, and can also "island" if need be to keep running during grid outages, particularly relevant for extreme weather conditions and damages to distribution lines; these interconnected battery systems would outperform any existing generators in the market as first responders to the national grid, providing short-duration emergency power to arrest a frequency decline, allowing larger generators to ramp up and do the heavy lifting.
- Building resilience on demand—the grid can interact with the microgrid in an IoT environment, such that consumer loads are better managed.

However, in order for such an approach to become a reality, innovation in energy policy needs to take place. While it makes little sense that a small SHS feeds into the national grid (not economically justifiable), if there was a way (Fig. 4) to bundle those systems (e.g., in the hundreds) both technically—through a point of common coupling—and contractually—through a community power purchasing agreement (CPPA), this provides an opportunity for both to address poverty and reduce inequalities as well as a way to promote the realization of countries' NDC goals or acceleration of co-benefits of climate change through special CPPAs [27, 28]. A PPA is usually a long-term contract to buy the electricity generated by a renewable energy installation. The contracts are arranged between the buyer (an electricity supply company) and the seller (the owner or developer) of the installation. In the case of the CPPA, the bundled SHS households in a microgrid would be rewarded for stepping up from their familiar roles as passive power suppliers during daylight hours to battery-enabled energy service providers to the national grid, available at any hour. Under such a scheme, a "poor" household can be a net importer of electricity

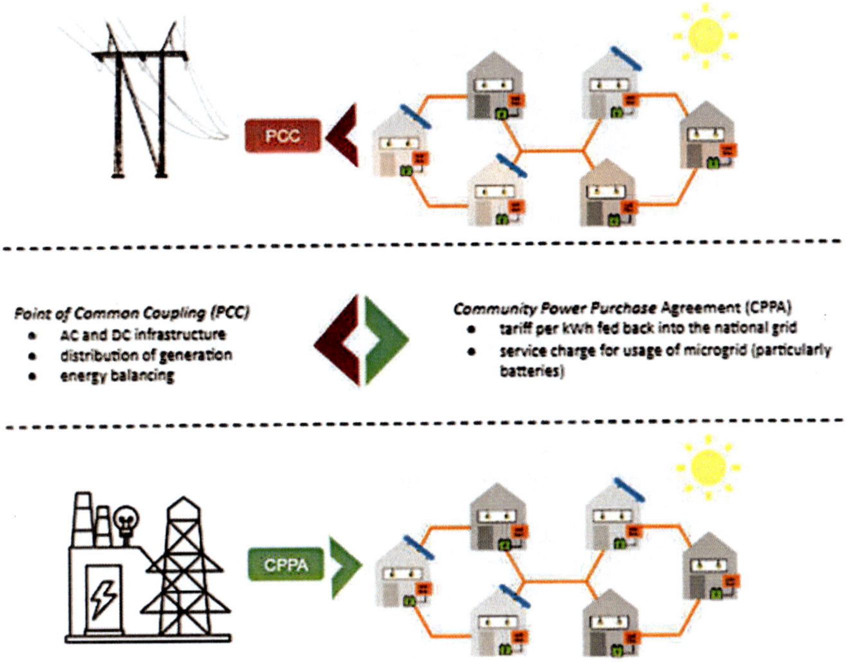

Fig. 4 PCC and CPPA layout. *Source* [11]

while still being a net earner given a fair evaluation of her timed renewable power contribution at the edge of the grid and improving her payback time for the battery installation (as part of the SHS).

2.3 4IR Tools

At the heart of the technical concept is the interconnection of distributed energy resources, such as the millions of isolated SHS into a peer-to-peer grid facilitating electricity exchange. This requires artificial intelligence (AI), edge computing, and the IoT. Each node in such a peer-to-peer grid is an IoT device, a so-called bi-directional smart electricity meter. The setup allows for dynamic pricing signals that strengthen the business case of the peer-to-peer grids [11, 29]. This group of solar home systems is then again interconnected with the national utility grid in the PCC, which in turn allows bi-directional electricity exchange between private households and the national grid, fully automated. In the backend, the settlement is executed automatically based on each individual meter's electricity flows which are directly connected to a mobile wallet allowing real-time settlement through mobile money. For the public utility though, in this case, BREB, the group of many individual nodes/SHS, is perceived as one single node or off-taker which eases the process significantly. Each meter learns from previous transactions which inform the efficiency of trading algorithms to allow for a more effective allocation of the available electricity, a minimization of electricity bills for net consumers, and profit maximization for net producers. The PCC will further learn when the BREB gird gets under pressure, can store up power for those instances, and thus effectively counter low-voltage issues in the regional grid. On a cautionary note, this will require an effective layer of cyber security to avoid manipulation from third parties. With smart financial incentives, this layer of distributed storage and generation will have the ability to stabilize the grid as it grows. This is a critical puzzle piece in the quest for a smarter national grid that is able to accommodate larger amounts of intermittent renewable energy.

3 Conclusion

In essence, we suggest an energy policy innovation promoting a greener and at the same time more inclusive recovery from the ongoing C-19 pandemic. A policy innovation with a special CPPA tariff that is a multiple of the actual national grid tariff earmarked for owners of SHS from the base of the pyramid. This will result in multiple co-benefits while fighting climate change: in direct financial support at low-income levels through pro-poor policy and in adding renewable energy generation and storage to the grid which otherwise would be largely underutilized and supports grid stabilization at the edge of the grid. Several of those claims require a thorough assessment of their actual impact, therefore, as a next step, we propose to test for the practical implementation of the proposed scheme based on a regulatory sandbox with strict data monitoring. This could be a reference for other developing countries seeking smart ways for integrated electrification efforts.

References

1. Bangladesh—gross domestic product (GDP) growth rate 2026 | Statista. https://www.statista.com/statistics/438214/gross-domestic-product-gdp-growth-rate-in-bangladesh/. Accessed Sept. 02, 2021
2. Bangladesh Population—Worldometer (2020). https://www.worldometers.info/world-population/bangladesh-population/. Accessed Apr. 28, 2020
3. Least Developed Country Category: Bangladesh Profile | Department of Economic and Social Affairs. https://www.un.org/development/desa/dpad/least-developed-country-category-bangladesh.html. Accessed Sept. 02, 2021
4. Bangladesh Poor Experience 80pc Drop in Income for Coronavirus—BRAC Institute of Governance and Development. https://bigd.bracu.ac.bd/news/bangladesh-poor-experience-80pc-drop-in-income-for-coronavirus/. Accessed Sept. 02, 2021

5. Cyclone Amphan: 10m people suffer from power outage across Bangladesh, *Dhaka Tribune*, May 21, 2020. https://www.dhakatribune.com/bangladesh/power-energy/2020/05/21/cyclone-amphan-10m-people-suffer-from-power-outage-across-bangladesh. Accessed May 23, 2020
6. Eckstein, D., Hutfils, M.-L., Winges, M., Germanwatch: Global Climate Risk Index 2019 Who Suffers Most From Extreme Weather Events? Weather-related Loss Events in 2017 and 1998 to 2017 (2018)
7. International Energy Agency (IEA), IRENA, United Nations Statistics Division (UNSD), World Bank Group (WB), and World Health Organization (WHO), Tracking SDG 7: The Energy Progress Report 2019, Washington DC, 2019. https://sustainabledevelopment.un.org/content/documents/2019_Tracking_SDG7_Report.pdf
8. Pachauri, S., Mueller, A., Kemmler, A., Spreng, D.: On measuring energy poverty in Indian households. World Dev. **32**(12), 2083–2104 (2004). https://doi.org/10.1016/j.worlddev.2004.08.005
9. Groh, S.: The role of energy in development processes—the energy poverty penalty: Case study of Arequipa (Peru). Energy Sustain. Dev. **18**, 83–99 (2014). https://doi.org/10.1016/j.esd.2013.12.002
10. Nasrul Hamid: Bangladesh close to 100% electricity coverage | Dhaka Tribune. https://www.dhakatribune.com/bangladesh/2021/08/20/bangladesh-close-to-100-electricity-coverage. Accessed Sept. 02, 2021
11. Dumitrescu, R., Lüth, A., Weibezahn, J., Groh, S.: Prosumer empowerment through community power purchase agreements: a market design for peer-to-peer swarm grids. Econ Energy Environ Policy (EEEP) **1**(1), 127–144 (2022). https://doi.org/10.5547/2160-5890.11.1.rdum
12. Heinemann, G., Dumitrescu, R., Von Hirschhausen, C., Kebir, N., Philipp, D.: Lessons from Deploying Large-Scale Solar Electrification in Bangladesh: Can the Last Mile Become the First?. Coimbra, Portugal, pp. 75–86 (2019). https://doi.org/10.2495/ESUS190071
13. Taft, J.: Electric Grid Resilience and Reliability for Grid Architecture. U.S. Department of Energy (2018)
14. Power cuts make households, industries suffer in Bangladesh. New Age | The Most Popular Outspoken English Daily in Bangladesh. https://www.newagebd.net/article/101874/power-cuts-make-households-industries-suffer-in-bangladesh. Accessed May 06, 2020
15. P. 29 J. 2020 | 16:00 GMT, Bangladesh Scrambles to Deliver Electricity to Its 160 Million Residents in 2021—IEEE Spectrum. IEEE Spectrum: Technology, Engineering, and Science News. https://spectrum.ieee.org/energy/fossil-fuels/bangladesh-scrambles-to-deliver-electricity-to-its-160-million-residents-in-2021. Accessed May 06, 2020
16. Amin, S., Mansur, A.H., Kamal, S.M., Groh, S.: Powering up a country into the middle-income club—the story of Bangladesh. In: Groh, S., Hirschhausen, C., Karplus, V. (eds.) Electricity Access, Decarbonization, and Integration of Renewables—Insights and Lessons from the Energy Transformation in Bangladesh, South Asia, and Sub-Sahara Africa. Springer Nature, Forthcoming
17. Chowdhury, S.A., Aziz, S., Groh, S., Kirchhoff, H., Leal Filho, W.: Off-grid rural area electrification through solar-diesel hybrid minigrids in Bangladesh: resource-efficient design principles in practice. J. Cleaner Prod., 95, 194–202 (2015). https://doi.org/10.1016/j.jclepro.2015.02.062
18. Groh, S., Straeten, J., Edlefsen Lasch, B., Gershenson, D., Leal Filho, W., Kammen, D.M.: Decentralized Solutions for Developing Economies Addressing Energy Poverty Through Innovation. Springer International Publishing, Cham (2015). Accessed May 27, 2020. https://doi.org/10.1007/978-3-319-15964-5
19. Mondal, A.H., Kamp, L.M., Pachova, N.I.: Drivers, barriers, and strategies for implementation of renewable energy technologies in rural areas in Bangladesh—An innovation system analysis. Ener. Policy 38(8), 4626–4634 (2010). https://doi.org/10.1016/j.enpol.2010.04.018
20. Palit, D., Chaurey, A.: Off-grid rural electrification experiences from South Asia: Status and best practices. Ener. Sustain. Dev. **15**(3), 266–276 (2011). https://doi.org/10.1016/j.esd.2011.07.004
21. Groh, S., Philipp, D., Lasch, B.E., Kirchhoff, H.: Swarm electrification: investigating a paradigm shift through the building of microgrids bottom-up. In: Groh, S., van der Straeten, J.,

Edlefsen Lasch, B., Gershenson, D., Leal Filho, W., Kammen, D.M. (eds.) Decentralized Solutions for Developing Economies. Springer International Publishing, Cham, pp. 3–22 (2015). https://doi.org/10.1007/978-3-319-15964-5_1
22. Groh, S.: Rural Bangladesh has already embraced renewable energy. Here's what the rest of the world can learn, *World Economic Forum*, Davos (2018).
23. Groh, S., Hirschhausen, C., Karplus, V.: Decentral electrification, network interconnection, and local power markets—an introduction. Econ Energy Environ Policy **11**(1) (2022)
24. Comello, S.D., Reichelstein, S.J., Sahoo, A., Schmidt, T.S.: Enabling mini-grid development in rural India. World Dev. **93**, 94–107 (2017). https://doi.org/10.1016/j.worlddev.2016.12.029
25. Sustainable Energy for All SEforALL, "Integrated Electrification Pathways for Universal Access to Electricity: A Primer," SEforALL, May 2019. https://www.seforall.org/sites/default/files/2019-06/SEforALL_IEP_2019.pdf
26. "Mujib: A Nation Rises," Centre for Research and Information, Aug. 2020. Accessed Nov. 29, 2021. https://cri.org.bd/publication/Mujib/A-Nation-Rises.pdf
27. Campagnolo, L., Davide, M.: Can the Paris deal boost SDGs achievement? An assessment of climate mitigation co-benefits or side-effects on poverty and inequality. World Dev. **122**, 96–109 (2019). https://doi.org/10.1016/j.worlddev.2019.05.015
28. Lüth, A., Weibezahn, J., Zepter, J.M.: On distributional effects in local electricity market designs—evidence from a German case study. Energies **13**(8), 1993 (2020). https://doi.org/10.3390/en13081993
29. Groh, S., Waris, E., Werth, A.: Analytics on pricing signals in peer-to-peer solar microgrids in Bangladesh. Econ Energy Environ Policy (EEP) **11**(1) (2022). https://doi.org/10.5547/2160-5890.11.1.sgro

Healthcare

Denial-of-Service (DoS) Threat Detection Using Supervised Machine Learning Algorithms on CICIDS2018 Dataset

Abu Jafar Mohammad Saleh and Nasim Adnan

Abstract Identification and protection against ever-growing cyber threats is a constantly evolving challenge for traditional Intrusion Detection Systems. Thus, extensive research on implementation of data mining, machine learning, and deep learning methods in conjunction with traditional intrusion detection systems have been carried out to achieve better accuracy, faster execution time, and better utilization of resources. This paper examines the application of a few widely used data mining and machine learning algorithms into intelligent detection of Denial-of-Service (DoS) attacks. For this purpose, this study utilized one of the most recent intrusion detection datasets—CICIDS2018 dataset. The experimental outcome using Naïve Bayes, Random Forest, and J48 classifiers for detecting DoS attacks exhibits promising results with respect to accuracy, precision, recall, or sensitivity as well as F-measure on this specific dataset.

Keywords Intrusion detection · Machine learning · Denial-of-Service · Feature selection · Naïve Bayes · Random forest · J48

1 Introduction

Through the advent of modern technology and digitalization, computer and the internet has become an integral part of everyday communication. As people rely more on network services, malicious intrusion incidents have become more prevalent. This has placed the availability, confidentiality, and integrity of sensitive data at ever higher peril. Kaspersky reported that during the months between November 2019 and October 2020, at least one Malware-class attack was detected on 10.18% of internet consumer computers, and Ransomware attacks were thwarted on 549,301 unique

A. J. M. Saleh (✉)
Bangladesh University of Professionals, Dhaka, Bangladesh
e-mail: saleh@bup.edu.bd

N. Adnan
Jashore University of Science and Technology, Jashore, Bangladesh
e-mail: nasim.adnan@just.edu.bd

S. Hossain et al. (eds.), *Proceedings of International Conference on Fourth Industrial Revolution and Beyond 2021*, Lecture Notes in Networks and Systems 437,
https://doi.org/10.1007/978-981-19-2445-3_36

computers around the world [1]. Therefore, fortifying networks, as well as user devices against malware intrusion and malicious threats, are of utmost importance.

To secure critical data/services from malevolent activities, all network infrastructure and connected devices must use more than one security mechanism and tool which include anti-virus, user authentication, firewall, methods of encryption, intrusion detection system (IDS), etc. Hence, an IDS which is a program or hardware that detects network traffic and looks for suspected threats as well as unusual or malicious behavior can complement a firewall by analyzing threat data and resolving security issues [2].

According to the method of identifying potential threats, Signature-based (SIDS) and Anomaly-based (AIDS) are two of the most common types of IDSs [3]. SIDSs use pattern-matching techniques to detect proven attacks. However, due to reliance on database of previously known signatures, SIDSs are unable to identify previously unseen attacks and Zero-day attacks. On the contrary, an AIDS examines normal activities and constructs a model of normal behavior. It then monitors network traffic, compares it with the "Normal Model" and detects any deviation or new or unknown threat. Nevertheless, an AIDS may identify previously unknown yet benign activities as malicious and thus generate high false positive rates [3].

Beside traditional methods of detection, researchers have proposed numerous intelligent IDS models which are built on Data Mining (DM), Machine Learning (ML), Deep Learning (DL), and other novel techniques to enhance accuracy and eliminate false alerts. Despite the successful application of DM, ML, and DL techniques in other fields such as recommender systems, spam detection systems, etc., little success has been achieved so far towards the implementation of DM, ML, DL, or combination of these techniques towards the development of an efficient and accurate IDS that works on devices as well as network systems. In terms of cyber defense, an effective ML, DL, or DM algorithm and their implementation which can reduce false alarm rates while maintaining high accuracy may be revolutionary [4].

The focus of this study is on effective detection of Denial-of-Service (DoS) threats using ML techniques as this belongs to the most prevalent types of threats. In subsequent sections, this paper examines prior works on relevant fields, elaborates on experimental setup, and discusses outcomes and observations from the experimentation. At the end, the paper concludes by focusing on major contributions and future research scopes.

2 Related Works

DM is the use of complex algorithms for identifying patterns from data, whereas ML works as a compilation of rules or methods that focuses on inference or prediction through learning from given training data. However, in many cases, researchers have used these terms interchangeably and implemented similar methods in their research. DL, which is a subset of ML, is a computational construct that tries to mimic organic neural networks found in the human brain. Typically, DL models perform exceedingly better than ML models with large datasets [5]. In the last few decades, a lot of researches were conducted into the use of ML and DM in intrusion detection and relevant fields and many of those researchers have also used DL techniques to improve detection accuracy, speed of detection, and overall practical applicability. Several surveys and reviews, such as [3, 6, 7] and so on, were conducted over the years to compare various methods and propose taxonomy of evasion techniques, intrusion detection methods, dataset issues, and relevant challenges in the field. ML methods are categorized into three broad groups, such as supervised learning, unsupervised learning, and reinforcement learning [6]. Labeled or categorized training data are used while building and training Supervised learning-based models. On the contrary, variables and data patterns are not labeled in unsupervised learning-based ML methods. Instead, the machine must use unsupervised learning techniques to reveal hidden relationships and label data accordingly. Unlike supervised and unsupervised learning, reinforcement learning strengthens the model over time by incorporating input from previous iterations.

2.1 Applications of DM, ML, and DL in IDSs

In literature, a vast range of supervised learning methods, for example, Decision Trees (DT), Support Vector Machines (SVM), Neural Networks (NN), Naïve Bayes (NB), Nearest Neighbor, etc., have been investigated, highlighting their own sets of advantages and disadvantages. In general, DTs are easier to understand and simpler to implement than some other algorithms such as SVMs or NN [3]. In [7], 54 of the most widely used classifiers from different groups were applied to NSLKDD, ISCXIDS2012, and CICIDS2017 datasets and evaluated on 16 performance measures. Though the J48Consolidated algorithm achieved excellent performances in the class imbalance instances of the CICIDS2017 dataset, considering the Receiver Operating Characteristic (ROC) values, the J48 exhibits superior results than the J48Consolidated. Hussain et al., in their study [8], applied NB, RF, and J48 algorithms to the NSL-KDD dataset and observed that RF and J48 achieved greater accuracy but NB trained and tested in much less time. In [9], 15 classifiers (Bayes Net, NB Multinomial Text, NB Updateable, NB, Simple and Logistic Regression (LR), Voted Perception, ZeroR, JRip, PART, OneR, Decision Table, Random Tree (RT), J48, and Random Forest) were implemented on NSL-KDD dataset. Evaluation of its performance was done on four parameters, which are the rate of false discovery, the root mean square error, the model building time, and the average accuracy. RT algorithm performed best with low model building time and therefore, was recommended for deployment in real-time network IDSs.

In [10], linear-SVM algorithm was compared against Multi-Layer Perception (MLP) and Radial Bias Function (RBF) with their application on KDD'99 dataset. The experiment showed that linear-SVM more efficiently detected various attacks such as DoS, Probe, etc. In [11], seven classifiers such as DT, RF, NB, and LR including some gradient boosting methods (CatBoost, XGBoost, and Light Gradient Boosting Machine—LightGBM) were applied to CICIDS2018 dataset. This study performed ensemble feature selection during pre-processing stage before applying desired methods. Obtained results showed that feature selection can deliver better performance in terms of lower computation time and resources. Researchers of this study also concluded that the choice of classifiers can impact model performance and accuracy significantly. In studies [12–14], some enhanced algorithms such as Forest by Continuously Excluding Root Node (ForestCERN), Forest by Penalizing Attributes (ForestPA), and Balanced Decision Forest (BDF) algorithms were developed by enhancing decision forest algorithm. Researchers have demonstrated that these algorithms hold potential for better performance and applicability in intrusion detection systems.

Some of the unsupervised learning methods such as hierarchical clustering, Density-Based Spatial Clustering of Applications with Noise (DBSCAN), K-means clustering, etc., are used in intrusion detection. In [15], hierarchical clustering and K-means clustering (unsupervised learning methods) were, respectively, used on unlabeled data with good success. Q-Learning, Deep Q-networks, and Monte Carlo Learning are a few of the often-used algorithms in reinforcement learning category [2].

2.2 *Intrusion Detection Datasets*

To implement an intrusion detection model, it must be trained on an evaluation dataset. In [16], available public datasets were examined and classified into seven broad categories. Among these categories, network traffic-based datasets comprise the largest category. During our study, we found that KDD Cup 99 dataset [17], though one of the earliest publicly available intrusion detection datasets, is still being used by many researchers for intrusion detection applications. However, due to too many redundant records and a lack of modern attack types in KDD'99, Tavallaee et al. [18] later developed NSL-KDD dataset to mitigate earlier issues. In 2015, another widely used dataset called UNSW-NB15 [19] containing more recent attack types was developed in the University of New South Wales, Australia. Some of the other major datasets are CAIDA datasets, ISCX2012 datasets, CICIDS2017, CICIDS2018 [20] datasets, etc. As a more recent dataset, CICIDS2017 and CICIDS2018 contain labeled real-world data which includes a wide variety of attack types and different protocols. In a few analyses, such as [21, 22], CICIDS2017 and CICIDS2018 datasets were studied and found to be more relevant with a wider range of real-world attack instances. CICIDS2018 dataset was generated in an experimental network of attack and target machines and contains total 16,233,002 instances in ten CSV files. In general, six broad classes of threats which comprises 17% of all instances are represented in the dataset. More recent studies, for example [23, 24], tend to use these datasets to depict a current and wider attack scenario during their experimentations.

3 Applied ML Algorithms, Dataset, and Evaluation Metrics

For this research work, CICIDS2018 dataset was used. Since this dataset is labeled, supervised learning techniques were applied. The algorithms as well as evaluation metrics used in this study are described in this section.

3.1 Algorithms

We applied NB, RF, and J48 algorithms to this experimentation. NB is one of the simplest Bayesian models that applies Bayes' theorem with the assumptions that features are strongly (naïve) independent. Many practical uses of NB have been demonstrated, including text categorization and medical diagnosis, etc. [25]. RF is a powerful classification algorithm that implements an ensemble learning technique. It creates multiple deep decision trees that use different features from same dataset to make decisions [25]. J48 is a statistical classifier that generates a decision tree utilizing C4.5 (an extension of ID3) [26]. In many relevant studies, such as [3, 7, 9], it was found that algorithms from DT groups performed better than that of other groups. In [7], researchers observed that J48 performed best in terms of accuracy and ROC values. Again, RF algorithm was found to perform better when applied to NSL-KDD datasets [8], however, NB showed that it can train and test the model in much lesser time in comparison to RF and J48 algorithms [7, 8]. In our study, we applied RF and J48 because we wanted to determine their performances on a more current dataset (CICIDS2018). We also applied NB to understand the trade-off in accuracy while achieving faster model building time. Therefore, these algorithms were chosen for this study.

3.2 Evaluation Metrics

Only accuracy results of a classification technique often are not enough to predict the suitability of an ML model. Therefore, in this study, accuracy, as well as precision, recall, and F-measures, have been chosen to evaluate applied ML models. These values are derived from True Positive (TP), False Positive (FP), True Negative (TN), and False Negative (FN) measures and are explained in Table 1.

Accuracy. Accuracy is the ratio between the correct number of detections and the total number of detected instances.

Table 1 Factors used for evaluation metrics

	Actual positive	Actual negative
Predicted positive	TP	FP
Predicted negative	FN	TN

$$\text{Accuracy} = \frac{\text{TP} + \text{TN}}{\text{TP} + \text{FN} + \text{FP} + \text{TN}} \quad (1)$$

Precision. Precision refers to the ratio between true positive detections and the total number of detected threats.

$$\text{Precision} = \frac{\text{TP}}{\text{TP} + \text{FP}} \quad (2)$$

Recall. Recall indicates the ratio between the detected true positive instances and the actual number of threat instances. It is often referred to as Sensitivity or True Positive Rate (TPR).

$$\text{Recall} = \frac{\text{TP}}{\text{TP} + \text{FN}} \quad (3)$$

F-Measure. F-Measure, often referred to as F-Score or F1-Score, corresponds to an overall balanced view of precision and recall.

$$\text{F1Score} = \frac{2 * \text{Precision} * \text{Recall}}{(\text{Precision} + \text{Recall})} \quad (4)$$

Above Eqs. (1–4) were used to calculate the evaluation metrics.

4 Experimental Setup and Methodology

This section outlines the hardware and tools used as well as the steps followed for model implementation and validation process. The study was done on a local machine with the following specifications: Intel® Core™ i7-10,510 CPU @ 1.80–2.30 GHz CPU, 8 GB 3200 MHz Memory, Windows 10, and Pro (v.20H2) with Weka version 3.8.5 [27].

4.1 Data Collection

For this experiment, we applied ML techniques to a subset of CICIDS2018 dataset. This dataset is one of the most recent datasets available for intrusion detection testing. As this study dealt with DoS threats, data files containing only DoS traffic were experimented on. Traffic data was collected on 16 February (Friday) 2018 containing Dos-Hulk and DoS-SlowHTTPTest attacks and on 15 February (Thursday) 2018 containing DoS-GoldenEye and DoS-Slowloris attacks along with benign instances. These DoS attacks were simulated using Hulk, SlowHTTPTest, GoldenEye, and

Slowloris tools with an attacking infrastructure of 50 machines [20]. The file generated on 16 February 2018 with 1,048,574 instances and 80 attributes of data was primarily used for model building and experimentation.

4.2 Pre-Processing

Dataset Analysis. Analysis of the dataset is carried out for a better understanding of the dataset structure. Table 2 shows attack types, number of benign, and attack instances in the subset.

From the above table, it has been observed that the training datasets used were slightly imbalanced. However, there was a very large number of instances from each type of clas to adequately train the model. This slight class imbalance might have impacted the mathematical results which we plan to address in our future research.

Data Cleaning. In the data cleaning stage, removing certain fields of data was our first step. We removed 1 (one) record from the 16 February (Friday) 2018 file that was a repetition of the header row. There was no invalid ("NaN" and "infinity") data in this data file. As Destination Port and Protocol fields are redundant, we dropped the Protocol field from the file. We removed the Timestamp field since we didn't want the learners to be able to tell the difference between attack predictions based on time. Then, the following ten fields that had values of "Zero" in all instances were removed: BwdPSH Flags, FwdURG Flags, BwdURG Flags, CWEFlag Count, FwdBytes/b Avg, FwdPkts/b Avg, FwdBlk Rate Avg, BwdBytes/b Avg, BwdPkts/b Avg, and BwdBlk Rate Avg. We used this cleaned dataset with 68 features as one of the datasets to apply selected algorithms.

Feature Selection. To reduce the number of features without affecting the accuracy of the model, we applied feature selection technique. For feature selection purpose, we utilized Feature Correlation (Pearson's) in Weka. The term "Feature Correlation" refers to the degree of correlation among features. In such cases, general understanding implies that if we find a collection of features that are significantly correlated to each other in a particular set, then only one of the features in the set may be used without compromising accuracy of the model. In this experiment, correlation attribute evaluation along with ranker search method was applied with default settings of Weka. Following Table 3 shows the list of 10 attributes with the highest correlation (their correlation with label attribute) which were finally selected for model building purpose.

Table 2 Distribution of instances

Label	Number
Benign	446,772
DoS attacks-SlowHTTPTest	139,890
DoS attacks-Hulk	461,912

Table 3 Selected 10 features after feature selection

Label	Correlation	Label	Correlation
Fwd Pkt Len Max	0.79323	Pkt Len Max	0.75574
TotLen Fwd Pkts	0.79072	Pkt Len Std	0.75419
Fwd Pkt Len Std	0.79004	Pkt Len Var	0.74953
Dst Port	0.78723	Pkt Len Mean	0.74927
Fwd Pkt Len Mean	0.78359	Pkt Size Avg	0.74839

Normalization and K-fold Validation. In the normalization stage, Weka default setting of scale = 1.0 and translation = 0.0 was utilized to obtain values within [0, 1] to compute normalization intervals. In this experimentation, we utilized k-fold cross-validation for experimentation to ensure that we can repeat the experiments in Weka to obtain same results each time. We have a very large number of instances in our dataset; therefore, we chose k = 5 as with this value of k, each segment would be representative of the different attack types available while reducing computation time.

Experimentation. Two different scenarios were implemented to evaluate the performances of selected ML algorithms on target dataset. The first scenario was to assess the subset of CICIDS2018 dataset (data file generated on 16 February 2018 with benign, DoS-SlowHTTPTest, and DoS-Hulk attack types) with all the selected algorithms. In this scenario, three ML models were applied to the same dataset with a fivefold validation process. In the other scenario, each trained model was used against another subset of data that was created taking 30% instances from the 15 February (Thursday) 2018 file that contained Benign, DoS-GoldenEye, and DoS-Slowloris attacks. The purpose of this scenario was to train all the models with DoS attacks available in one data file (16 February 2018) and then test those trained models to detect different types of DoS attacks from another smaller data file (15 February 2018). This process was conducted to evaluate whether trained ML models can successfully identify novel DoS attacks from untrained data.

5 Results and Findings

Following section discusses the results and findings obtained from our experimentation.

5.1 Performance Analysis with K-fold Cross-Validation Data

As described in Sect. 3, selected ML algorithms were applied to data contained in "Friday, 16 February 2018" file to build and test suitable learning models. Figure 1 depicts the comparison among model building time (in seconds) applying fivefold

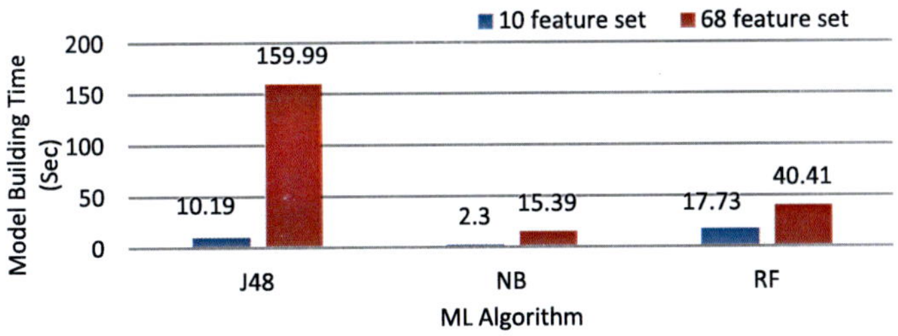

Fig. 1 Model building time for applied algorithms

cross-validation for RF (numIterations = 5, maxDepth = 6 [m]), NB (default setting in Weka), and J48 (default settings in Weka) on three classes of data instances: Benign, DoS-SlowHTTPTest, and DoS-Hulk. Required time for building models with 10 features and 68 features dataset is shown in the figure as well. It was observed that NB took the shortest time in both cases. On the contrary, while building models with 10 features and with 68 features, RF and J48 took the longest time, respectively.

Tables 4 and 5 summarize precision, recall, and F-measure values obtained after applying ML algorithms on datasets with 10 features and 68 features, respectively. In both cases, all three algorithms performed excellently. Models with J48 algorithm achieved the best result with perfect precision, recall, and F-measure values. RF algorithm delivered the next best result with ten feature-dataset. However, one interesting observation was that RF performed relatively better when applied to the dataset with 10 features than that of the dataset with 68 features. This phenomenon might be caused by overfitting or addition of more noises with larger number of features. In case of NB, it was observed that, while it scored relatively lower in comparison with other two methods, it still achieved very high values. In Table 6, we have compared our precision, recall, and F-measure values with those obtained by Sharafaldin et al. [20]. However, as we did not have any information about the hardware platform used

Table 4 Precision, Recall, and F measure values with 10 feature datasets

ML	Attack types	10 feature set		
		Precision	Recall	F-measure
J48	Benign	1.00	1.00	1.00
	DoS-SlowHTTPTest	1.00	1.00	1.00
	DoS-Hulk	1.00	1.00	1.00
NB	Benign	0.978	0.990	0.984
	DoS-SlowHTTPTest	0.970	1.00	0.985
	DoS-Hulk	1.00	0.979	0.982
RF	Benign	1.00	1.00	1.00
	DoS-SlowHTTPTest	1.00	1.00	1.00
	DoS-Hulk	1.00	1.00	1.00

Table 5 Precision, Recall, and F measure values with 68 feature datasets

ML	Attack types	68 feature set		
		Precision	Recall	F-measure
J48	Benign	1.00	1.00	1.00
	DoS-SlowHTTPTest	1.00	1.00	1.00
	DoS-Hulk	1.00	1.00	1.00
NB	Benign	0.997	0.991	0.989
	DoS-SlowHTTPTest	0.997	1.00	0.998
	DoS-Hulk	0.991	0.996	0.988
RF	Benign	1.00	0.998	0.999
	DoS-SlowHTTPTest	1.00	1.00	1.00
	DoS-Hulk	0.998	1.00	0.999

Table 6 Performance comparison between our study and a similar study

ML algorithm	Our study (10 features, fivefold, avg)			Sharafaldin et al. [20]		
	Precision	Recall	F-measure	Precision	Recall	F-measure
NB	0.982	0.9896	0.9836	0.88	0.84	0.84
RF	1.00	1.00	1.00	0.98	0.97	0.97
J48	1.00	1.00	1.00	Not applied		

by Sharafaldin et al. [20], we did not compare model building time. In the comparison we found that in our study, both NB and RF algorithms attained better precision, recall, and F-measure values. However, we could not compare the results of J48 as researchers did not apply this algorithm in [20].

Tables 7 and 8 show confusion matrices for all three models. Both J48 and RF methods successfully detected nearly all Benign and other DoS attacks. In case of J48, performance with both 68 and 10 feature sets were the same. However, in case of RF, it performed better when applied to the dataset with 10 features than the dataset with 68 features. In the case of NB, though, performance was slightly poorer than the other two algorithms. Another finding was that all the classifiers could identify DoS-SlowHTTPTest-type attacks with perfect accuracy but could not detect all DoS-Hulk attacks. Overall, J48 performed best among these three algorithms when applied with fivefold cross-validation testing.

5.2 *Performance Analysis on Untrained Data*

For this experimentation, during the training phase, both DoS-SlowHTTPTest and DoS-Hulk attack types were renamed as DoS. So, the training dataset contained two types of instances: benign and DoS. The test dataset was generated by randomly

Table 7 Confusion matrix with 10 features

		Predicted Benign	Predicted DoS-SlowHTTPTest	Predicted DoS-Hulk
J48	Actual Benign	446,767	0	5
	Actual DoS-SlowHTTPTest	0	139,890	0
	Actual DoS-Hulk	0	0	461,912
NB	Actual Benign	442,372	4367	33
	Actual DoS-SlowHTTPTest	0	139,890	0
	Actual DoS-Hulk	9855	0	452,057
RF	Actual Benign	446,766	0	6
	Actual DoS-SlowHTTPTest	0	139,890	0
	Actual DoS-Hulk	0	0	461,912

Table 8 Confusion matrix with 68 features

		Predicted Benign	Predicted DoS-SlowHTTPTest	Predicted DoS-Hulk
J48	Actual Benign	446,767	0	5
	Actual DoS-SlowHTTPTest	0	139,890	0
	Actual DoS-Hulk	0	0	461,912
NB	Actual Benign	442,532	57	4183
	Actual DoS-SlowHTTPTest	0	139,890	0
	Actual DoS-Hulk	1543	424	459,945
RF	Actual Benign	446,766	0	947
	Actual DoS-SlowHTTPTest	0	139,890	0
	Actual DoS-Hulk	0	0	461,912

selecting 30% of the instances from "15 February, (Thursday) 2018" file and renaming all DoS-GoldenEye and DoS-Slowloris instances as DoS. Thus, the test dataset contained only two types of instances: benign and DoS. Then this test dataset was tested with two models built with the previous 10 features dataset and 68 features dataset separately.

Table 9 exhibits the result of these two tests. It was generally observed that all three models struggled to identify attack instances from untrained data. Following Figs. 2 and 3 shows the corresponding accuracy levels of testing on same dataset (with fivefold cross-validation) and testing on untrained data. Figure 2 (trained with 68 features) and Fig. 3 (trained with 10 features) show that classifier performance

Table 9. Confusion matrix with 10 and 68 features against untrained data

		10 feature set		68 feature set	
		Predicted Benign	Predicted DoS	Predicted Benign	Predicted DoS
J48	Actual Benign	169,534	129,235	254,819	43,950
	Actual DoS	2823	12,980	6403	9400
NB	Actual Benign	3734	295,035	76126	222,643
	Actual DoS	0	15,803	3046	12,757
RF	Actual Benign	170,535	128,234	239,482	59,287
	Actual DoS	15,255	548	6940	8863

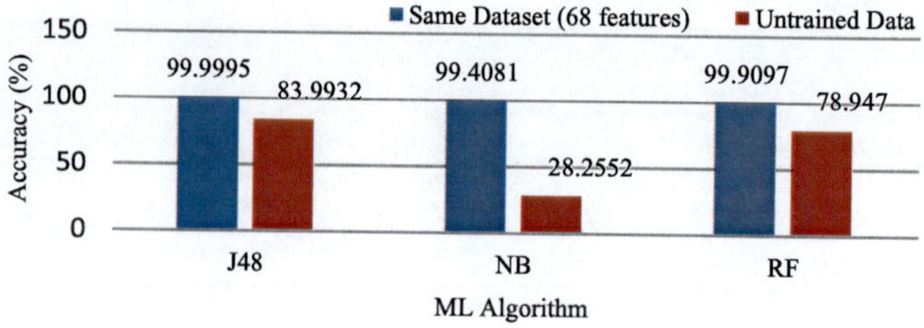

Fig. 2 Accuracy levels with fivefold cross-validation (68 features) and untrained data

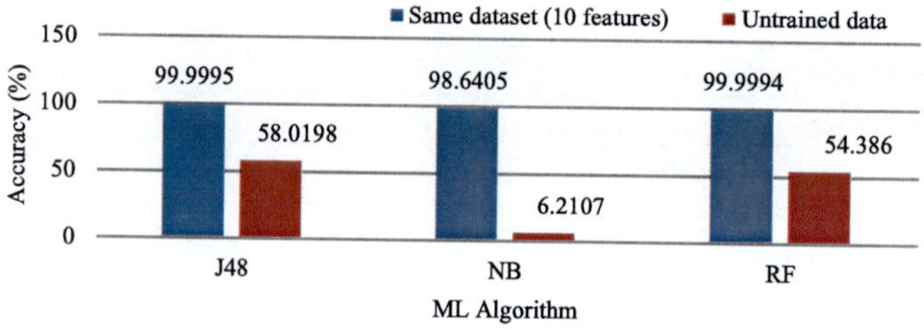

Fig. 3 Accuracy levels with fivefold cross-validation (10 features) and untrained data

was not as accurate as with fivefold validation. In this second experimentation, when the models were trained with 68 and 10 feature datasets, J48 attained accuracy of 83.9932% and 58.0198% and RF attained 78.947% and 54.386%, respectively. In case of NB, accuracy was much lower at 28.2552% and 6.2107%, respectively.

In case of fivefold validation, even if attribute independence assumption of NB algorithm was not maintained properly, NB still performed exceedingly well. But when it came to identifying attack instances from untrained data, NB showed very poor performance. This pointed to the fact that though similar in nature, different DoS attacks exhibited different attributes and therefore were very difficult to generalize. Dip in performance of other two algorithms might be attributed to this fact as well. Slightly better results with large feature sets show that models which are trained on a greater number of feature sets, tend to perform better while testing on untrained data. Though the models performed relatively better when trained on dataset with more feature sets, their performance still falls way short of any real-world suitability. This phenomenon points to the fact that it is very difficult to generalize attack patterns and hence very difficult to successfully identify novel attacks. However, in this experiment, we only trained the models on two types of DoS attacks. Training the models on more varied types of attacks may improve their accuracy to detect novel attacks from untrained data.

6 Conclusions and Future Works

Our experiment studies the performances of three ML techniques which are known to be efficient and therefore, widely used. These algorithms were applied to a subset of a very recent dataset (CICIDS2018) to analyze their suitability to identify several types of DoS attacks. Obtained results showed that efficiency of applied algorithms with k-fold cross-validation to identify DoS attacks with trained data was very high. However, while applying the same algorithms on similar but untrained data-subset, result was not as high and left room for improvement to be utilized in the real world.

Since this study is limited to only a few ML algorithms, we have an opportunity to further study the application of other ML and DL techniques. We trained the models only on various DoS attack types before applying them to untrained data. Future works can train the model on a varied number of attack types and then apply those models to untrained data which may yield better success. Similarly, research on the application of ensemble learning techniques such as Forest CERN, Forest PA, and BDF may also be carried out to successfully detect previously undiscovered attacks.

References

1. Kaspersky Security Bulletin 2020, pp. 6–6. https://go.kaspersky.com/rs/802-JN-240/images/KSB_statistics_2020_en.pdf. Accessed Apr. 21, 2021
2. Sarker, I.H., Kayes, A.S.M., Badsha, S. et al.: Cybersecurity data science: an overview from machine learning perspective. J. Big Data 7(41) (2020). https://doi.org/10.1186/s40537-020-00318-5
3. Buczak, A.L., Guven, E.: A survey of data mining and machine learning methods for cyber security intrusion detection. IEEE Commun. Surv. Tutor. 18(2), 1153–1176 (2016). https://doi.org/10.1109/COMST.2015.2494502
4. Sommer, R., Paxson, V.: Outside the closed world: on using machine learning for network intrusion detection. In: 2010 IEEE Symposium on Security and Privacy, Oakland, CA, USA, pp. 305–316. IEEE (2010). https://doi.org/10.1109/SP.2010.25
5. Aldweesh, A., Derhab, A., Emam, A.Z.: Deep learning approaches for anomaly-based intrusion detection systems: a survey, taxonomy, and open issues. Knowl.-Based Syst. 189 (2020). https://doi.org/10.1016/j.knosys.2019.105124

6. Leevy, J.L., Khoshgoftaar, T.M.: A survey and analysis of intrusion detection models based on CSE-CIC-IDS2018 Big Data. J. Big Data **7**, 104 (2020). https://doi.org/10.1186/s40537-020-00382-x
7. Panigrahi, R., Borah, S., Bhoi, A.K., Ijaz, M.F., Pramanik, M., Jhaveri, R.H., Chowdhary, C.L.: Performance assessment of supervised classifiers for designing intrusion detection systems: a comprehensive review and recommendations for future research. Mathematics **9**, 690 (2021). https://doi.org/10.3390/math9060690
8. Hussain, J., Lalmuanawma, S.: Feature analysis, evaluation and comparisons of classification algorithms based on noisy intrusion dataset. Proc. Comput. Sci. **92**, 188–198 (2016). https://doi.org/10.1016/j.procs.2016.07.345
9. Chaudhary, P.: Usage of machine learning for intrusion detection in a network. Int. J. Comput. Netw. Appl. (IJCNA) 3(6), 139–147 (2016). https://doi.org/10.22247/ijcna/2016/41278.
10. Keshta, I.M.: Intelligent intrusion detection system based on MLP, RBF and SVM classification algorithms: a comparative study. Int. J. Comput. Sci. Inf. Secur. (IJCSIS) 16(5) (2018)
11. Leevy, J.L., Hancock, J., Zuech, R., Khoshgoftaar, T.M.: Detecting cybersecurity attacks across different network features and learners. J. Big Data **8**, 38 (2021). https://doi.org/10.1186/s40537-021-00426-w
12. Adnan, M.N., Islam, M.Z.: Forest CERN: a new decision forest building technique. In: Bailey, J., Khan, L., Washio, T., Dobbie, G., Huang, J., Wang, R. (eds.) Advances in Knowledge Discovery and Data Mining. PAKDD. Lecture Notes in Computer Science, vol 9651. Springer, Cham (2016). https://doi.org/10.1007/978-3-319-31753-3_25
13. Adnan, M.N., Islam, M.Z.: Forest PA: constructing a decision forest by penalizing attributes used in previous trees. Exp. Syst. Appl. 89, 389–403 (2017). https://doi.org/10.1016/j.eswa.2017.08.002
14. Adnan, M.N., Ip, H.L., Bewong, M., Islam, Z.: BDF: a new decision forest algorithm. Inf. Sci. **569**, 687–705 (2021)
15. Annachhatre, C., Austin, T.H., Stamp, M.: Hidden Markov models for malware classification. J. Comput. Virol. Hacking Tech. **11**, 59–73 (2015). https://doi.org/10.1007/s11416-014-0215-x
16. Ferrag, M.A., Maglaras, L., Moschoyiannis, S., Janicke, H.: Deep learning for cyber security intrusion detection: approaches, datasets, and comparative study. J. Inf. Secur. Appl. 50 (2020). https://doi.org/10.1016/j.jisa.2019.102419
17. KDD Cup 1999 Data, University of California, Irvine (2021). http://kdd.ics.uci.edu/databases/kddcup99/kddcup99.html. Accessed Apr. 04, 2021
18. Tavallaee, M., Bagheri, E., Lu, W., Ghorbani, A.A.: A detailed analysis of the KDD CUP 99 data set. In: 2009 IEEE Symposium on Computational Intelligence for Security and Defense Applications, pp. 1–6. Ontario, Canada (2009). https://doi.org/10.1109/CISDA.2009.5356528
19. UNSW-NB15 Dataset, University of New South Wales. https://cloudstor.aarnet.edu.au/plus/index.php/s/2DhnLGDdEECo4ys. Accessed Apr. 12, 2021
20. Sharafaldin, I., Lashkari, A.H., Ghorbani, A.A.: Toward generating a new intrusion de- tection dataset and intrusion traffic characterization. In: 4th International Conference on Information Systems Security and Privacy (ICISSP), pp. 108–116. SciTePress, Portugal (2018). https://www.scitepress.org/Papers/2018/66398/66398.pdf
21. Ankit, T., Ritika, L.: A review of the advancement in intrusion detection datasets. Proc. Comput. Sci. **167**, 636–645 (2020). https://doi.org/10.1016/j.procs.2020.03.330
22. Dwibedi, S., Pujari, M., Sun, W.: A comparative study on contemporary intrusion detection datasets for machine learning research. In: IEEE ISI 2020: International Conference on Intelligence and Security Informatics. IEEE, Arlington, VA, USA (2020). https://doi.org/10.1109/ISI49825.2020.9280519
23. Kshirsagar, D., Kumar, S.: An efficient feature reduction method for the detection of DoS attack. ICT Express **7**(3), 371–375 (2021). https://doi.org/10.1016/j.icte.2020.12.006
24. Rios, A.L.G., Li, Z., Bekshentayeva, K., Trajkovic´, L.: Detection of denial of service attacks in communication networks. In: IEEE International Symposium on Circuits and Systems (ISCAS), pp. 1–5. IEEE, Seville, Spain (2020)
25. Mitchell, T.M.: Machine Learning. McGraw-Hill, New York (1997)

26. Sahu, S., Mehtre, B.M.: Network intrusion detection system using J48 decision tree. In: International Conference on Advances in Computing, Communications and Informatics (ICACCI), pp. 2023–2026. IEEE, Kochi, India (2015)
27. Weka-The workbench for machine learning, University of Waikato, New Zealand, https://www.cs.waikato.ac.nz/ml/weka/. Accessed Apr. 21, 2021

Implementation of a Digital Healthcare Service Model for Ensuring Preventive and Primary Health Care in Rural Bangladesh

Md. Mashiar Rahman, Moinul H. Chowdhury, Rubaiyat Alim Hridhee, Tanvir Islam, Mazharul Islam Leon, Mithila Faruque, Mohammad Badruddozza Mia, Farhana Sarker, Md. Jashim Uddin, and Khondaker A. Mamun

Abstract Primary Health Care (PHC) is a core component of the universal healthcare system and is considered the first line of defense against health problems. The establishment of a comprehensive PHC service ensures better management of health care needs of the population. However, the current primary healthcare system in Bangladesh is far behind international standards in terms of both accessibility and quality. People in rural areas often don't receive quality primary care and end up with avoidable health complications. This causes a major burden on the healthcare system, increases out-of-pocket expenditure, and untimely deaths. To address this, "CMED Health" has developed a cloud-based digital healthcare platform which creates digital health accounts for personal health records and uses IoT-enabled smart medical devices and AI to collect health data from individuals and households and assess health risks for referral. CMED Health addresses the needs of rural people by introducing digital healthcare services utilizing the platform and community health workers to complement the existing healthcare system. Accordingly, CMED Health has partnered with Palli Karma-Sahayak Foundation (PKSF) to pilot this model where

Md. M. Rahman · Md. J. Uddin
Palli Karma-Sahayak Foundation, Dhaka, Bangladesh

M. H. Chowdhury · R. A. Hridhee · T. Islam · M. I. Leon · M. Faruque · F. Sarker · K. A. Mamun (✉)
CMED Health Limited, Dhaka, Bangladesh
e-mail: mamun@cse.uiu.ac.bd

M. Faruque
Department of NCD, Bangladesh University of Health Sciences, Dhaka, Bangladesh

M. B. Mia
School of Business and Economics, United International University, Dhaka, Bangladesh

F. Sarker
Department of CSE, University of Liberal Arts Bangladesh, Dhaka, Bangladesh

K. A. Mamun
AIMS Lab, Department of CSE, United International University, Dhaka, Bangladesh

S. Hossain et al. (eds.), *Proceedings of International Conference on Fourth Industrial Revolution and Beyond 2021*, Lecture Notes in Networks and Systems 437,
https://doi.org/10.1007/978-981-19-2445-3_37

trained health workers went door to door, provided preventive and primary healthcare service, and collected other socio-demographic data. The pilot was conducted in one union where 31,868 people received monthly doorstep digital healthcare service for 6 months. The outcome of the pilot demonstrates the benefits of the data-driven digital platform to increase the effectiveness of primary healthcare service delivery by establishing health records, structured referral systems, risk assessment for early intervention, and enhancing equity in rural areas. Moreover, the scale-up of this model will help to realize Digital Bangladesh by creating digital health inclusion and achieving universal health coverage of SDG.

Keywords Digital health inclusion · Preventive and primary healthcare · Digital health platform · mHealth · Internet of Things (IoT) · Artificial intelligence

1 Introduction

1.1 Overview

The United Nations' Sustainable Development Goal (SDG) target 3 lists global targets in order to achieve a healthy life and promote well-being for all ages [1]. These goals are designed to work as a blueprint for global development in the future. Although all countries are focused on achieving these goals, the developing countries are struggling more than the developed countries. One reason for this is that there has been a major shift in disease burden from communicable diseases to non-communicable and chronic diseases. The reduction in mortality due to communicable diseases has increased average life expectancy, but on the other hand, deaths due to non-communicable diseases (NCDs) have increased drastically. In 2017, 73.4% of total deaths were due to NCDs, while 18.6% were due to communicable, maternal, neonatal, and nutritional (CMNN) causes [2]. Although this is an alarming issue for the entire world, study shows that the majority of these deaths occur in countries with lower economic status. According to WHO facts sheets, about 77% of these deaths are in low- and middle-income countries [3]. This is a major concern for these groups of countries because they have primarily focused their healthcare efforts towards preventing communicable, child-maternal, and nutritional diseases.

Being a country with poor economy, these statistics are really concerning for Bangladesh. According to the World Bank, Bangladesh is classified as a lower middle-income country. The condition is worsened further by the population of Bangladesh. As of the year 2020, Bangladesh has a total population of 164 million [4] which is projected to be 186 million by the year 2030 [5]. This huge population creates a massive burden on hospitals because the current healthcare system in Bangladesh is not well organized. It is supported by insufficient resources, apoor management system, and a highly centralized service delivery model [6]. The entire healthcare system in Bangladesh is governed using a top-down approach which has created issues in terms of monitoring and reporting [7]. This system is not easily expandable due to infrastructure and personnel costs making it especially troubling for people living in rural areas because they can't get fast access to primary-level medical services.

One solution to this problem is to use digital health services or mobile health applications to increase the accessibility of healthcare to rural people. Mobile health (mHealth) is a concept which utilizes the power of mobile phones to address this issue. Its primary objective is moving the focal point of health care to a more patient-driven model with the help of mobile technology, allowing service mobility, timely intervention, and increased service reach. These technologies can help reduce system loss, improve treatment, collect, and store patients' data in accessible manner and even advance research [8]. MHealth is sometimes supported by digital medical devices and the Internet of Things (IoT). IoT features combined with medical devices are labeled as m-IoT [9]. The contribution of IoT devices in healthcare settings is automatic data collection and storage, data processing, health monitoring, and overall ease in the healthcare system [9]. The data collection feature of m-IoT devices combined with the power of cloud computing provides us with a powerful feature for managing large amounts of data.

Although there are many digital health apps and mHealth systems currently in place, their accessibility to rural people has been somewhat ignored. Due to the lack of proper education, most of the people living in villages don't know how to use these systems for their benefits. CMED's cloud-based medical framework is focused on rural households in developing countries and addresses their needs by providing them with comprehensive preventive and primary healthcare service at their doorstep through the use of IoT-enabled smart medical devices, cloud-based data storage, and artificial intelligence (AI)-driven clinical decision support systems. [10].

In order to assess its capability to deliver primary health service in rural areas, CMED has partnered with Palli Karma-Sahayak Foundation (PKSF) to digitize their rural health program using CMED digital health service. PKSF is a pinnacle advancement association, also a non-governmental organization (NGO) that works for the rural peoples in Bangladesh [11]. CMED Health has conducted pilot activities from 14 June 2018 to 31 December 2018 in the Somvag union of Bangladesh to digitize primary healthcare activities on the basis of a Memorandum of Understanding (MoU) with PKSF. CMED Health has built a digital platform that combines digital systems and smart medical devices according to the scope of PKSF's activities. The specific contributions of this paper are

- Implement digital framework to provide comprehensive preventive and primary healthcare service at doorstep in the Somvag union of Bangladesh.
- Implement a digital primary healthcare model in Bangladesh.
- Conduct regular health assessments using digital and IoT devices and raise health awareness through digitalization of primary healthcare in Bangladesh.
- Establish referral to health facilities using data-driven clinical decision support system (CDSS) for rural areas of Bangladesh.
- Implement digital health records for people in rural areas of Bangladesh.

1.2 Background Studies

There are a lot of mHealth systems currently in operation in Bangladesh. Ahmed et al. in 2014 identified and reported 26 e-health and mHealth services [12]. These include both government and private initiatives. Recently, there have been a lot of new digital health services which cover various ranges of healthcare aspects from simple text messaging systems to medical consultation services which is also known as telemedicine service. Jeeon (https://www.jeeon.co/#services) provides mHealth services in Bangladesh as well as telemedicine services in rural areas [13]. But recently, they have stopped their telemedicine service after facing challenges [14]. Digital Healthcare Solutions (https://dh.health/OurServices) provides medical consultancy services, micro-health insurance services as well as other health programs on diabetes, communicable diseases, and maternal and child care [15]. Digital Hospital (https://digitalhospital.health/) is a social brand of Digital Healthcare Solutions which provides telemedicine services as well as medicine home delivery [16]. Although they are providing remote medical consultancy services, their access to the rural people is not fully assessed. They also do not have any IoT devices for patients to measure health vitals. Daktarbhai (https://www.daktarbhai.com/) is another telemedicine service in Bangladesh consisting of general practitioners as well as specialist doctors [17]. They also provide pathology tests at doorstep upon request. Their system is effective in secondary- and tertiary-level health services but does not provide a comprehensive primary-level screening. Another similar e-health service provider is Doctorola (https://doctorola.com/) which provides online consultancy services from specialized doctors, home test sample collection, report delivery, medicine home delivery as well as insurance coverage [18]. Dhaka University Telemedicine Programme (DUTP) (https://bmpt.du.ac.bd/telemedicine/) developed a telemedicine system for rural areas focusing on the primary- and secondary-level of healthcare [19]. They developed four low-cost telemedicine devices, namely, ECG, Stethoscope, digital microscope, and an X-ray view box which support remote doctors to view these data and provide remote consultation to patients. The ECG system developed by them supports real-time data transmission which can be more beneficial and informative for medical practitioners to make important medical decisions. Their system, however, does not take environmental, economic, and other socio-demographic factors into consideration when providing service.

Although 62.6% of the population in Bangladesh live in rural areas [20], there is a significant shortage of qualified health workers in these areas [21]. Rural health service is dominated by drug-sellers and local doctors who do not have formal training to practice modern medicine. mHealth services can be useful for ensuring quality health services in rural areas. CMED was developed as a ful cloud-based medical framework focused on the primary-level healthcare of rural people [10]. Faruque et al. addressed the usability and efficacy of the system by conducting a feasibility study of CMED digital health service to provide prevention against NCDs [22]. They concluded that the service is capable of providing preventive care and raising awareness against NCDs. They also reported the ease and increased speed in data collection of health workers while delivering service.

2 Methodology

2.1 *System Overview*

CMED started the program with 14 health visitors and 2 health officers in the Somvag union of Dhamrai Upazila of Dhaka district. Figure 1 shows the working procedure and referral structure of the system. The health visitors provide primary health services through monthly household visits and provide referral to health officers after health risk assessment. The health officers provide limited primary health service and provide referrals to general practitioners or health facilities for further intervention. A pilot study was conducted among 31,868 individuals belonging to 7081 households. The specific segments of the system are described below:

Health visitor and health officer. PKSF dispatched 14 health visitors (HV) for field-level data collection and 2 health officers/medical assistants to supervise them. The health visitors are equipped with a CMED Health Kit. Their main objectives are to visit each "khana" or "household", provide primary health service, and collect data regarding the khana member's health as well as other household information. Using the digital platform, the workload of the health visitor decreased substantially and they were able to collect data much faster than the health visitors using analog methods. Figure 2 shows khana profile creation and editing view of the mobile app.

Primary health monitoring. Primary health monitoring is the core service of CMED Health. Health visitors use the CMED smart health monitoring kit (Fig. 3) to record the health data of members. The members can see their health data on the screen and the health visitors interpret it to them and give them necessary advice. This data is

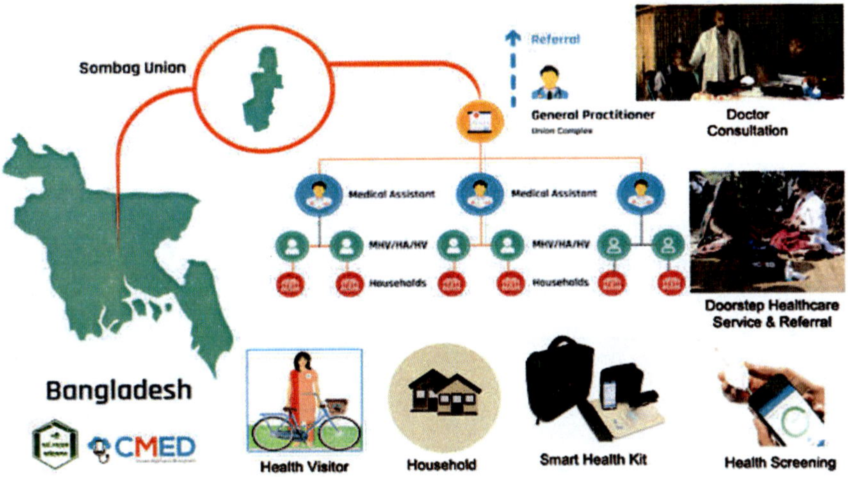

Fig. 1 Overall working procedure of CMED digital health service and referral system

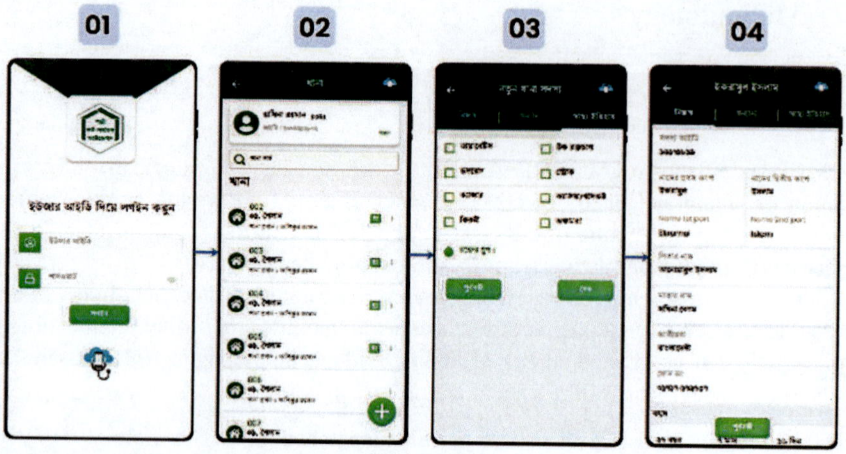

Fig. 2 Profile, measurement, and household survey application view for mobile application

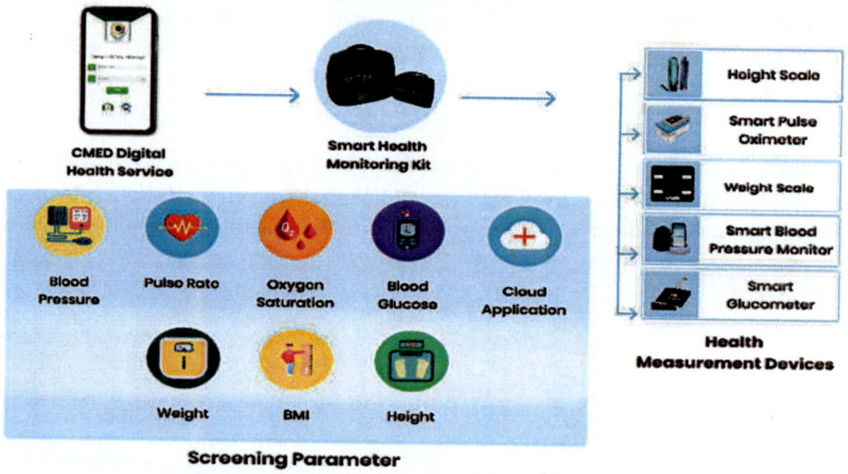

Fig. 3 IoT-based smart medical devices provided by CMED digital health service

automatically uploaded and stored in the cloud server, which acts as an electronic health record. The smart health monitoring kit contains a smart pulse oximeter, smart blood pressure monitor, smart glucometer, analog weight scale, and analog height scale.

Dashboard visualization. Figure 4 shows the web dashboard which is used to get an overall view of the system. The interactive dashboard supports searching household information by household id. It also enhances transparency by conceptualizing data and improving visibility. The dashboard is also used to monitor the progress of each health visitor.

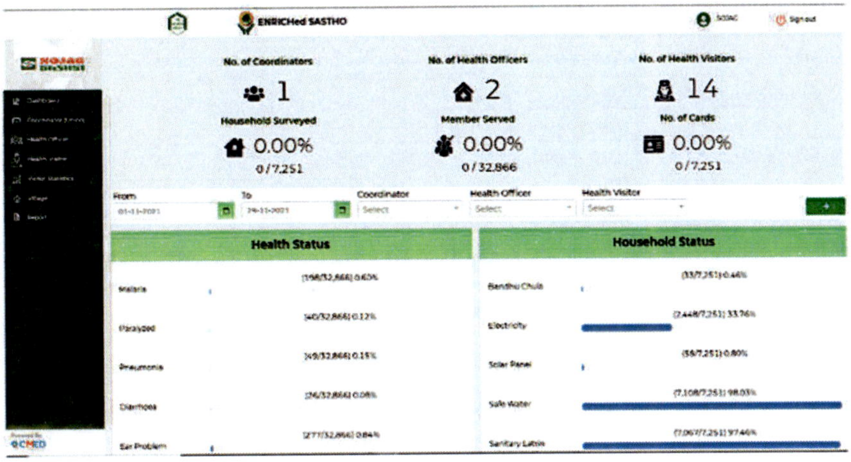

Fig. 4 Web dashboard to view the overall activity of health visitor

Health measurement devices and cloud storage. Health visitors use manual, digital, or IoT-enabled devices to collect health vitals from households. The data obtained from the manual and digital devices are entered into the app by health visitors, whereas IoT devices automatically enter the data through Bluetooth technology. These devices are interconnected through the mobile phone and store the data on the cloud server. These combination devices help to accelerate data collection ensuring minimum human error.

Clinical decision support system (CDSS). The health visitors and health officers provide referrals to the people using a clinical decision support system. This feature uses artificial intelligence to analyze patients' conditions and makes complex decisions regarding requirements for referral.

Other features. CMED has continuously updated and improved the system throughout the operation of the service and updates were introduced to the health visitors/officers through training and a user manual document. The user interface is compatible with both Bangla and English, providing language flexibility to the user. To avoid dependency on internet availability, CMED has introduced an offline synchronization system. This is a very crucial feature because continuous internet connectivity is a problem in the rural areas of Bangladesh.

2.2 Data Collection

Each Health Visitor has a catchment area and visits each khana at least once every month. During their visit, they collect socio-demographic, environmental, economic, educational, and health measurement data. They also collect personal data such as

Table 1 Data collection categories and their attributes

Category	Attributes
Socio-demographic	Name, age, national id, village, date of birth, marital status, education, occupation
Environmental	Safe drinking water, sanitary latrine, bondhu chula, solar system
Economic	Health card, monthly income, monthly medicine cost, monthly health other cost
Health Measurements	Diabetes, hypertension, cardiovascular diseases (CVD), stroke, cancer, chronic obstructive pulmonary disease (COPD), relative's health measurement, family diabetes, family hypertension, physical disability, mental disability, eye disability, ear disability, other disability, paralyzed, malaria, diarrhea, hepatitis, leprosy, tuberculosis, blood pressure, malnutrition, Oxygen Saturation (SPO2), Body Mass Index (BMI), Mid-Upper Arm Circumference (MUAC)

id, name, family information, birthdate, mobile number, age, household head, etc. The segments and the data are listed in Table 1.

Each health visitor is assigned a specific number of households to visit for the month. They generally have about 20 households a day in order to achieve their monthly goal. For health measurements, they use IoT devices and other measurement devices.

3 Results

During the period June 2018–December 2018, CMED Health provided health measurement services to 17,521 adults and collected socio-demographic data of a total 31,868 people using the CMED digital health service. Most of these people had not previously have their health data measured and thus were not aware of their health conditions. Due to the initiative of CMED, their data have been collected using CMED's smart health kit, and the data is stored in the cloud. Not only that, the health visitors gave advice to the individuals regarding their health in cases where the data showed the risk of a certain health problem. This suggestion system is built-in to the health app of CMED.

Table 2 shows the population distribution based on different age groups separated into male and female. One observation from the data is there are more females measured than males. Over 60% of the measured subjects were female. This is because the visit was during the day and most male members were at workplaces during that time. It is also observed from the data that there are more people of ages 18–49 than other groups.

From Table 3, it is also observed that about 21% of the people have high blood sugar. This is a matter of concern because this is a large portion of the population who did not know they may have diabetes. Due to CMED's intervention, they can

Table 2 Distribution of overall measured population based on different age groups

Characteristics	No. of people/denominator (%)		
	Male	Female	Total
Total population	6497 (37.08%)	11,024 (62.91%)	17,521
>=18	4772 (34.37%)	9111 (65.63%)	13,883
18–29	876 (18.36%)	2506 (27.51%)	3382 (24.36%)
30–39	1085 (22.74%)	2344 (25.73%)	3429 (24.7%)
40–49	1028 (21.54%)	1776 (19.49%)	2804 (20.2%)
50–59	744 (15.59%)	1187 (13.03%)	1931 (13.9%)
60–69	632 (13.24%)	879 (9.65%)	1511 (10.88%)
>=70	407 (8.53%)	419 (4.60%)	826 (5.95%)

now be more cautious about their health and get proper medical treatment at health facilities. CMED's cloud storage system also stores their data for future use and this works as a medical history of a patient which doctors can get access to in order to make an informed decision regarding the patient's treatment.

After the primary screening, health visitors regularly visited a household once a month and measured blood pressure, blood glucose level, body mass index, SPO2, and MUAC (for children under 6 years old). A total of 17,521 people had at least one measurement taken. For people of ages 18 or older, blood pressure was measured at 13,427, BMI 12,062, blood sugar 1323, and SPO2 for 13,430 individuals. For children under 18, blood pressure was measured at 485, BMI 2807, blood sugar 35, and 3143 children were measured for SPO2. MUAC was measured for only children under 6 years of age. Among 666 children whose MUAC measurement was taken, 14.56% were malnourished (Table 4).

4 Discussion

Primary Health Care is an integral part of a community which addresses the most important health problems of that community and provides preventive solutions to those problems [23]. Currently, the PHC system in Bangladesh is unorganized, inefficient, understaffed, and relies on analog systems for operation. The use of digital technologies in the field of primary healthcare service presents us with an opportunity to transform the landscape of the healthcare industry. The introduction of smart technologies to the healthcare system has the potential to bring revolutionary changes in terms of diagnosis, disease prevention, risk monitoring, health management, etc. [24].

To improve the primary healthcare services in our country and to mitigate disease burden of secondary- and tertiary-level health facilities in Bangladesh, CMED digital health service is an effective solution. During the period of service from June 2018 to December 2018, a total of 17,521 people measured their health vitals using this

Table 3 Distribution of overall measured population over 18 based on BP, BMI, SPO2, and blood glucose

Characteristics	No. of people/denominator (%)		
	Male	Female	Total
>=18	4772 (34.37%)	9111 (65.63%)	13,883
Blood pressure	4448 (33.13%)	8979 (66.87%)	13,427
Low	187 (4.20%)	691 (7.70%)	878 (6.54%)
Normal	3347 (75.25%)	6520 (72.61%)	9867 (73.49%)
Mild high	362 (8.14%)	725 (8.07%)	1087 (8.10%)
Moderate high	72 (1.62%)	202 (2.25%)	274 (2.04%)
Severe high	30 (0.67%)	42 (0.47%)	72 (0.54%)
Pre-hypertension	450 (10.12%)	799 (8.90%)	1249 (9.30%)
BMI	3538 (29.33%)	8524 (70.67%)	12,062
Underweight	246 (6.95%)	498 (6.95%)	744 (6.17%)
Normal	2588 (73.15%	5366 (62.95%)	7954 (65.94%)
Overweight	618 (17.47%)	2127 (24.95%)	2745 (22.76%)
Obesity	81 (2.29%)	456 (5.35%)	537 (4.45%)
Highly obesity	5 (0.14%)	65 (0.76%)	70 (0.58%)
Morbid obesity	0 (0%)	12 (0.14%)	12 (0.1%)
Glucose level	479 (36.20%)	844 (63.80%)	1323
Normal	353 (73.70%)	640 (75.83%)	993 (75.06%)
Diabetic (need confirmation)	19 (3.97%)	15 (1.78%)	34 (3.42%)
High	77 (16.08%)	136 (16.11%)	213 (21.45%)
High (borderline)	16 (3.34%)	27 (3.30%)	43 (4.33%)
Pre-diabetic	9 (1.88%)	9 (1.07%)	18 1.81%)
Low (Hypoglycemia)	5 (1.04%)	17 (2.01%)	22 (2.21%)
SPO2	4441 (33.07%)	8989 (66.93%)	13,430
Normal	4377 (98.56%)	8951 (99.58%)	13,328 (99.24%)
Low	54 (1.22%)	29 (0.32%)	83 (0.62%)
Very low	10 (0.23%)	9 (0.10%)	19 (0.14%)

service among which, 13,883 (79.23%) are adults. Among the adult population, the percentages of measurements are—BP 61.12%, BMI 54.91%, blood sugar 6.02%, and SPO2 61.14%. It is also observed that there are 14.56% malnourished children under the age of 6. Table 5 shows the total number of health measurements during the operation period. A noteworthy observation is that the total measurement of glucose is very low (1765 for adults). One reason for this is that people had to pay a small amount (30tk) per blood glucose measurement. This caused low enthusiasm for glucose measurement.

Table 4 Distribution of overall measured population under 18 based on BP, BMI, SPO2, blood glucose, and MUAC for children under 6

Characteristics	No of people/denominator (%)		
	Male	Female	Total
<18	1725 (47.42%)	1913 (57.75%)	3638
Blood pressure	162 (33.40%)	323 (66.60%)	485
Low	75 (46.30%)	133 (41.18%)	208 (42.87%)
Normal	75 (46.30%)	159 (49.29%)	234 (48.25%)
High	10 (6.17%)	23 (7.12%)	33 (6.80%)
Mild high	0 (0)	1 (0.31%)	1 (0.21%)
Pre-hypertension	2 (1.23%)	7 (2.17%)	9 (1.86%)
BMI	1309 (46.63%)	1498 (53.37%)	2807
Underweight	364 (27.81%)	370 (24.70%)	734 (26.15%)
Normal	779 (59.51%)	943 (62.95%)	779 (27.75%)
Overweight	94 (7.18%)	124 (8.28%)	94 (3.35%)
Obesity	72 (5.50%)	59 (3.94%)	131 (4.67%)
Highly obesity	0 (0)	2 (0.13%)	2 (0.07%)
Morbid obesity	0 (0)	0 (0)	0 (0)
Glucose level	14 (40%)	21 (60%)	35
Normal	11 (78.57%)	15 (71.43%)	26 (74.29%)
Low	3 (21.43%)	5 (23.81%)	8 (22.86%)
Low (Hypoglycemia)	0 (0)	1 (4.76%)	1 (2.56%)
SPO2	1460 (46.45%)	1683 (53.54%)	3143
Very low	5 (0.24%)	4 (0.34%)	9 (0.29%)
Low	7 (0.36%)	6 (0.48%)	13 (0.41%)
Normal	1448 (99.18%)	1673 (99.41%)	3121 (99.3%)
MUAC (<=5)	340 (51.05%)	326 (48.95%)	666
Nourished	291 (85.89%)	266 (81.60%)	557 (83.63%)
Malnutrition	44 (12.94%)	53 (16.26%)	97 (14.56%)
Severe malnutrition	5 (1.47%)	7 (2.15%)	12 (1.80%)

Table 5 Overall health measurements across age categories

Measurement type	Age >= 18	Age < 18	Age <= 5
BMI	72,176	8454	–
Blood pressure	81,750	1764	–
Glucose	1765	43	–
SPO2	69,284	4728	–
MUAC	–	–	942

Majority of the people who were served using the CMED digital health service did not previously know their health vitals and none of them had a proper health record. Due to this service, they have an understanding of their health vitals and how it could affect their health. The health workers gave each individual valuable advice based on their health measurements and for individuals with extreme health measurements, they advised them to seek treatment from specialized doctors and referred them to upper level health facilities. Thus, each person under the service of this program has a digital health record and knowledge on how to properly maintain good health based on their measurement. By doing so, the platform establishes a primary screening process which ensures preventive health care for all people. The system is available for all people irrespective of any socio-economic or ethnic background which ensures health inclusion.

4.1 Strengths and Limitations

The strength of the system comes from its combined feature of collecting and storing data using smart IoT health measuring devices, ability to make intelligent decisions, and provide medical advice based on the individual's data. IoT-enabled data collection and cloud storage also help to keep medical records of a patient's treatment thus making the system able to scale up as necessary.

Health visitors also take other data besides only health vitals measurements. These include data related to living environment, economic condition of a household, education and occupational-related data, availability of proper sanitation facility, health card usage, etc. These data can be used to classify people of different income levels and education groups, and this also helps understand the overall condition of a household.

The suggestion system of the app helps identify people at health risk and forward them to continue further check-ups at a designated health facility. CMED also provides a comprehensive dashboard to view the overall condition of a household and it also helps monitor the activity of health visitors.

This system has a few limitations and CMED is working towards overcoming these challenges. Table 2 shows that over 60% of the measured people are women. The reason behind it is that the men are at their workplaces during the daytime when the measurements were taken. CMED plans to overcome this challenge by conducting multiple visits to the same household.

People cannot confirm their diabetes status through this system as it requires biochemical tests in the laboratory. However, this system identifies the suspected cases for confirmatory tests in the diagnostic laboratory.

4.2 Future Work

Currently, there are many improvement projects undertaken by CMED. A GPS tracking system is being developed to track the real-time progress of health visitors. The collected data can be used to determine economic status and create an insurance scheme which will help people get universal health coverage. Another project currently in the works is to identify people living in extreme poverty who cannot afford the basic needs. By doing this it is possible to direct government help to those who really need it. CMED intends to educate rural people about their health through a user app.

The location of the pilot being limited to only one location is a drawback as the results from it may not translate to other areas. CMED plans to extend the service to other varying demographic locations to strengthen the research and to better understand the necessities of the population.

Tracking progress is an integral part of any development. Currently, SDGs are used to monitor global development. CMED plans to track SDG-related data in order to attain its effectiveness in the field. By doing so, they can make evident-based decisions which will help Bangladesh achieve these goals by concentrating resources in areas where it is really necessary.

5 Conclusion

Most of the people living in rural areas of Bangladesh are not aware of their health conditions and only visit health facilities when they are sick. On one hand, this puts their life in danger because many of these diseases could have been prevented by only a balanced lifestyle; on the other hand, this puts a burden on the current secondary- and tertiary-level health facilities. As a result, the individual out-of-pocket expenditure increases and which causes suffering to livelihood, especially for the people of lower economic statuses.

The primary goal of CMED Health is to provide comprehensive preventive and primary healthcare service to the people of Bangladesh with special attention to underprivileged rural and urban populations. CMED has demonstrated the advantage of a patient-centered comprehensive digital primary healthcare service over a hospital-centered healthcare system. The use of doorstep health visits, m-IoT devices, cloud storage, and data-driven clinical decision support systems yield an end-to-end digital primary health service for rural areas, which can be extended to other parts of Bangladesh. Through effective deployment of the service, it is possible to provide comprehensive preventive and primary health service to the people of Bangladesh at an affordable cost.

A smart and modern health monitoring system is long overdue in Bangladesh and CMED has proven the capability of its system to deliver exactly that. CMED is able to provide this service at a low cost, making it affordable for people of various socio-economic demographics which ensures health inclusion. By expanding the service to other areas, it is possible to improve the healthcare system throughout the nation, which is key to improving management of health needs of the population, achieving SDG-3 goals, and realizing the goals of Digital Bangladesh.

Acknowledgements The authors are grateful to Enhancing Resources and Increasing Capacities of Poor Households Towards Elimination of their Poverty (ENRICH) program of PKSF; PKSF management; Partner Organization SOJAG; and United International University, Bangladesh for their support and cooperation to successfully implement the pilot program.

References

1. Sustainable Development Goals | United Nations Development Programme. https://www.undp.org/sustainable-development-goals#good-health. Accessed Sept. 07, 2021
2. Roth: Global, regional, and national age-sex-specific mortality for 282 causes of death in 195 countries and territories, 1980–2017: a systematic analysis for the Global Burden of Disease Study 2017. Lancet 392, 1736–1788 (2018). https://doi.org/10.1016/S0140-6736(18)32203-7
3. Siddiqi, K.: Non-communicable diseases. In: Public Health: An action guide to improving health (2010). https://doi.org/10.1093/acprof:oso/9780199238934.003.15
4. CountryProfile | World Development Indicators. https://databank.worldbank.org/views/reports/reportwidget.aspx?Report_Name=CountryProfile&Id=b450fd57&tbar=y&dd=y&inf=n&zm=n&country=BGD. Accessed Sept. 06, 2021
5. Bangladesh Population 2021 (Demographics, Maps, Graphs). https://worldpopulationreview.com/countries/bangladesh-population. Accessed Sept. 12, 2021
6. Islam, A., Biswas, T.: Chronic non-communicable diseases and the healthcare system in Bangladesh: current status and way forward. Chronic Dis. Int. **1**, 1007–1012 (2014)
7. Karim, S.M.T., Alam, S.: The health care system in Bangladesh: an insight into health policy, law and governance. Aust. J. Asian Law. **20**, 1–19 (2020)
8. Mitchell, M., Kan, L.: Digital technology and the future of health systems. Heal. Syst. Reform. **5**, 113–120 (2019). https://doi.org/10.1080/23288604.2019.1583040
9. Yu, Y., Wang, Q., Liu, J.: Pervasive health monitor and analysis based on multi-parameter smart armband. In: Proceedings of Annual *International Conference* of the *IEEE* Engineering in Medicine and *Biology* Society (EMBS), pp. 5493–5496 (2015). https://doi.org/10.1109/EMBC.2015.7319635
10. Sailunaz, et al.: CMED: Cloud based medical system framework for rural health monitoring in developing countries. Comput. Electr. Eng. **53**, 469–481 (2016). https://doi.org/10.1016/j.compeleceng.2016.02.005
11. Gonzalez-vega, C.: Palli Karma Sahayak Foundation (PKSF) (1998)
12. Ahmed, T., Bloom, G., Iqbal, M., Lucas, H., Rasheed, S., Waldman, L., Khan, A.S., Islam, R., Bhuiya, A.: E-health and M-Health in Bangladesh: opportunities and challenges. Evid. Rep. No. 60; Horiz. scanning (Tomorrow Today) (2014)
13. Jeeon—Making Well-being Universal. https://www.jeeon.co/#services. Accessed Sept. 29, 2021
14. Khan, D.M.M.: IoT Based Smart Healthcare Services for Rural Unprivileged People in Bangladesh: Current Situation and Challenges, p. 7535 (2021). https://doi.org/10.3390/asec2020-07535
15. Our Services | Digital Healthcare Solutions. https://dh.health/OurServices. Accessed Sept. 29, 2021

16. Digital Hospital. https://digitalhospital.health/?&lang=en. Accessed Sept. 29, 2021
17. Your Health, Take Care | Daktarbhai. https://www.daktarbhai.com/. Accessed Sept. 29, 2021
18. Doctorola—Book Appointment With a Doctor Near You !. https://doctorola.com/. Accessed Sept. 29, 2021
19. Rabbani et al.: Dhaka University Telemedicine Programme, Targeting Healthcare-Deprived Rural Population of Bangladesh and Other Low Resource Countries. Lecture Notes Computer Science (including Subser. Lect. Notes Artif. Intell. Lect. Notes Bioinformatics). 11786 LNCS, pp. 580–598 (2019). https://doi.org/10.1007/978-3-030-30033-3_45
20. Statista: Bangladesh: share of rural population 2010–2019 | Statista. https://www.statista.com/statistics/760934/bangladesh-share-of-rural-population/. Accessed Oct. 05, 2021
21. Khatun, F., Hanifi, S.M.A., Iqbal, M., Rasheed, S., Rahman, M.S., Ahmed, T., Hoque, S., Sharmin, T., Khan, N.U.Z., Mahmood, S.S., Peters, D.H., Bhuiya, A.: Prospects of mHealth services in Bangladesh: Recent evidence from Chakaria. PLoS One. 9, (2014). https://doi.org/10.1371/journal.pone.0111413
22. Faruque et al.: Feasibility of digital health services for educating the community people regarding lifestyle modification combating noncommunicable diseases. Lecture Notes on Computer Science (including Subser. Lect. Notes Artif. Intell. Lect. Notes Bioinformatics). 11587 LNCS, pp. 333–345 (2019). https://doi.org/10.1007/978-3-030-21935-2_25
23. Bryant, J.H., Richmond, J.B.: Alma-Ata and primary health care: an evolving story. Int. Encycl. Public Heal. 83–102 (2016). https://doi.org/10.1016/B978-0-12-803678-5.00017-5
24. Tian, S., et al.: Smart healthcare: making medical care more intelligent. J. Glob. Health. **3**, 62–65 (2019). https://doi.org/10.1016/j.glohj.2019.07.001

An LSTM-Based Forecast Of COVID-19 For Bangladesh

Ashifur Rahman, Md. Akbar Hossain, and Mohasina Jannat Moon

Abstract Preoperative events can be predicted using deep learning-based forecasting techniques. It can help to improve future decision-making. Deep learning has traditionally been used to identify and evaluate adverse risks in a variety of major applications. Numerous prediction approaches are commonly applied to deal with forecasting challenges. The number of infected people, as well as the mortality rate of COVID-19, is increasing every day. Many countries, including India, Brazil, and the United States, were severely affected; however, since the very first case was identified, the transmission rate has decreased dramatically after a set time period. Bangladesh, on the other hand, was unable to keep the rate of infection low. In this situation, several methods have been developed to forecast the number of affected, time to recover, and the number of deaths. This research illustrates the ability of DL models to forecast the number of affected and dead people as a result of COVID-19, which is now regarded as a possible threat to humanity. As part of this study, we developed an LSTM based method to predict the next 100 days of death and newly identified COVID-19 cases in Bangladesh. To do this experiment we collect data on death and newly detected COVID-19 cases through Bangladesh's national COVID-19 help desk website. After collecting data we processed it to make a dataset for training our LSTM model. After completing the training, we predict our model with the test dataset. The result of our model is very robust on the basis of the training and testing dataset. Finally, we forecast the subsequent 100 days of deaths and newly infected COVID-19 cases in Bangladesh.

Keywords LSTM · Deep learning · Who · Affected · Death · Coronavirus · Covid-19

A. Rahman (✉) · Md. A. Hossain · M. J. Moon
Bangladesh University of Business and Technology (BUBT), Dhaka, Bangladesh
e-mail: ashifurrahman.bubt@gmail.com

S. Hossain et al. (eds.), *Proceedings of International Conference on Fourth Industrial Revolution and Beyond 2021*, Lecture Notes in Networks and Systems 437,
https://doi.org/10.1007/978-981-19-2445-3_38

1 Introduction

Coronavirus disease 2019, also known as COVID-19, was first found in Wuhan, the capital of China's Hubei province in Dec 2019 and quickly spread all over the world. There is now a global pandemic affecting all countries. The worries and anxiety that occur as a result of the growing spread of afflicted cases and deaths around the world are difficult to manage [19, 22]. Currently, more than 211 million people have been infected and more than 4.79 million have died as a result of the virus spreading worldwide. This is largely thanks to the amount of verified infected cases growing exponentially every day. It is alarming to see the number of deaths and affected people in Bangladesh in COVID-19. In Bangladesh, this virus has caused more than 28 thousand deaths so far, and the affected number is more than 1.55 million people. Coronavirus disease is an RNA virus with a complicated nature and variable genomic structure; these characteristics make it much more difficult to develop a vaccine against it. Predicting an epidemic and ensuring social distance and isolation are the safest ways to combat the virus. To develop a standardized epidemiology model, scientists are using ML, DL, and AI algorithms. Due to the severity of the infection, more risky places are identified and isolated [3, 18, 25]. This modeling technique is related to the overall amount of deaths and recorded incidents in a given location. In addition, it is related to the age of infected individuals, their sex and medical record, travel history, contacts tom other patients, and other factors [5, 23, 24].

There is no doubt that Bangladesh, one of the world's fastest-growing countries, is on the verge of experiencing the devastating effects of this virus, which has already resulted in a huge amount of fatalities and a rapid increase in infections. As a result, it is critical to develop a comprehensive model to forecast the spread of the virus and allow the government to take the requirements needed to prevent it [1, 17, 31].

The SIR [4] and SEIR [21] models are frequently used to identify epidemics and pandemics, although they have certain drawbacks. Machine Learning (ML) techniques are now frequently used to built a model to predict viral infection like as Ebola, Zika, H1N1, and others and have shown to be effective [16, 40]. Researchers are already utilizing AI and Machine Learning to diagnose COVID-19 by evaluating chest x-rays and screening [26], estimate the effectiveness of antimicrobial drugs [35], forecast, and determine the virus's genetic code [8], and analyze economical impacts due to the epidemic [30], among others.

In this article, we offer a revolutionary LSTM-based technique for predicting death and newly infected COVID-19 patients in Bangladesh over the next 100 days. To conduct this study, we used the Bangladesh national COVID support desk website to collect data on deaths and newly infected COVID-19 cases. We processed the data after collecting it to create a dataset for training our LSTM model. Models are predicted using the test dataset after we have completed the training. Our model is extremely accurate when it comes to the training and testing datasets. Finally, we estimated how many deaths and newly infected COVID-19 cases occur in Bangladesh in the next 100 days.

2 Related Work

Recent advances in Deep Learning (DL) research, along with applications in fields such as viral epidemiology, are growing. Deep Learning and Machine Learning is a cutting-edge method for detecting early virus infection, developing medicines and vaccinations, risk analysis, and recognizing virus protein molecules, among other uses. Many of the most common uses of Machine Learning and Deep Learning in the field of viral epidemiology. This is to develop a statistical framework or prediction method that has been shown to be beneficial in the transmission of diseases like Zika, Ebola, Dengue, Malaria, and others.

To anticipate potential dengue outbreaks, Yuhanis Yusof and Zuriani Mustaffa [39] proposed algorithms based on the LS-SVM (Least Squares Support Vector Machine). For this experiment, the dataset included data on dengue cases and rainfall levels in Selangor, Malaysia's local government regions. A total of 520 data points were processed, with 70% of the information used for training and 30% reserved for evaluation. Throughout the preprocessing process, input data was normalized to decimal points and output data was standardized to decimal points. It achieved an overall accuracy rate of 86.84%, significant higher than the 65.58% accuracy rate of the Neural Network Model. In order to track the Zika outbreak, Dong Jiang et al. [16] implemented Backward-Propagation-Neural-Network (BPNN), Gradient-Boosting-Machine (GBM), and Random-Forest (RF) methods. Focusing just on the virus's spread, the analysis identified high-risk zones. BPNN achieved an AUC score of 0.966 for both the training and testing datasets, which was higher than GMB and RF (0.964, 0.963). It was also discovered that the prediction uncertainty for BPNN was much more severe than for the others.

Naiyar Iqbal and Mohammad Islam [15] used 7 distinct machine learning techniques to predict dengue outbreaks. The dataset they gathered from patients' individual test results consisted of 75 patients; among them, 36 of them were dengue negative, and the rest of them were dengue infected. The LogitBoost Ensemble model outperformed all other models, with a 92% accuracy rate. The LogitBoost model also beat the other model with a calculated sensitivity of 90% and a specificity of 94%. To anticipate the outbreak of malaria in Ethiopia, Chekol et al. [10] conducted a mathematical model of malaria using SVR and ANFIS. A malaria dataset covering the years 2013–2017 was selected depending on altitude, moisture, temperature, the lag case of malaria, and a positive case of malaria in the most recent month. The model was trained on 7516 data and tested on 3765 data. The study demonstrates that ANFIS beat SVR based on the correlation coefficient and root-mean-square-error (RMSE).

Colubri et al. [12] developed a prognosis model for the Ebola epidemic based on Multivariate Logistic Regression. The dataset used in this research is EVD. The dataset consists of 470 EVD confirmed cases in 5 distinct places in Sierra Leone and Liberia, with 264 cases from two separate datasets used only for validation. By achieving AUC ranges of 0.70–0.79 and accuracy ranges of 0.64–0.74, the prognosis model was very efficient in identifying and contaminating the pandemic. On sow farms, Gustavo et al. [12] used machine learning methods to anticipate the emergence of the Porcine Epidemic Diarrhea Virus (PEDV). RF, SVM, and GBM were the ML algorithms utilized in this research to forecast the pandemic. The MSHMP provided the dataset, with 80% of it being utilized to train the model and 20% being used to test the model. The model's performance has been measured based on a variety of criteria, including AUC scores, accuracy, sensitivity, and specificity. According to the RF model, accuracy is 83.0%, AUC is 96%, sensitivity is 65.2%, and specificity is 95%.

Considering COVID-19 is a global pandemic, researchers are striving to develop an epidemiological model that can anticipate the pandemic's spread. Sina et al. [2] developed a model that used MLP and ANFIS techniques. In accordance with a sample of processing data, different scenarios for expanding the focus of the research were presented (on a weekly and daily basis). In terms of MLP, the number of hidden neurons was changed, and in the case of ANFIS, various kinds of classifiers were utilized. For time series epidemics in multiple nations, the test results demonstrated that MLP performed much better than ANFIS. Pandey et al. [28] created a model that uses the SEIR and Regression model to forecast the spread of COVID-19 over the next two weeks. The SEIR model has an estimated RMSE of 1.52, while the Regression model has an RMSE of 1.75 in this investigation.

Tuli et al. [34] presented an ML based models, like the Gaussian prediction model and the Robust Weibull model. These ML models were used for the Azure cloud platform, on which the FogBus [33] architecture was used for faster processing. The Robust Weibull model has a lower calculated MSE than that of the Gaussian model, while the Robust Weibull model has a greater MAPE than the Gaussian model. To estimate COVID-19 epidemic expansion in China, Yang et al. [38] developed a customized SEIR epidemiology model as well as an LSTM model. Integrating population mobility data into the SEIR formula, the classic SEIR equation is altered. The enhanced SEIR model can forecast pandemic transmission extremely well, and the LSTM model, which was trained on earlier SARS datasets, performed well as well. The outbreak peaked in February, according to the research, and then diminished by the end of April.

Pinter et al. [29] developed a prediction model to assess COVID-19 utilizing a hybrid Machine Learning approach that includes ANFIS and MPICA to develop an analytical model to assess COVID-19 utilizing records of infected patients and death (MLP-ICA). Data for this study were compiled from frequently reported incidents, as well as Hungary's mortality rate. The RMSE of MLP-obtained ICA for total cases is 187.88, while ANFIS's is 194.10, and the RMSE of MLP-obtained ICA for total cases stands at 8.32, compared to 15.25 for ANFIS, meaning MLP-ICA performed better than ANFIS. In a survey performed in Japan by Kuniya [20], the SEIR model was utilized to predict COVID-19. The analysis predicted when the outbreak would reach its height. The fundamental repetition numbers were estimated using the least square-based method and Poisson noise in this study. It has been demonstrated that perhaps the SEIR model can accurately anticipate the scale of a pandemic. It can also analyze the steps that must be taken to enhance the health service.

3 Dataset

COVID-19 transmission is the main concern of this research, which focuses on newly infected persons in Bangladesh and the death toll from COVID-19. The dataset used in this experiment is collected from the Bangladesh National COVID help desk website. We prepared the data into a comma-separated values (CSV) file. The file contains three columns (named Date, New_Case, and Death) and 365 rows. Only the last 365 days' data are collected for this experiment.

Table 1 shows some features and elements that have in our experiments dataset. We plotted all the values of our dataset separately. Both graphs in Fig. 1 are represented on the x-axis by the number of days, and on the y-axis by the number of newly infected cases per day (for a) and deaths per day (for b).

Due to the recent emergence of COVID-19, there are only a few widely accepted datasets are available. We know that this is a small amount of data but our model

Table 1 Dataset of newly infected and death cases of COVID-19 is given below. The first five rows of the dataset are given in this table

No:	Date	New_Case	Death
1	9-27-2020	1275	32
2	9-28-2020	1407	32
3	9-29-2020	1488	26
4	9-30-2020	1436	32
–	–	–	–
364	9-26-2021	980	21
365	9-27-2021	1212	25

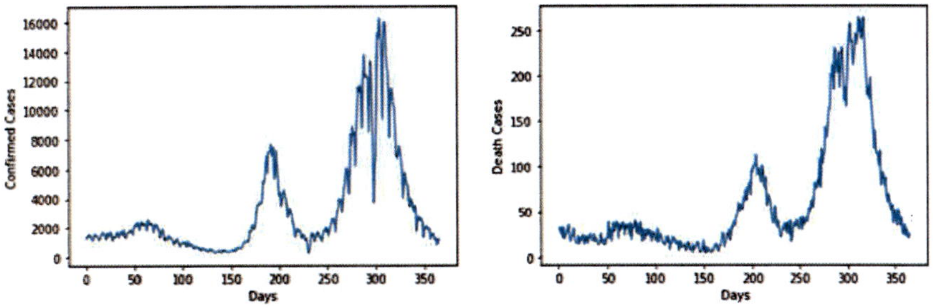

Fig. 1 Here we plot all the values that given in the dataset

gives a very good result on this small dataset. We also assure that if we increased the volume of data then the model gives much better accuracy than now.

4 Methodology

The related work section describes several methods to improve the method for predicting different epidemics and pandemics. We presented a novel pandemic outbreak prediction method based on stacked-LSTMs [6] in this paper. Figure 2 describes all the processes of our experiment.

In this research work at first, we load the COVID-19 dataset. After loading the whole dataset we processed it to train our model. In the processing step, we converted all the death and newly infected numbers between 0 and 1. To convert the dataset

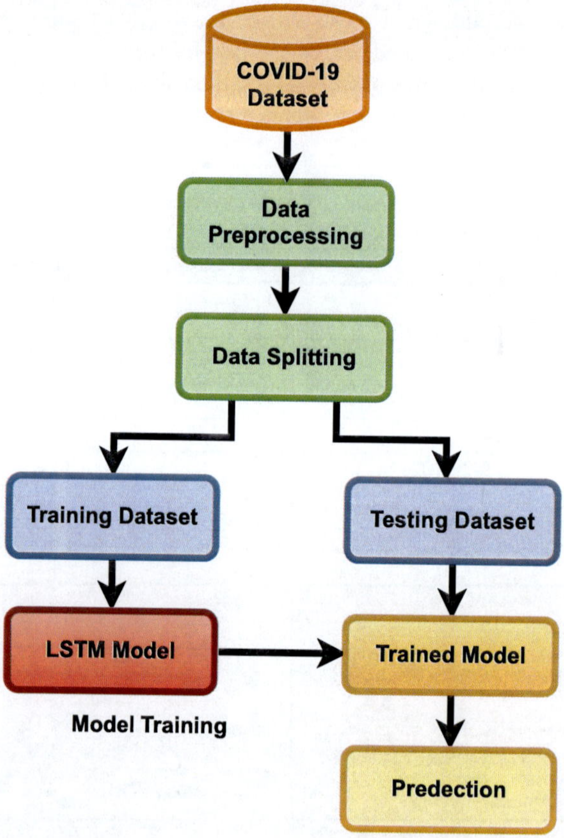

Fig. 2 This figure narrates the system architecture of our LSTM model. The whole process is divided into three main parts (1) Data preprocessing, (2) model training, and (3) Prediction

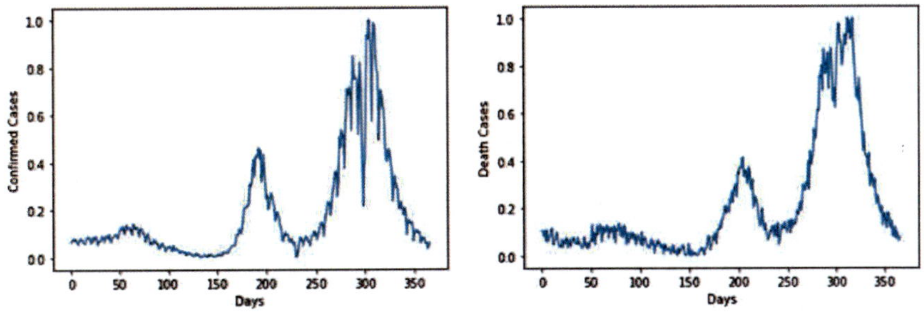

Fig. 3 This figure illustrates how the dataset was converted from 0 and 1

between 0 and 1 we use the MinMaxScaler [32] function. The converted dataset is shown in Fig. 3.

After converting the dataset, we split our COVID-19 dataset into two-part names, train and test. For training and testing, we split the dataset into 65/35. After building the train and test dataset we are ready to train and predict our model performance. We build a Stack LSTM [6] model that contains three Long Short-Term Memory (LSTM) [14] layers and one dense layer which is also the output layer. The output dimension of all three LSTM layers is 50, but on the other hand, the output dimension of dense layers is 1. We used mean_squared_error [36] as loss function and adam [41] as optimizer. We train our model into 200 epochs with batch size 32. After completing training we test our model through a test dataset. The testing result is given in the result section. To evaluate our model we use MSE, RMSE, and MAE.

$$MSE = \frac{1}{n}\Sigma_{i=1}^{n}\left(y_d - y_c\right)^2 \quad (1)$$

$$RMSE = \sqrt{\frac{1}{n}\Sigma_{i=1}^{n}\left(y_d - y_c\right)^2} \quad (2)$$

$$MAE = \frac{1}{n}\Sigma_{i=1}^{n}\left|y_d - y_c\right| \quad (3)$$

After predicting the model through the test dataset, we are ready to forecast the next 100 days of death and infection. To do this we consider the past 100 days' death and infection values to estimate the 101st days' result. After predicting the 101st-day result we are ready to predict the 102nd 102nd-day result. Here we again consider the past 100 days' death and infection values to forecast the 102nd day's result. We continue this procedure until we reach 100 days' prediction.

5 Result

Transmission of Coronavirus is unpredictable and forecasting its result is also difficult. Only public knowledge and government actions can have an impact on COVID-19's transmission. Lockdown, social separation in crowded locations, testing procedures, and hospitalization are all policies that the government is constantly modifying. Using our model, the government will be able to track the outbreak's spread in real-time and adjust policies to keep it under control.

Figure 4 depicts the result of our model in the train and test dataset. Our model gives very accurate results in both train and test datasets. The result is almost similar to the original dataset values.

Figure 5 depicts our model prediction of the next 100 days of COVID-19 affected and death cases in Bangladesh. Here the prediction is made by analyzing the past days of COVID-19 affected and death results.

In Fig. 6 we added the whole dataset with the next 100 days predicting the result. This clearly depicts how well our model forecast the next 100 days infected and death result of COVID-19 in Bangladesh. We evaluated our model through three evaluation matrices MSE, RMSE, and MAE. That result is given in Table 2.

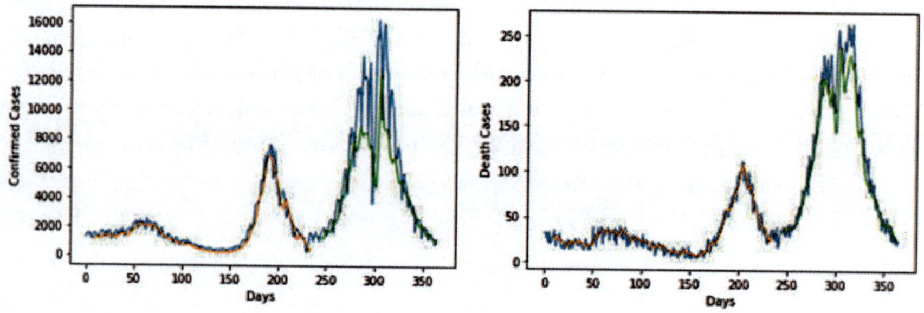

Fig. 4 Figure depicts the result of our model in train and test dataset. Here blue color line represents the original dataset, yellow, and green colors represent the train and testing data predictive values respectively

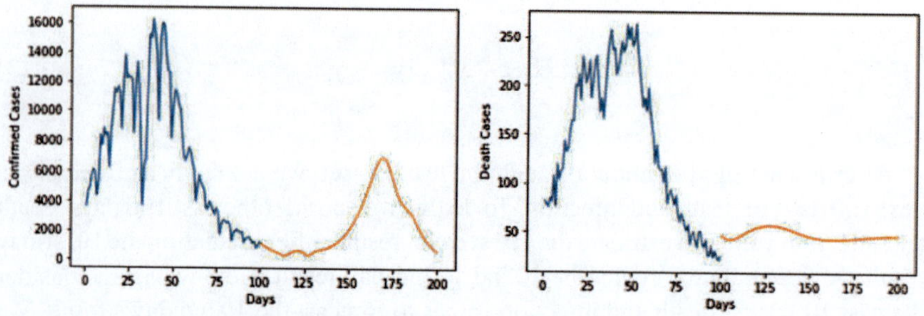

Fig. 5 Here we plot the next 100 days infected and death prediction with the past 100 days infected and death result

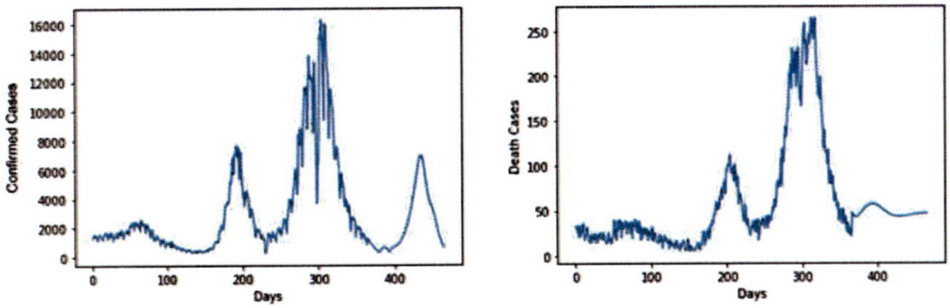

Fig. 6 Here we plot the whole dataset and next 100 days predicting result

Table 2 Here given our model performance on test data. We evaluated the model through MSE, RMSE, and MAE

Evaluation matrix	Confirmed cases	Death
MSE	4674052.3834	2754.8013
RMSE	2161.9557	52.4862
MAE	2120.7637	52.4458

6 Conclusion

As a result of the COVID-19 pandemic's precariousness, the potential for worldwide catastrophe has increased. Various governmental and scientific organizations around the world are concerned that the pandemic will impact a large section of the world's population [13]. The presented work is to predict the next 100 days of the COVID-19 outbreak in Bangladesh using the stacked LSTM model. In this approach, deep learning algorithms are used to analyze a dataset that includes daily previous data and make predictions for the 100 days ahead. The LSTM method did quite well in terms of prediction accuracy for the models when it came to calculating MSE values.

Deep Learning models could be utilized in place of epidemiological models that do not require previous beliefs by following the transmission of the disease and determining the actual affected, and fatalities, our proposed work can help the policymakers of Bangladesh to fight against COVID-19.The model can be improved in future research by merging it with other methods like Gated Recurrent Units(GRUs) [11] and adopting cloud services such as Azure [9, 37] and Amazon Web Service (AWS) [27]. In addition to predicting how the Coronavirus will impact our socio-economic sectors, we can use the model to develop new legislation to mitigate the virus' effects. Besides examining COVID-19's socio-economic effects we can also analyze share market price, education, psychological health, garments sector, remittance, and so on by this model [7].

References

1. Afsana, F., Asif-Ur-Rahman, M., Ahmed, M.R., Mahmud, M., Kaiser, M.S.: An energy conserving routing scheme for wireless body sensor nanonetwork communication. IEEE Access **6**, 9186–9200 (2018)
2. Ardabili, S.F., Mosavi, A., Ghamisi, P., Ferdinand, F., Varkonyi-Koczy, A.R., Reuter, U., Rabczuk, T., Atkinson, P.M.: Covid-19 outbreak prediction with machine learning. Algorithms **13**(10), 249 (2020)
3. Asif-Ur-Rahman, M., Afsana, F., Mahmud, M., Kaiser, M.S., Ahmed, M.R., Kaiwartya, O., James-Taylor, A.: Toward a heterogeneous mist, fog, and cloud-based framework for the internet of healthcare things. IEEE Internet Things J. **6**(3), 4049–4062 (2018)
4. Beretta, E., Takeuchi, Y.: Global stability of an sir epidemic model with time delays. J. Math. Biol. **33**(3), 250–260 (1995)
5. Biswas, S., et al.: Cloud based healthcare application architecture and electronic medical record mining: an integrated approach to improve healthcare system. In: ICCIT, pp. 286–291 (2014)
6. Breuel, T.M.: Benchmarking of LSTM networks. arXiv preprint arXiv:1508.02774 (2015)
7. Buheji, M., da Costa Cunha, K., Beka, G., Mavric, B., De Souza, Y., da Costa Silva, S.S., Hanafi, M., Yein, T.C.: The extent of covid-19 pandemic socio-economic impact on global poverty. a global integrative multidisciplinary review. Am. J. Econ. **10**(4), 213–224 (2020)
8. Bullock, J., Luccioni, A., Pham, K.H., Lam, C.S.N., Luengo-Oroz, M.: Mapping the landscape of artificial intelligence applications against covid-19. J. Artif. Intell. Res. **69**, 807–845 (2020)
9. Calder, B., Wang, J., Ogus, A., Nilakantan, N., Skjolsvold, A., McKelvie, S., Xu, Y., Srivastav, S., Wu, J., Simitci, H., et al.: Windows azure storage: a highly available cloud storage service with strong consistency. In: Proceedings of the Twenty-Third ACM Symposium on Operating Systems Principles, pp. 143–157 (2011)
10. Chekol, B.E., Hagras, H.: Employing machine learning techniques for the malaria epidemic prediction in Ethiopia. In: 2018 10th Computer Science and Electronic Engineering (CEEC), pp. 89–94. IEEE (2018)
11. Chung, J., Gulcehre, C., Cho, K., Bengio, Y.: Empirical evaluation of gated recurrent neural networks on sequence modeling. arXiv preprint arXiv:1412.3555 (2014)
12. Colubri, A., Hartley, M.A., Siakor, M., Wolfman, V., Felix, A., Sesay, T., Shaffer, J.G., Garry, R.F., Grant, D.S., Levine, A.C., et al.: Machine-learning prognostic models from the 2014–16 ebola outbreak: data-harmonization challenges, validation strategies, and mhealth applications. EClinical Med. **11**, 54–64 (2019)
13. Govt, B.: Harvard professor sounds alarm on 'likely' coronavirus pandemic: 40% to 70% of world could be infected this year.0. https://www.mediaite.com/news/harvardprofessor-sounds-alarm-on-likely-coronavirus-pandemic-40-to-70-ofworld-could-be-infected-this-year/ (2020). Accessed 15 Oct 2021
14. Graves, A., Fernández, S., Schmidhuber, J.: Bidirectional LSTM networks for improved phoneme classification and recognition. In: International conference on artificial neural networks, pp. 799–804. Springer (2005)
15. Iqbal, N., Islam, M.: Machine learning for dengue outbreak prediction: a performance evaluation of different prominent classifiers. Informatica **43**(3) (2019)
16. Jiang, D., Hao, M., Ding, F., Fu, J., Li, M.: Mapping the transmission risk of zika virus using machine learning models. Acta Tropica **185**, 391–399 (2018)
17. Kaiser, M.S., et al.: Advances in crowd analysis for urban applications through urban event detection. IEEE Trans. Intell. Transp. Syst. **19**(10), 3092–3112 (2018)
18. Kaiser, M.S., Chowdhury, Z.I., Al Mamun, S., Hussain, A., Mahmud, M.: A neuro-fuzzy control system based on feature extraction of surface electromyogram signal for solar-powered wheelchair. Cognit. Comput. **8**(5), 946–954 (2016)
19. Kaiser, M.S., Mahmud, M., Noor, M.B.T., Zenia, N.Z., Mamun, S.A., Mahmud, K.M.A., Azad, S., Aradhya, V.N.M., Stephan, P., Stephan, T., Kannan, R., Hanif, M., Sharmeen, T., Chen, T., Hussain, A.: iworksafe: towards healthy workplaces during covid-19 with an intelligent phealth app for industrial settings. IEEE Access **9**, 13814–13828 (2021). https://doi.org/10.1109/ACCESS.2021.3050193

20. Kuniya, T.: Prediction of the epidemic peak of coronavirus disease in Japan, 2020. J. Clin. Med. **9**(3), 789 (2020)
21. Li, M.Y., Smith, H.L., Wang, L.: Global dynamics of an seir epidemic model with vertical transmission. SIAM J. Appl. Math. **62**(1), 58–69 (2001)
22. Lum, L.H.W., Tambyah, P.A.: Outbreak of covid-19-an urgent need for good science to silence our fears? Singapore Med. J. **61**(2), 55 (2020)
23. Mahmud, M., Kaiser, M.S., Hussain, A.: Deep learning in mining biological data. arXiv preprint arXiv:2003.00108 (2020)
24. Mahmud, M., et al.: A brain-inspired trust management model to assure security in a cloud based IoT framework for neuroscience applications. Cognit. Comput. **10**(5), 864–873 (2018)
25. Makhataeva, Z., Varol, H.A.: Augmented reality for robotics: a review. Robotics **9**(2), 21 (2020)
26. Narin, A., Kaya, C., Pamuk, Z.: Automatic detection of coronavirus disease (covid-19) using x-ray images and deep convolutional neural networks. Pattern Anal. Appl. 1–14 (2021)
27. Narula, S., Jain, A., et al.: Cloud computing security: Amazon web service. In: 2015 Fifth International Conference on Advanced Computing & Communication Technologies. pp. 501–505. IEEE (2015)
28. Pandey, G., Chaudhary, P., Gupta, R., Pal, S.: Seir and regression model based covid-19 outbreak predictions in India. arXiv preprint arXiv:2004.00958 (2020)
29. Pinter, G., Felde, I., Mosavi, A., Ghamisi, P., Gloaguen, R.: Covid-19 pandemic prediction for Hungary; a hybrid machine learning approach. Mathematics **8**(6), 890 (2020)
30. Qiu, Y., Chen, X., Shi, W.: Impacts of social and economic factors on the transmission of coronavirus disease 2019 (covid-19) in China. J. Popul. Econ. **33**(4), 1127–1172 (2020)
31. Rahman, S., Al Mamun, S., Ahmed, M.U., Kaiser, M.S.: Phy/mac layer attack detection system using neuro-fuzzy algorithm for IoT network. In: ICEEOT, pp. 2531–2536. IEEE (2016)
32. Shaheen, H., Agarwal, S., Ranjan, P.: Minmaxscaler binary PSO for feature selection. In: First International Conference on Sustainable Technologies for Computational Intelligence, pp. 705–716. Springer (2020)
33. Tuli, S., Mahmud, R., Tuli, S., Buyya, R.: Fogbus: a blockchain-based lightweight framework for edge and fog computing. J. Syst. Softw. **154**, 22–36 (2019)
34. Tuli, S., Tuli, S., Tuli, R., Gill, S.S.: Predicting the growth and trend of covid-19 pandemic using machine learning and cloud computing. Internet Things **11**, 100222 (2020)
35. Wang, J.: Fast identification of possible drug treatment of coronavirus disease-19 (covid-19) through computational drug repurposing study. J. Chem. Inf. Model. **60**(6), 3277–3286 (2020)
36. Wang, Z., Bovik, A.C.: Mean squared error: love it or leave it? a new look at signal fidelity measures. IEEE Signal Process. Mag. **26**(1), 98–117 (2009)
37. Wilder, B.: Cloud architecture patterns: using Microsoft azure. O'Reilly Media, Inc. (2012)
38. Yang, Z., Zeng, Z., Wang, K., Wong, S.S., Liang, W., Zanin, M., Liu, P., Cao, X., Gao, Z., Mai, Z., et al.: Modified SEIR and AI prediction of the epidemics trend of covid-19 in china under public health interventions. J. Thoracic Disease **12**(3), 165 (2020)
39. Yusof, Y., Mustaffa, Z.: Dengue outbreak prediction: a least squares support vector machines approach. Int. J. Comput. Theory Eng. **3**(4), 489 (2011)
40. Zhang, P., Chen, B., Ma, L., Li, Z., Song, Z., Duan, W., Qiu, X.: The large scale machine learning in an artificial society: prediction of the Ebola outbreak in Beijing. Comput. Intell. Neurosci. **2015** (2015)
41. Zhang, Z.: Improved Adam optimizer for deep neural networks. In: 2018 IEEE/ACM 26th International Symposium on Quality of Service (IWQoS), pp. 1–2. IEEE (2018)

Smart Home Surveillance Based on IoT

Lutfun Nahar, Mohammad Shahadat Hossain, Nusrat Jahan, Mayesha Tasnim, Karl Andersson, and Md. Sazzad Hossain

Abstract Internet of things (IoT) is a process of inter-networking of constitutional design, vehicles (also referred to as "connected devices" and "smart devices"), and various objects—included with software, electronics, actuators, sensors, and inter-network connectivity which is used to gather and interchange of data and information. IoT can control objective and gather information from distance across the network. Recently smart home systems attained more popularity to increase the quality of life where smartphone application plays an important role to monitor home apparatus using wireless communication. The target of this research is to developed an IoT-based smart home to control specific devices and monitoring house using android mobile devices. Featured that are used here is temperature, gas detection, and door lock. A channel, alarm system, door lock can be visualized by the users from any mobile device. If any change happened in the monitored data, the user will be notified.

Keywords IoT · Smart home · Temperature · Gas · Door lock · Smart bin

1 Introduction

IoT is a process of interconnected counting instruments, digital instruments, object, animals, and people that are giving the power to change data over an interconnection without demanding interaction between human and computer. In IoT, Internet with

L. Nahar (✉) · N. Jahan · M. Tasnim
International Islamic University Chittagong, Chittagong, Bangladesh
e-mail: lutfacsecu@gmail.com

M. S. Hossain
University of Chittagong, Chittagong, Bangladesh

K. Andersson
Lulea University of Technology, Skelleftea, Sweden

Md. S. Hossain
University of Liberal Arts Bangladesh, Dhaka 1209, Bangladesh
e-mail: sazzad.hossain@ulab.edu.bd

S. Hossain et al. (eds.), *Proceedings of International Conference on Fourth Industrial Revolution and Beyond 2021*, Lecture Notes in Networks and Systems 437,
https://doi.org/10.1007/978-981-19-2445-3_39

smart technology is used to establish communication between devices using actuators, and wireless sensors that has been controlled remotely and also monitored by the people for automation is mentioned as Internet of Things. The term "Things" is widely used for physical instruments like sensors, microcontroller, cameras, and other devices.

A smart home is a concept that uses devices which is connected through Internet to enable the remote monitoring and management. It associates everything in the network with Internet. In this term, we can monitor homey apparatus like fluorescence, air cooling, ventilator, TV, Fridge, automatic washer, safety technique, etc. It can also be self-discipline and advised itself by using artificial intelligence. These cases additionally aid improvement home's vigor usefulness, which can minimize vigor bill.

The basic most essential objectives of home automation are to save users time and making livelihood common and standard. This research focuses on the implementation of smart house that can monitor house air condition by pressing a key. On the other hand, there is an android app for controlling the home apparatus.

This research integrates all features of home apparatus so that one can easily operate their home apparatus from distance by using smart device.

The remaining part of the paper is prepared as follows. Relevant works on smart home automation system describe in Sect. 2. Section 3 outlines the system description. Section 4 shows the results and comparison is illustrated in Sect. 5; while Sect. 6 gives the conclusion and discussion on further research.

2 Literature review

There are several works related to IoT. In this paper [1] they gathered a lot of knowledge about IoT. They implemented keyless entry and motion detection and smartphone alert system.

In [2] they used Raspberry Pi and smartphone application users to control a series of home appliances and sensors.

In[3] The planned method not only instruct the visual data, like thermal reading, oratory, fluorescence, displacement transducers, but also incites a method conforming to the necessity of light on if it gets gloomy. It also saves the sensor limitations in the cloud (Gmail) in a sharp method. This co-operated the user to explore the situation of varied restrictions in the house at any moment.

In [4, 6, 13] IoT-Home Automation is introduced where an ongoing home automation apparatus with short and obvious view was focused. The benefit of home automation, such as alleviate installation value, system durability, simple augmentation, esthetical advantages, integration of mobiles devices are practiced here.

In [5] they implement a prototype for house automation at peak hour where they use WSN devices. The current research of USA on wireless sensor, integration of artificial intelligence and IoT, nanotechnology, and RFID provides us with the idea of RFID and IoT. This research focused on the implementation of intelligence office and house system for the foreigners which gives automatic information time to time. Household, medical, and environmental appliances are also declared here.

In [9–12] new age house automation is elaborately discussed. In this method they permitted the user to monitor apparatus and radiances in their house from a Cell Phone and multiprocessor from any place in the society by a cyberspace relation. It also permitted the user to monitor their instruments systems within their house from the house server using GUI. The home server GUI will rule over the method if neither the Cell phone nor multiprocessor is capable to rule the device entities in the house.

This is huge informative and work worthy paper based on IoT smart home [7]. In this assessment paper, they explored and appreciated context-aware computing investigation endeavors to feel how the protest in the plot of context-aware computing has been intercepted in personal computer, web, cellular phone, device webs, and diffusive computing instance. A huge number of authorizations subsist in terms of methods, multivendor, petitions, technic, and examples proposed by the searcher to resolve various protests in context-aware computing. Some of the directions in the plot that recognize by the audit. The consequences fairly reveal the worth of factors sensation in the IoT instance.

In this study [8] a narrative planning for small tariff and gentle house monitor and controlling method using Android formed cellular phone is mentioned and executed. They implemented a web services for transmitting data from distance. Any Android cellular mobile with WIFI can monitor the gadgets of house. In addition without Wi-Fi, cellular phone webbing 4G can also be permitted to go into the structure. Further research will give attention on implementing a wireless network using ZIGBEE.

In [14, 15, 24, 26] IP-based wireless sensor networks are used to take decision and IP-based WSN is used for smart irrigation system, predicting natural disaster, and water quality monitoring [17, 19, 20, 23].

In this study [25] IoT-based system is developed for agriculture. In [21, 22] author implemented agro-sensor communication smart system and i-voting smart system so there is a clear indication that IoT-based smart system plays an important role in our ever day life to make life more easier and comfortable and save time.

Therefore, in this research, IoT-based smart home system is implemented where dock lock, gas detection, and smart been system are introduced. Our proposed system architecture will be described in the following section.

3 System Description

3.1 System Architecture

Smart home defines a system of network, governable device which can work altogether to make our home more relax, customized, proficient, and invulnerable. It makes our life simple. In view of this system come into being five major portions

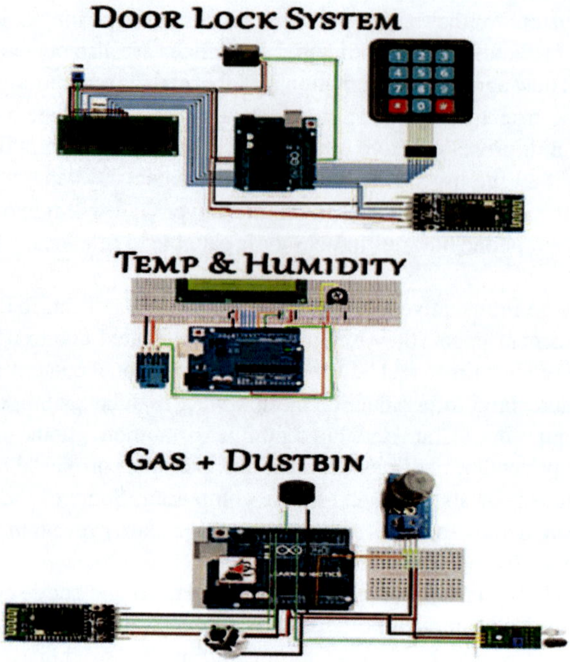

Fig. 1 System Architecture

Arduino, Relay drivers, Bluetooth module, android application, and step down transformer. At first we have to give power to the step down transformer, in step down the input voltage and provide to the Arduino with VIN pin. The Bluetooth module is also needed to associate with Arduino to Rx and Tx pin that gives the news to the microcontroller. Microcontroller reads the facts and dispatches to transfer the drivers which exertion as control In Arduino we transmit the program as per necessity then it carries some mathematical and rational functioning to monitor the relay drivers. System architecture is given in Fig. 1.

3.2 Block Diagram

Here is the block diagram of this system where we showed the overall process of our work.

3.3 Hardware

Arduino Uno: It is a microcontroller fume depending on the Atmega328 (datasheet) with 14 I/O pins, in which 6 pins can be turned to as yields, 6 pins are set as simple information sources. It has 16 MHz clay resonator, a USB association, a power jack, and a reset button. The microcontroller has 32kB of ISP flash memory, 2kB RAM, and 1kB EEPROM. The board provides serial indication competence via UART, SPI, and 12C. Cause of well scheme in the structure of Arduino it is simple to feel. In Arduino we conduct high level of programming languages like C language, C++ language, etc. It is simple to feel and uses amicable language. It has exceedingly precedence like multitasking, mechanization, time domain, etc. Arduino Uno is given below.

Bluetooth Module: We are using android app to connect the project and control it for remote connection between app and Arduino we are using Bluetooth module (master or slave at a time). HC-05 Bluetooth module is conducted to an android application with the microcontroller. Bluetooth is sent to the microcontroller and receives the information from the user (Arduino). It is easy to conduct Bluetooth Serial Port Protocol (SSP), planned as transmission sequential connection established. The Bluetooth of serial port module is Advanced Bluetooth v2.0+ Enhanced data Rate at 3 Mbps modulation with 2.4 GHz radio receiver with BB (base band). The Bluetooth of Rx and Tx pins are joined to the Arduino pins of Tx and Rx in some respect. To utilize Bluetooth SPP (Serial Port Protocol) module HC-05 module is simple and resolves for directly distant pursuant association setup. It puts CSR Blue canter 04-External single fume Bluetooth framework with CMOS innovation and with AFH (Adaptive Frequency Hopping Feature).

IR sensor: It is an electronic instrument, which can send in apropos sense various perspectives of the adjacent. An IR sensor can magnitude the warmth of an device like finds the motion. Only infrared radiation can control this type of radiation, more than transmitting it especially called a passive IR sensor. Habitually, in the electromagnetic spectrum, all the purposes diffuse several structures of thermal radiation. We use IR sensor in our project in the dustbin module. The dustbin module in our project sense the objects located in front of it. Here IR sensor is used to sense any obstacle and object thus we come to know that the dustbin is full.

Potentiometer: It is a three-terminal instrument panel with a turning contact or sliding which can set up an constant voltage divider. If only two terminals are used, one end and the wiper, it acts as a changeable resistor. In our work we use this to monitor the contrast and brightness of LCD display.

PCB Board: It electrically Joins electronic materials using conductive pads, tracks and other components inscribe from one or more over sheet layers of copper laminated onto or/and among sheet layers of a non-conductive substrate. Elements are usually sell onto the PCB to both mechanically fasten and electrically join them to it. Here in our project there are a lot of cables and circuits to integrate the spread circuits in one circuit and minimize the cable connection we used PCB board.

Gas sensor: A gas sensor is an instrument that can find the concentration or appearance of gases in the environment. Directed on the concentration of the gas sensor creates a analogous potential difference by switching the resistance of the material within the sensor, which can be adequate as output voltage. Based on this voltage standard the type and concentration of the gas can be calculated. Our project also alarms the user if it finds any gas around it. So we used gas sensor to sense the gas.

Buzzer: The buzzer is a sounding instrument that can change audio gesture into sound gestures. It is commonly powered by DC voltage. It is broadly used in dread, computers, printers, and other electronic manufactures as sound instrument. It is basically parted into electromagnetic buzzer and piezoelectric buzzer described by the letter "HA" or "H" in the circuit. According to different uses and designs, the buzzer can transmit different sounds such as alarm, electric bell, music, buzzer, and siren. To create different alarms for different scenarios we used a couple of buzzer in our buzzer.

Servo Motor: A servomotor is a linear actuator that allows for linear or velocity, monitor of angular and stimulation. It forms a compatible motor coupled to a sensor for position recompose. In our project we use a servo motor to control the automated door. In our project we used servo motor that rotates around a predetermined angle.

3.4 *Software*

Here Arduino IDE is used for implementing this smart home idea. For connecting the feature with the user we need to implement the module through proper program. We have to set program in various parts. We implement door lock system, gas sensor, temperature sensor, and smart dustbin separately.

Figure 2 describes the main components and Fig. 3 illustrates processing steps of our proposed system.

Arduino, Bluetooth Device, Sensor, Buzzer, PCB Board, and Potentiometer are used for the hardware implementation. The collision buttons of home or loads apparatus are an interface to 8051 series microcontroller by the command signals dispatch through RF transmitter after encoding order signals. The RF transmitter end is a RF Remote which can be used by the user as a distant controller for operating home apparatus. The receiver end structure of the RF receiver circuit form a decoder to decode the encoded order signals received from the transmitter. The decoded signals are brought up to the microcontroller and the orders are sent to the conduct loads through Opto-Isolator.

Here we use an android APP named Arduino Bluetooth Controller. Via this APP we can connect with that Bluetooth which we placed in our hardware component. Once our smartphones' Bluetooth is connected with the Bluetooth module of our smart home then it can transfer data and can happen to transmit between them to run the project properly and to have the expected result from the project. This is a one-way transmission at a time the module can work as a sender or a receiver.

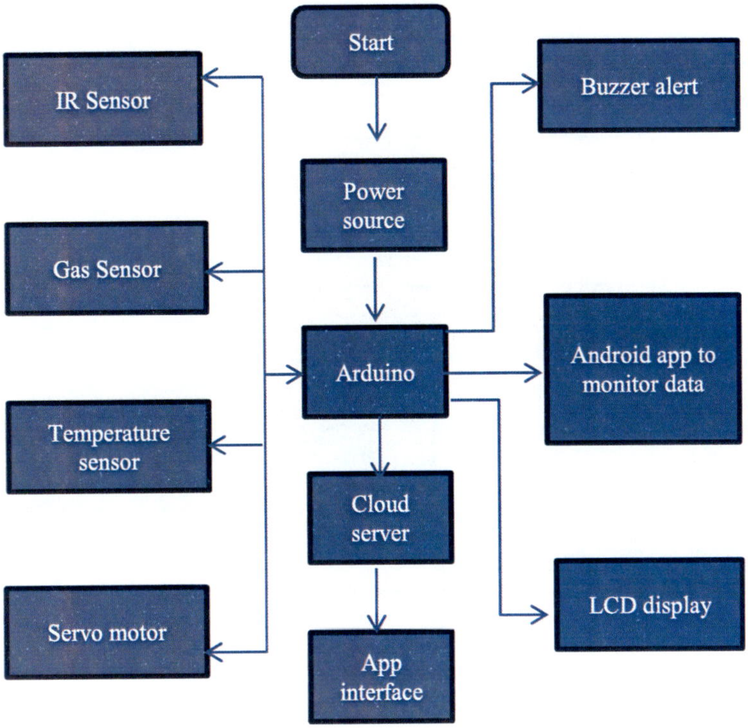

Fig. 2 Block diagram

4 Implementation and Results

A cloud server account has been opened to store information and then an API data fetching framework is implemented in the local mobile app system to store data in the cloud. Data are updated from time to time in the cloud and it is monitored from the smart device and also a notification has been sent to the mobile device if gas is detected. In addition an LCD display is added to show bin status and buzzer alert system is implemented for gas detection. The experimental results for all features have been shown below.

4.1 Door Lock System

The door of the smart home can automatically open or close by Servo Motor by giving password. Using Arduino control app, we can open or close house door. Figure 4 illustrated the door lock system.

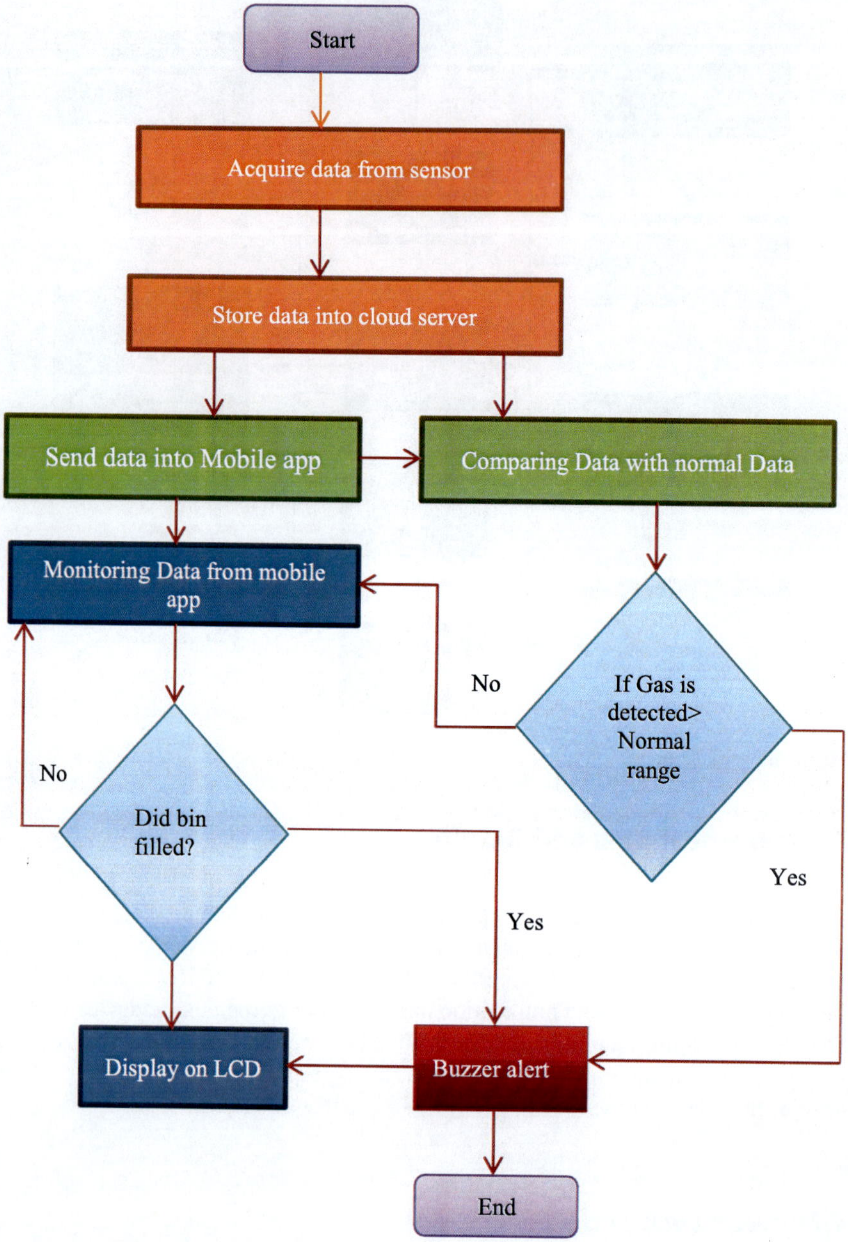

Fig. 3 Work flow diagram

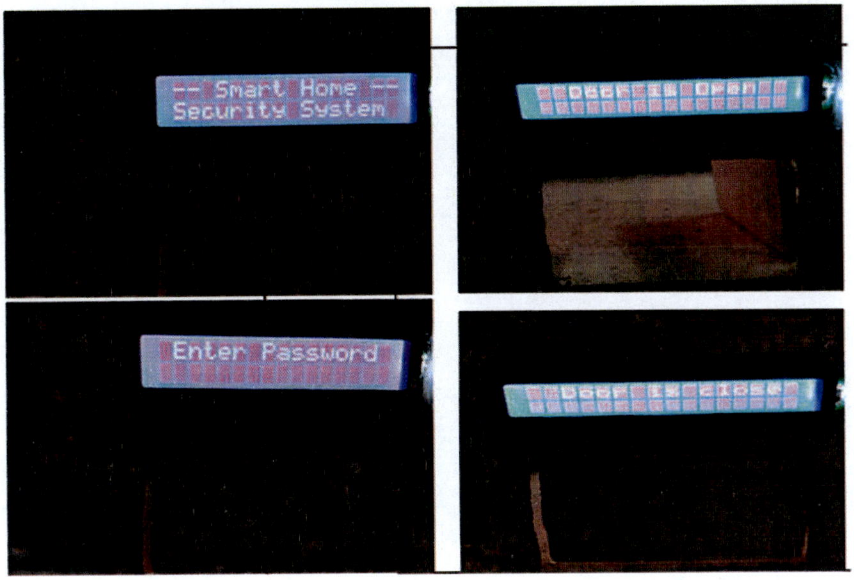

Fig. 4 Door lock system

4.2 *Temperature and Humidity System*

Dht11 sensor can sense the temperature and humidity and show it on LCD display. Temperature is shown in Fig. 5.

4.3 *Gas Detection and Smart Bin System*

Smart dustbin is also an element of our smart home which is on the buzzer on the user's phone if it can sense any dirt around it. Gas sensor can detect gas around of sensor then giving siren in buzzer. Smart bin alert system and gas detection alert system are shown in Fig. 6.

5 Data Analysis

To validate our proposed system we compare real-time data and system data which is found from Dht11 sensor. The comparisons are shown in Fig. 7.

Fig. 5 Temperature sensor

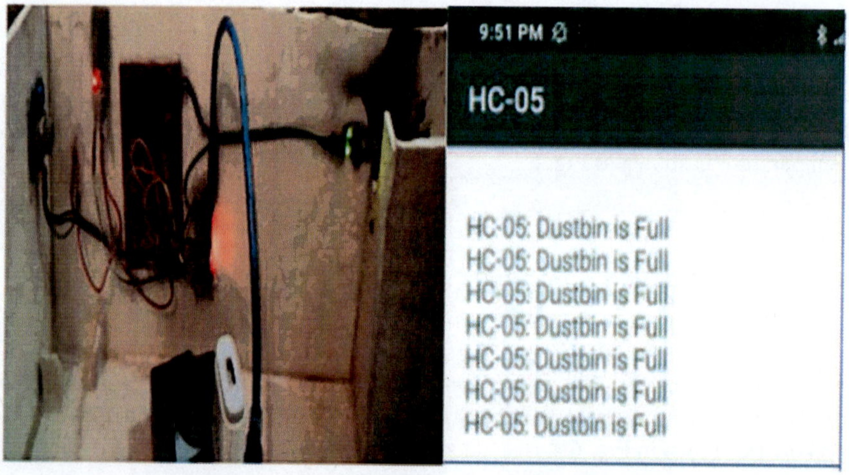

Fig. 6 Gas sensor and smart bin

6 Conclusion and Future work

The goal of this research is to implement a smart home system which can monitor home apparatus with a touch of a finger. An android app is implemented for controlling the home apparatus. Servo motor is used for door locking system and Dht11 sensor is used for temperature, humidity, and gas detection. Moreover an alert system

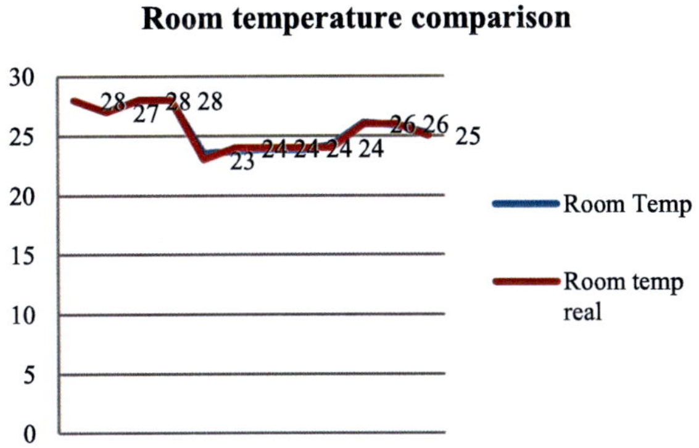

Fig. 7 Comparison of the real data and experimental data

is implemented to detect gas when it is leaked into the house. We make a fully functioning house mechanization technique with the facility to keep eye on apparatus and make our life easy and safe.

Further research concerning the integration of the Belief Rule Base Expert system (BRBES) with IoT [16, 18, 27]. It is broadly used to handle uncertain knowledge and accomplish the task of evidential reasoning.

References

1. Stolojescu-Crisan, C., Crisan, C., Butunoi, B.: An IoT-based smart home automation system. Sensors **21**, 3784 (2021)
2. Mustafa, B., Iqbal, M.W., Saeed, M., Shafqat, A.R., Sajjad, H., Naqvi, M.R.: IOT based low-cost smart home automation system. In: 2021 3rd International Congress on Human-Computer Interaction, Optimization and Robotic Applications (HORA), pp. 1–6. IEEE (2021)
3. Vinay, S.K.N., Kusuma, S.M.: Home automation using internet of things. Int. Res. J. Eng. Technol. (IRJET). 02, 1965–1970 (2015)
4. Li, B., Yu, J.: Research and application on the smart home based on component technologies and internet of things. Proc. Eng. **15**, 2087–2092 (2011)
5. Praveena, B., Rajganesh, K., Vinitha, B., Kandasamy, S., Chandraman, M.: Home security system for controlling theft using android ADK. Int. J. Eng. Res. Technol. 3 (2015)
6. Javale, D., Mohsin, M., Nandanwar, S., Shingate, M.: Home automation and security system using android ADK. Int. J. Electr. Commun. Comput. Technol. (IJECCT). 3, 382–385 (2013)
7. Perera, C., Zaslavsky, A., Christen, P., Georgakopoulos, D.: Context aware computing for the internet of things: a survey. IEEE Commun. Surv. Tutor.16, 414–454 (2014)
8. Mohammad El-Basioni, B.M., El-kader, S.M.A., Fakhreldin M.A.: Smart home design using wireless sensor network and biometric technologies. Int. J. Appl. Innov. Eng. Manage. (IJAIEM) 2, 413–429 (2013)
9. Jain, S., Vinoth Kumar, N., Paventhan, A., Kumar Chinnaiyan, V., Arnachalam, V., Pradish, M.: Survey on smart grid technologies- smart metering, IoT and EMS. In: 2014 IEEE Students Conference on Electrical, Electronics and Computer Science, pp. 1–6. IEEE (2014)
10. Gaikwad, P.P., Gabhane, J.P., Golait, S.S.: A survey based on smart homes system using internet-of-things. In: 2015 International Conference on Computation of Power, Energy, Information and Communication (ICCPEIC), pp. 0330–0335, IEEE (2015)

11. Dickey, N., Banks, D., Sukittanon, S.: Home automation using cloud network and mobile devices. In: 2012 Proceedings of IEEE Southeastcon, pp. 1–4. IEEE (2012)
12. Pavithra, D., Balakrishnan, R.: IoT based monitoring and control system for home automation. In: 2015 Global Conference on Communication Technologies (GCCT), pp. 169–173. IEEE (2015)
13. da Cruz, M.A.A., Rodrigues, J.J.P.C., Al-Muhtadi, J., Korotaev, V.V., de Albuquerque, V.H.C.: A reference model for internet of things middleware. IEEE Int. Things J. 5, 871–883 (2018)
14. Thombre, S., Islam, R.U., Andersson, K., Hossain, M.S.: IP based wireless sensor networks: performance analysis using simulations and experiments. J. Wirel. Mob. Netw. Ubiquit. Comput. Dependable Appl. 7(3), 53–76 (2016)
15. Andersson, K., Hossain, M.S.: Heterogeneous wireless sensor networks for flood prediction decision support systems. In: 2015 IEEE Conference on Computer Communications Workshops (INFOCOM WKSHPS) (2015)
16. Andersson, K., Hossain, M.S.: Smart risk assessment systems using belief-rule-based DSS and WSN technologies. In: 2014 4th International Conference on Wireless Communications, Vehicular Technology, Information Theory and Aerospace & Electronic Systems (VITAE) (2014)
17. Abedin, Z., Chowdhury, A., Shahadat Hossain, M.S., Andersson, K., Karim, R.: An interoperable IP based WSN for smart irrigation system. In: 2017 14th IEEE Annual Consumer Communications & Networking Conference (CCNC) (2017)
18. Alam, M., Kaiser, M., Hossain, M.S., Andersson, K.: An IoT-belief rule base smart system to assess autism. In: 2018 4th International Conference on Electrical Engineering and Information & Communication Technology (iCEEiCT) (2018)
19. Abedin, Z., Paul, S., Akhter, S., Siddiquee, K., Hossain, M.S., Andersson, K.: Selection of energy efficient routing protocol for irrigation enabled by wireless sensor network. In: 2017 IEEE 42nd Conference on Local Computer Networks Workshops (LCN Workshops) (2017)
20. .Islam, R., Andersson, K., Hossain, M.S.: Heterogeneous wireless sensor networks using CoAP and SMS to predict natural disasters. In: 2017 IEEE Conference on Computer Communications Workshops (INFOCOM WKSHPS) (2017)
21. Alam Siddiquee, K., Khan, F., Andersson, K., Hossain, M.S: Optimal dynamic routing protocols for agro-sensor communication in MANETs. In: 2017 14th IEEE Annual Consumer Communications & Networking Conference (CCNC) (2017)
22. Alam Siddiquee, K., Andersson, K., Khan, F., Hossain, M.S.: A scalable and secure MANET for an i-voting system. J. Wirel. Mob. Netw. Ubiquit. Comput. Dependable Appl. 8(3), 1–17 (2017)
23. Chowdury, M.S.U., Emran, T.B., Ghosh, S., Pathak, A., Alam, M.M., Absar, N., Andersson, K., Hossain, M.S.: IoT based real-time river water quality monitoring system. Proc. Comput. Sci. 155, 161–168 (2019)
24. Thombre, S., Ul Islam, R., Andersson, K., Hossain, M.S.: Performance analysis of an IP based protocol stack for WSNs. In: 2016 IEEE Conference on Computer Communications Workshops (INFOCOM WKSHPS) (2016)
25. Pathak, A., AmazUddin, M., Abedin, M.J., Andersson, K., Mustafa, R., Hossain, M.S.: IoT based smart system to support agricultural parameters: a case study. Proc. Comput. Sci. 155, 648–653 (2019)
26. Noor, K., Siddiquee, E., Sarma, D., Nandi, A., Akhter, S., Hossain, S., Andersson, K., Hossain, M.S.: Performance analysis of a surveillance system to detect and track vehicles using Haar cascaded classifiers and optical flow method. In: 2017 12th IEEE Conference on Industrial Electronics and Applications (ICIEA) (2017)
27. Monrat, A., Ul Islam, R., Hossain, M.S., Andersson, K.: A belief rule based flood risk assessment expert system using real time sensor data streaming. In: 2018 IEEE 43rd Conference on Local Computer Networks Workshops (LCN Workshops) (2018)

Brain Tumor Classification using Transfer Learning from MRI Images

Nasrin Akter, Jubair Ahmed Junjun, Nazmun Nahar, Mohammad Shahadat Hossain, Karl Andersson, and Md. Sazzad Hoassain

Abstract One of the most vital parts of medical image analysis is the classification of brain tumors. Because tumors are thought to be origins to cancer, accurate brain tumor classification can save lives. As a result, CNN (Convolutional Neural Network)-based techniques for classifying brain cancers are frequently employed. However, there is a problem: CNNs are exposed to vast amounts of training data in order to produce good performance. This is where transfer learning enters into the picture. We present a 4-class transfer learning approach for categorizing Glioma, Meningioma, and Pituitary tumors and non-tumors in this study. The three most prevalent types of brain tumors are glioma, meningioma, and pituitary tumors. Our presented method, which employs the theory of transfer learning, utilizes a pre-trained InceptionResnetV1 method for classifying brain MRI images by extracting features from them using the softmax classifier method. The proposed approach outperforms all prior techniques with a mean classification accuracy of 93.95%. For the evaluation of our method we use kaggle dataset. Precision, recall, and F-score are one of the key performance metrics employed in this study.

Keywords Brain tumor · Transfer learning · MRI · InceptionResNetV2

N. Akter · J. A. Junjun · N. Nahar
BGC Trust University Bangladesh, Bidyanagar, Chandanaish, Bangladesh
e-mail: nazmun@bgctub.ac.bd

M. S. Hossain (✉)
University of Chittagong, University-4331, Chittagong, Bangladesh
e-mail: hossain_ms@cu.ac.bd

K. Andersson
Lulea University of Technology, SE-931 87, Skellefteå, Sweden
e-mail: Karl.andersson@ltu.se

Md. S. Hoassain
University of Liberal Arts Bangladesh, Dhaka 1209, Bangladesh

S. Hossain et al. (eds.), *Proceedings of International Conference on Fourth Industrial Revolution and Beyond 2021*, Lecture Notes in Networks and Systems 437,
https://doi.org/10.1007/978-981-19-2445-3_40

1 Introduction

According to statistics [27] brain tumors are one of the deadliest and most common types of cancer in both children and adults. Brain tumor categorization into subtypes is a difficult topic since brain tumors vary greatly in shape and size [10]. Furthermore, different forms of brain tumors may have similar features, making categorization more difficult [11]. Three subtypes of brain tumors, Meningioma, Pituitary, and Glioma are the most common and can be approximately likened to three stages of cancer [30]. As a result, detecting and classifying brain tumors early and accurately is critical and can save many lives. That is why healthcare image processing is putting a lot of effort into improving brain tumor categorization tools.

MRI (Magnetic Resonance Imaging) is a well-known healthcare equipment that can be used to detect and study a variety of disorders, including brain tumors, neurological ailments, and epilepsy, among others. Typically, a system that is entirely processed by hardware/computer aids in the automation of this process in order to produce correct and timely results. MRI was found to be more sensitive, but not more specific, in several trials comparing CT and MRI for brain evaluation [17]. The benefits of an MRI facility are that it can discover abnormalities that are not detectable or only shown poorly on CT. When only a generalized group effect may be seen on CT scans, an MRI can be used to detect the tumor's exact scope and location. Furthermore, MRI can better detect the specific location of the lesion in relation to important neuroanatomical structures because of its enhanced contrast discrimination and capacity to capture images at many levels. As a result, in this work, we compare the MIR data set with other deep learning models.

Existing research on the classification of brain tumors into meningioma, glioma, and pituitary tumors has the following limitations. The performance of state-of-the-art algorithms is insufficient given the healthcare significance of the categorization challenge. Prior to classification, earlier approaches depended on manually drawn tumor sections. They were unable to be fully computerized as a result of this. Automatic algorithms based on CNN and its derivatives were unable to achieve a significant development in performance. Additionally, existing methods were tested on the unbalanced figshare dataset (with respect to tumor classes). As a result, performance analysis through metrics other than accuracy becomes important.

We provide in this study an accurate and automatic categorization system for three pathogenic forms of brain tumor (glioma, meningioma, and pituitary tumor). For the categorization of brain tumors, we use a deep transfer learning CNN model called InceptionResNetV2. The proposed system is then thoroughly evaluated. When tested on the available dataset from kaggle, the suggested system outperformed all other related works in classification results. This paper's important contributions are given below.

- For the first time, deep transfer learning is used to a three-class brain tumor classification task and one non-tumor classification task.
- The proposed method had the highest classification accuracy when compared to state-of-the-art approaches.

- We have also compared our approach to other recent research.

Remainder of the paper is organized into many sections. Part ii discusses the related work. The methodology is discussed in part iii. The results of the experiment are described in part iv. The final part discusses the conclusion and future work.

2 Related Work

Recent developments in computer-aided medical diagnostics have resulted in enhanced performance due to the introduction of deep learning ideas.

Cheng et al. [10] provided a technique for extracting characteristics from the region of interest of T-1 MRI images based on manually specified tumor borders, and the best result was achieved with an SVM model that used bag of words (BOW) features. This has become one of the first analyses on the figshare brain MRI image dataset.

Qadar et al. [16] explored a feature extraction method which is Gabor filter and discrete wavelet transform, followed by a multi-layer perceptron algorithm for classification. The disadvantage of these systems is that they both rely on semi-manual procedures for feature extraction. CNNs have an obvious advantage here because they do not need to be manually segmented and can extract relevant features on their own.

Swati et al. [29] developed a VGG-19 model that has been pre-trained to diagnose brain tumors from a figshare brain MRI image dataset, employing transfer learning technique. The same transfer learning strategy was utilized by Deepak et al. [12] to deploy an updated GoogLeNet architecture on the same figshare dataset.

Justin Ker et al. [18] used the InceptionV3 model for brain histology classification in the past, whereas histology slices from brain and breast tissues were classified.

The challenge of brain tumor classification [1] was effectively solved using a CNN-based deep learning network. CNN-based classifier systems have the advantage of eliminating the need for manually segmented tumor regions and providing a completely automated classifier.

A CNN architecture was designed by Pashaei et al. [23] to extract attributes from brain MRI samples. The model included five learnable layers with $3x3$ filter sizes across all of them. The model contained five learnable layers with $3x3$ filter sizes across all of them. The classification accuracy of the CNN model was 81%. The performance was improved by combining CNN features with an extreme learning machine classifier model (ELM). In this research, the recall measures for pituitary tumors were fairly high, while those for meningioma were quite poor. This indicates that the classifier's discrimination capacity is limited.

Afshar et al. [3] developed a customized CNN model described as a capsule network (CapsNet) to classify brain tumors. The tumor's spatial interaction with its surrounding tissues was used by CapsNet. Even so, there was only a minor boost in performance.

In their work on brain abnormality classification, Talo et al. [32] used deep transfer learning to obtain outstanding classification results. The researchers employed ResNet-34 in their studies, which included training of modified dense layers, data augmentation in training, and fine tuning of a transfer learnt model. The results of the experiments showed that a deep transfer learning model may be used to classify medical images with minimal pre-processing.

Yang et al. [34] utilized AlexNet and GoogLeNet to grade gliomas from MRI data in their research. For the task, GoogLeNet outscored AlexNet on the observed performance measures.

In content-based image retrieval (CBIR) [30], transfer learning was applied to detect brain cancers. The experiment was run using a publicly available dataset, and the results were encouraging.

3 Methodology

This section outlines the strategy we suggest for detecting brain tumors. The methodology is divided into various parts, the first of which is the collecting of a training dataset, the second one is data preprocessing, data augmentation, develop a transfer learning model, and classification of brain tumor. Figure 1 shows the methodology of our proposed model.

3.1 Dataset Description

The data was obtained from kaggle [8], an open source resource. The dataset contains 3264 T2 weighted contrast images. The dataset contains data of three types of brain tumor which are Meningioma, Glioma, and Pituitary tumor and without brain tumor. The MRI scans of the brain are different in their axial, coronal, and sagittal aspects. The images in the dataset are divided into four sub file. The MRI images are in matrix

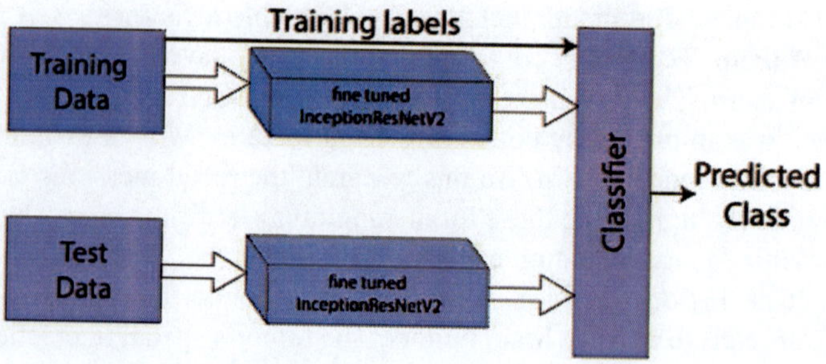

Fig. 1 Overview of our proposed model

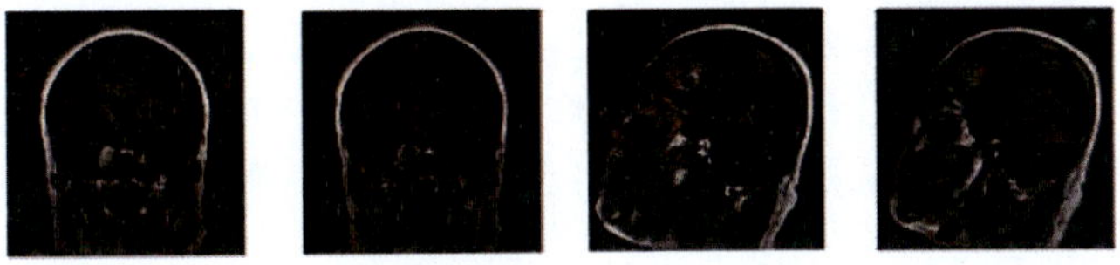

Fig. 2 Meningioma tumor class of MRI images

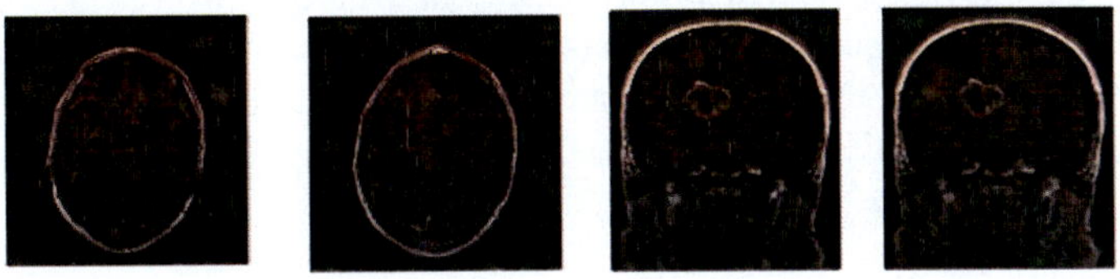

Fig. 3 Glioma tumor class of MRI images

Fig. 4 Data preprocessing stage

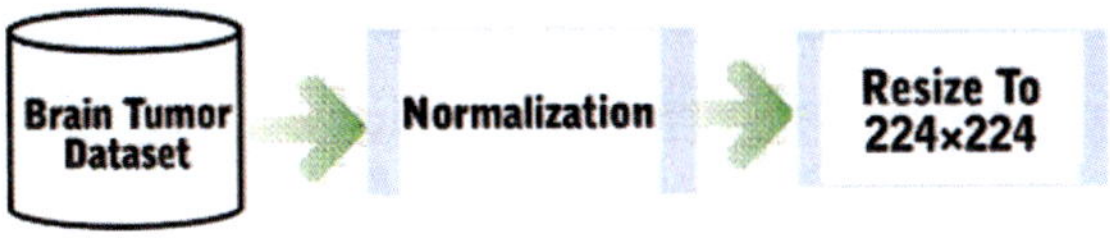

format with a resolution of 512 × 512 pixels. Some sample of our data set is given below.

3.2 Data Preprocessing

The collected dataset contains 512 × 512 brain MRI images. However, preprocessing is required because InceptionResnetV2 is created to receive colored images with an input layer size of 224 × 224 × 3. The intensity levels of the images were first standardized using Z-score normalization so that the value will be within 0–1, for each image, normalization was carried out by Eq. 1.

$$z = \frac{x - \sigma}{s} \tag{1}$$

where x denotes a training sample and s indicates the training sample's standard deviation. The images were then scaled to 224 × 224 pixels.

3.3 Data Augmentation

A limitation of training data is a major hindrance to developing successful deep learning models, such as the InceptionResNetV2 transfer learning model for brain tumor classification. More data is required to assure the neural network-based model's stability and to broaden the model's functional variety. To do this, we use image data augmentation [26] to subtly distort the dataset. This improved data enables the model to generalize more successfully. We have several possibilities for selecting values from various sizes for image augmentation, such as horizontal flip, rotation range, shear range, and zooming, among others. The image augmentation settings used in our study are shown in Table 1.

3.4 Transfer Learning

Transfer learning is a method for learning a new set of data using the knowledge gained from a previously trained model[33]. Provided an input space (Ds) and a training task (Ts) in the input space, a target domain (Dt), and the associated training task (Tt), the goal of transfer learning is to increase learning in Dt by using information from Ds and T. Depending on the type of task and the nature of the data available at the source and destination domains, different transfer learning settings are created. The transfer learning approach is known as inductive transfer learning [22] when labelled data is available in the source and target domains for a classification job. In this case, the domain is D = (xi, yi) where xi denotes the feature vector for the ith training set and yi denotes the classifier for the ith training set.

The InceptionResNetV2 is a 164-layer network and has 24 million trainable parameters [31]. For training and optimization, this type of deep model needed a large dataset and that is why InceptionResNetV2 was trained on the ImageNet dataset [13] and there are about 1.2 million images in 1000 distinct categories in this database. As a result, for smaller datasets such as the brain MRI dataset, the model is likely to suffer from overfitting. It is at this point when transfer learning comes into the picture. We use pre-trained weights to create the model, and then fine-tune

Table 1 Data augmentation parameter

Parameters	Value
Rotation range	40
Width shift range	0.2
Height shift range	0.2
Shear range	0.2
Zoom range	0.2
Horizontal	flip True

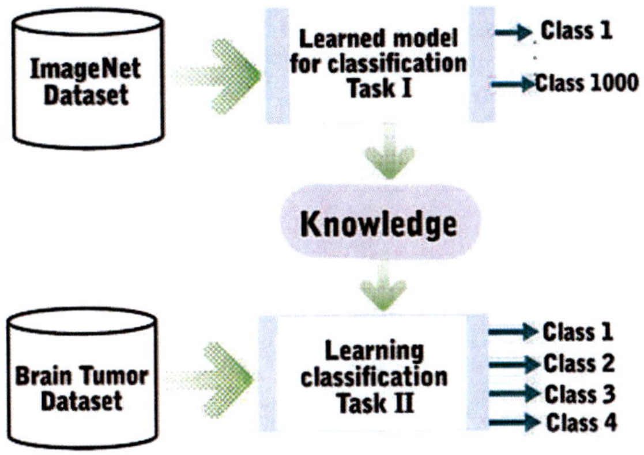

Fig. 5 Transfer learning settings

it to make it capable of doing the desired task, which in our scenario is categorizing brain tumors, rather than training the model from beginning for smaller datasets like the brain tumor dataset.

Because the InceptionResNetV2 model was originally built for a different objective, significant structural modifications are required for classifying brain tumors. The InceptionResnetV2 model's last three layers were adjusted in order to adapt it to the destined purpose. The original model's average pooling layer was substituted with a flatten layer, and InceptionResNetV2's fully connected layer, which was expected to categorize 1000 different classes, was also eliminated and replaced with a new FC layer with a four-output size. The softmax activation layer, which came after the FC layer, was also changed and substituted by a new one.

4 Experimental Result

This section contains the outcomes of our model's experiments.

4.1 System Configuration

The transfer learning model was implemented using a GPU with tensor-flow support. Because the neural network performs many matrix multiplication operations, a sufficient GPU is essential for deep learning model training. The proposed pretained architecture is designed for training on the Google Colaboratory cloud server because to the constraints of a CPU's processing capacity. The GPU and the Jupyter Notebook environment are used in the Google Colab. The Google Colab was created specifically to assist the machine classifier in overcoming the processing unit diffi-

Table 2 Training, validation, and test datasets

Class	Training images	Validation images	Testing images
Meningioma tumor	468	50	138
Glioma tumor	800	160	105
Pituitary tumor	800	150	100
No tumor	350	30	105
Total	2418	390	448

Table 3 Hyperparameters settings

Hyper-parameters	Value
Loss function	Categorical cross-entropy
Batch size	32
Epochs	50
Learning rate	0.001
Optimizer	Adam

culty. Google Colab was utilized to train and evaluate the deep learning technique we presented.

4.2 Training, Validation and Test Datasets

We examined 3256 RGB colored 512 × 512 sized images of brain tumor. We use 60% of the data for training, 20% for validation, and 20% for testing operations. These data is shown in table. Validation data is employed to obtain ideal hyper-parameters to avoid the model from overfitting.

4.3 Hyperparameters of Our Transfer Learning Model

While training our model, we employ categorical crossentropy as the loss function. We train our model for a total of 50 epochs because no further changes in training and validation accuracies are seen. The loss function is optimized by using Adam's optimizer. Table 4 displays the most optimally configured hyper-parameters in our model. In our experiment, the number of epochs and batch size are set at 50 and 32 respectively.

Table 4 Accuracy and Loss of InceptionResNetV2 (transfer learning)

Model part	Accuracy	Loss
Training	98.15%	0.098
Validation	95.08%	0.230
Testing	93.75%	0.407

4.4 Performance Matrices

Several performance measures are provided for the standard evaluation of a classifier. Classification accuracy is the most commonly used performance metric. The proportion of properly identified observations to the total number of data observations is used to measure accuracy in classification. Precision and recall (or sensitivity) are important metrics that can be determined utilizing the equations shown in following.

$$ACC = \frac{TP + TN}{TP + FP + FN + TN} \tag{2}$$

$$Precision = \frac{TP}{TP + FP} \tag{3}$$

$$Recall = \frac{TP}{TP + FN} \tag{4}$$

Here TP, FP, TN, and FN represent the number of classified true positives, false positives, true negatives, and false negatives. The harmonic mean of precision and recall yields the F-score, which is an essential statistical measure of classification. The equation of F1-score is given below

$$F1 - Score = 2\frac{Precision + Recall}{precision * Recall} \tag{5}$$

4.5 Result

The classification training, validation, and testing accuracy and loss of our proposed model are shown in Table 2. From the table we can see training accuracy and loss are 98.15% and 0.098 respectively. Validation and testing accuracy are 95.08% and 93.75% respectively and validation and testing loss are 0.230 and 0.407 respectively.

Classification accuracy is a useful metric for measuring performance when the test dataset contains an equal number of samples from each class. However, the dataset used to solve the mentioned classification problem is imbalanced. This needs

Table 5 Performance comparision of our model

Model	Accuracy	Precision	Recall	F1-sscore	AUC
InceptionResNetV2 (Transfer Learning)	93.75%	0.875	0.875	0.875	0.921
VGG19	76.56%	0.533	0.500	0.500	0.767
DenseNet	85.16%	0.783	0.562	0.656	0.847
ResNet50	81.82%	0.723	0.512	0.643	0.801

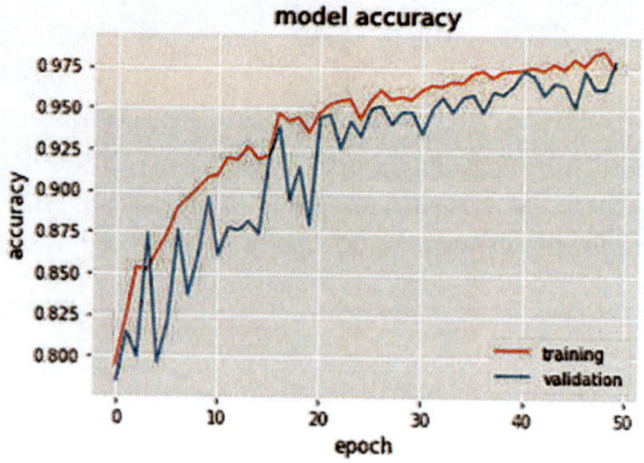

Fig. 6 Training and validation accuracy of transfer learning model

a more comprehensive analysis of the proposed system using additional performance metrics.

Table 3 shows the Precision, Recall, F1-score, and AUC of our proposed transfer learning method, and we also compare our approach with other method like VGG19, DenseNet, and ResNet50. Precision, Recall, F1-score, and AUC of our proposed approach are 0.875, 0.875, 0.875, and 0.921 respectively. From the table we can also see that our model performs better than the other three pre-trained model.

Under default parameters, the proposed CNN-based transfer learning model is trained with a numerous number of epochs (up to 50). Because there are no more changes in training and validation accuracies for our model, the best set epoch is 50. Figures 6 and 7 show the model accuracy and loss. In Fig. 7 at the first epoch the training and validation loss is same that is 0.450. In Fig. 6 at the initial epoch the train and validation accuracy is almost same. Accuracy is increasing and loss is decreasing when the number of epoch is increased.

We have also compare our proposed transfer learning method with state-of-arts techniques. Table 4 shows the comparison. According to the table, our proposed approach has a better accuracy than the other.

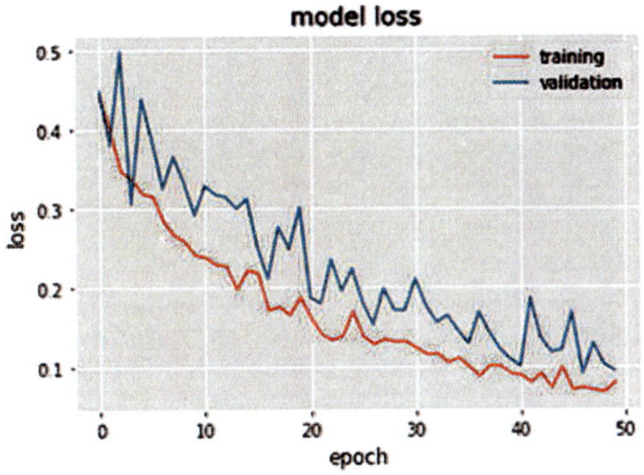

Fig. 7 Training and validation accuracy of transfer learning model

Table 6 Accuracy and loss of InceptionResNetV2 (transfer learning)

Author	Algorithm	Performence matrics
Cheng [10]	BoW-SVM algorithm	91.28% Accuracy
Abiwinanda [1]	CNN model	84.19% Accuracy
Afsar [3]	CapsNet Model	90.89% Accuracy
Proposed method	InceptionResNetV2 (transfer learning)	93.75%

5 Conclusion and Future Work

In this paper, we built a transfer learning-based classification model for brain tumor classification using MRI data. When compared to related publications, our suggested model had the highest classification accuracy with the least amount of data preprocessing. When tested against various performance metrics, our suggested model outperformed all state-of-the-art techniques on the same dataset. In this paper we can see that our proposed pretrained method work better than the other method. Here, we classify four types brain tumor using our pretrained model. That is the main strength our work. The main weakness our paper is that the data is limited. Although the accomplishments discussed in this study, various modifications are still achievable. There are some limitation in our paper like we see the overfitting problem. In future we will try to collect more data of four brain tumor. We also apply some technique for avoiding the overfitting problem. We also try to apply some other method [2, 4–7, 9, 14, 15, 19–21, 24, 25, 28] to classify brain tumor.

References

1. Abiwinanda, N., Hanif, M., Hesaputra, S.T., Handayani, A., Mengko, T.R.: Brain tumor classification using convolutional neural network. In: World Congress on Medical Physics and Biomedical Engineering 2018, pp. 183–189. Springer (2019)
2. Afroze, T., Akther, S., Chowdhury, M.A., Hossain, E., Hossain, M.S., Andersson, K.: Glaucoma detection using inception convolutional neural network v3. In: International Conference on Applied Intelligence and Informatics, pp. 17–28. Springer (2021)
3. Afshar, P., Plataniotis, K.N., Mohammadi, A.: Capsule networks for brain tumor classification based on MRI images and coarse tumor boundaries. In: ICASSP 2019-2019 IEEE International Conference on Acoustics, Speech and Signal Processing (ICASSP), pp. 1368–1372. IEEE (2019)
4. Ahmed, F., Hossain, M.S., Islam, R.U., Andersson, K.: An evolutionary belief rule-based clinical decision support system to predict Covid-19 severity under uncertainty. Appl. Sci. **11**(13), 5810 (2021)
5. Basnin, N., Nahar, L., Hossain, M.S.: An integrated CNN-LSTM model for micro hand gesture recognition. In: International Conference on Intelligent Computing & Optimization, pp. 379–392. Springer (2020)
6. Basnin, N., Nahar, N., Anika, F.A., Hossain, M.S., Andersson, K.: Deep learning approach to classify Parkinson's disease from MRI samples. In: International Conference on Brain Informatics, pp. 536–547. Springer (2021)
7. Basnin, N., Sumi, T.A., Hossain, M.S., Andersson, K.: Early detection of Parkinson's disease from micrographic static hand drawings. In: International Conference on Brain Informatics, pp. 433–447. Springer (2021)
8. Bhuvaji, S.: Brain Tumor Classification(MRI). https://www.kaggle.com/sartajbhuvaji/brain-tumor-classification-mri. Accessed 20 September 2021
9. Biswas, M., Chowdhury, S.U., Nahar, N., Hossain, M.S., Andersson, K.: A belief rule base expert system for staging non-small cell lung cancer under uncertainty. In: 2019 IEEE International Conference on Biomedical Engineering, Computer and Information Technology for Health (BECITHCON), pp. 47–52. IEEE (2019)
10. Cheng, J., Huang, W., Cao, S., Yang, R., Yang, W., Yun, Z., Wang, Z., Feng, Q.: Enhanced performance of brain tumor classification via tumor region augmentation and partition. PLoS ONE **10**(10), e0140381 (2015)
11. Cheng, J., Yang, W., Huang, M., Huang, W., Jiang, J., Zhou, Y., Yang, R., Zhao, J., Feng, Y., Feng, Q., et al.: Retrieval of brain tumors by adaptive spatial pooling and fisher vector representation. PLoS ONE **11**(6), e0157112 (2016)
12. Deepak, S., Ameer, P.: Brain tumor classification using deep CNN features via transfer learning. Comput. Biol. Med. **111**, 103345 (2019)
13. Deng, J., Dong, W., Socher, R., Li, L.J., Li, K., Fei-Fei, L.: Imagenet: A large-scale hierarchical image database. In: 2009 IEEE conference on computer vision and pattern recognition. pp. 248–255. IEEE (2009)
14. Gosh, S., Nahar, N., Wahab, M.A., Biswas, M., Hossain, M.S., Andersson, K.: Recommendation system for e-commerce using alternating least squares (ALS) on apache spark. In: International Conference on Intelligent Computing & Optimization, pp. 880–893. Springer (2020)
15. Islam, R.U., Ruci, X., Hossain, M.S., Andersson, K., Kor, A.L.: Capacity management of hyperscale data centers using predictive modelling. Energies **12**(18), 3438 (2019)
16. Ismael, M.R., Abdel-Qader, I.: Brain tumor classification via statistical features and back-propagation neural network. In: 2018 IEEE International Conference on Electro/information Technology (EIT), pp. 0252–0257. IEEE (2018)
17. Iv, M., Telischak, N., Feng, D., Holdsworth, S.J., Yeom, K.W., Daldrup-Link, H.E.: Clinical applications of iron oxide nanoparticles for magnetic resonance imaging of brain tumors. Nanomedicine **10**(6), 993–1018 (2015)

18. Ker, J., Bai, Y., Lee, H.Y., Rao, J., Wang, L.: Automated brain histology classification using machine learning. J. Clin. Neurosci. **66**, 239–245 (2019)
19. Nahar, N., Ara, F., Neloy, M., Istiek, A., Biswas, A., Hossain, M.S., Andersson, K.: Feature selection based machine learning to improve prediction of Parkinson disease. In: International Conference on Brain Informatics, pp. 496–508. Springer (2021)
20. Nahar, N., Ara, F., Neloy, M.A.I., Barua, V., Hossain, M.S., Andersson, K.: A comparative analysis of the ensemble method for liver disease prediction. In: 2019 2nd International Conference on Innovation in Engineering and Technology (ICIET), pp. 1–6. IEEE (2019)
21. Nahar, N., Hossain, M.S., Andersson, K.: A machine learning based fall detection for elderly people with neurodegenerative disorders. In: International Conference on Brain Informatics, pp. 194–203. Springer (2020)
22. Pan, S.J., Yang, Q.: A survey on transfer learning. IEEE Trans. Knowl. Data Eng. **22**(10), 1345–1359 (2009)
23. Pashaei, A., Sajedi, H., Jazayeri, N.: Brain tumor classification via convolutional neural network and extreme learning machines. In: 2018 8th International Conference on Computer and Knowledge Engineering (ICCKE), pp. 314–319. IEEE (2018)
24. Pathan, R.K., Uddin, M.A., Nahar, N., Ara, F., Hossain, M.S., Andersson, K.: Gender classification from inertial sensor-based gait dataset. In: International Conference on Intelligent Computing & Optimization, pp. 583–596. Springer (2020)
25. Pathan, R.K., Uddin, M.A., Nahar, N., Ara, F., Hossain, M.S., Andersson, K.: Human age estimation using deep learning from gait data. In: International Conference on Applied Intelligence and Informatics, pp. 281–294. Springer (2021)
26. Shorten, C., Khoshgoftaar, T.M.: A survey on image data augmentation for deep learning. J. Big Data **6**(1), 1–48 (2019)
27. Siegel, R.L., Miller, K.D., Jemal, A.: Cancer statistics, 2019. CA Cancer J. Clin. **69**(1), 7–34 (2019)
28. Sultana, Z., Nahar, L., Basnin, N., Hossain, M.S.: Inference and learning methodology of belief rule based expert system to assess chikungunya. In: International Conference on Applied Intelligence and Informatics, pp. 3–16. Springer (2021)
29. Swati, Z.N.K., Zhao, Q., Kabir, M., Ali, F., Ali, Z., Ahmed, S., Lu, J.: Brain tumor classification for MR images using transfer learning and fine-tuning. Comput. Med. Imaging Graph. **75**, 34–46 (2019)
30. Swati, Z.N.K., Zhao, Q., Kabir, M., Ali, F., Ali, Z., Ahmed, S., Lu, J.: Content-based brain tumor retrieval for MR images using transfer learning. IEEE Access **7**, 17809–17822 (2019)
31. Szegedy, C., Ioffe, S., Vanhoucke, V., Alemi, A.A.: Inception-v4, inception-resnet and the impact of residual connections on learning. In: Thirty-First AAAI Conference on Artificial Intelligence (2017)
32. Talo, M., Baloglu, U.B., Yıldırım, Ö., Acharya, U.R.: Application of deep transfer learning for automated brain abnormality classification using MR images. Cogn. Syst. Res. **54**, 176–188 (2019)
33. Tan, C., Sun, F., Kong, T., Zhang, W., Yang, C., Liu, C.: A survey on deep transfer learning. In: International Conference on Artificial Neural Networks, pp. 270–279. Springer (2018)
34. Yang, Y., Yan, L.F., Zhang, X., Han, Y., Nan, H.Y., Hu, Y.C., Hu, B., Yan, S.L., Zhang, J., Cheng, D.L., et al.: Glioma grading on conventional MR images: a deep learning study with transfer learning. Front. Neurosci. **12**, 804 (2018)

Detection of Parkinson's Disease Using Multimodal Dataset

Md. Rawnak Saif Adib and M. Shamim Kaiser

Abstract Parkinson's disease (PD) is a neurological condition that causes tremors, stiffness, and difficulty walking, balancing, and coordinating. There is no cure for PD, and that is why it is important to look for early signs so that it can be mitigated right from the beginning. Previous works have used different data to detect PD, but we have focused on using multimodal data, Speech, and Image datasets to predict PD with various Machine Learning (ML) algorithms. We have used seven different ML algorithms to compare the speech dataset and DenseNet201 for the image dataset. Our image data consisted of Spiral and Wave hand-drawings of PD patients and healthy control. In comparison with other authors' work, our model has shown promising results with the highest accuracy of 94.87% for speech data. The accuracy for Spiral and Wave images were 73% and 93%, respectively. The novelty of our work lies in the fact that we have used Speech data, Spiral, and Wave hand-drawn images all together in one work to detect PD. Although one particular work used Speech data and Spiral images, but no dataset for wave images was used there.

Keywords Parkinson's disease · Multimodal data · Speech dataset · Spiral dataset · Wave dataset · Random forest · XGBoost · DenseNet201

1 Introduction

Parkinson's disease (PD) is a progressive neurological disorder in which a portion of the brain weakens over time. It is a degenerative central nervous system condition that affects the elderly for a lengthy time (60–80). When Parkinson's disease strikes,

Md. R. Saif Adib (✉)
Department of ICT, Bangladesh University of Professionals, Mirpur Cantonment, Dhaka, Bangladesh
e-mail: saifadib24@gmail.com; 17511024@student.bup.edu.bd

M. S. Kaiser
Institute of Information Technology, Jahangirnagar University, Savar, Dhaka 1342, Bangladesh
e-mail: mskaiser@juniv.edu

S. Hossain et al. (eds.), *Proceedings of International Conference on Fourth Industrial Revolution and Beyond 2021*, Lecture Notes in Networks and Systems 437,
https://doi.org/10.1007/978-981-19-2445-3_41

a patient's ability to maintain dopamine-producing neurons is harmed, resulting in tremor or trembling in the hands, arms, legs, jaw, and face, rigidity or stiffness of the limbs and trunk, bradykinesia (slowness of movement), and postural uncertainty (impaired balance and coordination). Tremors, sluggish movement, sleep problems, postural instability, and depression are just a few of the motor and non-motor symptoms of this illness. From 1999 to 2017, the age-adjusted death rate for Parkinson's disease among 65 rose from 41.7 to 65.3 per 100,000. Men's mortality rates increased from 65.2 per 100,000 in 1999 to 97.9 per 100,000 in 2017. Women's rates had climbed from 28.4 per 100,000 in 1999 to 43.0 per 100,000 in 2017. PD affects almost one million persons in the United States alone. Researchers discovered that of all the data, speech data had the best accuracy in predicting whether a patient would be diagnosed with Parkinson's disease or not. They also discovered that drawing spirals and waves by hand helped them identify PD patients.

This research focuses on a multimodal approach to detecting Parkinson's disease utilizing various ML techniques. A voice dataset and a hand drawing image collection, including spiral and wave images were used in this study. The paper is arranged in the following way. Section 2 contains the related works that were done previously. Section 3 contains our multimodal approach using the Speech and Image dataset and the models we used in the prediction. Section 4 contains the results of our model and comparative analysis. Finally, Sect. 5 concludes our paper.

2 Related Work

Bi et al. [1] have created a feasible multimodal feature fusion approach to identify correlations between the brain areas and genes in sequence information and extract more recognized features. To examine the fusion characteristics of samples, a new neural network ensemble model called CERNNE is proposed. This method uses an evolution strategy to optimize the classification outcomes of ensemble learners. They have used Functional Magnetic Resonance Imaging (fMRI) data and Single Nucleotide Polymorphism (SNP) data.

Oungl et al. [2] acquired data from healthy subjects utilizing triaxial wearable on-body sensors (accelerometer, gyroscope, and magnetometer). A to-fold cross-validation approach evaluated the system efficiency. The results reveal that People with Parkinson's (PWP) can be differentiated from healthy controls with the highest average accuracy, sensitivity, specificity, and Receiver Operating Characteristic (ROC) of over 88% from the proposed machine learning algorithms. The acceleration data were transferred via a Bluetooth link to an online data processing wearable computer system. They have used Probabilistic Neural Network (PNN) and Support Vector Machine (SVM) for classification.

Vasquez-Correa et al. [3] proposed a methodology approach to deep learning to assess the multimodal information of PD patients. Speech, handwriting, and gait data were considered. Patients' neurological status, particular deficits in the lower and top limbs and muscles, and the condition's impact on speech were assessed.

When patients start/stop vibrating their vocal folds, speech transitions are noticed. Handwriting transitions are detected when the patient holds the pen in the air and places it on the tablet's surface, and gait transitions are identified when the patient walks. The deep learning approach was based on Convolutional Neural Network (CNN). When information from three bio-signals is integrated, the system's precision improves up to 97.3%.

Pham et al. [4] have used multimodal dataset Voice and Spiral image test with different ML methods. Classification for voice data set is done using Pairwise correlation and k-means clustering techniques. They have also proposed incorporating the multimodal technique into a touch-enabled smartphone-based application that can be used to conduct early PD tests without the requirement for extra medical staff or specific equipment. They conducted three tests which are Dynamic Spiral Test (DST), Static Spiral Test (SST), and Stability Test on Certain Point (STCP). In DST, k-Nearest Neighbors (KNN) classifier gave the highest accuracy of 99.6%. The Logistic Regression classifier and the Adaptive Boosting classifier on the SST and STCP gave 98.8% and 94.9% accuracy, respectively.

Freezing of gait (FOG) detection in Parkinson's disease was done using EEG and accelerometer signals by Wang et al. [5]. The model employing motion data had a sensitivity of 65% and a specificity of 54% for single-modal models.

Non-motor features such as Rapid sleep Behaviour Disorder (RBD) and olfactory loss, Cerebrospinal fluid (CSF) markers, and dopaminergic SPECT imaging markers (Neuroimaging markers) were obtained from the Parkinson's Progression Markers Initiative (PPMI) database, which were combined in this paper to detect PD diagnosis by Prashantha et al. [6]. The classifiers such as Naive Bayes, SVM, Boosted Trees, and Random Forests were utilized to distinguish early PD patients from healthy controls. Among all these, SVM gave the best performance, 96.40% accuracy, 97.03% sensitivity, and 95.01% specificity.

Heidarivincheh et al. [7] described a multimodal deep learning strategy for distinguishing between patients with and without Parkinson's disease. Their proposed architecture MCPD-Net, trains variational autoencoder (VAE) models using two data modalities obtained from vision and accelerometer sensors in a home setting. They have used CNN, unimodal VAE, RF, and Long Short-Term Memory (LSTM) models for classification.

Iakovakis et al. [8] evaluated triaxial accelerometer, voice- and keystroke data recordings acquired from 70 PD patients. They have processed keystroke pipelines to measure the severity of stiffness and bradykinesia in subject typing sessions by considering 8 keystrokes from Hold-time (HT) and intra keystroke flight time (FT) sequences that are split into time windows from where statistical features are extracted and used as input for an ML model.

Bind et al. [9] have presented a comprehensive review of the prediction of PD by using ML based approaches. They have provided a detailed overview of techniques for Parkinson's disease prediction. There are 7 Machine learning methods mentioned in this paper. These are Artificial Neural Networks (ANN), KNN, SVM, Naïve Bayesian Classifier, Random Forest, Bagging, and Boosting.

3 Materials and Methods

This section contains the data collection and pre-processing of our multimodal data and proposed speech and image data model.

3.1 Data Collection

In our proposed work, we examined two datasets: the Speech and Image datasets. The following sections provide detailed information on both datasets' data collection.

3.1.1 Speech Data

This dataset contains a variety of biological voice measurements from 31 individuals, 23 of whom have PD. Each column in the table refers to a specific voice measure, and each row corresponds to one of the 195 voice recordings made by these people ("name" column). The primary goal of the data is to distinguish healthy persons from those with Parkinson's disease using the "status" column, which is set to 0 for healthy people and 1 for those with PD. The data is stored in ASCII CSV format. Each row in the CSV file corresponds to a single voice recording instance. There are around six recordings per patient, with the patient's name in the first column. Max Little of the University of Oxford generated the dataset with the National Center for Voice and Speech in Denver, Colorado, which recorded the speech signals [10]. Table 1 lists the features of the dataset that was used. We can see from the table that the features are divided into six categories, namely, Frequency Parameters, Pulse Parameters, Amplitude Parameters, Voicing Parameters, Pitch Parameters, and Harmonicity Parameters.

3.1.2 Image Data

The dataset is owned by Zham et al. [11], and it consists of Spiral and Wave hand-drawn drawings. Total 55 people were recruited and assessed for Unified Parkinson's Disease Rating Scale (UPDRS) to draw the images, among which, 28 are healthy control, and 27 are PD patients. There are 204 photos in the collection, with Spiral and Wave each comprising 102 samples. Furthermore, each kind is divided into the Testing and Training categories, with 15 healthy and 15 Parkinson's drawing samples in the testing section and 36 healthy and 36 Parkinson's drawing samples in the training section. Figure 1 shows examples of spiral and wave drawings.

Table 1 Features of speech data

Features	Category
Jitter (local) Jitter (local, absolute) Jitter (rap) Jitter (ppq) Jitter (ddp)	Frequency parameters
Number of pulses Number of periods Mean period Standard dev. of period	Pulse parameters
Shimmer (local) Shimmer (local, dB) Shimmer (apq) Shimmer (apq3) Shimmer (apq5) Shimmer (dda)	Amplitude parameters
Fraction of locally unvoiced frames Number of voice breaks Degree of voice breaks	Voicing parameters
Median pitch Mean pitch Standard deviation Minimum pitch Maximum pitch	Pitch Parameters
Autocorrelation Noise-to-harmonic Harmonic-to-noise	Harmonicity parameters

Fig. 1 Spiral and wave drawings

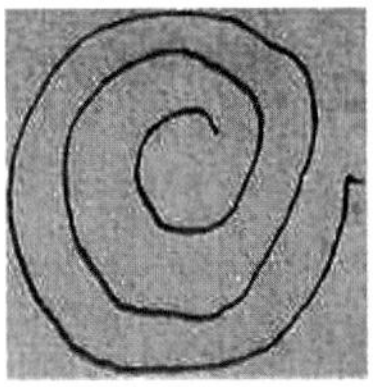

(a) Spriral Sample.

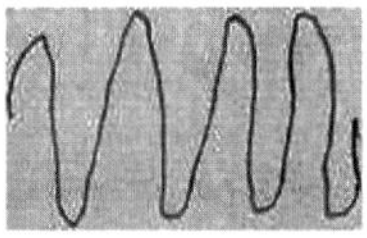

(b) Wave Sample.

3.2 *Data Pre-processing*

3.2.1 Speech Data

After obtaining the dataset, we have pre-processed the data by dropping the columns "name" and "status". Among 24 attributes of the dataset, the datatype object is assigned to the "name" attribute, while the datatype integer is assigned to the "status" attribute. The datatype of the remaining attributes is float. There are 147 PD positive records and 48 healthy control records in the collection. We also obtain statistical information and all the features' mean values based on the "status" column. Then we divided the data into train test split by 20% and standardized the data. Among 195 data records, the model has 156 training records and 39 testing record data. We have defined the status as 1 for Parkinson's Positive and 0 for Healthy. We also obtained the mean, standard deviation, minimum, and maximum values for all the features.

3.2.2 Image Data

We have used the TensorFlow library of python in our code. We directed the spiral-test, spiral-train, wave-test, and wave-train path directories and obtained the classes in which we will predict the result. We loaded every image from those paths, set the color mode to RGB, and converted every image into an array. In this dataset, we divided the data into train test split by 20%.

3.3 *Methods*

3.3.1 Speech Data

For the detection of voice data, we deployed seven distinct methods. These are SVM, Decision Tree, Random Forest, KNN, Nave Bayes, XGBoost, and AdaBoost. The reason for choosing these algorithms is that they are more accurate than others. We fitted both the training and testing data into the model after normalizing and converting the data. The training data's accuracy score was then anticipated and calculated. Following that, we projected our testing results and calculated the accuracy. We are also developing a prediction system that can analyze an individual's medical history and forecast whether or not they have Parkinson's disease. We did the above tasks in all seven models. Figure 2 shows the workflow diagram for our speech dataset.

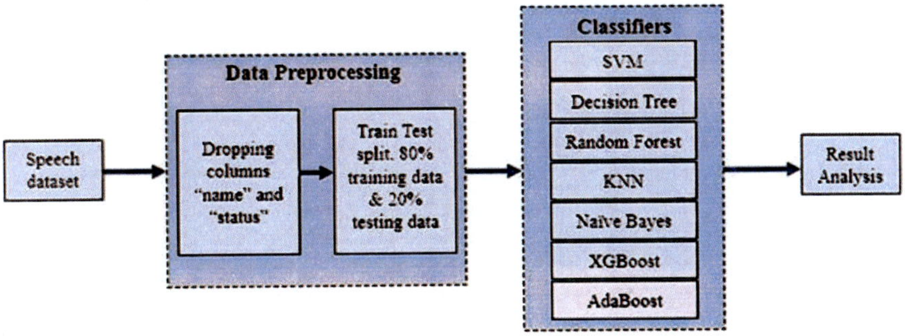

Fig. 2 Workflow diagram for speech dataset

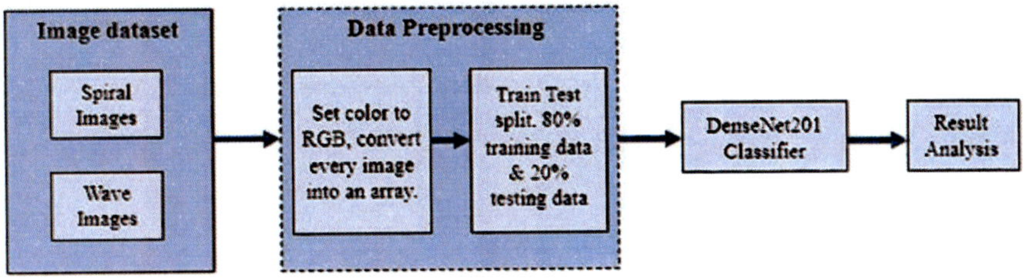

Fig. 3 Workflow diagram for image dataset

3.3.2 Image Data

We have used the DenseNet201 model for the image dataset. It is a Convolutional Neural Network that is 201 layers deep. Image Data Generator has been used to generate batches of image data later used to fit the samples into our model. Every batch size was defined with a size of 32. We also obtained precision, recall, f1-score, and support for both Spiral and Wave images. And finally, we have built a predictive system for both Spiral and Wave images that takes a single sample image as an input, load it, and predicts whether a patient draws that image with PD or by a healthy person. It shows the output "Prediction is healthy" if drawn by a healthy person and shows "Prediction is Parkinson's" if a PD patient draws the image. Figure 3 shows the workflow diagram for our image dataset.

4 Results

Speech Data

We have tried to show a comparative analysis among different machine learning models about which one gives us the best result. And from Fig. 4, we can see that Random Forest and XGBoost achieved the best accuracy, 94.87%. Among the others, SVM has

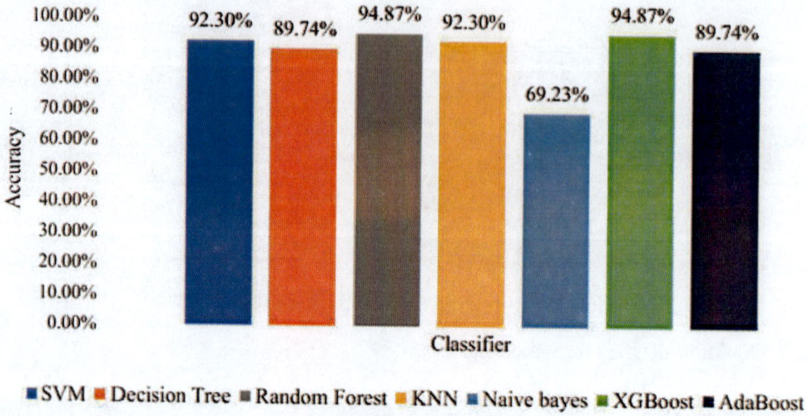

Fig. 4 Accuracy of the models

Table 2 A Comparison among the work of other authors

Authors	Classifier used	Best Accuracy
Das et al. [12]	Neural Network	92.90%
Kadam et al. [13]	Deep Neural Network	93.84%
Olanrewaju et al. [14]	K-means clustering	80%
Andr'e	Gravitational Search Algorithm	87.55%
Spadoto et al. [15]	Optimum Path Forest (GSA-OPF)	
Our Proposed model	Random Forest, XGBoost	94.87%

achieved 92.3%, Decision Tree achieved 89.74%, KNN achieved 92.3%, Naïve Bayes achieved the lowest accuracy 69.23%, and finally, AdaBoost has achieved 89.74%. Table 2 shows a comparative analysis of our work with some previous works for the speech dataset. And it shows that our work gives the best result than the existing ones.

Image Data:

Tables 3 and 4 have shown precision, recall, f1-score, and support for both Spiral and Wave along with the accuracy, macro average, and weighted average. The maximum training and testing (validation) accuracy and loss are obtained from our model, which we have shown in Figs. 5, 6, 7 and 8. The graphs show that the training and testing accuracy for spiral images are 98% and 73% (approx.), respectively, and that of the wave images are 100% and 93%, respectively. Thus it concludes that wave images give better accuracy. The training and testing (validation) loss for spiral images has ranged from 1.6 to 0.62 (maximum to minimum) and from 1.28 to 0.1

Table 3 Precision, Recall, F1-score and Support of Spiral Images

	Precision	Recall	F1-score	Support
Healthy	0.50	0.33	0.40	6
Parkinson's	0.64	0.78	0.70	9
Accuracy			0.60	15
Macro Avg	0.57	0.56	0.55	15
Weighted Avg	0.58	0.60	0.58	15

Table 4 Precision, Recall, F1-score and Support of Wave Images

	Precision	Recall	F1-score	Support
Healthy	0.00	0.00	0.00	6
Parkinson's	0.40	1.00	0.57	9
Accuracy			0.40	15
Macro Avg	0.20	0.50	0.29	15
Weighted Avg	0.16	0.40	0.23	15

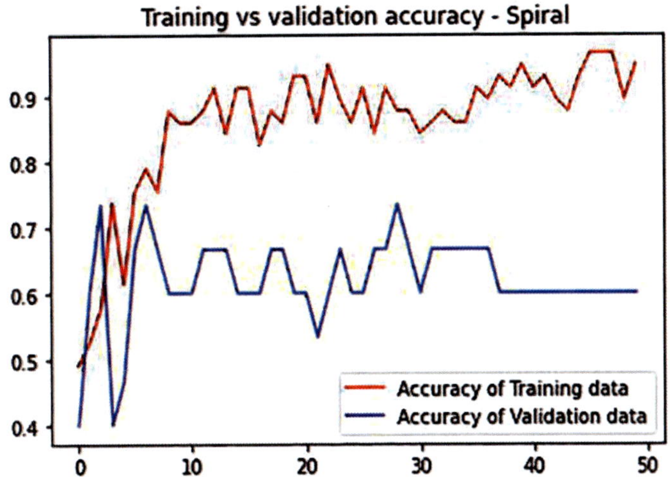

Fig. 5 Training versus validation accuracy-spiral

(maximum to minimum), respectively, and that of the wave images has ranged from 1.3 to 0.18 (maximum to minimum) and from 1.0 to 0.23 (maximum to minimum) respectively.

Muralikrishna et al. [16] achieved an accuracy of 80% and 90% for Spiral and Wave images respectively using Resnet34, Resnet50, and VGG19 classifiers. Shaban et al. [17] utilized Vgg19 and achieved an accuracy of 88.5% and 88% using tenfold cross-validation for Spiral and Wave images respectively. Chakraborty et al. [18]

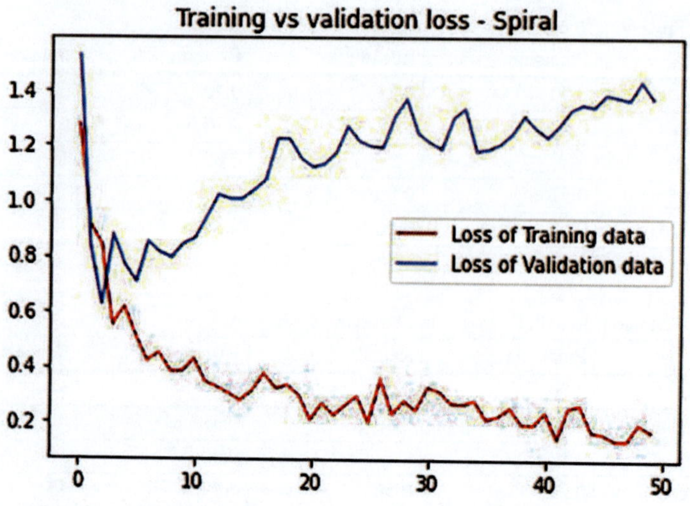

Fig. 6 Training versus validation loss-spiral

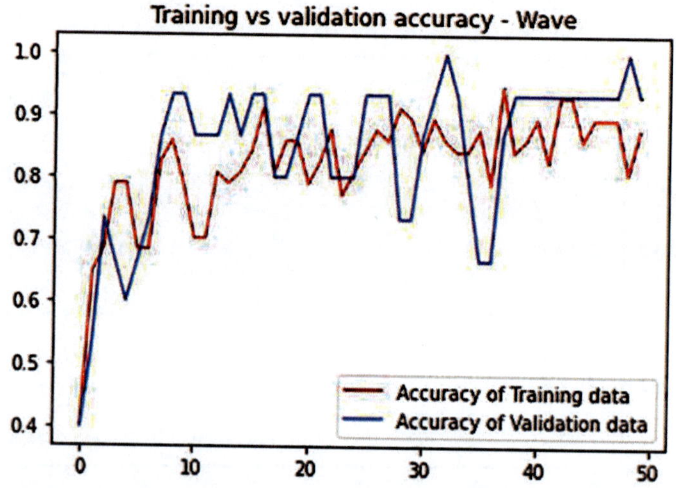

Fig. 7 Training versus validation accuracy-wave

achieved 93.9% accuracy using CNN with 1024 layers and 5 fold cross-validation. Vatsaraj et al. [19] combined the histogram of gradients (HOG) feature extractor with SVM and KNN to achieve 73.05% and 73.33% accuracy for Spiral and Wave images respectively. And for feature extraction, neural networks such as Xception, NASNet, and InceptionV3 with pretrained ImageNet weights and Fully Connected Neural Networks were utilized, with an accuracy of 87.22%, 89.52%, and 92.84%,

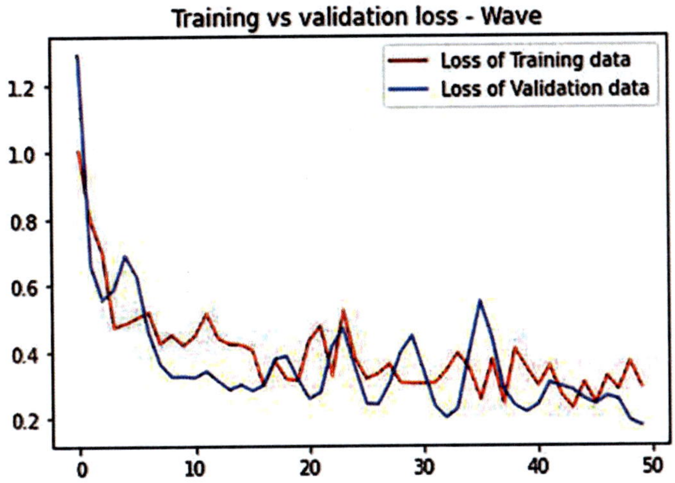

Fig. 8 Training versus validation loss-wave

respectively. The above examples demonstrate that despite providing good results, our model has a few limitations, which can be improved by using more advanced and improved algorithms and classifiers.

5 Conclusion

The detection of PD using various ML algorithms for multimodal data is proposed in this paper. We used speech and image data to detect PD, with the models SVM, Decision Tree, Random Forest, KNN, Nave Bayes, XGBoost, and AdaBoost were used for speech DenseNet201 used for image data. In the case of the speech data, Random Forest and XGBoost achieved the highest accuracy of 94.8%, while Nave Bayes achieved the lowest accuracy of 69.2%. In the case of image data, Spiral images had a 73% accuracy, while Wave images had a 93% accuracy. Though our model produced excellent results for the speech and wave image datasets, the output for our spiral images was slightly lower than the existing results. However, the accuracy of our model for spiral images is still below acceptable levels. If a larger dataset and better classifiers had been used, the accuracy of the spiral data could have been improved. In the future, we intend to use the Decision Fusion Model to combine the results of Speech and Image data with improving our work's efficiency.

References

1. Bi, X., Xi, H., Xie, Y., Hao, W.: A novel CERNNE approach for predicting Parkinson's Disease-associated genes and brain regions based on multimodal imaging genetics data. Med. Image Anal. **67**, 101830 (2021)
2. Oung, W., Hariharan, M., Lee, H.L., Basah, S.N., Sarillee, M., Lee, C.H.: Wearable multimodal sensors for evaluation of patients with Parkinson Disease, p. 6 (2015)
3. Vasquez-Correa, J.C., Arias-Vergara, T., Orozco-Arroyave, J.R., Eskofier, B., Klucken, J., Noth, E.: Multimodal assessment of Parkinson's Disease: a deep learning approach. IEEE J. Biomed. Health Inf. **23**(4), 1618–1630 (2019)
4. Pham, H.N., Do, T.T.T., Chan, K.Y.J., Sen, G., Han, A.Y.K., Lim, P., Cheng, T.S.L., Nguyen, Q.H., Nguyen, B.P., Chua, M.C.H.: Multimodal detection of Parkinson Disease based on vocal and improved spiral test. In: 2019 International Conference on System Science and Engineering (ICSSE), pp. 279–284. Dong Hoi, IEEE, Vietnam (2019)
5. Wang, Y., Beuving, F., Nonnekes, J., Cohen, M.X., Long, X., Aarts, R.M., van Wezel, R.: Freezing of gait detection in Parkinson's disease via multimodal analysis of EEG and accelerometer signals. In: 2020 42nd Annual International Conference of the IEEE Engineering in Medicine & Biology Society (EMBC), pp. 847–850. IEEE, Montreal, QC, Canada (2020)
6. Prashanth, R., Roy, S.D., Mandal, P.K., Ghosh, S.: High-accuracy detection of early Parkinson's Disease through multimodal features and machine learning. Int. J. Med. Inf. **90**, 13–21 (2016)
7. Heidarivincheh, F., McConville, R., Morgan, C., McNaney, R., Masullo, A., Mirmehdi, M., Whone, A.L., Craddock, I.: Multimodal classification of Parkinson's Disease in home environments with resiliency to missing modalities. Sensors **21**(12), 4133 (2021)
8. Hawkes, C.H., Del Tredici, K., Braak, H.: A timeline for Parkinson's disease. Parkinsonism Relat. Disord. **16**(2), 79–84 (2010)
9. Bind, S., Tiwari, A.K., Sahani, A.K.: A survey of machine learning based approaches for parkinson disease prediction **6**, 8 (2015)
10. Little, M.A., McSharry, P.E., Roberts, S.J., Costello, D.A.E., Moroz, I.M.: Exploiting nonlinear recurrence and fractal scaling properties for voice disorder detection. BioMed. Eng. Online **6**(1), 23 (2007)
11. Zham, P., Kumar, D.K., Dabnichki, P., Arjunan, S.P., Raghav, S.: Distinguishing different stages of Parkinson's Disease using composite index of speed and pen-pressure of sketching a spiral. Front. Neurol. **8**, 435 (2017)
12. Das, R.: A comparison of multiple classification methods for diagnosis of Parkinson disease. Expert Syst. Appl. **37**(2), 1568–1572 (2010)
13. Kadam, V.J., Jadhav, S.M.: Feature ensemble learning based on sparse autoencoders for diagnosis of Parkinson's Disease. In: Iyer, B., Nalbalwar, S.L., Pathak, N.P. (eds.) Computing, communication and signal processing, vol. 810, pp. 567–581. Series Title: Advances in Intelligent Systems and Computing. Springer, Singapore (2019)
14. Olanrewaju, R.F., Sahari, N.S., Musa, A.A., Hakiem, N.: Application of neural networks in early detection and diagnosis of Parkinson's disease. In: 2014 International Conference on Cyber and IT Service Management (CITSM), pp. 78–82. IEEE, South Tangerang, Indonesia (2014)
15. Spadoto, A.A., Guido, R.C., Carnevali, F.L., Pagnin, A.F., Falcao, A.X., Papa, J.P.: Improving Parkinson's disease identification through evolutionary-based feature selection. In: 2011 Annual International Conference of the IEEE Engineering in Medicine and Biology Society, pp. 7857–7860. IEEE, Boston, MA (2011)
16. Muralikrishna, A.B.: Efficient detection of Parkinson Disease using multiple machine learning techniques, p. 23
17. Shaban, M.: Deep convolutional neural network for Parkinson's Disease based handwriting screening. In: 2020 IEEE 17th International Symposium on Biomedical Imaging Workshops (ISBI Workshops), pp. 1–4. IEEE, Iowa City, IA, USA (2020)

18. Chakraborty, S., Aich, S., Sim, J.S., Han, E., Park, J., Kim, H.C.: Parkinson's Disease detection from spiral and wave drawings using convolutional neural networks: a multistage classifier approach. In: 2020 22nd International Conference on Advanced Communication Technology (ICACT), pp. 298–303. IEEE, Phoenix Park, Pyeong Chang, Korea (South) (2020)
19. Vatsaraj, I., Nagare, G.: Early detection of Parkinson's Disease using contrast enhancement techniques and CNN. Int. J. Eng. Res. **10**(05), 4

Neurophysiological Feature Based Stress Classification Using Unsupervised Machine Learning Technique

Moumita Bhowmik, Naim Ibna Khadem Al Bhuyain, Md. Rokonuzzaman Reza, Nafiz Imtiaz Khan, and Muhammad Nazrul Islam

Abstract Mental stress is the primary concern of increasing mental health problems and other medical problems like strokes, heart attacks, and ulcers. Thus, identifying and classifying stress at an early age is the prime requirement to avoid such diseases. Although a number of studies focused to predict and classify stress based on neurophysiological (brain wave and heart rate) data and used the supervised machine learning technique, but a little attention has been paid to explore the performances of unsupervised learning techniques in stress prediction. In this article, an unsupervised machine learning approach is proposed to classify mental stress into three categories: acute (low stress), episodic acute (moderate stress), and severe (high stress). The K-means clustering algorithm was used in the proposed methodology to create three different clusters, which depicts the aforementioned stress levels. The goodness of the clustering technique was evaluated by Silhouette Coefficient, and a standard fitness score of 0.76 was achieved.

Keywords Mental stress · Brain signal · Heart rate · Machine learning

1 Introduction

Nowadays a number of studies have focused on assessing a person's mental stress while performing cognitively demanding tasks. Mental stress can be defined as the result of a heavy workload and tight deadlines that can impair a person's ability to cope with a potential threat to their physical, emotional, or psychological well-being [22]. According to Gallup's 2019 Global Emotions Report, approximately a third of the world's population encountered stress and stress-related negative emotions in

M. Bhowmik · N. I. K. Al Bhuyain · Md. R. Reza · N. I. Khan · M. N. Islam (✉)
Department of Computer Science and Engineering, Military Institute of Science and Technology (MIST), Mirpur Cantonment, Dhaka 1216, Bangladesh
e-mail: nazrul@mist.cse.ac.bd

M. Bhowmik
e-mail: bhowmikmoumita99@gmail.com

S. Hossain et al. (eds.), *Proceedings of International Conference on Fourth Industrial Revolution and Beyond 2021*, Lecture Notes in Networks and Systems 437,
https://doi.org/10.1007/978-981-19-2445-3_42

2018 [25]. As a result, prevention-oriented stress management is essential for both individual health and societal well-being [12]. Again the negative moods such as depression, anger, and anxiety caused by stress result in increased right-prefrontal activity. Therefore, the asymmetric analysis of the frequency-band powers in the Electroencephalography (EEG) measured at the prefrontal cortex has been generally applied in previous stress studies [1]. As stress is associated with a range of behavioural, cognitive, neurovascular, cardiovascular, and molecular effects, in such a context, the simultaneous analysis of both the HRV and EEG is likely to provide a more precise assessment of stress [4, 6].

Although mental pressure cannot be precisely measured, various technologies have been developed in recent years to acquire data that can reliably predict a person's mental stress. Pupil diameter (PD), head-worn eye-trackers, electroencephalography (EEG), heart rate (HR), galvanic skin response (GSR), skin temperature, and respiration rate are just a few examples of measurements in this domain. These methods each have their own set of strengths and weaknesses, but all have been used individually to detect stress through the help of machine learning estimating task difficulty which occurs due to measurement noise such as eye blinks [2, 5, 14, 21]; head motion [14, 16]; various lighting effects [8, 14], and other environmental conditions [2, 5, 8, 21]. Unfortunately, these studies only examined stress symptoms after a person has gotten stressed, and offered subjective answers, which causes treatment to be delayed. Subjective feelings have no bearing on the severity of cognitive impairment. Furthermore, they may be misjudged due to a person's hesitancy to admit that they are stressed [11]. As a result, objective diagnostic methods for quantifying stress employing neuroimaging techniques are required to ensure adequate treatment.

Therefore, the objectives of this research are to measure and classify stress using neurophysiological features and an unsupervised ML technique. Stress is classified into three categories which are acute stress or low, episodic acute stress or moderate and severe stress or highly stressed.

The rest of the paper is organized as follows: Sect. 2 describes the overview of the prior related studies. The methodological overview is given in Sect. 3 followed by the result and analysis in Sect. 4. This research is concluded in Sect. 5.

2 Literature Review

People may feel low levels of self-esteem, high levels of depression, frustration because of cognitive workload which can lead to acute, episodic acute, and chronic stress. Therefore, in recent years, many different technologies have been used to gather data for providing a person's mental workload that includes: eye gaze as well as pupil diameter (PD), electroencephalography (EEG), heart rate variability (HRV), skin conductance (GSR), skin and facial temperature, etc. using machine learning [15]. Most of the studies have been focused on (a) developing a system for monitoring stress using ML [8] and (b) motivated people to evaluate mental workload on the basis of neurological features and frequency characteristics [2, 14, 16].

Despite technological progress in biological signal acquisition, this neurological and neurophysiological signal analysis is still a major challenge.[13] Because in case of measuring stress the diameter of the pupil may vary in different environments for different people under various lighting conditions and cognitive work whereas other neurophysiological methods use extra sensors which may be stressful in itself [10]. For example, Bates and Instance [3] found that EGT systems including PD analysis for computer cursor control are sometimes considered “uncomfortable” and “not easy to use” by their users because of the lack of pointing accuracy.

Most previous studies on mental stress detection fall into three broad categories: (a) Find correlation between cognitive load and psychological stress; (b) detect mental stress using machine learning framework with great accuracy and (c) Measure stress into various levels for medical treatments of psychological diseases like anxiety, depression, anger, post-traumatic disorder even cardiovascular and neurological diseases. In the first category, in order to prove the relation between cognitive workload and stress, researchers have used several mental tasks like Montreal Imaging Stress Task (MIST), Stroop Color and Word Test (SCWT), mental arithmetic tasks, etc. [9]. Some have given driving tasks, public speaking tasks, questionnaires, or have shown music videos and emotional movies to induce cognitive pressure [9]. Using these types of stressors, researchers have developed various methodologies to detect stress into different levels through levelling the tasks. Unfortunately, these levels of stress may not represent the ideal classifications of stress because different persons feel cognitive pressure on different levels which may result in a varied range of stress classification. Again, in the case of measuring stress, after analysing the cost, efficiency, feasibility, and accuracy of different techniques, researchers have recommended EEG and HR as the more feasible technique for stress detection. Some have even made a fusion of EEG and HR for better accuracy and found that the model incorporating both EEG and HRV variables performed better, demonstrating that simultaneous EEG and HRV measurements improve stress estimates [1]. Several classifiers like Support vector machines (SVM), Linear Discriminant Analysis (LDA), K-Nearest Neighbour (KNN), Multilayer Perceptron (MLP), Naive Bayes (NB), Random Forest (RF), etc.[9] have achieved better performance. However, labelling stress through supervised learning is endorsed by all but in this research, a different approach has been taken where unlabelled data is clustered using the k-means clustering algorithm.

Moreover, many others have concentrated on identifying emotions in various contexts. Zaki et al. [24], for example, have embarked on a machine learning-based study to discover user’s emotional experiences through neurological and physiological measures when playing serious games which will aid interactive computing and entertainment systems to grow in accordance with user’s satisfaction. On the other hand, Kaklauskas et al. [8] proposed an intelligent tutoring system for teachers to discover a student’s weaknesses by detecting their stress and making appropriate recommendations to enhance teaching techniques. In this context, all of them have looked into the relationship between stress and cognitive work, as well as how they can be quantified utilizing the technologies mentioned above. Some have found the best method to detect stress and some have increased the accuracy of any of the methods. Besides, various examinations took a gander at the viability of stress-soothing measures, like huggable correspondence contraptions [18] and patients holding a pressure ball during activity [7, 23]. The EEG also relates to pressure alleviation methodologies, for example, paying attention to traditional music [10] and reflection VR encounters [20] have additionally been concentrated in a few exploration.

In sum, several supervised machine learning techniques have been proposed in the existing literature to find the stress by using neurophysiological studies. But finding the stress using unsupervised learning could be a new step for classifying stress in the neurophysiological field. Therefore, this research concentrated on detecting stress levels in three different sections (acute, episodic acute, and severe) on the basis of unsupervised learning conditions.

3 Methodological Overview

The overview of the study methodology is presented in Fig. 1.

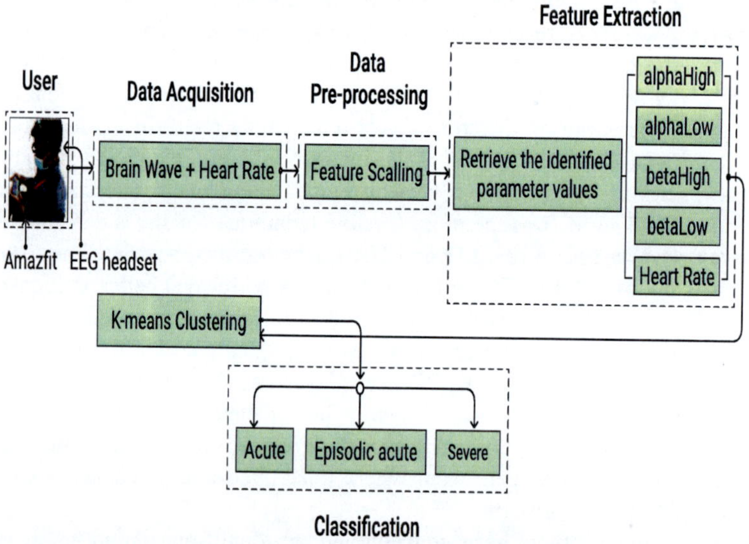

Fig. 1 Overview of the research methodology

In the experimental process, user tasks were defined by time pressure and cognitive load. At the data acquisition phase, neurophysiological data is collected from the EEG headset Neurosky,[1] eegID[2] mobile application and Amazfit[3] respectively while users

[1] https://store.neurosky.com/pages/mindwave.

[2] https://play.google.com/store/apps/details?id=com.isomerprogramming.application.eegID.

[3] https://www.amazfit.com/en/gtr.

were given to play the game "River Crossing IQ-IQ Test" on the smartphone. This game is available in Google Play Store containing 24 levels. Four individuals, all of whom were 22 years old and had prior experience playing mobile games, were recruited to participate in the experiment. None of them had played this selected game before. They were briefed regarding the goal of this experiment and assured that the hooked gadgets would not do them any bodily harm. In addition, before beginning the game, players were informed of all of the game's regulations, and due to time limitations, players were instructed to complete as many levels as they could in 20 minutes without using coins for assistance. As stress can be detected from brain wave signals and heart rate (beat per 15 s) data, authors focused on those particular features.

The adaptive least mean square technique was used for filtering out noise from brain wave signals and a digital filter is used to overcome the noises in the heart rate dataset. Feature scaling was used for scaling the features for training the model. After getting the processed data, alphaHigh, alphaLow, betaHigh, betaLow signals (brain wave signal), and heart rate data were extracted for training the machine learning model. With the k-means clustering technique, selected features were employed for clustering. As a consequence of the method used, the dataset was divided into three clusters and they denote three different levels of mental stress.

4 Result and Analysis

The proposed system has been made in three stages: Brain signal acquisition, heart rate acquisition, and user stress classification.

4.1 Data Acquisition

During the experiment, each participant was given the EEG headset to wear on their heads and the smartwatch to wear on their hands. The examiner was on standby to aid in the wearing of the devices, and their data was initially measured in order to check the gadgets. After 5 minutes, they were briefed appropriately and ensured that they would not be judged during the experiment. They were then given exactly 20 minutes to play the game (River Crossing IQ-IQ Test) while their brain waves and heart rates were continuously recorded utilizing the equipment.

To gather brain wave data, Electroencephalography (EEG) is used as it is a non-invasive way of acquiring brain waves. Neurosky Mindwave Mobile 2 Headset records a single channel of EEG from a dry electrode implanted in the frontal lobe of the brain (FP1) and EEG was collected using an electrode inserted on the ear lobe. Using the eegID application brain wave signals are stored in a CSV file. From all the sets of brain wave signals, only the alphaHigh, alphaLow, betaHigh, and betaLow signals were extracted. The eegID program reads the numbers while keeping a particular time period of 1 sec in mind. Lastly, the CSV file is processed and used in the proposed system containing total of 450 samples as input. Brainwave signals are converted into 15 sec timestamp depicted in Table 1 which shows the max, min, median, mean, and std deviation of alpha high, alpha low, beta high and beta low respectively.

For collecting heart rate, the Amazfit GTR smartwatch has been used. It includes a biological tracking optical sensor and a Huami-developed algorithm that provides advanced full-scale heart-rate monitoring and data was collected using the android application named ZEPP.[4] The heart rate was captured every 1 min and stored in a CSV file. The reading heart rate value is also converted into 15-sec timestamps and declared now as beat per 15 sec in Table 1. It also has the value of the same measure as brainwave signals.

Again a box plot is generated in Fig. 2 that provides a summary of the key features used in this research. The box plot shows the comparison between the median values of the features.

Table 1 Statistical analysis of data

Measures	Alpha low	Alpha high	Beta low	Beta high	Beat per 15 sec
Average	3295710.68	3208987.85	2978722.06	2029593.52	20.32
Mean	3295711	3208988	2978722	2029594	20.32
Median	3362796	3363427	2836225	1172053	20.25
Max	11180073.50	11177797	8948767	6728239	24.75
Min	9952.07	11033.67	6441.93	3223.60	13.50
Std deviation	1979058	1917721	1916186	1783309	2.91

[4] https://play.google.com/store/apps/details?id=com.huami.watch.hmwatchmanager.

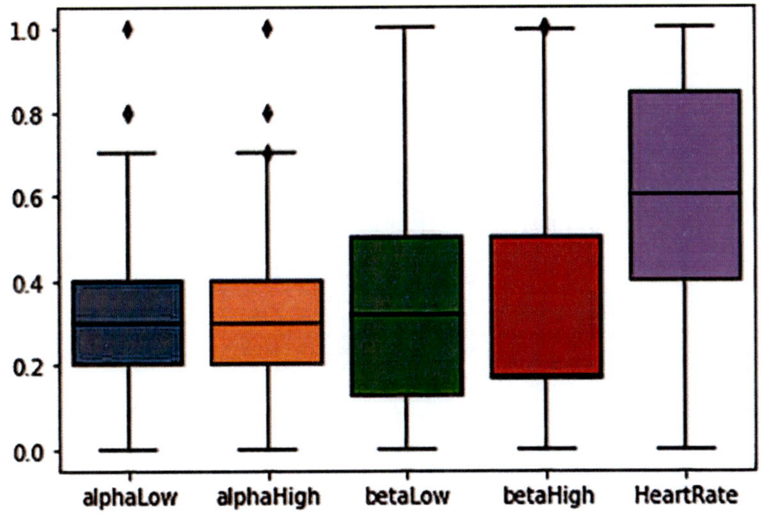

Fig. 2 Box plot of dataset

4.2 Stress Classification

Machine learning was used to classify stress levels from brain signal and heart rate data, with the objective of employing unsupervised learning for stress categorization. K-means clustering algorithm is used to divide the cluster into three levels. There is a total of five features in the dataset namely alphaHigh, alphaLow, betaHigh, betaLow and heartrate, while values were scaled from 0 to 1 using MinMaxScaler. Then K-means clustering was applied, and three different clusters were found from the dataset. This clustering algorithm's main goal is to minimize the sum of distances between data points and their respective clusters. The algorithm takes an unlabelled dataset as input, separates it into a k-number (predefined) of clusters, and continues the procedure until no better clusters are found. Next to evaluate the consistency of the clusters, Silhouette Score has been calculated using the mean intra-cluster distance denoted by a and the mean nearest-cluster distance denoted by b for each sample where the Silhouette Coefficient for a sample is $(b - a) / max(a, b)$. The score has been found 0.76 between 1 to -1 where the best value is 1 and the worst value is -1, and values near 0 indicate overlapping clusters. Thus, the score of this research shows an acceptable cluster range. Again to identify which cluster denotes which level of mental stress some statistical analysis was performed and some significant intuition has been found (see Table 2).

From prior literature [17, 19] it is found that alphaLow and betaHigh contributed mostly to mental stress whereas heart rate is proportional to the stress level. The average and median values of alphaLow, betaHigh, and heart rate is considered from Table 2, and a comparison has been made among these metrics (see Fig. 3).

Table 2 Statistical analysis of data after clustering

Clusters	Measures	Alpha low	Alpha high	Beta low	Beta high	Beat per 15 sec
Cluster 0	Average	2939078.55	2439398.64	1750333.34	4147640.43	22.36
	Median	2259741.53	2247141.20	1134551.67	3374096.87	22.50
	Max	11180073.47	11177796.73	5594991.27	4505291.27	20.75
	Min	9952.07	11033.67	6441.93	3223.60	13.50
	Std deviation	2055660.849	2037498.04	1720675.52	1290250.95	3.82
Cluster 1	Average	3894500.18	4240669.35	3796407.84	1058204.78	17.20
	Median	4479715.53	4488222.20	3374547.80	1128901.80	17.75
	Max	8953923.40	8950492.80	8948767.40	6728239.27	24.75
	Min	28755.80	1133760.93	29902.67	2245908.53	15.25
	Std deviation	2047539.17	1873502.86	1969284.05	1359669.34	2.46
Cluster 2	Average	3147187.58	3076116.51	3453190.46	1186262.89	21.22
	Median	3362625.27	3363002.10	3369183.40	1133619.53	21.75
	Max	5620244.07	6715417.60	7826318.07	2279556.27	24.75
	Min	13634.93	14806.07	9340.13	7874.87	18.25
	Std deviation	1734918.52	1481755.30	1746222.92	846870.13	1.53

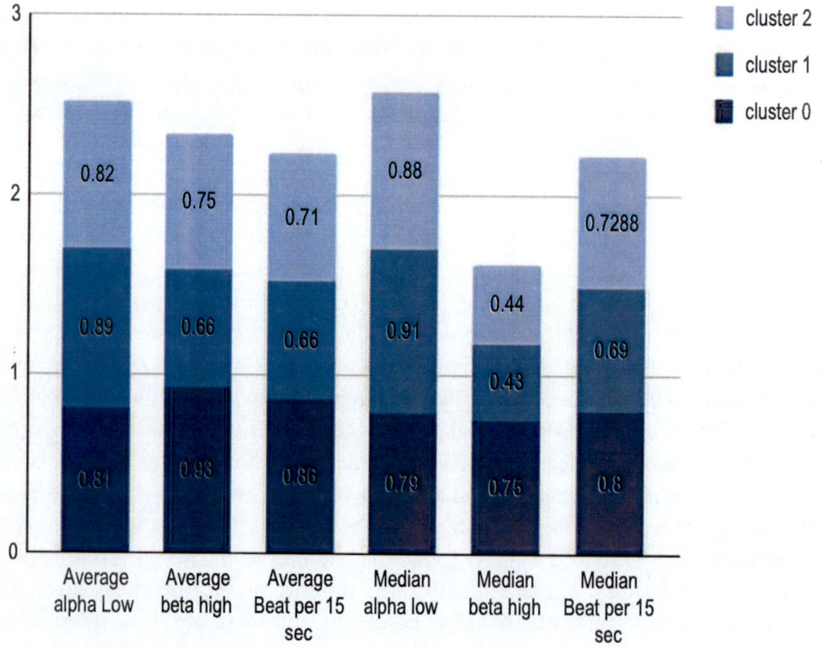

Fig. 3 Comparison of clusters

Table 3 Comparison pattern obtained from clusters

Metrics	Features	Comparison pattern
Average and Median	alphaLow	cluster 0 <cluster 2 <cluster 1
	betaHigh	cluster 0 >cluster 2 >cluster 1
	Beat per 15 s (heart rate)	cluster 0 >cluster 2 >cluster 1

Therefore, a significant pattern is noticed among the metrics as presented in Table 3. Here, the alphaLow values shows minimum, medium and maximum in cluster 0, 2 and 1 respectively while the betaHigh values shows a reverse result for these clusters and heart rate is found minimum, medium and maximum in cluster 0, 2 and 1. Existing studies [17, 19] showed that in highly stressed condition, alphaLow is lowest, the betaHigh is highest and the heart rate behaves proportional to the stress level. As such in this research cluster 0 can be represented as severe stress, cluster 2 as episodic acute and cluster 1 as acute stress respectively.

In addition, a violin plot in Fig. 4 confirms the above findings. In severe stress, alphaLow is minimum, in episodic acute it is medium and maximum in acute stress. In another violin plot in Fig. 5 shows the reverse order in terms of betaHigh. In sum it can be said that this study provides proper classification of stress through using the unsupervised approach.

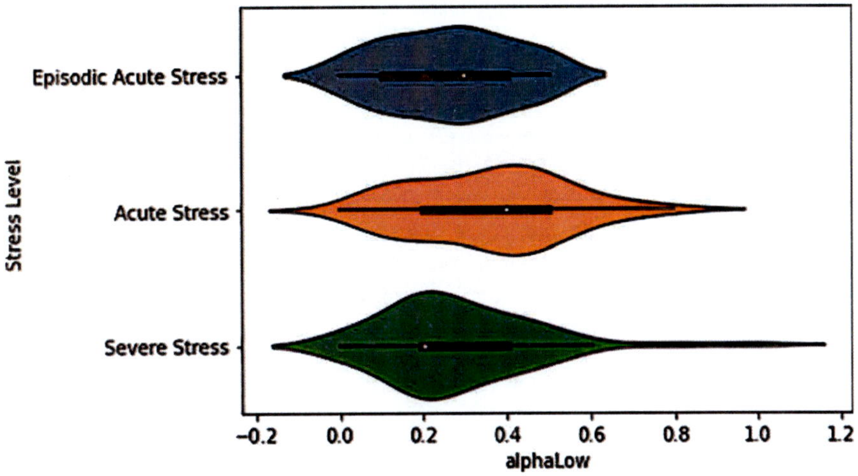

Fig. 4 Comparison of stress levels according to alphaLow

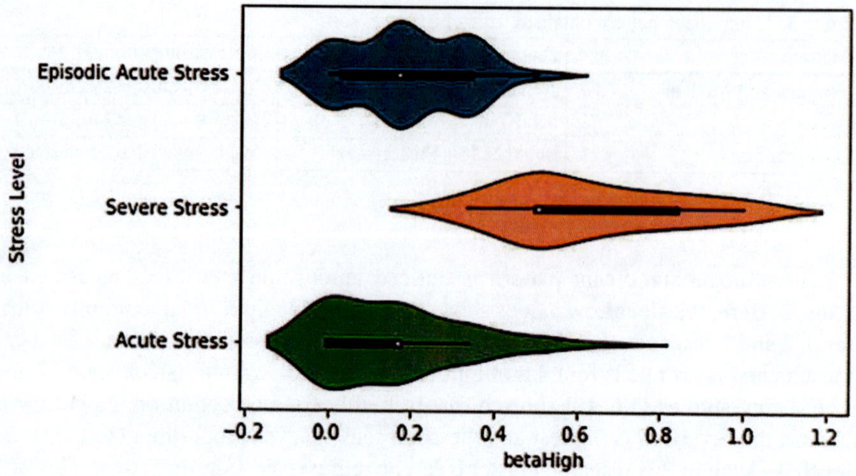

Fig. 5 Comparison of stress levels according to betaHigh

5 Conclusions

This study presents an unsupervised machine learning approach for stress classification into three categories. Existing studies used a supervised approach and labelled the stress beforehand while this study takes neurophysiological features into account and finds out stress level using an unsupervised approach. The study found that cluster 0, cluster 2, cluster 1 represent severe stress, episodic acute, and acute stress respectively. Nonetheless, in the future, other kinds of measures like pupil diameter can be included for more precise stress classification with other present features and the best-performed classifier can be explored using different machine learning techniques. The unsupervised learning technique, adopted in this study is a new approach to understand the stress level that can be helpful for taking measures to prevent mental disorders. The goodness of the clustering algorithm was evaluated by Silhouette Coefficient, and a standard fitness score of 0.76 was achieved. The performance obtained by the proposed approach showed the efficacy of the study.

References

1. Ahn, J.W., Ku, Y., Kim, H.C.: A novel wearable EEG and ECG recording system for stress assessment. Sensors **19**(9), 1991 (2019)
2. Baltaci, S., Gokcay, D.: Stress detection in human-computer interaction: fusion of pupil dilation and facial temperature features. Int. J. Hum. Comput. Interact. **32**(12), 956–966 (2016)
3. Barreto, A., Gao, Y., Adjouadi, M.: Pupil diameter measurements: Untapped potential to enhance computer interaction for eye tracker users? In: Proceedings of the 10th International ACM SIGACCESS Conference on Computers and Accessibility, pp. 269–270 (2008)

4. Carneiro, D., Castillo, J.C., Novais, P., Fernández-Caballero, A., Neves, J.: Multimodal behavioral analysis for non-invasive stress detection. Expert Syst. Appl. **39**(18), 13376–13389 (2012)
5. Gao, Y., Barreto, A., Zhai, J., Rishe, N.: Digital filtering of pupil diameter variations for the detection of stress in computer users. In: Proceedings of the 11th World Multi-conference on Systemics, Cybernetics and Informatics, Orlando, pp. 30–35 (2007)
6. Gärtner, M., Rohde-Liebenau, L., Grimm, S., Bajbouj, M.: Working memory-related frontal theta activity is decreased under acute stress. Psychoneuroendocrinology **43**, 105–113 (2014)
7. Hudson, B., Ogden, J., Whiteley, M.: Randomized controlled trial to compare the effect of simple distraction interventions on pain and anxiety experienced during conscious surgery. Eur. J. Pain **19**(10), 1447–1455 (2015)
8. Kaklauskas, A., Vlasenko, A., Raudonis, V., Zavadskas, E.K., Gudauskas, R., Seniut, M., Juozapaitis, A., Jackute, I., Kanapeckiene, L., Rimkuviene, S., et al.: Student progress assessment with the help of an intelligent pupil analysis system. Eng. Appl. Artif. Intell. **26**(1), 35–50 (2013)
9. Katmah, R., Al-Shargie, F., Tariq, U., Babiloni, F., Al-Mughairbi, F., Al-Nashash, H.: A review on mental stress assessment methods using EEG signals. Sensors **21**(15), 5043 (2021)
10. Khan, M.A., Chennafi, M., Li, G., Wang, G.: Electroencephalogram-based comparative study of music effect on mental stress relief. In: 2018 11th International Congress on Image and Signal Processing, BioMedical Engineering and Informatics (CISP-BMEI), pp. 1–5. IEEE (2018)
11. Koh, K.B., Park, J.K., Kim, C.H., Cho, S.: Development of the stress response inventory and its application in clinical practice. Psychosom. Med. **63**(4), 668–678 (2001)
12. Manning, M.R., Jackson, C.N., Fusilier, M.R.: Occupational stress, social support, and the costs of health care. Acad. Manag. J. **39**(3), 738–750 (1996)
13. Pedrotti, M., Mirzaei, M.A., Tedesco, A., Chardonnet, J.R., Mérienne, F., Benedetto, S., Baccino, T.: Automatic stress classification with pupil diameter analysis. Int. J. Hum. Comput. Interact. **30**(3), 220–236 (2014)
14. Pfleging, B., Fekety, D.K., Schmidt, A., Kun, A.L.: A model relating pupil diameter to mental workload and lighting conditions. In: Proceedings of the 2016 CHI Conference on Human Factors in Computing Systems, pp. 5776–5788 (2016)
15. Ren, P., Barreto, A., Gao, Y., Adjouadi, M.: Affective assessment by digital processing of the pupil diameter. IEEE Trans. Affect. Comput. **4**(1), 2–14 (2012)
16. Sakamoto, K., Aoyama, S., Asahara, S., Mizushina, H., Kaneko, H.: Relationship between emotional state and pupil diameter variability under various types of workload stress. In: International Conference on Ergonomics and Health Aspects of Work with Computers, pp. 177–185. Springer (2009)
17. Seo, S.H., Lee, J.T.: Stress and EEG. Converg. Hybrid Inf. Technol. **27** (2010)
18. Sumioka, H., Nakae, A., Kanai, R., Ishiguro, H.: Huggable communication medium decreases cortisol levels. Sci. Rep. **3**(1), 1–6 (2013)
19. Taelman, J., Vandeput, S., Spaepen, A., Van Huffel, S.: Influence of mental stress on heart rate and heart rate variability. In: 4th European Conference of the International Federation for Medical and Biological Engineering, pp. 1366–1369. Springer (2009)
20. Tarrant, J., Viczko, J., Cope, H.: Virtual reality for anxiety reduction demonstrated by quantitative EEG: a pilot study. Front. Psychol. **9**, 1280 (2018)
21. Torres-Salomao, L., Mahfouf, M., El-Samahy, E.: Pupil diameter size marker for incremental mental stress detection. In: 2015 17th International Conference on e-health Networking, Application & Services (HealthCom), pp. 286–291. IEEE (2015)
22. Xia, L., Malik, A.S., Subhani, A.R.: A physiological signal-based method for early mental-stress detection. Biomed. Signal Process. Control **46**, 18–32 (2018)
23. Yanes, A.F., Weil, A., Furlan, K.C., Poon, E., Alam, M.: Effect of stress ball use or hand-holding on anxiety during skin cancer excision: a randomized clinical trial. JAMA Dermatol. **154**(9), 1045–1049 (2018)

24. Zaki, T., Khan, N.I., Islam, M.N.: Evaluation of user's emotional experience through neurological and physiological measures in playing serious games. In: International Conference on Intelligent Systems Design and Applications, pp. 1039–1050. Springer (2020)
25. Zhang, Y., Wang, Q., Chin, Z.Y., Ang, K.K.: Investigating different stress-relief methods using electroencephalogram (EEG). In: 2020 42nd Annual International Conference of the IEEE Engineering in Medicine & Biology Society (EMBC), pp. 2999–3002. IEEE (2020)

A Cyber-Physical Fusion System for Stress Detection Using Multimodal and Social Media Data

Jasiya Fairiz Raisa, Sobhana Jahan, and M. Shamim Kaiser

Abstract Stress is identified as one of the most common human responses to physical, mental or emotional pressure. Long-term stress can cause cardiovascular diseases, depression, anxiety and even death. Stress can be recognized by observing physiological activity data and social media posts of individuals. This explorative study is performed to find the effect of fusion of physiological measurement with social media textual posts for classifying stress. The proposed model implements Heart Rate Variability (HRV) datasets as physiological stress datasets and social media post dataset as textual dataset. At first the datasets were individually implemented with different machine learning models to find the best fit model. It is shown that Random Forest showed the best classification result with an accuracy of 99.85% for the HRV data and the Logistic Regression model performed best for the social media data with an accuracy of 96.4%. The two models are combined using fuzzy fusion technique with an accuracy of 98%. To our knowledge, the fuzzy fusion technique for combining physiological and textual data is a novel approach for stress detection with significant applicability.

Keywords Stress detection · Fusion · Artificial intelligence · Heart rate variability · Natural language processing

J. F. Raisa (✉) · S. Jahan
Information and Communication Technology, Bangladesh University of Professionals, Dhaka 1216, Bangladesh
e-mail: jasiyafairiz@gmail.com

M. S. Kaiser
Institute of Information Technology, Jahangirnagar University, Savar, Dhaka 1342, Bangladesh
e-mail: kaiser@juniv.edu

S. Hossain et al. (eds.), *Proceedings of International Conference on Fourth Industrial Revolution and Beyond 2021*, Lecture Notes in Networks and Systems 437,
https://doi.org/10.1007/978-981-19-2445-3_43

1 Introduction

Stress has become a buzzword of present times, affecting a large number of people worldwide. Given the coronavirus pandemic and lockdown situation worldwide, more people are at risk of stress now, irrespective of gender, age group or work profile. In the long run, stress can lead to exhaustion of the neural, hormonal, cardiac, and muscle functions due to a lack of time for recovery and repair. Furthermore, stress has been shown to be one of the leading causes of major illnesses and, as a result, an earlier death [1].

Early detection of stress can prevent long-term disorders such as depression and anxiety. Researchers are working to find stress detection, measurement and management systems and find different factors causing stress.

Stress can be detected using different analysis approaches using psychological questionnaires, physiological activity data and social media [2]. However, the psychological questionnaires can be affected by personal convictions, hence providing false-positive results.

The majority of the studies for stress detection using physiological data have used cardiovascular assessments like Pulse Rate (PR), HR, BP, and Heart Rate Variability (HRV). Among these assessments, HRV has been approved as one of the most valuable and necessary cardiovascular responsive bio markers [3].

Due to the rapid development of the social media, people now use these platforms vastly than anytime before. The vast amount of data consisting of people's opinions, emotions and activities, makes it possible to leverage social media data for stress detection. It is found that textual data is the most available and popular form of expression in social media.

A combination of different types of data can provide a better classification of stress. Recently, several research works have been performed, focusing on fusing different data sources for stress classification. However, not many reliable studies are performed that fuse physiological data with social media data for stress classification.

This work aims to develop a fusion model to detect stress from physiological and social media datasets to develop a more accurate stress classification system. The proposed work is structured to answer the following set of research questions. **RQ1:** Does adding social media data for stress detection increase the reliability of the standard physiological data-based stress detection models? **RQ2:** How is the proposed ensemble model validated for detecting stress?
The focal contribution of the proposed system can be pointed as follows:

1. Development and validation of stress detection models individually for physiological and social media data.
2. Development of a novel fusion model for classifying stress based on social media and physiological datasets.

2 Related Works

At present, a significant number of research works are found that are developed using machine learning-based models to detect stress from multimodal datasets. More recently, ensemble and fusion models for stress detection have been proposed to improve stress classification accuracy.

Indikawati et al. [4] presented a novel system to detect stress from multimodal data. Here, multimodal data were collected using wrist-worn sensors named Empatica E4. The authors had performed four-class classification and using random forest, logistic regression, and decision tree algorithms. Random Forest was found to be the best performing algorithm with a promising accuracy rate of 88–99%.

Bobade et al. [5] proposed various ML and DL methods for detecting stressful emotion using multimodal datasets captured from wearable physical and body movement sensors. The authors used the WESAD dataset and analysed the results on three-class primary conditions with the accuracy of 81.65% for SVM and 84.32% for ANN. While for two-class conditions, the accuracy was 93.20% for SVM and 95.21%for ANN.

Radhika et al. [6] presented a multimodal fusion model for detecting stress using a convolutional layer. The proposed method is validated on two datasets, CLAS and ASCERTAIN, and the outputs indicate that middle fusion on the convolutional layer produces better results. ASCERTAIN dataset implemented with the deep multimodal model achieved the highest accuracy score of 75.5%.

Mohan et al. [7] proposed a work that uses a Photoplethysmography (PPG) optical sensor to collect heart rate variability data to identify a person's stress level. The authors divided the heart rate into shallow frequency, higher frequency, and lower frequency. However, the model cannot give a precise stress level prediction because it only measures one physiological parameter.

Cheng et al. [8] proposed a novel sentiment assessment model based on DL models to obtain emotion from social media. They used LSTM, GRU, and BI-LSTM algorithms and found that the Bi-LSTM was most promising, with an accuracy of 87.17%, a precision of 85.80%, recall of 88.89%, specificity of 82.88%, and F1-score of 87.29%.

Vashishtha et al. [9] used a new set of fuzzy rules involving numerous lexicons and datasets to calculate the emotion of social media posts. The authors conducted a comparative analysis of their method using nine public Twitter datasets, three sentiment lexicons, four state-of-the-art strategies for unsupervised sentiment classification, and one state-of-the-art strategy for supervised ML.

3 Key Concepts

The proposed model for stress detection uses the HRV dataset and social media dataset. For classification purposes, several machine learning algorithms are used. The key concepts behind the proposed model are discussed in this section.

3.1 Heart Rate Variability (HRV)

HRV is a measure of the time difference between each heartbeat. This variety is maintained by the autonomic nervous system, a fundamental part of the autonomic nervous system (ANS). It performs regardless of our willingness and regulates, among many other factors, our heart rate, respiration, digestion, and blood pressure. HRV should rise during relaxing activities and fall naturally during stressful situations [10].

3.2 Support Vector Machine (SVM)

SVM is an algorithm which is used to solve regression and classification tasks. SVM is based on a fundamental idea: the algorithm divides information into classes by drawing a line or a hyperplane. It applies to any vector that encodes any data. This implies that texts must be converted into vectors to take advantage of the SVM text classification [11].

3.3 K-Nearest Neighbour (K-NN)

K-NN is a straightforward ML algorithm that employs supervised learning. This method assumes similarity between the novel data and established cases and assigns the new case to the most similar group to the existing categories. K-NN can be used for both classification and regression tasks, but it is most widely used for classification tasks. Using the nltk library, one can create synonyms and use similarity measures between texts for text classification in K-NN [12].

3.4 Random Forest (RF)

Random Forest refers to an ensemble learning technique that makes predictions based on the integration of various decision trees. It essentially fits an amount of decision tree classifiers to different small-samples of the set of data. Furthermore, every tree in the forest is selected according to a random best subset of the features. RF classification methods are well-suited to involve high dimensional noisy data in classification tasks [13].

3.5 Logistic Regression (LR)

The Logistic Regression classifier takes the weighted combination of the input features and transmits them into a sigmoid function, which accepts any real-valued counts and maps them to a range between 0 and 1. Logistic regression is an excellent algorithm for text-related classification [14].

3.6 Naive Bayes (NB)

Naive Bayes is a powerful and easy-to-use algorithm mainly used as text classification learning method [15]. The Multinomial Naive Bayes (MultinomialNB) classifier is a probabilistic algorithm that is commonly used in natural language processing. It can be used to classify data with discrete features, for example, word counts for classifying text data [16].

3.7 Passive-Aggressive (PA)

Passive-aggressive algorithms are widely used for massive learning. It is one of the very few "online-learning methodologies". Online machine learning algorithms process data in a sequential manner and update the machine learning model in a series of steps. This is especially useful to deal with large amounts of data. This algorithm is appropriate for text classification in NLP [17].

3.8 Adaptive Boosting (AdaBoost)

AdaBoost is a meta-algorithm that is dependent on the statistical classification that can be used to enhance the performance and productivity of any machine learning algorithm. On a classification problem, this model can achieve accuracy above the incidental chance. Decision tree is one of the most appropriate and widely used technique to work with AdaBoost [18].

3.9 Multi-Layer Perceptron (MLP)

A Neural Network is interconnected to a MLP classifier. Unlike most other classifiers, such as NB or SVM, the MLP classifier uses a fundamental Neural Network to determine the class. The MLP is a supervised machine learning algorithm that learns a function by training on a dataset with the same number of input data dimensions as output data dimensions [19].

3.10 Data Fusion Technique

Data fusion is the process of merging information from multiple sources to create a much more strong, specialized, and comprehensive representation of a method that has been witnessed [20]. Fusion of any particular information can be performed generally in three types of fusion:

1. **Data-level fusion**: Data-level fusion occurs when same type of raw data from various sources is merged.
2. **Feature-level fusion**: The features have been extracted from raw data in various modalities are merged in a feature-level fusion.
3. **Decision-level fusion**: The final outcome of various classifiers are merged in the decision level fusion. This fusion type is of the least computational complexity [21].

Since we have implemented separate machine learning models with two different types of datasets, the decision-level fusion technique is implemented for achieving more accurate classification.

4 System Architecture and Methodology

The system can be divided into three main parts: (1) Data acquisition and preparation, (2) Development and validation of individual stress detection model, and (3) Decision-level fusion. The proposed model is a fusion of two individual stress detection models, as shown in Fig. 1.

4.1 Data Acquisition and Preparation

For the physiological data-based stress detection model, we have primarily used an HRV dataset publicly available dataset, WESAD. For validation purposes, we have implemented another publicly available dataset, SWELL. WESAD is a pub-

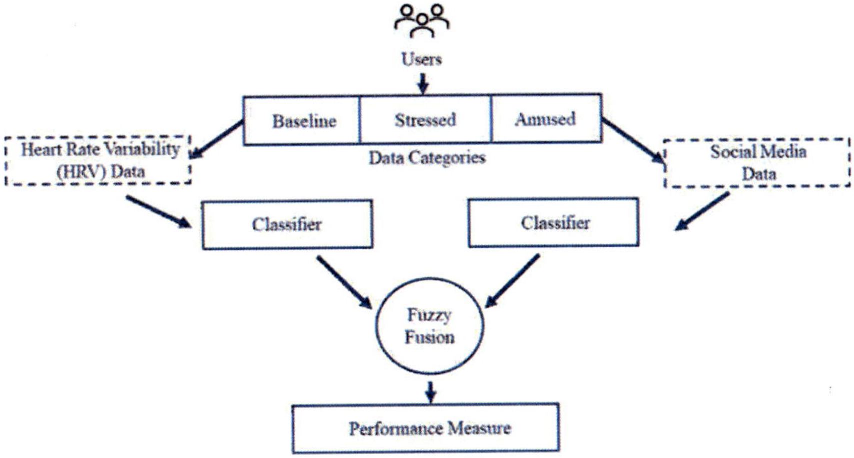

Fig. 1 Block diagram of the proposed stress detection model

licly accessible dataset for detecting wearable stress and its impact. This multi-modal dataset contains physiological and movement data collected from 15 patients throughout a lab study using both chest-worn sensor (RespiBAN) and wrist-worn sensor (Empatica E4) [22].

SWELL dataset is made up of heart rate variability (HRV) indices, collected using Video, Kinect 3D sensors and body sensors [23]. We have split the datasets into train-test sets for the implementation of WESAD and SWELL datasets, where the test carries 33% of the total dataset.

For the social media data-based model, we have merged two publicly available datasets, Sentiment140 and Dreaddit. The Sentiment140 dataset contains 1.6 million tweets collected using the Tweeter API and categorized into three types of emotions (positive, negative, and neutral) [24].

The Dreaddit dataset includes 190,000 posts from five various Reddit groups. This dataset contains longer social media posts divided into five categories, each of which has non-stressful and stressful messages and various ways to describe stress [25].

For stress detection using textual data, we have merged these two datasets and combined the emotion label of the datasets. For developing the corpus from the raw data, a set of preprocessing operations in the merged dataset are performed. We have removed the stopwords, reduced inflection in various words to the core forms developed a corpus. A bag of words is created where the top five thousand English words are taken and our corpus is mapped with them. We have split our dataset into train-test sets, where the test carries 33% of the total dataset. Figure 2 provides a clear view of the Workflow for data acquisition and preparation.

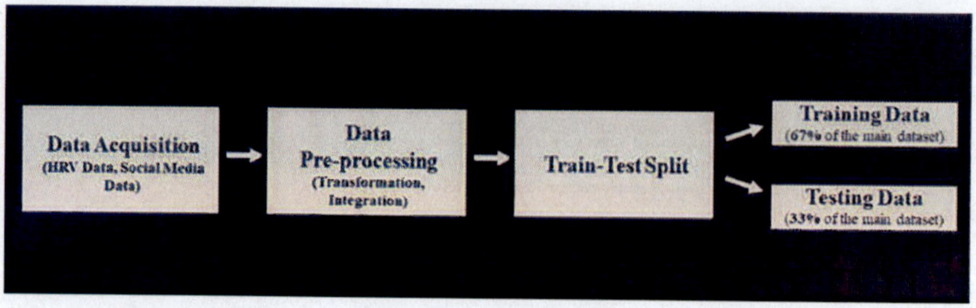

Fig. 2 Workflow diagram for data acquisition and preparation

4.2 Development and Validation of Individual Stress Detection Systems

To develop a physiological data-based model for stress detection, we have performed SVM, KNN, Random forest and MLP methods with the WESAD dataset. The model performed best with the Random Forest classifier. For the validation of the model, we have used the SWELL dataset. Since this dataset is extensive, we have implemented a feature selection approach before classification. The model with the random forest classifier performed best with the SWELL dataset. Therefore, the random forest model for stress detection with physiological data is the best performing model.

We have trained MultinomialNB, Passive-aggressive, SVM, LR, RF, and Adaboost models individually for the social media-based model. Among the models we experimented with, SVM was used as the base estimator in AdaBoost. We discovered that Logistic Regression performed better than the other models.

4.3 Development of the Fusion Model

After developing and validating the individual models, we perform a decision-level fusion of these models. By investigating the complementarities of source and pattern of two modalities, it has been illustrated that decisions from various classifiers can really be combined to achieve better classification results [26]. The Mamdani Fuzzy Inference System (FIS) is the central component of a fuzzy logic system that performs decision-making. It draws essential decision rules using the "IF...THEN" rules and uses "OR" or "AND" connectors. The linguistic terms used in this FIS are baseline, stressed, and amused (for HRV dataset) and neutral, stressed, and positive (for social media dataset). The FIS uses those linguistic terms as input 1 and input 2. The "IF...THEN" rule of our proposed FIS is visualized in Fig. 4.

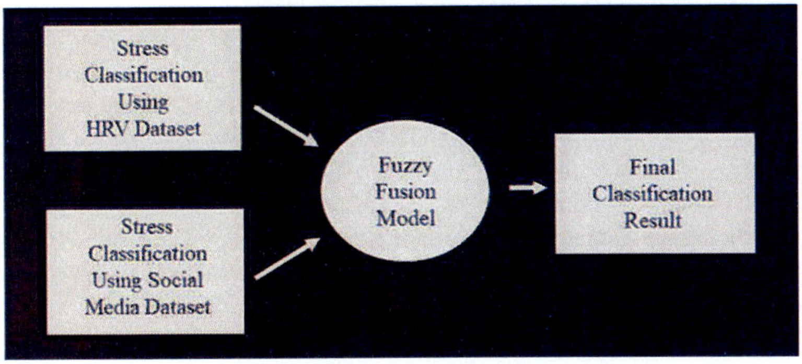

Fig. 3 Workflow diagram for decision fusion

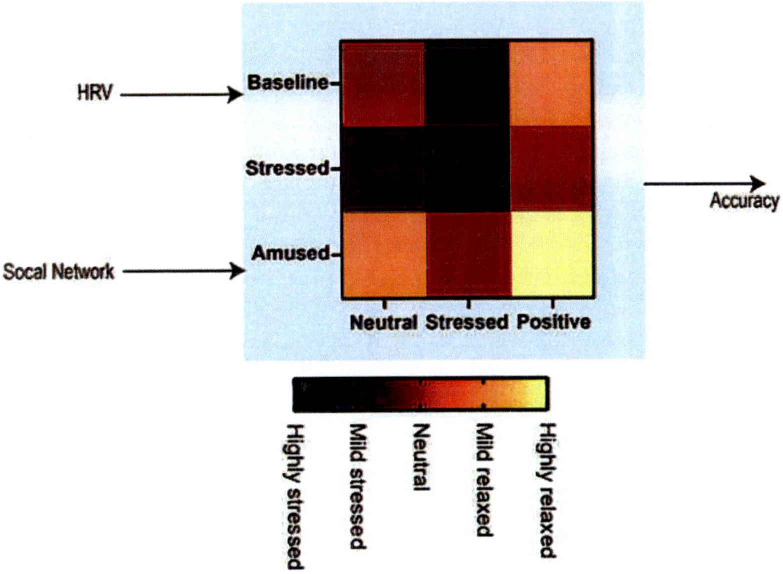

Fig. 4 Rule base for the proposed fuzzy model

Figure 3 shows the workflow for fuzzification of the individual models. The proposed and executed fusion module combines the scores products of sub-systems using fuzzy logic methodological approaches. For the proposed fuzzy model, we have developed a rule base. Figure 4 visualizes the rule base for the model.

5 Performance Analysis

For both physiological data-based model and social media textual data-based model, each data is classified using three conditions: stressed, amused, and baseline, after completing the model training.

For the machine learning models implemented using the HRV and social media datasets, the accuracy of each algorithm is listed in Table 1. The result presented in the table shows that the Random Forest and Logistic Regression classifiers performs with the highest accuracy for HRV and textual data, respectively.

The fusion of the two machine learning models is achieved using the fuzzy fusion technique. It provides a decision level fusion of the models. The result shows that the fusion model performs with an accuracy of 98%, which is significantly better than the two individual models.

For three linguistic terms, the accuracy of this fuzzy model is 95%. Whereas, this accuracy increases to 98% by using five linguistic terms as output. Those five linguistic terms are highly relaxed (HR), mild relaxed (MR), neutral (N), mild stressed (MS), and highly stressed (HS). In both cases, for HRV data three inputs are baseline, stressed, and amused and for social media data three inputs are neutral, stressed, and positive. Figure 4 comes with a view of the input and output linguistic term using a rule base. Figure 5 gives us a clear view of result of this fuzzy model. In Fig. 5, (a) shows the degree of membership for HRV dataset, (b) presents the degree of membership for social media dataset, (c) tells us the degree of membership after the fusion, and (d) presents the view of the surface of the accuracy of this fusion model.

The experimental result of the proposed model answers the research questions that we set previously. Since both physiological and social media data are implemented, the proposed model considers physical and mental reactions to the stressors. It reduces the chance of achieving false-positive results significantly, making the model more reliable for stress detection. This answers the first research question.

The two individual models using HRV data and social media data implements two different datasets each. Therefore, the models can be said to be free from data dependency and validated for stress detection. Since we perform a decision-level fusion of the models, where the models were previously validated with more than

Table 1 Result comparison using HRV and social media dataset

Physiological data		Textual data	
Classifier	Accuracy	Classifier	Accuracy
KNN	52%	MultinomialNB	88.60%
SVM	53%	PA	96%
MLP	97.80%	RF	94.40%
RF	99.85%	SVM	94.80%
		Adaboost	93.70%
		LR	96.40%

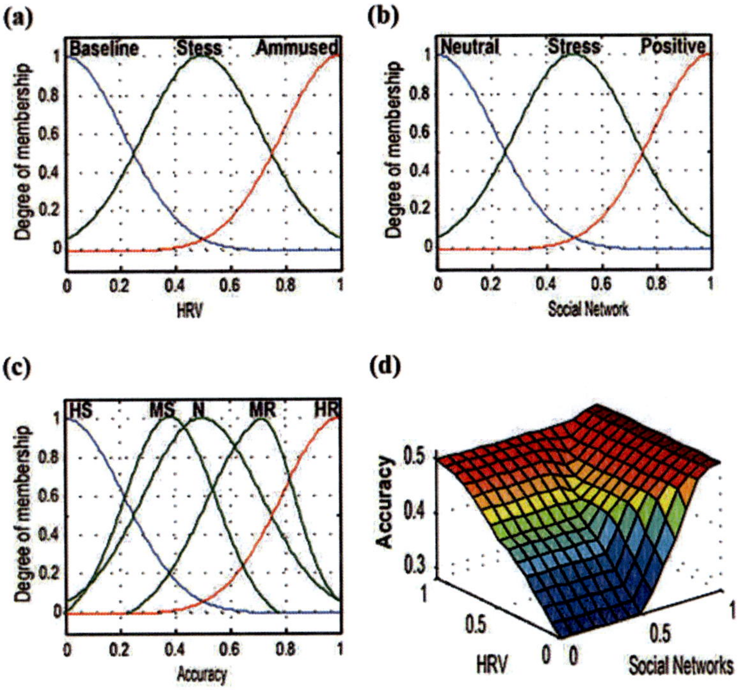

Fig. 5 Final result of the proposed fusion model. **a** the degree of membership for HRV dataset, **b** the degree of membership for social media dataset, **c** the degree of membership after the fusion, and **d** accuracy (98%) of the model (surface is plotted)

one dataset, the fusion model can also be said to be validated for stress detection. The result analysis shows that the fuzzy fusion model performs higher accuracy than the two individual models. This answers the second research question.

6 Conclusion

Stress can be identified through individuals' physiological activity data and social media posts. In this paper, with the help of these two types of data, we have developed a fuzzy fusion model for stress classification. Two separate models are developed based on physiological and social media data, with 96.4 and 99.85% accuracy for stress detection. Using the decisions from the models, we performed decision level fusion using a Fuzzy algorithm. The accuracy of our fusion model is 98%, which is significant to show model applicability in future works. In future, the proposed work can be advanced to detect stress using physiological data and social media data of the same individuals. To do so, we have planned to create our dataset beforehand.

Moreover, fusion in data-level and feature-level can also be performed to predict people's stress levels. Comparison among different level-wise fusion models can be performed to find the more suitable fusion technique for stress detection.

References

1. Schneiderman, N., Ironson, G., Siegel, S.D.: Stress and health: psychological, behavioral, and biological determinants. Annu. Rev. Clin. Psychol. **1**, 607–628 (2005)
2. Palmer, S., Dryden, W., Hickey, L.: Counselling for stress problems. J. Psychosom. Res. **5**(40), 553 (1996)
3. Aimie-Salleh, N., Malarvili, M.B., Whittaker, A.C.: Fusion of heart rate variability and salivary cortisol for stress response identification based on adverse childhood experience. Med. Biol. Eng. Comput. **57**(6), 1229–1245 (2019)
4. Indikawati, F.I., Winiarti, S.: Stress Detection from Multimodal Wearable Sensor Data, vol. 771, p. 012028. IOP Publishing (2020)
5. Bobade, P., Vani, M.: Stress detection with machine learning and deep learning using multi-modal physiological data. In: 2020 Second International Conference on Inventive Research in Computing Applications (ICIRCA), pp. 51–57 (2020)
6. Radhika, K., Oruganti, V.R.M.: Deep multimodal fusion for subject-independent stress detection. In: 2021 11th International Conference on Cloud Computing, Data Science Engineering (Confluence), pp. 105–109 (2021)
7. Mohan, P.M., Nagarajan, V., Das, S.R.: Stress measurement from wearable photoplethysmographic sensor using heart rate variability data. In: 2016 International Conference on Communication and Signal Processing (ICCSP), pp. 1141–1144 (2016)
8. Cheng, L.C., Tsai, S.L.: Deep learning for automated sentiment analysis of social media. In: Proceedings of the 2019 IEEE/ACM International Conference on Advances in Social Networks Analysis and Mining, ASONAM '19, pp. 1001–1004. Association for Computing Machinery. New York, NY, USA (2019)
9. Vashishtha, S., Susan, S.: Fuzzy rule based unsupervised sentiment analysis from social media posts. Expert Syst. Appl. **138**, 112834 (2019)
10. Shaffer, F., Ginsberg, J.P.: An overview of heart rate variability metrics and norms. Front. Public Health **5**, 258 (2017)
11. Srivastava, D., Bhambhu, L.: Data classification using support vector machine. J. Theor. Appl. Inf. Technol. **12**, 1–7 (2010)
12. Li, B., Yu, S., Lu, Q.: An improved k-nearest neighbor algorithm for text categorization (2003)
13. Ali, J., Khan, R., Ahmad, N., Maqsood, I.: Random forests and decision trees. Int. J. Comput. Sci. Issues(IJCSI) **9** (2012)
14. Peng, J., Lee, K., Ingersoll, G.: An introduction to logistic regression analysis and reporting. J. Educ. Res. **96**, 3–14 (2002)
15. Rish, I.: An empirical study of the naïve Bayes classifier. In: IJCAI 2001 Work Empir Methods Artif Intell, Vol. 3 (2001)
16. Abbas, M., Ali, K., Memon, S., Jamali, A., Memon, S., Ahmed, A.: Multinomial naive Bayes classification model for sentiment analysis (2019)
17. Crammer, K., Dekel, O., Keshet, J., Shalev-Shwartz, S., Singer, Y.: Online passive-aggressive algorithms. J. Mach. Learn. Res. **7**, 551–585 (2006)
18. Chengsheng, T., Huacheng, L., Bing, X.: AdaBoost typical Algorithm and its application research. In: MATEC Web of Conferences , vol. 139, p. 00222 (2017)
19. Nazzal, J., El-Emary, I., Najim, S.: Multilayer Perceptron Neural Network (MLPs) for analyzing the properties of Jordan Oil Shale. World Appl. Sci. J. **5** (2008)
20. Raol, J.R.: Multi-Sensor Data Fusion with MATLAB. CRC Press, Boca Raton (2009)

21. Liu, K.: Dual-sensor approaches for real-time robust hand gesture recognition. PhD thesis (2015)
22. Schmidt, P., Reiss, A., Duerichen, R., Marberger, C., Van Laerhoven, K.: Introducing WESAD, a multimodal dataset for wearable stress and affect detection. In: Proceedings of the 20th ACM International Conference on Multimodal Interaction, ICMI '18, pp. 400–408. Association for Computing Machinery. New York, NY, USA (2018)
23. Koldijk, S., Sappelli, M., Verberne, S., Neerincx, M.A., Kraaij, W.: The SWELL knowledge work dataset for stress and user modeling research. In: Proceedings of the 16th International Conference on Multimodal Interaction, ICMI '14, pp. 291–298. Association for Computing Machinery. New York, NY, USA (2014)
24. Go, A., Bhayani, R., Huang, L.: Twitter sentiment classification using distant supervision. Processing 150 (2009)
25. Turcan, E., McKeown, K.: Dreaddit: a reddit dataset for stress analysis in social media, 107p (2019)
26. Dasarathy, B.V.: Decision fusion. IEEE Computer Society Press, Los Alamitos, Calif (1994). OCLC, p. 28332530

Signals and NLP

Bangla Image Caption Generation Through CNN-Transformer Based Encoder-Decoder Network

Md. Aminul Haque Palash, Md. Abdullah Al Nasim, Sourav Saha, Faria Afrin, Raisa Mallik, and Sathishkumar Samiappan

Abstract Automatic Image Captioning is the never-ending effort of creating syntactically and validating the accuracy of textual descriptions of an image in natural language with context. The encoder-decoder structure used throughout existing Bengali Image Captioning (BIC) research utilized abstract image feature vectors as the encoder's input. We propose a novel transformer-based architecture with an attention mechanism with a pre-trained ResNet-101 model image encoder for feature extraction from images. Experiments demonstrate that the language decoder in our technique captures fine-grained information in the caption and, then paired with image features, produces accurate and diverse captions on the BanglaLekhaImage-Captions dataset. Our approach outperforms all existing Bengali Image Captioning work and sets a new benchmark by scoring 0.694 on BLEU-1, 0.630 on BLEU-2, 0.582 on BLEU-3, and 0.337 on METEOR.

Keywords Bengali Image Captioning · ResNet-101 · Transformer

1 Introduction

Depiction aims to describe an image's substance in text. Image subtitling in studies has been focused on a translation technique, including a visual encoder as well as a language decoder [11]. The method of producing image titles may be used for automating driving automobiles, implementing face recognition systems, describing persons with visual impairment, improving the quality of the photo query, etc. At the junction of computer vision and natural language interpretation, the challenge of producing natural language descriptions of the data of an image resides.

Md. A. H. Palash · Md. A. A. Nasim (✉) · S. Saha · F. Afrin · R. Mallik
Pioneer Alpha, Dhaka, Bangladesh
e-mail: info@pioneeralpha.com

S. Samiappan
Mississippi State University, Starkville, USA

S. Hossain et al. (eds.), *Proceedings of International Conference on Fourth Industrial Revolution and Beyond 2021*, Lecture Notes in Networks and Systems 437,
https://doi.org/10.1007/978-981-19-2445-3_44

Image captioning seems to be the function in the natural language to describe the visual data. An algorithm is required to comprehend and represent the links underlying visual and textual information and to create a series of the text output. Studies are based on the encoder-decoding framework [11] which is quite close to the machine transcription sequence-to-sequence paradigm. The framework contains coded language information and visual characterize complete transformer connectivity in the Caption Transformer (CPTR) [11] replaces a CNN with a fully convolution-free transformer encoder throughout the encoder portion. We directly sequence raw pictures as an input when compared to the traditional subtitled models that take the feature derived by CNN or feature detector as the source. Influenced by the studies above, we explore a new sequence-to-sequence approach for the resolution of the picture subtitling issue. Statistics are then fused into one multi-modal layer which predicts the term of the title by term [1].

Deep sequence model methodologies have produced amazing results. It was demonstrated recently that exposure differentiation and non-differentiable task metric questions may be solved by the incorporation of reinforcement learning approaches. At the WMT 2014, BLEU will be delivered to the model [10]. Elements and other prominent areas of the image are a lot better natural foundation of attention to create more human-like descriptions and inquiry responses. Researchers offer a joint visual recognition mechanism from the bottom up to the top down. The findings on the MSCOCO [10] production servers provide a new cutting edge for the captioning challenge. To summarize, the paper will contribute to these factors proceeding ahead:

- We propose a novel architecture of transformer based implementation for image captioning with attention mechanism. It will be able to tailor the complex image region structure of image captioning [10].
- The model architecture will be designed using the ResNet-101 [7] for feature extraction from images.
- We will perform the image depiction by using the train set and validation set of the BanglaLekhaImageCaptions dataset [12].
- On the BanglaLekhaImageCaptions dataset, we evaluate our approach both qualitatively and quantitatively.

2 Related Works

A rudimentary method to creating the image caption is to anticipate and connect words from image areas. Researchers built one of the earliest systems of picture subtitling in 1999. The underlying job of image translation was subsequently re-translated in 2002 which fails to correctly map an image to a word that entirely misses all kinds of interactions between the elements. With its introduction, most subtitling systems used RNNs as speech models for encoding visual input on one or more levels of a CNN [8]. Self-attention would be a technique that relates several places in a specific setting to calculate a sequence presentation. It is the backbone for the Extended Neural models GPU, ByteNet, and ConvS2S [10], all used as fundamental building blocks for deep convolutional neural network models. Dropping a constant stage, the average attention-weighted places are lowered at the cost of such effective resolution describe the steps.

In the field of image subscription, a search-based method was prominent. One proposed approach is to integrate the pictures and text in the same place in the vector to facilitate searches. Similar efforts include using a recurring visual representation of a photograph and using deep reinforcement education [23] for picture submission. Image extraction is the second collection of data in the BanglaLekha series [13] focused on pictures of solitary characteristics in Bangla. Specific direct techniques were also suggested, including the query extension method Yagcioglu et al. to cope with picture description issues [2]. They have the problem of not intuitively observing the items or activities in the image. In this study, we examine the process of developing techniques of picture description in past years and describe the essential framework and improvements. The first is the injection architecture, where the RNN is employed as a generator for the caption conditioning of the picture characteristics. The task of the decoder is to need future words with the absolute most excellent chance from all vocabulary words within the previously created title. Based on these fusion model accomplishments and the NLP CNN [7] success story, we propose an encoder-decoder-based model for Bengali Image Captioning based on the fusion architecture [22]. Contrary to the CNN used in our study, a pooling layer is used to capture essential and relevant information. We also compared our CNN with qualitative and quantitative data mixture model, and thus the CNN-LSTM [10] data-based mixture model was proposed. The CNN-based language encoder is used to improve the proposed model's overall performance. Introducing compassionate models in which the recurring connection for self-attention is discontinued gives unparalleled set-up and sequence modeling possibilities. Multi-modal structure of the sub-titling of images for specific architectures, differing from the need for a single modality. M2 Transformer with Meshed-Memory features [5] consists of self attention-based encoder. Tested on the COCO reference point, the "Karpathy" testing data [5], the model attained a state-of-the-art method in both individual and set configurations. Significantly, it exceeds current online test server suggestions and ranks first between the methods provided.

3 Methodology

We employed an Encoder-Decoder approach where the image extraction process was carried out in this research utilizing the ResNet-101 model. In earlier models, ResNets executed other functions, including object detection [13] and semantical picture segmentation [2]. In computer vision, experts have progressively substituted the conventional extractors for ResNet-101. The underlying mechanism for the efficiency of ResNets remains to be defined, though. ResNets are initially used to remove great

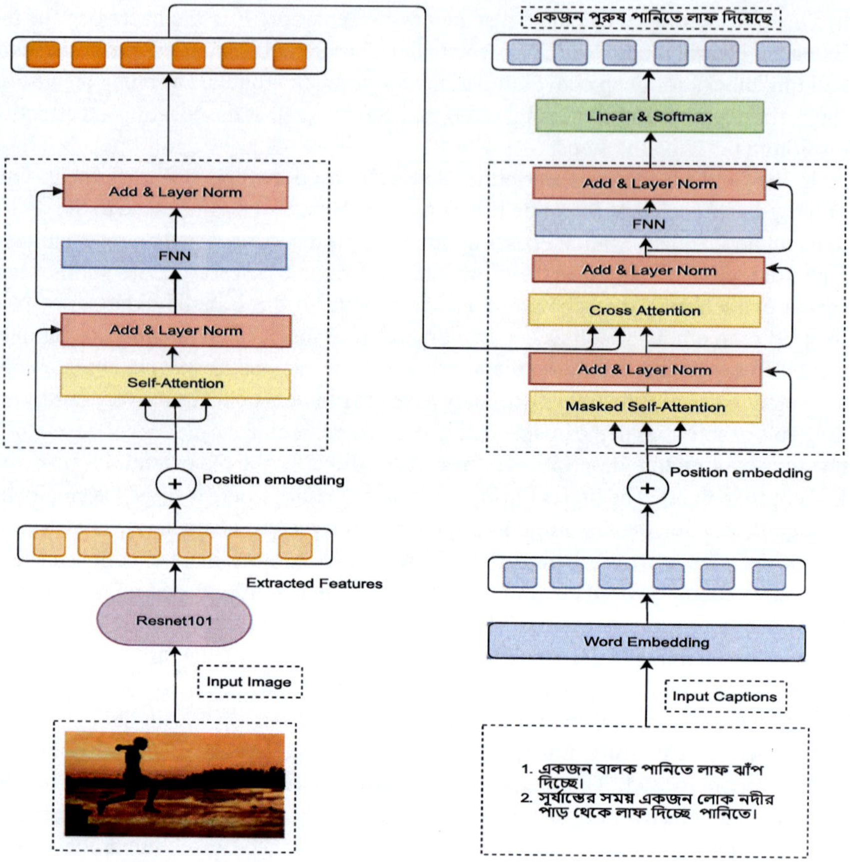

Fig. 1 The overview of the framework of our proposed model

features from the input data. Last but not least, the state sequence is employed to recognize the picture elements. The ResNet encoder comprises Ne stacked equivalent layers, each consisting of the sublayer and the positional forward-feed substitute. Then Positional encoding is being used to retain location coordinates as a shared component. The basic Transformer encoding methodology, [21] employs sinusoidal oscillations to fill out the weighted network model of positioning encoding. An attention layer ties all places to a predetermined sequential number of stages, whereas a recurring layer involves sequential operations. Auto-attention might be limited to investigate the input data focused on the position of the respective component. The proposed image captioning algorithm is illustrated in Fig. 1 as an overview. It is comprised of two parts: an encoder and a decoder.

3.1 Encoder

All input sequences must be translated into a consistent abstract representation that comprises all the learned information for this type of sequence by the Encoder layer. The Encoder consists of two sub-modules: multi-headed attention and another is a completely linked network. Layer normalization creates multiple residual connections between the two sublayers (see Fig. 2).

3.1.1 Feature Extraction

As an image feature extractor, we utilized the pre-trained ResNet-101 [3]. The ImageNet dataset [18] was used to train it. Pattern recognition has always been a strong suit for neural networks with several layers. There are overfitting problems, and they are difficult to optimize. Residual CNNs are built with fast connections between layers in mind. Identity mapping is performed via these relationships. ResNets are simple to tune, and their implementation improves as the network depth grows. We discarded the ResNet-101's final output layer, which contains the picture classification output, and only used the encoded image characteristics created by the hidden layers.

3.1.2 Positional Encoding

We must provide data about the relative or absolute positions of the tokens in the series for the model to use the sequence order because ours lacks recurrence and convolution. To accomplish so, we add "positional threescore encodings" to the bottoms of the input embeddings of the encoder and decoder stacks. There are a variety of learned and fixed positional encodings from which to choose. We utilize sine and cosine functions of various frequencies in this work: Let t stand for the

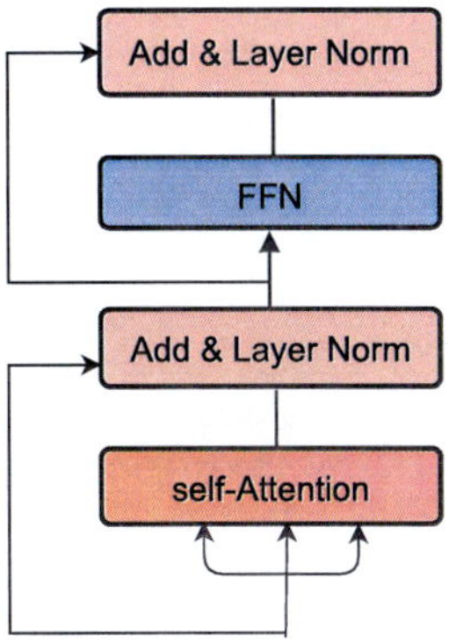

Fig. 2 Encoder architecture

desired location in an input sentence, be the encoding that corresponds to it, and d stand for the encoding dimension, $\overrightarrow{P_t} \in \mathbb{R}^d$ be its corresponding encoding, and d be the encoding dimension (where $d \equiv 20$) then $f : \mathbb{N} \rightarrow \mathbb{R}^d$ will be the function that produces the output vector $\overrightarrow{P_t}$ and it defined as follows:

$$\overrightarrow{P_t}^{(i)} = f(t)^{(i)} \quad (1)$$

Here, the limit is between $\sin(w_k.t)$, **if** $i = 2k$ *and* $cos(w_k.t)$, **if** $i = 2k + 1$ where

$$\omega_k = \frac{1}{10000^{2k/d}} \quad (2)$$

The frequencies decrease along the vector dimension, as can be seen from the function description. As a result, the wavelengths follow a geometric progression from 2π to $10000{\cdot}2\pi$ on the wavelengths. The positional embedding pt may alternatively be thought of as a vector containing pairs of sines and cosines positional $embedding \rightarrow pt$ for each frequency (note that d is divisible by 2).

3.1.3 Multi-Headed Attention

Multi-headed attention in the encoder uses a unique attention technique called self-attention model. Self-attention, in our case, allows models to connect each pixel in the input three-three to other pixels (see Fig. 3). Where the Query, Key, and Value Vectors are Q, K, and V, respectively. To achieve self-attention, we feed the input into three distinct, ultimately linked layers to produce the query, key, and value vectors.

Fig. 3 Multi-Headed attention mechanism

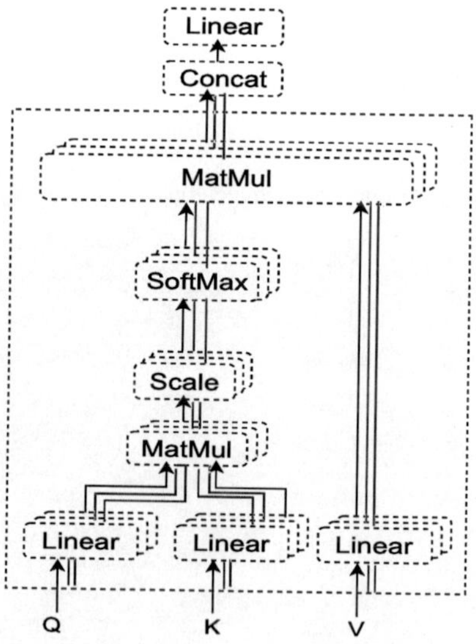

3.1.4 Dot Product of Query and Key

The questions and keys are multiplied by a dot product matrix to produce a scoring matrix after passing the query, key, and value vectors through a linear layer. The scoring matrix indicates how much importance a pixel should obtain in comparison to other pixels. As a result, each pixel in the time step will be assigned a score related to the scores of other pixels in the time step. The higher the score, the greater the concentration.

3.1.5 Scaling Down the Attention Scores

After that, the scores are scaled-down by dividing them by the square root of the query and critical dimensions. This score allows for more steady gradients because multiplying numbers can have explosive results. After that, the scores are scaled-down by dividing them by the square root of the query and critical dimensions. This score allows for more steady gradients because multiplying numbers can have explosive results.

3.1.6 Softmax of the Scaled Scores Multiply Softmax Output with Value Vector

The softmax of the scaled score is then used to determine the attention weights, yielding probability values ranging from 0 to 1. Your higher scores rise while your lower scores decline when you do a softmax. It makes choosing which pixel to focus on for the model much more accessible.

$$Z = softmax(\frac{Q * K^T}{\sqrt{d_k}}) * V \tag{3}$$

where Z is the self-attention matrix, Q is the query matrix, K is the fundamental matrix, V is the value matrix, and dk is the primary matrix's dimension. [21] improved the self-attention layer by incorporating a "Multi-Head" attention mechanism. The self-attention computation is implemented eight times with various weight matrices in Multi-Head attention.

3.1.7 The Residual Connections, Layer Normalization, and Feed Forward Network

The residual connection is formed when the multi-headed attention output vector is cumulated to the original positional input embedding. The output of the residual link is normalized at the layer level before being projected for further processing using a pointwise feed-forward network. A ReLU activation separates two linear layers in the pointwise feed-forward network. The output is then normalized by adding it to the input of the pointwise feed-forward network. By allowing gradients to pass straight through the networks, the remaining links improve network training. The grid is stabilized using layer normalizations, resulting in a considerable reduction in training time. The pointwise feed-forward layer is used to project better possible attention outcomes.

$$FFN(x) = FC_2(Dropout(ReLU(FC_1(x)))) \tag{4}$$

A sublayer connection in each sublayer is made up of a residual link proceeded by layer normalization.

$$x^{out} = LayerNorm(x^{in} + Sublayer(x^{in})) \tag{5}$$

where x^{in}, x^{out} are the input and output of one sublayer, feed-forward, and the sublayer can be the attention feed-forward-forward layer.

3.2 Decoder

On the decoder side, we combine the word embedding features with sinusoid positional embedding and use both the extra findings and encoder output features as input. The decoder is made up of Nd identical layers, each of which has a mask multi-head self-attention sublayer and a multi-head cross attention sublayer, as well as a positional feed-forward sublayer in that order. A linear layer with an output dimension identical to the vocabulary size uses the output feature throughout the last decoder layer to figure out the meaning of the next word. Given a ground truth sentence $y^*_{1:T}$ and the prediction y^*_t of the captioning model with parameters Θ, we minimize the subsequent cross-entropy loss:

$$L_{XE}(\Theta) = -\sum_{t=1}^{T} log(p_\Theta(y^*_t | y^*_{1:t-1})) \tag{6}$$

We used the sahajBERT model from the Hugging Face library for word embedding in the decoding stage. Like other encoding processes, we have used multi-head self-attention sublayer, multi-head cross attention sublayer, and positional feed-forward sublayer in the decoding stage to finetune our model.

4 Experiments

4.1 Datasets

This part mainly focuses on methods for analyzing open-source datasets and generated sentences in this domain, where a good dataset might help an algorithm or model perform better. We have tested and trained our model on the BanglaLekhaImageCaptions dataset [12].There are 9,154 images in the dataset presented are collected from the public domain. Besides, it is relevant to Bengali culture to some extent. While this collected data is small in comparison to other datasets where the prominent image captioning datasets such as MSCOCO [4], PASCAL 1K [17], Flickr8k [9], Flickr30k [15] contain a strong western cultural bias, with annotations written in English. To train our model, we split the data set into three parts: train, validation, and test where we have used 2 captions per image. We used 7154 x 2 images for the training. 1000 images for validation. And the remaining 1000 images were used during testing.

4.2 Data Preprocessinng

A considerable amount of human bias exists in the BanglaLekhaImageCaptions dataset [12]. Any model's capacity to describe non-human subjects is hampered because of this bias. In certain circumstances, the captions are not as descriptive as they should be, resulting in inaccurate training and evaluation of any model. For this reason, we have done some preprocessing to ensure that our dataset is noise-free and that it can perform well. Even though the Hugging Face library has a large number of pre-trained Bangla models and for preprocessing we have used sahajBERT [19] which is a pre-trained language model that consists of

- A tokenizer created specifically for Bengali and
- An ALBERT architecture that has been cooperatively pre-trained on a dump of Wikipedia in Bengali and the Bengali section of OSCAR.

By fine-tuning their pre-trained models three times on two downstream tasks in Bengali. Besides, we assess the quality of the sahajBERT model and two other model benchmarks (XLM-R-large and IndicBert).

5 Evaluation

5.1 Quantitative Analysis

For Quantitative results, there are five evaluation indexes, namely, BLEU, METEOR, ROUGE, CIDEr, and SPICE to evaluate the models' predicted caption. Among all

Table 1 Comparing the performance of various models quantitatively

Model	BLEU-1	BLEU-2	BLEU-3	BLEU-4	METEOR
Our proposed model	**0.694**	**0.580**	**0.505**	**2.22e-308**	**0.337**
CNN + ResNet-50 [merged] [10]	0.651	0.426	0.278	0.175	0.297
CNN + LSTM [mixture] [20]	0.632	0.414	0.269	0.168	0.291
CNN + Bi-LSTM [inject] [16]	0.619	0.403	0.261	0.163	0.296

five indicators, BLEU [14] is the most often used criterion for evaluation. BLEU considers translation length, word order, and word choice when determining how near a machine translation is to a human reference translation. It is used for image captioning, machine translation, etc. METEOR [6] is also used to evaluate machine translation, in which the model's generated translation is compared to the reference translation. It is intended to address some of the issues with BLEU.

Here, higher metric results suggested superior performance in the BLEU and METEOR indexes. In Table 1, the scores have been highlighted with the highest accuracy in boldface where a comparative study among our model, CNN language encoder with a merged-based architecture [10], LSTM language encoder with a mixture-based architecture [20], and Bi-LSTM language encoder with an inject-based architecture [16] can be found here. Above all models used the same BanglaLekhaImageCaptions dataset [12] through their presented methods which were different compared to our proposed model.

As can be seen, our proposed model achieved better results than the traditional architecture. It is also worth mentioning that the BLEU and METEOR scores are higher, indicating that our captions are of higher quality.

5.2 *Qualitative Analysis*

Figure 4 depicts a qualitative example of our model's performance. While the assessment metrics provide a numerical indication of the validity of the captions, any result can be misinterpreted. Detecting subtle differences in the generated captions compared to the original human language description can be evaluated via qualitative analysis. This concept will be bolstered even more by the qualitative assessment offered below.

For non-native Bengali speakers, we provide the predicted Bengali caption and the associated English Translated caption. Our algorithm outperformed the competitive

192.168.0.0
Original caption 1(Bangla): দুইজন মানুষ সাইকেল চালাচ্ছে।
(Two people are cycling.)
Original caption 2(Bangla): দুইজন ছেলে দুইটা সাইকেল চালিয়ে রাস্তা দিয়ে যাচ্ছে।
(Two boys are riding two bicycles on the road.)
Predicted caption(Bangla):দুইজন পুরুষ সাইকেল চালিয়ে যাচ্ছে।
(Two men are riding bicycles.)

BLEU 1: 0.882
BLEU 2: 0.833
BLEU 3: 0.795
BLEU 4: 1.508e-231

192.168.0.0
Original caption 1(Bangla): বৃষ্টিতে একজন পুরুষ রিক্সা টানছে ।
(A man pulls a rickshaw in the rain.)
Original caption 2(Bangla): বৃষ্টির মাঝে পুরুষ যাত্রী নিয়ে রিক্সা চালিয়ে যাচ্ছেন একজন পুরুষ।
(A man is driving a rickshaw with male a passenger in the rain.)
Predicted caption(Bangla):একজন পুরুষ রিক্সা চালিয়ে যাচ্ছেন।
(A man is driving a rickshaw)

BLEU 1: 0.969
BLEU 2: 0.906
BLEU 3: 0.870
BLEU 4: 2.225e-308

192.168.0.0
Original caption 1(Bangla): পানিতে একটি শিশু আছে।
(There is a baby in the water.)
Original caption 2(Bangla): লাল-সবুজ শাড়ি পরে হাতে একটা ফুলের ঝুড়ি নিয়ে একটা বাচ্চা মেয়ে পানিতে বসে হাত ভেজাচ্ছে।
(A baby girl in a red-green sari with a flower basket in her hand is sitting in the water and getting her hands wet.)
Predicted caption(Bangla):একজন ছোট মেয়ে পানিতে দাড়িয়ে পানি ছিটা দিচ্ছেন।
(A little girl is standing in the water and sprinkling water.)

BLEU 1: 0.755
BLEU 2: 0.395
BLEU 3: 0.215
BLEU 4: 2.225e-308

Original caption 1(Bangla): একজন রিকশাচালক একটা দেয়াল লিখন পড়তে পড়তে যাচ্ছে।
(A rickshaw puller is going and reading graffiti.)
Original caption 2(Bangla): একজন পুরুষ আছে। দেয়ালে ছবি আঁকা আছে।
(There is a man. There are pictures on the wall.)
Predicted caption(Bangla): একজন পুরুষ দাড়িয়ে দেয়ালে চিত্রকর্ম দেখছে।
(A man is standing and looking at the paintings on the wall.)

BLEU 1: 0.664
BLEU 2: 0.382
BLEU 3: 0.282
BLEU 4: 2.225e-308

Original caption 1(Bangla): গাছের ডালে ডালে অনেকগুলো পাখি আছে।
(There are many birds on the branches of the tree.)
Original caption 2(Bangla): একটা গাছের ডালের মধ্যে এক ঝাঁক কাক বসে আছে।
(A flock of crows is sitting in the branches of a tree.)
Predicted caption(Bangla): গাছে কয়েকটি পাখি আছে।
(There are a few birds in the tree.)

BLEU 1: 0.526
BLEU 2: 0.441
BLEU 3: 0.289
BLEU 4: 2.225e-308

Fig. 4 Qualitative examples of Bengali captions generated by our proposed model with untouched captions and BLEU score edge not only in terms of scoring but also in terms of caption quality. In comparison to other models, our model correctly identified the most appropriate caption as well as very well-detailed.

6 Limitation and Discussion

In our work, we use CNN-transformer base encoder-decoder architecture for image captioning. However, we are using CNN for feature extraction but the main architecture is a transformer. As the mechanism of the transformer, it requires a high volume of datasets. The amount of our dataset is good enough but not perfect for transformer architecture. As we mentioned, we have collected our dataset from [12]. This is an open-source dataset, and the caption quality is not that good as compared to the English [4] dataset available online. As a result our model sometimes faces barriers to counting objects from images. Second, we were planning to add more combinations to our model, such as multi Head, multi Layer. But due to GPU resources limitations, we were unable to add that combination to our model. Nevertheless, in our next work, we will try to combine our current dataset with the English [4] dataset. Afterward, we will try to increase our GPU resource power and will add more combinations to generate more accurate results.

7 Conclusion and Future Work

This paper introduces a novel Transformer-based architecture using a Deep Neural Network for automatically developing captions in Bengali from an input image, with a comprehensive test on the BanglaLekhaImageCaptions dataset, which yielded excellent results, justifies the feasibility of our proposed approach. Besides, the BLEU and METEOR scores were used to evaluate the proposed architecture's performance and the results, which were analyzed using both quantitative and qualitative methods. The result shows that image captioning in the Bengali language can be improved shortly. Furthermore, we will aim to adapt the visual attention and transformer models instead of ResNet-101 for improved feature extraction and more precise captions. We anticipate that our approach will encourage others to develop more efficient transformer-based architectures to aid image captioning as well as other computer vision tasks requiring relational attention.

References

1. Anderson, P., He, X., Buehler, C., Teney, D., Johnson, M., Gould, S., Zhang, L.: Bottom-up and top-down attention for image captioning and visual question answering. In: Proceedings of the IEEE Conference on Computer Vision and Pattern Recognition, pp. 6077–6086 (2018)
2. Bernardi, R., Cakici, R., Elliott, D., Erdem, A., Erdem, E., Ikizler-Cinbis, N., Keller, F., Muscat, A., Plank, B.: Automatic description generation from images: a survey of models, datasets, and evaluation measures. J. Artif. Intell. Res. **55**, 409–442 (2016)
3. Chen, H., Wang, Y., Guo, T., Xu, C., Deng, Y., Liu, Z., Ma, S., Xu, C., Xu, C., Gao, W.: Pre-trained image processing transformer. In: Proceedings of the IEEE/CVF Conference on Computer Vision and Pattern Recognition, pp. 12299–12310 (2021)
4. Chen, X., Fang, H., Lin, T.Y., Vedantam, R., Gupta, S., Dollár, P., Zitnick, C.L.: Microsoft coco captions: data collection and evaluation server. arXiv preprint arXiv:1504.00325 (2015)
5. Cornia, M., Stefanini, M., Baraldi, L., Cucchiara, R.: Meshed-memory transformer for image captioning. In: Proceedings of the IEEE/CVF Conference on Computer Vision and Pattern Recognition, pp. 10578–10587 (2020)
6. Denkowski, M., Lavie, A.: Meteor universal: language specific translation evaluation for any target language. In: Proceedings of the Ninth Workshop on Statistical Machine Translation, pp. 376–380 (2014)
7. Ghosal, P., Nandanwar, L., Kanchan, S., Bhadra, A., Chakraborty, J., Nandi, D.: Brain tumor classification using resnet-101 based squeeze and excitation deep neural network. In: 2019 Second International Conference on Advanced Computational and Communication Paradigms (ICACCP), pp. 1–6. IEEE (2019)
8. Herdade, S., Kappeler, A., Boakye, K., Soares, J.: Image captioning: transforming objects into words. arXiv preprint arXiv:1906.05963 (2019)
9. Hodosh, M., Young, P., Hockenmaier, J.: Framing image description as a ranking task: data, models and evaluation metrics. J. Artif. Intell. Res. **47**, 853–899 (2013)
10. Khan, M.F., Sadiq-Ur-Rahman, S., Islam, M.S.: Improved bengali image captioning via deep convolutional neural network based encoder-decoder model. In: Proceedings of International Joint Conference on Advances in Computational Intelligence, pp. 217–229. Springer (2021)
11. Liu, W., Chen, S., Guo, L., Zhu, X., Liu, J.: Cptr: Full transformer network for image captioning. arXiv preprint arXiv:2101.10804 (2021)
12. Mansoor, N., Kamal, A.H., Mohammed, N., Momen, S., Rahman, M.M.: Banglalekha image captions, Mendeley Data, V2 (2019). https://doi.org/10.17632/rxxch9vw59.2
13. Naim, F.A.: Bangla handwritten digit recognition based on different pixel matrices. In: International Conference on Innovative Computing and Communications, pp. 325–341. Springer (2022)
14. Papineni, K., Roukos, S., Ward, T., Zhu, W.J.: Bleu: a method for automatic evaluation of machine translation. In: Proceedings of the 40th annual meeting of the Association for Computational Linguistics, pp. 311–318 (2002)
15. Plummer, B.A., Wang, L., Cervantes, C.M., Caicedo, J.C., Hockenmaier, J., Lazebnik, S.: Flickr30k entities: collecting region-to-phrase correspondences for richer image-to-sentence models. In: Proceedings of the IEEE International Conference on Computer Vision, pp. 2641–2649 (2015)
16. Rahman, M., Mohammed, N., Mansoor, N., Momen, S.: Chittron: an automatic bangla image captioning system. Proc. Comput. Sci. **154**, 636–642 (2019)
17. Rashtchian, C., Young, P., Hodosh, M., Hockenmaier, J.: Collecting image annotations using amazon's mechanical turk. In: Proceedings of the NAACL HLT 2010 Workshop on Creating Speech and Language Data with Amazon's Mechanical Turk, pp. 139–147 (2010)
18. Rennie, S.J., Marcheret, E., Mroueh, Y., Ross, J., Goel, V.: Self-critical sequence training for image captioning. In: Proceedings of the IEEE Conference on Computer Vision and Pattern Recognition, pp. 7008–7024 (2017)
19. Saifullah, K.: neuropark/sahajbert (2021), https://github.com/khalidsaifullaah/BERTify

20. Tanti, M., Gatt, A., Camilleri, K.P.: What is the role of recurrent neural networks (rnns) in an image caption generator? arXiv preprint arXiv:1708.02043 (2017)
21. Vaswani, A., Shazeer, N., Parmar, N., Uszkoreit, J., Jones, L., Gomez, A.N., Kaiser, Ł., Polosukhin, I.: Attention is all you need. In: Advances in Neural Information Processing Systems, pp. 5998–6008 (2017)
22. Wang, H., Zhang, Y., Yu, X.: An overview of image caption generation methods. Comput. Intell. Neurosci. **2020** (2020)
23. Wu, X., Jackson, D.J., Chen, H.C.: Novel fractal image-encoding algorithm based on a full-binary-tree searchless iterated function system. Opt. Eng. **44**(10), 107002 (2005)

A Decision Support System (DSS) for Interview-Based Personnel Selection Using Fuzzy TOPSIS Method

Md. Muzahidul Islam Rahi, A. K. M. Amanat Ullah, and Dr. Md. Golam Rabiul Alam

Abstract Throughout human history, we have looked for ways to enrich our judgment. Human nature has always been biased, especially when it comes to the selection process involving multiple decision-makers. This becomes even more complex when it is about an interview and also for personnel selection or personnel ranking. Due to the fast expansion, we require skillful, professional human resources, and the ability to demolish nepotism or any form of biasedness. This paper presents a modified Fuzzy TOPSIS approach to interviewing mechanism where we could select or rank personnel from multiple decision-makers for Multi-Criteria Decision-Making (MCDM) Mechanism. The idea of using multiple decision-makers solves the problems of decision-making from various opinions. Finally, we collect our own dataset from scholars and lecturers from reputed universities in Bangladesh and Fuzzy TOPSIS is applied to the linguistic set to derive a set with a triangular fuzzy score for ranking between the responses. We apply the intuitionistic fuzzy set mechanism to form a triangular number for our member function.

Keywords Multi-criteria decision-making (MCDM) · Fuzzy TOPSIS · Linguistic set · Triangular fuzzy score · Decision-maker · Personnel selection · personnel ranking · Intuitionistic fuzzy sets · Member function

Md. M. I. Rahi (✉) · Dr. Md. G. R. Alam
Brac University, Dhaka, Bangladesh
e-mail: md.muzahidul.islam.rahi@g.bracu.ac.bd

A. K. M. A. Ullah
University of British, Okanagan, BC, Canada

S. Hossain et al. (eds.), *Proceedings of International Conference on Fourth Industrial Revolution and Beyond 2021*, Lecture Notes in Networks and Systems 437,
https://doi.org/10.1007/978-981-19-2445-3_45

1 Introduction

1.1 Research Goals

The aim of this paper is to build a grading mechanism for interviews that can be used for personnel selection using Fuzzy TOPSIS. This will take a fuzzy linguistic set and by applying the Fuzzy Technique of Order Preference Similarity to the Ideal Solution (TOPSIS) mechanism, we would like to rank the interviewees or alternatives according to specific criteria. We see how these methods can be used in evaluating, assessing, selecting, and ranking alternatives across diverse platforms [20].

1.2 Intuitionistic Fuzzy Set (IFS)

We used the concept of intuitionistic fuzzy set (IFS) [2] to determine the linguistic set to fuzzy set. An interval-valued intuitionistic fuzzy set (IVIFS) is used to determine the weights of linguistic terms, later described in Fig. 4. We used the Triangular Fuzzy Number (TFN) for computing the linguistic values of IVIFS for the linguistic term [21].

1.3 Fuzzy Weights Estimation

We have seen how the Fuzzy weights estimation method based on the linguistic expression of criteria relevance is used in [4]. They have used Bellman's approach [14] to a fuzzy solution problem to estimate the membership function in a triangular fuzzy score [6]. In this paper, we also come across multiple objective decision-making procedures where they have used different computationally effective approaches [7].

1.4 Hierarchical Structure of the Methodology

Initially, we would have *n* number of decision-makers or in this case interviewers. There will be certain criteria, upon which students will be judged (e.g., Analytical Skills, Time taken to solve, etc.). The interviewee or student can also vary. The fuzzy set will be comprised of levels such as **Low, Very Low, Average, High, Very High**. Figure 1 demonstrates the block diagram of our System model. The following steps need to be followed:

1. Convert the Linguistic Fuzzy set by applying weights to a triangular Fuzzy Score.
2. Form a Combined Decision Matrix by assigning weights to Judge criteria.

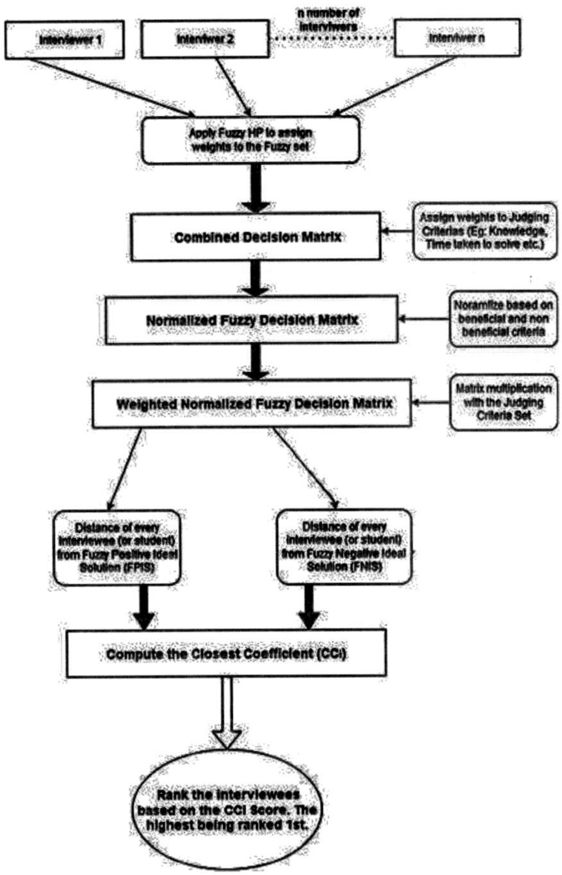

Fig. 1 Block diagram of system model

3. Form the Normalized Fuzzy Decision Matrix by normalizing based on beneficial and non-beneficial (cost) criteria.
4. Form the Weighted Normalized Fuzzy Decision Matrix to multiply the whole matrix of step 3 with the Judging Criteria Set.
5. Calculate the distance of every interviewee from the Fuzzy Positive Ideal Solution (FPIS) and Fuzzy Negative Ideal Solution (FNIS).
6. Compute the Closest Coefficient (CC_i) for every interviewee.
7. Rank the alternatives according to (CC_i). The highest being ranked first.

2 Related Works

We see Multi-Criteria Decision-Making (MCDM) being used to determine a solution between different responses. When implemented, TOPSIS may be a very active MCDM mechanism that was used to choose vehicles and solve vehicle routing problems (VRP) [1]. This paper acts as a shooting point to use TOPSIS and also to use TOPSIS in other areas. Another paper describes all the key points of using TOPSIS and also applies it as a linguistic set [2]. One of the main benefits of choosing a fuzzy set over a numeric set is that it is more readily available and has a much simpler architecture. It is usually more accurate and given the fact that we use linguistic languages everywhere, it is easier to collect data and to manipulate it as well. We have used the triangular fuzzy scoring of linguistic set [4]. Furthermore, we see a comparable method of interview selection in Fathi [3], but our system has employed additional criteria, and our model may be superior than Aghaie [5] in terms of majority opinion-based agreement.

Our motivation for this paper comes from the VRP discussed in [1]. They used a type 2 fuzzy logic [8], where they provide additional degrees of freedom to represent real-life scenarios [9]. Our selection process is quite similar to the VRP [1] method, which uses fuzzy type-2 logic to generalize the commonly used type -1 fuzzy logic [10]. We also see the use of MCDM problems being discussed and its application for finding the best alternative being very complex [18, 19]. The selection procedure used on [3] uses similar Fuzzy TOPSIS methods based on interval-valued fuzzy sets [11]. However, they have used only 12 criteria and do not classify their minimally-based weight selection [12]. We also see that they have used triangular fuzzy based on alpha level sets for risk management [13]. As we have seen, the majority opinion aggregator can be used in MCDM problems [5]. This is relevant to our research work because they select under group decision-making which allows multiple decision-makers [15]. The multiple attributes in decision-making help the MCDM model to remain optimal and efficient [16, 17].

In Freeman [23], we see a selection model where the index system was used. However, this TOPSIS system was limited to Green Supply selection and did not include the opportunity to allow for multiple decision-makers. A similar approach was also visualized in Doka [24], where a fuzzy set was also introduced while a DSS was designed to select Human Resources. Subsequently, this also did not include multiple decision-makers. Previously, we see in Yue [25], the impact of using decision-makers has been discussed, but their model was limited to only non Fuzzy sets.

3 Dataset Collection

3.1 Methods of Dataset Collection

We collected this dataset with the help of Google forms. In this dataset, we see our dataset is divided into three major parts:

1. Interviewee Section: This section consisted of a google form for interviewees (or students). All of these interviewees were university graduates from reputed universities across Bangladesh who have faced many interviewees and are accustomed to the set of questions. For the purpose of fairness, it was decided that all of these interviewees will face the same interviewers. Keeping in mind the interviewer's attention span and lack of resources, the number of interviewees was decided to be 4. In this way, we get better evaluation performance from the interviewers, as increasing the number of interviewees meant they needed to go through more questions which might reduce their uniformity while evaluating. The interviewees were given 7 questions to evaluate.

2. Interviewer Section: In this part of the collection, we passed another Google form to the interviewers which included responses from all 4 interviewees. The interviewers were asked to evaluate 7 questions based on the criteria. For validating our research, all the interviewers are lecturers or faculty members from BRAC University with experience as an interviewer. All of the decision-makers (interviewers) were given equal weight. There were 4 decision-makers for this dataset.

3. Rating Section: Another Google form was made to rate or have weights between the 18 criteria. The decision-makers and also others (lecturer or faculty members), with interviewing experience were asked to determine the weights between the criteria. An average was done to determine the weight for the 18 criteria.

3.2 Questions and Criteria

Each responder (interviewee) was given the seven questions, which were then utilized to assess 18 criteria. These are further described in Fig. 2. The questions were:

1. **[Analytical Skills]** If you had to complete a task, with no given information and you need to choose between two options. How did/will you do it? How do you measure the pros and cons?
2. **[Responsibility, Biasedness]** During a live presentation to key stakeholders, you spot a mistake in your manager's report, but your manager is not at the presentation. How do you handle this?
3. **[Team Spirit, Flexibility, Openness to Criticism]** According to your thinking, your manager gives you an impossible task to complete. How would you do it, given that one of your team members is also sick?
4. **[Problem Solving Skills, Reacting to Stress]** How do you know when to solve a problem on your own or to ask for help?
5. **[Leadership, Teamwork, Communication, Time taken]** How would you react if a team leader encouraged competition between team members instead of collaboration? Given that team members are not cooperating and the deadline is approaching.
6. **[Decision-Making, Organizational Skills, Managerial Skills, Work Ethics]** If you discovered your supervisor was breaking the company's code of conduct, what would you do?
7. **[Integrity, Social Responsibility]** Have you ever faced an ethical dilemma at work? If so, what was the issue, and what did you do?

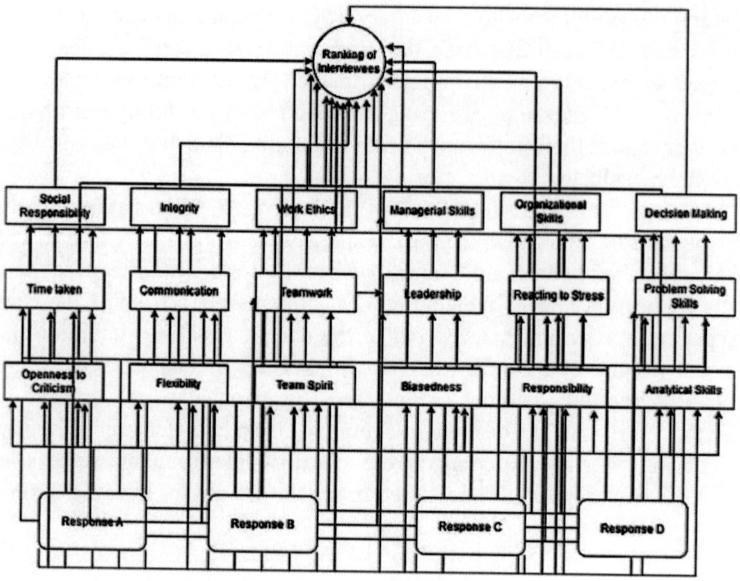

Fig. 2 Integration of the 18 criteria to define our model. Response A, B, C, D are the 4 interviewees. As you can see every criterion has incoming arrows from the responses, and while ranking, we see all the 18 arrows linking the criteria and the ranking process

3.3 Dataset

The interviewers ranged their evaluation of each criterion from a range of:

- Very High
- High
- Average
- Low
- Very Low

Figure 3 shows the output of our initial data collection in linguistic form from all 4 decision-makers.

4 Conversion to a Fuzzy Set

4.1 Applying Weights to Criteria

As we have already mentioned that a Google form was assigned to put weight to the criteria. The assigned weights are discussed in Table 1.

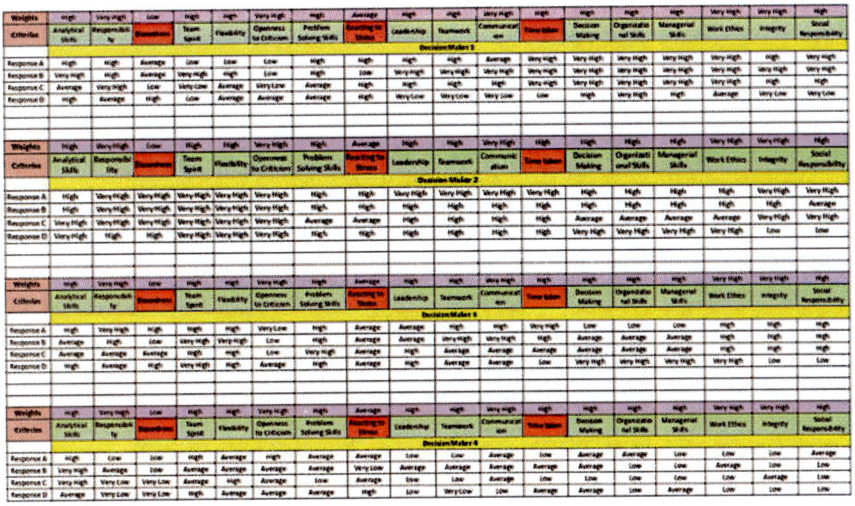

Fig. 3 Our dataset consists of the 4 decision-makers evaluating 4 responses from a range of Very High to Very Low based on the given 18 criteria in a linguistic set

Table 1 The criteria and the associated linguistic priorities

Criteria	Priority
Analytical skills	High
Responsibility	Very high
Biasedness	Low
Team spirit	High
Flexibility	High
Openness to criticism	Very high
Problem solving skills	High
Reacting to stress	Average
Leadership	High
Teamwork	High
Communication	Very high
Time taken	High
Decision making	High
Organizational	Skills high
Work ethics	Very high
Integrity	Very high
Social responsibility	High

4.2 Applying Triangular Fuzzy Score and It's Membership Function

The range of the indicated criteria on which the interviewees will be evaluated is provided in linguistic form (Very High to Very Low). To use it in TOPSIS, we apply a triangular fuzzy score to these. Furthermore, in Table 2, we see the numerical weights assigned to our range in intuitionistic fuzzy triangular format from a range of 0–1 [22]. The Triangular Fuzzy Number is defined by the Membership Function, $x(t)$, which is derived in (1) [2]. This is further described in the member function diagram in Fig. 4.

$$x(t) = \begin{cases} 0 & \text{if} \quad t < a \\ \frac{t-a}{b-a} & \text{if} \ a \le t < b \\ \frac{c-t}{c-b} & \text{if} \ b \le t < c \\ 0 & \text{if} \quad t > c \end{cases}, \quad \text{where } a \le b \le c \tag{1}$$

Hence, we find the numeric fuzzy set after we apply the weights to our previous linguistic set.

Table 2 Numerical fuzzy triangular weights assigned linguistic range

Range	Triangular fuzzy number
Very low	0.1, 0.1, 0.3
Low	0.1,0.3,0.5
Average	0.3,0.5,0.7
High	0.5,0.7,0.9
Very high	0.7,0.9,0.9

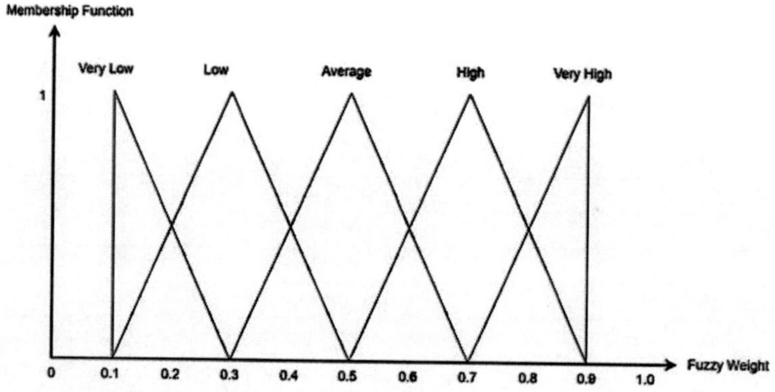

Fig. 4 Member function of the triangular fuzzy weights

5 Fuzzy TOPSIS Analysis and Results

Previously, we have used triangular fuzzy scoring to put weight in our linguistic dataset [2]. Our fuzzy score can be determined in (2)

$$\tilde{x}_{ij} = \left(a_{ij}, b_{ij}, c_{ij}\right) \tag{2}$$

5.1 Combined Matrix

As we had 4 decision-makers, we need to combine all those matrices into 1 combined matrix. Equation (3) explains how we came to have our combined matrix. According to Fuzzy TOPSIS, we take min of the first element, an average of the second element, and the maximum of the third. Figure 5 shows our combined matrix.

$$a_{ij} = \min_k \left\{a_{ij}^k\right\}, b_{ij} = \frac{1}{K}\sum_{k=1}^{K} b_{ij}^k, c_{ij} = \max_k \left\{c_{ij}^k\right\} \tag{3}$$

5.2 Normalized Fuzzy Matrix

We normalized our dataset to have a fixed range. Normalization was achieved by (4). Among our 18 criteria, there were three cost criteria: **Biasedness, Reacting to Stress, and Time Taken**. The other 15 criteria were beneficial.

$$\begin{aligned} \tilde{r}_{ij} &= \left(\frac{a_{ij}}{c_j^*}, \frac{b_{ij}}{c_j^*}, \frac{c_{ij}}{c_j^*}\right) \text{ and } c_j^* = \max_i \left\{c_{ij}\right\} \text{ (benefit criteria)} \\ \bar{r}_{ij} &= \left(\frac{a_j^-}{c_{ij}}, \frac{a_j^-}{b_{ij}}, \frac{a_j^-}{a_{ij}}\right) \text{ and } a_j^- = \min_i \left\{a_{ij}\right\} \text{ (cost criteria)} \end{aligned} \tag{4}$$

5.3 Weighted Normalized Fuzzy Matrix

As our criteria have different priorities, so we multiply each element of our normalized matrix with our criteria matrix. This is displayed in (5).

$$\widetilde{A_1} \otimes \widetilde{A_2} = (a_1, b_1, c_1) \otimes (a_2, b_2, c_2) = (a_1 \times a_2, b_1 \cdot b_2, c_1 * c_2) \tag{5}$$

Weights	0.5	0.7	0.9	0.7	0.9	0.9	0.1	0.3	0.5
Criteria	Analytical Skills			Responsibility			Biasedness		
Response A	0.277778	0.544444	0.9	0.077778	0.7	0.9	0.011111	0.05	0.5
Response B	0.166667	0.583333	0.9	0.233333	0.7	0.9	0.011111	0.06	0.5
Response C	0.166667	0.544444	0.9	0.077778	0.6	0.9	0.011111	0.066667	0.5
Response D	0.166667	0.544444	0.9	0.077778	0.45	0.9	0.011111	0.054545	0.5

	0.5	0.7	0.9	0.5	0.7	0.9	0.7	0.9	0.9
	Team Spirit			Flexibility			Openness to Criticism		
	0.055556	0.505556	0.9	0.055556	0.466667	0.9	0.077778	0.5	0.9
	0.166667	0.622222	0.9	0.166667	0.583333	0.9	0.077778	0.5	0.9
	0.055556	0.427778	0.9	0.166667	0.544444	0.9	0.077778	0.45	0.9
	0.055556	0.544444	0.9	0.166667	0.505556	0.9	0.233333	0.6	0.9

	0.5	0.7	0.9	0.3	0.5	0.7	0.5	0.7	0.9
	Problem Solving Skills			Reacting to Stress			Leadership		
	0.166667	0.505556	0.9	0.033333	0.083333	0.233333	0.055556	0.427778	0.9
	0.166667	0.505556	0.9	0.033333	0.125	0.7	0.166667	0.505556	0.9
	0.055556	0.427778	0.9	0.033333	0.090909	0.233333	0.055556	0.466667	0.9
	0.166667	0.466667	0.9	0.033333	0.076923	0.233333	0.055556	0.35	0.9

	0.5	0.7	0.9	0.7	0.9	0.9	0.5	0.7	0.9
	Teamwork			Communication			Time taken		
	0.055556	0.505556	0.9	0.233333	0.65	0.9	0.055556	0.093333	0.9
	0.166667	0.583333	0.9	0.233333	0.75	0.9	0.055556	0.1	0.3
	0.055556	0.427778	0.9	0.233333	0.6	0.9	0.055556	0.116667	0.9
	0.055556	0.272222	0.9	0.077778	0.4	0.9	0.055556	0.155556	0.9

	0.5	0.7	0.9	0.5	0.7	0.9	0.5	0.7	0.9
	Decision Making			Organizational Skills			Managerial Skills		
	0.055556	0.466667	0.9	0.055556	0.466667	0.9	0.055556	0.427778	0.9
	0.166667	0.505556	0.9	0.055556	0.466667	0.9	0.055556	0.466667	0.9
	0.055556	0.427778	0.9	0.055556	0.427778	0.9	0.055556	0.427778	0.9
	0.166667	0.583333	0.9	0.055556	0.583333	0.9	0.166667	0.583333	0.9

	0.7	0.9	0.9	0.7	0.9	0.9	0.5	0.7	0.9
	Work Ethics			Integrity			Social Responsibility		
	0.077778	0.7	0.9	0.077778	0.65	0.9	0.166667	0.583333	0.9
	0.233333	0.7	0.9	0.077778	0.65	0.9	0.055556	0.466667	0.9
	0.077778	0.6	0.9	0.233333	0.7	0.9	0.055556	0.505556	0.9
	0.077778	0.65	0.9	0.077778	0.25	0.5	0.055556	0.194444	0.5

Fig. 5 Combining the 4 decision-makers into one combined dataset

5.4 *Calculating A* for Fuzzy Positive Ideal Solution (FPIS) and A^- for Fuzzy Negative Ideal Solution (FNIS)*

For calculating the Fuzzy Positive Ideal Solution (FPIS), we need to find A*, which is a Euclidean distance from the maximum value. For calculating the Fuzzy Negative Ideal Solution (FNIS), we need to find A^-, which is a Euclidean distance from the minimum value. These are described in Eqs. (6) and (7), which demonstrate how to determine the Euclidean distance.

$$A^* = \left(\tilde{v}_1^*, \tilde{v}_2^*, \cdots, \tilde{v}_n^*\right), \text{ where } \tilde{v}_j^* = \max_i \left\{v_{ij}\right\} \\ A^- = \left(\tilde{v}_1^-, \tilde{v}_2^-, \cdots, \tilde{v}_n^-\right), \text{ where } \tilde{v}_j^- = \min_i \left\{v_{ij}\right\} \tag{6}$$

$$d(\tilde{x}, \tilde{y}) = \sqrt{\frac{1}{3}\left[(a_1 - a_2)^2 + (b_1 - b_2)^2 + (c_1 - c_2)^2\right]} \tag{7}$$

5.5 Ranking the Alternatives Based on Closest Coefficient (CC_i)

For the last part of our Fuzzy TOPSIS, we calculate the Closest Coefficient (CC_i), shown in (8). The results are displayed in Fig. 6. The higher the CC_i, the better ranking. In our model, Response B was ranked first among the four interviewees.

$$CC_i = \frac{d_i^-}{d_i^- + d_i^*} \tag{8}$$

5.6 Discussion

The mentioned model allows us to select personnel based on:

- A modified TOPSIS approach to interviewing mechanism
- Inclusion of Fuzzy set to allow for more diverse use
- Multiple decision-makers (DM) with different weights can be allocated
- This model can also be used for ranking or priority determination
- A Multi-Criteria Decision-Making (MCDM) Mechanism which can work for any other selection dataset.

Fig. 6 Ranking using the closest coefficient (CC_i)

	CcI	Rank
Response A	2.256974926	2
Response B	5.151507636	1
Response C	1.348702414	3
Response D	1.228828521	4

6 Limitations and Future Works

6.1 Limitations

Initially, we wanted to use our model in an existing dataset which would make our model credible. As we could not find such a model, we decided to work on our own acquired dataset. Furthermore, as there was no funding, we had to depend on the voluntary work of faculties and graduate students from BRAC University, Bangladesh. Hence, our data size is small and our response was restricted to 4.

6.2 Future Works

One thing we want to do is use type 2 fuzzy logic, which is a generalization of customary fuzzy logic (type 1) as invulnerability is not just restricted to the linguistic factors yet, in addition, is available in the meaning of the membership functions [10]. Also, we could use MCDM methods such as Vikor, AHP, etc., or even machine learning methods such as AdaBoost or GradientBoost in the case of decision-makers. Also, we can feed our system to a neural network or use transfer learning mechanisms once our dataset is significantly large and split into tests and trained to evaluate our system's accuracy.

7 Conclusion

The aim of this survey was to build an interview system that will select the most skillful, professional, and efficient workers. All the questions asked were descriptive and unique to everyone which meant there was no scope of biasedness from the decision-makers. We valued the 7 questions asked into 18 criteria to determine the linguistic set evaluated by 4 decision-makers. Moreover, we applied the concept of the intuitionistic fuzzy set to apply a triangular fuzzy number to our linguistic variables and hence determine the membership function. Furthermore, we used an empirical method to apply weights to our linguistic dataset. Eventually, we applied the Fuzzy TOPSIS mechanism to rank and select our personnel. In our ranking, Response B was ranked first.

References

1. Mehlawat, M.K., Gupta, P., Khaitan, A., Pedrycz, W.: A hybrid intelligent approach to integrated fuzzy multiple depot capacitated green vehicle routing problem with split delivery and

vehicle selection. IEEE Trans. Fuzzy Syst. **28**(6), 1155–1166 (2020). https://doi.org/10.1109/TFUZZ.2019.2946110
2. Nădăban, S., Dzitac, S., Dzitac, I.: Fuzzy topsis: a general view. Proc. Comput. Sci. **91**, 823–831 (2016)
3. Fathi, M.R., Matin, H.Z., Zarchi, M.K., Azizollahi, S.: The application of fuzzy TOPSIS approach to personnel selection for Padir Company Iran. J. Manag. Res. **3**(2), 1–14 (2011)
4. Borovička, A. (2014). Fuzzy weights estimation method based on the linguistic expression of criteria relevance
5. Aghaie, H., Shafieezadeh, S., Moshiri, B.: A new modified fuzzy TOPSIS for group decision making using fuzzy majority opinion based aggregation. In: 2011 19th Iranian Conference on Electrical Engineering, (pp. 1–6). IEEE (2011)
6. Gupta, M., Bhattacharjee, D.: Multi-objective problem in a fuzzy environment where resources are triangular fuzzy number. Eur. J. Sci. Res. **46**(1), 99–106 (2010)
7. Lai, Y.J., Hwang, C.L.: Fuzzy multiple objective decision making. In: Fuzzy Multiple Objective Decision Making, pp. 139–262. Springer, Berlin, Heidelberg (1994)
8. Bera, A.K., Jana, D.K., Banerjee, D., et al.: Supplier selection using extended IT2 fuzzy TOPSIS and IT2 fuzzy MOORA considering subjective and objective factors. Soft. Comput. **24**, 8899–8915 (2020). https://doi.org/10.1007/s00500-019-04419-z
9. Chen, S.M., Lee, L.W.: Fuzzy multiple attributes group decision-making based on the interval type-2 TOPSIS method. Expert Syst. Appl. **37**(4), 2790–2798 (2010)
10. Castillo, O., Melin, P.: 3 Type-2 fuzzy logic. In: Type-2 Fuzzy Logic: Theory and Applications. Studies in Fuzziness and Soft Computing, vol. 223. Springer, Berlin, Heidelberg (2007). https://doi.org/10.1007/978-3-540-76284-3-3
11. Ashtiani, B., Haghighirad, F., Makui, A., Ali Montazer, G.: Extension of fuzzy TOPSIS method based on interval-valued fuzzy sets. Appl. Soft Comput. **9**(2), 457–461 (2009)
12. Jessop, A.: Minimally biased weight determination in personnel selection. Eur. J. Oper. Res. **153**(2), 433–444 (2004)
13. Wang, Y.M., Elhag, T.M.: Fuzzy TOPSIS method based on alpha level sets with an application to bridge risk assessment. Expert Syst. Appl. **31**(2), 309–319 (2006)
14. Bellman, R.E., Zadeh, L.A.: Decision-making in a fuzzy environment. Manag. Sci. **17**(4), B–141 (1970)
15. Chen, C.T.: Extensions of the TOPSIS for group decision-making under fuzzy environment. Fuzzy Sets Syst. **114**(1), 1–9 (2000)
16. Tzeng, G.H., Huang, J.J.: Multiple Attribute Decision Making: Methods and Applications. CRC Press (2011)
17. Pasi, G., Yager, R.R.: Modelling the concept of fuzzy majority opinion. In: International Fuzzy Systems Association World Congress, pp. 143–150. Springer, Berlin, Heidelberg (2003)
18. Abdullah, L.: Fuzzy multi-criteria decision making and its applications: a brief review of category. Procedia. Soc. Behav. Sci. **97**, 131–136 (2013)
19. Aruldoss, M., Lakshmi, T.M., Venkatesan, V.P.: A survey on multi-criteria decision making methods and its applications. Am. J. Inf. Syst. **1**(1), 31–43 (2013)
20. Behzadian, M., Otaghsara, S.K., Yazdani, M., Ignatius, J.: A state-of-the-art survey of TOPSIS applications. Expert Syst. Appl. **39**(17), 13051–13069 (2012)
21. Zavadskas, E.K., Antucheviciene, J., Hajiagha, S.H.R., Hashemi, S.S.: Extension of weighted aggregated sum product assessment with interval-valued intuitionistic fuzzy numbers (WASPAS-IVIF). Appl. Soft Comput. **24**, 1013–1021 (2014)
22. Yang, T., Hung, C.C.: Multiple-attribute decision-making methods for plant layout design problem. Robot. Comput. Integr. Manufact. **23**(1), 126–137 (2007)
23. Freeman, J., Chen, T.: Green supplier selection using an AHP-Entropy-TOPSIS framework. Int. J. Supply Chain Manag. (2015)
24. Doka, K.M., Ahmad, F., Shamsuddin, S.N.W., Awang, W.S.W., Ghazali, N.: Integrated decision support system for human resource selection using topsis based models. Appl. Math. Sci. **9**(129), 6403–6414 (2015)
25. Yue, Z.: A method for group decision-making based on determining weights of decision makers using TOPSIS. Appl. Math. Model. **35**(4), 1926–1936 (2011)

Performance Analysis and Implementation of Pre-trained Model Using Transfer Learning on Bangla Document Clustering

Arefin Niam, Avijit Das, and Summit Haque

Abstract Natural Language Processing has ushered in a revolutionary improvement in making intelligent systems for various purposes using machine learning techniques to process human language. NLP tasks like machine translation, summarization, sentiment analysis, etc. have helped make these applications possible. The techniques and algorithms behind these tasks are changing fast with each improvement. Furthermore, scholars from all over the world use cutting-edge methodologies and approaches in a variety of languages. In this research, we explore the state-of-the-art NLP framework BERT that has been pre-trained in several languages including Bangla. We demonstrate the importance of contextual language representation in Bangla, analyze performance and implement the state-of-the-art BERT pre-trained language model on the Bangla Document Clustering task accompanied by a web app to demonstrate the aforementioned implementation.

Keywords Transfer learning · Natural language processing · Document clustering · Unsupervised learning · Document similarity · Document embedding · Topic modelling

1 Introduction

1.1 Background

In recent years, there has been unprecedented development in the innovation of computer technologies. The rapid development in internet and cloud computing has facilitated more efficient ways of transmitting and processing huge amounts of data [11], and data is at the heart of machine learning.

A. Niam (✉) · A. Das · S. Haque
Department of Computer Science and Engineering, Shahjalal University of Science and Technology, Sylhet, Bangladesh
e-mail: arefin003@student.sust.edu

S. Hossain et al. (eds.), *Proceedings of International Conference on Fourth Industrial Revolution and Beyond 2021*, Lecture Notes in Networks and Systems 437,
https://doi.org/10.1007/978-981-19-2445-3_46

The heavy usage of internet applications has resulted in the generation of huge amounts of data from every user every day. A machine learning algorithm works by training itself on a huge amount of data. If given a certain number of conditions where one has to make out an outcome of a situation, the more the amount of data that dictates the conditions to a certain pattern, the more accurate is the outcome. We can predict an outcome from a human-comprehensible amount of data manually. But large amounts of data are impossible to be carried out with strictly human dictations, that's where machine learning comes in.

1.2 Motivation

There have been numerous works in Natural Language Processing in English and other major languages of the world. Bangla is a language that is spoken by more than 210 million people in the world as their first or second language [1]. It is a very widely used language for literature, communication and social networks. The widespread use of Bangla in modern platforms is resulting in the creation of massive amounts of data that can be used to conduct various machine learning tasks. Without large- scale incorporation of NLP tasks, it cannot be interpreted to computers for developing intelligent systems and applications that operate in Bangla. A simple search on Google Scholar for the term "Natural Language Processing" returns about 4,070,000 as of the time of writing of this paper. And a search for the term "Bangla Natural Language Processing" and "Bengali Natural Language Processing" return 9,250 and 19,200 results, respectively. This provides a loosely vague image of the contemporary state of research in Bangla Natural Language Processing.

1.3 Objective

The term Transfer Learning represents the task of unravelling separate training and learning structure of a particular task and application of the acquired knowledge to solve a similar related task. In traditional learning methods, the learning process is separate for every task independent of their relativity, whereas in transfer learning, the knowledge acquired from previous tasks (weight, feature, etc.) can be applied similarly to achieve faster learning even with a relatively smaller dataset. In October 2018, a paper by Jacob Devlin, Ming-Wei Chang, Kenton Lee, Kristina Toutanova [9] introduced BERT, which is a language model for NLP. BERT or Bidirectional Encoder Representations from Transformers is a state-of-the-art method of pre-training language representation that can achieve remarkable results on a large number of NLP tasks. Even though mainly trained for NLP tasks in English, the BERT multilingual model released by Google covers quite a lot of Bangla text. This allows us to incorporate this newly adopted state-of-the-art methodology for the training of language models for task-specific implementation in Bangla NLP.

In this research, we perform the task of document clustering using the pre-trained BERT model and perform subsequent evaluation and comparison. We also develop a web application that implements the machine learning model with its application demonstration by performing clustering tasks on Bangla news articles.

2 Background Study

2.1 *Related Works*

The amount of work in incorporating transfer learning in Bangla NLP is very low compared to other languages. There have been only one relevant work in Bangla document clustering which used WMD and Word2Vec [2]. The most relevant work in work embeddings performance analysis has been done in [33]. A detailed review of the current trends and challenges in Bangla NLP and Document clustering has been demonstrated in our previous research [27]. Performance analysis of tri-gram [34]and n-gram [17] has been demonstrated in Bangla language modelling. But with time, newer methods are being adopted and it demands subsequent implementation in Bangla. Among laudable works is an implementation of AWD-LSTM architecture and various other techniques [19]. BERT is the latest methodology in language modelling and its adoption will open doors for more research in Bangla using transfer learning.

2.2 *Overview of Transfer Learning*

What is Transfer Learning In traditional ML algorithms, the conduction of similar tasks is separate despite their proximity. If two tasks are similar in nature, there is no traditional way to use the knowledge from training one particular task in a relative secondary task.

But in transfer learning, the outcome of training in a task is used in a relative target task. This allows for training a general task that helps conduct other tasks based on that general task which helps faster and efficient convergence of the relative deep learning models.

Therefore, in various learning tasks, we see that the use of this method has evolved from the application of fine-tuning deep neural networks that are trained on [8] for mostly image and computer vision-related machine learning tasks. It is now feasible to perform natural language processing using this method.

Transfer Learning in NLP Traditional NLP models were trained using randomized initialization of the parameters that are used to construct the NLP model which can be described as weights. But transfer learning has led to the generalization of tasks and

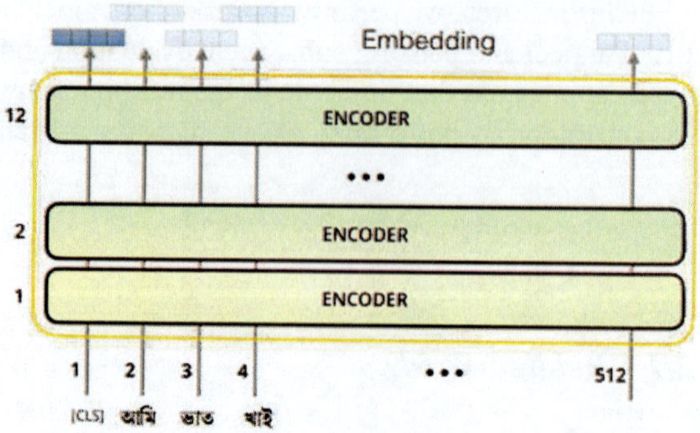

Fig. 1 Input Format for the Encoders in BERT

has helped alleviate the necessity of going through the arduous process of specific dataset fine-tuning.

Due to the very nature of transfer learning's transfer of inductive bias, this method is highly applicable in cases where there can be a lack of large correctly annotated training dataset [22].

2.3 *BERT*

A combination of recurrent neural network and convolutional neural network is utilized for solving NLP problems like sequence transduction problems. The traditional models while exploring the contexts can be either unidirectional, meaning they read the context from the input sequence from left to right, right to left or a combination of both utilizing architecture like LSTM and its variations [12][14]. But with the introduction of BERT, it is possible to read the context of the sequence in both directions at the same time [9]. The basic architecture of BERT is based on the transformer [38] by utilizing the multiple layers that are available in the encoder.

Input Representation The problem with mainstream language models is that they are constructed to be trained with fixed input parameters [16], whereas BERT allows for input representation with variable input parameters. For example, the input can either be an ordered pair of textual data or directly a sequence of textual data.

In every input sequence of BERT, the first token is always defined by *[CLS]* token which is termed as the classification token, and for separation task, the token *SEP* is always used. In completion of the task, based on the estimation that is output from the task with each iteration, the tokens are placed next to each other based on their feasibility [36]. This procedure is demonstrated along with the encoders in 1.

In Fig. 1, a brief visualization of the Bangla words being input as tokens into the encoders is provided. At the end of the 12 layer encoders, the embeddings are generated for the Bangla tokens.

2.4 *Embeddings*

Word embeddings started from distributional semantics [13] despite it not being one of the driving forces in its application. In a survey exploring the applicability of distributional semantics, its applications were demonstrated in [37]. Consequently, research on performance analysis of distributional semantics modelling on various tasks [23] was conducted. Eventually, *Word2Vec* was introduced which solidified the use of word embeddings in NLP [24]. In word2vec using dense vector representations of words from the text, a corpus is produced by **CBOW**(Continuous Bag of Words) and SG(Skip Gram) models. In [28], Global Vectors for Word Representation is introduced which is commonly termed as GloVe. In contrast to the earlier word2vec model, this model attempts to minimize the dimensionality of the co-occurrence matrix.

Word Embeddings to Sentence Embeddings One of the most impactful innovations in information retrieval tasks is Term-Frequency-Inverse Document Frequency (TF-IDF) [29]. Various search engines have been extensively used in this method for better extraction of information from very large amounts of data. Another approach to represent a large volume of textual data is Doc2Vec, which is also known for its ability to generate vectors from paragraphs in articles [21]. Though there are two approaches, the latter performs better in most cases [20]. This was made widely popular by its integration in Genism using the later approach of DBOW [30].

The current state-of-the-art approach to generating sentence embeddings is Sentence-BERT, which utilizes BERT to create the embeddings and is far superior to other approaches due to its deep contextualization of the corpus [31].

2.5 *Document Clustering*

Text mining is a branch of NLP that has helped other academic branches like health, medical science, social sciences, etc. One of the exemplary works of text mining has been demonstrated in [7] where the technique is used for exploring the link between genetic attributes and breast cancer. In the health sector, this technique has also been used in understanding patients for more efficient diagnosis [26]. And also in cases like fraudulent activity exploitation [18] and social media interactions understanding [3].

One of the major elements in the text mining field is document clustering. The task of grouping objects that are similar in nature in one group and vice versa is called clustering or cluster analysis [10]. For identifying and clustering documents in different groups, the word frequencies from the previous stage are used to incorporate cluster algorithms. Some words are, therefore, carefully selected for clustering the documents [25].

3 Methodology

3.1 Data Collection and Preprocessing

Data Source and collection For our task, we needed Bangla news articles collected from various web portals that had a variety in the collection of the articles. For tasks like sentiment analysis, document classification, text summarization similar datasets are prepared. Therefore, we spared the strenuous task of building a news article dataset from scratch. Instead, we searched for the term "Bengali news article dataset" on various dataset repositories available on the internet that host the publicly available dataset. The dataset available in [15] was selected as the training dataset for the training of the model and for evaluation a python scraper was developed to form around 1000 test sets.

Data Processing we processed the dataset in the input format of sentence transformer architecture for generating sentence embeddings. The standard format of input for sentence embedding generation is a sequence of strings that represent each sentence which can also be used as documents.

The dataset was in CSV file format and we use the pandas data frame library to process the dataset for our model. The contents of the news articles were processed for feeding into the encoder of the BERT model.

3.2 Using Pre-Trained Model

Bangla BERT base that has been pre-trained in [35] is used in forming the embeddings using the sentence transformer model [32]. The pre-trained model that we are using in this project is available on Hugging face.

4 Generating Embeddings

After the pre-trained model has been initialized and trained, we connect it with the sentence transformer interface for the training of sentence representations from the pre-trained BERT model. For this task, we use the tokenizer available in the Bangla BERT Base model and the masked language modelling language technique for evaluating the context of a word in a sentence. The process of tokenization and subsequent masked language modelling has been described in Fig. 2.

In the process that has been shown in Fig. 2, the core idea of BERT is visualized where a word from the sequence is masked and then predicted based on the context provided in the sequence. The score corresponding to the sentences is the metric of feasibility of the predicted words.

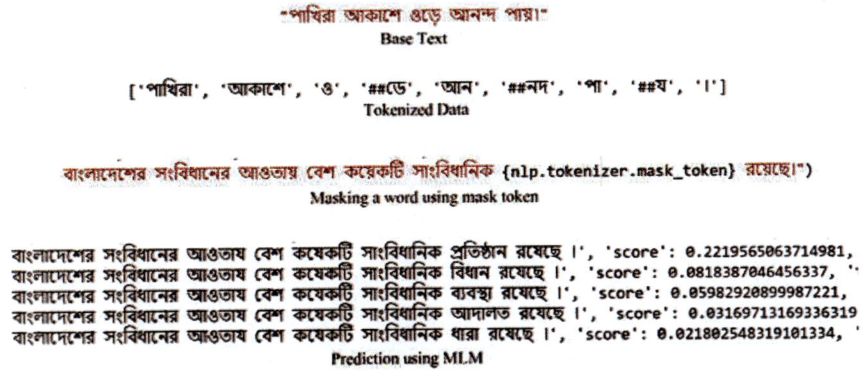

Fig. 2 Tokenization and masked language modelling

4.1 Clustering

For clustering, we select two of the most common clustering algorithms used in the field of document clustering. Despite being old and simple, these clustering algorithms are considered to be the benchmark techniques of cluster analysis [4]. We settled on using the elbow method for efficiently estimating the number of clusters [5].

In this process, we used the k-means library that is available on scikit-learn and ran 15 iterations of k-means clustering. We chose k = 6 as the optimum value of k in for clusters. The fit_transform of scikit-learn is used to join the steps for the initial fitting of the parameters on the training set *x*. At the same time, it also returns the transformation denoted as *x'* as described in the following equations:

$$x' = \frac{x - \mu}{\sigma} \tag{1}$$

5 Result and Analysis

5.1 Embeddings Analysis

For the generation of the embeddings, PyTorch interface for BERT by Hugging Face was utilized. Most of the training tasks and generative tasks were performed using Google Colaboratory TPU resources.

For performing the embedding generation tasks we use *SentenceTransformers* [32]. As Bangla was not available on the multilingual models hosted in the framework we performed pre-training using [35] to prepare a sentence transformer model for generating document embeddings.

text1	text2	distance
উত্তর	উত্তর	1.000000
উত্তর	হিমালয় বাংলাদেশের উত্তরে অবস্থিত।	0.434111
উত্তর	চুপ করে আছ কেন, উত্তর দাও।	0.446134
উত্তর	স্বাধীনতা উত্তর বাংলাদেশে নাটকের প্রভূত উন্নতি...	0.403095
উত্তর	দোয়া করি উত্তর কালে তুমি একজন দেশবরেণ্য নেতা হও।	0.419401
উত্তর	শব্দের উত্তরে প্রত্যয় ব্যবহৃত হয়।	0.401441

	text1	text2	distance
12	চুপ করে আছ কেন, উত্তর দাও।	উত্তর	0.446134
13	চুপ করে আছ কেন, উত্তর দাও।	হিমালয় বাংলাদেশের উত্তরে অবস্থিত।	0.532551
14	চুপ করে আছ কেন, উত্তর দাও।	চুপ করে আছ কেন, উত্তর দাও।	1.000000
15	চুপ করে আছ কেন, উত্তর দাও।	স্বাধীনতা উত্তর বাংলাদেশে নাটকের প্রভূত উন্নতি...	0.492490
16	চুপ করে আছ কেন, উত্তর দাও।	দোয়া করি উত্তর কালে তুমি একজন দেশবরেণ্য নেতা হও।	0.593660
17	চুপ করে আছ কেন, উত্তর দাও।	শব্দের উত্তরে প্রত্যয় ব্যবহৃত হয়।	0.529438

Fig. 3 Measuring Distance between contextual embeddings and context-free embeddings for the same word used in different meanings in different sentences

For the generation of embeddings the sentence transformer model was initialized with the number of layers = 13 (initial embeddings + 12 BERT layers), and the number of hidden units = 768. Taking a look at the range of the values for a given layer and the tokens in the input, we find that the range is generally similar for a given layer and token.

After generating the embeddings, we evaluate them by using the document clustering process. For that, we measure the distance between the embeddings of the same word used in various meanings in different sentences as described in Fig. 3. As described in Fig. 3, we generate a context free embedding for the term by generating embedding by only feeding the word in the encoder and then we perform the same for other sentences containing the same word.

Then we measure the distance of the contextualized word in a sentence from other sentences containing the same word in different meanings.

The values are demonstrated in Fig. 3. These values represent the Euclidean distance of the sentence from the sentence or word token as described in the figure. In this figure, we can see the same word being used in a different context in each of the sentences. It gives us an idea of the variation of distance based on the context. When all of the generated embeddings are averaged together, a context-average embeddings value is generated for understanding the average meaning of the word from all of the trained contexts of the word. Distance between the context-free and context averaged embeddings in the experiments was found to be **0.6759477266455169**.

5.2 Clustering Evaluation

We use UMAP for dimensionality reduction. For visualizing the clusters in a two-dimensional graph, we further reduce the dimensionality to 2 and the outliers in the implementation are represented as grey points. This gives a brief overview of the distribution of the data points in the cluster formations (Tables 1 and 2).

For creating the representation of the topics that were created by clustering, the representatives of the clusters were selected by top 50 words from each cluster. A class-based TF-IDF variant is used to measure the uniqueness of the articles in comparison to each other using the following formula:

$$c - TF - IDF_i = \frac{t_i}{w_i} \times \log \frac{m}{\sum_j^n t_j} \quad (2)$$

With the same formed embeddings, we perform the K-Means Clustering with the value of k set to 6. .

For accuracy of the measurement of our cluster analysis, we selected around 200 news articles from various Bangla news sources to be analyzed in our clustering

Table 1 Generated cluster and frequency of topics using K-means

	Topic	Size
5	5	14620
3	3	9573
4	4	8348
1	1	7545
2	2	5274
0	0	3318

Table 2 Generated cluster and frequency of topics using HDBSCAN

	Topic	Size
0	−1	22264
1150	1149	536
271	270	494
284	283	456
109	108	400
940	939	344
544	543	312
1025	1024	217
1170	1169	214
1177	1176	201

No	Ground Truth		K-Means			HSBSCAN		
	Label Topic	No. of items	Correct	Incorrect	Undefined	Correct	Incorrect	Undefined
1	আটক ও শাস্তি সম্পর্কিত	14	23	1	0	3	0	11
2	অপরাধমূলক। হতাহতের ঘটনা	24	24	0	0	10	2	12
3	খেলাধুলা	22	22	0	0	17	3	2
4	বিনোদনমূলক	35	31	4	0	11	3	21
5	আইন ও রাজনীতি	19	19	0	0	7	2	10
6	সরকারি ও দাপ্তরিক	40	38	2	0	10	7	23

Fig. 4 Comparison of the clustering task with ground truth

model. We manually determine the similarity and dissimilarity among the news articles and generate a ground truth table to test news articles and perform a comparison with our employed clustering algorithms in Fig. 4.

We perform the external criteria evaluation of the clusters to measure the quality of the clusters in our experiment. We particularly use the rand index and F-measure of our clustering task.

For evaluation of True Positive and True Negative decisions assignment, we take two similar news articles being in the same cluster and different cluster, respectively. And for the assignment of False Positive and False Negative, we take two different articles being in the same cluster and similar news articles being in different clusters, respectively. Thus, measuring these criteria, we find the precision and recall and calculation of the F-measure of our cluster analysis Fig. 3.

Our k-means implementation on the contextual embeddings performed the best out of the two employed methods.

6 Conclusion

There has been unprecedented advancement in natural language processing in recent years. The state-of-the-art models like GPT-3 [6] have shown huge potential in NLP tasks. A huge surge in the usage and application of internet communication and

Table 3 Evaluation of the implemented clustering techniques

	K-Means	HDBSCAN
Number of clusters	6	7
True positive	117	42
True negative	145	168
False positive	9	23
False negative	53	88
Rand index (accuracy)	0.8086	0.654
Precision	0.9285	0.6461
Recall	0.688	0.323
F-Score	0.7903	0.4305

mobile phones are generating a huge amount of data every instance and are being utilized to develop intelligent applications.

In this research, we explored transfer learning and developed a model for performing the clustering task using contextual embeddings generated using the state-of-the-art model BERT [9]. We showed various implications of embeddings and the impact of their contextualization and performed the clustering task with the generated embeddings. We analyzed the embeddings and clustering subsequently. In future, we plan to further explore the usage of contextualized embeddings using transfer learning in other Bangla NLP tasks.

Finally, we built a web application using the python-based Django framework to show a brief demonstration of the applicability of Bangla Document Clustering. The source code is publicly available https://github.com/IcarusX60/Bangla-Document-Clustering.git here.

References

1. Encyclopedia Britannica (2019)
2. Ahmad, A., Amin, M.R., Chowdhury, F.: Bengali document clustering using word movers distance. In: 2018 International Conference on Bangla Speech and Language Processing (ICBSLP), pp. 1–6. IEEE (2018)
3. Al-Daihani, S.M., Abrahams, A.: A text mining analysis of academic libraries' tweets. J. Acad. Libr. **42**(2), 135–143 (2016)
4. Andrews, N.O., Fox, E.A.: Recent developments in document clustering. Tech. rep., Department of Computer Science, Virginia Polytechnic Institute & State ... (2007)
5. Bholowalia, P., Kumar, A.: Ebk-means: A clustering technique based on elbow method and k-means in WSN. Int. J. Comput. Appl. **105**(9) (2014)
6. Brown, T.B., Mann, B., Ryder, N., Subbiah, M., Kaplan, J., Dhariwal, P., Neelakantan, A., Shyam, P., Sastry, G., Askell, A., Agarwal, S., Herbert-Voss, A., Krueger, G., Henighan, T., Child, R., Ramesh, A., Ziegler, D.M., Wu, J., Winter, C., Hesse, C., Chen, M., Sigler, E., Litwin, M., Gray, S., Chess, B., Clark, J., Berner, C., McCandlish, S., Radford, A., Sutskever, I., Amodei, D.: Language models are few-shot learners (2020)

7. Cohen, K.B., Hunter, L.: Getting started in text mining. PLoS Comput. Biol. **4**(1), e20 (2008)
8. Deng, J., Dong, W., Socher, R., Li, L.J., Li, K., Fei-Fei, L.: Imagenet: A large-scale hierarchical image database. In: 2009 IEEE Conference on Computer Vision and Pattern Recognition, pp. 248–255. IEEE (2009)
9. Devlin, J., Chang, M.W., Lee, K., Toutanova, K.: Bert: Pre-training of deep bidirectional transformers for language understanding. arXiv preprint arXiv:1810.04805 (2018)
10. Erisoglu, M., Calis, N., Sakallioglu, S.: A new algorithm for initial cluster centers in k-means algorithm. Pattern Recognit. Lett. **32**(14), 1701–1705 (2011)
11. Gadducci, F., Tavosanis, M.: History and Philosophy of Computing. Springer (2016)
12. Ghosh, S., Vinyals, O., Strope, B., Roy, S., Dean, T., Heck, L.: Contextual lstm (clstm) models for large scale nlp tasks. arXiv preprint arXiv:1602.06291 (2016)
13. Griffiths, T.L., Steyvers, M., Tenenbaum, J.B.: Topics in semantic representation. Psychol. Rev. **114**(2), 211 (2007)
14. Hochreiter, S., Schmidhuber, J.: Long short-term memory. Neural Comput. **9**(8), 1735–1780 (1997)
15. Hossain, M.Z., Rahman, M.A., Islam, M.S., Kar, S.: BanFakeNews: A dataset for detecting fake news in Bangla. In: Proceedings of the 12th Language Resources and Evaluation Conference. pp. 2862–2871. European Language Resources Association, Marseille, France (May 2020), https://aclanthology.org/2020.lrec-1.349
16. Houlsby, N., Giurgiu, A., Jastrzebski, S., Morrone, B., De Laroussilhe, Q., Gesmundo, A., Attariyan, M., Gelly, S.: Parameter-efficient transfer learning for nlp. In: International Conference on Machine Learning. pp. 2790–2799. PMLR (2019)
17. Ismail, S., Rahman, M.S.: Bangla word clustering based on n-gram language model. In: 2014 International Conference on Electrical Engineering and Information & Communication Technology, pp. 1–5. IEEE (2014)
18. Kamaruddin, S., Hamdan, A., Bakar, A.A., Nor, F.M.: Outlier detection in financial statements: a text mining method. In: Data Mining X: Data Mining, Protection, Detection and Other Security Technologies, Vol. 42, p. 71 (2009)
19. Khatun, A., Rahman, A., Chowdhury, H.A., Islam, M.S., Tasnim, A.: A subword level language model for bangla language. In: Proceedings of International Joint Conference on Computational Intelligence, pp. 385–396. Springer (2020)
20. Lau, J.H., Baldwin, T.: An empirical evaluation of doc2vec with practical insights into document embedding generation. arXiv preprint arXiv:1607.05368 (2016)
21. Le, Q., Mikolov, T.: Distributed representations of sentences and documents. In: International Conference on machine learning, pp. 1188–1196. PMLR (2014)
22. Malte, A., Ratadiya, P.: Evolution of transfer learning in natural language processing. arXiv preprint arXiv:1910.07370 (2019)
23. McNamara, D.S.: Computational methods to extract meaning from text and advance theories of human cognition. Top. Cogn. Sci. **3**(1), 3–17 (2011)
24. Mikolov, T., Chen, K., Corrado, G., Dean, J.: Efficient estimation of word representations in vector space. arXiv preprint arXiv:1301.3781 (2013)
25. Müllner, D.: Modern hierarchical, agglomerative clustering algorithms. arXiv preprint arXiv:1109.2378 (2011)
26. Neustein, A., Imambi, S.S., Rodrigues, M., Teixeira, A., Ferreira, L.: 1. application of text mining to biomedical knowledge extraction: analyzing clinical narratives and medical literature. In: Text Mining of Web-based Medical Content, pp. 3–32. De Gruyter (2014)
27. Niam, A., Das, A., Chowdhury, M.S., Al Mumin, M.A.: A literature review of bangla document clustering. Int. J. Comput. Appl.**975**, 8887
28. Pennington, J., Socher, R., Manning, C.D.: Glove: Global vectors for word representation. In: Proceedings of the 2014 conference on empirical methods in natural language processing (EMNLP), pp. 1532–1543 (2014)
29. Ramos, J., et al.: Using tf-idf to determine word relevance in document queries. In: Proceedings of the First Instructional Conference on Machine Learning, vol. 242, pp. 29–48. Citeseer (2003)

30. Řehřek, R., Sojka, P., et al.: Gensim-statistical semantics in python. Retrieved from genism.org (2011)
31. Reimers, N., Gurevych, I.: Sentence-bert: Sentence embeddings using siamese bert-networks. arXiv preprint arXiv:1908.10084 (2019)
32. Reimers, N., Gurevych, I.: Making monolingual sentence embeddings multilingual using knowledge distillation. In: Proceedings of the 2020 Conference on Empirical Methods in Natural Language Processing. Association for Computational Linguistics (11 2020), https://arxiv.org/abs/2004.09813
33. Ritu, Z.S., Nowshin, N., Nahid, M.M.H., Ismail, S.: Performance analysis of different word embedding models on bangla language. In: 2018 International Conference on Bangla Speech and Language Processing (ICBSLP). pp. 1–5. IEEE (2018)
34. Saha, D., Hossain, M.S., Islam, M., Ismail, S.: Bangla word clustering based on tri-gram, 4-gram and 5-gram language model. arXiv preprint arXiv:1701.08702 (2017)
35. Sarker, S.: Banglabert: Bengali mask language model for Bengali language understanding (2020), https://github.com/sagorbrur/bangla-bert
36. Sennrich, R., Haddow, B., Birch, A.: Neural machine translation of rare words with subword units. arXiv preprint arXiv:1508.07909 (2015)
37. Turney, P.D., Pantel, P.: From frequency to meaning: Vector space models of semantics. J. Artif. Intell. Res. **37**, 141–188 (2010)
38. Vaswani, A., Shazeer, N., Parmar, N., Uszkoreit, J., Jones, L., Gomez, A.N., Kaiser, Ł., Polosukhin, I.: Attention is all you need. In: Advances in Neural Information Processing Systems, pp. 5998–6008 (2017)

Bangla Speech-To-Braille Interaction Device for Visual and Hearing Impaired

Rafia Rizwana Rahim, Atik Ishrak Al Nahian, Rakiba Benta Kalam, Atanu Sen Gupta, and Nusrat Binta Nizam

Abstract This research project proposes a prototype design for a refreshable braille display to convert Bangla Speech to Braille output. It aims to be a communication and learning module for deaf-blind or blind people to pursue inclusive education. First, we took speech input from a transducer, and speech recognition was done by DeepSpeech, a Natural Language Processing framework developed by Mozilla. Then, the converted text output was linked with its corresponding braille pins and was shown in a refreshable braille display. Finally, the entire computation was carried away by a Raspberry Pi. We have proposed the usage of our prototype as a special ICT device in the journey towards inclusive education in Bangladesh. In addition, future development has been suggested to broaden the periphery of the project.

Keywords DeepSpeech · Bangla natural language processing · Refreshable Braille display

1 Introduction

The goal of this research project is to create a refreshable braille display that accepts Bangla speech as input and outputs active braille pins. It is primarily intended to serve as a communication and learning tool for deaf-blind persons, particularly those aged 5–18. When we saw that there were few technologies for blind students and none for deaf-blind pupils, we were inspired to create such a product. There are 285 million visually impaired (VI) people in the world [1], where 90% of them are from developing countries [1]. Nearly 19 million children below 15 age group are VI [1].

R. R. Rahim · A. I. Al Nahian · R. B. Kalam · A. S. Gupta
Department of Mechanical Engineering, Bangladesh University of Engineering and Technology, Dhaka 1205, Bangladesh
e-mail: 1710045@me.buet.ac.bd

N. B. Nizam (✉)
Department of Biomedical Engineering, Bangladesh University of Engineering and Technology, Dhaka 1205, Bangladesh
e-mail: nusratbintanizam@bme.buet.ac.bd

S. Hossain et al. (eds.), *Proceedings of International Conference on Fourth Industrial Revolution and Beyond 2021*, Lecture Notes in Networks and Systems 437,
https://doi.org/10.1007/978-981-19-2445-3_47

Among them, 90% children do not have access of education in developing countries [2]. Bangladesh is a developing country which has 341,819 VI children in the 6–11 age group [5]. Also, a significant portion is illiterate. The literacy rate is 57.7%, which demands ensuring inclusive education, where the system will make education affordable for all, irrespective of their background or disabilities [6]. Implementation of special Information and Communication Technology (ICT) devices in education can help to develop the quality of life and empower the VI and deaf-blind people [7]. But the accessibility and availability of ICT educational tools are a major issue because these tools can help to ensure equal opportunities for all students across the world.

2 Literature Review

Even while Braille Displays are pretty famous, adding a language-specific speech to the braille module is the real issue. It is a pity that Bangla, the world's seventh most spoken language, lacks a fully developed speech-to-text module that we may use in our project. We found the closest potential work that can assist us in creating our voice recognition module when searching for a good speech recognition approach. We found the following suggested approaches and theoretical results for Bangla Speech to Text Conversion Linear Predictive Coding (LPC) and Cepstral Coefficients for feature extraction and a Multilayer Perceptron (MLP) approach for recognition [9], Mel Frequency Cepstrum [10], Unsuper vised learning by recognizing repeated patterns [11], tokenizing the input speech [12], unsupervised term discovery to cluster repeated patterns [13], classification of unlabeled data [14]. More recent studies used CNN, RNN, and LSTM-based models for Bangla speech recognition purpose [17]. Also, augmented dataset has given closer mnemonics and better results than other studies [18]. Hannun et al. [15] showed a deep recurrent neural network architecture to recognize continuous speech. Experimentation showed that it gave highly accurate Transcriptions to continuous speech. With data of 600hours, the model yields 90% accuracy for Mandarin Chinese and English. They published as Mozilla's open-source framework DeepSpeech, which can be used to model any language for speech recognition and, therefore, has been used in our research. To support the requirements of DeepSpeech, we searched for speech to text dataset and found that the dataset provided by the openslr platform is the most suitable for our purpose [19]. Previous studies have also been done using the same dataset. For Example Sultana et al. (2012) used the corresponding dataset in their SAPI-based speech recognition module [16]. Therefore, this dataset is a proven one to be used in our research.

3 Dataset Preparation

A huge preprocessed dataset is the primary need of any deep learning frame work. DeepSpeech also requires over 100,000 audio and text samples and numerous other auxiliary files to begin its training. The steps of dataset preprocessing can be described sequentially.

3.1 Dataset Selection and Download

We looked for an open-source Bangla audio dataset and found 250 thousand audio of 347 h and corresponding text data in the openslr platform [19], where it contained 229 h of unique speech to text data and the rest of the dataset was either repeated or augmented. The dataset was in a very raw format and needed preprocessing.

3.2 .wav Conversion, Riff Code Addition and Audio Data Debug

DeepSpeech only accepts.wav audio files with a specified riff code attached before the audio file. Using AudioConverter 0.3.0 [22], we converted the .flac files to.wav files with riff code attached. Then we built ftwdebug.py [21], a debugger program to see if any audio files were missing the riff code. Furthermore, DeepSpeech will not accept audio input with a duration of fewer than 3 s. As a result, we built a new debugger code [21] to erase and exclude all audio with a duration of fewer than 3 s.

3.3 Corresponding Text Manual Correction and Unicode Debug

DeepSpeech requires audio files to be accompanied by a text file that contains no numerical characters or punctuation marks. As a result, all of these had to be manually erased. The openslr dataset included a .tsv file containing file names (without extensions), anonymous ids, and matching text for each audio file. As a result, we initially copied it into Excel. Next, we divided the spreadsheet file into 250 separate excel files, each with 1000 items, to make human corrections easier. All punctuation marks and emoji were then manually removed. In this case, the number of characters was replaced with Bangla text. Unfortunately, the text contained an unexplained error, and DeepSpeech could not locate Unicode for the Bangla script. So we built xldebug.py [21], a debugging program to see if each of our characters has a UTF+8 code associated with it.

3.4 .csv File Generation

DeepSpeech requires the following format for audio text in.csv: filename (with extension), file size (in Kilobytes), text. To create such a .csv file, we created the file name size.py script [21], which can create .csv files containing the file name and size. Our audio files were divided into 14 folders, each with 14 subfolders. We put all audio files in one folder and made a .csv file with the file name and size.

3.5 Text Input In .csv Files

Now, we had two files one having the file name and corresponding texts, the other having file name and size. We needed to merge these two excel sheets concerning file names. When we did so, we saw that several audio files did not have written text with them. Also, some of the audio files did not exist in our dataset, but their file name and text were written in the excel file. So we needed to delete those entries manually. Here is our sequential operation we copied the file names in a text file for each segmented excel file. Then we run a code File Copy Reading txt to Folder and Check.py [21] that will copy the file reading the .txt from our audio folder. It will also show which files do not exist. So we can remove those entries manually. This same process had to be repeated 250 times. Then we can merge text with file name and size .csv file.

3.6 Training, Development, and Testing Dataset

DeepSpeech requires three folders having audio files and .csv files containing the file name, size, and text. They are called training (train), development (dev), test folders. In each training cycle, train, dev, and test contain 70, 20, and 10% audio data. For example, we trained our model 5 times, each with 40–80 thousand data. For this, we created test, train, dev folders for each training using the above-stated manner.

3.7 Generating alphabet.txt File

We needed a code to generate all alphabet of Bangla in a text file. Moreover, we needed it to tell the training framework that we are using a different alphabet set than English, and it is Unicode based. So, we created a python code alphabet.py [21] that generates an alphabet of Bengali language and saves it in a text file.

3.8 Generating a Scorer File

A scorer is a language model, and DeepSpeech uses it to improve the accuracy of the transcription. A language model predicts which words are more likely to come after one another. The scorer determines the likelihood of words appearing together. DeepSpeech's default scorer is trained using the LibriSpeech dataset. The LibriSpeech dataset is based on LibriVox, an open repository of public domain and out-of-copyright works. Therefore, we needed scorer training DeepSpeech in another language, training a speech recognition model for a particular domain. In our case, it is day-to-day formal Bangla speeches mostly available in news reports and to improve the accuracy of the transcription.

DeepSpeech supports the optional use of an external scorer. We collected many of the sentences from Bangla Wikipedia through a module named banglaWIKI corpus, which allowed us to scrap through Wikipedia and download the Bengali sentences, which we will use to build the scorer. It supports our DeepSpeech model with predefined sentences that we regularly use. If a model does not understand a voice fully, it tries to match its word with the scorer file and shows outputs accordingly.

4 Design and Methodology

4.1 Bangla Automatic Speech Recognition

A microphone should capture the sound waves, convert them to a digital signal, and store them in memory. The sound must be linked to the corresponding test. To find the appropriate text for the sound, we need pre-requisites of 2 types a large dataset with a variety of audio and text and a framework having an Acoustic Model, Pronunciation Model, and Language Model (Fig. 1).

Preparation for DeepSpeech Framework We have used Google's cloud-based machine learning platform, Google Colab, to train our model. To start training in Google Colab, first, we need to set up an environment of DeepSpeech in the following manner. Installation of DeepSpeech, mounting Google Colab with Google Drive to train the data from the drive directly, enabling GPU, and installing other packages such as an appropriate version of TensorFlow have been done, respectively. After preparing the train and dev folder, we need to start the training by calling flags on DeepSpeech. Our ultimate target is to decrease the value of the loss. The validation part is the development part using the dev folder.

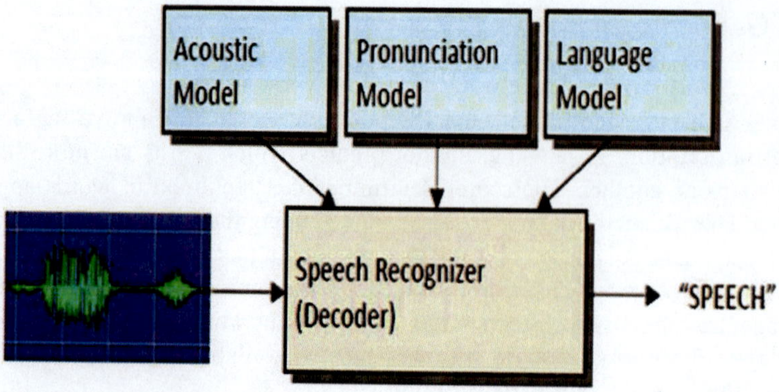

Fig. 1 Speech recognition flow chart

4.2 Bangla Text to Braille (TTB) Conversion

Every character of the Bangla language has a unique Unicode. For every Unicode, a character has a unique combination of braille pins. Braille is a language where letters are shown via the combination of 6 pins in a braille cell numbered vertically. Most of the letters in Bangla need one set of braille pins to resemble the character. Nevertheless, prefixes and suffixes are incorporated when it comes down to words, joint letters, or sentences. We have followed all the rules of Bangla Braille to Convert Bangla Text to Braille pins. The following rules were considered while setting conditional loops in the TTB code, btb.py. Our code (btb.py) [21] first splits every word in the text obtained from DeepSpeech. If there is any joint letter, it is detected by Unicode of hosonto between those letters. Then the braille pins are detected according to the rules of Bangla Braille Adaption [27]. The numbers of braille pins are the input of our refreshable braille display. The display is designed so that the output is shown cell by cell, one cell at an instant (Figs. 2, 3 and 4).

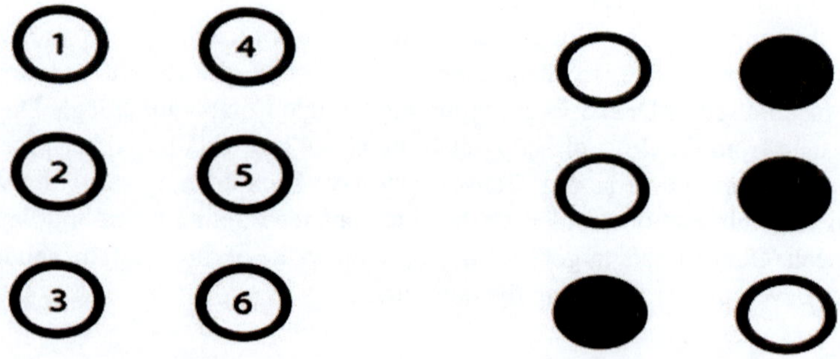

(a) Braille Cell Numbering **(b) Braille Cell Representing 'আ'(345)**

Fig. 2 Braille cell representations

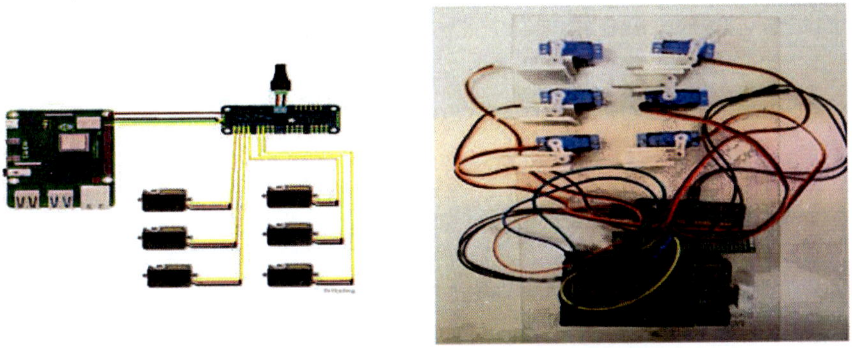

Fig. 3 Electrical circuit simulation using fritzing and electrical circuit

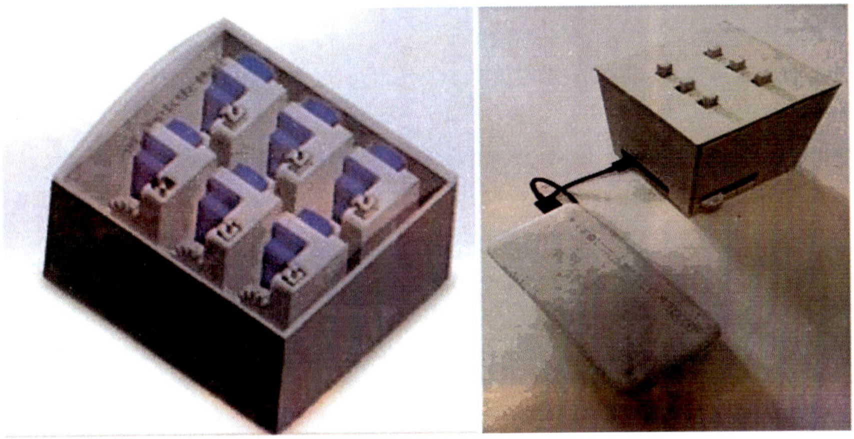

Fig. 4 Solid works design and final assembly

4.3 Refreshable Braille Display

Electrical Circuit Design: We have used six independent servo motors with a linear actuation system to show the braille pins obtained. Here the servo has been controlled by a Pulse Width Modulation (PWM) controller named Adafruit PCA9685. This PWM controller takes input from Raspberry Pi and sends corresponding signals to the servos. The servos' precise motion helps to step up the braille pin when necessary. We have done our circuit design using Fritzing [24]. Once our circuit was complete, we combined all of our programs into a single file called Speech to Braille Hardware.py. [21] While operating, this program requests the user's input audio stream, transforms it to text, and then generates the matching Braille pin numbers and servos. The electrical simulation was also used to create a printed circuit board (PCB).

Mechanical Part Design: We intend to use a rack and pinion gear system as our primary hardware and designed the device accordingly in SolidWorks [25]. However, because of the lockout in the Covid-19 issue, we were unable to 3D print our rack and pinions. As a result, we devised a less expensive alternative: the reverse crank mechanism. The servo shaft will be attached to a crank here. The connecting rod will be linked to the other end of the crank. The connecting rod is connected to a piston (braille pin) that can reciprocate with a predetermined slot. As a result, the reverse crank mechanism acts as a linear actuator, converting the servo's rotating motion to the linear motion of the braille pin [26].

After trailing for several weeks, we saw that our prototype was working fine. So we did our final assembly. Here the circuit is mounted in a 10cm*10cm*10cm box. The microphone adapter is directed outward. The raspberry pi is getting power from a power bank.

5 Results

Here src means the actual source text of the speech, and res means predicted text by DeepSpeech. There are some critical features of our testing data (Fig. 5). First, the word count per sentence and letter count per word is almost ac- curate. Second, common verbs and words are predicted correctly. Finally, the mnemonics of the words were predicted correctly. That is, the position of vowel and consonant modifiers and silent letters are in the correct position. Bangla is segmented into groups of letters according to the way they sound. Unfortunately, our model could not predict the exact sound, and the place is taken over by the closed possible sound in that group. Here WER means Word error Rate, and CER means character Error Rate. In our best WER, we can see that our model recognized 77% of characters in a sentence and

```
WER: 0.333333, CER: 0.133333, loss: 15.634764
 - wav: file://../drive/MyDrive/Datal/test/195590bd5b.wav
 - src: "অনেক দামের সুতা"
 - res: "সনের দামের সুতা"
------------------------------------------------------------
WER: 0.500000, CER: 0.384615, loss: 23.318504
 - wav: file://../drive/MyDrive/Datal/test/19e034d0al.wav
 - src: "ইউনিবিজয় নামে"
 - res: "সমিনিজল নামে"
------------------------------------------------------------
WER: 0.500000, CER: 0.263158, loss: 21.237675
 - wav: file://../drive/MyDrive/Datal/test/199c157a10.wav
 - src: "হত্যা করা হয়েছে বলে"
 - res: "সতা করা হরিছে বলে"
```

Fig. 5 Example of ASR test result

67% of the word in a sentence. So as we can see from Figs. 6 and 7, we have trained our model, but we still have a lot of WER and CER errors. Therefore, to overcome those errors, we need to improve our language model.

Our TTB code gives 100% accurate results while converting text to braille (Figs. 8 and 9).

```
Best WER:
--------------------------------------------------------------------------------
WER: 0.333333, CER: 0.230769, loss: 10.730057
 - wav: file://../drive/Shareddrives/SynthiaSoft.com-Unlimited-03/Data_for_Atik/Data_for_Atik_4/testwav/918e8af019.wav
 - src: "মিত্র মিত্র বাল"
 - res: "মি মিত্র বাল"
--------------------------------------------------------------------------------
WER: 0.500000, CER: 0.480000, loss: 56.481258
 - wav: file://../drive/Shareddrives/SynthiaSoft.com-Unlimited-03/Data_for_Atik/Data_for_Atik_4/testwav/915a7dd10c.wav
 - src: "অর্থমন্ত্রণালয়কে জানিয়েছে"
 - res: "সত্তাজবাকে জানিয়েছে"
--------------------------------------------------------------------------------
WER: 0.500000, CER: 0.388889, loss: 25.654314
 - wav: file://../drive/Shareddrives/SynthiaSoft.com-Unlimited-03/Data_for_Atik/Data_for_Atik_4/testwav/91370868fc.wav
 - src: "সমাজের পূর্বাকিত"
 - res: "সমাজের পমাধিত"
--------------------------------------------------------------------------------
WER: 0.500000, CER: 0.157895, loss: 20.535526
 - wav: file://../drive/Shareddrives/SynthiaSoft.com-Unlimited-03/Data_for_Atik/Data_for_Atik_4/testwav/90cd3a27bc.wav
 - src: "ঝামেলা না করার জন্য"
 - res: "সামেলা না করার জন্ল"
--------------------------------------------------------------------------------
WER: 0.500000, CER: 0.555556, loss: 18.539881
 - wav: file://../drive/Shareddrives/SynthiaSoft.com-Unlimited-03/Data_for_Atik/Data_for_Atik_4/testwav/91a97a499d.wav
 - src: "আমার মিল"
 - res: "ম মিল"
```

Fig. 6 Best case scenario test result

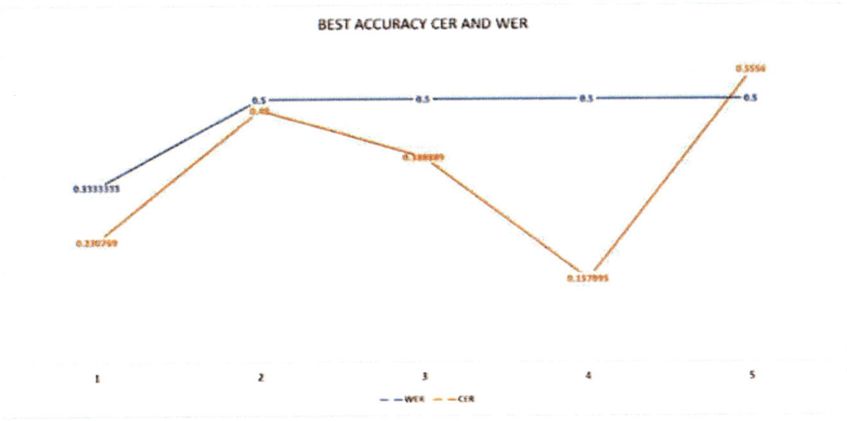

Fig. 7 Accuracy of CER and WER for best case scenario

```
Worst WER:
--------------------------------------------------------------------------------
WER: 2.000000, CER: 0.545455, loss: 16.047106
 - wav: file://../drive/Shareddrives/SynthiaSoft.com-Unlimited-03/Data_for_Atik/Data_for_Atik_4/testwav/90a5ce2ba8.wav
 - src: "বেচারামবাবু"
 - res: "সিচানাম বব"
--------------------------------------------------------------------------------
WER: 2.000000, CER: 1.250000, loss: 13.515151
 - wav: file://../drive/Shareddrives/SynthiaSoft.com-Unlimited-03/Data_for_Atik/Data_for_Atik_4/testwav/90cd3119a6.wav
 - src: "কর্ণ"
 - res: "সন্দল স"
--------------------------------------------------------------------------------
WER: 2.000000, CER: 0.333333, loss: 9.219383
 - wav: file://../drive/Shareddrives/SynthiaSoft.com-Unlimited-03/Data_for_Atik/Data_for_Atik_4/testwav/91358a8456.wav
 - src: "তারপরও"
 - res: "সার পরও"
--------------------------------------------------------------------------------
WER: 3.000000, CER: 1.000000, loss: 21.332855
 - wav: file://../drive/Shareddrives/SynthiaSoft.com-Unlimited-03/Data_for_Atik/Data_for_Atik_4/testwav/914cc6a1c5.wav
 - src: "কানাইঘাট"
 - res: "সনাই দা তান"
--------------------------------------------------------------------------------
WER: 3.000000, CER: 0.750000, loss: 13.563680
 - wav: file://../drive/Shareddrives/SynthiaSoft.com-Unlimited-03/Data_for_Atik/Data_for_Atik_4/testwav/91867b3a36.wav
 - src: "পলাশবাড়ী"
 - res: "সনা শ বারি"
```

Fig. 8 Worst case scenario test result

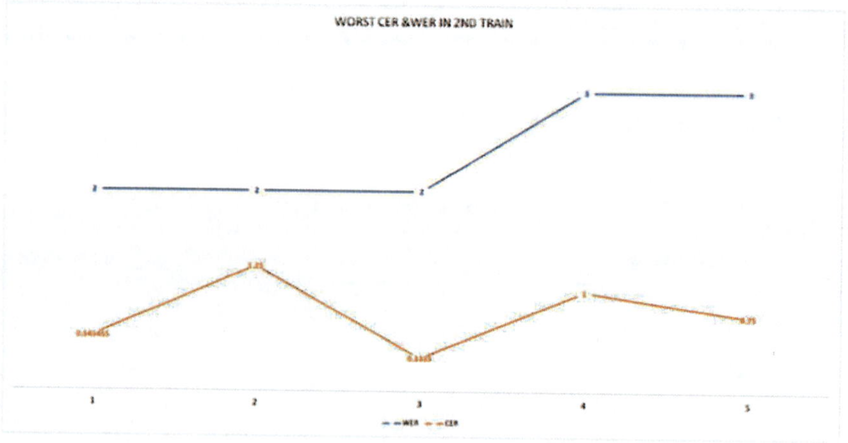

Fig. 9 Accuracy of CER and WER for worst case scenario

6 Discussion

The ASR results described in the figures are self-explanatory to understand how our ASR model is dealing with test data. The model can successfully identify mnemonics of common verbs, vowels, consonant, and vowel modifiers. More data can reduce the loss in this model and therefore give a better accuracy. We have realized that perfecting any speech recognition module could take months. Furthermore, it is a one-way communication medium; a person may convert speech to braille but not vice versa using the same instruments. How- ever, we do not consider all of this a weakness; instead, we believe that it will only open the doors to a plethora of opportunities soon. It is self-evident that the accuracy of any ASR model improves as the dataset size grows. We used a dataset that was created by chopping up news snippets for our project. As a result, it is unable to distinguish the dialect of a specific Bangladeshi district. Its vocabulary has to be expanded in order to recognize more complex works. As a result, the production of datasets is our key priority as we upgrade our prototype. Our prototype is a one-way communication medium. We intend to incorporate Braille to Speech using the same devices. So when a deaf-blind person is asked something using our device, he/she can get an immediate response. Such de- vices can work as their eyes and voice. This complete package can help millions of people to demolish the barriers around them and add them to mainstream education.

7 Conclusion

Our project has a broad spectrum of usage when it comes down to inclusive education. It can be used as a communication and learning device not only for deaf-blind people but also for blind people. Our equipment is simple to use for school-aged youngsters who are deaf-blind. Because it records audio and subsequently translates it to braille, the gadget can be used as a note-taking device for them and a great aid while listening to lessons. As a result, implementing our initiative following the Bangladesh Education Ministry's objective can illuminate the path to inclusive education on its own.

Acknowledgements At first, we would like to thank our family members, especially our parents. Without their hospitality, the project would not be possible in the lockdown due to the Covid-19 situation. Next, we are ever grateful to Satyaki Banik and Nusrat Binte Nizam, who greatly encouraged us to take up a project to bridge the gap between Bangla Natural Language Processing (NLP) and Mechanical Engineering. We want to thank Apurba Sarker for his initial involvement in preparing DeepSpeech Framework, Rasman Mubtasim Swargo, and Aniruddha Ganguli for their knowledgeable suggestions regarding NLP. Finally, we are grateful to Partha Pratim Das for the Google Colab Pro Support and Shakti Banik for his contribution in conceptualizing the design.

References

1. WHO Media Centre. Visual impairment and blindness. http://www.who.int/mediacentre/factsheets/fs282/en/ (2013). Accessed 15 March 2014
2. United Nations. Factsheet on Persons with Disabilities. http://www.un.org/disabilities/default.asp?id=18 (2013). Accessed 25 May 2013
3. Mitra, S., Posarac, A., Vick, B.: Disability and poverty in developing countries: a snapshsot from the World Health Survey. (World Bank Sp discussion paper (1109)), The World Bank, Washington, DC, USA (2011)
4. Rahman, K.F.: Blindness, 'Vision 2020 and Bangladesh. Finan. Express **20**(436), 10 (2012)

5. Das, A.: Inclusion of student with disabilities in mainstream primary education of Bangladesh. J. Int. Dev. Cooper. **17**(2), 1–10 (2011)
6. EFA Global Monitoring Report 2010. Reaching the Marginalized. United Nations Educational, Scientific and Cultural Organization. Oxford University Press, Oxford, UK
7. Hirayama, M.J.: Digital talking book contents of beading handicraft design diagram for visually impaired persons. In: 4th International Conference on Interaction Sciences (ICIS), pp. 96–101, Busan, South Korea (2011)
8. CDD. https://cdd.org.bd/deaf-blind-support/
9. Paul, A.K., Das, D., Kamal, M.M.: Bangla speech recognition system using lpc and ann. In: 2009 Seventh International Conference on Advances in Pattern Recognition, pp. 171–174, February 2009.
10. Ahammad, K., Rahman, M.M.: Connected Bangla speech recognition using artificial neural network. Int. J. Comput. Appl. **149**(9), 38–41 (2016)
11. Park, A.S., Glass, J.R.: Unsupervised pattern discovery in speech. IEEE Trans. Audio Speech Language Process. **16**(1), 186–197 (2008)
12. Kamper, H., Jansen, A., Goldwater, S.: Unsupervised word segmentation and lexicon discovery using acoustic word embeddings. IEEE/ACM Trans. Audio Speech Language Process. (TASLP), **24**(4), 669–679 (2016)
13. Bansal, S., Kamper, H., Lopez, A., Goldwater, S.: Towards speech-to-text translation without speech recognition. arXiv preprint arXiv:1702.03856 (2017)
14. Honglak Lee, Peter Pham, Yan Largman, and Andrew Y. Ng. Unsupervised fea- ture learning for audio classification using convolutional deep belief networks. In Y. Bengio, D. Schuurmans, J. D. Lafferty, C. K. I. Williams, and A. Culotta, editors, Advances in Neural Information Processing Systems 22, pages 1096–1104. Curran Associates, Inc., 2009.
15. Hannun, A. et al.: Deep speech: scaling up end-to-end speech recognition. Baidu Research – Silicon Valley AI. arXiv:1412.5567v2. [cs.CL], 19 December 2014
16. Sultana, S., Akhand, M.A.H., Das, P.K., Rahman, M.H.: Bangla speech-to-text conversion using SAPI. In: 2012 International Conference on Computer and Communication Engineering (ICCCE), pp. 385–390 (2012)
17. Sabab, M., Chowdhury, M.A.R., Nirjhor, S.M., Uddin, J.: Bangla speech recognition using 1D-CNN and LSTM with different dimension reduction techniques. In: International Conference for Emerging Technologies in Computing, pp. 158–169. Springer, Cham (2020).
18. Maruf, M., Faruque, M., Muhtasim, M., Nelima, N.N., Mahmood, S., Riad, M., Uddin, M.: Comparative study of effective augmentation method for Bangla ASR using convolutional neural network." In: Emerging Technologies in Data Mining and Information Security, pp. 203–213. Springer, Singapore (2021)
19. Large Bengali ASR training data set. https://tinyurl.com/Open-SLR
20. Bangla Text to Braille code (2021). https://tinyurl.com/Bangla-Text2Brail
21. Shobdo-Kolpo-Drum/Project-codes (2021). https://github.com/Shobdo-Kolpo- Drum/Project-codes
22. AudioConverter 0.3.0. https://pypi.org/project/AudioConverter/
23. Mozilla/DeepSpeech. https://github.com/mozilla/DeepSpeech
24. Simulated circuit design.fzz (2021). https://tinyurl.com/Circuit-Design-A25
25. Shobdo Koplo Drum (2021). https://tinyurl.com/Shobdo-kolpo-drum
26. Group A25 Demo (2021). https://tinyurl.com/Group-A25
27. Hossain, S.A. et al.: Bangla braille adaptation. In: Technical Challenges and Design Issues in Bangla Language Processing, Chapter 2 (2013)

Particle Swarm Optimized DWT-SVD Watermarking Scheme for Securing EEG Data

Akash Kumar Gupta, Chinmay Chakraborty, and Bharat Gupta

Abstract A watermark embedding provides a way to secure the audio, video, image, text, and other types of data by which data authentication, ownership, visual quality, etc. can be maintained. In this paper particle swarm optimized watermarking algorithm is proposed for securing the EEG data. This method optimizes the scaling factor to achieve higher imperceptibility and robustness. This proposed watermarking method is based on the singular value decomposition (SVD) in discrete wavelet transform (DWT) domain. Performance of proposed algorithm analyzed against various attacks like rotation, scaling, noise, etc. The experimental result shows that proposed algorithm performs better as compared to already existing techniques in terms of peak signal to noise ratio (PSNR) and Normalized correlation coefficient (NCC). The procured PSNR value is 55.2 dB and NCC value is 0.95.

Keywords DWT · Imperceptibility · PSO · Robustness · SVD · Watermarking

1 Introduction

Today all data is accessible in the computerized form on the web. This accessibility of information on the web permits user to share and access all the information and data digitally and encroaches the law of copyright possession. The data embedded with watermark can be utilized for sharing in between sender and authenticated receiver only. Digital watermarking permits the user to add information like a logo or image

A. K. Gupta (✉) · C. Chakraborty
Department of Electronics and Communication Engineering, Birla Institute of Technology, Mesra, Jharkhand, India
e-mail: akgupta@bitmesra.ac.in

C. Chakraborty
e-mail: cchakrabarty@bitmesra.ac.in

B. Gupta
Department of Electronics and Communication Engineering, National Institute of Technology Patna, Patna, Bihar, India
e-mail: bharat@nitp.ac.in

S. Hossain et al. (eds.), *Proceedings of International Conference on Fourth Industrial Revolution and Beyond 2021*, Lecture Notes in Networks and Systems 437,
https://doi.org/10.1007/978-981-19-2445-3_48

in a message to demonstrate ownership, authenticity, and duplicity control. Process of watermarking has different applications like copyright security, fingerprinting, information authentication, and a lot more [1]. The watermarking algorithm must keep up the nature of the original image along with watermark embedding that highlights imperceptibility and robustness against different attacks like rotation, scaling, translation, etc. [2].

In this paper, DWT-SVD-PSO-based watermarking algorithm has been proposed to perform secured watermarking over EEG data so that the privacy of patients and their information must maintain. Here, firefly algorithm has been utilized to optimize scaling factor and watermark logo successfully implanted and extracted at the time of requirement. Watermarking of digital information can be performed either in spatial or in transform domain. Pixel values of image can directly be utilized in spatial domain, whereas it needs to be converted in frequency coefficients at the time of transform domain watermarking. In spatial domain, to perform watermarking is easier but efficiency can be lower [3]. To perform watermarking in transform domain, there are a number of methods like DWT, (Discrete Fourier Transform) DFT, (Discrete Cosine Transform) DCT, (Redundant Discrete Wavelet Transform) RDWT, (Lifting Wavelet Transform) LWT, etc. These schemes provide higher efficiency and flexibility, but we need to be careful as watermarking in lower frequency components may change the original image, whereas in high-frequency region it may affect during noise attacks like low pass filtering by which watermark shows floppiness. There is huge research going on in wavelet domain, as it shows higher efficiency and faster response as compared to other available techniques.

To solve issues of security, a robust watermarking technique has been proposed by author in [4] that utilizes DWT-SVD with optimization algorithm has been proposed. To enhance the performance in terms of imperceptibility and robustness author uses dual watermarking scheme over medical image. Author has achieved more robustness, imperceptible and good capacity simultaneously. Author in [5] has proposed DWT-DCT-SVD along with chaotic encryption, to provide security of medical data in watermark perspective. A robust SVD-based hybrid watermarking scheme for color images has been proposed by author in [6]. This paper also utilizes optimization based on particle swarm to decide optimize value of scaling factor. Using Lifting Wavelet Transform (LWT) and SVD, a watermarking scheme for color image has been proposed by author in [7], which applies Arnold Transform (AT) to enhance the security of watermark. Author has also used Grey Wolf Optimization (GWO) method to enhance the performance of proposed scheme. Using Firefly algorithm, to optimize multiple scaling factors (MSF), A DWT-SVD-based watermarking scheme has been proposed by author in [8]. Here, singular values of binary watermark have been embedded into level three sub-bands of image data. The objective function proposed by the author has imperceptibility and robustness as a component, which are linearly related and shows good (Peak Signal to Noise Ratio) PSNR, (Normalized Correlation Coefficient) NCC values. As we know watermarking provides a way to secure information copyright. Based on different embedding strength, a watermarking method has been proposed by the author in [9]. Author has utilized discrete cosine transform for selecting a block from an image with higher variance. A scrambled binary watermark bits using AT has been inserted into image's DCT coefficient. Proposed scheme by author has shown higher imperceptibility and achieved good quality watermarked image.

To provide security for digital data, a watermarking algorithm has been proposed by author in [10], which uses DWT-SVD technique and is optimized by Genetic Algorithm (GA), Particle Swarm Optimization (PSO), and Artificial Bee Colony (ABC). and DCT. Author in [11] proposed a color image watermarking which use the concept of DCT and further optimized by PSO for the need of good imperceptibility and robustness. Another paper based on PSO optimization proposed by author in [12] to protect ownership of data, which utilizes the DWT-DCT watermarking principle. Author in [13] proposed a robust watermarking technique for digital multimedia based on DWT-SVD-PSO.

The very first introduction section has information regarding watermarking and its application along with literature survey of recent research in the field of proposed work area. The rest of the paper is arranged into 5 sections. Section 2 includes 4 subsections, in which the first three subsections describe watermarking methodology, whereas its subsection four elaborates PSO algorithm. Section 3 discusses EEG data set and its short-time Fourier transform, which is utilized further, for the purpose of watermarking. In Sect. 4, we have presented results and discussion on performance of proposed DWT-SVD-FA-based watermarking methodology. Finally, Sect. 5 concludes our work of successful implementation with improved performance.

2 Watermarking

2.1 DWT Domain Watermarking

DWT decomposition provides four sub-bands, namely, approximation image of lower resolution (LL) along with its horizontal (HL), vertical (LH), and diagonal (HH) sub-bands [8].

After decomposing first level sub-bands, according to requirement, we can further decompose them to second, third, or higher level. As LL region is lower frequency sub-band, henceforth we usually prefer to embed the watermark in higher frequency sub-bands (HL, LH, and HH), for the purpose of quality maintenance of image data. In LL region during embedding of watermark it can alter the information, thus watermark as useful and important information embedded in higher level sub-bands [14]. This method alone is not fruitful to achieve good robustness. There are various hybrid methodologies that can be utilized for achieving good performance.

2.2 Watermarking with the Use of SVD

SVD is a decomposition method-based linear algebra and utilized for matrix diagonalization. It basically factorizes the digital data available in form of matrix of size P by Q into three numbers of matrices, namely, U and V as unitary matrices of size P by P and Q by Q, respectively, and matrix S is termed as diagonal matrix, also known as singular matrix of size P by Q [15, 16]. Singular matrix plays the crucial role as it is used for embedding. As we know, it is a diagonal matrix; its elements are arranged in diagonal pattern. By making any change in message/data with the use of singular matrix, its singular values won't be changed significantly. SVD method can be applied with various combinations or in different forms. Its singular values either directly can be applied into data or in a modified way, which allows us to embed watermark of any type. Against different types of attacks, this SVD method shows good robustness and with the use in different transform domain, it also represents transparency and imperceptibility. High stability and algebraic property of singular values of image/digital data obtained through SVD are the keys to utilizing it.

2.3 Proposed DWT-SVD-Based Watermarking Technique

In this paper, we have proposed DWT-SVD-based watermarking technique. To analyze the scaling factor effect over PSNR and NCC for the study of imperceptibility and robustness of proposed algorithm. Further, scaling factor has been optimized by firefly algorithm. Presented DWT-SVD watermarking method basically constitutes following steps, namely, watermark embedding and extraction [17, 18]. Block diagram for watermark embedding and extraction is presented in Fig. 1.

Watermark embedding steps:

- We have taken two-dimensional EEG data as host image (H_I) and second-level DWT has been applied over it.

$$\text{DWT}\,(\text{H\_I},\ \text{`HAAR'}) = [\text{LL1},\ \text{HL1},\ \text{LH1},\ \text{HH1}] \tag{1}$$

$$\text{DWT}\,(\text{HL1},\ \text{`HAAR'}) = [\text{LL2},\ \text{HL2},\ \text{LH2},\ \text{HH2}] \tag{2}$$

- Then, we have applied SVD on HL2 sub-band coefficient and watermark logo (W_L).

$$\text{SVD}\,(\text{HL2}) = \left[\text{U}_{\text{W\_I}},\ \text{S}_{\text{W\_I}},\ \text{V}_{\text{W\_I}}\right] \tag{3}$$

$$\text{SVD}\,(\text{W\_L}) = \left[\text{U}_{\text{W\_L}},\ \text{S}_{\text{W\_L}},\ \text{V}_{\text{W\_L}}\right] \tag{4}$$

- Updated singular value of HL2 sub-band coefficient

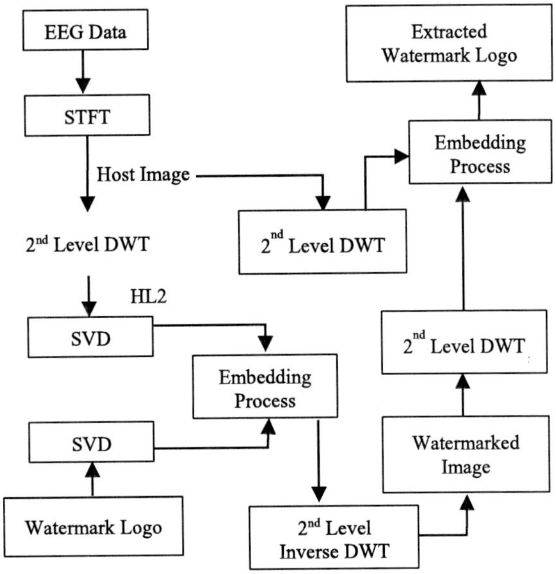

Fig. 1 Block diagram for watermark embedding and extraction

$$S_{W\_I'} = S_W\_I + (\alpha \cdot S_{W\_L}) \quad (5)$$

- Modified HL2′ sub-band coefficients

$$HL2' = U_{W\_I} * S_{W\_I'} * V_{W\_I}{}^T \quad (6)$$

- Finally, second-level IDWT needs to be applied, to get watermarked image.

$$W_M\_I' = IDWT(LL2, HL2', LH2, HH2) \quad (7)$$

$$W_M\_I = IDWT(LL1, W_M\_I', LH1, HH1) \quad (8)$$

Watermark extraction steps:

- For extraction second-level DWT need to be applied on host image (H_I) and watermarked image (W_M_I), that helps to receive HL2 and HL2′ sub-band coefficients.
- Further, we have applied SVD on both of these coefficients.

$$SVD\ (HL2) = [U_{W\_I}, S_{W\_I}, V_{W\_I}] \quad (9)$$

$$SVD(HL2') = [U_{W\_I'}, S_{W\_I'}, V_{W\_I'}] \quad (10)$$

- To extract the watermark logo, singular value of it needs to evaluate

$$S_{W\_L'} = \left(S_{W\_I'} - S_{W\_I}\right)/\alpha \tag{11}$$

- Further extracted watermark logo can be recovered

$$W\_L' = U_{W\_L} * S_{W\_L'} * V_{W\_L}{}^{T} \tag{12}$$

2.4 Particle Swarm Optimization (PSO) to Optimize the Scaling Factor

To optimize scaling factor, we have chosen here PSO method [10–13]. It is a population-based bio-inspired technique which is based on the behavior of particles (birds, fish's swarms, etc.) when they are in group. To find the optimum solution of any problem, it follows the way as they move for searching space and find their best position and velocity either individually (L_{best}) or in group (G_{best}). As it is an iterative process of finding optimum solution, after each iteration particles updates it best value in term of position and velocity. To implement this optimization method random population (n), inertia weight of particle (W), no. of iterations and acceleration coefficient need to take care (A1, A2). Particles update their position (P) and velocity (V) by following equations:

$$V_i(t+1) = W \cdot V_i(t) + A1 \cdot R1 \cdot (L_{best} - P_i(t)) + A2 \cdot R2 \cdot (G_{best} - P_i(t)) \tag{13}$$

$$P_i(t+1) = P_i(t) + V_i(t) \tag{14}$$

Here, R1 and R2 are the random vectors, values belong to (0, 1). To execute this optimization method, we have selected A1 and A2 as 1 and 2, respectively. Considered value of weight is taken 1. The numbers of particle populations taken as 20 with no. of iteration varies from 20 to 60. These updating steps will repeat itself up to the completion of number of iterations.

To find the optimum solution for scaling factor fitness function can be evaluated using the following equation:

$$f\_f = \left|\mathrm{PSNR} - \mathrm{PSNR}_{\mathrm{optimum}}\right| + \left(1 - \frac{1}{N}\sum_{1}^{N} \mathrm{NCC}_i\right) \tag{15}$$

N	number of attacks
PSNR	PSNR value after embedding watermark
$\mathrm{PSNR}_{\mathrm{optimum}}$	Ideal/Preset value
NCC_i	Normalized Correlation Coefficient of ith attack

In this paper, we have selected $\mathrm{PSNR}_{\mathrm{optimum}}$ value as 50. As we know, here we are optimizing our algorithm performance with scaling factor optimization, it is minimization technique. Henceforth, corresponding to decreasing value of f_f, PSNR value reaches toward optimum value and NCC toward 1.

3 Short-Time Fourier Transform of EEG Data

In this paper, publically accessible dataset introduced in [19], given by University of Bonn has been utilized for performing watermark in EEG data. This accessible information is already preprocessed; consequently without utilization of any noise elimination strategy dataset has been utilized as it is.

STFT has been utilized to carry out time-frequency domain conversion of EEG information [20]. We have used Z-class EEG among all accessible. In this work, hamming window of size odd (N/4) is utilized. STFT of message signal x (t) utilizing small length hamming window h (t) is given by the Equation

$$\mathrm{STFT(t, f)} = \mathrm{x(t, f)} = \int_{-\infty}^{\infty} \mathrm{x}(\tau)\mathrm{h}(\tau - \mathrm{t})\mathrm{e}^{(-if\tau)}d\tau \tag{16}$$

4 Results and Discussion

In this paper, a watermark logo is successfully embedded into EEG data obtained after STFT. Here, non-linear EEG data firstly converted into two-dimensional image form of EEG using STFT and used as a host image. An image alphabet “A” is used as a watermark logo.

After embedding watermark logo into host image, we have successfully extracted and recovered the logo. The performance of algorithm has been evaluated through parameters like PSNR and NCC, which indicates imperceptibility and robustness of proposed algorithm. Results are analyzed corresponding to without attack and different kind of attacks a well. In this paper, the results presented are corresponding to Z-class of EEG data. EEG data. Its STFT as host image and watermark logo has been presented in Fig. 2.

Imperceptibility:

With the help of performance parameter PSNR, quality of watermarked image can be visualized and given by the equation:

$$\mathrm{PSNR} = 20\ \log\ 10 \frac{(\text{Maximum Intensity of Image})^2}{\text{MSE}} \tag{17}$$

Fig. 2 **a** Z-class EEG data, **b** STFT, **c** Watermark logo

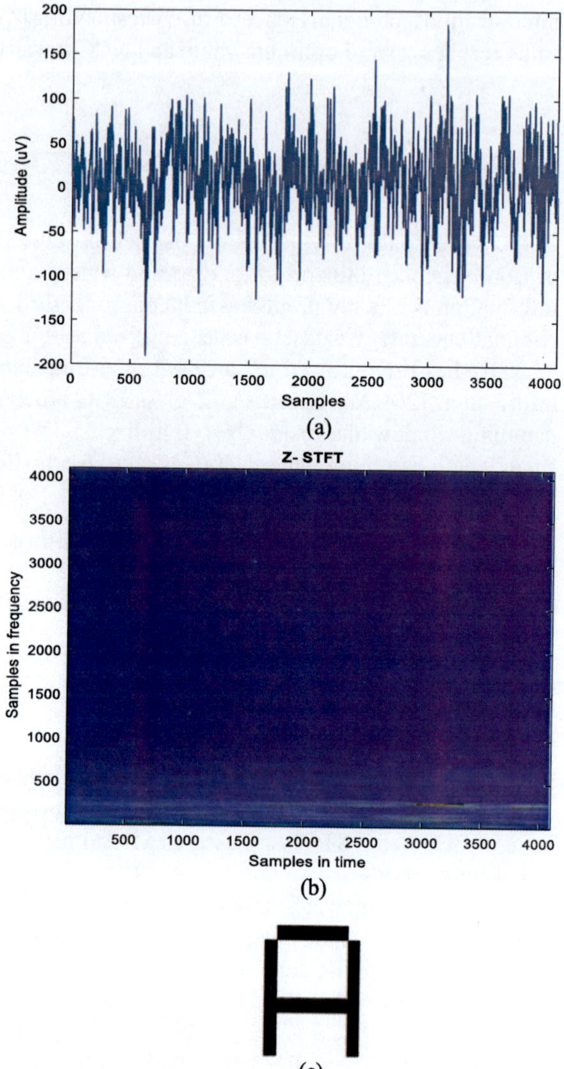

Robustness:

Robustness is an important parameter to observe the performance of proposed algorithm. In this paper, robustness is analyzed against different kinds of attacks. Scaling Rotation, JPEG compression, Gaussian noise (density 0.001), and Low pass filtering are the few of them, against which performance have been presented over here in terms of parameter NCC.

Normalized Correlation Coefficient (NCC) is given by following equation:

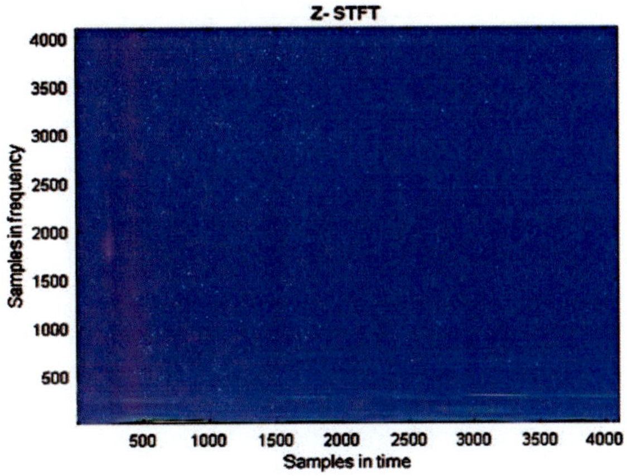

Fig. 3 Watermarked Z-class EEG data

$$\mathrm{NCC}(\mathrm{W\_L}, \mathrm{W\_L}') = \frac{\left(\sum_{i=0}^{R}\sum_{j=0}^{S}\left[\mathrm{W\_L}(\mathrm{I}, \mathrm{j}) \cdot \mathrm{W\_L}'(\mathrm{I}, \mathrm{j})\right]\right)}{\sum_{i=0}^{R}\sum_{j=0}^{S}\left[\mathrm{W\_L}(\mathrm{I}, \mathrm{j})\right]^2} b \quad (18)$$

R by S is the Size of watermark logo and extracted watermark logo. After embedding watermark on EEG data, watermarked image is presented in Fig. 3.

Here, we have presented a comparison table of our work with recent previous work of various researchers that shows performance of our proposed DWT-SVD-PSO-based watermarking method is comparable and highlight good imperceptibility and robustness. Table 1 provides comparative result of proposed algorithm, whereas Table 2 highlights values of PSNR and NCC against various attacks. Extracted watermark logo after different attacks listed in Table 2 is presented in Fig. 4.

NCC gives us the idea about robustness, with Table 2, we can say that our proposed algorithm shows good robustness. Here, extracted watermark logo is very much similar to the applied one. With this proposed watermarking algorithm, we have achieved the good value of PSNR and NCC against each attack that we have considered here.

5 Conclusion

In this paper, a watermarking methodology based on DWT-SVD-PSO is proposed for enhancing the data security. To embed the watermark, second-level intermediate frequency HL sub-band obtained through DWT of STFT version EEG data has been utilized. We have achieved good quality watermarked image and successfully extracted watermark logo from it, which shows good value of PSNR and NCC as a result. Our best result is shown as 55.92 dB of PSNR value and 0.95 of NCC value against JPEG compression.

Table 1 Comparison of Proposed DWT-SVD-FA-based watermarking methodology with already existing FA based methods

Refs.	Method	Logo	Host Image	Results
[10]	A DWT-SVD-based watermarking optimized by PSO	An image of size 64 *64	An image of size 256 *256	PSNR 45.88 dB NCC 0.91
[11]	optimized (DCT-PSO) color image watermarking	Watermark logo	Standard images like lena, baboon, tiger, peepers	PSNR 41–42 dB NCC 0.80–0.99
[12]	DWT-DCT with PSO optimization	Watermark logo	image of a man and cameraman	PSNR 26.54–27.12 dB NCC 0.89–0.93
[13]	DWT, SVD and PSO based digital image watermarking	Watermark logo	Standard lena image	PSNR 48–56 dB NCC 0.90–0.99
This work	DWT-SVD-PSO based watermarking	Watermark logo	Bonn's EEG dataset	PSNR 55.92 dB NCC 0.87–0.95

Table 2 Values of PSNR and NCC against different attacks for S class EEG data

S. no	Type of attack	PSNR (dB)	NCC
1.	No attack	56.22	0.97
2.	Scaling	51.83	0.92
3.	Rotation clockwise by 5°	53.12	0.90
4.	JPEG compression by 50%	55.92	0.95
5.	Gaussian noise (density 0.001)	49.72	0.87
6.	Low pass filtering of window size 2 by 2	51.65	0.93

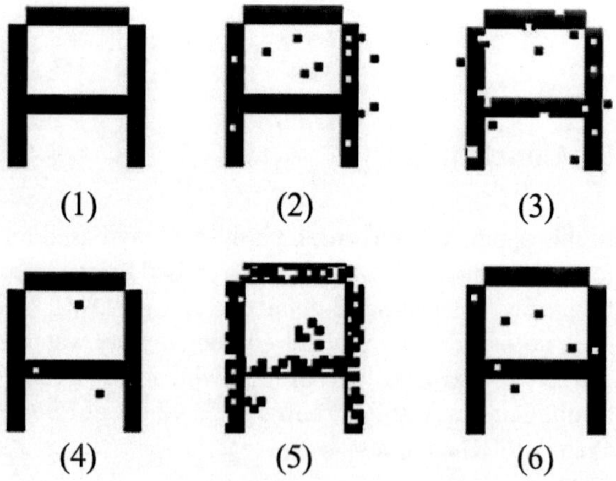

Fig. 4 Extracted watermark logo from watermarked image after (1) No attack, (2) Scaling, (3) Rotation, (4) JPEG compression, (5) Gaussian noise, (6) Low pass filtering

Acknowledgements Authors are grateful to R. G. Andrzejak, who has made the EEG database publically available.

References

1. Garg, P., Kishore, R.: Performance comparison of various watermarking techniques. Multimed. Tools Appl. **79**, 25921–25967 (2020)
2. Priya, S., Santhi, B., Swaminathan, P., Mohan, J.R.: Hybrid transform based reversible watermarking technique for medical images in telemedicine applications. Optik Int. J. Light Electron Opt. **145**, 655–671 (2017)
3. Kazemivash, B., Moghaddam, M.E.: A robust digital image watermarking technique using lifting wavelet transform and firefly algorithm. Multimed. Tools Appl. **76**, 20499–20524 (2017)
4. Amrit, P., Anand, A., Kumar, S., Singh, A.K.: Robust transmission of medical records using dual watermarking and optimization algorithm. In: Journal of Physics: Conference Series, 2020 International E- Conference on Data Analytics, Intelligent Systems and Information Security (ICDIIS), vol. 1767, pp. 1–10 (2021)
5. Thakur, S., Singh, A.K., Ghrera, S.P., Elhoseny, M.: Multi-layer security of medical data through watermarking and chaotic encryption for tele-health applications. Multimed. Tools Appl. **78**(3), 3457–3470 (2019)
6. Baba Ahmadi, S.B., Zhang, G., Jelodar, H.: A robust hybrid SVD-based image watermarking scheme for color images. In: IEEE 10th Annual Information Technology, Electronics and Mobile Communication Conference, IEMCON, pp. 682–688 (2019)
7. Pandey, M.K., Parmar, G., Gupta, R., Sikander, A.: Lossless robust color image watermarking using lifting scheme and GWO. Int. J. Syst. Assur. Eng. Manag. **11**(2), 320–331 (2020)
8. Agarwal, C., Mishra, A., Sharma, A., Bedi, P.: Optimized gray-scale image watermarking using DWT-SVD and Firefly Algorithm. Expert Syst. Appl. **41**(17), 7858–7867 (2014)
9. Ariatmanto, D., Ernawan, F.: An improved robust image watermarking by using different embedding strengths. Multimed. Tools Appl. **79**, 12041–12067 (2020)
10. Kulkarni, B.P., Sai, K.S., Meenakshi, K., Padmavathi, K., Swaraja, K.: Performance analysis of optimization algorithms GA, PSO, and ABC based on DWT-SVD watermarking in OpenCV Python environment. In: 2020 International Conference for Emerging Technology (INCET), pp. 1–5 (2020)
11. Garg, P., Kishore, R.: Optimized color image watermarking through watermark strength optimization using particle swarm optimization technique. J. Inf. Optim. Sci. **41**, 1–14 (2020)
12. Yadav, N., Rajpoot, D., Dhakad, S.K.: Optimization of watermarking in image by using particle swarm optimization algorithm. In: 2020 International Conference on Signal Processing and Communication (ICSC), pp. 85–90 (2020)
13. Sanku, D., Kiran, S., Takore, T.T., Kumar, P.R.: Digital image watermarking in RGB Host Using DWT, SVD, and PSO techniques. In: Lecture Notes in Electrical Engineering, pp. 333–342 (2018)
14. Araghi, T.K., Manaf, A.A., Araghi, S.K.: A secure blind discrete wavelet transform based watermarking scheme using two-level singular value decomposition. Expert Syst. Appl. **112**, 208–228 (2018)
15. Alshoura, W.H., Zainol, Z., Teh, J.S., Alawida, M., Alabdulatif, A.: Hybrid SVD-based image watermarking schemes: a review. IEEE Access **9**, 32931–32968 (2021)
16. Abdelhakim, A.M., Saad, M.H., Sayed, M., Saleh, H.I.: Optimized SVD-based robust watermarking in the fractional Fourier domain. Multimed. Tools Appl. **77**, 27895–27917 (2018)
17. Anand, A., Singh, A.K.: An improved DWT-SVD domain watermarking for medical information security. Comput. Commun. **152**, 72–80 (2020)

18. Zermi, N., Khaldi, A., Kafi, R., Kahlessenane, F., Euschi, S.: A DWT-SVD based robust digital watermarking for medical image security. Forensic Sci. Int. 320 (2021)
19. Andrzejak, R.G., Lehnertz, K., Mormann, F., Rieke, C., David, P., Elger, C.E.: Indications of nonlinear deterministic and finite-dimensional structures in time series of brain electrical activity: dependence on recording region and brain state. Phys. Rev. E Stat. Nonlin. Soft. Matter Phys. **64**(6), 1–8 (2001)
20. Sameer, M., Gupta, B.: ROC analysis of EEG subbands for epileptic seizure detection using Naive Bayes classifier. J. Mob. Multimed **17**(1), 299–310 (2021)

Urgent Text Detection in Bengali Language Based on Boosting Techniques

Rafsan Rahman, Tamanna Nazmin, Noor Nafeur Rahman, Miyad Bhuiyan, G. M. Shahariar, and Faisal Muhammad Shah

Abstract This paper presents a learning approach on a unique dataset formulated by authors that detects urgent texts from the posts on social media platforms in Bengali language. It is difficult to keep track of every information we go through social media. In the collision of numerous posts, it is easy to miss information that is urgent. In this advanced era of machine learning, detecting urgent texts among thousands of posts would be much easier if we can implement a model that can filter the urgent text out of them. Therefore, we propose an approach that can identify any type of urgent texts from public posts by leveraging a manually constructed dataset that is fully human annotated. Apart from traditional machine learning classifiers, we applied boosting algorithms in our proposed method in addition. Experimentally, a significant increase in accuracy has been noticed by boosting weak learners. Support Vector Machine (SVM) achieved 80.9% accuracy where gradient boosting outperformed the traditional approach with 82% accuracy while detecting urgent texts in Bengali language.

Rafsan Rahman, Tamanna Nazmin, Noor Nafeur Rahman, Miyad Bhuiyan—These authors contributed equally to this work.

R. Rahman (✉) · T. Nazmin · N. N. Rahman · M. Bhuiyan · G. M. Shahariar · F. M. Shah
Department of Computer Science and Engineering, Ahsanullah University of Science and Technology, Dhaka, Bangladesh
e-mail: rafsanrahman35549@gmail.com

T. Nazmin
e-mail: tamanna.naz98@gmail.com

N. N. Rahman
e-mail: nafirahman27@gmail.com

M. Bhuiyan
e-mail: miyadbhuiyan@gmail.com

G. M. Shahariar
e-mail: shahariar_shibli.cse@aust.edu

F. M. Shah
e-mail: faisal.cse@aust.edu

S. Hossain et al. (eds.), *Proceedings of International Conference on Fourth Industrial Revolution and Beyond 2021*, Lecture Notes in Networks and Systems 437,
https://doi.org/10.1007/978-981-19-2445-3_49

Keywords Urgent text · Boosting algorithms · Machine learning · Classifiers

1 Introduction

With the growing usage of technology, we are getting more dependent on social media constantly. We convey our thoughts, socialize with everyone, and seek help through social media platforms. The number of posts and information we go through every day is huge. It is difficult to keep track of each post, unnecessary to some extent. Nevertheless, some of them are very important for us to know. We may experience miserable consequences for missing them out. Among them, some texts are called urgent. Urgency can be defined as something that requires rapid action. When a situation of grave danger requires us to act now, then that is a form of urgency. In our day-to-day life, we confront urgent situations that need our fastest response possible. Posting an urgent text on social media has two advantages. Firstly, it connects us with a wider range of people. Secondly, we can get responses faster.

We increased our scrutiny in this matter and gained a huge motivation to work on the detection of urgent texts. But this is a narrow area of research in NLP. The volume of literature published in this field is not significant. Most of them focused on urgency detection from social media posts in crises like disasters in paper [1–5]. All of the researches are conducted in English language. A few works explored some other languages like Kayi et al. [2]. To date, there has been no work in Bengali language in this area. Moreover, researches conducted in other languages focused only on detecting urgent texts, posted in a specific situation like disaster and crisis. But urgency can be raised in any situation. These are the reasons we come up with an idea that can detect any type of urgent text in Bengali language. Regardless of the progress in the area of NLP, the major problem of limited essential resources in Bengali language persists. As Bengali is the sixth most spoken language in the world, our interest thrived even more to work on this topic in Bengali. A fully human-annotated dataset has been formulated for that purpose.

Our approach is completely unique for the following aspects.

1.1 Human-Annotated Real-Life Dataset

For our research purpose, we have constructed a unique dataset that is human annotated. For that purpose, we have searched on popular social media platforms in Bangladesh. We have found that there are approximately 88% of Facebook users and 3% of Twitter users in 2021. These two platforms are optimal sources of data to achieve our target. Therefore, 3,372 Facebook posts and Tweets have been taken to detect urgency. Finally, a balanced dataset containing 3,034 data has been constructed for this research purpose.

1.2 Versatility in Urgent Text Detection

In previous works on the detection of urgent text in various languages, the definition of "urgent" texts was enclosed within a particular situation. In [1–5], authors have established their approaches to detect urgency status from crises. The authors of [6] worked on detecting email containing requests for action. Again in [7], detection of real-time events in earthquake has been observed. To our best knowledge, there is no prior work that has considered every type of urgent text in any language. Therefore, our research is implemented on a unique idea. For example, হার্ট সার্জারির জন্য বি পজেটিভ রক্ত দরকার। আজ দুপুর ২টায় অপারেশন করা হবে। (B-positive blood is needed for heart surgery. The operation will be performed at 2 pm today.) বাংলাদশে সরকার কর্তক ঘোষিত লকডাউনের প্রেক্ষিতে বাংলাদেশের সকল ভারতীয় ভিসা আবেদন কেন্দ্র পরবর্তী বিজ্ঞপ্তি না দেয়া পর্যন্ত বন্ধ থাকবে।(Due to the lockdown announced by the Government of Bangladesh, all Indian Visa Application Centers in Bangladesh will remain closed until further notice.)

These texts are also very urgent in our everyday life. Missing them out can cost us badly. Detection of such urgency is equally important as detection of crisis posts.

1.3 Implementation of Boosting Algorithm

In this paper, we have introduced an approach that uses boosting techniques to detect urgent text detection in Bengali besides the use of traditional machine learning classifiers. We have shown how boosting techniques beat the accuracy of machine learning classifiers. No work in the area of NLP that deals with the detection of urgent text in Bengali language so far.

2 Related Works

Recent studies have shown the versatility of crisis-related data on social media for calamity response. Several approaches have been developed to classify, extract, and summarize crisis-relevant information from social media. Kayi et al. [2] released two datasets in English and low-resource languages and implemented cross-lingual embedding models that are zero-shot learners to detect urgent tweets in natural crises. Traditional machine learning and deep learning algorithms such as SVM, Random Forest, CNN, and MLP were adopted. Urgency varies in context and situation. Giving the importance to the context Alshehri [8] experimented on official emails to detect important and urgent texts. The author labeled the dataset in various categories depending on the available time and level of importance of a particular task assigned through emails. He implemented ordinal regression, Decision Tree j48, and neural network on his model. 93.33%, 95.55%, and 96.66% accuracies were gained, respectively. Guo et al. [9] exploited a hybrid neural network combining CNN, LSTM, and Char-CNN to gain both word-level representation and character-level representation to detect urgent posts from the MOOC forum. Kejriwal and Zhou [3] demonstrated low supervision and transfer learning methodologies to detect urgency in short crisis texts. Transfer learning outperformed others with a 69.4% F-measure. Lampert et al. [6] applied a supervised machine learning algorithm including and excluding N-gram features and zone classifiers. Comparing the approaches, the inclusion of n-gram features and zone classifier gained more accuracy. Several non-lexical features and lexical features had been used. Nguyen et al. [4] experimented with both binary classification and multi-class models to fetch informative tweets from crisis-related tweets. The binary classification model progressed with 78% accuracy whereas the multi-label classification model escalated up to 60% accuracy. CNN was applied as a deep learning algorithm in this research. It is a known fact that texts of different contexts and subjects exhibit distinctive accuracies in particular models. In order to identify the most responsive traits, Aipe et al. [1] explored multi-label classification of crisis-related tweets. Researchers categorized seven types of tweets and applied seven separate neural network classifiers to train the model. Word2Vec model was used to generate a matrix representation of tweet text. Imran et al. [5] implemented Naive Bayesian classifier to classify tweets into many categories of information to extract information nuggets from disaster-related texts. Various features were used for different categories. Among them, unigrams, bigrams, POS tags, and the Verbnet classes were common in all the features. Caragea et al. [10] established a comparison between different algorithms such as Naive Bayes, SVM, ANN, and CNN to identify informative texts from tweets regarding floods. CNN outperformed all the other algorithms with 85.52% accuracy. It is clearly noticeable that the papers on this branch focus on a particular type of urgency detection, e.g., urgency in natural disasters. To our best knowledge, there is no research that has been performed on detecting urgent texts of numerous types involving our day-to-day life in any language. That being the case, researchers achieved the inspiration to work on this important topic.

3 Dataset Construction

3.1 Data Collection

We have constructed a new dataset for our research purpose as there is no previous work on urgent text detection in Bengali language. Data for our study were retrospectively collected from two eminently popular social media platforms: Facebook and Twitter. Assorted Facebook pages and tweets that may contain urgent posts were our main focus. These include popular blood banks, renowned newspapers, news channels, and education. The dataset was collected manually by professional data collectors and completely preprocessed and annotated by the authors.

3.2 Annotation Process

For the sake of accurate annotation, we have classified the collected data into four categories. The annotation process comprises the following four levels depending on the level of urgency:

- **Immensely Urgent**: The aspect of the post refers to extremely urgent and unignorable. This category is annotated as "3" ইমারজেন্সি অপারেশন এর জন্য আজ বিকালে ১ ব্যাগ ও পজেটিভ রক্ত প্রয়োজন (e.g., a bag of O positive blood is needed today afternoon for emergency operation).
- **Clearly Urgent**: The post conveys urgency certainly but the urgency level is not as high as an Extremely Urgent post. This category is annotated as "2" এখন কোন ভ্যাক্সিন দিচ্ছে? আজ গেলে কি ভ্যাক্সিন পাব? (e.g., is the vaccination process still ongoing? Will I get a vaccine today?).
- **Partially Urgent**: The post is skeptical of being urgent and most likely seems to be non-urgent. This category is annotated as "1" কোন রুট দিয়ে গেলে খরচ কম পড়বে? (e.g., which route will cost less?).
- **Not Urgent**: The post is definitely not urgent. This category is annotated as "0" বাংলাদেশের অনেকেই কমবেশি ফ্যাটি লিভার রোগে আক্রান্ত (e.g., many people from Bangladesh are suffering from fatty liver disease).

The dataset contains 3,372 Bengali posts that include 43 duplicate values and 2 blanks. The number of data after removing those rows is 3,327. The distribution of data is shown in Table 1.

All of the annotators were queried and everyone agreed to the fact that some data are extremely confusing to label directly as "non-urgent" or "urgent". This type of data seemed to belong in both categories. But if we set multiple levels of urgency status, it gets easier to decide which level of urgency a data belongs to. And later on all the levels were converted into binary set of tags. We merged the "Partially urgent" and "Not urgent" labels into "0" (non-urgent) as they are more likely to be less significant in the scale of urgency and the rest of them into "1" (urgent). The authors of [2] have applied the same approach to label their dataset and got benefited in the annotation process. After the conversion, we acquired 1,760 urgent and 1,567 non-urgent data, which still represents an imbalanced dataset. For the sake of building a balanced dataset, we finally removed a small amount of data. Eventually,

Table 1 Annotated data before the conversion into binary classification

Category	Amount
Immensely urgent	988
Clearly urgent	772
Partially urgent	326
Not urgent	1241

Table 2 Data preprocessing and classification of the preprocessed data

Raw text	Preprocessed text	Text classification
একজন প্রসূতি মায়ের অপারেশনের জন্য আর্জেন্ট রক্তের প্রয়োজন। রক্তের গ্রুপঃ এ + রক্তের পরিমাণঃ ১ ব্যাগ রক্তদানের স্থানঃ যশোর সদর হাসপাতাল যোগাযোগঃ ০১৩০৩ ৮৫৩০৫২	একজন প্রসূতি মায়ের অপারেশনের জন্য আর্জেন্ট রক্তের প্রয়োজন রক্তের গ্রুপ এ পজেটিভ রক্তের পরিমাণ ১ ব্যাগ	Urgent text
চুরি নিয়ে নির্মিত কয়েকটি সিনেমার নাম বলবেন।উল্লেখ, চুরির সিনের পাশাপাশি অনেক অ্যাকশনও থাকা লাগবে। অগ্রিম ধন্যবাদ।	চুরি নিয়ে নির্মিত কয়েকটি সিনেমার নাম বলবেন।উল্লেখ, চুরির সিনের পাশাপাশি অনেক অ্যাকশনও থাকা লাগবে। অগ্রিম ধন্যবাদ।	Non-urgent text

the concluding dataset consists of 3,034 data that possesses 1,517 urgent and 1,517 non-urgent data.

A detailed description of our dataset has been presented in this section. The dataset contains a total of 3,034 Bengali posts. After the conversion to binary labeling, we were able to pull out a balanced dataset consisting of 1,517 urgent and 1517 non-urgent data.

Table 2 represents an overview of our labeled dataset construction after preprocessing.

4 Methodology

In this section, we will be discussing our methodology of urgent text detection broadly. Figure 1 represents our proposed approach. We have distributed our proposed approach into five phases. At the onset, our first phase contains dataset preparation. The second phase involves dataset preprocessing by utilizing a few natural language preprocessing techniques. Subsequently, the third phase consolidates the feature selection process. This process includes Term Frequency-Inverse Document Frequency (TF-IDF) and N-grams for both traditional machine learning models and boosting algorithms. The fourth phase is the preeminent phase of our proposed method that is the appliance of multiple machine learning models and boosting techniques to classify text as urgent or non-urgent. At last, we have examined the performance of each algorithm and presented the best performed one in the last phase. In the following subsections, we have discussed all the phases briefly.

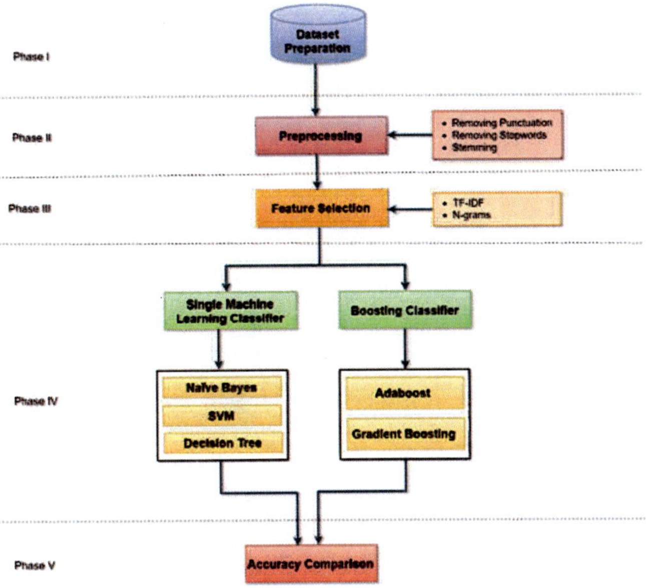

Fig. 1 Proposed model for urgent text detection

4.1 Phase I-Dataset Preparation

We have gathered 3,034 data from Facebook and Twitter written in Bengali. The overall dataset construction process is described in Sect. 3.

4.2 Phase II-Preprocessing

In the field of NLP, data preprocessing is the first step of building a model. Depending on how well the data has been preprocessed, the results are seen. It transforms text into a more digestible form so that machine learning algorithms can perform better. The foremost problems are the fact that the posts and tweets from Facebook and Twitter comprise abounding unnecessary information, symbols, and errors. In order to clean the dataset, we have gone through the following steps of preprocessing.

Removing and altering unnecessary elements There are countless errors in the public posts of social media. Facebook posts are very noisy and mostly contain errors. On the other hand, Twitter posts are very short and abbreviated that it is difficult to understand the meaning of a post sometimes. For the sake of a decent dataset, we applied the following steps:

- Converting significant English words into Bengali: Conversion of English words into Bengali was completely terminated by the authors manually. The authors faced two situations while doing it.

(1) Some English words may have several meanings. It was ensured that the exact semantic meaning was extracted according to the context while converting a particular English word into Bengali, e.g., "**Fair**" ভাবে খেলুন। অতি আক্রমনাত্নক খেললে অনেক দুর্ঘটনা ঘটতে পারে . In English, "Fair" describes someone who has light skin or hair. It also means a gathering of people for a variety of entertainment or commercial activities. But in this sentence, it means playing without cheating or trying to achieve an unjust advantage. That is why it has been converted to ন্যায্য ভাবে খেলুন। অতি আক্রমনাত্নক খেললে অনেক দুর্ঘটনা ঘটতে পারে।

.

(2) Some English words are used directly in the Bengali language. The Bangla meaning of such words is not popularly used. Instead of converting them directly into Bengali we simply represented the English spelling in Bangla letters, e.g., "৫ বছরের শিশুর জন্য এক ব্যাগ **O-negative** রক্ত লাগবে।" . Here, the word negative means "নেতিবাচক" which is not widely used in the given context. Therefore, it has been converted to "৫ বছরের শিশুর জন্য এক ব্যাগ ও-নেগেটিভ রক্ত লাগবে।" . We did not keep any English words in our dataset because an enormous amount of English words may ruin the significance of learning Bengali in the model.

- Altering unintentional spelling mistakes: Spelling mistakes are fully corrected by human annotators manually. Four annotators were engaged to check the spelling of the complete dataset.
- Removing names, contacts, addresses, duplicate data, links and URLs, emoticons, hashtags and underscores and unnecessary signs convert them if necessary.

Stop word removal Stop words are a set of commonly used words of less significance in a language. The set of words contained in "stopwords" from corpus class of BNLP (Bengali Natural Language Processing) [11] toolkit are considered as stop words, e.g., "এই", "তবু", "যেন" etc. Stop words are commonly used in Text Mining and Natural Language Processing (NLP) to eliminate words that are so commonly used that they carry very little useful information.

Stemming Stemming is the process of reducing inflection in words to their root forms such as mapping a group of words to the same stem even if the stem itself is not a valid word in the Language. Stemming drastically improves the size of the index (or feature space).

It is not rational if all the steps stated for preprocessing are depleted with coding. The result will still enclose unnecessary elements for some of the steps. For a precisely preprocessed dataset, we have accomplished some of the steps manually by human annotators.

4.3 Phase-III Feature Selection

Feature selection is the third phase of our proposed model. We have used two feature extraction approaches: N-grams and TF-IDF. For traditional machine learning classifiers, we have used TF-IDF and N-grams methods.

N-gram An N-gram is an N-character slice of a longer string [12].

N-gram creates a document-term matrix where columns represent all columns of adjacent words of length n and cells represent count. It helps machines to get a better understanding of the meaning of the words in a particular context.

TF-IDF TF-IDF (term frequency-inverse document frequency) is a technique that measures how important a word is to a document in a collection of documents. Term Frequency (TF) measures how frequently a word occurs in a document. Suppose we have a number of N documents. Define fij to be the frequency of word i in document j. Then, the term frequency $TFij$ can be defined as follows:

$$TFij = \frac{f_{ij}}{max_k f_{kj}}$$

Inverse Document Frequency (IDF) measures how significant a word is. Suppose word i appears in ni of the N documents in the collection. Then, the Inverse Document Frequency (IDFi) can be defined [13] as follows:

$$IDF_i = log_2(N/n_i)$$

Each word or term has its own TF and IDF score. The TF and IDF product scores of a term are referred to the TF*IDF weight of that term. Simply we can state that the higher the TF*IDF score (weight) the rarer the term and vice versa.

4.4 Phase-IV Appliance of Different Algorithms

Traditional Machine Learning Classifiers We have implemented multiple machine learning classifiers to choose the most suitable classifier for our model. The classifiers are Naive Bayes (NB), Decision Tree (DT), and Support Vector Machine (SVM). Furthermore, we have applied Unigram, Bigram, and Trigram and also extracted the features using TF-IDF individually. Multiple combinations of N-grams and TF-IDF such as TF-IDF and Bigram, TF-IDF and Trigram were also used to examine the accuracy rate. Later, we have split the dataset into 70:30 and 80:20 training set-test set ratios to examine the performances. Additionally, we have run the model under fivefold and tenfold cross validation also.

Boosting algorithms Boosting is an ensemble learning strategy for minimizing training errors by combining a group of weak learners into a strong learner. A chunk of data is chosen, fitted with a model, and then trained progressively in boosting, that is, each model attempts to compensate for the shortcomings of its predecessor and generate a strong prediction rule. In this paper, our main goal is the appliance of Gradient Boosting and Adaptive boosting (AdaBoost) techniques.

The concept of transforming weak learners into strong learners with improved accuracy was initially proposed by the authors of [14]. The authors of [15, 16] then followed suit, making the boosting popular. They used it as a functional logistic regression approximation. “Gradient Boosting” is the name given to this method of issue solving. Gradient Boosting is a robust ensemble method which can deal well with the sparsity of dataset. It can optimize different loss function and provides several hyperparameter options that make the function fit with flexibility. This is why Gradient Boosting performs so well on binary classification models. We have also used AdaBoost algorithm [17] based on Decision Tree classifier for its performance with weak learners.

In our research, both the algorithms were trained and validated with the same train-test splits and validation sets after the feature selection process was conducted. The boosting algorithms used here generally have several hyperparameters. We used Grid Search to find the best hyperparameter set. The simulation procedure is the same as the traditional classification approach keeping the training data and feature selection techniques same for comparison purposes.

5 Experimental Result

In this section, we displayed the performances of all the algorithms with the combination of features that gives us the best accuracy individually. We conduct our observation by splitting the dataset into 80:20 and 70:30 train-test set ratios as well as implementing tenfold and fivefold cross validation for both machine learning classifiers and boosting algorithms. K-fold cross validation is used to estimate how precise the accuracy of a model on a limited data sample presents. We have evaluated the performance in both approaches of splitting and cross validating. Additionally, we have examined the different effects of N-gram features such as Unigram, Bigram, Trigram and some combinations like TF-IDF and Bigram, TF-IDF and Trigram. The simulation procedure is the same in both traditional classifiers and boosting algorithms approaches. We kept the training data and feature selection techniques same for the sake of ideal comparison.

In Tables 3 and 4, we have shown the accuracy, precision, recall, and F1-scores for each of the algorithms with the combination of features that gives the best performance individually. Table 3 specifies the performance record for 80:20 split and 70:30 split ratio. Table 4 specifies the performance record of each algorithms for both fivefold and tenfold cross validation.

We can observe that the Gradient Boosting algorithm on 80:20 ratio while using only Unigram feature gives us the best accuracy of 82%. Gradient Boosting also outperforms other algorithms in 70:30 training-test set ratio with 81.6% of accuracy. Among the traditional machine learning approach, SVM on tenfold cross validation

Table 3 Comparison of different algorithms using best feature performed with train-test ratio 80:20 and 70:30

Classifier	80:20					70:30				
	Technique	Accuracy	Precision	Recall	F1-score	Technique	Accuracy	Precision	Recall	F1-score
Naïve Bayes (NB)	Trigram+ TF-IDF	0.764	0.859	0.586	0.696	Trigram+ TF-IDF	0.760	0.881	0.568	0.691
SVM	TF-IDF	0.797	0.818	0.721	0.767	TF-IDF	0.802	0.833	0.729	0.777
Decision Tree (DT)	Bigram+ TF-IDF	0.759	0.796	0.643	0.711	Bigram	0.763	0.815	0.645	0.720
AdaBoost	Unigram	0.791	0.857	0.695	0.768	TF-IDF	0.805	0.864	0.718	0.784
Gradient Boosting	Unigram	**0.820**	0.930	0.696	0.796	Unigram	**0.816**	0.880	0.741	0.805

Table 4 Comparison of different algorithms using best feature performed with fivefold and tenfold cross validation

Classifier	5-fold					10-fold				
	Technique	Accuracy	Precision	Recall	F1-score	Technique	Accuracy	Precision	Recall	F1-score
Naïve Bayes (NB)	Bigram	0.659	0.611	0.873	0.719	Bigram	0.660	0.641	0.866	0.718
SVM	TF-IDF	**0.804**	0.584	0.857	0.695	TF-IDF	**0.809**	0.582	0.870	0.697
Decision Tree (DT)	Bigram	0.756	0.832	0.642	0.724	Bigram	0.756	0.823	0.651	0.727
AdaBoost	Unigram	0.800	0.872	0.704	0.778	Unigram	0.800	0.869	0.705	0.778
Gradient Boosting	Unigram	0.803	0.854	0.719	0.783	TF-IDF	0.804	0.860	0.717	0.783

shows the best performance with TF-IDF feature selection. From Table 4, it is clearly noticeable that SVM performed really well on tenfold cross validation while being trained with TF-IDF only. It shows 80.9% accuracy.

With selection of suitable feature, each algorithm shows a very close performance. It has already been mentioned that we have used grid search technique to find the best parameter tuning for boosting algorithms. On a final note, it can be concluded that Gradient Boosting is the best algorithm for our proposed methodology while the dataset is split into 80:20 ratio. Unigram is the suitable feature in this case.

6 Conclusion

In this research, we established a dataset that includes a variety of urgent texts that we convey in our daily life in Bengali language. All the data was collected from Facebook posts and Tweets. The annotation process of the full data was human conducted. After the dataset was fully ready and preprocessed precisely, we experimented it with different machine learning classifiers as well as boosting algorithms to detect whether a text is urgent or not. Finally, we established a comparison between the performances of those classifiers using different features. The result provides crucial insights to conclude that the Gradient Boosting algorithm provides the highest accuracy for our proposed approach.

Despite the dataset being difficult to construct in Bengali language, we tried to gather every type of urgent texts and annotated them on the level of their urgency. But if we could increase the amount of the data, the model could be trained more efficiently which will enhance the accuracy of our methodology hopefully. In future, our research will include the appliance of deep learning and word embedding models after we construct an adequate dataset as deep learning-based approaches require a huge amount of data to automatically extract complex representation. Additionally, we will advance the dataset with multi-label classification where machine can predict different categories of urgency.

References

1. Aipe, A., Mukuntha, N., Ekbal, A., Kurohashi, S.: Deep learning approach towards multi-label classification of crisis related tweets. In: Proceedings of the 15th ISCRAM Conference (2018)
2. Kayi, E.S., Nan, L., Qu, B., Diab, M., McKeown, K.: Detecting urgencystatus of crisis tweets: a transfer learning approach for low resource languages. In: Proceedings of the 28th International Conference on Computational Linguistics, pp. 4693–4703 (2020)
3. Kejriwal, M., Zhou, P.: Low-supervision urgency detection and transfer in short crisis messages In: Proceedings of the 2019 IEEE/ACM International Conference on Advances in Social Networks Analysis and Mining, pp. 353–356 (2019)
4. Nguyen, D.T., Joty, S., Imran, M., Sajjad, H., Mitra, P.: Applications of online deep learning for crisis response using social media information (2016). arXiv:1610.01030
5. Imran, M., Elbassuoni, S., Castillo, C., Diaz, F., Meier, P.: Extracting information nuggets from disaster-related messages in social media. In: ISCRAM (2013)
6. Lampert, A., Dale, R., Paris, C.: Detecting emails containing requests for action. In: Human Language Technologies: The 2010 Annual Conference of the North American Chapter of the Association for Computational Linguistics, pp. 984–992 (2010)
7. Sakaki, T., Okazaki, M., Matsuo, Y.: Earthquake shakes twitter users: real-timeevent detection by social sensors. In: Proceedings of the 19th International Conference on Worldwide Web, pp. 851–860 (2010)
8. Alshehri, Y.A.: Text mining for incoming tasks based on the urgency/importance factors andtask classification using machine learning tools. In: Proceedings of the 2020 the 4th International Conference on Compute and Data Analysis, pp. 183–189 (2020)
9. Guo, S.X., Sun, X., Wang, S.X., Gao, Y., Feng, J.: Attention-based character-word hybrid neural networks with semantic and structural information for identifying of urgent posts in MOOC discussion forums. IEEE Access **7**, 120522–120532 (2019)
10. Caragea, C., Silvescu, A., Tapia, A.H.: Identifying informative messages in disaster events using convolutional neural networks. In International Conference on Information Systems for Crisis Response and Management, pp. 137–147 (2016)
11. Sarker, S.: BNLP: Natural language processing toolkit for Bengali language (2021). arXiv:2102.00405
12. Cavnar, W.B., Trenkle, J.M., et al.: N-gram-based text categorization. In: Proceedings of SDAIR-94, 3rd Annual Symposium on Document Analysis and Information Retrieval, vol. 161175 (1994)
13. Rajaraman, A., Ullman, J.D.: Mining of Massive Datasets. Cambridge University Press, Cambridge (2011)

14. Schapire, R.E.: The strength of weak learnability. Mach. Learn. **5**(2), 197–227 (1990)
15. Friedman, J., Hastie, T., Tibshirani, R.: Additive logistic regression: a statistical view of boosting (with discussion and a rejoinder by the authors). Ann. Stat. **28**(2), 337–407 (2000)
16. Friedman, J.H.: Greedy function approximation: a gradient boosting machine. Ann. Stat. 1189–1232 (2001)
17. Freund, Y., Schapire, R.E.: A decision-theoretic generalization of on-line learning and an application to boosting. J. Comput. Syst. Sci. **55**(1), 119–139 (1997)

A Hybrid Deep Learning Approach to Detect Bangla Social Media Hate Speech

Tapotosh Ghosh, Ashraf Alam Khan Chowdhury, Md. Hasan Al Banna, Md. Jaber Al Nahian, M. Shamim Kaiser, and Mufti Mahmud

Abstract Social media has become an integral part of our day-to-day life. In our activities or posts on social media, the presence of hate speech written in the native language or English has increased significantly. It often leads to the spread of negativity, depression, or even sometimes considered cybercrime. In this paper, a hybrid deep learning approach has been taken to detect Bangla social media hate speech using fastText embedding, Long Short-Term Memory (LSTM), Gated Recurrent Unit (GRU), and Convolutional Neural Network. A publicly available dataset of 30000 samples has been used, and the proposed hybrid model achieved close to 90% accuracy with significant sensitivity and specificity in Bangla hate speech detection. Several related deep learning approaches were evaluated in this same dataset, but none of them performed better than the proposed model. The hybrid model also showed robustness which made it more suitable for this task.

Tapotosh Ghosh and Ashraf Alam Khan Chowdhury have contributed equally.

T. Ghosh (✉)
Department of Computer Science and Engineering, United International University, Dhaka, Bangladesh
e-mail: tapotoshghosh@gmail.com

A. A. K. Chowdhury
Northern University Bangladesh, Dhaka, Bangladesh

Md. H. A. Banna · Md. J. A. Nahian
Bangladesh University of Professionals, Dhaka, Bangladesh
e-mail: hasan.banna@bup.edu.bd

Md. J. A. Nahian
e-mail: jaber.nahian@bup.edu.bd

M. S. Kaiser
Jahangirnagar University, Savar, Dhaka, Bangladesh
e-mail: mskaiser@juniv.edu

M. Mahmud
Nottingham Trent University, Clifton Campus, Nottingham NG11 8NS, UK
e-mail: mufti.mahmud@ntu.ac.uk; mufti.mahmud@gmail.com

S. Hossain et al. (eds.), *Proceedings of International Conference on Fourth Industrial Revolution and Beyond 2021*, Lecture Notes in Networks and Systems 437,
https://doi.org/10.1007/978-981-19-2445-3_50

Keywords Hate speech · Negativity · LSTM · Hybrid architecture · Embedding

1 Introduction

Social media is one of the easiest and most popular means of communication and expressing emotions. People share their daily activities, feelings, emotions, and so through it. Nowadays, people use these means more than face-to-face communication. As it becomes popular, incidents of cyber-bullying hace also increased in a skyrocketing rate. According to a recent survey, around 49% of school-going students have gone through some form of cyber-bullying [30]. From another survey, it is found that the majority of the cybercrime victims are women [5]. Cyber-bullying leads to mental stress, depression, or even may turn into a possible cause of suicide.

Bangladesh is one of the most densely populated countries in the world, where most of the people speak Bangla as their first language. Around 45 million Bangladeshi use social media, and out of them [31], 79.13% of Bangladeshi social media users prefer to use Facebook [2]. Majority of them love to comment or post in their native language. Therefore, to reduce the impact of cyber-bullying or hate speech, the development of an automated hate speech detection system has become a necessity which can deal with Bangla at a high accuracy rate.

Machine learning and deep learning algorithms are being widely in various applications [6, 7, 14–16, 18]. Bangla is a low-resource language and the development of Bangla natural language processing is not up to the mark. Detecting sentiments or categorizing Bangla texts is difficult to the unavailability of reliable stemmer, lemmatizer, or pretrained embedding layers. Detecting Bangla social media hate speech is even more difficult due to the spelling mistakes, short forms, and emoticons in social media texts. There are a large number of dialects in Bangla that made it even more challenging as people find it comfortable to share their thoughts and views in their own dialect or in mixing Bangla-English. Several researches have been carried out to detect Bangla hate speech. Romim et al. [24, 25] proposed a Support Vector Machine (SVM)-based model to detect hate speech They created a new dataset of 30000 samples that were collected from YouTube and Facebook comments. Their contribution is making the dataset available in the field of Bangla abusive text detection. Das et al. [11] adopted an encoder–decoder architecture to classify the types of Bangla hate speech. The architectures were primarily built with Long Short-Term Memory (LSTM), Attention, and Gated Recurrent Unit (GRU) blocks. Their model was trained and tested on a dataset containing 7425 samples to classify hate speech which were further classified into 7 categories. Ahammed et al. [3] collected Bangla texts from Facebook and labeled them as hate or non-hate speech. They used classical machine learning algorithms to classify texts between hate and non-hate speech. Ishmam et al. [17] took a GRU-based approach to detect six different types of Bangla hate speech. Their proposed model performed better than most of the classical machine learning models. A Convolutional Neural Network (CNN)–LSTM-based approach was taken by Karim et al. [20] to classify different types of hate speech. They at first preprocessed the texts and used embedding before giving input to the models. They also tried to explain the reasons behind the models classifying a text as a specific type of hate. Ritu et al. [23] proposed a system that could filter abusive messages from radio live feeds. They used classical machine learning algorithms and a dataset of 45000 sentences to develop the system. The proposed system could successfully filter out abusive contents and showed only positive Bangla sentences. Sazzed et al. [28] developed a transliterated Bangla corpus of 3000 samples for abusive text detection. They evaluated several machine learning algorithms on their developed corpus. Among them, SVM performed the best. Chakraborty et al. [10] adopted a classical machine learning approach to detect Bangla abusive and threatening language. In their system, they considered unicode emoticon as valid input. Their proposed system performed the best while using SVM with linear kernel that provided better performance than the LSTM-based models. Cécillon et al. [9] proposed a combination of extracted features from graph and content that was further given input to classical machine learning algorithms to detect English abusive texts. Steimel et al. [29] proposed a sampling method which performed better than the Latent Dirichlet Allocation (LDA) approach. They tested their proposed approach in various languages. Akhter et al. [4] proposed a CNN-based approach to detect abusive Urdu texts that outperformed several classical machine learning and LSTM-based approaches.

In this paper, a Bangla hate speech detection system has been proposed using LSTM–GRU–CNN hybrid approach. For this, a publicly available dataset has been used to train and test the models. The main contribution of this paper is given below:

- A hybrid deep learning architecture has been proposed to detect Bangla hate speech. To the best of knowledge, a hybrid LSTM–GRU–CNN-based approach was not proposed in any existing literature.
- Robustness of the proposed model has been evaluated.
- Performance of the proposed model has been compared with eight other LSTM–GRU–CNN-based models and transformer-based models.
- Impact of different pre-trained embeddings has been evaluated.

This paper is organized as follows: Sect. 2 describes the methodology of this research work. The performance of the proposed model has been described in Sect. 3. Section 4 concludes the paper by discussing some research scopes in this field.

2 Methodology

In this paper, since social media text was used for hate speech detection, the usual procedure that is commonly practiced on text processing was followed. The hate speech detection dataset of Romim et al. [24] was used for this purpose. The dataset comprises of 30,000 comments from YouTube and Facebook comment section. These comments were on seven different categories of posts. The annotation procedure was carried out by crowdsourcing and 50 annotators labeled the whole dataset. Each

Table 1 Dataset properties

Total sample	30000
Hate speech	10000
Non hate speech	20000
Training samples	24000 (0:16000,1:8000)
Testing	6000 (0:4000,1:2000)
Total no of words before preprocessing	49372
Total no of words after preprocessing	37690

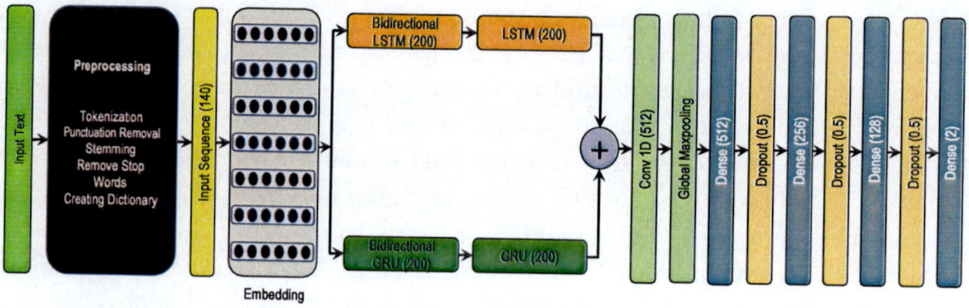

Fig. 1 Proposed model for hate speech detection

sample was evaluated three times and majority voting was considered for the final label. This dataset contains 10,000 hate speeches and 20,000 neutral comments. Table 1 provides the properties of the dataset (0 denotes non-hate speech and 1 denotes hate speech sample).

This dataset then went through some preprocessing stages. At first, tokenization was done to extract the words from the sentence. The Natural Language Toolkit tokenizer was used for this purpose. Then the punctuation was removed from the comments. The tokens were then converted to the basic form of the words using the Bangla Natural Language Processing Toolkit (BNLTK) stemmer. With the base form of the words, it becomes easy to convert the sentence into numeric form. A pool of Bangla 746 words was taken as stop words which adds little value to the sentence. Finally, the sentences were all converted to same length (140 characters) to create an input sequence.

To convert the text to numeric vector, word embeddings are used. GloVe, fastText, and Word2Vec are some of the most popular vectorizers for Bangla language. In this paper, for all the models, fastText was used, but performance was compared with GloVe and Word2Vec. fastText is trained with Common Crawl and Wikipedia data. This pretrained model helps to keep the internal relation among the words of a sentence though they are in numeric form. The output of this embedding layer is used as input to a bidirectional LSTM layer and a bidirectional GRU layer. Each of the layers has 200 neurons. These two layers are followed by an LSTM layer and a GRU layer, respectively, with 200 neurons in each layer. Output of the embedding

Table 2 Parameter and hyperparameter tuning range

Layer/Hyerparameter	Tuning range
Neuron in LSTM layer	50–350
Neuron in GRU layer	50–30
Filters in CNN layer	256–1024
Neuron in dense layer	50–500
Activation function in all layers	Tanh, sigmoid, ReLU
Dropout	0.25,0.5,0.75
Optimizer	Adadelta, Adagrad, Adam, RMSprop
Batch size	16, 32, 64
Learning rate	0.1,0.01,0.001,0.0001

layer was given as input to these two layers individually. These two parallel layers are then appended together and fed to a conv1D layer having 512 neurons. LSTM and GRU were chosen to capture long-term dependencies, where CNN was used to extract meaningful features from them. After the convolution operation was done in this layer, a global max-pooling layer was introduced. Three dense layers with 512, 256, and 128 neurons were used along with 50% dropouts. The dropout layers actually helped to provide a more generalized model. Finally, there was a dense layer with two neurons which worked as an output layer. For the output layer, softmax was used as an activation function and for the other layers, ReLU activation function was used. This model was trained for 20 epochs, where the Adam optimizer was used. The learning rate was set to 0.001 as it performed best to reach the goal. All the parameters and hyperparameters were found using Keras tuner that were searched in the range mentioned in Table. 2. Figure 1 shows the proposed model used in this paper.

For training purpose of this model, 24,000 social media comments were used, where 8000 samples were hate comments. For testing purpose, 6,000 samples were considered with 4,000 neutral comments and 2,000 hate comments. The accuracy, sensitivity, specificity, positive predictive value, and negative predictive value were calculated for evaluation of the performance of the model. The model was then compared with other Bangla NLP models. To check the robustness of the model, three different train-test sets were taken, and performance of each model was evaluated. Different combinations of CNN, LSTM, and GRU were taken, and their performances were compared with the proposed model. For each model, the time taken to complete each epoch was recorded to compare the efficiency of the compared models.

Table 3 Performance of the proposed model

Metrics	Proposed model
Accuracy	0.8985
True negative	3701
False positive	299
False negative	310
True positive	1690
Sensitivity	0.8450
Specificity	0.9253
P_0	0.9227
P_1	0.8497

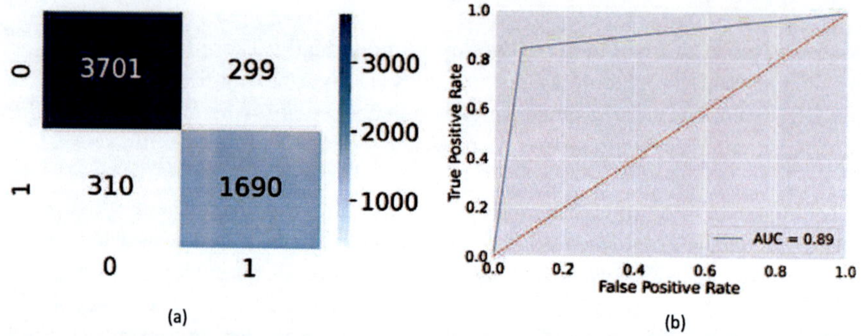

Fig. 2 **a** Confusion matrix **b** ROC curve of the proposed model. It correctly classified 84.50% of the hate text, and 92.53% of the non-hate text. AUC score of the proposed model was 0.89

3 Result Analysis and Discussion

3.1 Performance of the Proposed Model

The proposed LSTM–GRU–CNN hybrid model achieved 89.85% accuracy in Bangla social media hate speech detection using fastText embedding. It correctly classified 3701 out of 4000 non-hate texts and 1690 out of 2000 hate texts which are shown in Fig. 2. It could not detect 299 non-hate speech and 310 hate speech. The proposed model acquired 84.50% sensitivity and 92.53% specificity that was remarkable. Positive predictive value (P_1) was 0.8497, where the negative predictive (P_0) value was 0.9227. From the result, it is clear that the proposed model performed better in case of non-hate text detection. Table 3 provides the detailed result of the proposed model. Figure 2 illustrates the confusion matrix and ROC curve of the proposed model.

Table 4 Performance of the proposed model in different embeddings

Metrics	fastText	Word2Vec	GloVe
Accuracy	**0.8985**	0.8613	0.8683
True negative	3701	3433	3542
False positive	299	567	458
False negative	310	265	332
True positive	1690	1735	1668
Sensitivity	0.845	**0.8675**	0.834
Specificity	**0.9253**	0.85825	0.8855
P_0	0.9227	**0.9283**	0.9143
P_1	**0.8497**	0.7537	0.7846

3.2 Discussion on Embedding

In order to represent word and preserve the semantic and syntactic properties of words, embedding has been used in this work [13]. Three different embeddings (Word2Vec, fastText [19, 21, 32], and GloVe [22, 26]) have been tried out in this work. Table 4 shows the performance obtained by the proposed model while using three different types of embedding. Among these embeddings, the proposed model acquired the best accuracy (89.85%) and specificity (92.53%) while using fastText embedding. The best sensitivity (86.75%) was acquired by the proposed model while using Word2Vec embedding. The proposed model provided more generalized result in Word2Vec embedding. While using GloVe embedding, the proposed model acquired better accuracy than the Word2Vec embedding, but overall performance dropped significantly. Comparing all the parameters, it can be said that the proposed model performed the best while using fastText embedding. fastText breaks the words into multiple parts which helps to deal with misspelled and unknown words which are commonly seen in Bangla social media texts. This may be the reason behind the significant improvement in performance while using fastText embedding.

3.3 Robustness

To prove the robustness of the proposed model, the dataset was shuffled randomly and then divided into train and test parts. In this way, four different training–test sets were created. A 80:20 ratio was followed in all the sets while splitting data in training (24000 samples) and testing sets (6000 samples). Table 5 shows the result of the proposed in all the training–testing sets. In all the sets, the proposed model found over 89% accuracy. In all the cases sensitivity was greater than 84.40% and specificity was over 90%, which shows great consistency. By taking an average of four sets, the proposed model's average accuracy, sensitivity, and specificity were

Table 5 Performance of the proposed model in different training–test sets

SET	Accuracy	Sensitivity	Specificity
SET#1	0.8965	0.8737	0.9074
SET#2	0.8951	0.8446	0.9214
SET#3	0.8920	0.8657	0.9048
Considered Set	0.8985	0.8450	0.9253
Average	0.8955	0.8573	0.9147

Table 6 Performance of the transformer models

Model	Accuracy	Sensitivity	Specificity	Second/Epoch
Bangla electra [1]	0.8762	0.8390	0.8946	247
BERT-base-multilingual-cased [12]	0.8917	0.8435	0.9158	697
CSEBUETNLP/Bangla BERT [8]	**0.8993**	**0.8450**	**0.9265**	621
SagorSarkar/Bangla-BERT [27]	0.8860	0.8210	0.9185	665
Proposed model	0.8985	**0.8450**	0.9253	**151**

89.55%, 85.73%, and 91.47%, respectively. Therefore, the proposed model can be considered as robust.

3.4 *Performance Comparison with Transformer Models*

Transformer models are capable of achieving very high performance in natural language processing tasks. In this work, we have tried out five different pre-trained transformers [1, 8, 12, 27]. Table 6 provides the performance comparison between the proposed model and the transformers. Among them, CSEBUETNLP/Bangla-BERT [8] achieved better accuracy than the proposed model, where the sensitivity and specificity were quite similar. The proposed model performed quite similar to or better than all the transformer models. If the training time is taken into consideration, the proposed model can be denoted as the best performing as it took the least amount of time to train and test. It also contained the least number of parameters. Therefore, it can be said that the proposed model outperformed all the considered transformer models while taking time and evaluation metrics under consideration.

3.5 *Performance Comparison with Related Models*

To compare performance with related LSTM–GRU–CNN-based models, several models have been created. Figure 3 illustrates the architecture of the best-performing eight models. In all the models shown in Fig. 3, dense layers were same as the proposed model. In BiLSTM–LSTM–series-CNN model (Fig. 3a), a BiLSTM layer and an LSTM layer has been added sequentially. Output of this layer went through a convolutional layer. In the BiLSTM–LSTM–series architecture (Fig. 3b), no convolutional or max-pooling layers were added. Flatten layer replaced the max-pooling layer. Figure 3c, Fig. 3d illustrate BiGRU–GRU–series-CNN and BiGRU–GRU–series architecture, respectively. The only difference with previously discussed LSTM-based model was the replacement of LSTM layer with GRU layer. In LSTM–GRU–series-CNN (Fig. 3e) and LSTM-GRU-series (Fig. 3f) models, input at first goes through an LSTM layer with 200 neurons. Output of this layer went through GRU layers. The difference in these two models was the presence of a convolutional layer or flatten layer. Figure 3g, Fig. 3h show the architecture of BiLSTM–BiGRU–parallel-CNN and LSTM–GRU–parallel-CNN models. In both of the models, the input goes through LSTM and GRU layers individually. Output of these layers got concatenated, and the merged vector goes through a convolutional and max-pooling layers.

Figure 4 shows the performance comparison between the proposed model and the considered LSTM–GRU–CNN-based models. Among these models, BiGRU–GRU–seris model achieved the highest accuracy (88.10%). None of them could reach 89% mark in terms of accuracy. Sensitivity was below 83.50% in all the cases. The highest specificity achieved among these models was 94.65% by LSTM–GRU–series-CNN, but its sensitivity was not remarkable to consider. Therefore, the proposed model outperformed all these models in terms of overall performance.

3.6 Performance Comparison with Baseline Models

Romim et al. [24] developed this dataset, and also evaluated the performance of different deep learning and machine learning architectures in this dataset. Among them, SVM performed the best (accuracy: 87.5%). The proposed model achieved 89.55% average accuracy with 85.73% sensitivity, and 91.47% specificity (Table 5). In this comparison, we have considered the average achieved performance of different training-testing sets that helped to deal with the factors that may be arisen from a biased training-testing set. Therefore, it can be said that the proposed model performed better than its baseline models.

4 Conclusions

Due to the massive increase in social media usage, communication and expressing emotions through social media has become a daily routine in almost everyone's life. With this massive explosion of social media usage, cybercrime or cyber-bullying has increased a lot. Almost every social media users face bullying or social media hate

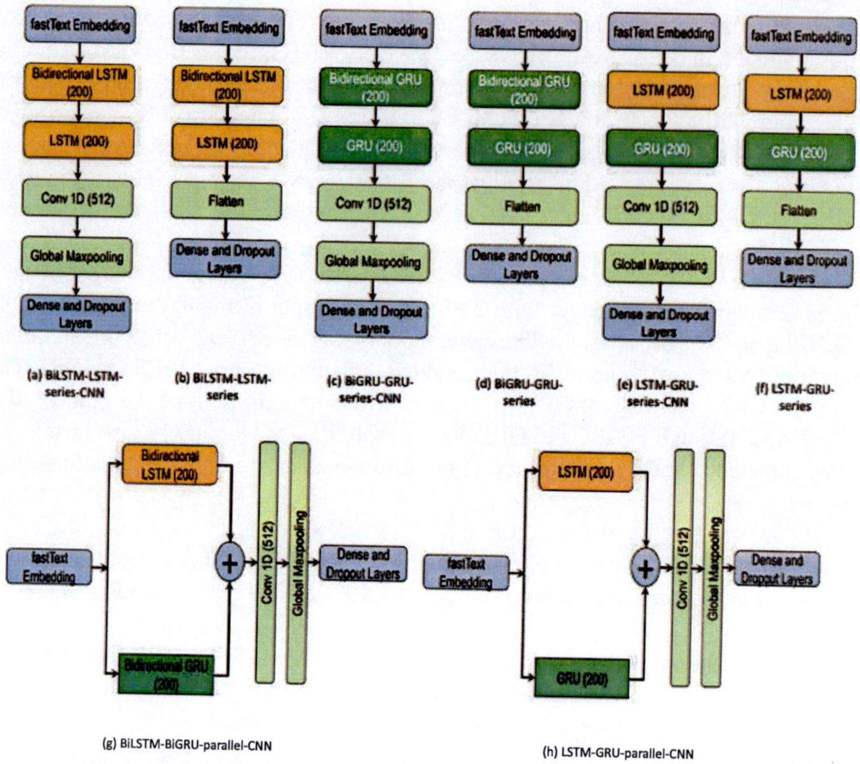

Fig. 3 Related models built using LSTM, GRU, and CNN layers. Dense layer of all the models was same as the proposed model

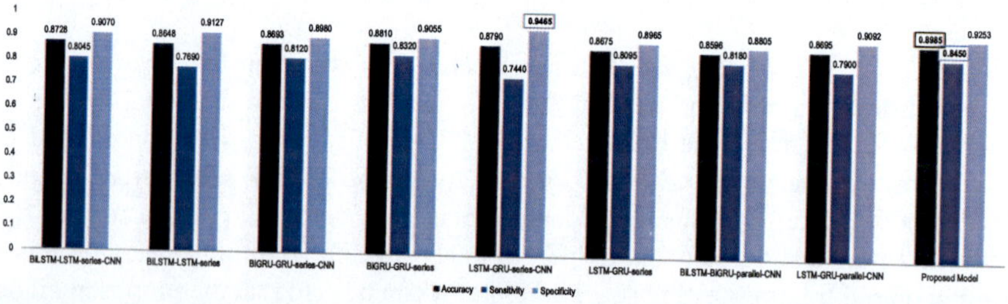

Fig. 4 Performance comparison with LSTM–GRU–CNN-based architectures. The proposed model outperformed all the considered models in terms of accuracy and sensitivity

speech at least once in their life. It sometimes reache to a severe extent and makes the user psychologically disturbed. Therefore, screening social media hate speech has become a necessity. In this paper, a deep learning-based hybrid approach has been taken to classify Bangla social media hate speech at a significant accuracy. Though the performance was satisfactory, it still requires improvement. Dataset contained around 30000 data that is not enough to examine a model's performance, especially in a low-resource and diverse language like Bangla. Larger dataset, more improved stemmer or lemmatizer, better embedding matrix, etc., need to be developed to enrich Bangla NLP and also achieve a satisfactory performance in Bangla hate speech detection.

References

1. monsoon-nlp/bangla-electra · hugging face. https://huggingface.co/monsoon-nlp/bangla-electra, (Accessed on 07/06/2021)
2. Social media stats bangladesh | statcounter global stats. https://gs.statcounter.com/social-media-stats/all/bangladesh(May 2021), (Accessed on 06/29/2021)
3. Ahammed, S., Rahman, M., Niloy, M.H., Chowdhury, S.M.H.: Implementation of machine learning to detect hate speech in bangla language. In: 2019 8th International Conference System Modeling and Advancement in Research Trends (SMART). pp. 317–320. IEEE (2019)
4. Akhter, M.P., Jiangbin, Z., Naqvi, I.R., AbdelMajeed, M., Zia, T.: Abusive language detection from social media comments using conventional machine learning and deep learning approaches. Multimed. Syst., 1–16 (2021)
5. Akter, F.: Cyber violence against women: the case of Bangladesh|genderit.org. https://www.genderit.org/articles/cyber-violence-against-women-case-Bangladesh(2018), (Accessed on 11/03/2021)
6. Al Banna, M.H., Ghosh, T., Al Nahian, M.J., Taher, K.A., Kaiser, M.S., Mahmud, M., Hossain, M.S., Andersson, K.: Attention-based bi-directional long-short term memory network for earthquake prediction. IEEE Access **9**, 56589–56603 (2021)
7. Al Nahian, M.J., Ghosh, T., Al Banna, M.H., Aseeri, M.A., Uddin, M.N., Ahmed, M.R., Mahmud, M., Kaiser, M.S.: Towards an accelerometer-based elderly fall detection system using cross-disciplinary time series features. IEEE Access **9**, 39413–39431 (2021)
8. Bhattacharjee, A., Hasan, T., Samin, K., Islam, M.S., Rahman, M.S., Iqbal, A., Shahriyar, R.: Banglabert: Combating embedding barrier in multilingual models for low-resource language understanding. CoRR **abs/2101.00204** (2021), https://arxiv.org/abs/2101.00204
9. Cecillon, N., Labatut, V., Dufour, R., Linarès, G.: Abusive language detection in online conversations by combining content-and graph-based features. Front. Big Data **2**, 8 (2019)
10. Chakraborty, P., Seddiqui, M.H.: Threat and abusive language detection on social media in bengali language. In: 2019 1st International Conference on Advances in Science, Engineering and Robotics Technology (ICASERT), pp. 1–6. IEEE (2019)
11. Das, A.K., Al Asif, A., Paul, A., Hossain, M.N.: Bangla hate speech detection on social media using attention-based recurrent neural network. J. Intell. Syst. **30**(1), 578–591 (2021)
12. Devlin, J., Chang, M., Lee, K., Toutanova, K.: BERT: pre-training of deep bidirectional transformers for language understanding. CoRR **abs/1810.04805** (2018), http://arxiv.org/abs/1810.04805
13. Ghannay, S., Favre, B., Esteve, Y., Camelin, N.: Word embedding evaluation and combination. In: Proceedings of the Tenth International Conference on Language Resources and Evaluation (LREC'16), pp. 300–305 (2016)
14. Ghosh, T., Al Banna, M.H., Al Nahian, M.J., Taher, K.A., Kaiser, M.S., Mahmud, M.: A hybrid deep learning model to predict the impact of Covid-19 on mental health form social media big data (2021)

15. Ghosh, T., Al Banna, M.H., Rahman, M.S., Kaiser, M.S., Mahmud, M., Hosen, A.S., Cho, G.H.: Artificial intelligence and internet of things in screening and management of autism spectrum disorder. Sustain. Cities Soc. **74**, 103189 (2021)
16. Ghosh, T., Banna, M., Al, H., Angona, T.M., Nahian, M., Al, J., Uddin, M.N., Kaiser, M.S., Mahmud, M.: An attention-based mood controlling framework for social media users. In: International Conference on Brain Informatics, pp. 245–256. Springer (2021)
17. Ishmam, A.M., Sharmin, S.: Hateful speech detection in public facebook pages for the bengali language. In: 2019 18th IEEE International Conference On Machine Learning And Applications (ICMLA), pp. 555–560. IEEE (2019)
18. Islam, T., Ahmed, N., Latif, S.: An evolutionary approach to comparative analysis of detecting Bangla abusive text. Bull. Electr. Eng. Inform. **10**(4), 2163–2169 (2021)
19. Joulin, A., Grave, E., Bojanowski, P., Douze, M., Jégou, H., Mikolov, T.: Fasttext. zip: Compressing text classification models. arXiv preprint arXiv:1612.03651 (2016)
20. Karim, M.R., Dey, S.K., Islam, T., Sarker, S., Menon, M.H., Hossain, K., Hossain, M.A., Decker, S.: Deephateexplainer: Explainable hate speech detection in under-resourced Bengali language. In: 2021 IEEE 8th International Conference on Data Science and Advanced Analytics (DSAA), pp. 1–10. IEEE (2021)
21. Liao, M., Shi, B., Bai, X., Wang, X., Liu, W.: Textboxes: A fast text detector with a single deep neural network. In: Thirty-first AAAI conference on artificial intelligence (2017)
22. Pennington, J., Socher, R., Manning, C.D.: Glove: Global vectors for word representation. In: Proceedings of the 2014 conference on empirical methods in natural language processing (EMNLP), pp. 1532–1543 (2014)
23. Ritu, S.S., Mondal, J., Mia, M.M., Al Marouf, A.: Bangla abusive language detection using machine learning on radio message gateway. In: 2021 6th International Conference on Communication and Electronics Systems (ICCES), pp. 1725–1729. IEEE (2021)
24. Romim, N., Ahmed, M., Talukder, H., Islam, M.S.: Hate speech detection in the Bengali language: A dataset and its baseline evaluation. CoRR **abs/2012.09686** (2020), https://arxiv.org/abs/2012.09686
25. Romim, N., Ahmed, M., Talukder, H., Islam, M.S.: Hate speech detection in the bengali language: A dataset and its baseline evaluation. In: Proceedings of International Joint Conference on Advances in Computational Intelligence, pp. 457–468. Springer (2021)
26. Sarker, S.: Github-sagorbrur/glove-Bengali: Bengali glove pretrained word vector. https://github.com/sagorbrur/GloVe-Bengali, (Accessed on 07/03/2021)
27. Sarker, S.: Banglabert: Bengali mask language model for Bengali language understanding (2020), https://github.com/sagorbrur/bangla-bert
28. Sazzed, S.: Abusive content detection in transliterated Bengali-English social media corpus. In: Proceedings of the Fifth Workshop on Computational Approaches to Linguistic Code-Switching, pp. 125–130 (2021)
29. Steimel, K., Dakota, D., Chen, Y., Kübler, S.: Investigating multilingual abusive language detection: A cautionary tale. In: Proceedings of the International Conference on Recent Advances in Natural Language Processing (RANLP 2019), pp. 1151–1160 (2019)
30. UNB: '49% Bangladeshi school pupils face cyberbullying' | the daily star. https://www.thedailystar.net/bytes/'49-bangladeshi-school-pupils-face-cyberbullying'-287209(2016), (Accessed on 11/03/2021)
31. UNB: Bangladesh charts 9m new social media users | dhaka tribune. https://www.dhakatribune.com/bangladesh/2021/04/26/bangladesh-charts-9m-new-social-media-users#: :text=A20study20has20demonstrated20the,February20by20We20Are20Social (2021), (Accessed on 11/03/2021)
32. Wu, S., Manber, U.: Fast text searching: allowing errors. Commun. ACM **35**(10), 83–91 (1992)

Approaches for Improving the Performance of Fake News Detection in Bangla: Imbalance Handling and Model Stacking

Md. Muzakker Hossain, Zahin Awosaf, Md. Salman Hossan Prottoy, Abu Saleh Muhammod Alvy, and Md. Kishor Morol

Abstract Imbalanced datasets can lead to biasness in the detection of fake news. In this work, we present several strategies for resolving the imbalance issue of fake news detection in Bangla with a comparative assessment of proposed methodologies. Additionally, we propose a technique for improving performance even when the dataset is imbalanced. We applied our proposed approaches to BanFakeNews (Hossain et al. in Banfakenews: A dataset for detecting fake news in bangla. arXiv preprint arXiv:2004.08789, 2020), a dataset developed for the purpose of detecting fake news in Bangla comprising 50K instances but is significantly skewed, with 97% of the majority instances. We obtained a 93.1% F1-score using data manipulation techniques such as SMOTE and a 79.1% F1-score using without data manipulation approaches such as Stacked Generalization. Without implementing these techniques, the F1-score would have been 67.6% for baseline models. We see this work as an important step towards paving the way of fake news detection in Bangla. By implementing these strategies the obstacles of imbalanced dataset can be removed and improvement in the performance can be achieved.

Keywords Text classification · Imbalanced data · Stacked generalization

The authors contributed equally to this paper.

Md. M. Hossain · Z. Awosaf · Md. S. H. Prottoy · A. S. M. Alvy · Md. K. Morol (✉)
American International University-Bangladesh, Dhaka, Bangladesh
e-mail: kishor@aiub.edu

Md. M. Hossain
e-mail: 18-37801-2@student.aiub.edu

Z. Awosaf
e-mail: 18-37064-1@student.aiub.edu

Md. S. H. Prottoy
e-mail: 18-36902-1@student.aiub.edu

A. S. M. Alvy
e-mail: 18-37487-1@student.aiub.edu

S. Hossain et al. (eds.), *Proceedings of International Conference on Fourth Industrial Revolution and Beyond 2021*, Lecture Notes in Networks and Systems 437,
https://doi.org/10.1007/978-981-19-2445-3_51

1 Introduction

The dissemination of misinformation, colloquially referred to as Fake News, is escalating as more people gain access to the internet. These information can be fed to us in the form of news, stories, or social media posts. These false information can be categorized into many forms such as Clickbait, Propaganda, Satire/Parody, Biased/Slanted News, and so on. These fabricated news can have a range of effects, from political polarization to communal violence. A study shows that people tend to share the falsified news more than the authentic ones on the mainstream social media [13]. Automated programs used to engage in social media or Social Bots that impersonate human behavior are more frequently used in popular social media. Separate reports have shown that between 9 and 15% of active Twitter accounts are Social Bots and about 60 million bots were found in Facebook [12]. And these automated accounts can be used to manipulate other users of these social platforms and spread misinformation. The impacts of falsified information are spread all over the world. In Bangladesh, many unfortunate incidents like communal violence occur because of some rumors, or false information shared on mainstream social media. Hence, it has become critical to identify these fake news.

A big hindrance to fake news detection can be improper or lack of datasets. Compared to prominent languages not many significant datasets can be found for Bangla. The only comprehensive dataset available in Bangla [4], has a big hindrance of data imbalance, meaning one class is significantly larger than the other. As a result, this is a significant setback in the route of research for detecting fake news in Bangla which can hinder the objective of tackling the risk of falsified information and fighting against rumors.

So in this work, we focus to remove the obstacle and pave the way for research in detecting fake news in Bangla.

Contributions. The bridge the current gap we-

- Propose several methods to handle the data imbalance
- Make a comparative study between the proposed models to find out the best-suited method for data imbalance handling
- Show an alternative way to optimize the performance in detecting fake news without any data manipulation, namely, Stacked Generalization

We strongly hope to contribute adequate information to overcome the limitations of the BanFakeNews dataset and make a significant contribution to the Bangla fake news detection system.

2 Related Work

The lack of a substantial dataset is the primary impediment to developing an automated system for detecting fake news in Bangla. Possessing a proper dataset can pave the way for potential research in this field. This issue is not apparent in popular languages such as English, which is why numerous substantial studies have been undertaken in this domain for English.

For example, [8] presents a dataset containing over 1 million samples of fake news in English. Another prominent dataset is [14] which is a news dataset for the purpose of investigating automated fake news detection using surface-level linguistic patterns.

Scarcity of datasets is one of the reasons why not so many researches in fake news detection in Bangla were conducted. To our best knowledge, [4] is the only proper dataset available for fake news detection purpose in Bangla. Although the number of authentic news is approximately 37.47 times higher than the number of fake news, it offers a significant amount of data for a language with scarce resources like Bangla. Since it is a comparatively new dataset in Bangla, not many researches have been conducted based on this dataset. The experimental analysis of [11]'s proposed model on real-world data demonstrates that the Passive–Aggressive Classifier and Support Vector Machine achieve 93.8% and 93.5% accuracy, respectively. And this research was conducted based on [4] dataset.

[10] used a wide range of resampling schemes to deal with the imbalanced dataset, including random oversampling, SMOTE, SMOTE-SVM, WESMOTE, and CSMOTE. In [1], a state-of-the-art oversampling technique called the Synthetic Minority Over-sampling Technique (SMOTE) was implemented. [3] implements the Adaptive Synthetic (ADASYN) oversampling method, which generates new synthetic instances based on difficult-to-learn instances of the minority class. An oversampling method to deal with imbalanced data was implemented in [9] which mainly concentrates on the borderline instances that are crucial for establishing decision boundary.

Another efficient technique for handling imbalanced datasets can be model stacking. While a single classifier is inefficient, stacking multiple generalizers can increase a model's performance by reducing the generalization error rate of one or more generalizers [15]. This approach is termed Stacked Generalization, as presented by [15]. The researchers demonstrated that by stacking models, the performance of each model can be significantly improved [7].

3 Background

In this section, we discuss the necessary background information related to our work.

3.1 Dataset

BanFakeNews [4], a publicly available annotated dataset consists of ≈50K Bangla news (Table 1). Owing to the lack of a well-established dataset for Bangla fake news, we used BanFakeNews to perform text classification in Bangla with the aim of detect-

Table 1 Features of BanFakeNews Dataset

Class name	Total rows	Percentage of total
Authentic	48,678	97.4%
Fake	1,299	≈2.6%

ing fake news. The dataset collected the authentic news from 22 most popular and mainstream trusted news portals in Bangladesh [4]. The fake news were commonly attributed to all kinds of satirical, clickbait, and falsified information.

As (Table 1) indicates, this dataset is utterly imbalanced containing only 2.6% of fake news. Therefore, in order to minimize this disparity, we utilized several strategies and compared the outcome to the baseline model. It's worth noting that the dataset comprises a variety of features, including the source, the source's domain, the headline, the date the article was published, and the article's category. However, we have used the article's text as our primary feature.

4 Methodology

In this section, we describe the data preprocessing, feature extraction from raw texts of dataset, and proposed techniques to handle the imbalance issue.

4.1 Data Preprocessing

As we are dealing with text classification, the dataset had to be prepared in a certain way before feeding to the models. Data preprocessing is a strategy for transforming unstructured data into a usable and effective format. Initially, the text contained many unnecessary emoticons, symbols, punctuation, digits, and foreign alphabets. So, we preprocessed the dataset by removing the unnecessary fragments at first, followed by stop words removal and stemming.

Standardization: The unnecessary fragments such as punctuation marks, symbols (e.g., @, #, $, %, &), English alphabets, Bangla and English digits, and emoticons have been omitted since they are irrelevant to text classification.

Stopwords Removal: Stop words are frequent but insignificant, these words do not deliver any necessary information and they make no contribution to the text classification process. So to prune off these words, we used a stop words library called "Stopwords Bengali (BN)" from GitHub,[1] which is a complete set of Bengali stopwords to make the process efficient. Some common stopwords are shown below:

[1] https://github.com/stopwords-iso/stopwords-bn.

এবং, ও, আর, কিন্তু, তবে, যেহেতু, etc.

Stemming: Stemming is the process of finding the root of a word. This makes the process even more efficient. We have used the python Bangla Stemmer[2] package to do the stemming.

4.2 Feature Extraction

Algorithms for machine learning are limited to numerical values. To implement machine learning over text, we should first transform our text data as vector values which are known as feature extraction or vectorization. We have used two of the most widely used feature extraction techniques for text classification—Count Vectorizer and TF-IDF Vectorizer, since these are not language-dependent and can simply be incorporated into machine learning models.

Count Vectorizer: Count Vectorizer creates a sparse matrix with columns for unique words and rows for text samples. The value of each cell corresponds to the number of occurring words in the text data.

TF-IDF Vectorizer: TF-IDF or Term Frequency-Inverse Document Frequency method computes a word's distinctiveness by comparing the number of times a word appears in a document to the total number of documents in which the word appears. Count Vectorizer counts the occurrences of a certain word in a document, whereas the TF-IDF also considers the number of documents in which the word appears. For our work, we have used the n-gram range (1, 2) which is a continuous sequence of n items extracted from a particular text or speech sample.

4.3 Structure of Models

In our work, initially, we developed the baseline model in order to establish a benchmark for performance measurement. Then three strategies, namely, Oversampling, Under-sampling, and Modifying the Class-Weight, were applied to handle the imbalanced data. Finally, an ensemble technique "Stacked Generalization" was further performed to improve the outcome of the models.

Baseline Model
Six classifiers (i.e., Logistic Regression, Support Vector Machine, Multinomial Naïve Bayes, Bernoulli Naïve Bayes, Random Forest Classifier, and Decision Tree Classifier) were chosen as baseline models in order to create a benchmark for further model testing.

Oversampling
As shown in (Table 1), the dataset included only 2.6% of the minority class. To com-

---

[2] https://pypi.org/project/bangla-stemmer/.

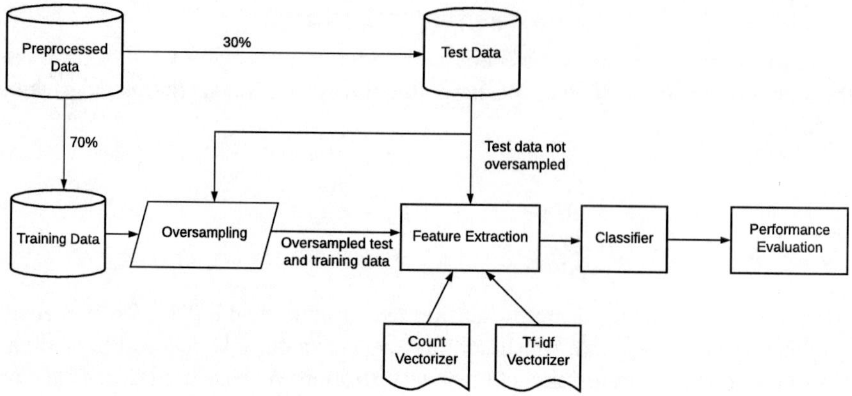

Fig. 1 Illustration of oversampling technique

Table 2 Oversampled data

	Before oversampling	After oversampling
Train data (authentic)	34,075	34,075
Train data (fake)	909	34,075
Test data (authentic)	14,603	14,603
Test data (fake)	390	14,603

pensate for the remaining 97.4%, the first technique we utilized was oversampling. Oversampling is a technique for resolving data imbalances that increases the sample size of minority class. We've performed three kinds of oversampling methods namely, Random Oversampling, SMOTE, and ADASYN.

Random Oversampling: Random Oversampling is a naive approach for generating new samples by randomly sampling current data samples with substitution.

SMOTE: Synthetic Minority Oversampling Technique (SMOTE) is an oversampling method that generates synthetic samples from the minority class. SMOTE begins by selecting random data from the minority class and then generates synthetic data using a k-nearest neighbor method.

ADASYN: Adaptive Synthetic Sampling Method or simply ADASYN, is a technique for adaptively producing minority data samples based on their distributions: more synthetic data is produced for minority class samples that are more difficult to learn than for minority class samples that are simpler to learn.

Prior to oversampling (Fig. 1), data for each technique were separated in a 70:30 ratio as train:test data to eliminate the chance of data contamination. Without doing so, the same data used for training and testing can be mixed together, resulting in a biased output. The overview of oversampled data is shown in (Table 2).

Undersampling

Undersampling is the reciprocal of oversampling, in which instances from the majority class are discarded in order to reduce the disparity between the majority and minor-

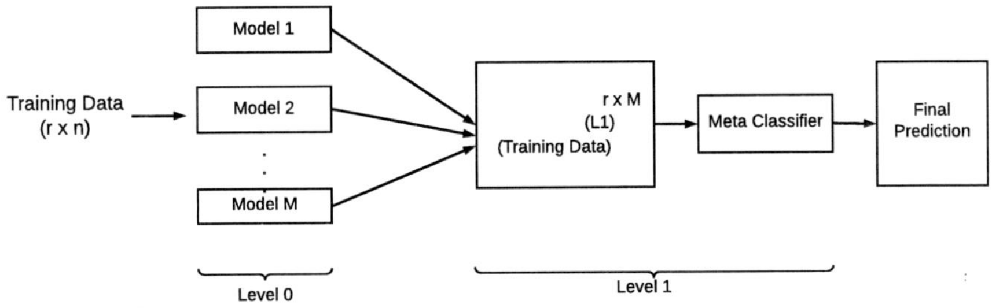

Fig. 2 Mathematical demonstration of model stacking

ity classes. We performed two different types of undersampling: Random Undersampling and NearMiss.

Random Undersampling: Random undersampling is the process of randomly choosing and discarding samples from the majority class from the training dataset.

NearMiss: This technique examines the distribution of the larger class and randomly discards samples from it. When two points in the distribution belong to different classes and are extremely close to each other, this technique eliminates the data points from the larger class, attempting to balance the distribution.

The total preprocessed data were first undersampled and then were split into 80:20 ratio for train and test data accordingly. By the following undersampling, the majority class was reduced to 1,299 rows, same as the minority class, and the data were split 80:20.

Modifying Class-Weight

The aim of modifying class weights is to penalize the minority class for misclassification by increasing class weights while decreasing weights for the majority class. In order to calculate class weight, we have used balanced as the argument for class weight function.

Model Stacking

Apart from resolving imbalances, another strategy that may be used to increase the model's performance is Model Stacking. Model stacking is a strategy of combining several machine learning models to have a better output. Model stacking can be completed by different ensemble techniques. In this work, we have implemented one of the most renowned ensemble techniques known as Stacked Generalization or stacking which employs a new model to discover the optimal way to combine predictions from a set of models trained on the same dataset. We executed the strategy by building a set of six models for level-0 and then picking one of them as the meta-classifier for level-1 (Fig. 2). This process was repeated six times. Initially, the training data contained r observations and n features. The different models were cross-validated (cv = 5) and each model produced outcome that was then cast into meta-classifier, meaning the M predictions became features (r x M) in the second level (Level-1). The second level model or meta-classifier was then trained again and tested for final prediction.

4.4 Experimental Setup

To begin with, we divide the preprocessed data (Sect. 4.1) into train and test sets using an 80:20 train-to-test split for all techniques except Oversampling which was a split of 70:30. Then, using the Count Vectorizer and the TF-IDF vectorizer, both the train and test data were transformed into vector forms. Following that, all six classifiers were fit to the train data and then evaluated against the test data.

The parameter list is as follows, for Logistic Regression the parameters are Solver: Saga, Multi-Class: Multinomial; for Support Vector Classification Kernel: rbf, Gamma: 1; for both of the Naïve Bayes method Alpha: 0.01; for Random Forest Classifier N-Estimators: 400, Max-Feature: sqrt; and for Decision Tree Classifier Criterion: Gini, Max-Depth: 6. Finally, in stacking: we used five-folds for cross-validation generator.

5 Results and Discussion

In this section, we describe an in-depth performance analysis of our proposed models. We calculate four performance indicators to evaluate the models' efficiency: Accuracy, Precision, Recall, and F1-score. Since the dataset was significantly imbalanced, we mainly focused on F1-score, which tends to better reflect the scenario of minority classes in an imbalanced dataset.

5.1 Overall Performance of All Methods

Here, (Table 3), (N) indicates that the test data for that specific method were not oversampled, only the test data obtained by 70:30 (train:test) split was used. The column "Classifier" denotes the classifier that produced the highest F1-score for the specified method.

A baseline model is a straightforward model that produces a decent outcome without any changes in the algorithm. As a result, it can serve as a benchmark for other models constructed using the same dataset. For the performance matrix, we focus on the F1-score because accuracy isn't a viable statistic for an imbalanced dataset [2]. And as for Precision and Recall, F1-score is the harmonic mean of both of these metrics and so F1-score is our ultimate statistic for evaluation. In our work, we've emphasized minority class while computing F1-Score. As there is a vast disparity between majority and minority classes and so the majority class has a nearly perfect F1-score.

As (Table 3) indicates, while undersampling methods (i.e., random undersampling and near-miss) generate better outcomes, we do not consider them to be the optimum method for dealing with data imbalance. Since undersampling is the pro-

Table 3 Overall performance of all methods

Method	Feature extraction	Classifier	F1-score
Baseline	Count	LR	0.604
	TF-IDF	LR	0.676
Random oversampling (N)	Count	LR	0.452
	TF-IDF	LR	0.703
Random oversampling	Count	MNB	0.845
	TF-IDF	LR	0.826
SMOTE (N)	Count	MNB	0.476
	TF-IDF	LR	0.696
SMOTE	Count	LR	**0.931**
	TF-IDF	LR	0.863
ADASYN (N)	Count	MNB	0.460
	TF-IDF	LR	0.695
ADASYN	Count	BNB	0.914
	TF-IDF	LR	0.866
Random undersampling	Count	BNB, RFC	0.893
	TF-IDF	MNB	0.911
Near-miss	Count	RFC	**0.943**
	TF-IDF	SVM	0.935
Modifying class-weight	Count	LR	0.493
	TF-IDF	LR	0.757
Model stacking	TF-IDF	RFC	0.791

cess of excluding data from the majority class, this might result in the omission of important instances from the dataset, hence influencing the desired outcome. Additionally, when previously unseen or new data are presented, undersampling models perform poorly.

Due to the random replication of the data in the case of random oversampling, the generated outcome may be biased, as the same instances are copied repeatedly and the prediction count is repeated for the same instances. However, this is not the case with SMOTE and ADASYN. SMOTE and ADASYN generate new synthetic data, and as shown in the table, when test data is oversampled with SMOTE and ADASYN, the results are significantly more accurate and consistent.

It is seen from (Table 3) that the modifying class weight is not appropriate for our dataset. It performed worse than the baseline models in the case of Count Vectorizer. The disparity between the two feature extraction strategies is especially evident in the case of class-weight modification, which is typically not apparent in other models.

Several studies have been conducted on fake news detection in Bangla but there's a scarcity of dataset. BanFakeNews [4] was published in 2020 and so not many works have been done utilizing this dataset. [4] proposed a benchmark for this dataset by using neural network models like CNN, LSTM, and BERT. For these models, they achieved F1-Score of 0.59, 0.53, and 0.68, respectively, for the minority classes, whereas we've achieved 0.931, 0.914, and 0.791 for methods like SMOTE, ADASYN, and Model Stacking, respectively. Other than [4], to our best knowledge only [11] had worked on this dataset before us. [11] had achieved an accuracy of 93.8% using Passive–Aggressive Classifier on this dataset. To overcome the imbalance issue, [11] worked with 3.5K authentic class instances and manually expanded the fake class instances by merging with another dataset, increasing the class instances to 2.3K. So, no proper correlation can be drawn to [11] with our work.

The other works that were conducted on Bangla fake news detection like [5] had achieved a 96.64% of accuracy for SVM on their own dataset consisted of around 2541 instances with real and fake being 60.92% and 39.08%, respectively. In other work, [6] had achieved an accuracy of 85% for Random Forest Classifier working on their own dataset consisting of 726 news articles. A common phenomenon is seen in these studies that their dataset is comparatively much smaller. Comparing to these studies, we've achieved the highest accuracy of 98.7% using Logistic Regression and our other models also outperform these studies.

5.2 *Performance of Model Stacking*

Data manipulation can always be a risky maneuver whether we apply oversampling or undersampling technique. Unlike prior strategies for resolving imbalances, model stacking does not need to manipulate the dataset. It ensembles the models and enhances the newly created model's performance. The particular stacking technique that we've employed is called Stacked Generalization. As it has been proven that the TF-IDF Vectorizer performs better than the Count Vectorizer as a feature extraction technique, we used the TF-IDF Vectorizer as our feature extraction technique for stacked generalization. We've used each of our six classifiers as meta-learner in the process. The result is shown on (Table 4).

The outcome of model stacking is much better than the baseline models with an F1-Score of 0.791 for Random Forest as the meta-classifier. Though the outcome is lower than the SMOTE or ADASYN, this method can be utilized for better comprehension and reliability of unseen data. The drawback of this technique is time and space complexity. As several models are stacked upon each other and the dataset is considerably large, it consumed a lot of memory and took a much longer time to fit the model and execute the outcome (Fig. 3).

Table 4 Performance of model stacking

	TF-IDF vectorizer			
	Accuracy	Precision	Recall	F1-score
LR	0.988	0.844	0.690	0.759
SVM	0.989	0.839	0.706	0.767
MNB	0.981	0.602	0.669	0.675
BNB	0.978	**0.923**	0.142	0.247
RFC	**0.990**	0.846	**0.742**	**0.791**
DTC	0.987	0.776	0.718	0.746

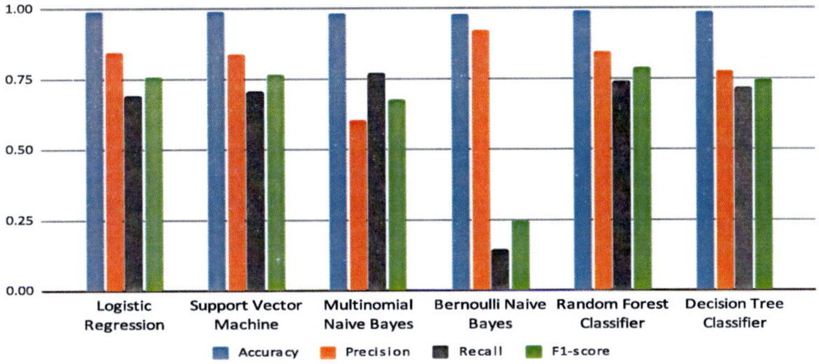

Fig. 3 Outcome of model stacking

6 Conclusion and Future Scope

Fake news detection in Bangla is still in its infancy. In previous studies related to this domain, we've seen researchers lack a proper dataset to accomplish their objective. Although we found [4] as our primary dataset that contains a sufficient amount of instances, the majority and minority classes are vastly imbalanced. Because of imbalance the expected result can be biased towards the majority class. Thus, we focused on addressing the issue and proposed numerous solutions to overcome it. To our best knowledge, we are the first to work on the full dataset of BanFakeNews [4]. We've not only addressed the imbalance issue of the dataset but also proposed numerous techniques to handle the disparity. Our proposed models successfully surpassed the existing works conducted in Bangla for the purpose of fake news detection in terms of performance metrics and have shown a comparative study between the models. We've shown that imbalance handling methods have their own pros and cons. Oversampling methods like SMOTE and ADASYN can be a good solution to reduce the discrepancy between majority and minority classes and the performance of the models also increase with comparison to baseline models (Table: 3). Model stacking can be an alternative to data manipulation. A better result can be achieved by using either of these models. Our work aimed to pave the way for expanding research on fake news detection in Bangla. We hope that future studies conducted on this dataset for the purpose of fake news detection, researchers can reliably use any of our proposed imbalance handling techniques.

It's worth noting that we focused exclusively on the text's content and omitted any other features. This is mostly because we have concentrated our efforts on resolving the data imbalance issue. Thus, in the near future, we'd like to work on other dataset features such as source, category, and headline. Additionally, we intend to experiment with different models, including deep learning models such as RNN, CNN, LSTM, and GRU.

References

1. Chawla, N.V., Bowyer, K.W., Hall, L.O., Kegelmeyer, W.P.: Smote: synthetic minority over-sampling technique. J. Artif. Intell. Res. **16**, 321–357 (2002)
2. Galar, M., Fernandez, A., Barrenechea, E., Bustince, H., Herrera, F.: A review on ensembles for the class imbalance problem: bagging-, boosting-, and hybrid-based approaches. In: IEEE Transactions on Systems, Man, and Cybernetics, Part C (Applications and Reviews) **42**(4), 463–484 (2011)
3. He, H., Bai, Y., Garcia, E.A., Li, S.: Adasyn: Adaptive synthetic sampling approach for imbalanced learning. In: 2008 IEEE International Joint Conference on Neural Networks (IEEE World Congress on Computational Intelligence), pp. 1322–1328. IEEE (2008)
4. Hossain, M.Z., Rahman, M.A., Islam, M.S., Kar, S.: Banfakenews: A dataset for detecting fake news in bangla. arXiv preprint arXiv:2004.08789 (2020)
5. Hussain, M.G., Hasan, M.R., Rahman, M., Protim, J., Al Hasan, S.: Detection of bangla fake news using mnb and svm classifier. In: 2020 International Conference on Computing, Electronics & Communications Engineering (iCCECE), pp. 81–85. IEEE (2020)
6. Islam, F., Alam, M.M., Hossain, S.S., Motaleb, A., Yeasmin, S., Hasan, M., Rahman, R.M.: Bengali fake news detection. In: 2020 IEEE 10th International Conference on Intelligent Systems (IS), pp. 281–287. IEEE (2020)
7. Jiang, T., Li, J.P., Haq, A.U., Saboor, A., Ali, A.: A novel stacking approach for accurate detection of fake news. IEEE Access **9**, 22626–22639 (2021)
8. Nakamura, K., Levy, S., Wang, W.Y.: r/fakeddit: A new multimodal benchmark dataset for fine-grained fake news detection. arXiv preprint arXiv:1911.03854 (2019)
9. Nguyen, H.M., Cooper, E.W., Kamei, K.: Borderline over-sampling for imbalanced data classification. Int. J. Knowl. Eng. Soft Data Parad. **3**(1), 4–21 (2011)
10. Padurariu, C., Breaban, M.E.: Dealing with data imbalance in text classification. Procedia Comput. Sci. **159**, 736–745 (2019)
11. Sraboni, T., Uddin, M., Shahriar, F., Rizon, R.A., Polock, S.I.S., et al.: FakeDetect: Bangla fake news detection model based on different machine learning classifiers. Ph.D. thesis, Brac University (2021)
12. Varol, O., Ferrara, E., Davis, C., Menczer, F., Flammini, A.: Online human-bot interactions: detection, estimation, and characterization. In: Proceedings of the International AAAI Conference on Web and Social Media, vol. 11 (2017)
13. Vosoughi, S., Roy, D., Aral, S.: The spread of true and false news online. Science **359**(6380), 1146–1151 (2018)
14. Wang, W.Y.: "liar, liar pants on fire": A new benchmark dataset for fake news detection. arXiv preprint arXiv:1705.00648 (2017)
15. Wolpert, D.H.: Stacked generalization. Neural Netw. **5**(2), 241–259 (1992)

Early Suicide Prevention: Depression Level Prediction Using Machine Learning and Deep Learning Techniques for Bangladeshi Facebook Users

Imran Hossen, Tohidul Islam, Md. Golam Rashed, and Dipankar Das

Abstract Depression is a crucial factor for deciding to suicide. However, few works have been done on depression analysis using the Bengali language based on social media data such as Facebook. In this paper, we propose a depression detection model for Facebook users using Logistic Regression and LSTM. This work aims to analyze the status updates from Facebook users within 2–3 years and evaluate them to detect whether the person is depressed or not. We collected data from 100 users' profiles from Facebook, containing on average 30 posts from each user, and proposed BenFED dataset. The proposed system considers sixteen emotional factors related to depression. Based on these emotions, Facebook users' statuses are labeled as four types, e.g., no, mild, moderate, and severe for determining the level of depression. We compared our proposed approach with other state-of-the-art approaches. It is revealed that our proposed approach outperformed most of the compared techniques for detecting emotions, depression levels, and depression statuses.

Keywords Social media · Facebook · Depression · Emotion · Suicide prevention · Classification · Logistic regression · LSTM

1 Introduction and Motivation

Every suicide is a tragedy, and to some extent, a mystery. Suicide often stems from a deep feeling of hopelessness. It is the tenth leading cause of death worldwide [1]. According to the World Health Organization (WHO), approximately 8,00,000 people die due to suicidal effects every year, that is one person every 40 s. While high-income countries had the highest incidence of 11.5 per 1,00,000.00, however,

I. Hossen · T. Islam · Md. G. Rashed (✉) · D. Das
Department of Information and Communication Engineering, University of Rajshahi, Rajshahi 6205, Bangladesh
e-mail: golamrashed@ru.ac.bd

D. Das
e-mail: dipankar@ru.ac.bd

S. Hossain et al. (eds.), *Proceedings of International Conference on Fourth Industrial Revolution and Beyond 2021*, Lecture Notes in Networks and Systems 437,
https://doi.org/10.1007/978-981-19-2445-3_52

79% of the world's suicides occurred in low- and middle-income countries. The rate of suicide occurrence is higher among men than women globally.

Depression is a key influence in suicide. Although Facebook is undoubtedly a good platform to communicate with each other to remain socially engaged and updated, Mamun and Griffiths inspect the association between Facebook addiction and depression. The study shows that 39.7% of Bangladeshi students are Facebook Addicted. In Bangladesh, 52.5% of undergraduate fresher students are prone to suffering from depression and 57.8% are suffering from anxiety problems [2]. These kinds of causes lead to suicide and suicidal behaviors among students in serious cases [3]. However, these statistics are all survey-based, and they didn't analyze the social media data which is the most used social platform in Bangladesh to express thoughts and daily life feelings.

Though depression analysis using paper-and-pencil-based techniques has been carried out for a long time, however, the acceptance of depression analysis through technology as a valid and reliable way is not yet established. There are only a few research works on depression analysis based on social media data available, e.g., [4–6]. However, most of these research works have taken solely the English language into consideration.

Ayah Zirikly et al. [6] proposed an assessment of a suicide risk system that relied on Reddit posts. Victor Leiva et al. [4] introduced early depression detection of social media users to prevent suicide occurrences. They used two time-aware classification measures: ERDE5 and ERDE50; they validate their results by applying different machine learning algorithms and combinations of them. However, in their study, they only considered two labels namely non-depressed or depressed.

Recently, in the Bengali language, A. A. Choudhury et al. utilized three machine learning algorithms to analyze depression among Bangladeshi Undergraduates [7]. However, their dataset was small (only 935 records and reduced to 577 after preprocessing); besides they only consider binary labels such as depressed or not, and the highest accuracy obtained was 75%. Masum Billah et al. introduced 'Depression detection' from Bangla Facebook status using machine learning approach [8]. They also considered binary classification such as depressed or non-depressed and achieved the highest 77.96% accuracy for SGDC with unigram + emoticons as a feature.

One recent study, however, applied Long Short-Term Memory (LSTM) Recurrent Neural Network technique to evaluate depression in the Bangla language [5]. They tried different hyperparameter tuning to extract the best working model; the best accuracy was 86.3% for five-layered LSTM with size 128, batch size 25, and learning rate 0.0001 over 20 epochs. They labeled their dataset, considering depressive, non-depressive, ambiguous, incomplete tweets, and the number of these tweets were 984, 2930, 699, and 387, respectively. However, finally, they removed ambiguous and incomplete tweets. Furthermore, they reduced the non-depressive tweets to 984. As a result, the final label was depressive and non-depressive, consisting of a total of 1968 tweets.

The researcher who developed Bangla depression detection system did not consider more than two labels for their datasets which is momentous for recommending prevention initiatives. Because, for example, if the tag of the dataset is only either depressed or non-depressed, then the users may get anti-depressive online resources or other deterrence measures which are only applicable at the extreme hour; although the users are not depressed, nor under extreme conditions. Therefore, there is a fine chance that the system might cause annoyance and the individuals might not be optimistic to use the system.

Thus, our research aim is to develop a depression detection model for Facebook users to predict suicidal risk in the Bengali language using machine learning, and deep learning techniques that overcome the limitation of previous studies. The challenge we face is to build the standard Bengali depressive dataset. Using automated tools like Facebook, API was not effective to collect potential depressed users. As a result, we manually collected 100 Facebook users' profiles in order to develop our dataset. This dataset is not big enough for real-world applications and even not enough for deep learning models. However, we were able to collect 3000 Facebook statuses from 100 users on an average of 30 statuses from each user. Furthermore, we performed plenty of experiments to determine the best model to be used for depression level prediction.

The contribution of this work is: we collected 100 users' Facebook profiles, containing 3000 posts on an average of 30 posts per user which are collected within 2–3 years time duration. We labeled the dataset considering 16 emotions, 4 depression levels, and depression statuses and proposed BenFED dataset. We performed plenty of ML and DL experiments on our dataset to recommend the best model to be used in the Bengali language for determining depression degree for early suicide prevention. Furthermore, we proposed the first state-of-the-art depression detection model in the Bengali language for predicting the level of depression to reduce the number of suicide activities. A comprehensive mental health action plan 2013–2020–2030 was adopted by the WHO to engage in a web-based consultation [9]. Therefore, the government or other private organizations can use our model and dataset for introducing web-based consultation programs, and other active and slow prevention measures with the help of our proposed model to assist Facebook users with their mental health and can also save from potential suicide risk. In the subsequent sections, methodology, result and discussion, ethical consideration, conclusion and future scope were described in Sects. 2, 3, 4, and 5, respectively.

2 Proposed Methodology

The soul of our dataset is Bengali Facebook posts. In fact, all the texts we have analyzed in this work are Facebook users' statues. After collecting the raw data from Facebook, a data pre-processing module was used to extract useful information and leave the insignificant portions. This way we made sure that our model gets noise-free information. Then, the dataset has been split into training (which contains 80%

of the dataset) and testing (which contains 20% of the dataset) data to evaluate the performance of the proposed model, considering random shuffling. We used tf-idf with unigram and bigram feature extraction for the machine learning model and word embedding for the deep learning model. We used different machine learning, and deep learning models, for classifying depression datasets. Thus, our model is trained and ready to make predictions.

2.1 BenFED: Proposed Dataset

We need a vast amount of data for depression analysis. Unfortunately, there is no standard dataset for depression analysis in the Bengali language. As Bangladeshi people are more active and willing to share their daily thoughts on Facebook, this is a good platform to get the desired datasets. We, firstly, identify the potential individuals from different Facebook groups, pages, and profiles where they ask for mental health support. Some people hide their identity in Facebook groups while asking for help. In that case, we carefully engaged in talks with them on a very personal level through chat. Upon explaining the nature of our works and the value it brings to the table, we humbly inquire whether they take an interest in disclosing their original Facebook profile for our research study. We also exchanged views in as many cases as possible going down extremely individual levels to rate our classified "emotion" and "depression" levels. The dataset is labeled manually by four different persons, having domain knowledge, and is further verified by two experts in the mental health domain.

We collected 100 user profiles manually from Facebook, consisting of 3000 posts, on an average of 30 posts per user, and introduced BenFED (Bengali Facebook Emotion Depression) dataset, for determining emotion and depression levels. Four human crawlers were assigned to collect the dataset. Identifying emotion is a great way to classify the potential depressed person. Therefore, we took 16 emotions for easily classifying depression levels. Based on the emotion, we group these into four depression levels such as no (e.g., joy, approval, love, amusement, surprise, neutral); mild (e.g., disgust, disapproval, anger, confusion); moderate (e.g., nervousness, remorse, embarrassment, fear, disappointment), severe (e.g., sadness). For suicide prevention, these four labels have been used for predicting depression levels.

2.2 Data Preprocessing

Before feeding the data to the ML and DL models, it is the pinnacle of importance to pre-process the data and get rid of the noises to score a favorable outcome. Accurately predicting depression level that requires several sophisticated tasks in the Bengali language one such task is cleaning (stop-word, emoticon, special word, non-Bengali

Original Facebook Status: জীবনে কখনো I love You শুনি নাই😫😔😔
Translation: (In this life, I did not hear "I love You")
Preprocessed text: জীবনে কখনো শুনি নাই

Fig. 1 Stop-word, special word, non-Bengali word, emoticon, are cleaned using python programming

word removal). Other tasks such as tokenization and normalization are performed, after cleaning the dataset (See Fig. 1.)

2.3 *Feature Extraction*

Feature extraction is a very important task to be done before performing text classification or sentiment analysis using machine learning or deep learning algorithm. We used unigram and bigram features containing at most 15,000 frequent words for machine learning (ML) models, and for deep learning models (DL), word embedding techniques were used. We used tf-idf (term frequency and inverse document frequency) as our feature extraction techniques for the ML models. Words occurring more frequently are more important than a less likely occurring word. However, some words such as we, you, they, a, an, the, etc., can be found more likely, however, they don't carry much weight in a document. Therefore, tf-idf is used to solve these conflicting scenarios. Term frequency [10] can be written as in Eq. (1)

$$tf_{t,d} = log_{10}(count(t, d) + 1) \tag{1}$$

The inverse document frequency [11] has the following form in Eq. (2):

$$\mathrm{idf}_t = \log_{10}\left(\frac{N}{\mathrm{df}_t}\right) \tag{2}$$

Here, N refers to the number of documents where the term t occurred. The tf-idf with value is the product of term frequency and inverse document frequency which is given by Eq. (3)

$$w_{t,d} = \mathrm{tf}_{t,d} \times \mathrm{idf}_t \tag{3}$$

3 Result and Discussion

In this section, we examine and present an extensive performance evaluation of several ML-, DNN-based models to reach a foregone conclusion regarding the emotional state of users in terms of Bengali Facebook statuses. The efficiency of the models is ranked on the basis of respective accuracy (Acc). For obtaining more insights into how models are yielding outputs, the precision (Pr), recall (Re), and F1-score metrics are also considered.

3.1 Machine Learning Approach

We began working on developing a depression level prediction system using machine learning algorithms. Logistic Regression (LR), Decision Tree (DT), Random Forest (RF), Naïve Bayes (NB), Support Vector Machine (SVM), K-Nearest Neighbors (KNN), and Adaptive Boosting (AdaBoost) are enacted for incorporating fresh perspectives into our work.

Effect of the number of features. We performed ten experiments for figuring out the effect of features on classification accuracy, among them eight are shown in Table 1. We concluded that maximum accuracy was obtained for a maximum of 15 thousand features. We observed that SVM works well in the lower number of features. Naïve byes work worst in the highest features.

In this stage, taking 15 thousand features into consideration, we performed several experiments, using different ML models, to confirm the best model to be used (see Table 2.). Among the ML techniques, the LR secured the topmost accuracy in all cases namely grouped emotions, tagging a mental state (depression levels and predicting whether a user is depressed or not. The accuracy rates are, respectively, 47.42%, 43.28%, and 22.93% while considering 6, 7, and 16 emotions; 46.64% in terms of detecting mental state and 73.11% in predicting whether depressed or not.

Table 1 Effect of features in F1 Accuracy of ML model. Here *15 K denotes that ambiguous emotion such as neutral, surprise, and confusion was discarded, becoming the total examples of 2702

Model	1 K	5 K	10 K	15 K	20 K	25 K	30 K	*15 K
LR	67.16	66.96	71.73	73.49	73.49	73.19	69.63	73.90
DT	64.40	66.22	62.63	65.03	65.25	65.93	61.92	69.92
RF	70.00	68.96	69.89	67.43	67.67	70.42	64.68	69.74
NB	66.19	67.46	65.10	65.10	65.10	65.19	61.59	68.87
SVM	77.17	69.46	68.66	68.06	68.06	68.12	62.71	70.59
KNN	64.32	63.94	64.89	65.19	65.19	70.87	61.87	71.75
Adabost	66.79	67.92	67.42	70.55	70.55	67.42	65.69	70.74

Table 2 F1-score for 16 emotions, 7 emotions, 6 emotions, 4 depression levels, and 2 sentiments either depressed or non-depressed are shown

Model	Six emotions	Seven emotions	Sixteen emotions	Four depression levels	Depressed or non-depressed
LR	47.42	43.28	22.93	46.64	73.11
DT	33.84	28.04	14.68	35.61	64.41
RF	41.39	33.44	16.37	34.26	69.03
NB	43.95	26.95	10.86	29.19	65.58
SVM	46.80	33.50	10.02	41.02	68.25
KNN	22.83	26.40	9.50	32.89	67.17
Adabost	36.54	26.19	8.30	32.47	69.69

However, NB and KNN are relatively poor performing; NB responded to the grouped emotions (43.95, 26.95, and 10.86%), and in figuring depressive individuals (65.58%) in the poorest manner when KNN showed the worst (67.17%) output in tracing depression state. It can be distinctively pointed out that the algorithms are more effective when they are needed to consider fewer labels. Hence, we chose the LR model as the desired model for early suicide prevention.

Table 3 shows the performance of our top model which results in an accuracy of 49% for six emotions. The model's proficiency (F1-score) is at its best with the acute sentimental emotions such as sadness (62%), anger (56%); which are also the most frequent emotions. However, the model functions poorly in terms of light emotions such as surprise (8%), which can be categorized as less frequent emotion. For four depression levels, the accuracy was decreased to 48%, as the total examples

Table 3 Six different emotions, four depression levels, and depression status

Class	Label	Train	Test	Pr(%)	Re(%)	F1(%)	Acc (%)
Emotionb	joy	469	98	31	35	33	49
	neutral	162	46	22	9	12	
	surprise	68	22	33	5	8	
	anger	530	132	52	61	56	
	fear	218	42	19	21	20	
	sadness	953	260	61	63	62	
Depression levels	no	450	117	40	34	37	48
	mild	540	122	49	64	55	
	moderate	348	98	30	21	25	
	severe	823	204	57	58	58	
Depression status	non-depressed	450	117	40	20	26	76
	debpress ed	1711	424	81	92	86	

were reduced to 2702 (ambiguous posts such as neutral, surprise, ad confusion were discarded). Finally, for two levels such as non-depressed, and depressed the accuracy was 76%. After performing an ensemble considering LR and SVM models for the two levels, the accuracy, precision, recall, and F1 were 78%, 72%, 78%, and 70%, respectively. Similarly, for four levels ensemble were 47%, 45%, 47%, and 45%, respectively.

3.2 *Deep Learning Approach*

DNN is a very useful tool for text categorization. Word embedding is used as a feature in the deep learning method. After performing several experiments, the optimized result was obtained for a bidirectional Long short-term Memory. The LSTM consists of an embedding layer (size of which 210×300), two bidirectional layers, and an output layer. In the hidden layer, we used 64 and 32 hidden units. We used sparse categorical cross-entropy as the loss function, and the learning rate is set to 0.001 and dropout 0.7. Adam optimizer was used for optimization, running the model up to 30 epochs.

We took 3000 Facebook users' posts, the average length of the post was 288, and after pre-processing, the length becomes 195. Then, we observed accuracy, considering different lengths with varying embedding sizes. We conclude that 288 lengths with 300 embedding sizes are the best fit for our model and dataset. See Table 4.

We have shown the accuracy and loss with respect to epoch. For four levels of depression (See Fig. 2.), we observed that after 22 epochs, the training accuracy is unchanged with increasing epochs. Furthermore, the validation loss also increases with increasing epochs. We conclude that there is no need to train the model after 22

Table 4 Effect of input words in the model, and word embedding size in model accuracy

Text length	Data label	Embedding size	Accuracy
250	no, mild, moderate, severe	300	46
275	no, mild, moderate, severe	300	44
288	no, mild, moderate, severe	300	47
300	no, mild, moderate, severe	300	45
330	no, mild, moderate, severe	300	47
288	no, mild, moderate, severe	500	46
288	no, mild, moderate, severe	1000	45
288	no, mild, moderate, severe	100	41
288	no, mild, moderate, severe	50	43
288	joy, neutral, surprise, anger, fear, sadness	300	48
288	depressed or nondepressed	300	80

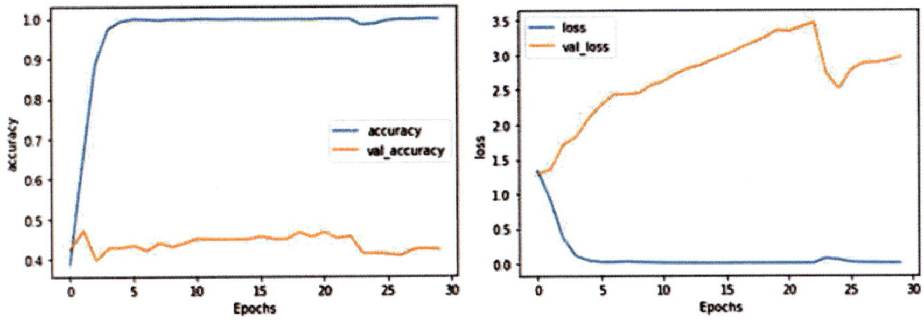

Fig. 2 Accuracy of the BiLSTM, having no, mild, moderate, severe depression level is presented on the left side; training and validation loss are presented on the right side

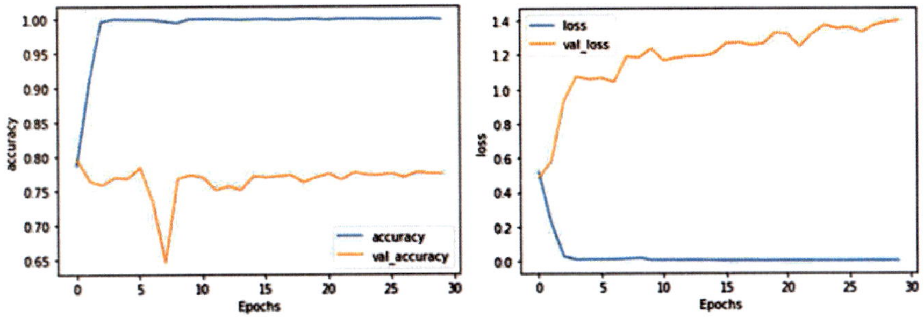

Fig. 3 Accuracy of the BiLSTM having depressed and non-depressed is presented on the left side; training and validation loss are presented on the right side

epochs. For two levels, i.e., non-depressed and depressed, shown in Fig. 3, the same results were observed.

3.3 Confusion Matrix

In this part, we present a thorough confusion matrix analysis. A class-wise proportion of the quantity of the predicted labels is demonstrated in Table 5. The findings reveal that a few data points were categorized erroneously. To illustrate, 25 out of 132 cases under the anger tag were mistakenly classified as sadness. In a similar manner, 34 out of 260 were grouped as anger with respect to sadness. Furthermore, among 46 data in neutral, 15 were misclassified as joy. It is observed that the trivial sum of data is wrongly labeled as a surprise, which is the least frequent emotion as well.

The class imbalance nature of data is the potential cause for such intricacy. In Bengali language, a good number of words and terms can be repeatedly used in several cases for different reasons. For example, expressions of disappointment can easily end up being anger or sadness feelings. On top of that, emotion classification

Table 5 Six emotions: confusion matrix

True/predicted	Anger	Fear	Joy	Neutral	Sadness	Surprise
Anger	80	10	14	3	25	0
Fear	2	9	7	1	22	1
Joy	26	5	34	2	31	0
Neutral	8	3	15	4	16	0
Sadness	34	18	36	6	165	1
Surprise	3	3	3	2	10	1

is extremely subjective, depending on the individual's viewpoint, and people may interpret a sentiment in a variety of ways [12]. Thus, erroneous predictions may be avoided to some extent by developing a balanced dataset containing diverse data.

3.4 Results Comparison

The results of our proposed approach are compared with other approaches to determine emotion and depression levels are shown in Table 6, where English and Bengali languages were used in social media platforms such as Twitter, Reddit, and Facebook. From Table 6, it is evident that the LR and the LSTM are comparatively better fitting for our dataset. It is also apparent that the work done in Bengali was not more than two labels. However, we consider 16 emotions, 4 depression, and finally 2 depression statues. In the case of English language, two labels annotation were addressed, alongside four levels of suicide risk were addressed for the Reddit dataset. However, the accuracy we obtained was far higher than that of them.

4 Ethical Considerations

Research conducted on mental health is evidently sensitive in nature, and when it particularly deals with complex areas such as suicide prevention, the entire process becomes oversensitive. User's data being collected and incorporated with technology in order to develop a model, inherently hints a collection of ambiguities, which prompt researchers to review additional ethical considerations. Benton et al. recommended considerate and detailed protocols in terms of ethical concerns on mental health researches based on social media platforms [17].

Table 6 Performance comparison analysis based on F1-score and Accuracy

Methods	Data Label	Language	Platform	F1	Accuracy
RF + 20 features, DASS 21- BV (Bangla),Beck scale [7]	depressed, non-depressed	–	Survey	60	75
LSTM(128, 5 layers, 50 batch size, 5 epoch [5]	depressed, non-depressed	Bengali	Twitter	–	67
LSTM (128 unit, 5 layers, 25 batch size, 20 epoch) [5]	depressed, non-depressed	Bengali	Twitter	–	86
SGDC with unigram + emoticons [8]	depressed, non-depressed	Bengali	Facebook	–	78
MNB + tf-idf (unigram, bigram), LR + tf-idf [13]	joy, surprise, anger, disgust, fear, sadness	Bengali	Online and Offline	49 61	54 (lowest) 61 (highest)
CNN + Word2Vec, BiLSTM + FastText [13]	joy, surprise, anger, disgust, fear, sadness	Bengali	Online and Offline	52 57	52 (lowest) 58 (highest)
SVM + tf-idf [4]	depressed, non-depressed	English	Twitter	58	–
CNN + GloVe [14]; LR + n-gram [15]; SVM + Word2Vec (skip-gram) [16]	no,mild, moderate, severe	English	Reddit	42 30 36	– – –
LR + tf-idf (proposed)	joy, neutral, suprise, anger, fear, sadness	Bengali	Facebook	48	49
LR + tf-idf (proposed)	no, mild, moderate, severe	Bengali	Facebook	76	48
LSTM + word embedding (proposed)	depressed, non-depressed	Bengali	Facebook	–	80

4.1 Participants and Research Oversight

Prioritizing user privacy and confidentiality issues and holding the question of their dignity high in regard, we collected user data predominantly undergoing two approaches. In the first approach, we compiled a good number of individual Facebook profiles going through mental disturbance, i.e., conducted a comprehensive search on different groups and pages on Facebook where user share their troubled feelings and experiences seeking assistance, reached out to enthusiasts working on this field and exploited our personal contacts with a view to identifying public profiles inclined with our research objectives. After that, we updated a google sheet that we used as our official dataset with information such as a link to the user ID, Sex, Age, Class,

Label, Text (Facebook user posts), and Comments. Among the collected 100 users' posts, there were 54 men, 46 women, and the average age was 23. Most of them were students, going to graduate schools. There are eight users who committed suicide after writing a post on Facebook.

4.2 Data Storing and Protection

Benton et al. indicated that even publicly available materials on social media sometimes contain information that needs to be safeguarded [17]. Depending on the posts we manually categorized the mental states of individuals, respectively. Therefore, we consider the labeled dataset classified despite the fact that they have been accumulated from multiple public forums. To keep collected data well-protected, safe, and secure, we ensured a robust preservation mechanism. We stored the dataset in the cloud and necessary technical securities have been enacted. Reliable security measures such as firewall and password have been appropriately administered and the trustworthiness of google drive as a cloud service provider has been properly investigated. It has been made sure that anyone who is not involved with this research work does not have access to the files.

5 Conclusion and Future Scope

Many Facebook user express their thoughts and feelings when they are at the worst time or before they commit suicide. When they are hopeless or down, they don't find anyone to express their hidden sorrows or maybe they are not eager to share sufferings. Hence, Logistic Regression and LSTM model for depression analysis will be the state-of-the-art techniques to predict their mental health condition. Based on the depression levels, online anti-depressive posts can be recommended for early suicide prevention.

In the future, we aim to collect 500 users' profiles, consisting of a total of 10,000 Facebook posts which will be labeled based on these 3000 records in a semi-supervised manner.

References

1. World Health Organization. Suicide Prevention. Retrieved from https://www.who.int/mental_health/prevention/suicide/suicideprevent/en
2. Al Mamun, M.A., Griffiths, M.D.: The association between Facebook addiction and depression: a pilot survey study among Bangladeshi students. Psychiatry Res. **271**, 628–633 (2019). https://doi.org/10.1016/j.psychres.2018.12.039

3. Arafat, S.M.Y., Al Mamun, M.A.: Repeated suicides in the University of Dhaka (November 2018): Strategies to identify risky individuals. Asian J. Psychiatry **39**, 84–85 (2019). https://doi.org/10.1016/j.ajp.2018.12.014
4. Leiva, V., Freire, A.: Towards suicide prevention: early detection of depression on social media. In: LNCS, pp. 428–436. Springer (2017). https://doi.org/10.1007/978-3-319-70284-1_34
5. Uddin, A.H., et al.: Depression analysis from social media data in Bangla language using long short term memory (LSTM) recurrent neural network technique. In: 5th IEEE IC4ME2, Rajshahi, BD (2019). https://doi.org/10.1109/IC4ME247184.2019.9036528
6. Zirikly, A., et al.: In: Proceedings of the 6th Workshop on Computational Linguistics and Clinical Psychology, Association for Computational Linguistics, USA, pp 24–33 (2019). https://doi.org/10.18653/v1/W19-300
7. Choudhury, A.A., et al.: Predicting depression in Bangladeshi undergraduates using machine learning. IEEE TENSYMP **2019**, 789–794 (2019). https://doi.org/10.1109/TENSYMP46218.2019.8971369
8. Billah, M., Hassan, E.: Depression detection from Bangla Facebook status using machine learning approach. Int. J. Comput. Appl. **178**(43), 9–14 (2019). https://doi.org/10.5120/ijca2019919314
9. World Health Organization. Mental Health Action Plan: 2013–2020–2030. Retrieved from https://www.who.int/mental_health/action_plan_2013/en/
10. Luhn, H.P.: A statistical approach to mechanized encoding and searching of literary information. IBM J. Res. Dev. **1**(4), 309–317 (2010). https://doi.org/10.1147/rd.14.0309
11. Jones, K.S.: A statistical interpretation of term specificity and its application in retrieval. J. Doc. 28, 11–21 (1972).https://doi.org/10.1108/eb026526
12. LeDoux, J.E., Hofmann, S.G.: The subjective experience of emotion: a fearful view. Curr. Opin. Behav. Sci. **19**, 67–72 (2018)
13. Das, A. et al.: Emotion classification in a resource constrained language using transformer-based approach, pp. 150–158 (2021). https://doi.org/10.18653/V1/2021.NAACL-SRW.19
14. Allen, K. et al.: ConvSent at CLPsych (2019) Task a: using post-level sentiment features for suicide risk prediction on Reddit. In: Proceedings of the *Sixth* Workshop on Computational Linguistics and Clinical Psychology, pp. 182–187. https://doi.org/10.18653/V1/W19-3024
15. Ríssola, E., et al.: Suicide Risk Assessment on Social Media: USI-UPF at the CLPsych Shared Task, pp. 167–171 (2019). https://doi.org/10.18653/V1/W19-3021
16. Morales, M., et al.: An investigation of deep learning systems for suicide risk assessment. In: Proceedings of the 6th Workshop on Computational Linguistics and Clinical Psychology, Association for Computational Linguistics, Stroudsburg, pp. 177–181 (2019). https://doi.org/10.18653/v1/w19-3023
17. Benton, A., et al.: Ethical research protocols for social media health research. In: Proceedings of the First ACL Workshop on Ethics in Natural Language Processing (2017)

Author Index

S. Hossain et al. (eds.), *Proceedings of International Conference on Fourth Industrial Revolution and Beyond 2021*, Lecture Notes in Networks and Systems 437,
https://doi.org/10.1007/978-981-19-2445-3